W9-BFC-709

THE TOTAL SYNTHESIS OF NATURAL PRODUCTS

The Total Synthesis of Natural Products

VOLUME 3

Edited by

John ApSimon

Department of Chemistry
Carleton University, Ottawa

A WILEY-INTERSCIENCE PUBLICATION

John Wiley & Sons, New York · London · Sydney · Toronto

Library of Congress Cataloging in Publication Data:

ApSimon, John.
The total synthesis of natural products.
Includes bibliographical references.

1. Chemistry, Organic–Synthesis. I. Title.

QD262.A68   547'.2   72-4075
ISBN 0-471-02392-2 (V. 3)

Printed in the United States of America

10 9 8 7 6 5 4 3 2

Contributors to Volume 3

T. Kametani, Tohoku University, Sendai, Japan
J. P. Kutney, University of British Columbia, Vancouver, Canada
R. V. Stevens, University of California, Los Angeles, California

Preface

Throughout the history of organic chemistry, we find that the study of natural products frequently has provided the impetus for great advances. This is certainly true in total synthesis, where the desire to construct intricate and complex molecules has led to the demonstration of the organic chemist's utmost ingenuity in the design of routes using established reactions or in the production of new methods in order to achieve a specific transformation.

These volumes draw together the reported total syntheses of various groups of natural products and commentary on the strategy involved with particular emphasis on any stereochemical control. No such compilation exists at present, and we hope that these books will act as a definitive source book of the successful synthetic approaches reported to date. As such, it will find use not only with the synthetic organic chemist but also perhaps with the organic chemist in general and the biochemist in his specific area of interest.

One of the most promising areas for the future development of organic chemistry is synthesis. The lessons learned from the synthetic challenges presented by various natural products can serve as a basis for this ever-developing area. It is hoped that these books will act as an inspiration for future challenges and outline the development of thought and concept in the area of organic synthesis.

The project started modestly with an experiment in literature searching by a group of graduate students about nine years ago. Each student prepared a summary in equation form of the reported total syntheses of various groups of natural products. It was my intention to collate this material and possibly publish it. During a sabbatical leave in Strasbourg in 1968-1969, I attempted to prepare a manuscript, but it soon became apparent that the task would take many years and I wanted to enjoy some of the other benefits of a sabbatical leave. Several colleagues suggested that the value of such a collection would be enhanced by commentary. The only way to encompass the amount of data collected and the inclusion of some words was to persuade experts in the various areas to contribute.

Volume 1 presented six chapters describing the total synthesis of a wide variety of natural products. The subject matter of Volume 2 was somewhat more related, being a description of some terpenoid and steroid syntheses. The present volume considers the syntheses of several classes of alkaloids. The authors originally provided me with their manucripts three years ago, and the delay in producing this volume is a result of a hope that another planned chapter would also appear in time for inclusion. Unfortunately, the author of that chapter has been unable to produce his contribution.

I have asked the authors of these chapters to provide wherever possible, an updating of their work by the use of supplementary references and addenda. The delay in producing the original work is in no way the fault of the present authors, and I apologize to them for this tardiness. However, I believe that their work is outstanding and well worth publishing. I hope the readers of this volume will find it useful as a reference work on total syntheses preformed in the alkaloid field.

I wish to express my thanks to Ms. Karen Bergenstein for preparing the index and to Karl Diedrich for preparing the illustrations to Chapter 2.

JOHN APSIMON

Ottawa, Canada
January 1977

Contents

The Total Syntheses of Isoquinoline Alkaloids

TETSUJI KAMETANI
Pharmaceutical Institute,
Tohoku University,
Aobayama, Sendai, Japan

*The author is deeply indebted to Dr. Keiichiro Fukumoto and Dr. Shiroshi Shibuya, Pharmaceutical Institute, Tohoku University, for many help for suggestions, as well as for help in the preparation of this review.

Isoquinoline or benzo[c]pyridine, an isomer of quinoline, was first obtained from coal tar by Hoogewerff and van Dorp in 1885 together with various alkylisoquinolines, and isoquinoline itself was synthesized by Gabriel in the same year. However, the natural occurrence of the isoquinoline ring system was first recognized in the opium alkaloid; papaverine, isolated as needles, m.p. 147°, $C_{20}H_{21}O_4N$, by Goldschmidt,[1] in one of the first structural determinations of alkaloids. Since Goldschmidt's recognition, efforts by chemists have been devoted to the chemistry of the alkaloids and by now about 1000 isoquinoline alkaloids are known.[2]

The numbering of isoquinoline ring system is shown as follow.

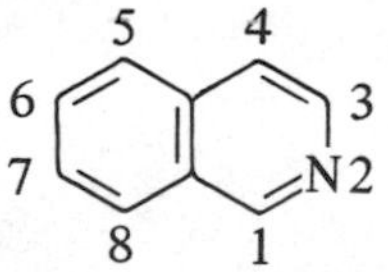

5 4
6 3
7 N2
8 1

Isoquinoline is obtained as hygroscopic colorless crystals, m.p. 24.6°, b.p.$_{760}$ 243.3°, b.p.$_{40}$ 142° with *p*Ka 5.14 in water at 20°. The odor of isoquinoline is almost the same as that of quinoline, but the former smells somewhat likes benzaldehyde. The basicity of isoquinoline is stronger than that of quinoline, which has *p*Ka 4.85 in water at 20°. Electronically, the chief difference between naphthalene and isoquinoline is due to the fact that the latter, isoquinoline has the "lone pair" at its nitrogen atom. Furthermore, the nitrogen attracts electron density from the carbon atoms so that these carbon atoms have a deficiency of the electron charge compared with the atoms in naphthalene.[3] Quantitatively, the charge on each atom can be calculated by the valence bond method or by the method of molecular orbitals.[3]

π-Electron Densities for Isoquinoline

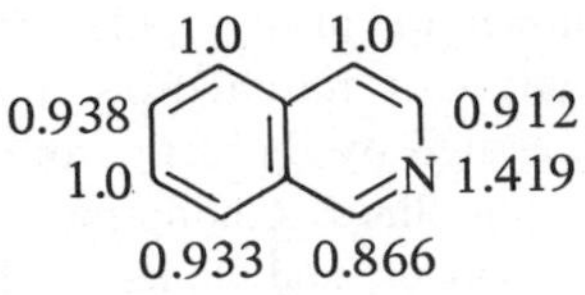

V. B. Method

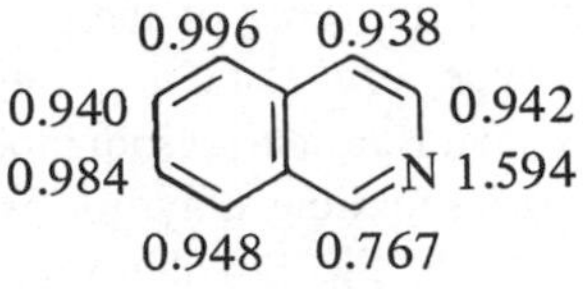

M. O. Method

In general, it would be expected that substitution with electrophilic reagents would occur at the carbon having the greatest π-electron density and that substitution with nucleophilic reagents would occur at the position having the smallest π-electron density.

1. GENERAL METHODS

A. Introduction

The methods for the synthesis of isoquinoline ring system can be classified systematically in five ways according to the mode of formation of the pyridine ring (Chart 1-1). The first type involves ring closure between the benzene ring and

Chart 1-1.

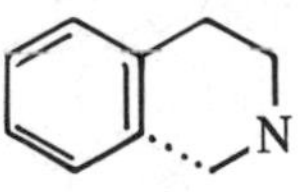

Type 1

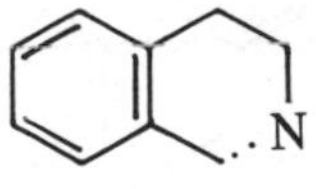

Type 2

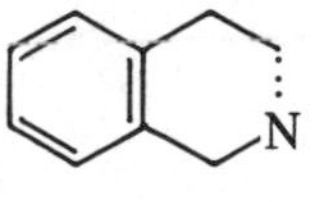

Type 3

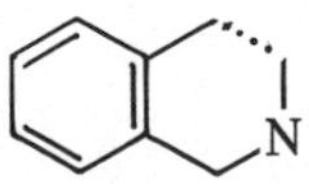

Type 4

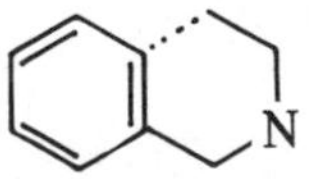

Type 5

the carbon atom, which forms the C_1-position of the resulting isoquinoline ring. The second type uses bond formation between the C_1-position and nitrogen, and the third type uses cyclization by the combination of nitrogen with the C_3-position. The fourth type is due to the formation of isoquinoline ring by ring closure between the C_3- and C_4-position. The fifth type necessitates ring closure between the benzene ring and C_4-position.

In the Chart 1-1, the dotted lines indicate the bond formation by cyclization. Although all the types of these reactions are known, the most popular reactions are the type of 1 and 5, giving usually dihydro- or tetrahydroisoquinoline derivatives and aromatic isoquinolines can be prepared by the dehydrogenation of the corresponding dihydro- or tetrahydroisoquinolines. Among reactions of type 1 and 5, the Bischler-Napieralski, Pictet-Spengler, and Pomeranz-Fritsch reactions are especially important.

Chart 1-2.

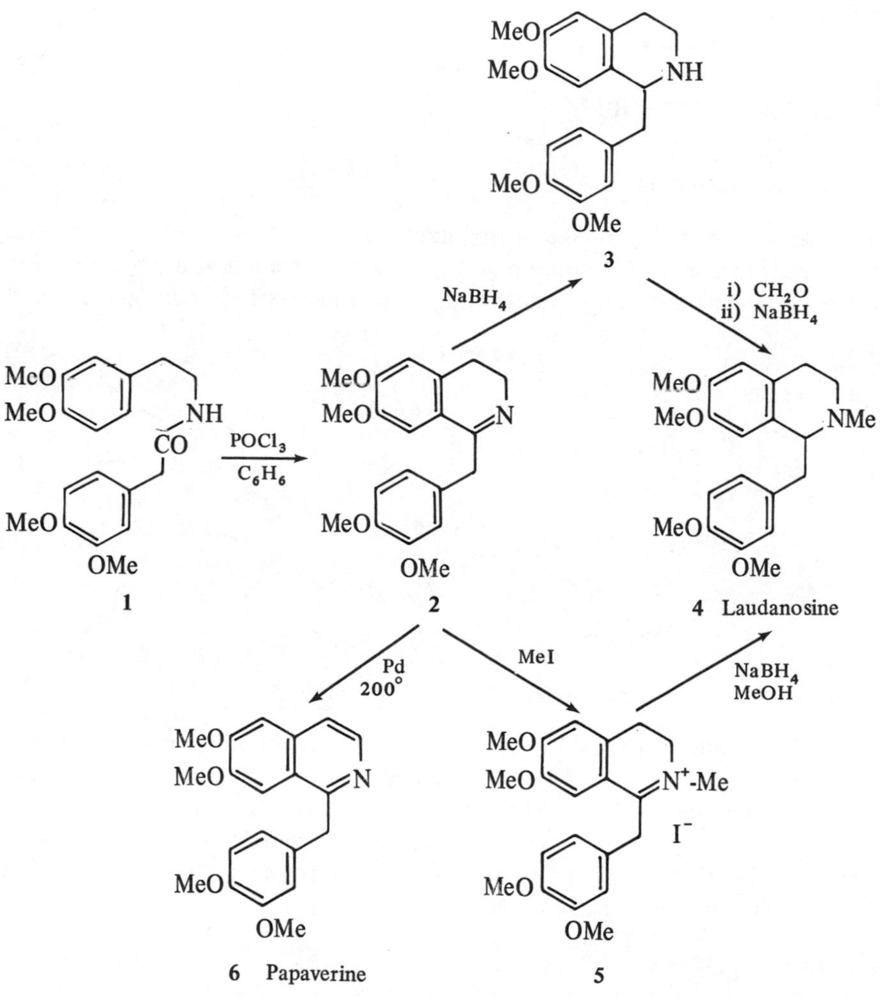

MeO
MeO
NH
MeO
OMe
3

$NaBH_4$

i) CH_2O
ii) $NaBH_4$

MeO
MeO
NH
CO
MeO
OMe
1

$POCl_3$
C_6H_6

MeO
MeO
N
MeO
OMe
2

MeO
MeO
NMe
MeO
OMe
4 Laudanosine

Pd
200°

MeI

$NaBH_4$
MeOH

MeO
MeO
N
MeO
OMe
6 Papaverine

MeO
MeO
N^+-Me
I^-
MeO
OMe
5

B. Type 1 Synthesis

Bischler-Napieralski Reaction[4] *(Chart 1-2)*

The Bischler-Napieralski route involves the cyclodehydration of an acyl derivative **1** of β-phenethylamine in the presence of a Lewis acid such as phosphoryl chloride or phosphorous pentoxide in an inert solvent to give a 3,4-dihydroisoquinoline **2**, which must be reduced to a 1,2,3,4-tetrahydroisoquinoline **3** since the isoquinoline alkaloids[5] exist as the tetrahydro derivatives in most cases. For this purpose, 3,4-dihydroisoquinoline hydrochloride can be directly reduced with sodium borohydride to give the tetrahydroisoquinoline derivative **3**.[6] When the *N*-methyl derivative **4** is desired, the Eschweiler-Clarke reaction of **3** with formalin and formic acid or sodium borohydride gives the expected *N*-methyl compound **4**.[7] Reduction of the methiodide **5** of a 3,4-dihydroisoquinoline with sodium borohydride to **4** is also recommended.[8] Recently, cinnamolaurine was synthesized by this method as shown Chart 1-3A.[8a] On the other hand, the mild dehydrogenation of a 3,4-dihydroisoquinoline **2** can be

Chart 1-3A.

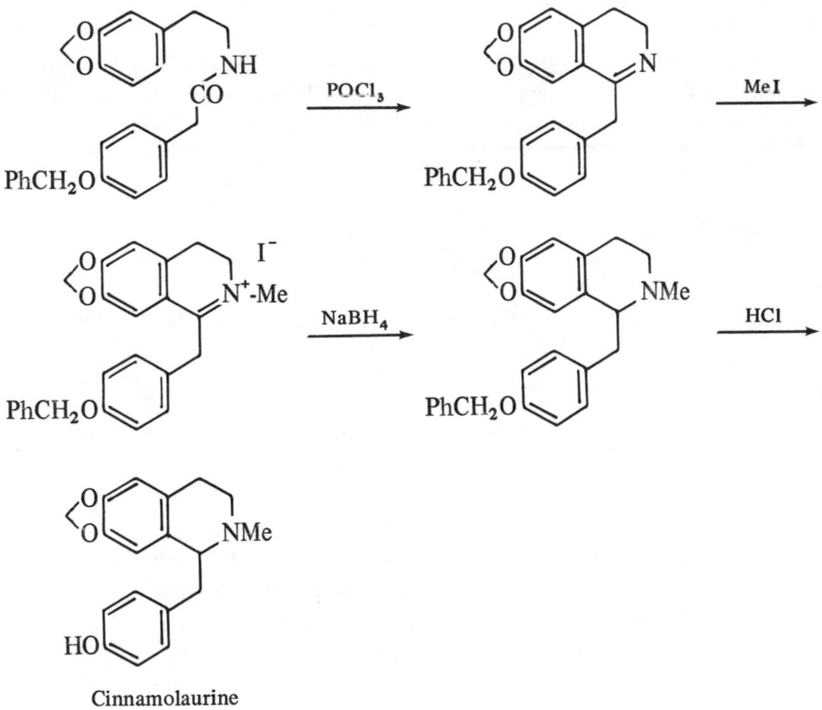

carried out to obtain the aromatic isoquinoline alkaloids such as papaverine **6**.[9]

One of the most important modifications of the Bischler-Napieralski reaction was introduced by Pictet and Gams.[10] This reaction gives the isoquinoline derivative instead of the 3,4-dihydro-compound by cyclization of a β-hydroxy-β-phenethylamide 7 with phosphorous pentoxide. For example, papaverine **6** was obtained directly from 7 (Chart 1-3).

Chart 1-3.

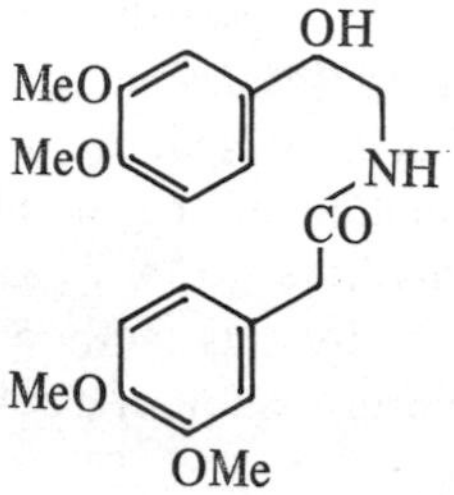

MeO, MeO, OH, NH, CO, MeO, OMe

7

Application of the Pictet-Gam's modification to β-methoxy-β-phenethylamide **7a** also gives the isoquinoline derivatives **6**, directly.[11] Therefore, the choice of the foregoing variation of the Bischler-Napieralski reaction should be made according to the availability of the starting amide (Chart 1-4).

Chart 1-4.

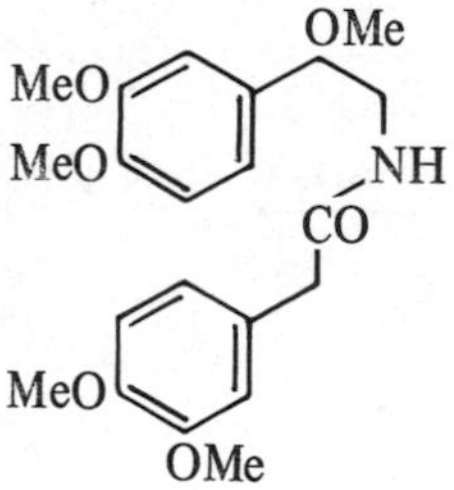

MeO, MeO, OMe, NH, CO, MeO, OMe

7a

Oximes **8**, which could lead to *N*-acyl-β-phenethylamide **9** by Beckmann rearrangement, are also useful as starting materials for the Bischler-Napieralski route, a method applied to the synthesis of carnegine **10**[12] (Chart 1-5).

These oximes are converted into the corresponding isoquinolines or 3,4-dihydroisoquinolines without isolation of the amides formed as intermediates.[13]

Chart 1-5.

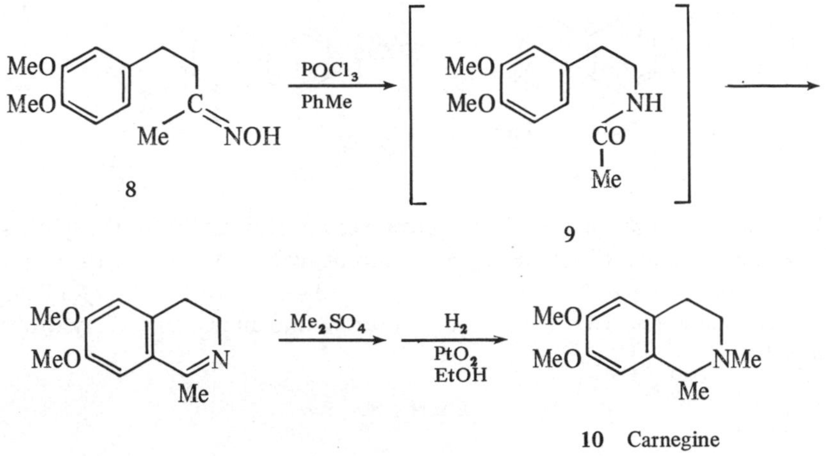

MeO, MeO; Me, C, NOH; 8 —$POCl_3$, PhMe→ [MeO, MeO; NH, CO, Me] 9 →

MeO, MeO; N, Me —Me_2SO_4→ —H_2, PtO_2, EtOH→ MeO, MeO; NMe, Me

10 Carnegine

The benzenesulfonyl ester **11** of an oxime undergoes cyclization to give the 3,4-dihydroisoquinoline derivative **12** by heating alone without any other reagent[14] (Chart 1-6). In some cases, the amidine, instead of the amide, is used

Chart 1-6.

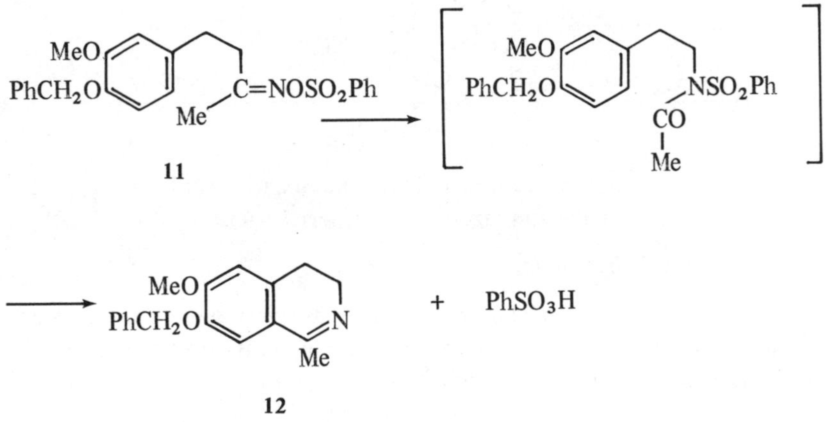

MeO, $PhCH_2O$; Me, $C{=}NOSO_2Ph$; 11 → [MeO, $PhCH_2O$; NSO_2Ph, CO, Me]

→ MeO, $PhCH_2O$; N, Me + $PhSO_3H$

12

for cyclization to give the phenanthridine derivatives in good yields.[15] Short and Brodrich[16] synthesized 3,4-dihydro-1-phenylisoquinoline **13** from amidine **14** by treatment with phosphoryl chloride (Chart 1-7).

Chart 1-7.

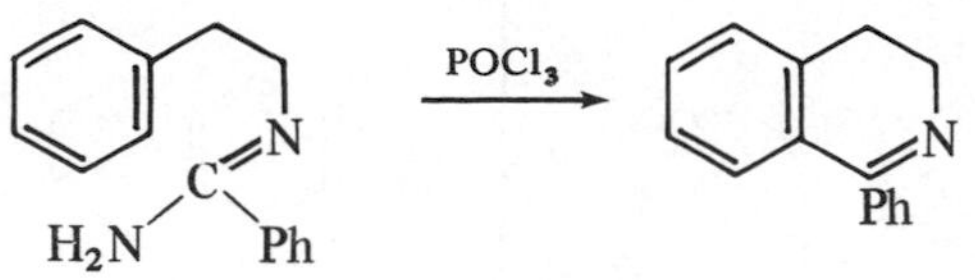

$POCl_3$

14 13

N-β-Phenylethylurea and urethane derivatives are also useful for the syntheses of 3,4-dihydroisoquinolines having an amino or hydroxyl group at the C_1-position.[17] For example, an urethane **15** yields 3,4-dihydro-1-hydroxy-6,7,8-trimethoxyisoquinoline **16**,[18] which was converted into anhalamine **17** by Brossi[18] (Chart 1-8).

Chart 1-8.

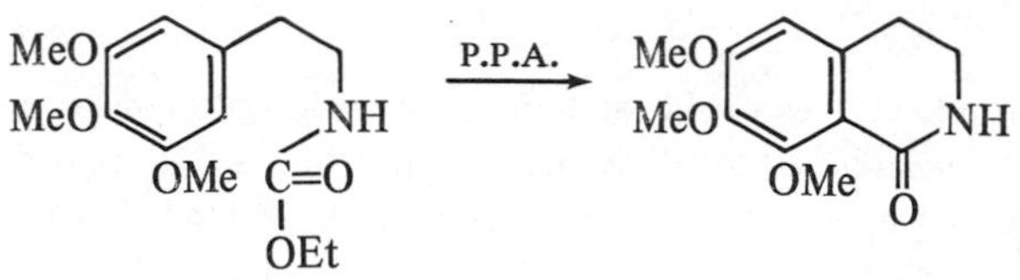

P.P.A.

15 16

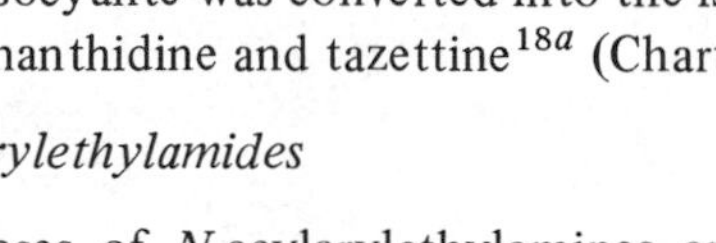

17 Anhalamine

Similarly, the isocyante was converted into the isocarbostyril, which was transformed into haemanthidine and tazettine[18a] (Chart 1-8A).

Syntheses of β-Arylethylamides

Since the syntheses of N-acylarylethylamines are very important as starting materials for Bischler-Napieralski reaction, and representative synthetic methods to the amides are described as follows.[19,20]

Schotten-Baumann Reaction (Chart 1-9). This reaction involves an acylation of amines by treatment with an acyl chloride under ice-cooling in dilute alkaline solution. In the case of substances labile to strong alkali, weaker alkaline reagents such as sodium carbonate, bicarbonate, or triethylamine can be used. In some cases, an excess of amine is used to remove the resulting hydrogen chloride

Chart 1-8A.

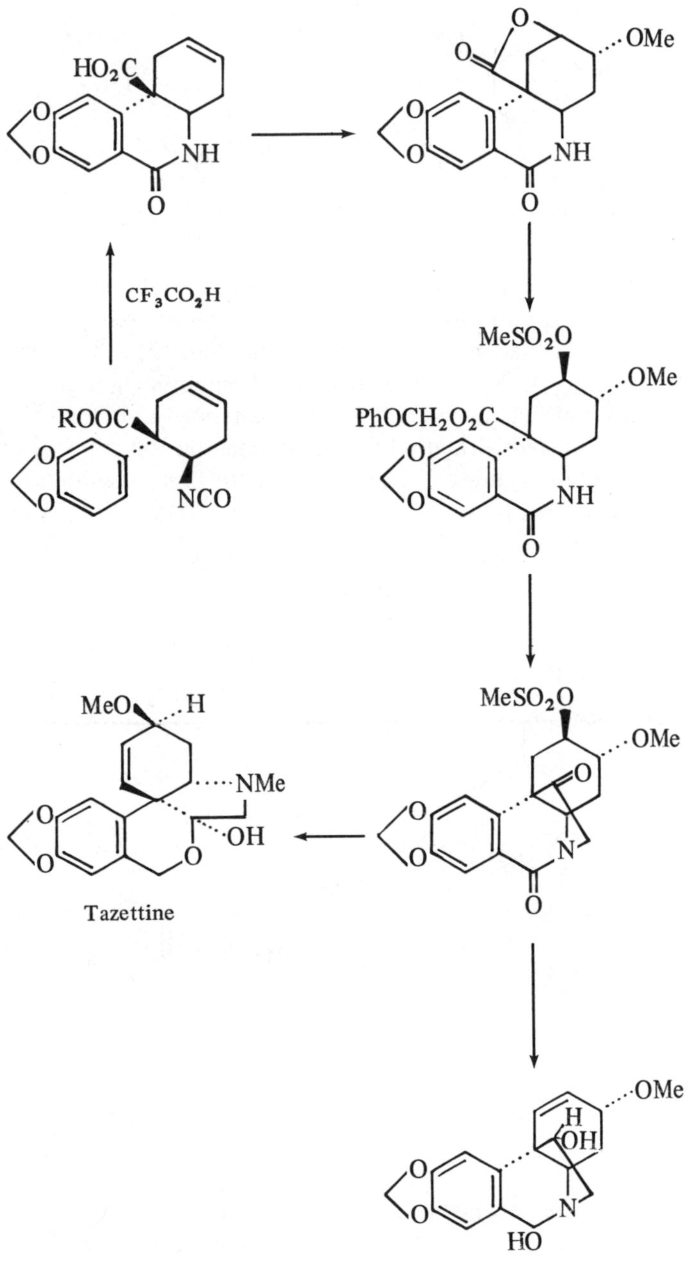

CF_3CO_2H

Tazettine

Haemanthidine

Chart 1-9.

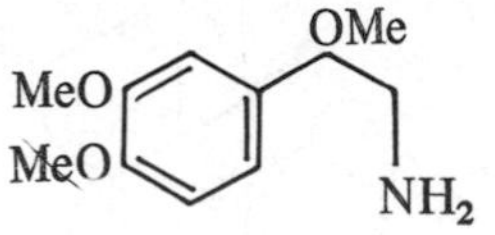

18 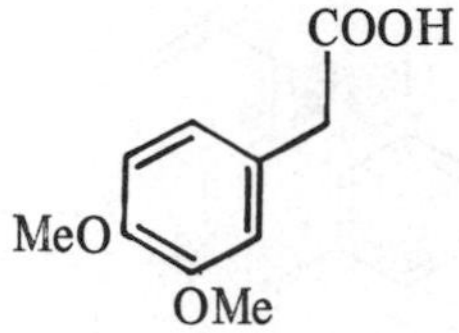 19

as its hydrochloride. Acylation with acid chloride in anhydrous pyridine also gives the amide in good yield. Sugasawa directly synthesized papaverine **6** by heating a mixture of the amine **18** and carboxylic acid **19**, without the isolation of the corresponding amide, in the presence of phosphoryl chloride.[21] The modification of this method was carried out by Battersby as follows.[21a] The carboxylic acid was treated with ethyl chlorocarbonate in the presence of triethylamine in dimethylformamide at $-5°$, and the resulting mixed anhydride, without isolation, was condensed with the homoveratrylamine at $-5\sim0°$ to afford the amide (Chart 1-10).

Chart 1-10.

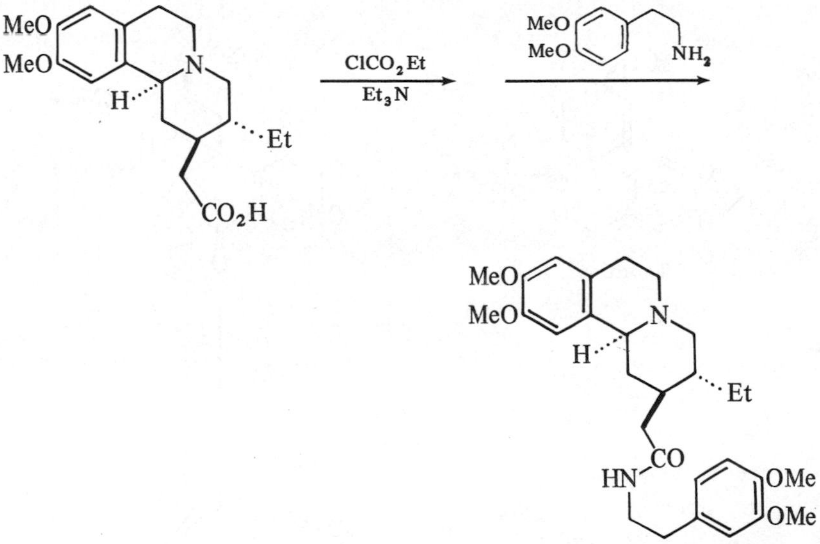

$ClCO_2Et$, Et_3N

The Condensation of a Carboxylic Acid with an Isocyanate (Chart 1-11). Amides, which are difficult to prepare by the Schotten-Baumann or other reactions, can often be obtained by the condensation of isocyanates with carboxylic acids. For instance, Sugasawa and Shigehara[22] prepared the amide **20** by the condensation of 2 moles of 3,4-dimethoxyhydrocinnamic acid azide **21** with γ,γ-ethoxycarbonylpimelic acid 22.

Chart 1-11.

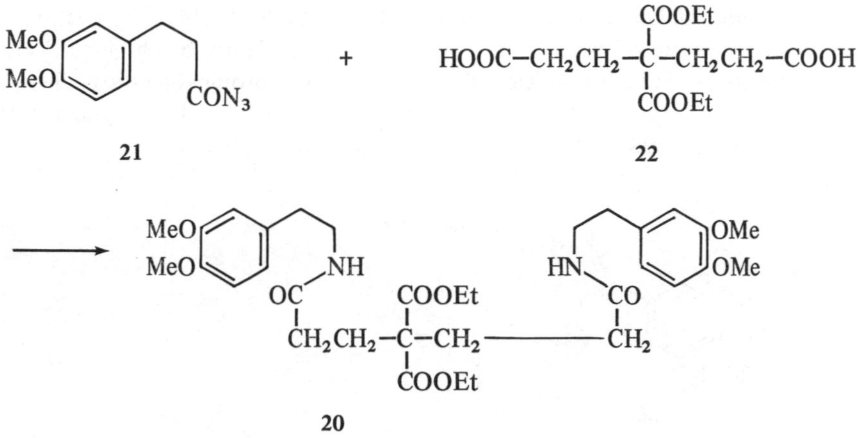

MeO
MeO
CON_3
+
COOEt
$HOOC-CH_2CH_2-C-CH_2CH_2-COOH$
COOEt

21 **22**

MeO
MeO
NH
OC
COOEt
$CH_2CH_2-C-CH_2$ —— CH_2
COOEt
HN
CO
OMe
OMe

20

In this reaction, the azide **21** can be converted into the corresponding isocyanate as an intermediate by heating in benzene solution, and the resulting isocyanate can also be used for the condensation, but normally a mixture of the azide **21** and dicarboxylic acid is heated in benzene to give the amide **20** directly. When the materials are not dried completely the urea derivative is formed as a by-product.

Chart 1-12.

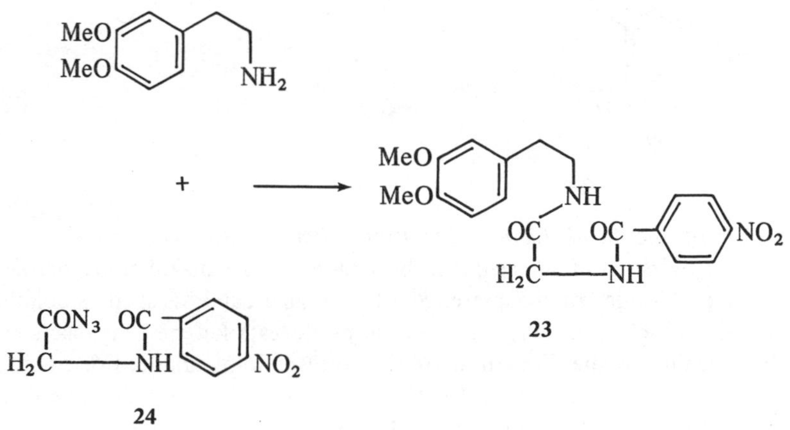

MeO
MeO
NH_2
+
CON_3 OC
H_2C —— NH
NO_2

24

MeO
MeO
NH
OC OC
H_2C —— NH
NO_2

23

The Condensation of an Amine with an Azide (Chart 1-12). This procedure uses the acylation of the amine with azide in cold solvent. Since this reaction is quite different from the preceding method, it must be carried out without formation of isocyanate from the azide. For example, 4-nitrohippuro-β-veratrylethylamide **23** was prepared by the condensation between 4-nitrohippuric acid azide **24** and homoveratrylamine.[23]

The Condensation Between an Ester and an Amine (Chart 1-12A). By heating or fusion of a mixture of an ester and an amine, the amide can be obtained easily. *N*-(3-Benzyloxy-4,5-dimethoxyphenethyl)-5-benzyloxy-2-bromo-4-methoxyphenylacetamide **25** was prepared from 3-benzyloxy-4,5-dimethoxyphenethylamine **26** with methyl 5-benzyloxy-2-bromo-4-methoxyphenylacetate **27**.[24]

Chart 1-12A.

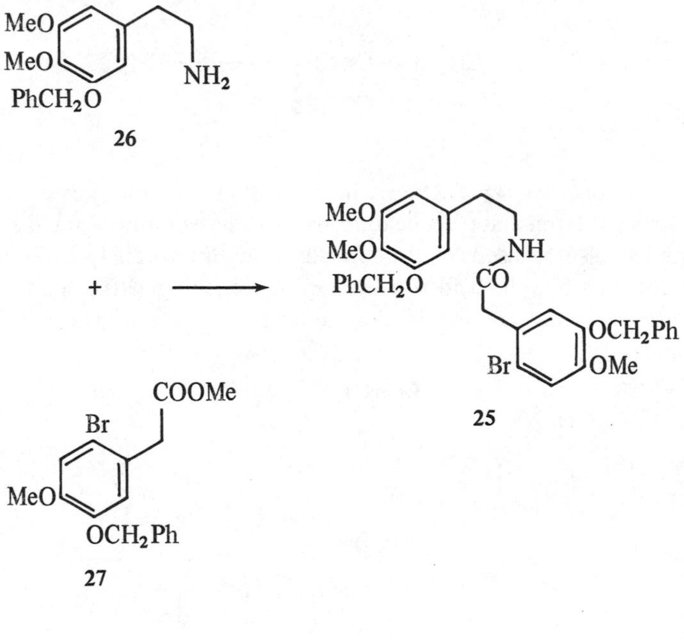

MeO
MeO
NH_2
$PhCH_2O$
26

\+ →

MeO
MeO
NH
$PhCH_2O$
CO
OCH_2Ph
Br
OMe
25

COOMe
Br
MeO
OCH_2Ph
27

Application of the Arndt-Eistert Reaction (Chart 1-13). This procedure involves the conversion of an acid to the amide via the diazoketone prepared from an acid chloride. In the presence of a suitable catalyst, such as colloidal silver, platinum, or copper, the diazoketone produces a ketene that reacts with an amine, leading to the formation of the corresponding amide. For example, Kametani prepared the amide **31** for the synthesis of dauricine **28**. The diazoketone **29** prepared from acid chloride **30** was reacted with homoveratrylamine

Chart 1-13.

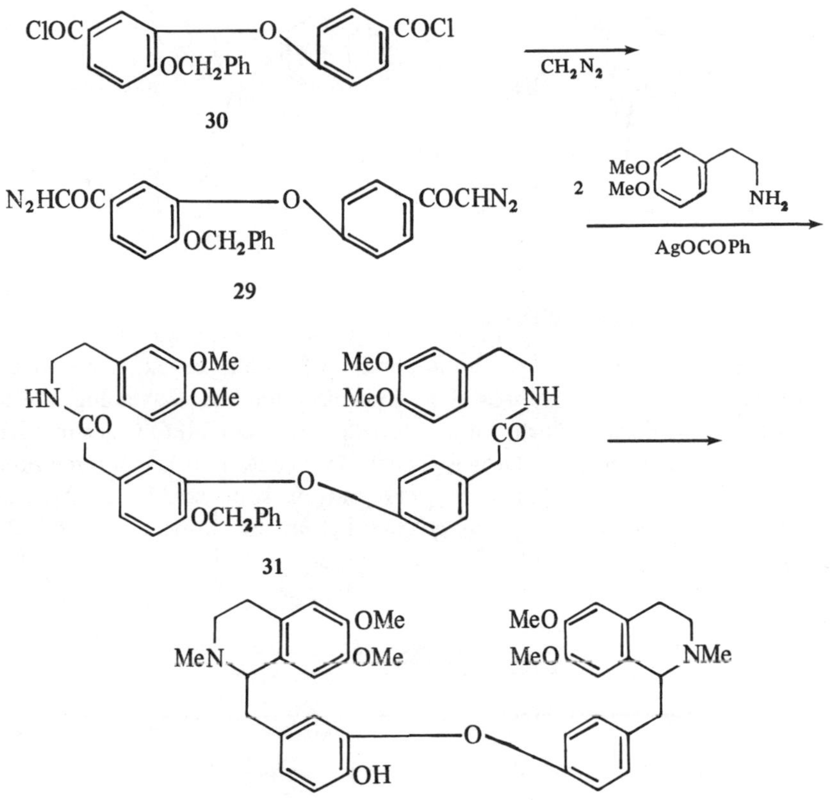

ClOC O COCl
OCH$_2$Ph
$\xrightarrow{CH_2N_2}$
30

N$_2$HCOC O COCHN$_2$
OCH$_2$Ph
2 MeO MeO NH$_2$
AgOCOPh
29

HN OMe OMe MeO MeO NH
CO CO
O
OCH$_2$Ph
31

MeN OMe OMe MeO MeO NMe
O
OH
28 Dauricine

to give the corresonding amide **31**.[25] This type of a double Bischler-Napieralski reaction is used for a synthesis of *O*-methyldauricine[25a] and cepharanthine.[25b]

Other Amide Syntheses. Beckmann rearrangement of an oxime can also give an amide, and, in many instances, further cyclization to isoquinoline is known to occur during the course of the foregoing reaction.[12] Furthermore, a mixture of arylethylamine and carboxylic acid gives the corresponding amide on heating to 170 to 180°.[26] On the other hand, Ritter and Murphy[27] obtained the 3,4-dihydro-3-methylisoquinoline from the nitrile and an allylbenzene in the presence of concentrated sulfuric acid, but not the amide. When the amine and/or carboxylic acid are susceptible to acid, base, or heat, a mixture of the amine and carboxylic acid is treated with dicyclohexylcarbodiimide as a condensation reagent in methylene dichloride at room temperature in order to obtain the corresponding amide (Chart 1-13A).[27b]

Chart 1-13A.

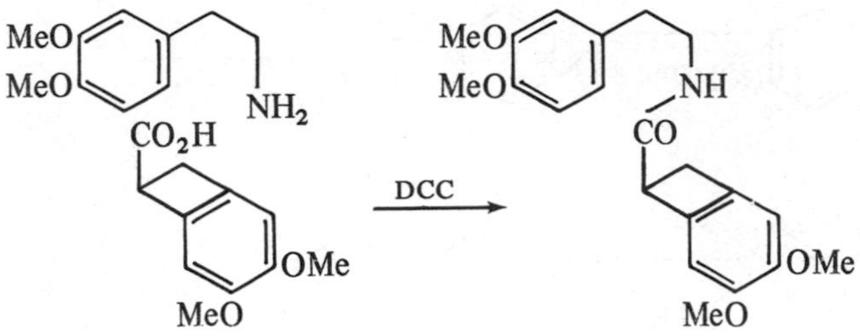

MeO MeO NH_2 CO_2H OMe MeO —DCC→ MeO MeO NH CO OMe MeO

Direction of Ring Closure (Chart 1-14)

Cyclization of *m*-methoxy-β-phenethylamide **32** would be expected to give either 6-methoxy- or 8-methoxy-3,4-dihydroisoquinoline depending on the direction of ring closure. When the *para* position to the methoxyl group has no substituent, cyclization preferentially occurs at the *para* to give a 6-methoxy-isoquinoline derivative **33**. When the *para* position is blocked, cyclization will proceed to the *ortho* position to the methoxyl group. For instance, *N*-acetyl-2,5-dimethoxyphenethylamine **34** was readily converted to 3,4-dihydro-5,8-dimethoxy-1-methylisoquinoline **35**[27a] (Chart 1-15).

Chart 1-14.

MeO NH CO R **32** → MeO N R **33** ⇸ N OMe R

If both available positions are activated to a similar extent, a mixture of both cyclized products is obtained, as in the case of cyclization of *N*-(3-benzyloxy-4,5-dimethoxyphenethyl)-4-benzyloxy-3-methoxyphenylacetamide **36** to the 8-benzyloxy-6,7-dimethoxy- **37** and 6-benzyloxy-7,8-dimethoxy-3,4-dihydroisoquinoline derivative **38**.[28] Moreover, the cyclic bisamide **39** also gave stebisimine

Chart 1-15.

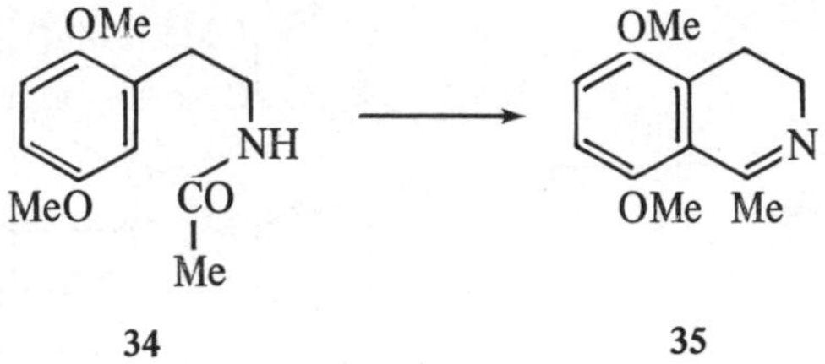

OMe
NH
MeO
CO
Me
OMe
N
OMe Me

34 35

Chart 1-16.

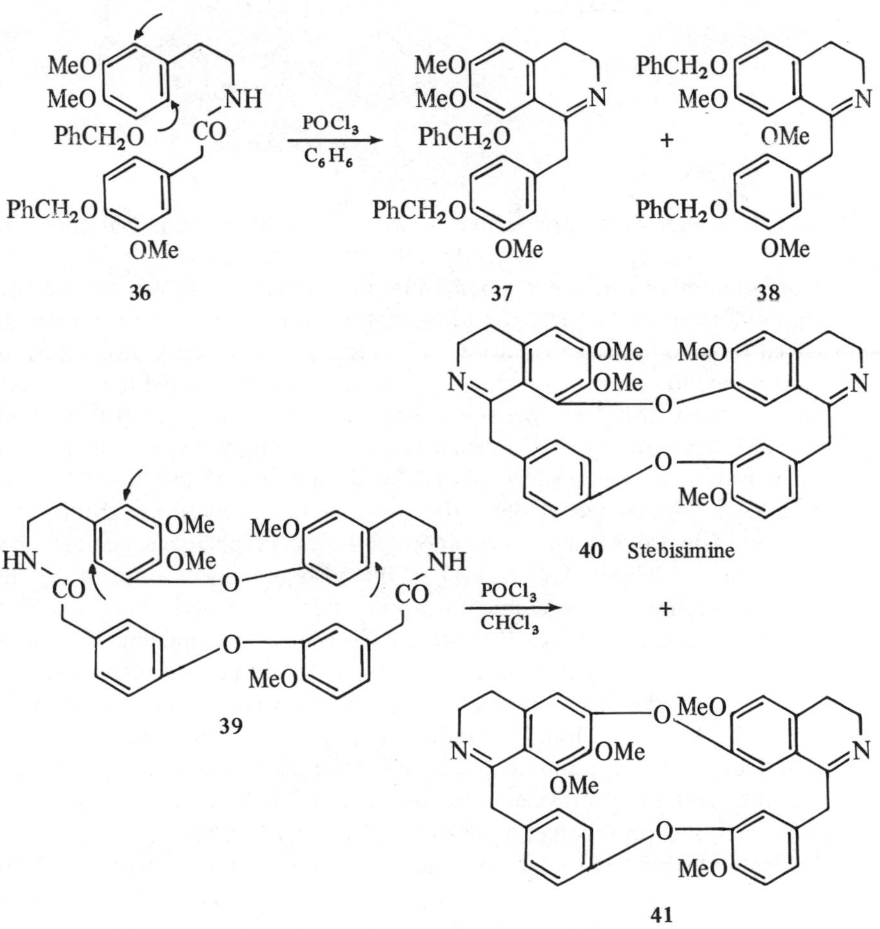

MeO
MeO
NH
PhCH$_2$O
CO
PhCH$_2$O
OMe

$\xrightarrow[C_6H_6]{POCl_3}$

MeO
MeO
N
PhCH$_2$O
PhCH$_2$O
OMe

+

PhCH$_2$O
MeO
N
OMe
PhCH$_2$O
OMe

36 37 38

OMe
OMe
MeO
N
N
O
O
MeO

40 Stebisimine

+

OMe
OMe
MeO
HN
NH
CO
O
CO
O
MeO

$\xrightarrow[CHCl_3]{POCl_3}$

39

O
MeO
N
N
OMe
OMe
O
MeO

41

Chart 1-16A.

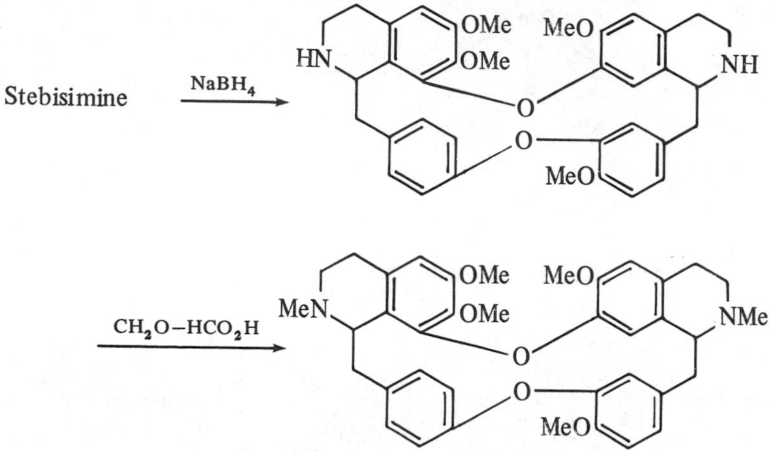

40 and its isomer 41 on cyclization[29] (Chart 1-16). Stebisimine was converted into obaberine by reduction and methylation[29a] (Chart 1-16A).

In an attempt to synthesize a berberine, the formamide 42 was treated with phosphoryl chloride to yield the bromine-free compound 43 rather than the expected bromodihydroberberine 43a. This result is a remarkable instance of the preferred direction of ring closure, a bromine atom being removed to allow cyclization to proceed at the *para* position to the electron-releasing group.[30] However, Kametani has assumed that the replacement of methoxyl by hydroxyl offsets the inactivation of the nucleus caused by the I-effect of the bromine atom, leading to the cyclization at the *ortho* position to the hydroxyl group. Thus N-(2-bromo-5-hydroxy-4-methoxyphenethyl)-4-methoxyphenylacetamide gave the 5-bromo-3,4-dihydro-8-hydroxy-7-methoxyisoquinoline derivative by the action of phosphoryl chloride in chloroform, which was converted into petaline 44 by the standard method. Recently, 8-oxygenated isoquinoline derivatives were obtained in the cyclization of *trans-N*-[2-(3-methoxyphenyl)cyclohexyl]-benzamide, but the main product was 6-methoxyisoquinoline derivative[30a] (Chart 1-17). Moreover, Tani[30b] achieved a cyclization of the formamide to the expected bromodihydroprotoberberine and succeeded in a synthesis of cheilanthifoline. This route provides a useful method for the total synthesis of the 9,10-disubstituted protoberberine alkaloids (Chart 1-17A).

This method is applied to the synthesis of caseadine-type compounds.[30c] Moreover, this problem was circumvented by using an ethoxycarbonylamino-β-phenethylamide in order to activate the *para* position and thus to effect the required

Chart 1-17.

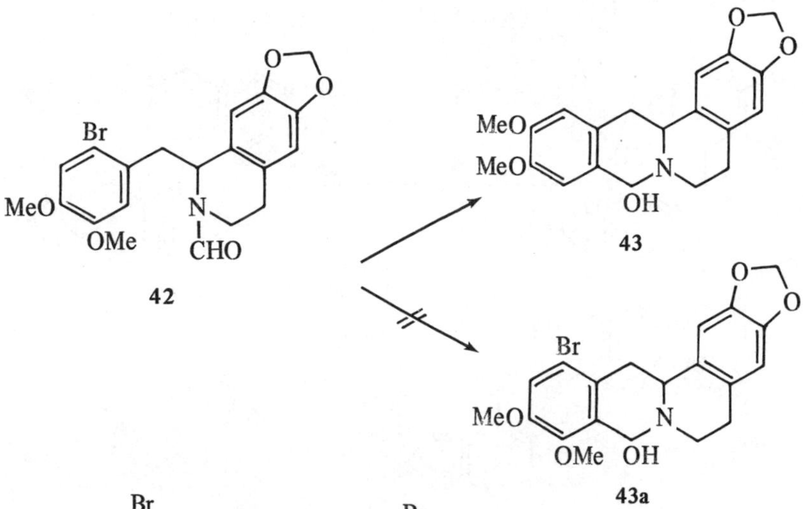

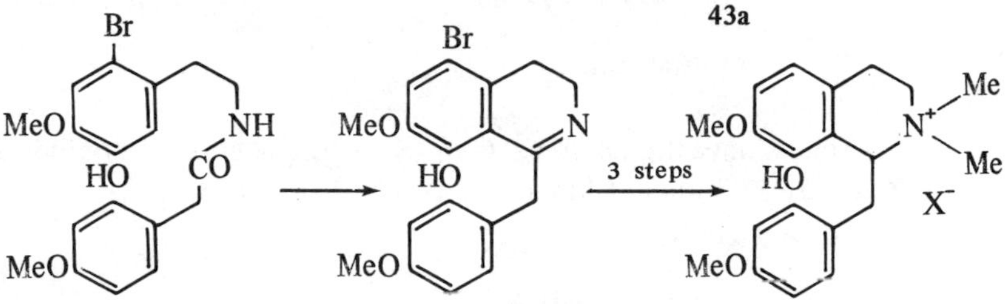

Br
MeO
OMe
N
CHO
O
O

42

MeO
MeO
N
OH
O
O

43

Br
MeO
OMe
OH
N
O
O

43a

Br
MeO
HO
NH
CO
MeO

Br
MeO
HO
N
MeO

3 steps

MeO
HO
N+
Me
Me
X−
MeO

44 Petaline

H
MeO
H
NH
CO

H
MeO
H
N

H
H
N
MeO

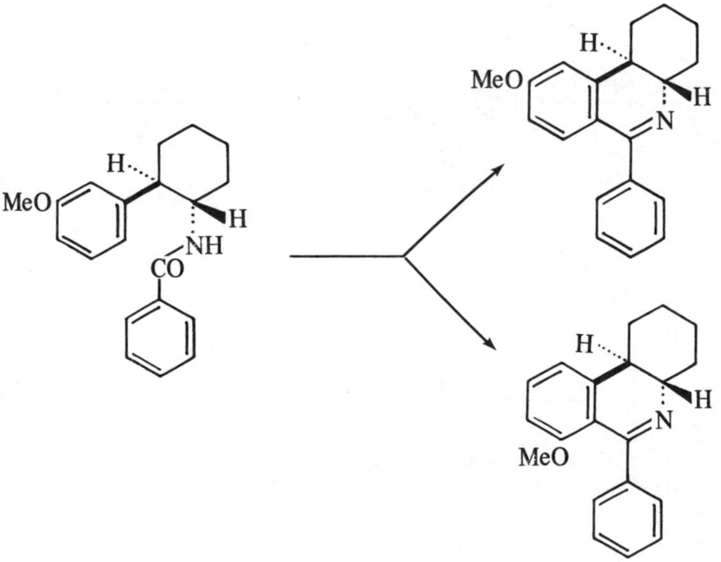

Chart 1-17A.

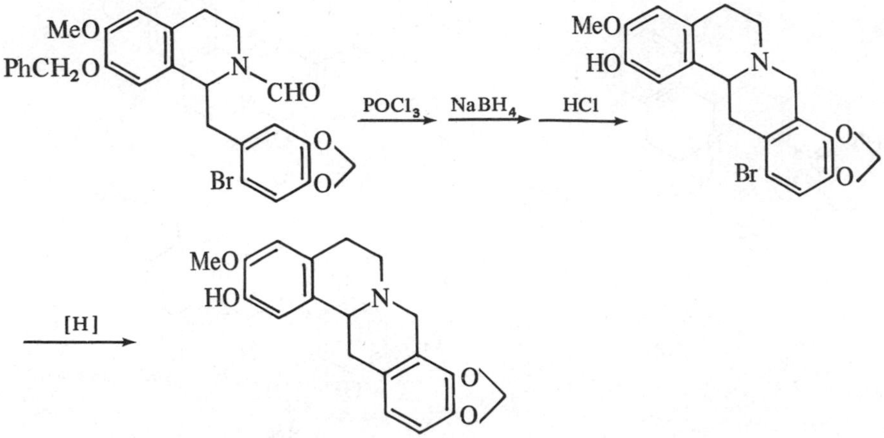

Cheilanthifoline

cyclization reaction. Conventional steps then led to the phenolic isoquinoline, which gave (±)-cularine[30d] under Ullmann reaction (Chart 1-17B).

Chart 1-17B.

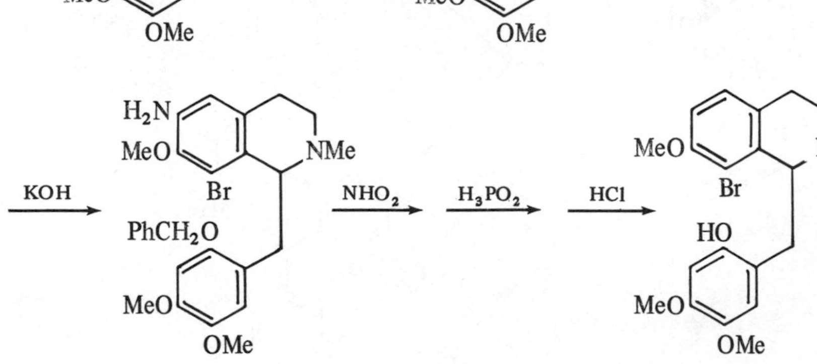

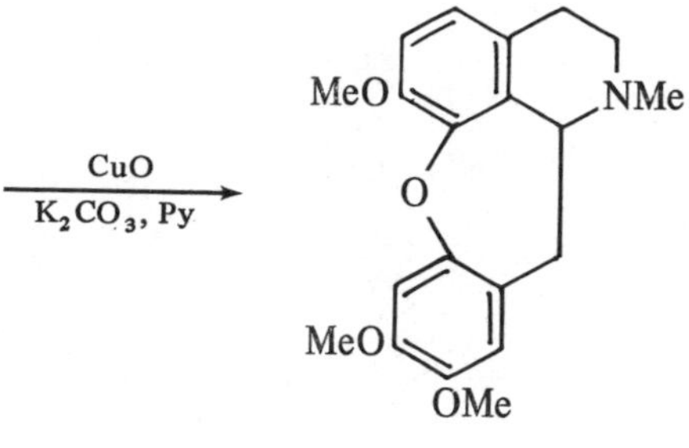

CuO
K_2CO_3, Py

MeO
NMe
O
MeO
OMe

Cularine

Position of the Double Bond

In most Bischler-Napieralski reactions, 3,4-dihydroisoquinolines are obtained; the double bond is formed between the carbonyl carbon and the nitrogen atom in the cyclodehydration. The presence of an active methylene group at the C_1-position in the compounds analogous to **45b** allows the double bond to become exocyclic in the free base, as in 1-benzal-1,2,3,4-tetrahydro-2-methylisoquinoline **45a**, whose color is yellow because of the extended conjugation. Cava synthesized an aporphine **46** by photooxidation of this type of isoquinoline **47**[31] (Chart 1-18).

Chart 1-18.

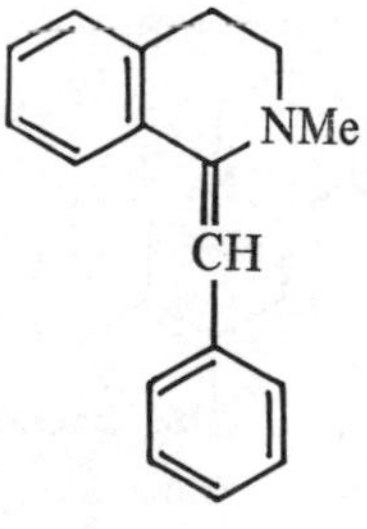

NMe
CH
45a

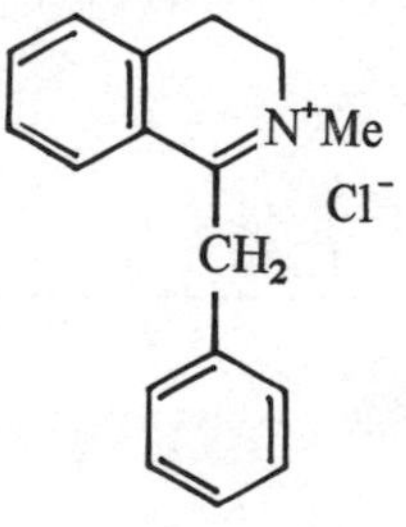

N^+Me
Cl^-
CH_2
45b

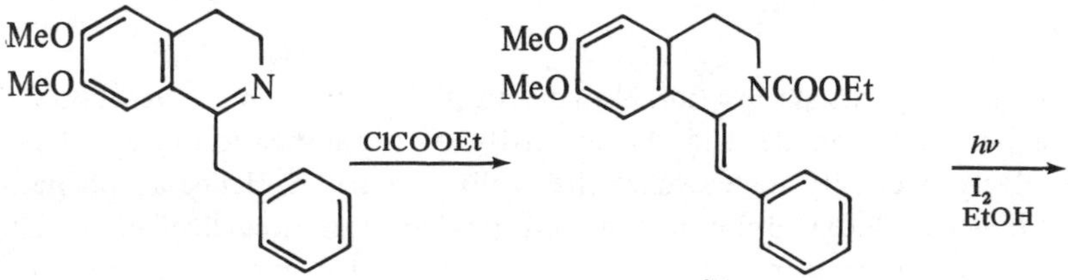

MeO
MeO
N

ClCOOEt

MeO
MeO
NCOOEt
47

$h\nu$
I_2
EtOH

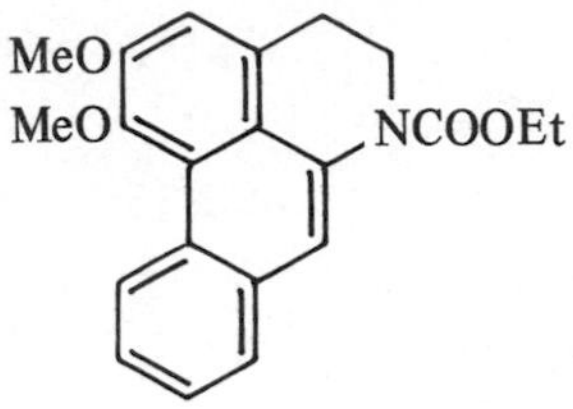

MeO
MeO
NCOOEt

46

Ninomiya also synthesized a protoberberine alkaloid, xylopinine, by a photolysis of the exo-methylene compound derived from 3,4-dihydro-6,7-dimethoxy-1-methylisoquinoline and veratric acid[31a] (Chart 1-18A).

Chart 1-18A.

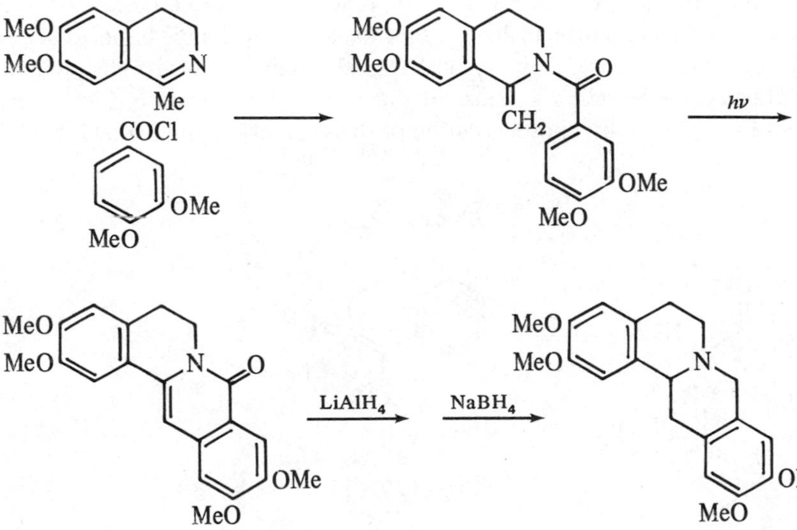

MeO MeO N Me COCl OMe MeO → MeO MeO N O CH₂ OMe MeO $\xrightarrow{h\nu}$

MeO MeO N O OMe MeO $\xrightarrow{LiAlH_4}$ $\xrightarrow{NaBH_4}$ MeO MeO N OMe MeO

Xylopinine

Peculiarities

The Bischler-Napieralski reaction is an electrophilic attack on the benzenoid ring of the β-phenethylamine, and the reactivity of the aromatic nucleus depends upon electron density increased at the cyclized position. Hence a β-phenethylamine that has alkoxyl group at the *meta* position can be cyclized easily, and it is clear that an electron-attracting group such as nitro group will inhibit this reaction. Nevertheless, the 3,4-dihydroisoquinoline derivative[32] **48** was prepared in 13% of the yield (Chart 1-19).

Chart 1-19.

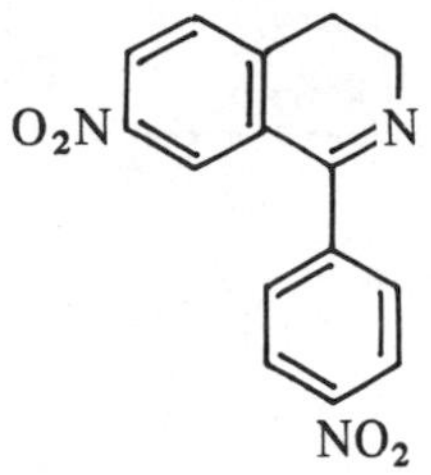

O_2N N NO_2

48

On the other hand, it is more obvious that an electron releasing group has an influence on the cyclization in the synthesis of 3,4-dihydroisoquinolines. For example, the yield of 3,4-dihydro-1-methyl-6,7-methylenedioxyisoquinoline[33] **49** is better than that of 3,4-dihydro-1-methylisoquinoline[34] **50** under the same conditions (Chart 1-20).

Chart 1-20.

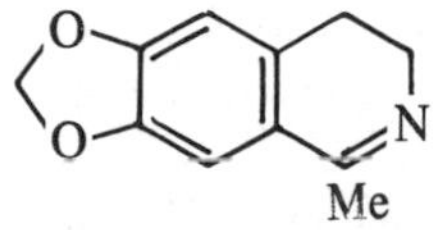

O O N Me

49

N Me

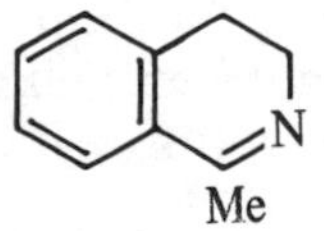

50

The influence of the other groups besides an alkoxyl group was examined in some cases. For example, 1-benzyl-6-ethoxycarbonylamino-3,4-dihydro-7-methoxyisoquinoline **51** has been prepared in good yield,[35] but the cyclization of *N*-(4-methoxyphenethyl)phenylacetamide was very difficult. Under the special condition using phosphorus pentoxide absorbed on Celite, cyclization of **52** gave the 3,4-dihydroisoquinoline in poor yield[35a] (Chart 1-21).

In general, the Bischler-Napieralski reaction is carried out by heating the appropriate amide with a dehydrating reagent in the presence of an inert and anhydrous solvent, such as chloroform, acetonitrile, benzene, toluene, xylene, nitrobenzene, or tetralin according to its boiling point. Cyclization is often carried out in the presence of an excess of phosphoryl chloride without solvent.

Chart 1-21.

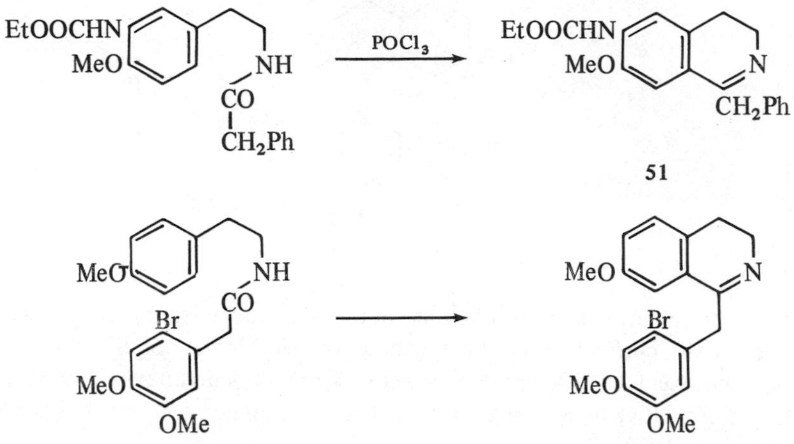

$$EtOOCHN,\ MeO\text{-}C_6H_2\text{-}CH_2CH_2NH\text{-}CO\text{-}CH_2Ph \xrightarrow{POCl_3} \mathbf{51}\ (1\text{-}CH_2Ph\text{-}3,4\text{-dihydroisoquinoline})$$

51

52

Phosphoryl chloride is the most popular dehydrating agent, but phosphorus pentoxide and pentachloride are also important in specific cases. Furthermore, various reagents such as polyphosphoric acid and its ester have been found to be useful.[35b]

Brossi and Teitel reported an improved synthesis of the phenolic benzylisoquinolines without protection of the hydroxy-group, in which Bischler-Napieralski reaction of the appropriate amide is achieved with phosphoryl chloride in chloroform or acetonitrile. In this fashion, coclaurine is obtained in 61% of the overall yield from the phenolic amine, isococlaurine and reticuline have been prepared in this way[35c] (Chart 1-21A).

Application of Bischler-Napieralski Reaction to the Total Synthesis of the Isoquinoline Alkaloids

Cherylline. Resolution of the β-(*p*-benzyloxyphenyl)homoveratrylamine with (−)-diacetone-5-keto-L-gulonic acid gave the (2R)-(+)- and (2S)-(−)-L-gulonate salts, which were transformed into the diastereomeric hydrobromides. The latter were formylated and subjected to Bischler-Napieralski reaction to give the 3,4-dihydroisoquinolines, which were debenzylated to yield the (4R)-(+)- and (4S)-(−)-dihydroisoquinoline hydrochlorides. The latter was selectively *O*-demethylated by 48% hydrobromic acid, quaternized with methyl iodide, and reduced with sodium borohydride to give (−)-cherylline[35d] (Chart 1-21B).

Phenylisoquinoline Alkaloids. (+)-Cryptostyline I, II, and III have been synthesized by the Bischler-Napieralski reaction and sodium borohydride reduction, followed by a resolution using (−)-diacetone-2-keto-L-gulonic acid and reductive *N*-methylation[35e] (Chart 1-21C).

Chart 1-21A.

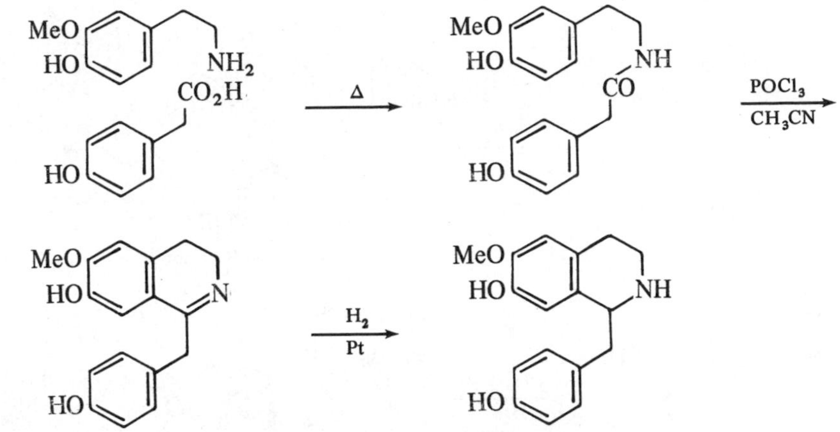

MeO, HO, NH_2, CO_2H, HO → (Δ) → MeO, HO, NH, CO, HO → ($POCl_3$, CH_3CN)

MeO, HO, N, HO → (H_2, Pt) → MeO, HO, NH, HO

Coclaurine

Chart 1-21B.

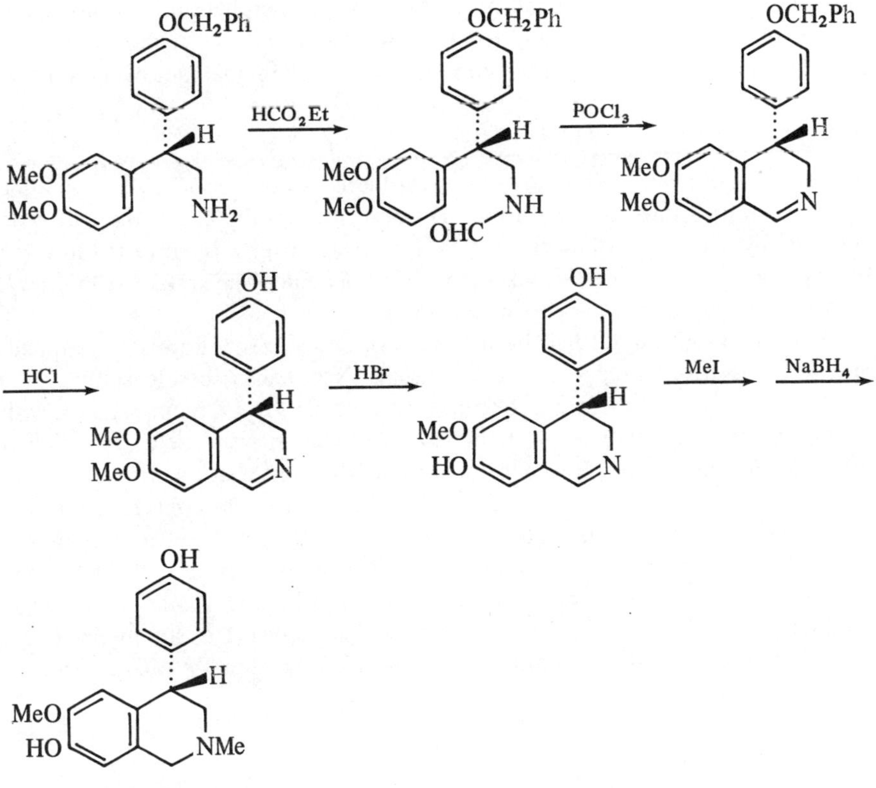

OCH_2Ph, H, MeO, MeO, NH_2 → (HCO_2Et) → OCH_2Ph, H, MeO, MeO, NH, OHC → ($POCl_3$) → OCH_2Ph, H, MeO, MeO, N

→ (HCl) → OH, H, MeO, MeO, N → (HBr) → OH, H, MeO, HO, N → (MeI) → ($NaBH_4$)

OH, H, MeO, HO, NMe

Cherylline

Chart 1-21C.

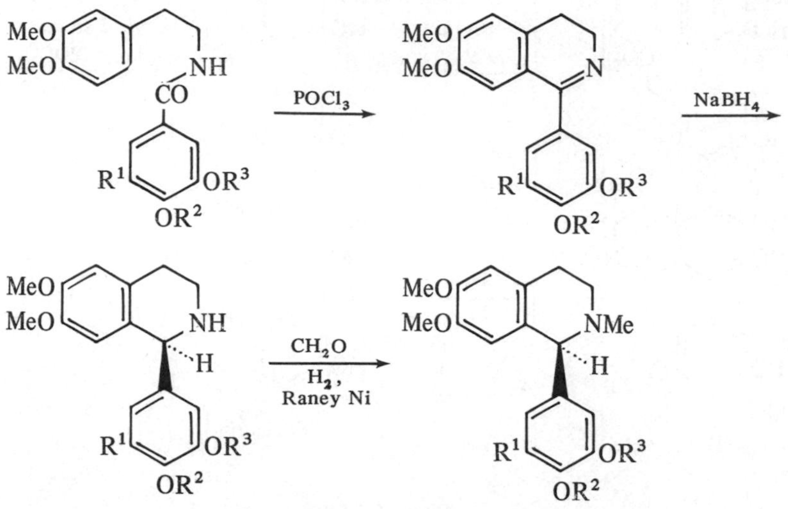

MeO MeO NH CO R[1] OR[3] OR[2] —$POCl_3$→ MeO MeO N R[1] OR[3] OR[2] —$NaBH_4$→

MeO MeO NH H R[1] OR[3] OR[2] —CH_2O / H_2, Raney Ni→ MeO MeO NMe H R[1] OR[3] OR[2]

Emetine. The lactone, obtained by a Mannich-type condensation of homoveratrylamine with keto triester and formalin, was subjected to Bischler-Napieralski reaction to give the benzoquinolizidine derivative. The remainder of the synthesis followed the usual way to afford the tricyclic ester, which had been converted into emetine[35f] (Chart 1-21D).

Bisbenzylisoquinoline Alkaloids. In a synthesis of the bisbenzylisoquinoline alkaloids, tetrandrine, isotetrandrine, and phaeanthine, the bases A and B were prepared by successive Ullmann reactions, and the latter was converted into the macrocyclic lactam, which was subjected to a Bischler-Napieralski reaction, *N*-methylation, and reduction[35g] (Chart 1-21E).

For example, *O*-methylthalicberine was synthesized as follow. The norlaudanidine derivative, prepared via a Bischler-Napieralski condensation, was resolved, and the (S)-(+)-chiral substance was subjected to Ullmann reaction with *N*-butoxycarbonyl-3-hydroxy-4-methoxyphenethylamine to yield a biphenyl ether. Hydrogenolysis, followed by a second Ullmann reaction with methyl *p*-bromophenylacetate, gave the *bis*-diphenyl ether, which was converted into the *p*-nitrophenyl ester, and the latter, after removal of the *t*-butoxycarbonyl group, was cyclized to a macrocyclic lactam. Bischler-Napieralski reaction of this followed by sodium borohydride reduction gave *O*-methyl-*N*-northalicberine as a single product, indicating that the latter reaction had proceeded stereoselectively. *N*-Methylation gave the naturally occurring alkaloid *O*-methylthalicberine[35h] (Chart 1-21F).

Chart 1-21D.

MeO, MeO, NH_2, MeO_2C, CH_2O, CO_2Me, Me, O, CO_2Me

t-BuOH, Δ

MeO, MeO, N, O, CO_2Me, Me, O, MeO_2C

$POCl_3$

MeO, MeO, N^+, Cl^-, CO_2Me, Me, O, CO_2Me

H_2 → HCl → MeOH, HCl

MeO, MeO, N, H, H, Me, H, O, CO_2Me

HS SH, BF_3–Et_2O → R–Ni, Δ

MeO, MeO, N, H, H, Et, H, CO_2Me

steps

MeO, MeO, N, H, H, Et, H, H, HN, OMe, OMe

Emetine

Chart 1-21E.

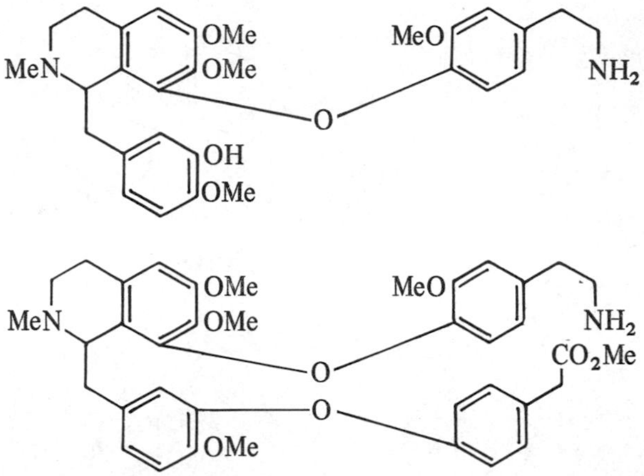

Tubocurarine iodide and its isomers have been prepared by a double Bischler-Napieralski reaction, followed by debenzylation, Ullmann reaction and quaternization[35i] (Chart 1-21G).

The occurrence of the dibenzo-*p*-dioxan unit in (±)-*O*-methyltiliacorine presented problems not encountered in the synthesis of other dimeric benzylisoquinoline alkaloids. Attempts to effect a double Bischler-Napieralski reaction of the bisamide under vigorous conditions were synthetically very inefficient, but milder conditions gave the monocyclized product. Possibly, as a result of less

Chart 1-21G.

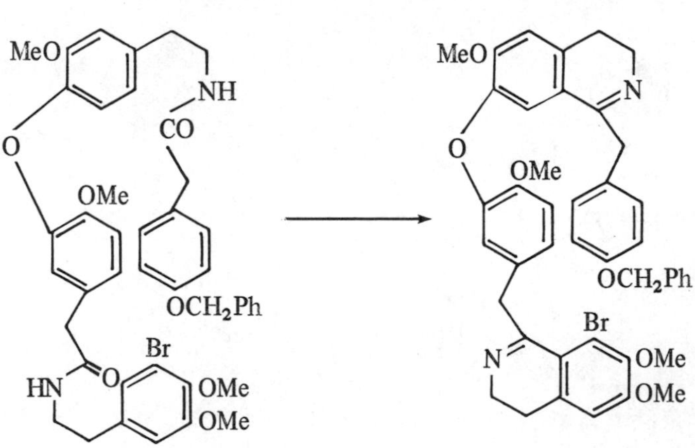

Chart 1-21F.

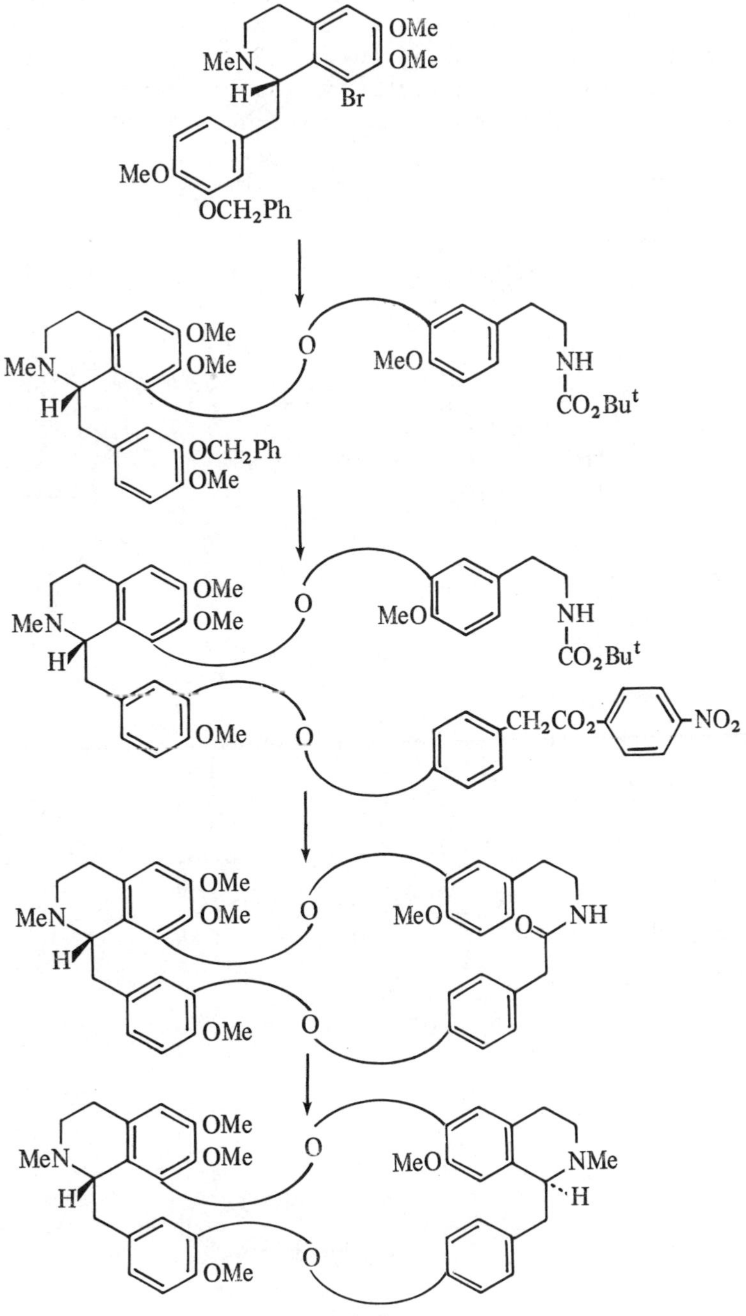

MeN, H, OMe, OMe, Br, MeO, OCH_2Ph

MeN, H, OMe, OMe, O, MeO, NH, CO_2Bu^t, OCH_2Ph, OMe

MeN, H, OMe, OMe, O, MeO, NH, CO_2Bu^t, OMe, O, CH_2CO_2, NO_2

MeN, H, OMe, OMe, O, MeO, O, NH, OMe, O

MeN, H, OMe, OMe, O, MeO, NMe, H, OMe, O

O-Methylthalicberine

molecular flexibility in the monocyclized compound in comparison to a starting bisamide, the second Bischler-Napieralski cyclization could now be effected under forcing conditions to yield the bicyclized product. Reduction and methylation yielded a diastereoisomeric mixture, from which (±)-*O*-methyltiliacorine could be isolated as the major product, thus completing the synthesis. Other alkaloids

Chart 1-21H.

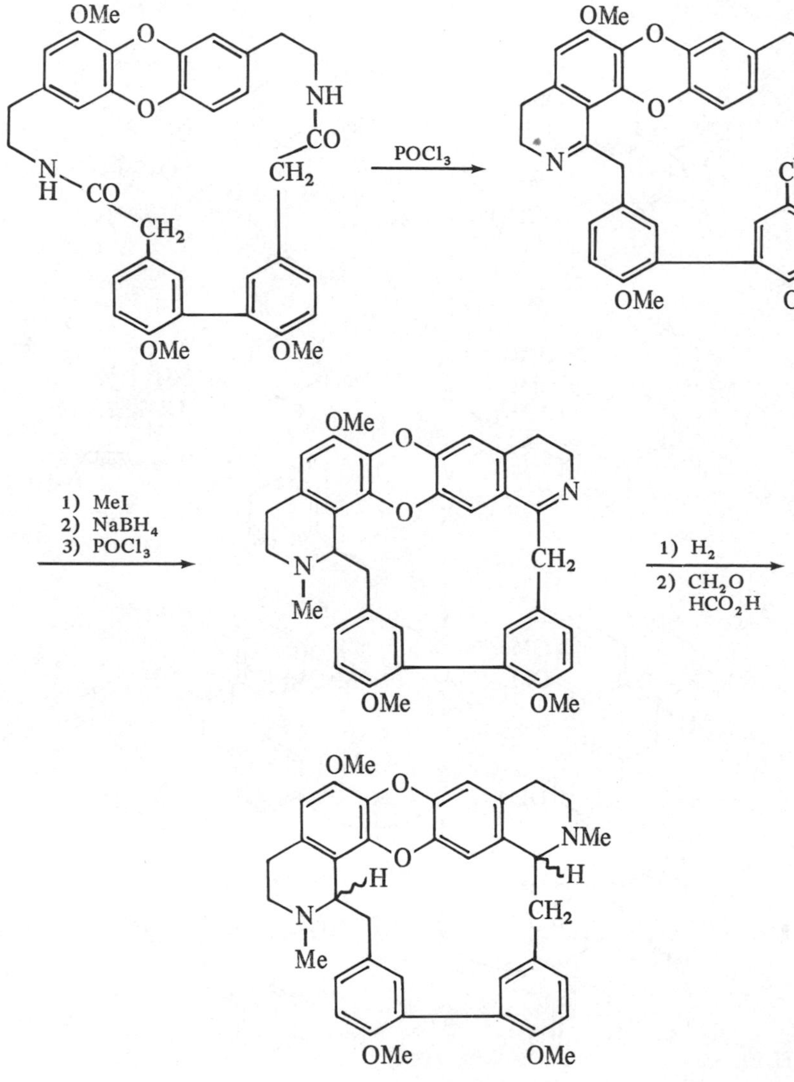

O-Methyltiliacorine

Chart 1-21I.

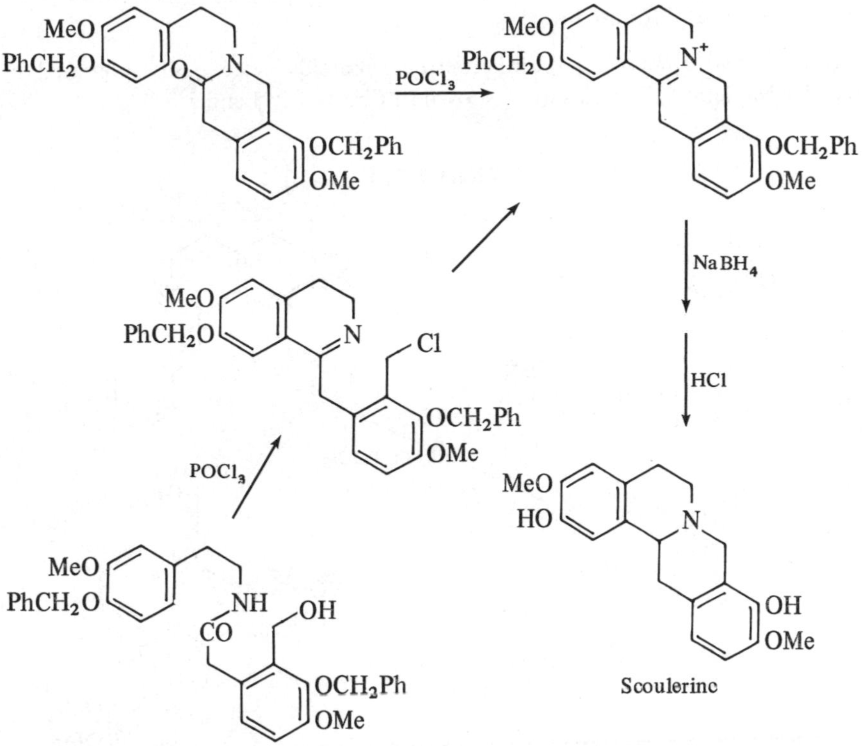

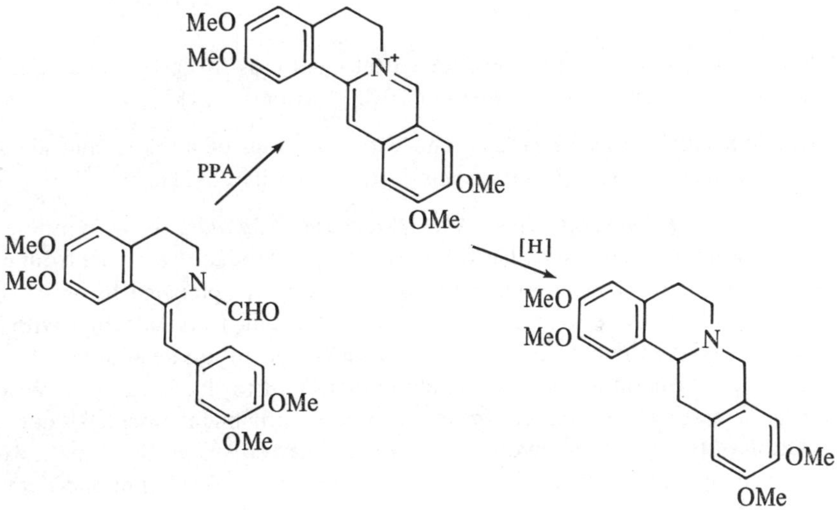

MeO
PhCH$_2$O
N
O
OCH$_2$Ph
OMe
POCl$_3$
MeO
PhCH$_2$O
N$^+$
OCH$_2$Ph
OMe
NaBH$_4$
HCl
MeO
PhCH$_2$O
N
Cl
OCH$_2$Ph
OMe
POCl$_3$
MeO
HO
N
OH
OMe
Scoulerine
MeO
PhCH$_2$O
NH
CO
OH
OCH$_2$Ph
OMe

MeO
MeO
N$^+$
OMe
OMe
PPA
[H]
MeO
MeO
N
CHO
OMe
OMe
MeO
MeO
N
OMe
OMe
Xylopinine

containing the dibenzo-*p*-dioxan system, which have been synthesized, were briefly reviewed[35j] (Chart 1-21H).

Protoberberine Alkaloids. Protoberberine alkaloids were easily prepared by Bischler-Napieralski ring closure as shown in Charts 1-21I and J.[35k-35n]

Chart 1-21J.

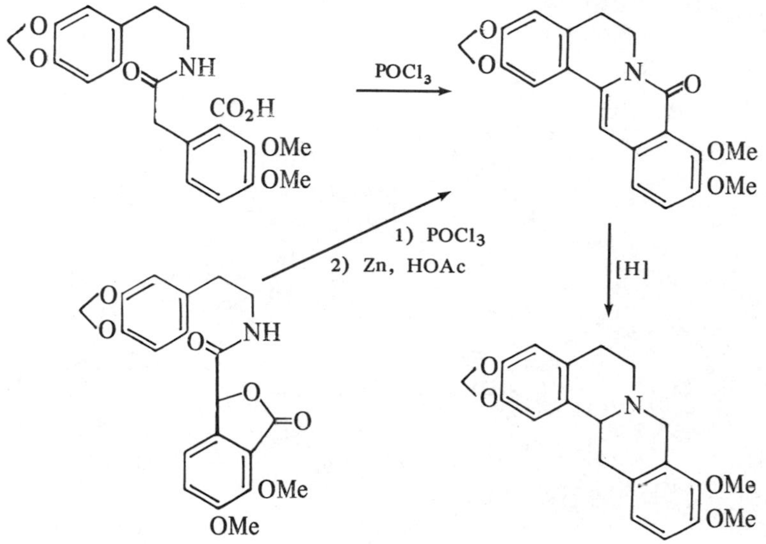

Canadine

Phthalideisoquinoline Alkaloids. A total synthesis of gnoscopine and (±)-narcotine, was reported by Kerekes in 1971[35o] (Chart 1-21K).

Benzophenanthridine Alkaloids. The total synthesis of nitidine chloride along the route developed by Robinson was reported in Chart 1-21L.[35p]

Tazettine and Haemanthidine, Amaryllidaceae *Alkaloids.* New syntheses of tazettine and haemanthidine has been reported by Tsuda.[35q] Starting from piperonyl cyanide, the key intermediate, keto lactam, was prepared by three steps, via the cyanopyruvate. Treatment of this with *N*-bromoacetamide, then with base gave the epoxyketone, whose treatment with boron trifluoride-etherate afforded the lactam. Lithium aluminum hydride reduction of the lactam gave stereospecifically a diol, which was converted into the formamide. Bischler-Napieralski reaction of this gave the 1-hydroxyisoquinoline derivative, which on methylation followed by alkaline treatment gave dihydrotazettine. Tosylation and detosylation of this product afforded tazettine. The oxidation of the 1-hydroxyisoquino-

Chart 1-21K

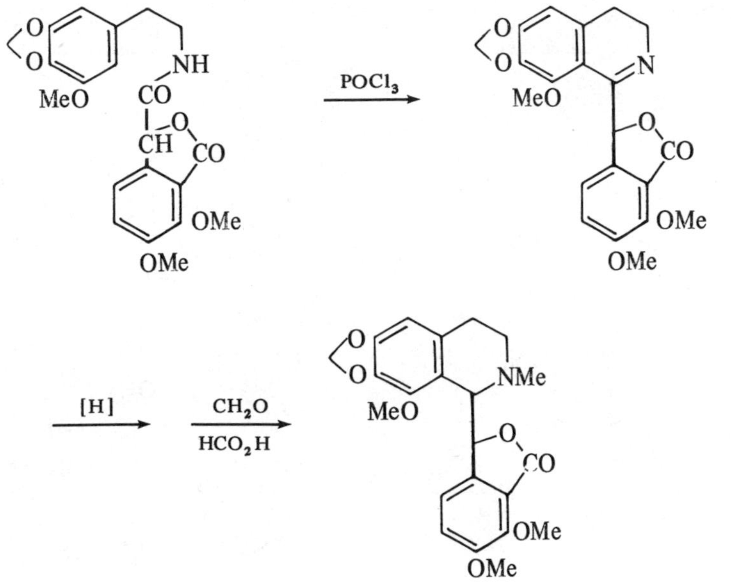
O
O
MeO
NH
CO
O
CH
CO
OMe
OMe
POCl$_3$
N
[H]
CH$_2$O
HCO$_2$H
NMe

Gnoscopine

Chart 1-21L.

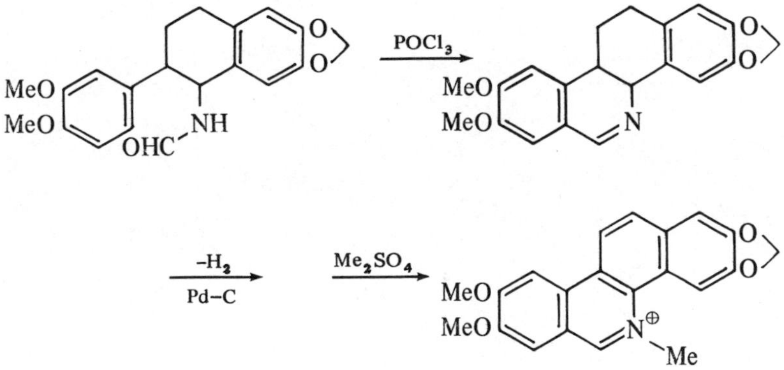
MeO
MeO
NH
OHC
O
O
POCl$_3$
N
-H$_2$
Pd–C
Me$_2$SO$_4$
N⊕
Me

Nitidine

line, followed by alkaline treatment and dehydration gave the lactam, which was reduced with lithium aluminum hydride to yield haemanthidine (Charts 1-21M and N).

Chart 1-21M.

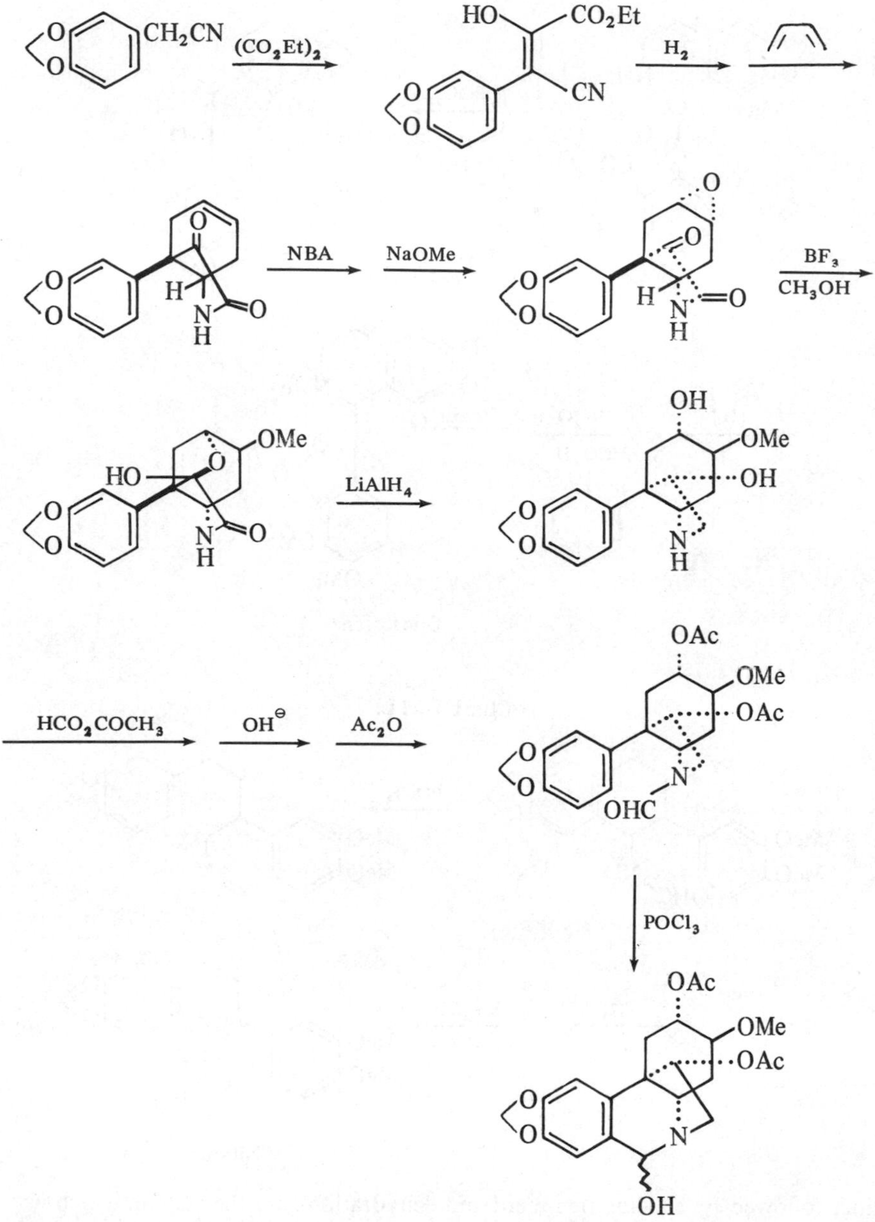

CH_2CN $\xrightarrow{(CO_2Et)_2}$ HO, CO_2Et, CN $\xrightarrow{H_2}$ $\longrightarrow$

$\xrightarrow{NBA}$ $\xrightarrow{NaOMe}$ $\xrightarrow[CH_3OH]{BF_3}$

HO, OMe, O, N, H, O $\xrightarrow{LiAlH_4}$ OH, OMe, OH, N, H

$\xrightarrow{HCO_2COCH_3}$ $\xrightarrow{OH^{\ominus}}$ $\xrightarrow{Ac_2O}$ OAc, OMe, OAc, N, OHC

$\downarrow POCl_3$

OAc, OMe, OAc, N, OH

Chart 1-21N.

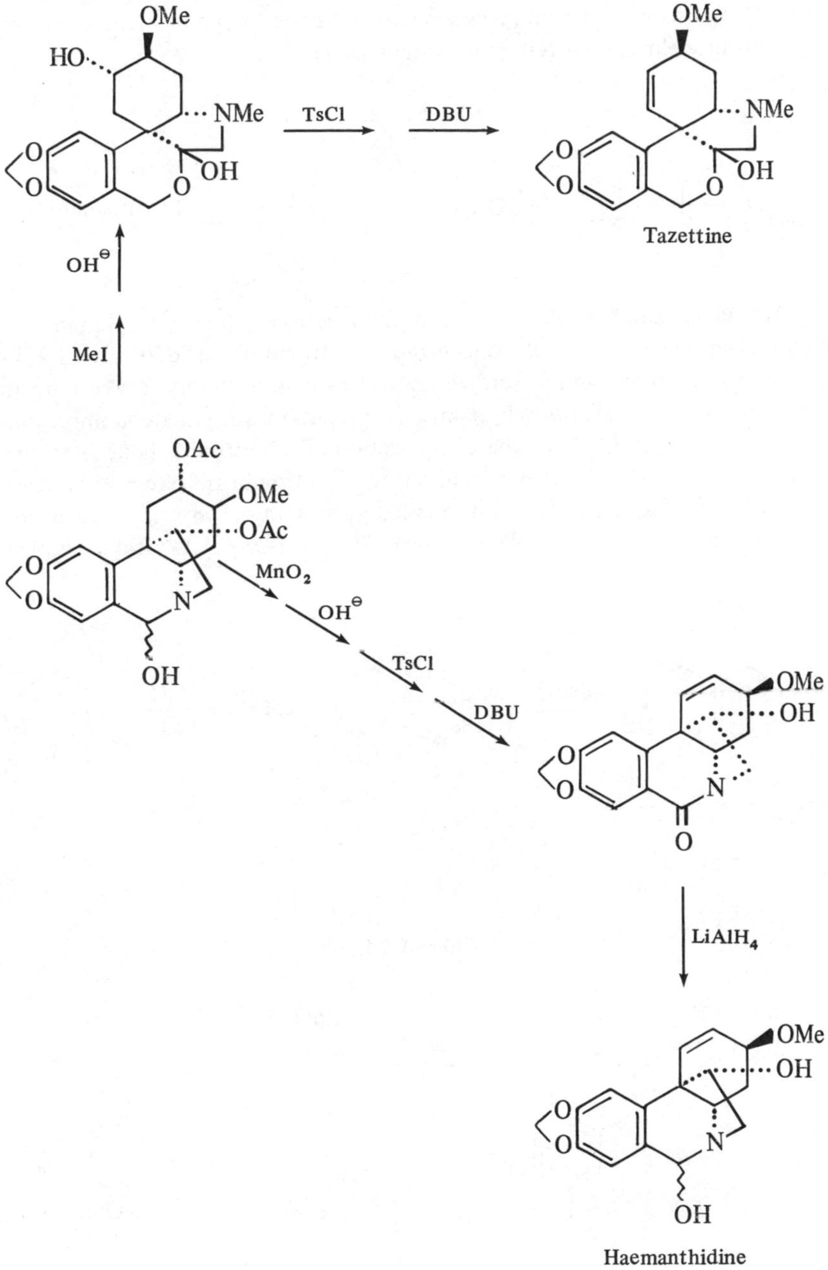

Tazettine

Haemanthidine

Pictet-Spengler Reaction[36] *(Chart 1-22)*[36a]

The Pictet-Spengler reaction, which is one of the special cases of the Mannich reaction,[37] consists in the condensation of a β-arylethylamine with a carbonyl compound to yield 1,2,3,4-tetrahydroisoquinoline **17**.

Chart 1-22.

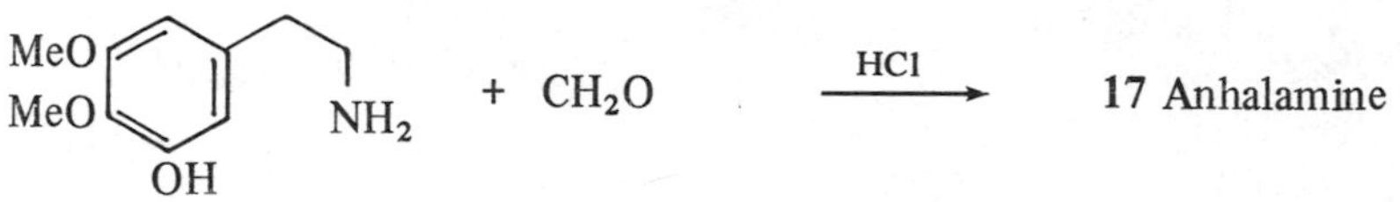

MeO MeO OH NH_2 + CH_2O —HCl→ **17** Anhalamine

In 1911, Pictet and Spengler[38] reported the reaction of β-phenethylamine with a methylal in the presence of concentrated hydrochloric acid to give 1,2,3,4-tetrahydroisoquinoline and described an experimental theory concerning the origin of isoquinoline alkaloids in plants. This reaction was extended immediately by Decker and Becker[39] to the condensation of substituted phenethylamines with various aldehydes. Decker carried out the reaction in two steps as indicated by the following equation. The intermediate azomethine **53** was often formed before the addition of the condensing agent (Chart 1-23). A typical example of

Chart 1-23.

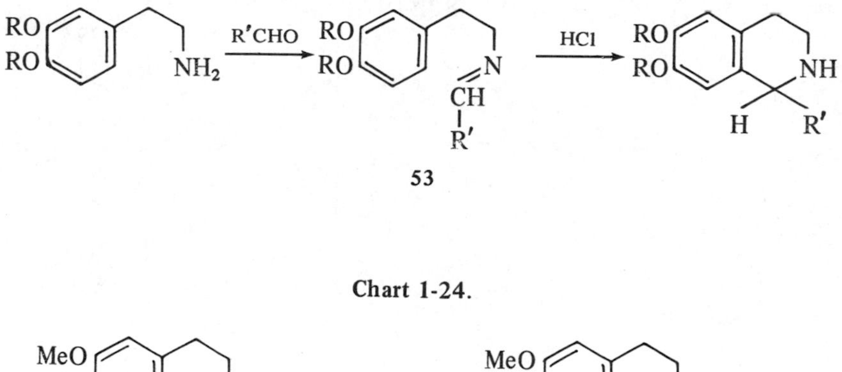

RO RO NH_2 —R′CHO→ RO RO N=CH–R′ (**53**) —HCl→ RO RO NH H R′

Chart 1-24.

MeO MeO NH OMe OMe —CH_2O, HCl→ MeO MeO N OMe OMe

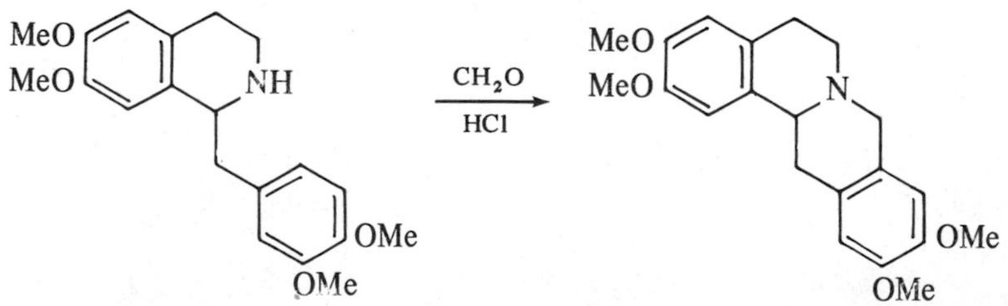

54 Xylopinine

further use is the synthesis of tetrahydro-2,3,10,11-tetramethoxy-5,6,13,13a-8H-dibenzo[a,g] quinolizine[40] **54** (Chart 1-24). Similarly, benzyldiscretine was synthesized from an appropriate isoquinoline and subjected to an optical resolution. The resulting (−)-base on debenzylation gave (−)-discretine[40a] (Chart 1-24A).

Chart 1-24A.

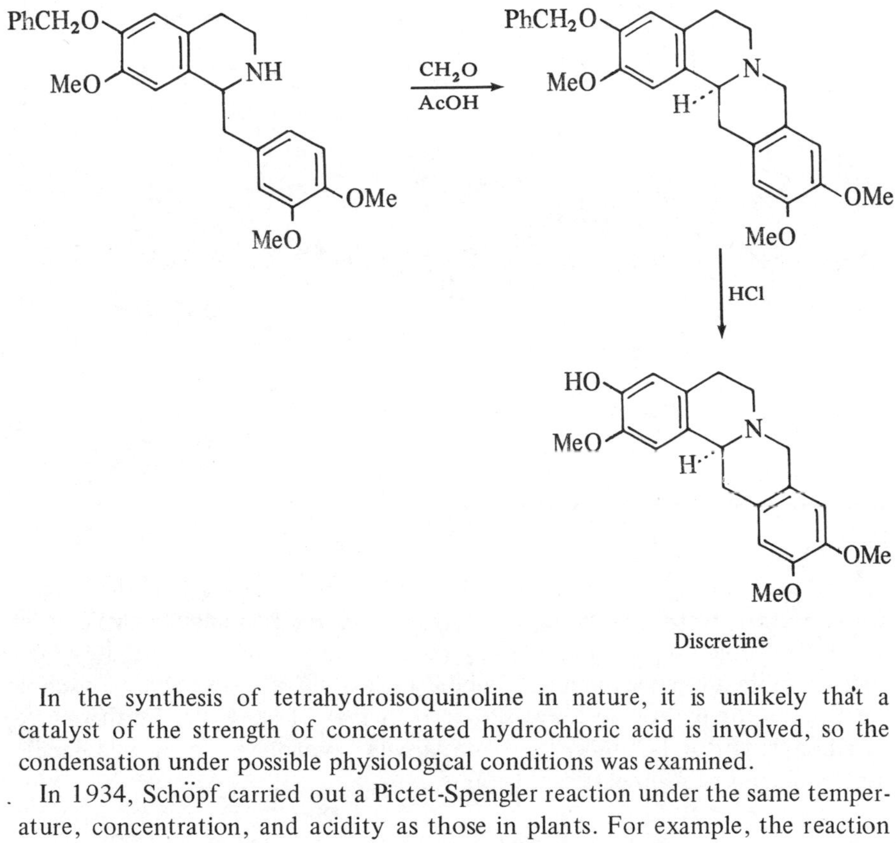

PhCH$_2$O, MeO, NH, OMe, MeO — CH_2O / AcOH → PhCH$_2$O, MeO, N, H, OMe, MeO — HCl → HO, MeO, N, H, OMe, MeO

Discretine

In the synthesis of tetrahydroisoquinoline in nature, it is unlikely that a catalyst of the strength of concentrated hydrochloric acid is involved, so the condensation under possible physiological conditions was examined.

In 1934, Schöpf carried out a Pictet-Spengler reaction under the same temperature, concentration, and acidity as those in plants. For example, the reaction of β-(3,4-dihydroxyphenyl)ethylamine with homopiperonal gave 1,2,3,4-tetrahydro-6,7-dihydroxy-1-piperonylisoquinoline[41] **55** at pH 6 at 25° (Chart 1-25).

Hahn and Schales proved that the ether derivative reacted in the same way as the phenolic base, but its reaction rate was found to be slower. For example, a mixture of homopiperonylamine and homopiperonal at pH 5 for 8 days at 25° gave a small amount of the corresponding isoquinoline base **56**[42,43] (Chart 1-26).

It is well known that naturally occurring phenylacetaldehyde probably is derived from its appropriate α-amino acid through the corresponding phenyl-

Chart 1-25.

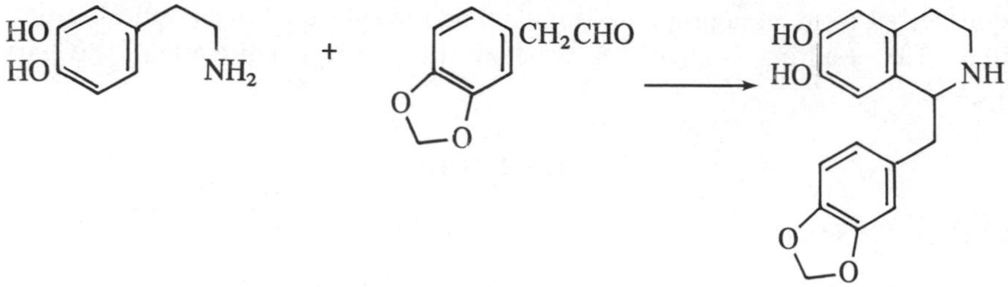

HO HO NH$_2$ + CH$_2$CHO O O → HO HO NH O O

55

Chart 1-26.

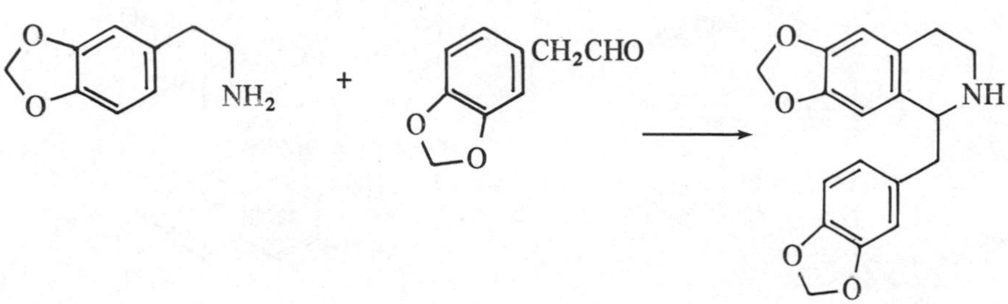

O O NH$_2$ + CH$_2$CHO O O → O O NH O O

56

pyruvic acid. Hahn[42] and Haun and Werner[43] have proposed that the α-keto acids are the actual precursors in the biogenesis of the isoquinoline alkaloids in nature. His suggestion was supported by the synthesis of 1-carboxy-1,2,3,4-tetrahydro-6,7-dihydroxyisoquinoline **57** under biologically plausible conditions, but the reaction of pyruvic acid was slower than that of aldehyde. Furthermore, decarboxylation of 1-carboxy-1,2,3,4-tetrahydroisoquinoline under mild conditions could not be realized (Chart 1-27).

Chart 1-27.

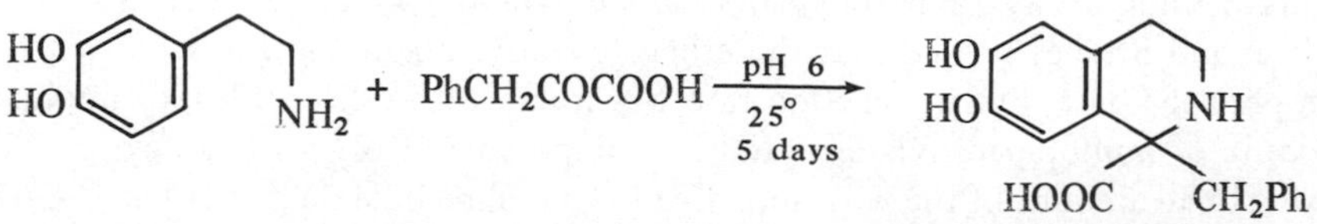

HO HO NH$_2$ + PhCH$_2$COCOOH $\xrightarrow[\text{25°, 5 days}]{\text{pH 6}}$ HO HO NH HOOC CH$_2$Ph

57

In support of a suggestion that the biosynthesis of isoquinoline alkaloids involves peptide chains, a model sequence designed to simulate this process has been investigated. The peptide analog was treated with the masked phenylpyruvate to give the diamide, which cyclized easily to a tetrahydroisoquinoline, the hydrolysis of which gave the amino acid. Presumably, if nature does indeed take a course analogous to the laboratory model, 1-benzyl-1-carboxyisoquinoline derivatives may well exist in benzylisoquinoline-producing plants[43a] (Chart 1-27A). Moreover, Yamada has reported a biogenetically patterned synthesis of (+)-laudanosine from (−)-dopa[43b] (Chart 1-27B).

Chart 1-27A.

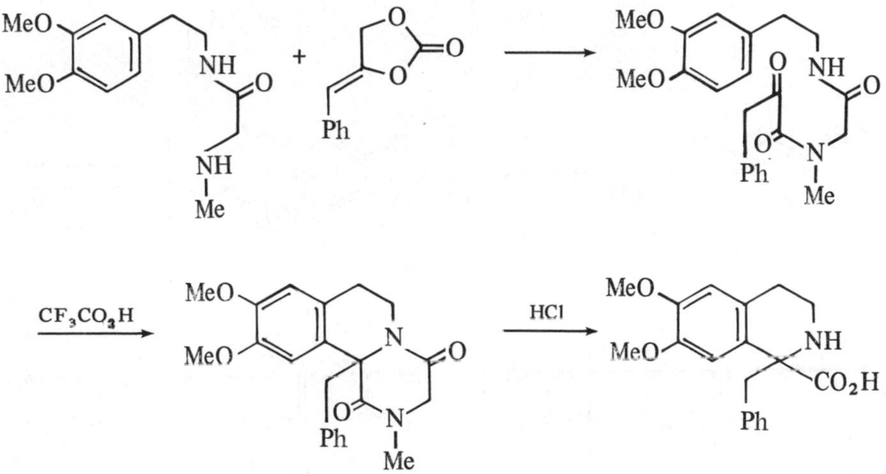

MeO MeO NH O NH Me + O O O Ph → MeO MeO O NH O O N Ph Me

CF_3CO_2H → MeO MeO N O O N Ph Me → HCl → MeO MeO NH CO_2H Ph

Chart 1-27B.

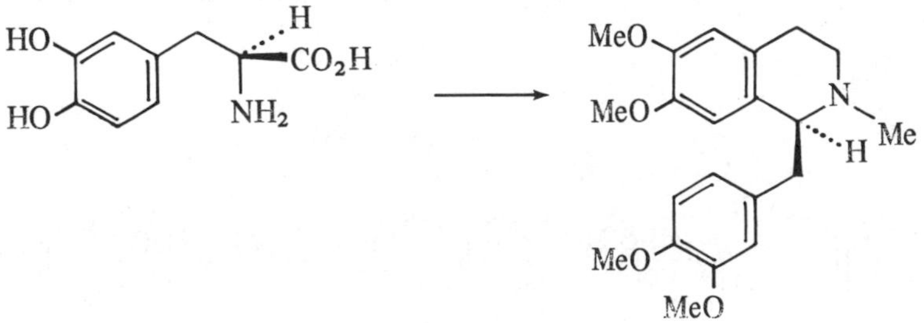

HO HO H CO_2H NH_2 → MeO MeO N Me H MeO MeO

Reaction Mechanism

The electronic mechanism of the Pictet-Spengler reaction has not yet been investigated completely, but it is well known that Schiff base **58** is isolated as an intermediate in some cases and can then be cyclized by acid. The formation of norhydrohydrastinine **59** from homopiperonylamine is illustrated as the following scheme (Chart 1-28).

Chart 1-28.

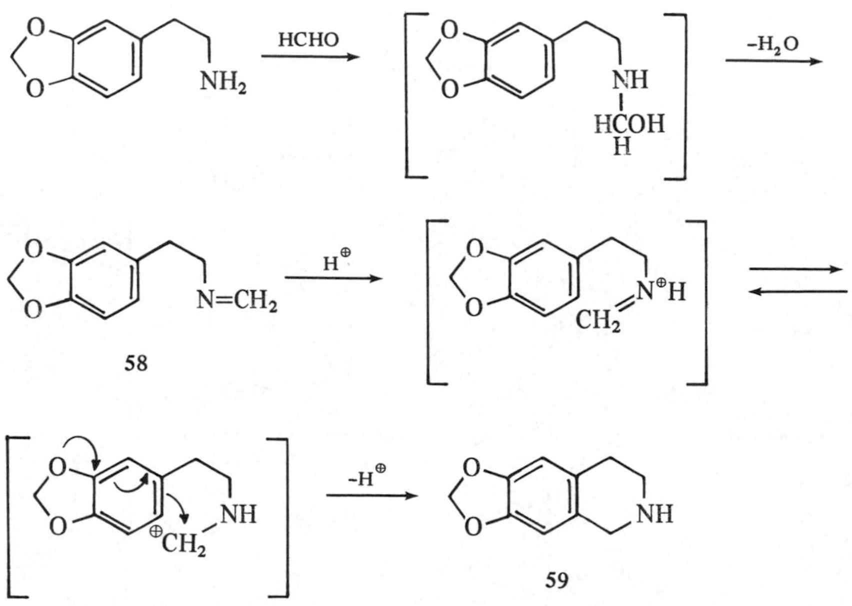

HCHO
NH_2
NH
HCOH
H
$-H_2O$
$N{=}CH_2$
58
$H^{\oplus}$
$N^{\oplus}H$
CH_2
NH
$^{\oplus}CH_2$
$-H^{\oplus}$
NH
59

As in the case of the Bischler-Napieralski reaction, this reaction also depends upon the extent of electron density at the *ortho* position in the ring closure. Thus, the reaction of phenethylamine with formaldehyde yields only 1,2,3,4-tetrahydro-6-methoxyisoquinoline **60** and no 8-methoxy-compound. This fact was proved by oxidation of the product to 4-methoxyphthalic acid[44] **61** (Chart 1-29).

Chart 1-29.

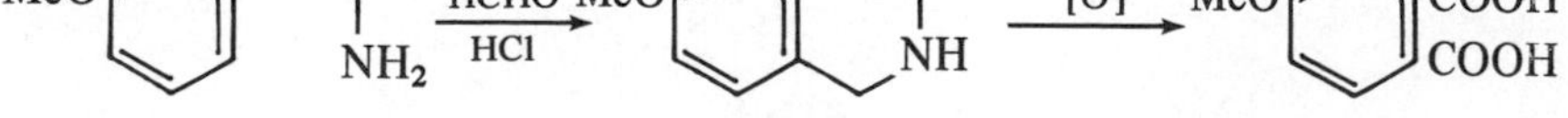

MeO
NH_2
HCHO
HCl
MeO
NH
60
[O]
MeO
COOH
COOH
61

The product obtained by cyclization of a 3,4-dialkoxy-β-phenethylamine is always the 6,7-dialkoxyl derivatives and 7,8-dialkoxyl derivatives are not formed. If both *ortho* positions are activated by a *m*-alkoxyl group, cyclization occurs in both directions to yield a mixture of the two possible tetrahydroisoquinoline derivatives. This fact was proved by the condensation of 1-(3-benzyloxy-4,5-dimethoxybenzyl)-1,2,3,4-tetrahydro-6,7-dimethoxyisoquinoline **62** with formaldehyde to give two products[45,45a] (Chart 1-30). Similarly, treatment of the

Chart 1-30.

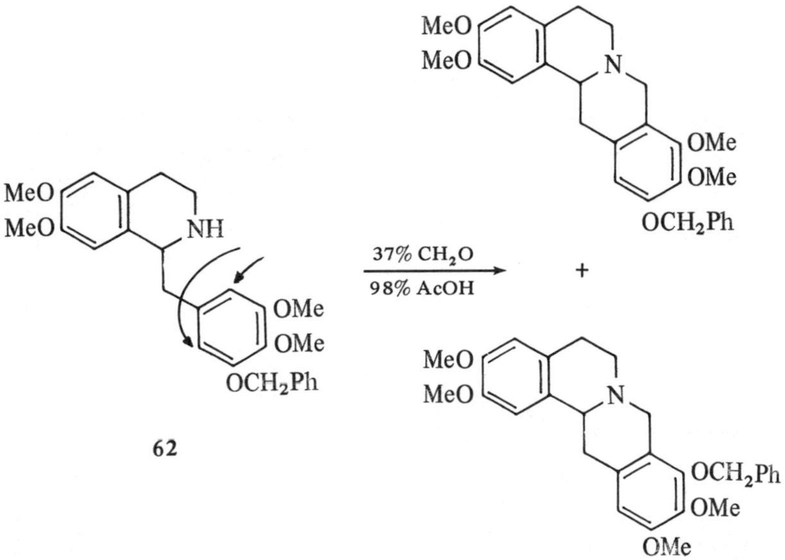

MeO
MeO
NH
OMe
OMe
OCH_2Ph
62
37% CH_2O
98% AcOH
MeO
MeO
N
OMe
OMe
OCH_2Ph
+
MeO
MeO
N
OCH_2Ph
OMe
OMe

1-(3-amino-4,5-dimethoxybenzyl)isoquinoline derivative with formaldehyde gave two isomeric products, either of which could be further transformed to deaminated or hydroxylated protoberberines as shown in Chart 1-30A.[45b]

Generally, cyclization has a tendency to occur at the *para* position to an alkoxyl group. For example, as a historical instance, a synthesis of tetrahydro-Φ-berberine **63** from 1-veratrylnorhydrohydrastinine can be mentioned. Pictet and Gams[46,47] emphasized the reaction product to be identical with natural tetrahydroberberine **64**, although the cyclization occurred at the less activated position. Subsequently, Haworth, Perkin, and Rankin[48] reported that this proposal was incorrect and tetrahydro-Φ-berberine **63** was the only reaction product that was different from the natural product (Chart 1-31).

On the other hand, Späth[49] revealed that if the alkoxyl groups are replaced by hydroxyl groups the orientation rule becomes invalid, and the ring closure

Chart 1-30A.

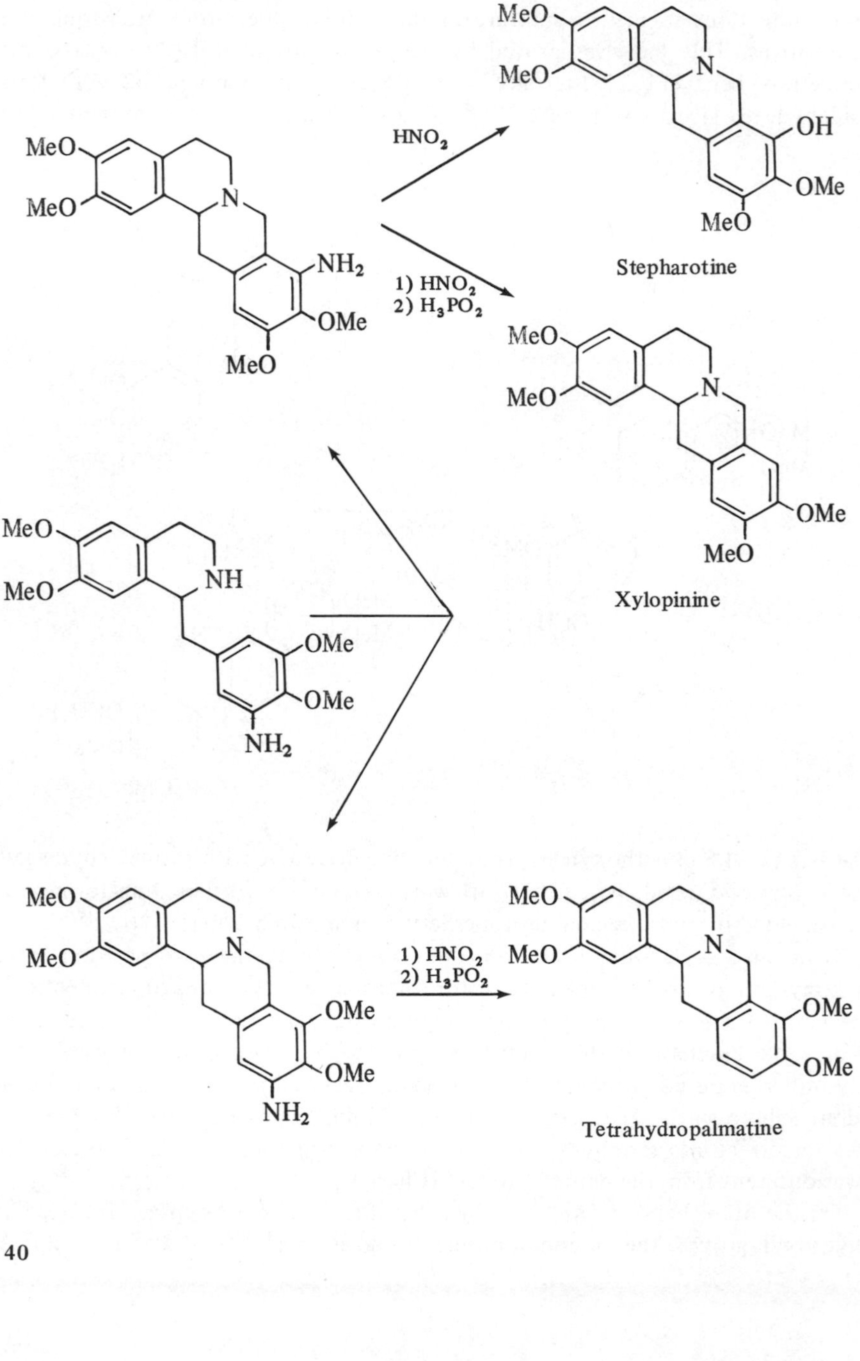

MeO MeO N NH$_2$ OMe MeO

HNO$_2$

1) HNO$_2$
2) H$_3$PO$_2$

Stepharotine

Xylopinine

MeO MeO NH OMe OMe NH$_2$

1) HNO$_2$
2) H$_3$PO$_2$

Tetrahydropalmatine

Chart 1-31.

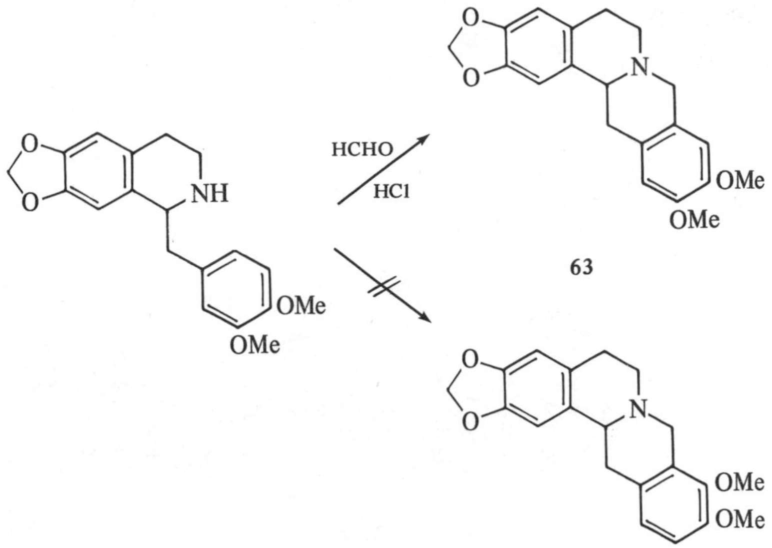

HCHO
HCl

63

64

proceeds to both *ortho* and *para* positions with nearly equal facility. Thus treatment of tetrahydropapaveroline **65** with formaldehyde afforded a mixture of products **66** and **67** in equal amounts. After conversion to the tetramethoxyl derivatives, xylopinine **54** and tetrahydropalmatine **68** were obtained (Chart 1-32). The same result was obtained in Pictet-Spengler reaction between norreticuline hydrochloride **69** and formalin at pH 6.3,[50] as well as in the synthesis of the protoberberine alkaloids, canadine,[50a] kikemanine,[50b] and capaurimine[50c] (Chart 1-33).

Schöpf[51] also obtained the compound **67** in 80% of the yield under physiological conditions. Apparently, the presence of free hydroxyl groups in the benzyl residue activates the *ortho* position. Most of the protoberberine alkaloids that occur in nature belong to the 2,3,9,10-oxygenated series, therefore, a number of attempts to synthesize these alkaloids by a Pictet-Spengler reaction have been investigated, but few successful examples have been reported by use of the usual method. Accordingly, Kametani decided that blocking of the usual cyclization position with bromine and replacement of methoxyl by hydroxyl to offset the inactivation of the nucleus caused by the I-effect of the bromine atom would be favorable for the promotion of the *ortho* coupling to the hydroxyl

Chart 1-32.

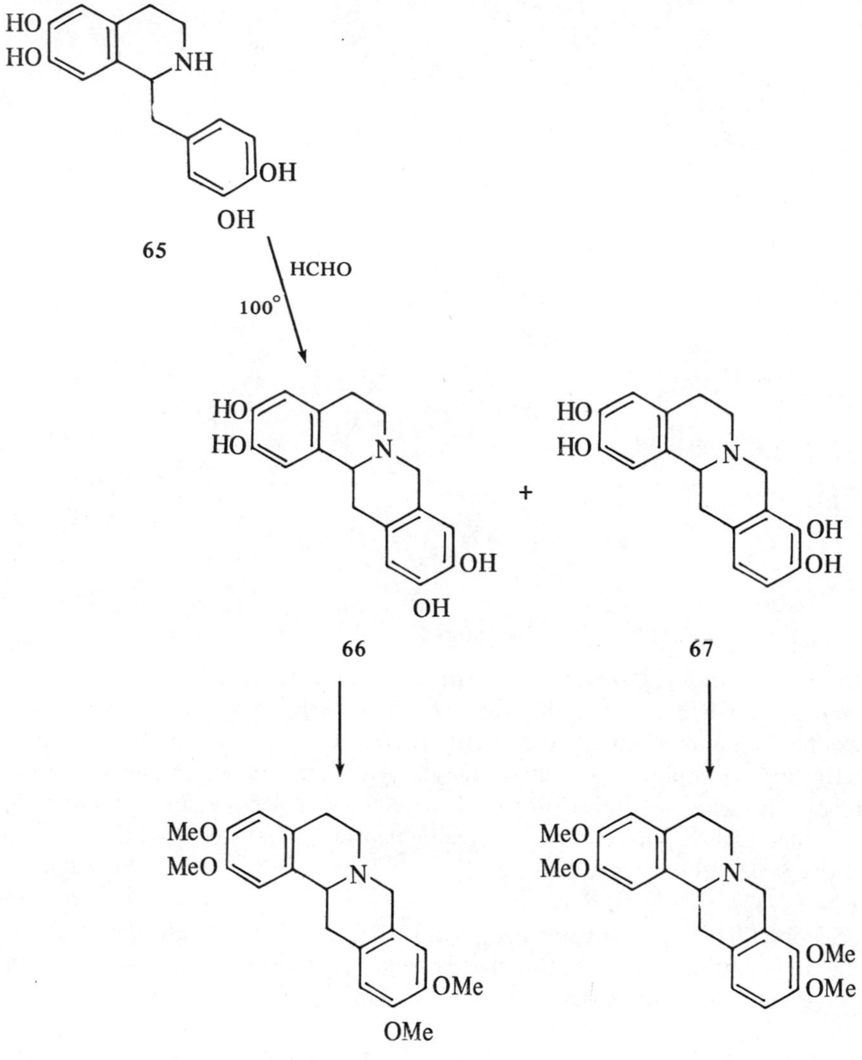

HO
HO
NH
OH
OH
65
HCHO
100°
HO
HO
N
OH
OH
66
+
HO
HO
N
OH
OH
67
MeO
MeO
N
OMe
OMe
54 Xylopinine
MeO
MeO
N
OMe
OMe
68 Tetrahydropalmatine

Chart 1-33.

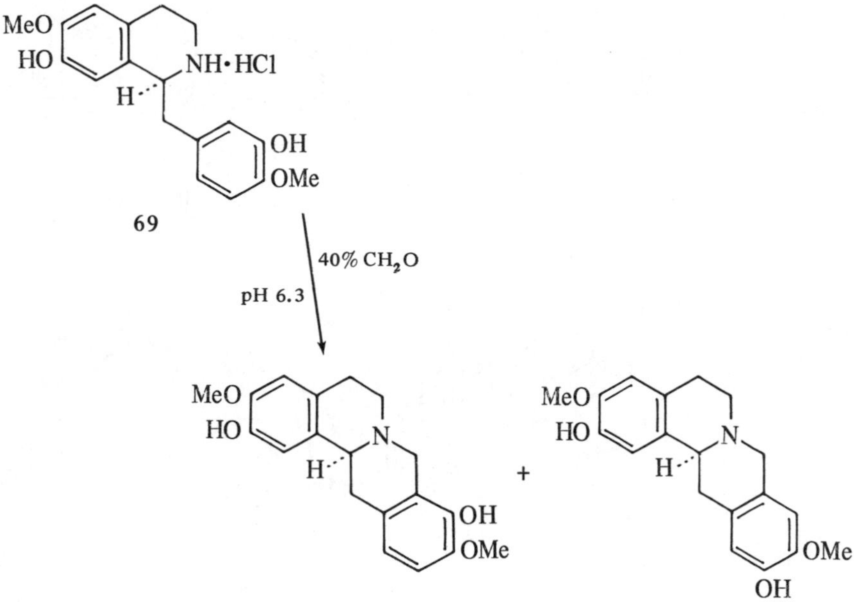

MeO
HO
NH•HCl
H
OH
OMe
69
40% CH_2O
pH 6.3
MeO
HO
N
H
OH
OMe
+
MeO
HO
N
H
OMe
OH

70 Scoulerine

71 Coreximine

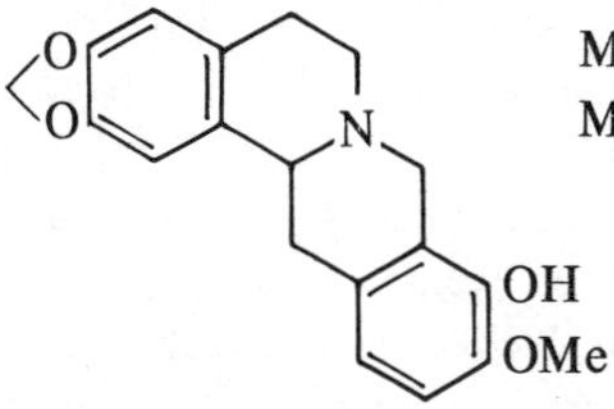

O
O
N
OH
OMe
MeO
MeO
N
OMe
OH

Canadine

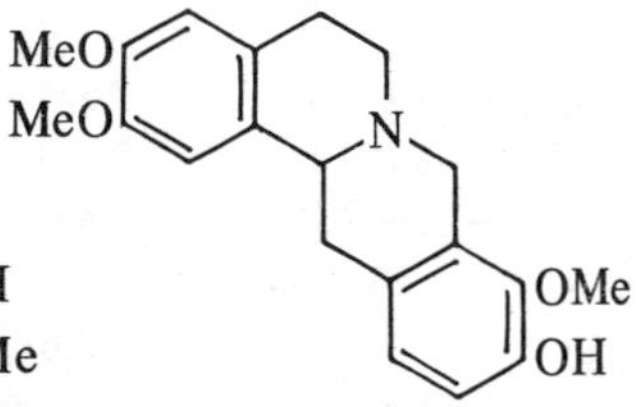

Kikemanine

group. Thus the reaction of the bromoisoquinoline **72** with formalin and hydrochloric acid gave the expected cyclized compound **73** in good yield which was converted into scoulerine **70** and tetrahydropalmatine **63**.[52] Similarly, capaurine was synthesized by the Mannich reaction of the 1-(2-bromo-4-methoxy-5-hydroxybenzyl)isoquinoline, followed by a reductive cleavage of the bromine[52a] (Chart 1-33A). Moreover, the condensation of 2-bromo-5-hydroxy-4-methoxyphenethylamine with homoanisaldehyde afforded 5-bromo-1,2,3,4-tetrahydro-8-hydroxy-7-methoxyisoquinoline derivative[30a] (Chart 1-34).

Chart 1-33A.

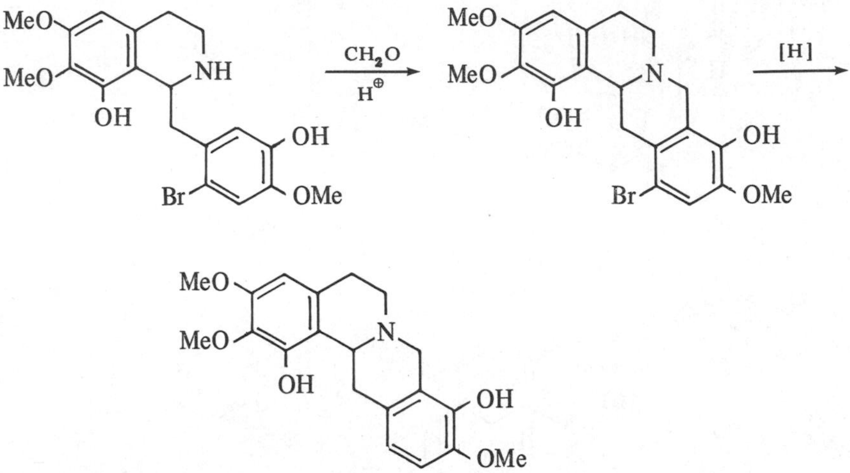

MeO MeO OH NH Br OH OMe $\xrightarrow[H^{\oplus}]{CH_2O}$ MeO MeO OH N Br OH OMe $\xrightarrow{[H]}$ MeO MeO OH N OH OMe

Factors Affecting the Ease of Cyclization

The reactivity of the aromatic nucleus of arylethylamines and the nature of the carbonyl component are important to the success of the Pictet-Spengler reaction. It has been considered that substituents on the side chain of the arylethylamine would have an influence on the ease of cyclization as in the case of the Bischler-Napieralski reaction, but the available data of checking the previous suggestion have not yet been sufficiently examined. The reaction is facilitated by increased electron density at the cyclized position in the case of the Pictet-Spengler reaction. Few phenethylamine, even if an alkoxyl or hydroxyl group *para* to the cyclized position would be lacked, can be cyclized. For example, β-phenethylamine and phenylalanine were converted to the corresponding tetrahydroisoquinolines in approximately 35% of the yield by treatment with methylal and hydrochloric acid.[38] However, this result has been disputed by Kondo and Ochiai,[53] who could obtain a trace of the product. Cyclization of the hydroxyl amine **74** to the hydroxytetrahydroisoquinoline **75** took place quantitatively,[54] and tyramine[55] and tyrosine[56] have also been cyclized in good yield, indicating that the reaction does not require a great activation (Chart 1-35). However, β-(2-ethoxyphenyl)ethylamine **76**[57] and 1,2,3,4-tetrahydro-6-methoxy-1-(4-methoxybenzyl)isoquinoline **77**[58] could not be cyclized (Chart 1-36). Furthermore, if a methoxyl group would be existed at the *para* position to the cyclized position, cyclization has not occurred under the ordinary conditions as in case of ω-aminoacetoveratrone.[59]

Chart 1-34.

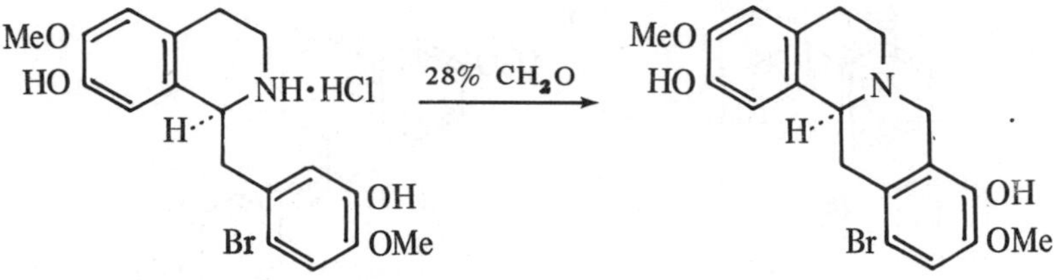

MeO
HO
NH·HCl
H
OH
Br
OMe
72

28% CH_2O

MeO
HO
N
H
OH
Br
OMe
73

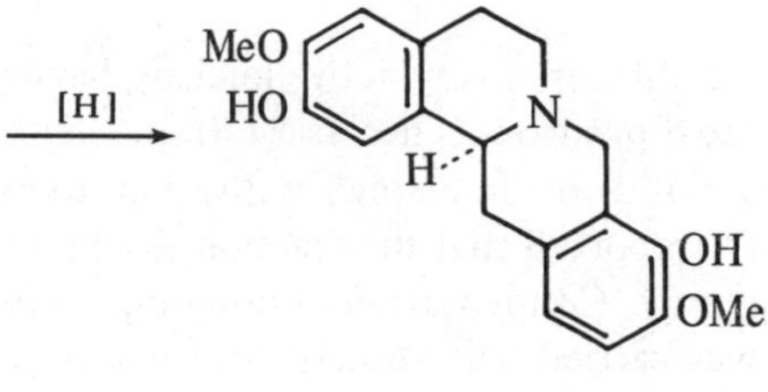

[H]

MeO
HO
N
H
OH
OMe

70 Scoulerine

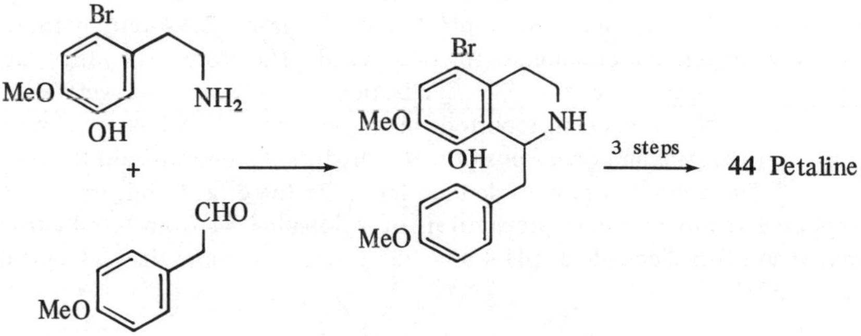

Br
MeO
NH_2
OH
+
CHO
MeO

Br
MeO
NH
OH
MeO

3 steps

44 Petaline

Chart 1-35.

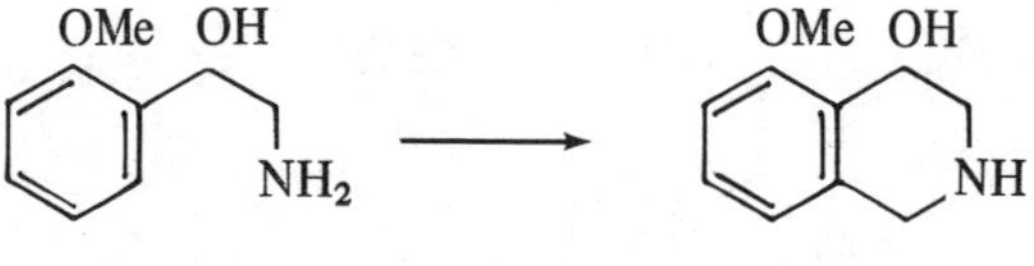

OMe OH
NH_2
74

OMe OH
NH
75

Chart 1-36.

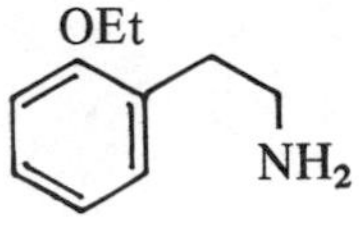

OEt

NH_2

76

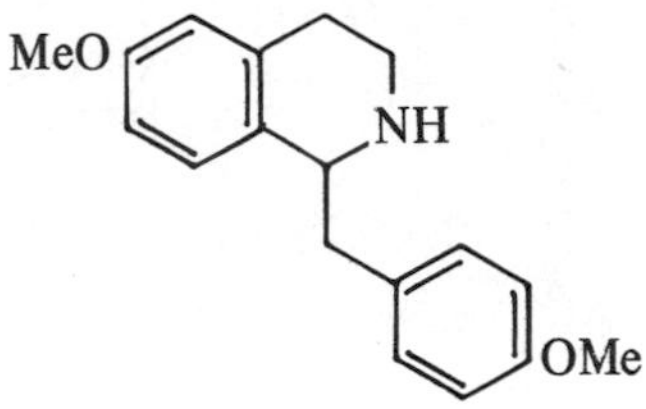

MeO

NH

OMe

77

In the case of physiological conditions, a very active nucleus, having an increased electron density at the cyclized position, is necessary. In the benzenoid series, even alkoxyl substituents do not furnish enough activation to promote the reaction satisfactorily. Schöpf[60] reported that the reaction would not proceed in the absence of free hydroxyl groups. Condensation of homopiperonylamine with piperonal at pH 5 and 25° was carried out successfully to afford the desired 1-piperonylnorhydrastinine in only 5% of the yield by Hahn,[61] but Späth disputed Hahn's theory.[62] In contrast, treatment of β-(3,4-dihydroxyphenyl)-ethylamine with homopiperonal at pH 6 and 25° gave 1,2,3,4-tetrahydro-6,7-dihydroxy-1-piperonylisoquinoline in 84% yield. Therefore, a hydroxylated benzene ring appears necessary for cyclization under quasi biological conditions.[43] Recently, Kametani reported many instances in which cyclization occurred at the *para* and *ortho* position to a hydroxyl group without the use of acids.[63,64] These results will be described later. On the other hand, there is one exceptional report regarding preparation of xylopinine **54** from tetrahydropapaverine and formaldehyde at pH 4 and 25° for 18 hr in more than 84% yield[65] (Chart 1-37).

Chart 1-37.

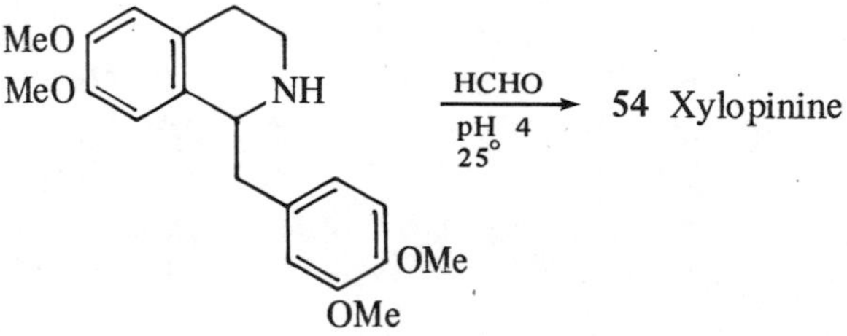

MeO

MeO

NH

OMe

OMe

HCHO

pH 4

25°

54 Xylopinine

Formaldehyde and methylal as carbonyl compounds have been employed most frequently in the original Pictet-Spengler reaction. Formaldehyde generally gives excellent yields in a number of instances and is preferable to methylal.[66,67] For instance, tetrahydropapaverine was cyclized to xylopinine **54** in 46% yield using methylal, whereas it was obtained with formaldehyde in 60% yield under the same conditions.[67] The poor yields in case of homopiperonal and homoveratraldehyde depend upon their instability in the presence of hydrochloric acid.[68,69]

Phenolic Cyclization[63,64]

The reaction of phenolic amines such as **78** with various carbonyl compounds without acidic catalyst gives 1,2,3,4-tetrahydroisoquinolines such as **79** and **80**. This reaction bears close similarity to the Pictet-Spengler reaction, but can be differentiated in that isoquinoline formation has occurred under nonacidic conditions. Therefore, Kametani has proposed that this type of nonacidic reaction should be called "phenolic cyclization" because the phenolic hydroxyl group on the benzene ring apparently plays an important role in this reaction[63] (Chart 1-38).

Chart 1-38.

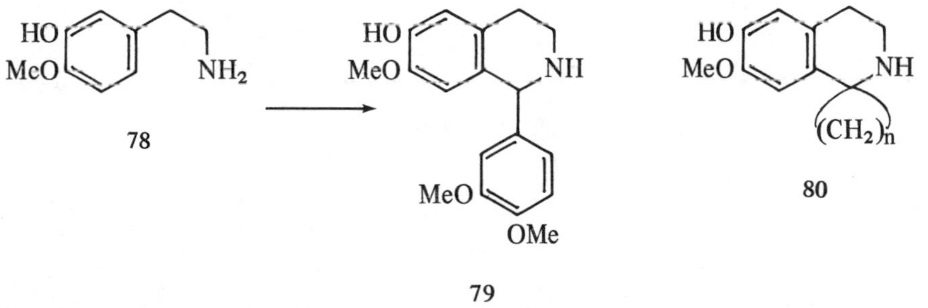

HO MeO NH$_2$ **78** → HO MeO NH MeO OMe **79** HO MeO NH $(CH_2)_n$ **80**

This phenolic cyclization is a useful method for the syntheses of 6-hydroxyisoquinoline type alkaloids and protoberberine. At first, syntheses of 1-phenethylisoquinoline derivatives as model experiments to the total syntheses of some alkaloids were successfully examined as follows[63,64] (Chart 1-39). The 1,1′-spiro compounds **81**, **82**, **83**, **84** were obtained by the reaction of the amines with cyclohexanone, a method that showed a useful route[63,70,71] (Chart 1-40).

The condensation of (±)-norreticuline **85** with 37% formalin in ethanol afforded (±)-coreximine **86** without using acid.[63] Interestingly, the reaction of *trans*-2-(3-hydroxyphenyl)cyclohexylamine with benzaldehyde gave two products

Chart 1-39.

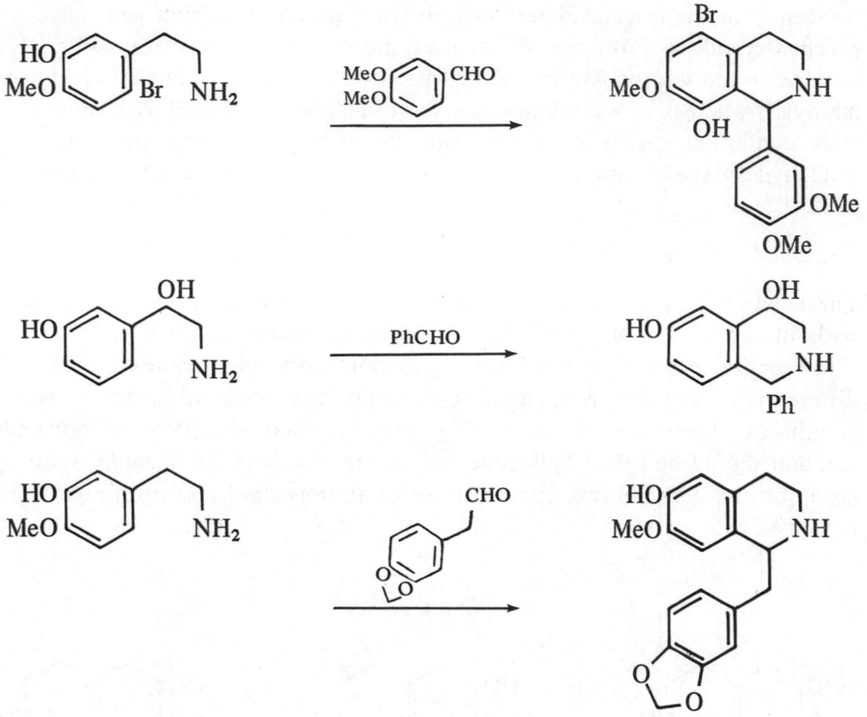

HO
MeO
Br
NH_2
MeO
MeO
CHO
Br
MeO
NH
OH
OMe
OMe

OH
HO
NH_2
PhCHO
OH
HO
NH
Ph

HO
MeO
NH_2
CHO
O
O
HO
MeO
NH
O
O

Chart 1-40.

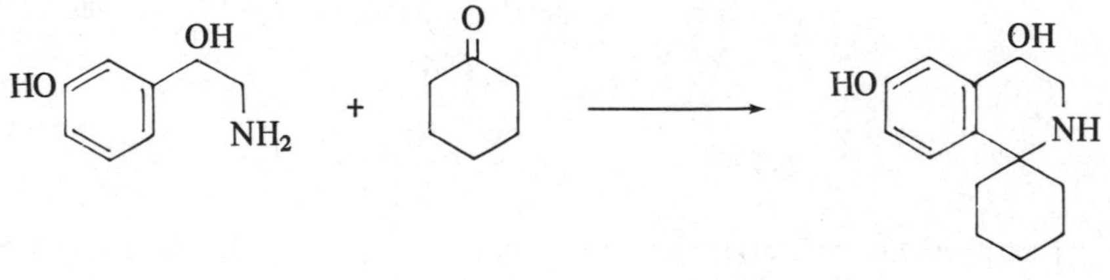

OH
HO
NH_2
+
O
OH
HO
NH

81

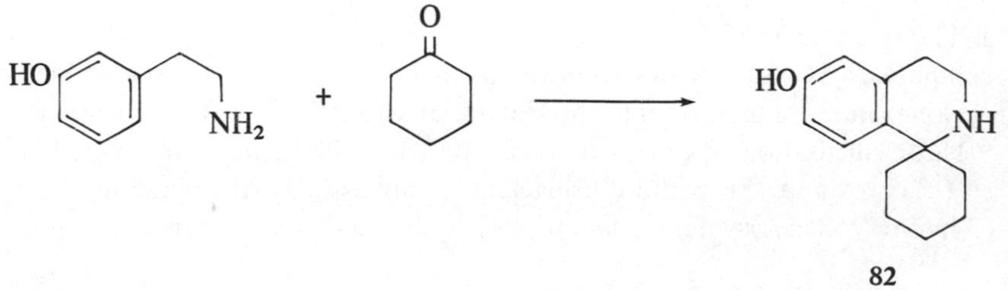

HO
NH_2
+
O
HO
NH

82

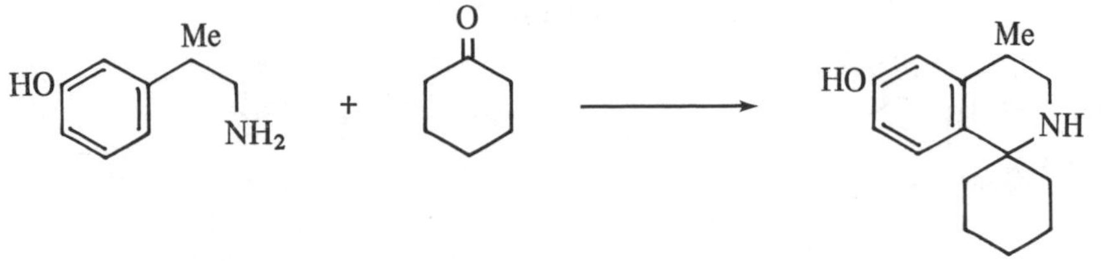

HO Me NH$_2$ + O → HO Me NH

83

HO NH$_2$ + O → HO NH

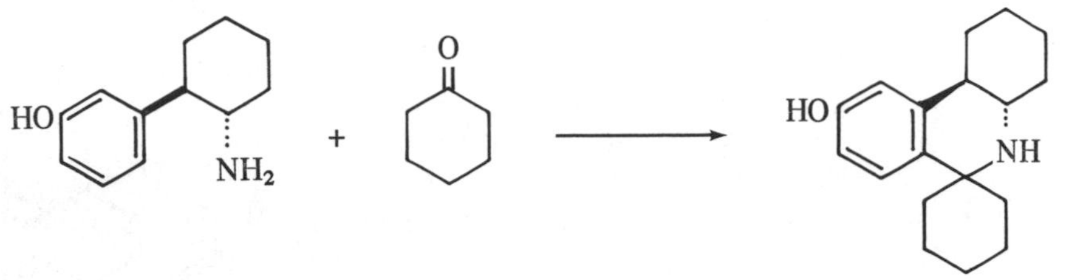

84

cyclized at the *ortho* and *para* position to the hydroxyl group. The formation of the former product would be an interaction of the phenolic hydroxyl group with π-electrons on the benzene ring by the approach of both aromatic ring in Schiff base[30b] (Chart 1-41).

The phenolic cyclization of the 1-phenethylisoquinoline **87** afforded the product **88** cyclized at the *ortho* position to the hydroxyl group, a reaction not observed in the 1-benzylisoquinoline series. On the other hand, Pictet-Spengler reaction gave the normal product **89** cyclized at the *para* position to the phenolic hydroxyl group.[72] The structure of **88** was identified by a standard synthesis from bromophenethylisoquinoline **90** (Chart 1-42). Isococlaurine was synthesized by this method[72a] (Chart 1-42A). Moreover, a key intermediate for the synthesis of petaline was also obtained by phenolic cyclization.[30a]

An Application of Pictet-Spengler Reaction to the Total Synthesis of the Isoquinoline Alkaloids

Ochotensine-type Alkaloids. Several syntheses of ochotensine and related bases have been achieved by a Pictet-Spengler reaction of 3-hydroxy-4-methoxyphenethylamine hydrohalide with the hydrindandione, followed by *O*-methylation, *N*-methylation, and the Wittig reaction.[72b] Protection of the hydroxyl group of the phenolic isoquinoline as the methoxymethyl ether, before the Wittig reaction, afforded a route to ochotensine[72c] (Chart 1-42B).

Three syntheses of ochrobirine by the indanone approach have appeared,[72d-e] and indanones have also been used in the preparation of fumaritine and fumariline[72f,g] (Chart 1-42C).

Chart 1-41.

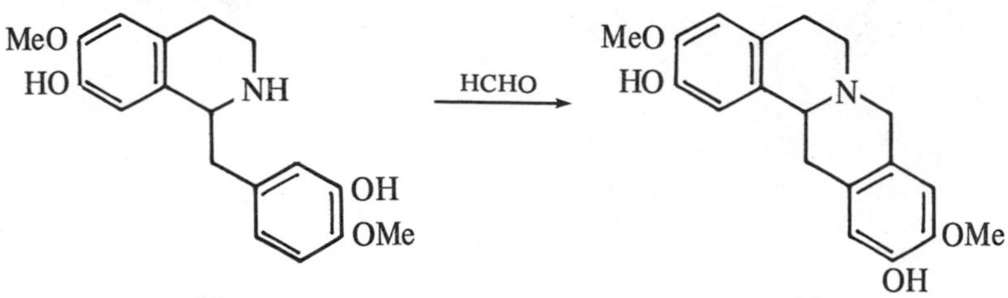

MeO HO NH OH OMe

85

HCHO

MeO HO N OMe OH

86
Corexímine

H HO H NH

H HO H NH HO

H HO H NH_2 CHO

H N=CH O–H H

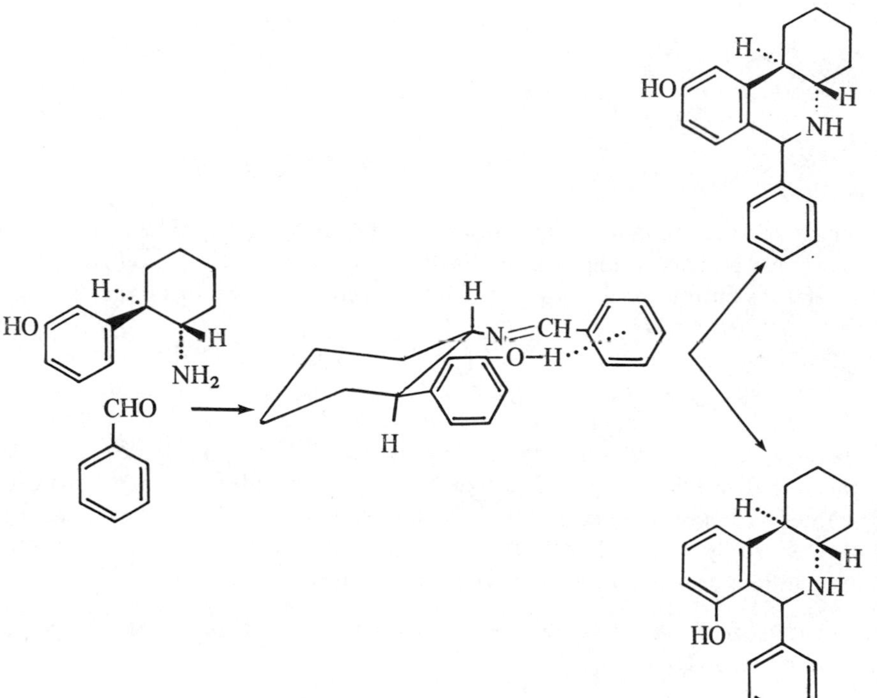

Benzophenanthridine Alkaloids. Benzophenanthridine ring, a basic system of nitidine and oxonitidine, has been prepared by a Pictet-Spengler reaction of the 1-amino-2-aryltetralone, and the product was converted into nitidine by a standard method[72h] (Chart 1-42D).

Tylophorine Alkaloids. The structures for two alkaloids isolated from *Cynanchum vincetoxicum* have been confirmed by total synthesis along the usual way. The 10-cyanophenanthrene, prepared by the Pschorr reaction and dehydrogena-

Chart 1-42.

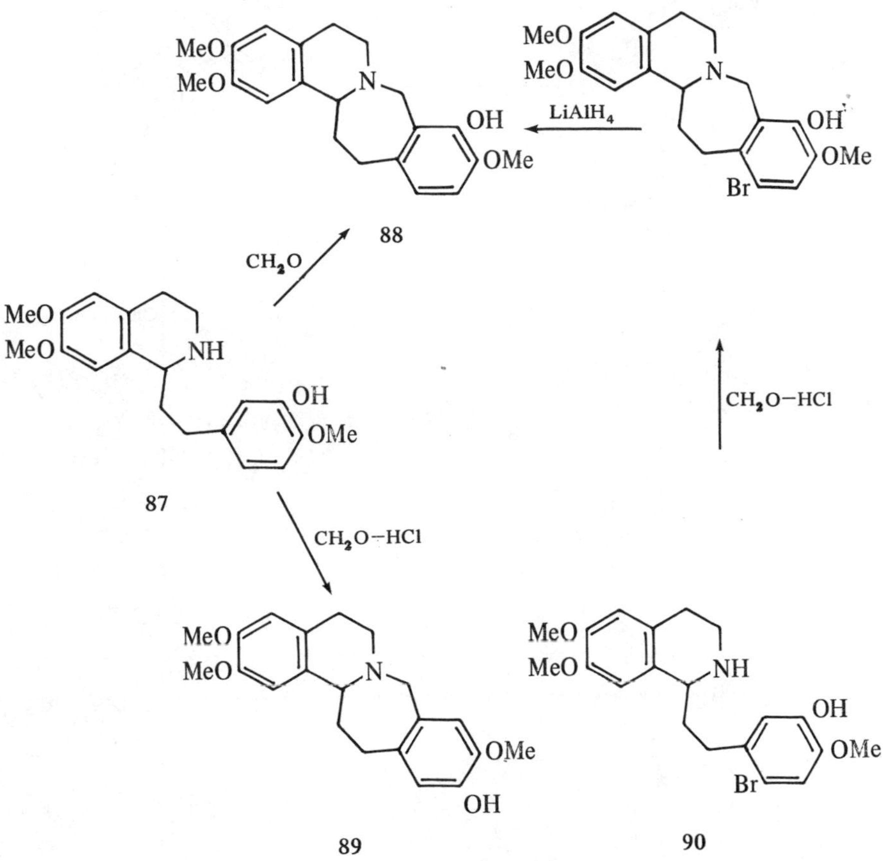

MeO
MeO
N
OH
OMe
88
LiAlH$_4$
MeO
MeO
N
OH
OMe
Br
CH$_2$O
MeO
MeO
NH
OH
OMe
87
CH$_2$O–HCl
CH$_2$O–HCl
MeO
MeO
N
OMe
OH
89
MeO
MeO
NH
OH
OMe
Br
90

Chart 1-42A.

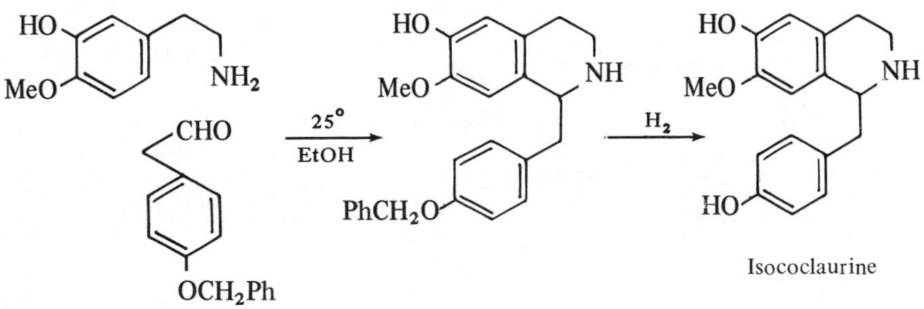

HO
MeO
NH$_2$
CHO
OCH$_2$Ph
25°
EtOH
HO
MeO
NH
PhCH$_2$O
H$_2$
HO
MeO
NH
HO
Isococlaurine

Chart 1-42B.

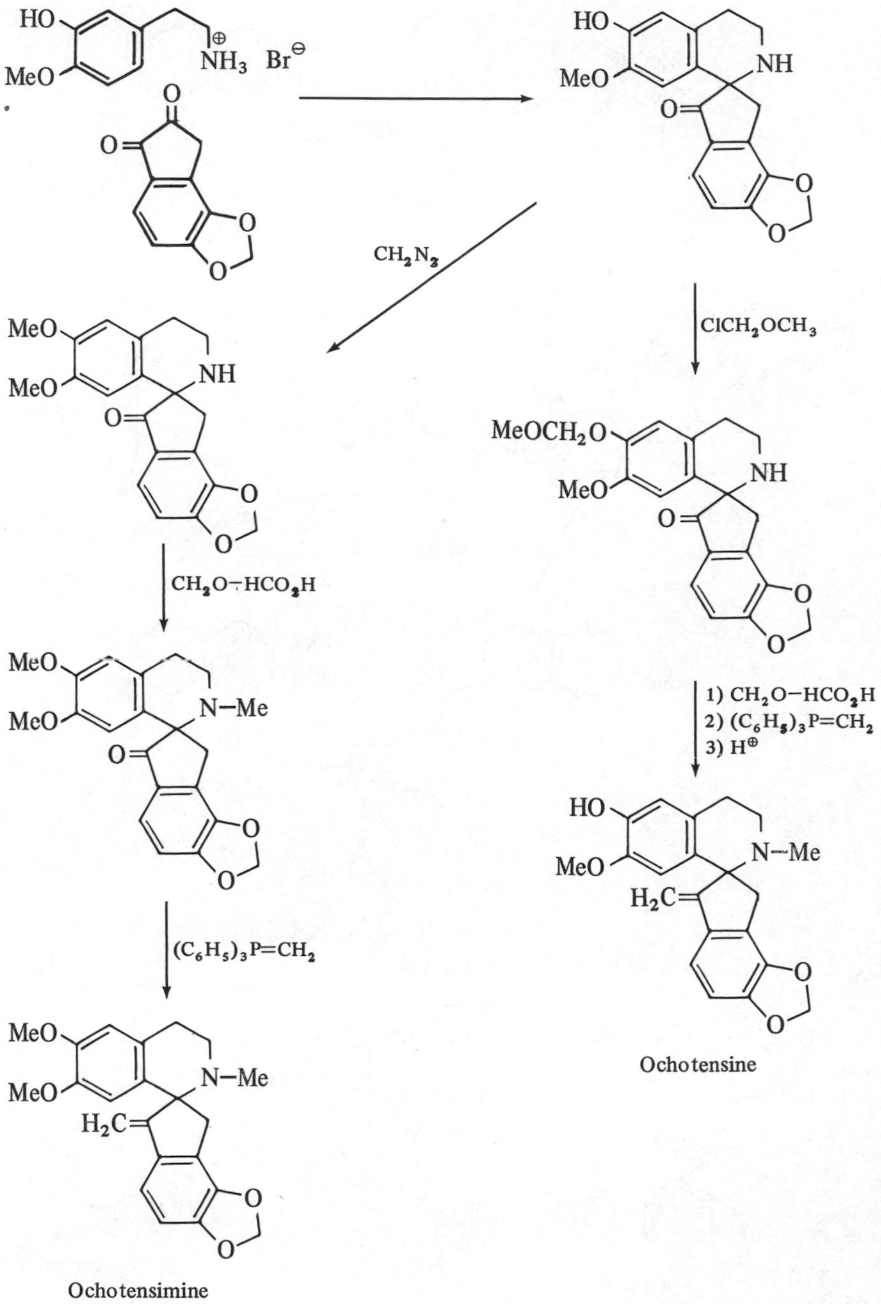

HO
$NH_3^{\oplus}$ $Br^{\ominus}$
MeO
O
O
O
O

NH
CH_2N_2
$ClCH_2OCH_3$

MeO
MeO
NH
CH_2O-HCO_2H

$MeOCH_2O$
MeO
NH

N–Me
$(C_6H_5)_3P=CH_2$

1) CH_2O-HCO_2H
2) $(C_6H_5)_3P=CH_2$
3) $H^{\oplus}$

HO
MeO
N–Me
$H_2C=$

Ochotensine

MeO
MeO
N–Me
$H_2C=$

Ochotensimine

Chart 1-42C.

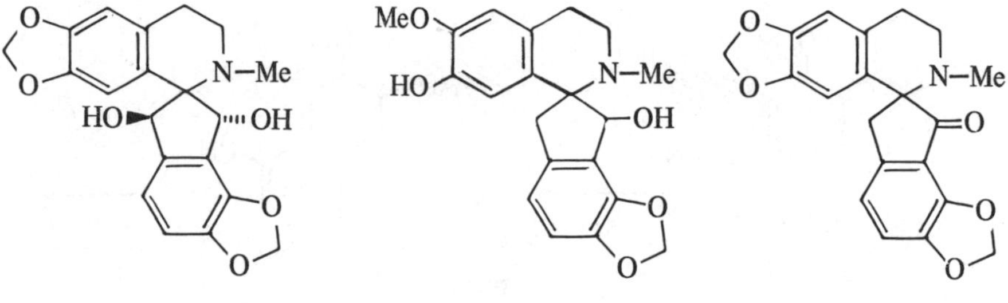

Ochrobirine

Fumaritine

Fumariline

Chart 1-42D.

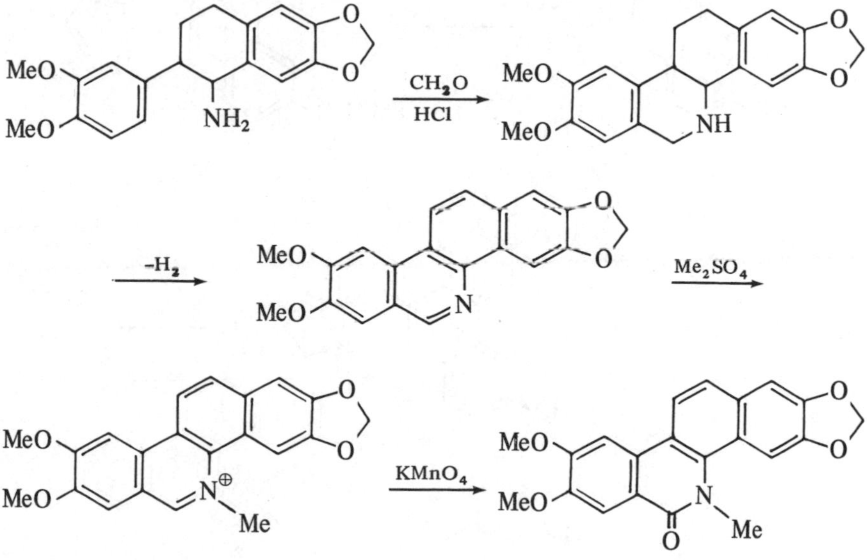

CH_2O
HCl

$-H_2$

Me_2SO_4

$KMnO_4$

Nitidine

Oxonitidine

tion, was converted by standard methods into chloromethylphenanthrene, which condensed with pyrrolemagnesium bromide to give the pyrrole derivative. Hydrogenolysis and hydrogenation afforded the phenolic pyrrolidine, which was formylated and cyclized with phosphoryl chloride. Finally, reduction with sodium borohydride gave alkaloid (C), which on methylation afforded alkaloid (A)[72i] (Chart 1-42E).

Chart 1-42E.

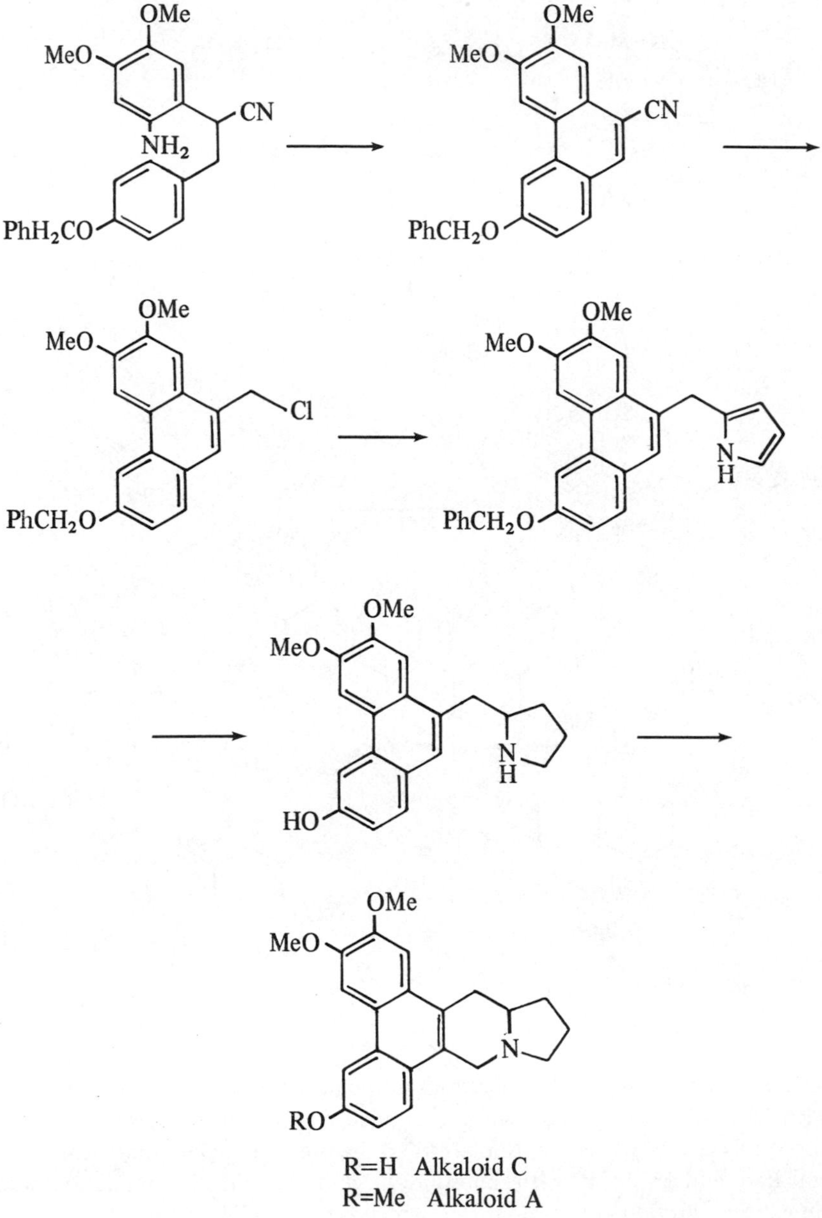

MeO, OMe, NH_2, CN, PhH_2CO → MeO, OMe, CN, $PhCH_2O$ →

MeO, OMe, Cl, $PhCH_2O$ → MeO, OMe, N H, $PhCH_2O$

→ MeO, OMe, N H, HO →

MeO, OMe, N, RO

R=H Alkaloid C

R=Me Alkaloid A

Amaryllidaceae Alkaloids. Elwesine, dihydrocrinine, has been synthesized by the following route, which may have generality for crinan alkaloid synthesis.[72j] Piperonyl cyanide was converted into the cyclopropyl imine in three steps, which was transformed via the enamine into the mesembrine-type compound. Successive reduction and debenzylation provided the alcohol that was subjected to a Pictet-Spengler reaction to give elwesine (Chart 1-42F).

Chart 1-42F.

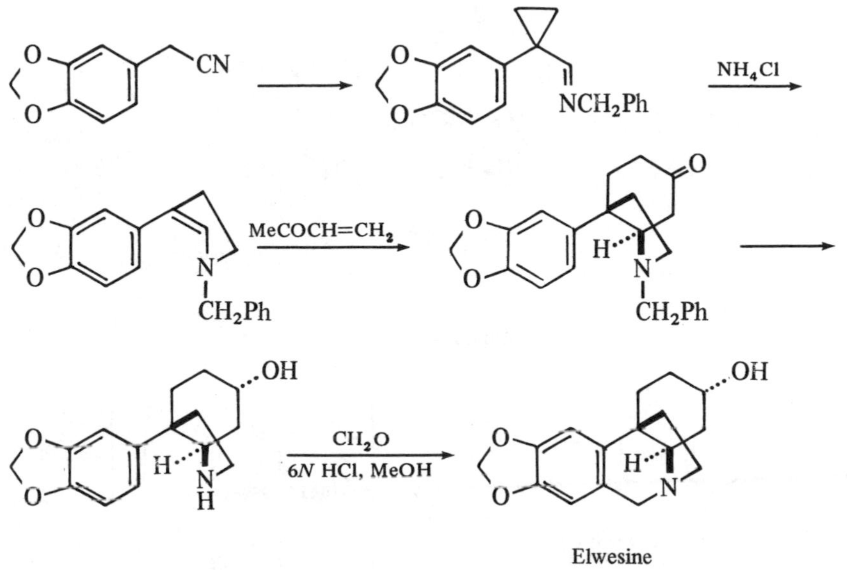

CN → NCH_2Ph —NH_4Cl→

CH_2Ph —$MeCOCH{=}CH_2$→ O, H, N, CH_2Ph →

OH, H, N, H —CH_2O, 6*N* HCl, MeOH→ OH, H, N

Elwesine

Stereoselective total synthesis of haemanthamine has been reported by Tsuda and Isobe.[72k] The Pictet-Spengler reaction of the diol, described in a synthesis of tazettine and haemanthidine by Bischler-Napieralski reaction,[35q] which was followed by acid treatment, yielded the ethanophenanthridine, which was converted into haemanthamine in two steps (Chart 1-42G).

Other Synthetic Reactions

Lora-Tamayo has developed a new synthesis of isoquinolines by a condensation of homoveratrylamine with homoveratronitrile-stannic chloride complex[72l] (Chart 1-42H). The tetrahydroisoquinoline synthesis by the sulfur dioxide dehydrative cyclization of *N,N*-dimethylphenethylamine *N*-oxide was reported by Norman[72m] (Chart 1-42I). Protoberberine alkaloids were prepared by the

Chart 1-42G.

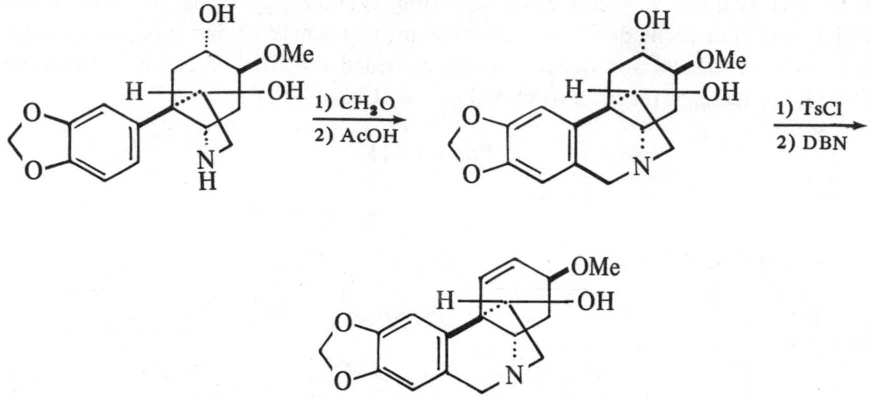

OH
OMe
H
OH
O
O
N
H
1) CH_2O
2) AcOH
OH
OMe
H
OH
O
O
N
1) TsCl
2) DBN
OMe
H
OH
O
O
N

Chart 1-42H.

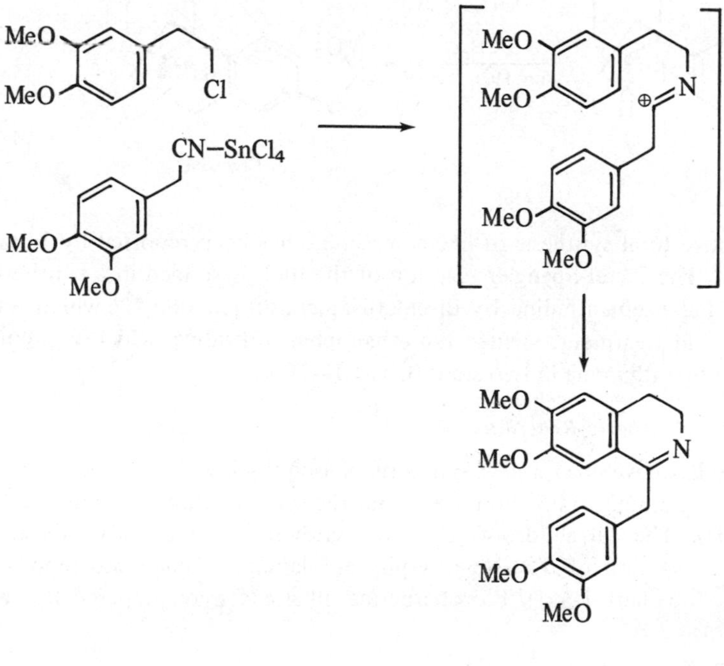

MeO
MeO
Cl
CN–$SnCl_4$
MeO
MeO
MeO
MeO
⊕
N
MeO
MeO
MeO
MeO
N
MeO
MeO

Chart 1-42I.

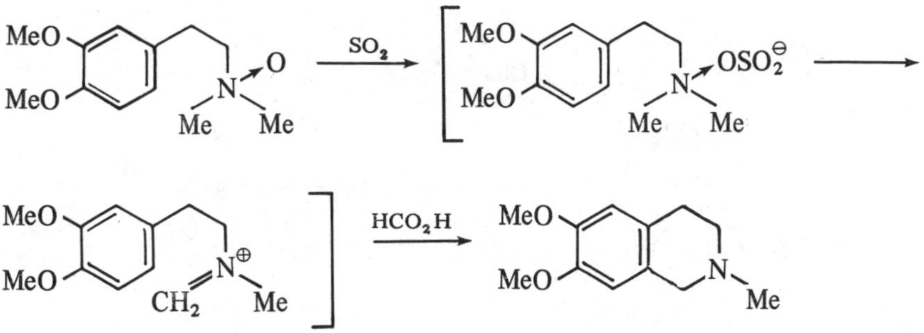

MeO, MeO, N→O, Me, Me —SO_2→ [MeO, MeO, N→$OSO_2^{\ominus}$, Me, Me] → [MeO, MeO, CH_2=N⊕, Me] —HCO_2H→ MeO, MeO, N, Me

acid-catalyzed ring closure from the 1,2-dihydroisoquinolines as in Chart 1-42J.[72n] Pavine-type alkaloids, bisnorargemonine, were also synthesized by the same mechanism of this xylopinine preparation. The key intermediate, prepared by Pictet-Gams reaction, was converted into methiodide, which was subjected to partial reduction to give the 1,2-dihydro-2-methylisoquinoline, and finally, acid-catalyzed cyclization afforded bisnorargemonine[72o] (Chart 1-42K). Interestingly, 1,2-dihydroisoquinoline derivatives do not give pavine-type compounds but rather undergo a variety of other reactions if excess acid is avoided. Pai synthe-

Chart 1-42J.

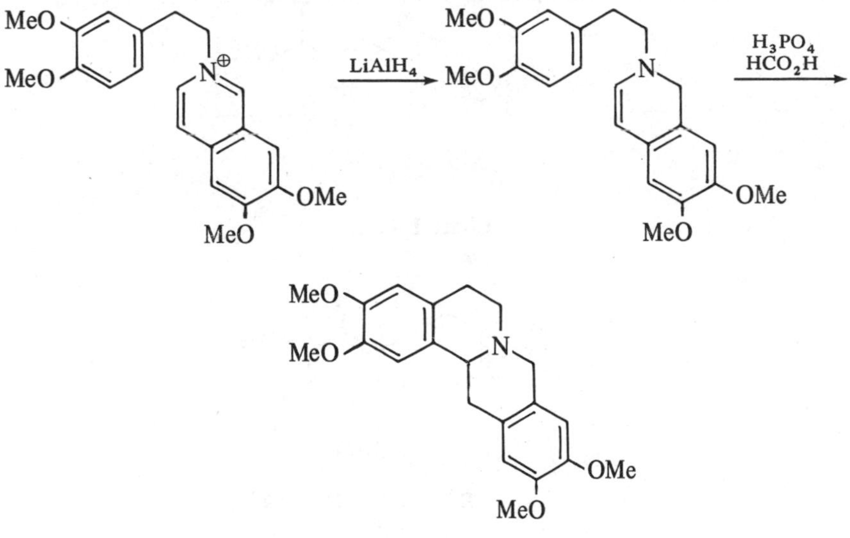

MeO, MeO, N⊕, OMe, MeO —$LiAlH_4$→ MeO, MeO, N, OMe, MeO —H_3PO_4, HCO_2H→

MeO, MeO, N, OMe, MeO

Xylopinine

sized caryachine by the same method,[72p] and platycerine was prepared from the appropriate 1,2-dihydroisoquinoline[72q] (Chart 1-42).

Chart 1-42K.

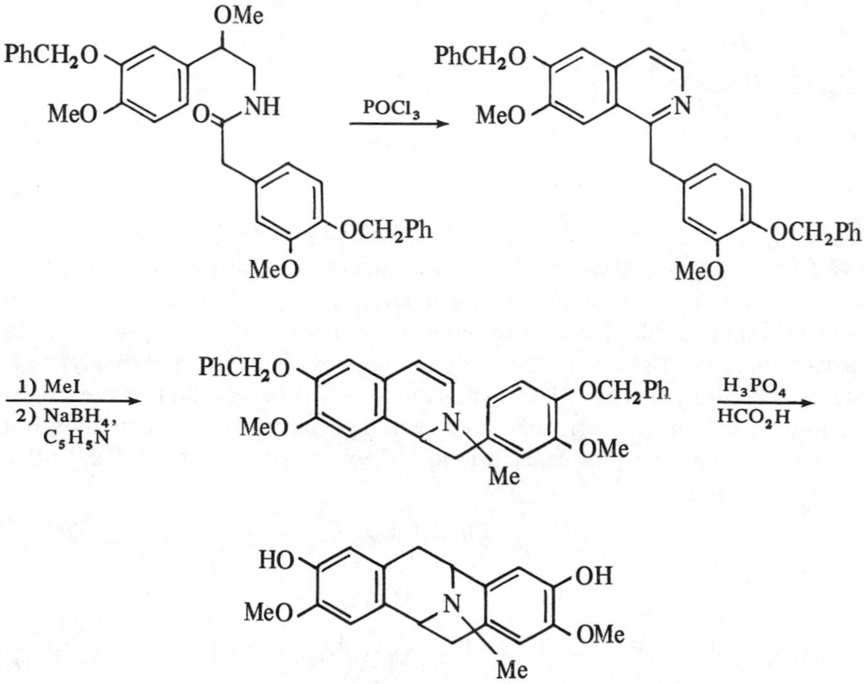

$\xrightarrow{POCl_3}$

$\xrightarrow[2)\ NaBH_4,\ C_5H_5N]{1)\ MeI}$

$\xrightarrow[HCO_2H]{H_3PO_4}$

Chart 1-42L.

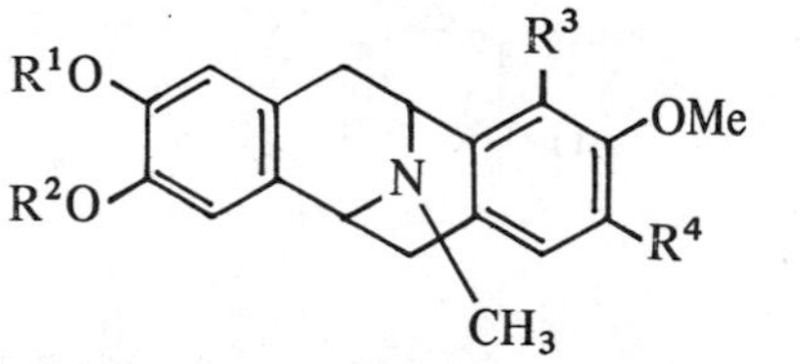

	R^1	R^2	R^3	R^4
Caryachine	$-CH_2-$		H	OH
Platycerine	Me	Me	OH	H

C. Type 2 Synthesis

Introduction

Among isoquinoline syntheses, the synthetic procedure belonging to Type 2 is not generally useful. β-Arylethylamines having active substituents at the *ortho* position give isoquinolines by intramolecular condensation. However, these starting materials are, in general, difficult to synthesize. In many cases, isocarbostyrils are obtained as an intermediate in both methods via isocoumarins and homophthalic acid.

Formation from Isocoumarins (Chart 1-43)

Chart 1-43.

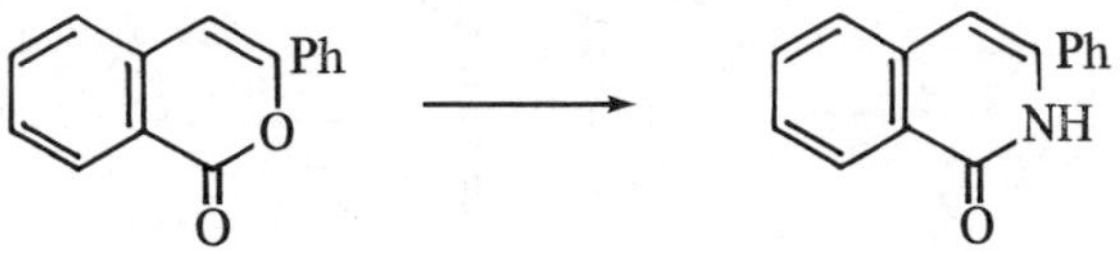

91 → **92**

Primary amines react with isocoumarins to afford the corresponding isoquinolines in many cases. Gabriel[73,74] reported the conversion of 3-phenylisocoumarin **91** into 3-phenylisocarbostyril **92** with ammonia. This reaction was applied to a synthesis of narciprimine isomer.[74a] Conversion of the amino-lactone into the corresponding phenol, followed by ammonolysis, gave the phenanthridinedione,

Chart 1-43A.

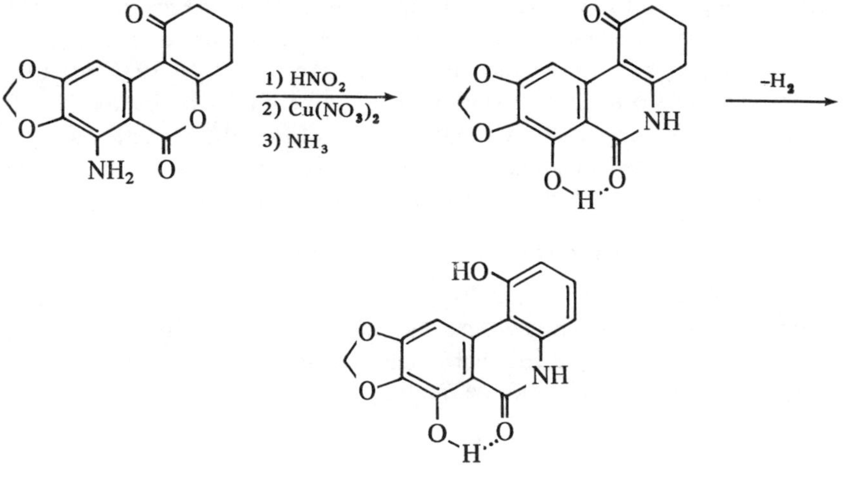

1) HNO_2
2) $Cu(NO_3)_2$
3) NH_3

$-H_2$

which on dehydrogenation yielded the phenanthridone, narciprimine isomer (Chart 1-43A).

Recently the total synthesis of cularimine **93** has been accomplished using this reaction.[75] Cyclization of the acid **94** with polyphosphoric acid gave the lactone **95**, which was converted into the lactam by ammonolysis with ethanolic ammonia. Catalytic hydrogenation of the lactam, followed by reduction with lithium aluminum hydride, gave (±)-cularimine **93**. A synthesis of *O*-ethylcularidine by the same method was also reported[75e] (Chart 1-44). The isoquinoline

Chart 1-44.

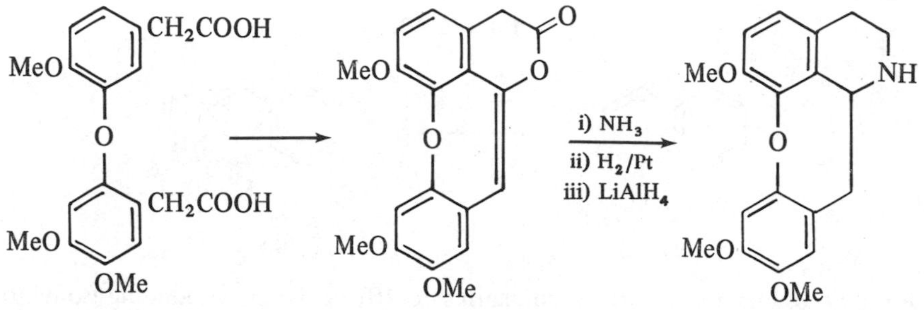

MeO CH$_2$COOH O MeO CH$_2$COOH OMe **94** → MeO O O MeO OMe **95** —i) NH$_3$; ii) H$_2$/Pt; iii) LiAlH$_4$→ MeO NH O MeO OMe **93**

synthesis by the similar routes was reported by many groups[75a-d] (Chart 1-44A). Moreover, this reaction was applied to synthesis of the benzophenanthridine alkaloid, chelerythrine[75f] (Chart 1-44B). However, these reactions have some disadvantages for the synthesis of isocoumarines. Zincke[76] and Bamberger and Kischelt[77] reported the synthesis of isoquinoline from isocarbostyril **96**, which was obtained from isocoumarin-carboxylic acid **97** via **98**. In this case, oxidation of β-naphthoquinone with calcium chloride gave the compound **99** (Chart 1-45).

In the foregoing reaction, the isoquinoline can be obtained directly from isocarbostyril carboxylic acid **98** without isolation of isocarbostyril **96**. Bamberger and Frew[78] converted isocoumarin and its carboxylic acid with ammonia into isocarbostyril and isocarbostyril-3-carboxylic acid, respectively.

Syntheses from Homophthalic Acid and Its Derivatives

Homophthalimide, namely, 1,2,3,4-tetrahydro-1,3-diketoisoquinoline, which was obtained from homophthalic acid **101** with ammonia and also by hydrolysis of 2-cyanobenzyl cyanide **102** with sulfuric acid,[79] was heated with zinc powder[80] or phosphoryl chloride and hydriodic acid to give isoquinoline (Chart 1-46).

Chart 1-44A.

MeO, MeO, CO_2H, PPA, CO_2H, O, OMe, MeO, Δ

MeO, MeO, O, O, OMe, MeO, $MeNH_2$, N, Me

1) H_2
2) B_2H_6

MeO, MeO, N, Me, OMe, MeO

Chart 1-44B.

O, O, O, O, O, O, NH_3, Δ, NH

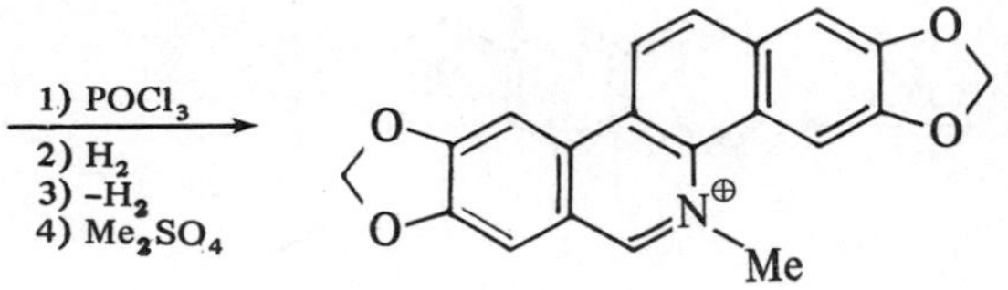

1) $POCl_3$
2) H_2
3) $-H_2$
4) Me_2SO_4

Chelerythrine

Chart 1-45.

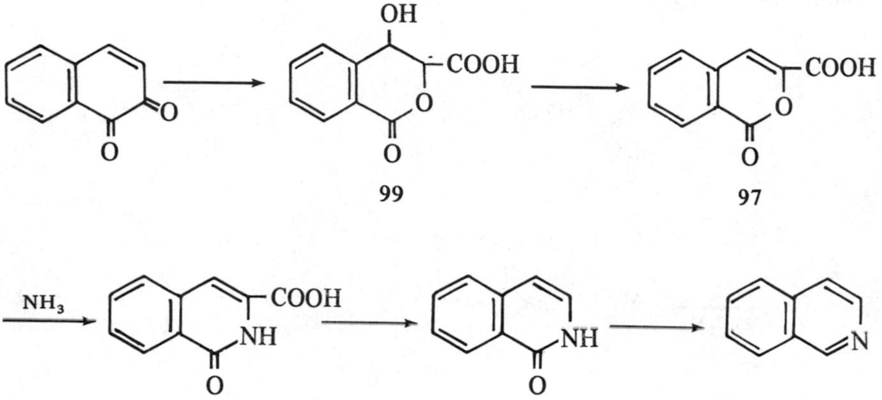

OH COOH O O 99

COOH O O 97

NH_3

COOH NH O 98

NH O 96

N

Chart 1-46.

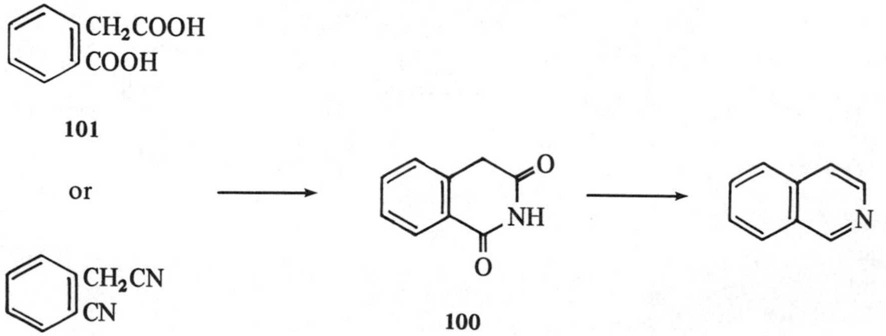

CH_2COOH COOH

101

or

CH_2CN CN

102

O NH O 100

N

Conversion of homophthalic acid into isoquinolines was reported by Grewe and Mondon.[81] Thus 5,6,7,8-tetrahydroisoquinolines, the key intermediate for the morphine skeleton, was prepared by the previous method. For example, treatment of the compound **103**, obtained from homophthalic acid derivative **104** and ammonia in 89% of the yield with phosphoryl chloride, gave the compound **105** in 95% of the yield. The latter was reduced in the presence of nickel catalyst at 80 atm in a current of hydrogen to give the tetrahydroisoquinoline **106** in 95% of the yield (Chart 1-47).

Chart 1-47.

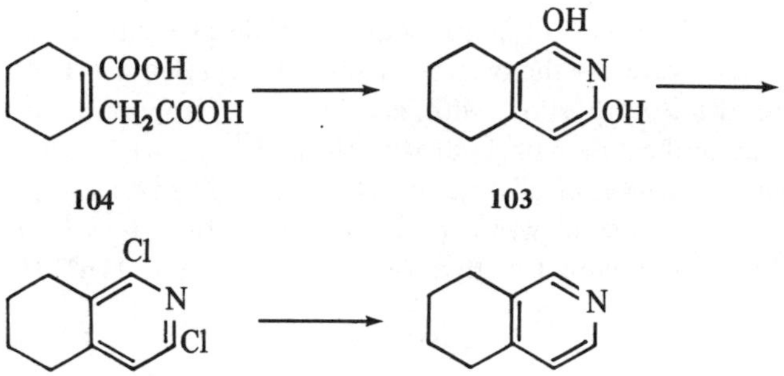

COOH
CH_2COOH
OH
N
OH
104 **103**
Cl
N
Cl
N
105 **106**

Syntheses from Homoxylene Derivatives

Brown and Zobel[82] reported the conversion of homoxylene dibromide **107** to 1,2,3,4-tetrahydro-2-phenylisoquinoline **108** with aniline. In this case, when dimethylamine was used the quaternary bromide of isoquinoline was obtained (Chart 1-48).

Chart 1-48.

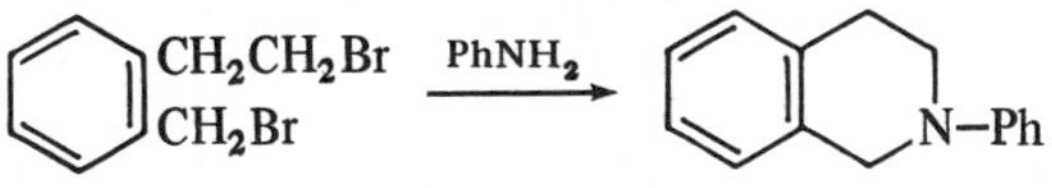

CH_2CH_2Br
CH_2Br
$PhNH_2$
N–Ph
107 **108**

In general, distillation of hydrochlorides of α,ω-diamino compounds, whose amino groups are located four or five methylene groups from each other, gives pyrrolodine or piperidine by elimination of ammonia. Helfer[83] applied this method to the synthesis of 1,2,3,4-tetrahydroisoquinoline **109** by distillation of the hydrochloride of β-(2-aminomethylphenyl)ethylamine **110** (Chart 1-49).

Chart 1-49.

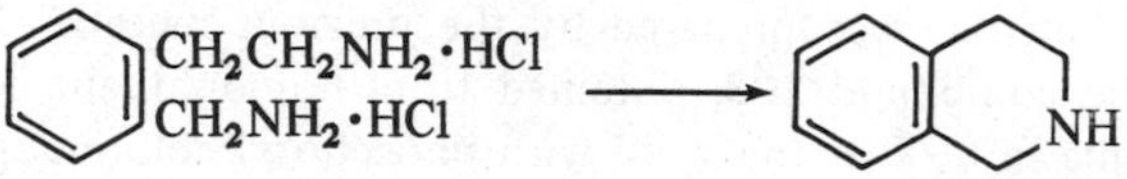

$CH_2CH_2NH_2 \cdot HCl$
$CH_2NH_2 \cdot HCl$ → NH

110 109

Other Type 2 Methods

Bain, Perkin, and Robinson[84] succeeded in the synthesis of 3-carboxy-1-hydroxyisoquinoline **111** by condensation of 2-carboxybenzaldehyde **112** with hippuric acid (Chart 1-50). Treatment of 2-cyanocinnamamide **113** with sodium hypochlorite in methanol gave *N*-ethoxycarbonyl-2-cyanostyrylamine **114**, which was converted to the isocarbostyril with elimination of ammonia, methanol, and carbon dioxide on heating with hydrochloric acid[85] (Chart 1-51). Furthermore, the reductive condensation of α-(2-acetylcyclohexyl)benzyl cyanide **115** with copper chromite in ethanol was carried out successfully to afford two stereoisomers of 2-alkyldecahydro-1-methyl-4-phenylisoquinoline **116**[86] (Chart 1-52).

Chart 1-50.

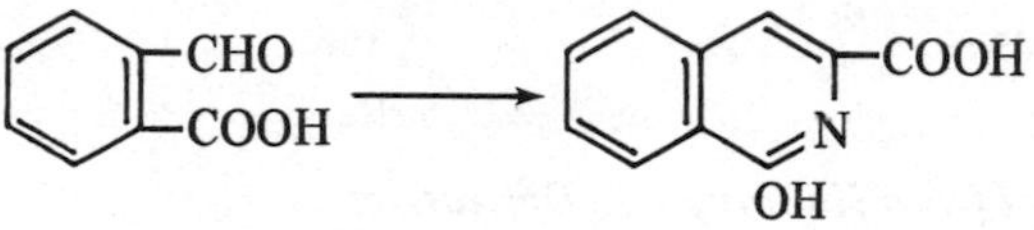

CHO
COOH → COOH
N
OH

112 111

Chart 1-51.

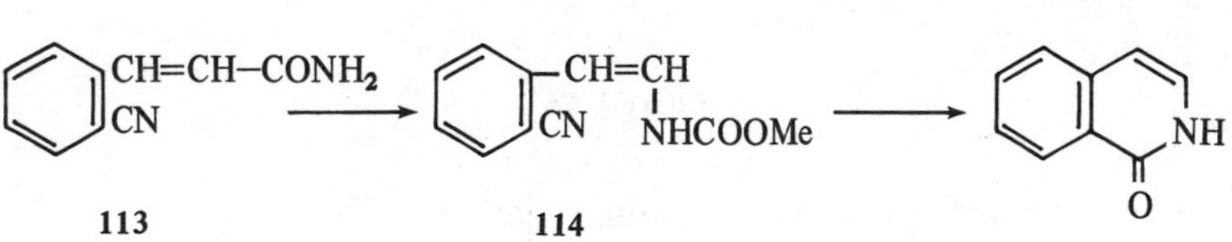

CH=CH–CONH$_2$
CN → CH=CH
CN NHCOOMe → NH
O

113 114

Chart 1-52.

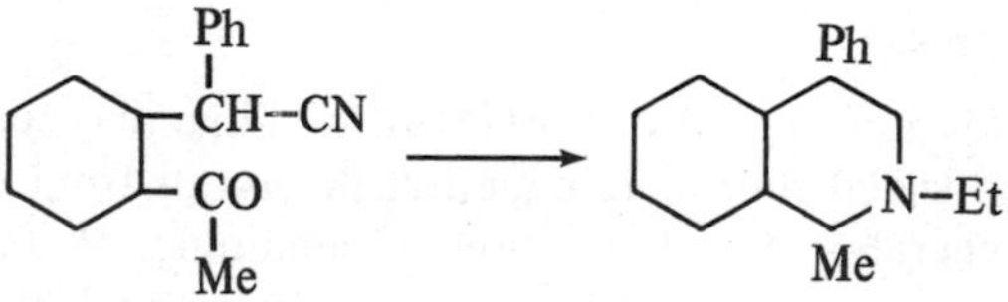

Ph
CH–CN
CO
Me → Ph
N–Et
Me

115 116

Chart 1-53.

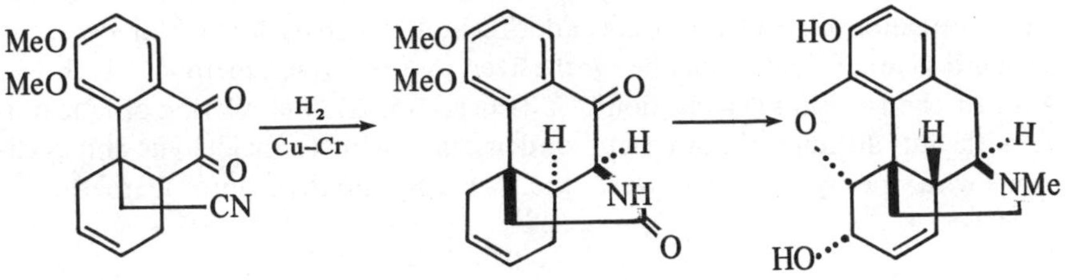

MeO MeO O O CN
H_2 Cu–Cr
MeO MeO O H H NH O
HO O H H NMe HO

117

118 Morphine

Chart 1-53A.

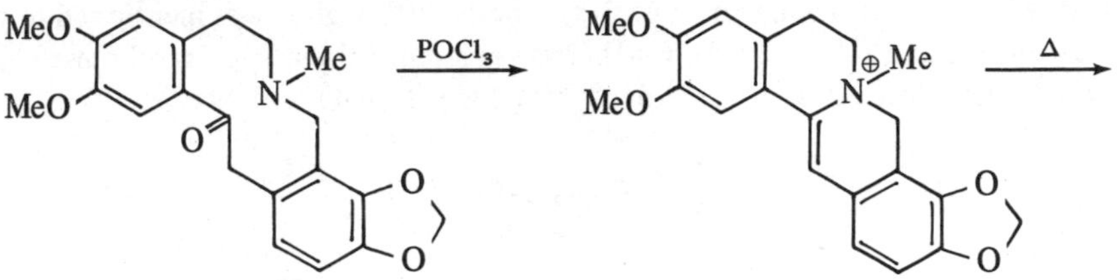

MeO MeO N Me O O O
$POCl_3$
MeO MeO N ⊕ Me O O
Δ

Cryptopine

MeO MeO N O O
Zn–HCl
MeO MeO N O O

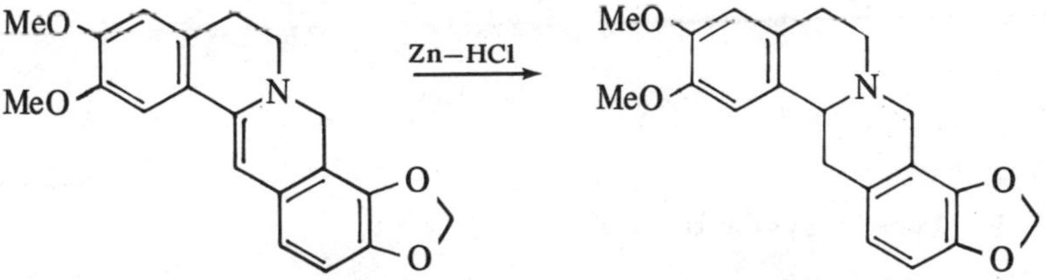

Sinactine

Chart 1-53B.

MeO MeO N Me OMe MeO
MeO MeO N ⊕ Me OMe OMe

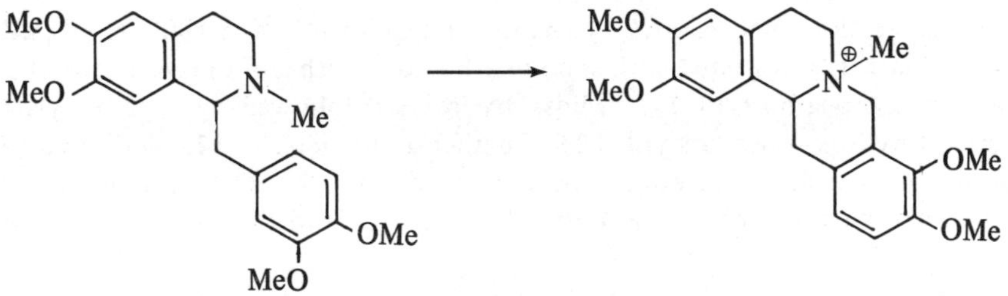

This method was applied to a synthesis of the key intermediate **117** to the synthesis of morphine **118** by Gates and Tschudi[87] (Chart 1-53). Moreover, the protoberberine alkaloids could be synthesized by the Type 2 method, which was a kind of the transannular reaction[87a, b] (Chart 1-53A). Methylation of laudanosine with butyllithium, followed by condensation with formaldehyde and cyclization with methyl chloride, gave the *N*-methyltetrahydroprotoberberine[87c] (Chart 1-53B).

D. Type 3 Synthesis

Type 3 synthesis, which consists of bond formation between the C_3-carbon and the 2-nitrogen atom, like Type 2 synthesis, has relatively few examples recorded. Reduction of the oxime ester **119** in methanol afforded a mixture of two stereoisomers of 1,2,3,4,4a,5,6,7,8,8a-decahydro-3-keto-1-methyl-4-phenylisoquinoline **120**, which might belong to Type 3 synthesis, although it was mentioned in the previous section (Chart 1-54). Furthermore, similar three type syntheses of the isoquinoline derivatives are recorded[88a, b] (Chart 1-54A).

Chart 1-54.

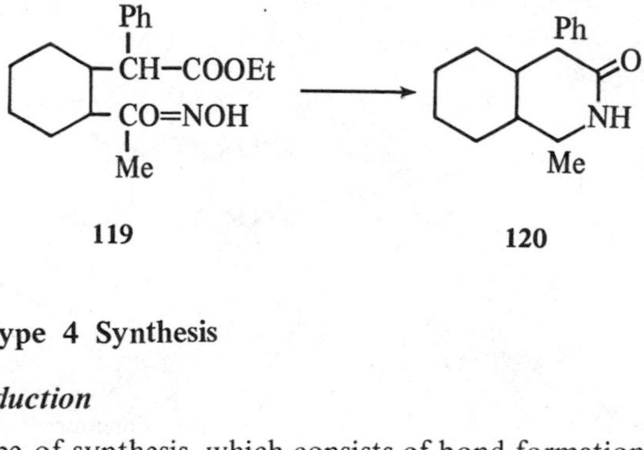

Ph
CH—COOEt
CO=NOH
Me

119

Ph
O
NH
Me

120

E. Type 4 Synthesis

Introduction

The fourth type of synthesis, which consists of bond formation between the C_3- and C_4-carbon atoms, includes a few limited examples. A typical example is reported in the rearrangement of phthalylglycine ester by Gabriel and Colman.

Gabriel-Colman Method

In this reaction, first reported by Gabriel and Colman[88, 89] in 1900, ethyl phthalylglycinate **121** is heated with sodium ethoxide in ethanol to give 3-carbethoxy-4-hydroxyisocarbostyril **122**, whose hydrolysis, followed by decarboxylation, gave 4-hydroxyisocarbostyril **123**. Further reduction of **123** with hydriodic acid and phosphorus gave the expected compound **124**. The isocarbostyril thus obtained is treated with phosphoryl chloride to afford the 1-chloroisoquinoline

Chart 1-54A.

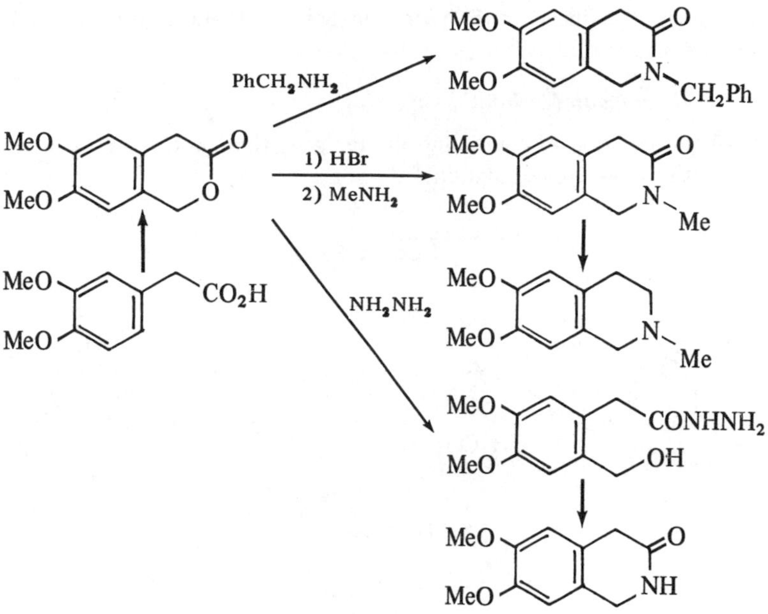

$PhCH_2NH_2$

1) HBr

2) $MeNH_2$

NH_2NH_2

Chart 1-55.

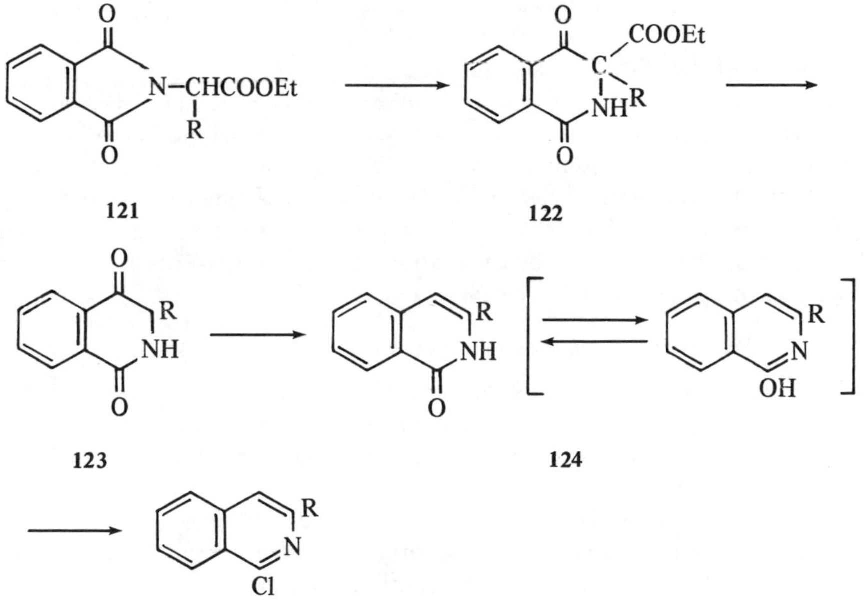

121

122

123

124

125

125, which is reduced with hydriodic acid and phosphorus to give the isoquinoline (Chart 1-55).

Dieckmann Method

Dieckmann reaction of the amino diester gave the β-keto ester, which was converted into the 4-ketoisoquinoline[89a] (Chart 1-55A).

Chart 1-55A.

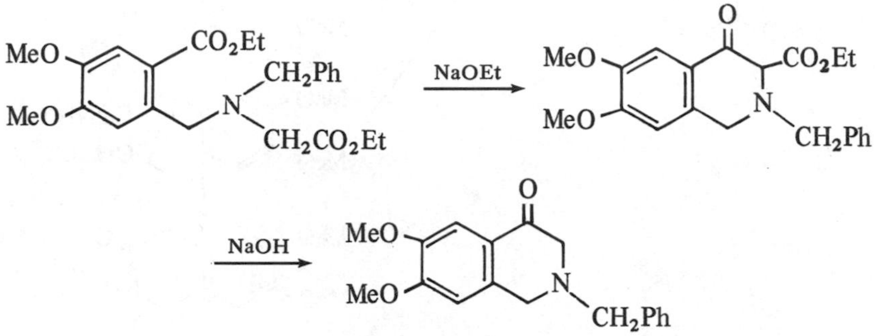

MeO, MeO, CO_2Et, CH_2Ph, N, CH_2CO_2Et —NaOEt→ MeO, MeO, O, CO_2Et, N, CH_2Ph —NaOH→ MeO, MeO, O, N, CH_2Ph

F. Type 5 Synthesis

Introduction

The fifth method consists of cyclization between the C_4- and C_{4a}-position in an aromatic ring and is similar to the modified Skraup reaction. Staub[90] has proposed the structural conditions necessary for the synthesis of isoquinolines; the cyclization occurs only in cases of a compound containing a conjugated double bond or compounds having an OR or OH group on the terminal carbon atom, which could be eliminated on cyclization. The Pomeranz-Fritsch reaction is a typical synthetic method for this type of synthesis.

Pomeranz-Fritsch Reaction[91] *(Chart 1-56)*

This reaction, first reported by Pomeranz[14] and Fritsch[92,93] has been utilized in the synthesis of a variety of isoquinoline derivatives. Cyclization of benzalaminoacetal **126** under acid-catalyzed conditions results in the formation of the expected isoquinoline.

This reaction proceeds in two stages; the first condensation gives the benzalaminoacetal, whose successive cyclization leads to the isoquinoline. In the first step, in which the Schiff base is formed by the reaction of an aromatic aldehyde and an aminoacetal, the reaction proceeds smoothly in good yield. Cyclization of the corresponding benzalaminoacetal is generally carried out with sulfuric

Chart 1-56.

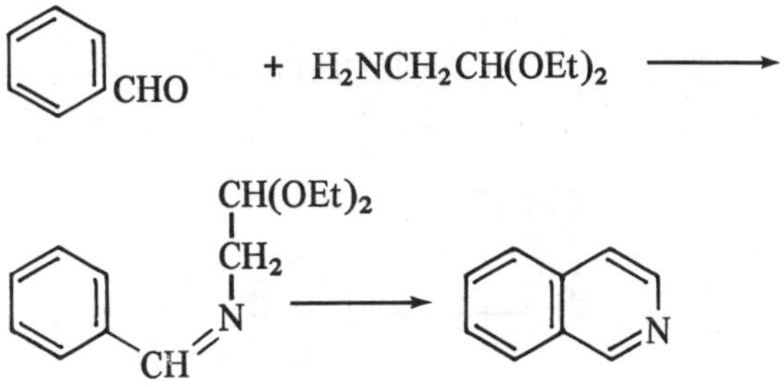

CHO + $H_2NCH_2CH(OEt)_2$ ⟶

$CH(OEt)_2$ — CH_2 — N = CH ⟶ N

126

acid or sulfuric acid containing other acidic reagents. The yield of isoquinolines varies widely according to the conditions. A modification of the Pomeranz-Fritsch reaction using a ketimine instead of aldimine, is suitable for the synthesis of 1-substituted isoquinolines, but the results hitherto reported are disappointing from the point of the yield.[94] The Pomeranz-Fritsch reaction offers the possibility of preparing isoquinolines with substituents, which would be difficult to obtain by the Bischler-Napieralski or the Pictet-Spengler reaction. Furthermore, the Pomeranz-Fritsch method yields a product as a fully aromatic isoquinoline, whereas the partially or fully hydrogenated isoquinolines are obtained in case of the above two reactions using phenethylamines.

Bradsher[95] has compared the Pomeranz-Fritsch cyclization and other aromatic cyclodehydration reactions. Certainly the use of strong acid in bond formation between the acetal carbon and the aromatic nucleus means an electrophilic process; thus the ease of cyclization will depend on the susceptibility of the benzene ring to electrophilic attack. Thus compounds having some groups at the *meta* position to carbonyl group will react under relatively mild conditions, whereas benzaldehyde and halogeno-substituted derivatives will react a higher temperature and in more acidic media and nitrobenzalaminoacetal having a lower activity in the nucleus due to the presence of nitro group does not react at all.

Formation of the Schiff Base

Condensation of aromatic aldehydes with aminoacetals occurs easily in good yield, and the product can be used in the cyclization either with or without purification. In general, the condensation proceeds smoothly when a mixture of aldehyde and aminoacetal is kept aside at room temperature or on the steam bath. An alternative method reported by Schlittler and Müller[94] is available in the reaction of benzylamine with glyoxal semiacetal. Cyclization of the product

Chart 1-57.

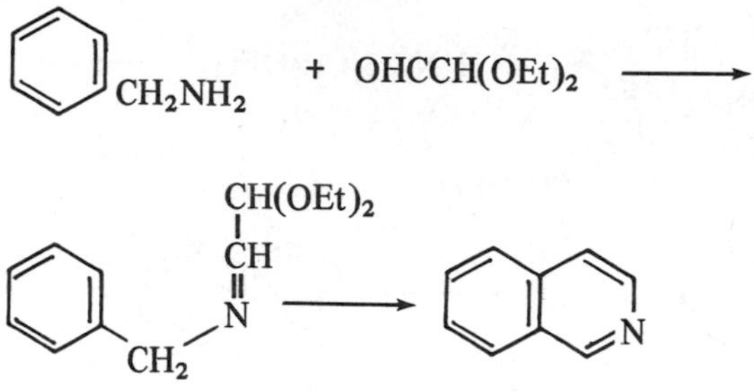

$C_6H_5CH_2NH_2$ + $OHCCH(OEt)_2$ ⟶

$C_6H_5CH_2N{=}CHCH(OEt)_2$ ⟶ **127**

so obtained **127** furnishes the same isoquinoline as that obtained from the Schiff base derived from the aromatic aldehyde and aminoacetal (Chart 1-57).

Cyclization

A variety of methods for cyclization with the use of sulfuric acid was reported. Sulfuric acid has been used in concentrations ranging from fuming acid to approximately 70% sulfuric acid or in admixture with such reagents as gaseous hydrogen chloride, acetic acid, phosphorous pentoxide, and phosphoryl chloride. Furthermore, the cyclization has been carried out at temperatures from 0° or below in case of alkoxy- or hydroxybenzalamines to 150-160° in case of halobenzalaminoacetals. The enol form of the benzoyl aminoacetal **128**, which involves a tautomeric double bond, is cyclized with sulfuric acid to afford the isocarbostyril[77] (Chart 1-58). Recently, Kametani[96] reported that the cyclization of the benzalaminoacetal derivative **129** with polyphosphoric acid at 65-70° for 3 hr afforded the isoquinoline **130** in good yield (Chart 1-59).

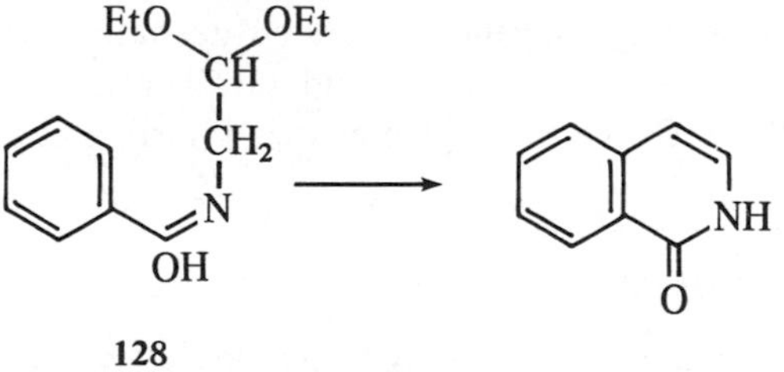

Chart 1-58.

$C_6H_5C(OH){=}NCH_2CH(OEt)_2$ (**128**) ⟶ isocarbostyril (NH, C=O)

Although the yields for Pomeranz-Fritsch method vary from zero to more than 80%, the yields are mostly below 50%. In case of 3-alkoxy-, 3-hydroxy-, and 3-

Chart 1-59.

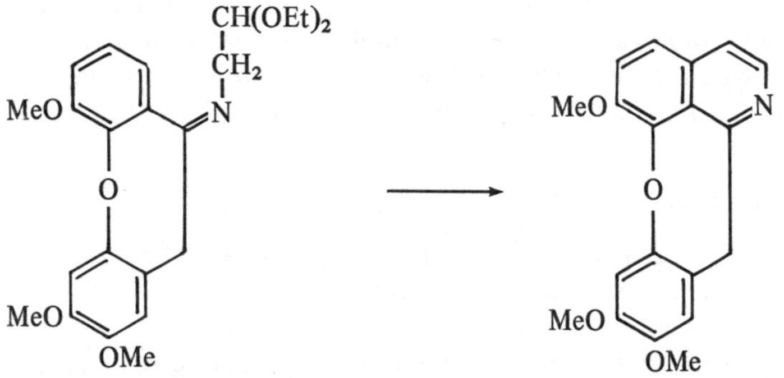

$$\text{129} \longrightarrow \text{130}$$

halobenzalaminoacetals, satisfactory results are obtained, whereas 2- or 4-alkoxyl (or hydroxyl) derivatives afford the isoquinolines in low yield or no products. The yields of isoquinolines are remarkably effected by the conditions used in cyclization and especially by the concentration of sulfuric acid, as illustrated by 3-ethoxybenzal-,[93] 3-hydroxybenzal-,[97] and 3,4-methylenedioxybenzalamino-acetal.[93] In case of 3-hydroxybenzalaminoacetal, the cyclization with 84, 82, 80, 78, 76, and 62% sulfuric acid affords the corresponding isoquinoline in 31, 44, 64, 59, 43 and 30% of the yield. A small deviation from the optimum acid concentration results in a remarkable decrease in yield. Variation of the yields with acid concentration may be attributed, at least in part, to the fact that competitive hydrolytic cleavage of the Schiff base may occur under conditions of cyclization.

Orientation of Cyclization

Cyclization of unsymmetrically substituted benzalaminoacetals, in which the two positions *ortho* to the aldimine group are not occupied, may lead to one or both of two isomeric isoquinolines. For instance, 3-ethoxybenzalaminoacetal affords 7-ethoxyisoquinoline in more than 80% of the yield.[93] 3-Hydroxybenzal-aminoacetal is transformed to a mixture of 7-hydroxyisoquinoline as a main product and 5-hydroxyisoquinoline as a by-product.[98] In case of 3,4-methylene-dioxybenzalaminoacetal, only 6,7-methylenedioxyisoquinoline is obtained, and 3,4-dimethoxybenzalaminoacetal yields 6,7-dimethoxyisoquinoline.[99,100] The application of the Pomeranz-Fritsch synthesis as a preparative method for isoquinolines is severely limited because of the yields. Actually, only 3-hydroxy- and 3-alkoxybenzaldehyde have been converted to the corresponding isoquinolines in more than 50% of the yield, and this synthetic method is more satisfactory than the synthesis of the corresponding 7-substituted 1,2,3,4-tetrahydroisoquinolines, which would be derived from the phenethylamine by Pictet-Spengler reaction. Without consideration of the yield, the Pomeranz-Fritsch synthesis is applicable to the preparation of a variety of substituted isoquinolines. For example, 8-substituted isoquinolines are obtained from the *ortho*-substituted benzaldehydes, whereas 8-substituted isoquinolines are generally not obtained from *meta*-substituted arylethylamines by the Bischler-Napieralski reaction.

Modification of the Pomeranz-Fritsch Reaction

When the ketone is used instead of the aromatic aldehyde in the Pomeranz-Fritsch reaction, the 1-substituted isoquinoline is obtained. For instance, 1-methylisoquinoline **131** is formed from acetophenone **132**, but the extension of this synthesis mostly results in poor yields, possibly due to difficulty in the condensation of the ketones with aminoacetals to yield Schiff bases. For the synthesis of 1-substituted isoquinolines, the Schlittler-Müller reaction of the Schiff bases is very useful. Compared with the difficulty of condensation of an aminoacetal with a ketone, a relatively facile formation of Schiff base is expected. α-Phenylethylamine **133** is first converted to the Schiff base with glyoxal semiacetal and then, on treatment with concentrated sulfuric acid at 160°, 1-methylisoquinoline is obtained.[94] The yield, which is given as 40%, is an improvement by comparison to the reaction between acetophenone and aminoacetal (Chart 1-60). Similarly, 7-methoxy-1-methylisoquinoline was obtained from α-(3-methoxyphenyl)ethylamine in 37.5% of the yield, whereas the yield from 3-methoxyacetophenone and aminoacetal was only 0.1%.[94]

Chart 1-60.

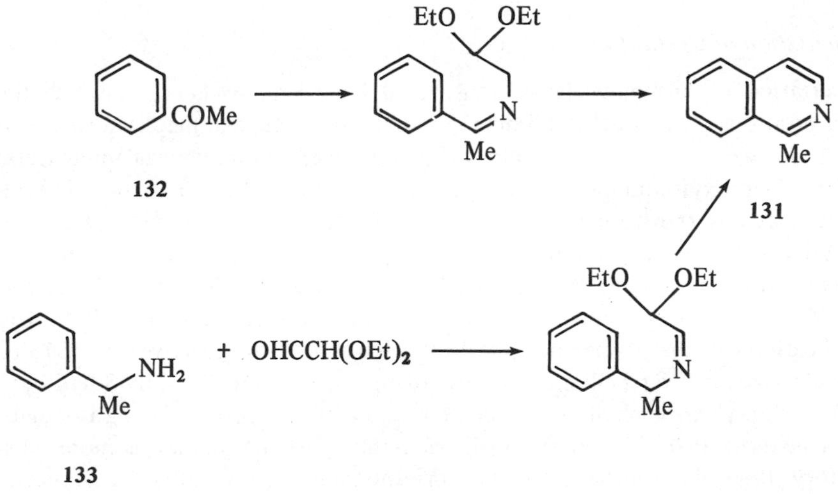

COMe

132

EtO OEt

N

Me

N

Me

131

EtO OEt

NH$_2$

Me

133

+ OHCCH(OEt)$_2$

N

Me

Recently, Bobbitt et al.[101] improved this reaction and obtained 1,2,3,4-tetrahydroisoquinoline derivatives in good yield. Thus, the Schiff base **134**, obtained from the aromatic aldehyde **135** and the aminoacetal, was reduced with hydrogen and platinum oxide to give a secondary amine **136**, which was cyclized with 6N hydrochloric acid. The resulting 1,2,3,4-tetrahydro-4-hydroxyisoquinoline **137** was hydrogenolyzed with hydrogen and 5% palladium-carbon to afford 1,2,3,4-tetrahydroisoquinoline **138** (Chart 1-61). This isoquinoline **138** was converted into petaline **144** by Grethe, Uskokovic, and Brossi.[102]

Chart 1-61.

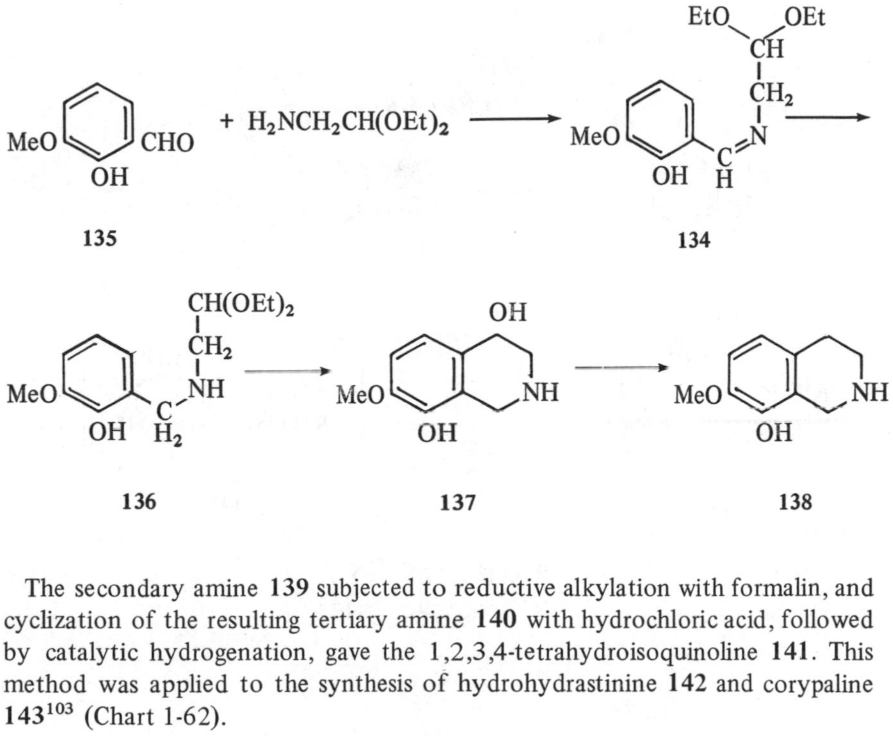

MeO, OH, CHO + $H_2NCH_2CH(OEt)_2$ → MeO, OH, EtO, OEt, CH, CH_2, N, C, H

135 **134**

MeO, OH, $CH(OEt)_2$, CH_2, NH, C, H_2 → MeO, OH, OH, NH → MeO, OH, NH

136 **137** **138**

The secondary amine **139** subjected to reductive alkylation with formalin, and cyclization of the resulting tertiary amine **140** with hydrochloric acid, followed by catalytic hydrogenation, gave the 1,2,3,4-tetrahydroisoquinoline **141**. This method was applied to the synthesis of hydrohydrastinine **142** and corypaline **143**[103] (Chart 1-62).

Reductive condensation of an aromatic aldehyde and an alkyl amine, followed by *N*-alkylation of the secondary amine **144** with glycerol, gave the tertiary amine **145**, which was converted into an unstable α-aminoaldehyde **146** with periodate. This was subjected to cyclization with hydrochloric acid and then catalytic hydrogenation afforded the 2-alkyl-1,2,3,4-tetrahydroisoquinolines **147**[103] (Chart 1-63).

The Bobbitt modification could be applied to a total synthesis of the simple isoquinoline alkaloids (anhalonidine, pellotine, tepenine, tehaunine, and *O*-methylgigantine)[103a,b] and benzylisoquinoline alkaloids (escholamine and

Chart 1-62.

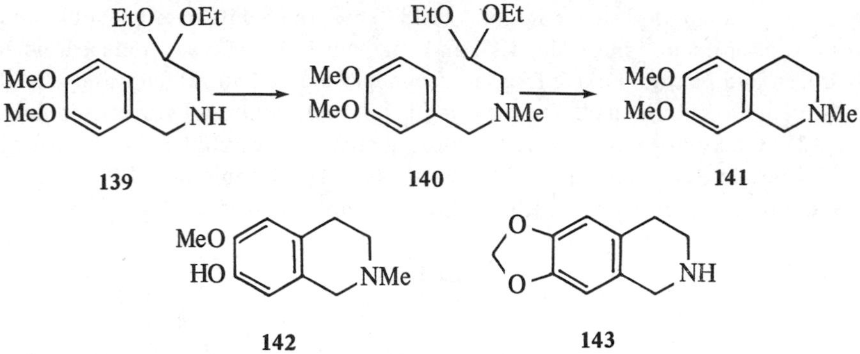

139 140 141 142 143

Chart 1-63.

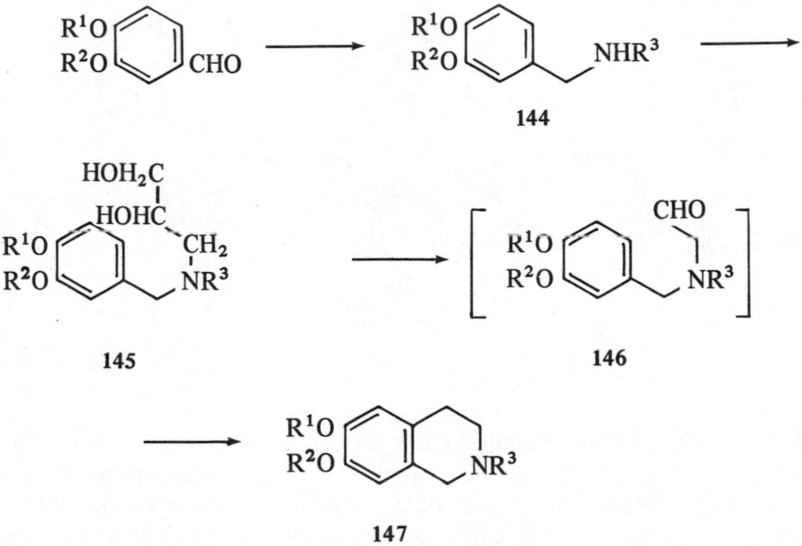

144 145 146 147

takatonine).[103c] Furthermore, the new synthetic routes for the ochotensine type[103d] and protoberberine type alkaloids[103e] were developed using this reaction (Chart 1-63A). This important intermediate for the benzophenanthridine alkaloid synthesis is obtained by Pictet-Spengler reaction[103f] (Chart 1-63B).

The Other Procedures

Grethe, Uskokovic, and Brossi generalized the following type of reaction[104] and synthesized so-called "gigantine" **148**.[105] The aldehydes or ketones were condensed with the amine to yield the Schiff-bases **149**, which were converted

Chart 1-63A.

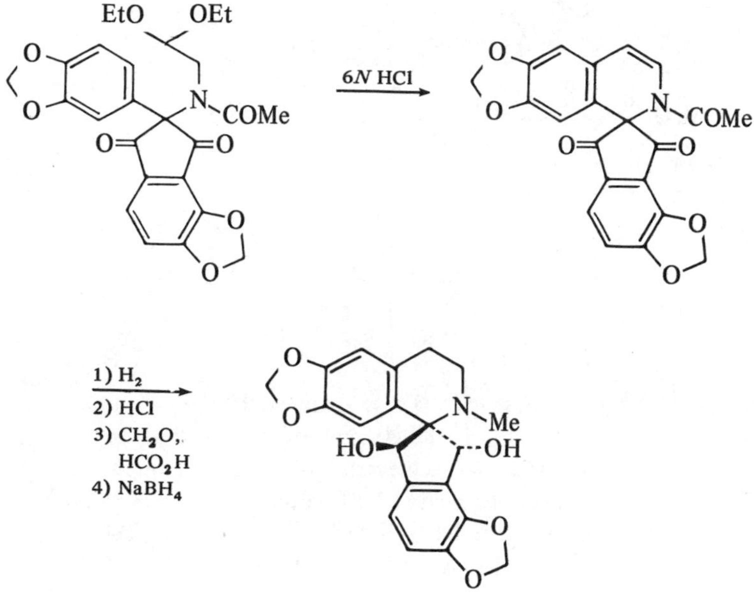

EtO OEt

N COMe

O O

O O

6*N* HCl

1) H_2
2) HCl
3) CH_2O, HCO_2H
4) $NaBH_4$

N Me

HO OH

Ochrobirine

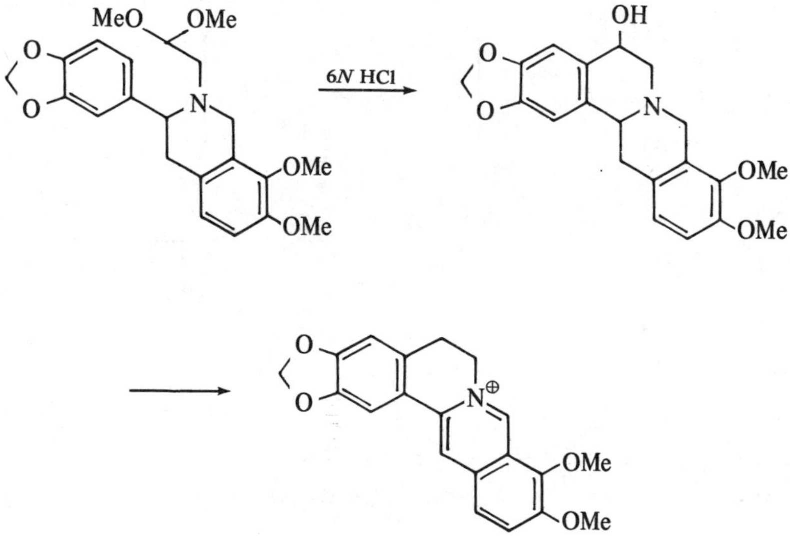

MeO OMe

N

OMe

OMe

6*N* HCl

OH

N

OMe

OMe

N⊕

OMe

OMe

Berberine

Chart 1-63B.

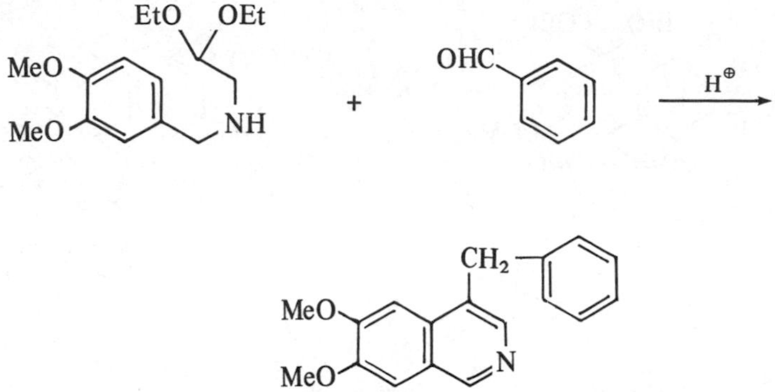

EtO OEt
MeO
MeO
NH
+
OHC
$H^{\oplus}$

CH_2
MeO
MeO
N

readily into benzylamines **150**. These amines **150**, on reaction with halo-acetates, were transformed into the glycine ester derivatives **151**. Formation of the isoquinoline ring **152** was achieved by cyclizing compounds **151** in the presence of 70 to 90% sulfuric acid at 100° (Chart 1-64). The 4-ketoisoquinoline was transformed into *O*-methylcherylline[105,105a] (Chart 1-64A). The cyclization of α-aminoether was applied to a synthesis of tlyphorine[105b,c] (Chart 1-64B).

Chart 1-64.

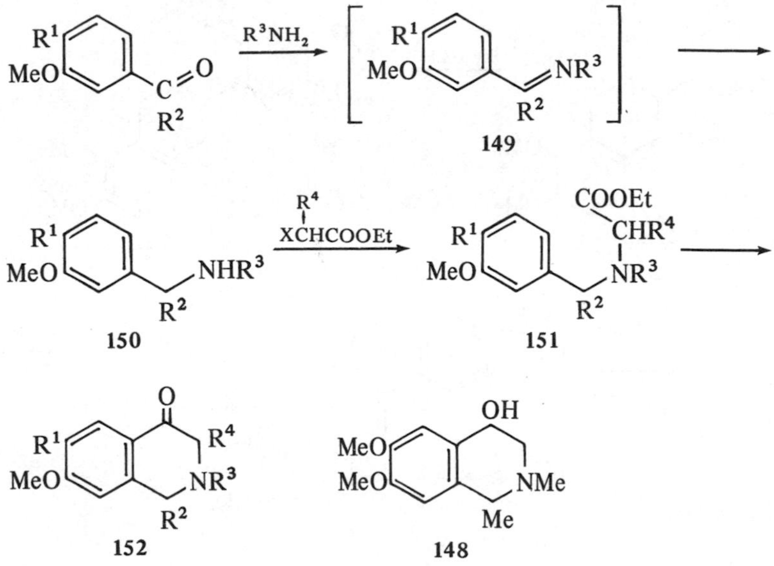

R^1 MeO C=O R^2 —R^3NH_2→ [R^1 MeO NR^3 R^2] **149** →

R^1 MeO NHR^3 R^2 **150** —R^4 XCHCOOEt→ R^1 MeO COOEt CHR^4 NR^3 R^2 **151** →

O R^1 MeO R^4 NR^3 R^2 **152**

OH MeO MeO NMe Me **148**

Chart 1-64A.

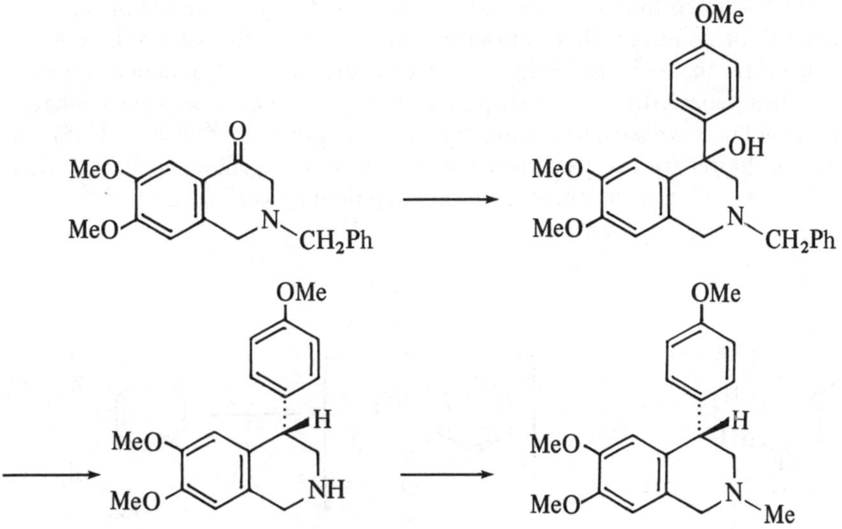

Chart 1-64B.

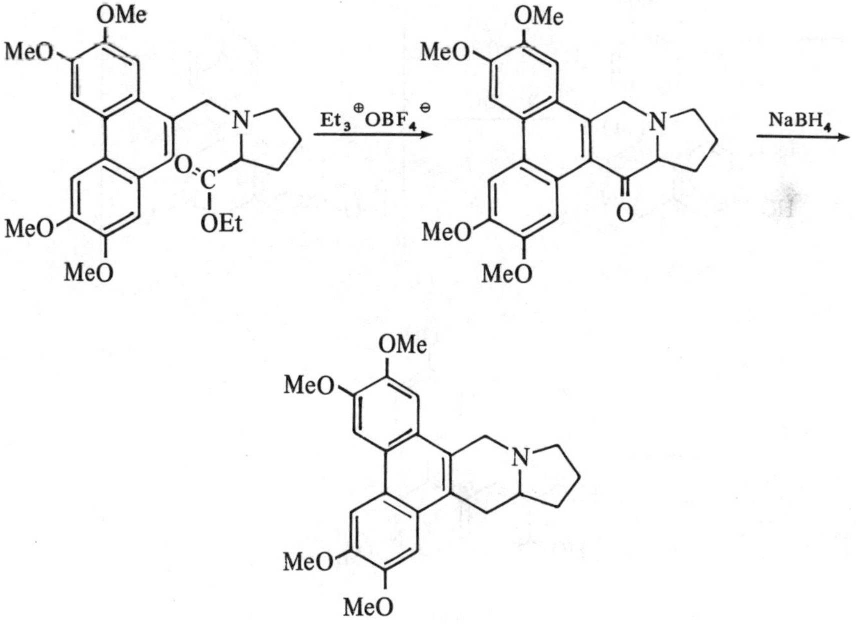

$Et_3^{\oplus}OBF_4^{\ominus}$

$NaBH_4$

Tylophorine

On the other hand, treatment of hippuric acid with phosphorous pentachloride afforded 1,4-dichloroisoquinoline **153**, though 4-hydroxyisocarbostyril would be assumed to be an intermediate. However, in a similar reaction of 3,4-dimethoxybenzalglycine, the expected isoquinoline derivative was not obtained. Furthermore, when benzalethylamine was passed through the hot tube, a small amount of isoquinoline was separated from the reaction products[106] (Chart 1-65). An interesting biogenetically patterned synthesis of (±)-cherylline has been carried out by a treatment of the norbelladine type compound with mild base[107] (Chart 1-65A).

Chart 1-65.

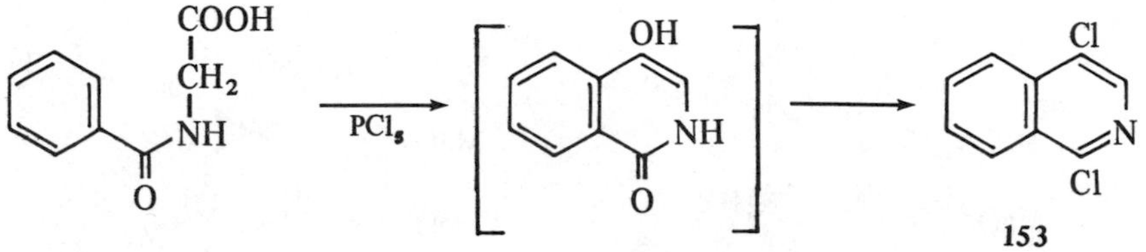

COOH CH₂ NH O —PCl₅→ [OH NH O] → Cl N Cl

153

Chart 1-65A.

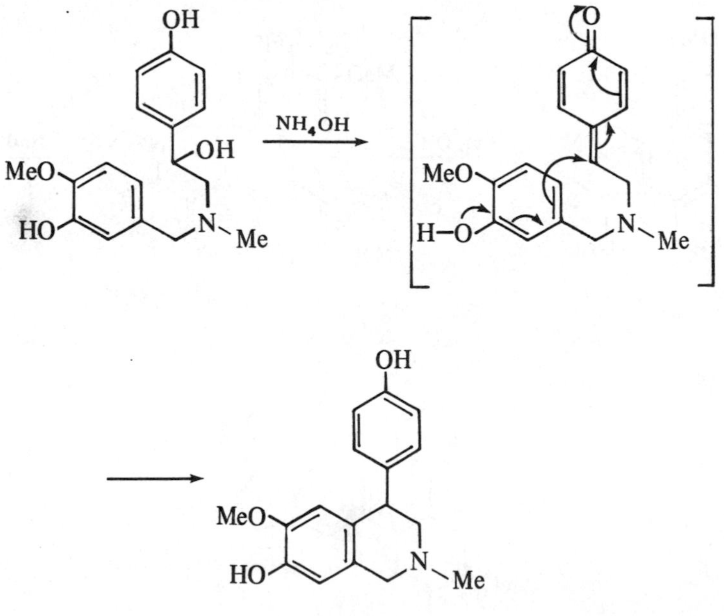

OH OH MeO HO N Me —NH_4OH→ [O MeO H–O N Me] →

OH MeO HO N Me

Cherylline

2. STEREOCHEMICAL PROBLEM IN THE SYNTHESIS OF ISOQUINOLINE ALKALOIDS

In this section the stereochemical problems in the total synthesis of isoquinoline alkaloids are described. The method for the synthesis of optically active isoquinoline alkaloids can be classified as follows: (A) optical resolution of racemic synthetic alkaloids, (B) syntheses of optically active isoquinolines by using optically active intermediates, and (C) stereospecific total synthesis.

Chart 2-1.

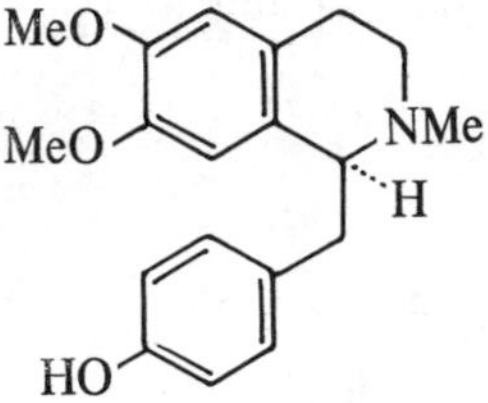

MeO MeO NMe H HO **1a**

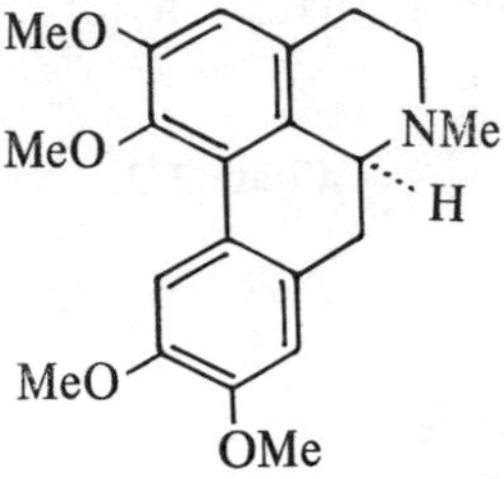

MeO MeO NMe H MeO OMe **2a**

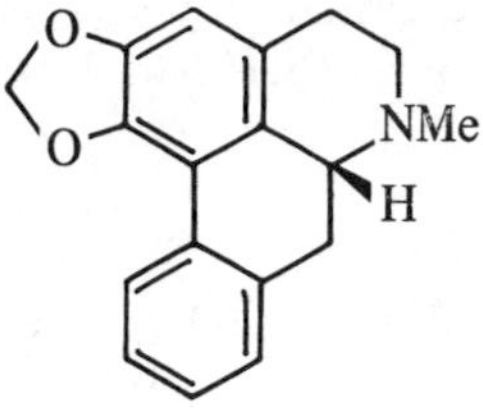

O O NMe H **3**

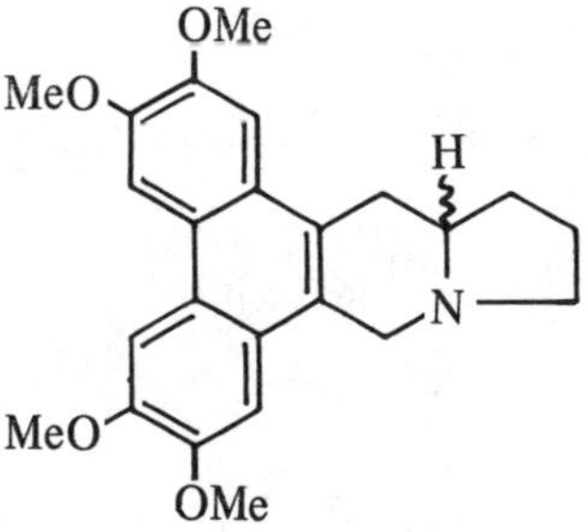

OMe MeO H N MeO OMe **4**

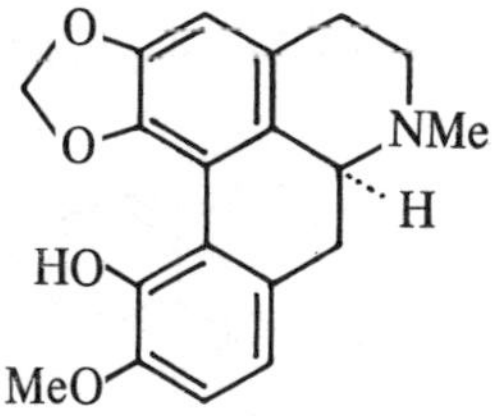

O O NMe H HO MeO **5**

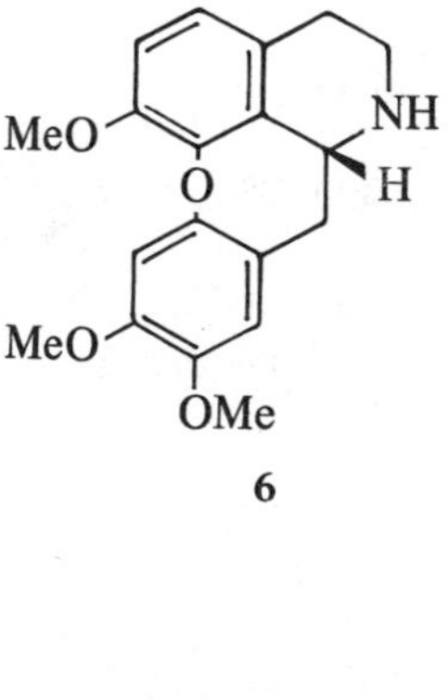

MeO O NH H MeO OMe **6**

A. Total Syntheses by Resolution of Racemate

At first we discuss some examples belonging to category (A), which has been used for the synthesis of isoquinolines having one asymmetric carbon atom. For instance, total syntheses of armepavine **1**, glaucine **2**, roemerine **3**, tylophorine **4**, bulbocapnine **5**, and cularimine **6** were accomplished by the optical resolution of these synthetic racemates (Chart 2-1).

Bischler-Napieralski reaction of the amide **7**, which was obtained by the Schotten-Baumann reaction of 3,4-dimethoxyphenethylamine with acid chloride **8**, gave 3,4-dihydroisoquinoline **9**, which was converted to *O*-benzoylarmepavine **10**. Hydrolysis of **10** gave racemic armepavine **1**, whose resolution with (+)-camphor-10-sulfonic acid afforded S(+)-armepavine **1a**[108] possessing the same configuration at the C_1-position as that of natural armepavine **1a** and its optical isomer **1b**, (Chart 2-2).

Chart 2-2.

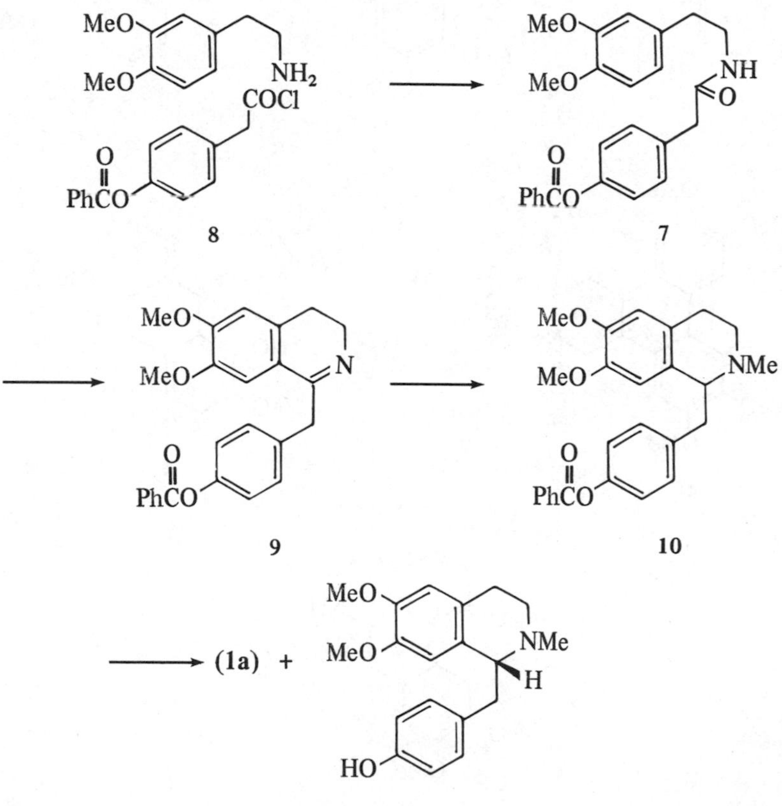

MeO MeO NH₂ COCl O PhCO **8** → MeO MeO NH O O PhCO **7**

→ MeO MeO N O PhCO **9** → MeO MeO NMe O PhCO **10**

→ (**1a**) + MeO MeO NMe H HO **1b**

Similarly, Pschorr reaction of 2′-aminolaudanosine **11**, which was obtained by reduction of the nitro derivative **13** of synthetic laudanosine **12**,[109] afforded (±)-glaucine **2**,[110] whose resolution with di-*p*-toluoyltartaric acid yielded (+)-glaucine **2a**. Pschorr reaction of 2′-amino derivative **14**, followed by resolution of (±)-roemerine, also gave (−)-isomer (roemerine **3a**).[111] (±)-Bulbocapnine **15** was also resolved to give (+)-bulbocapnine **16**,[112] which was confirmed by the X-ray analysis[113] (Charts 2-3 and 2-4).

Chart 2-3.

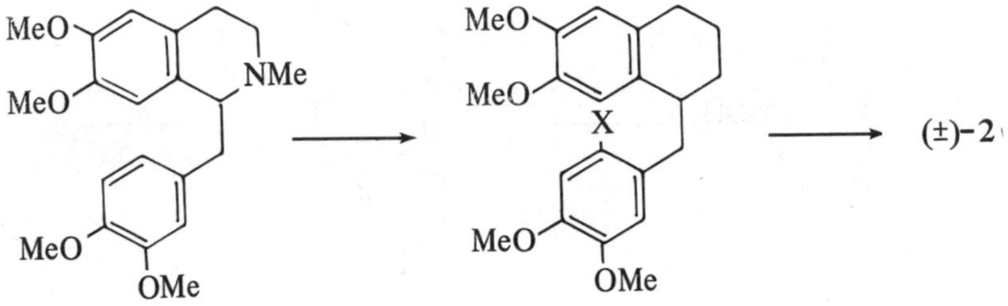

MeO, MeO, NMe, MeO, OMe → MeO, MeO, X, MeO, OMe → (±)-2

12

11 X = NH_2
13 X = NO_2

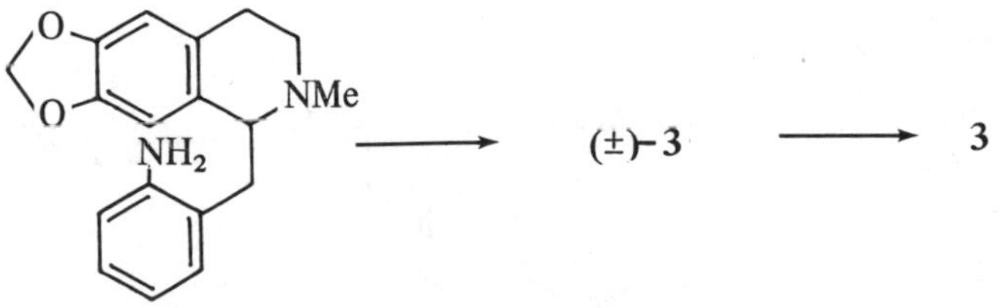

O, O, NMe, NH_2 → (±)-3 → 3

14

Chart 2-4.

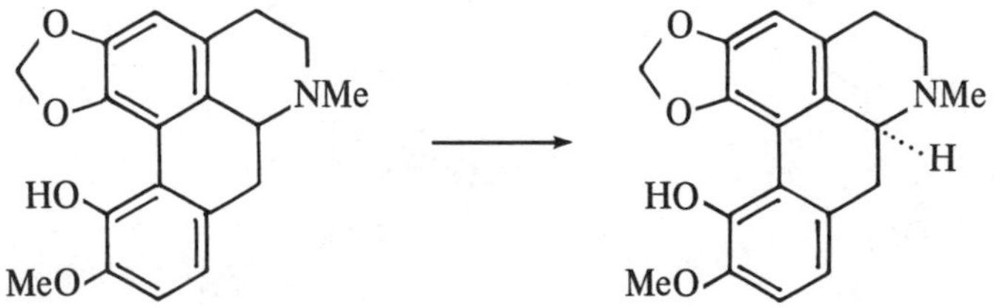

O, O, NMe, HO, MeO → O, O, NMe, H, HO, MeO

15 **16**

Cyclization of the dicarboxylic acid **17** with polyphosphoric acid gave the lactone **18**, which was converted into lactam **19** by ammonolysis. Catalytic hydrogenation of **19**, followed by reduction of **20** with lithium aluminum hydride, gave (±)-cularimine **21**,[114] whose optical resolution with tartaric acid or di-*p*-toluoyltartaric acid afforded (+)-cularimine **6** and (−)-isomer.[115] On the other hand, the stereochemistry of natural (+)-cularine was confirmed by X-ray analysis[116] as 22 (Chart 2-5).

Chart 2-5.

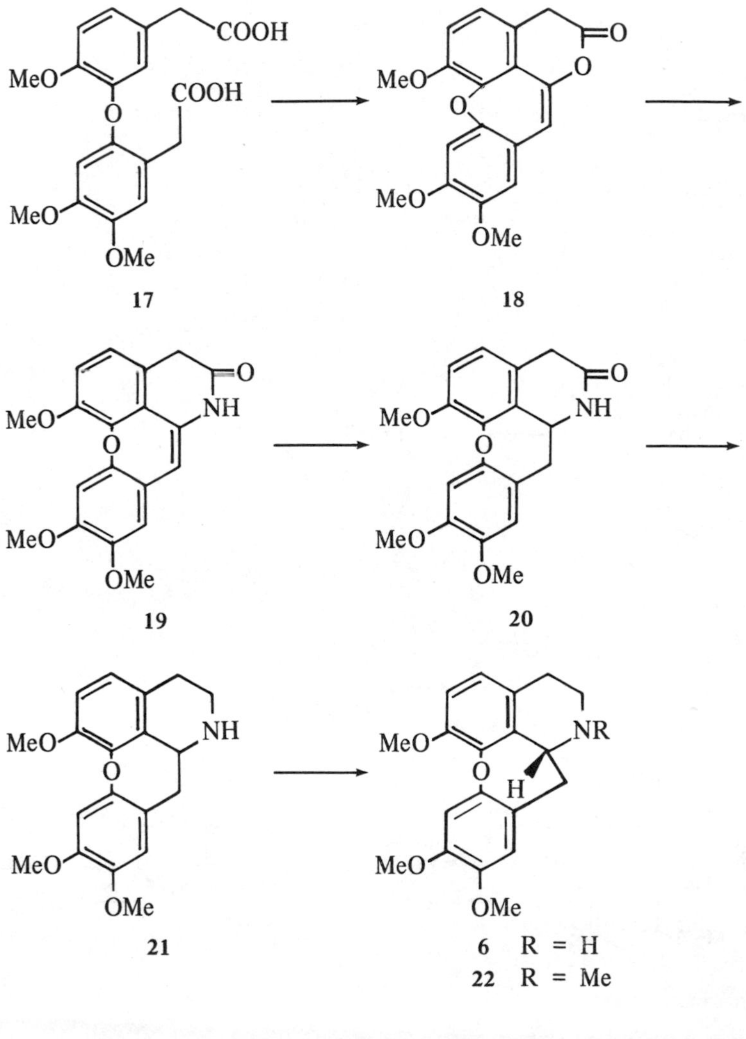

MeO, COOH, COOH, O, MeO, OMe

17

MeO, O, O, MeO, OMe

18

MeO, O, NH, O, MeO, OMe

19

MeO, O, NH, O, MeO, OMe

20

MeO, NH, O, MeO, OMe

21

MeO, NR, O, H, MeO, OMe

6 R = H
22 R = Me

Chart 2-6.

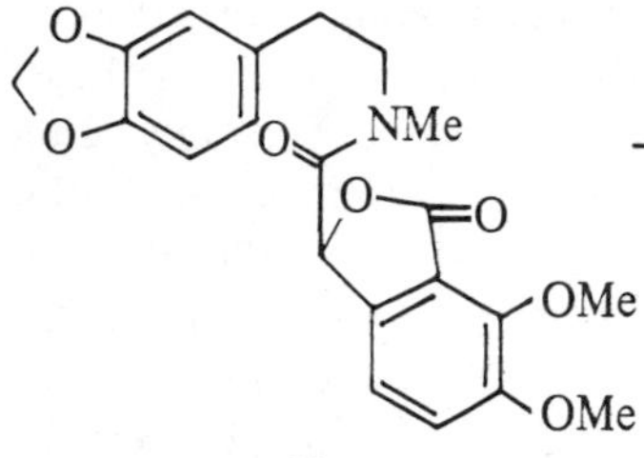

23

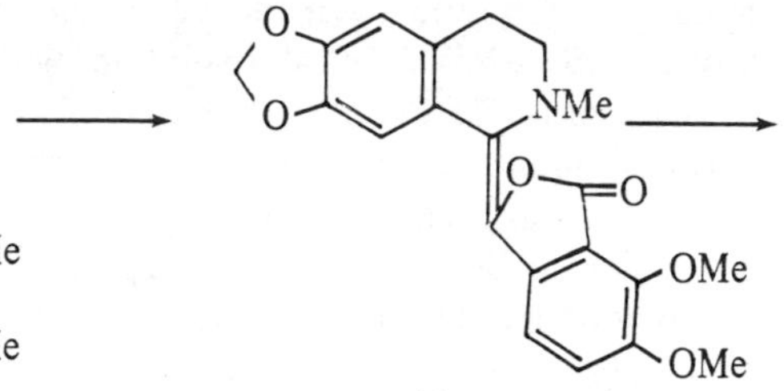

24

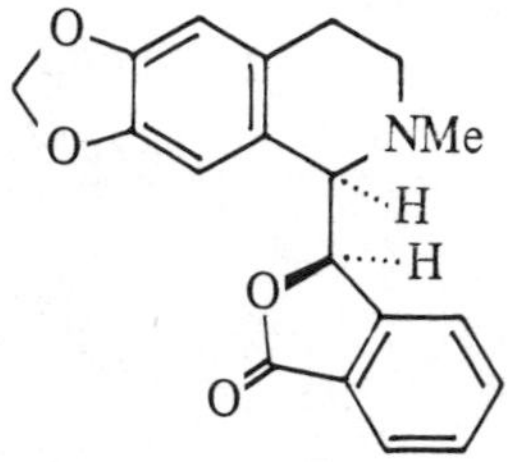

25 (±)-base
26 (−)-β-base

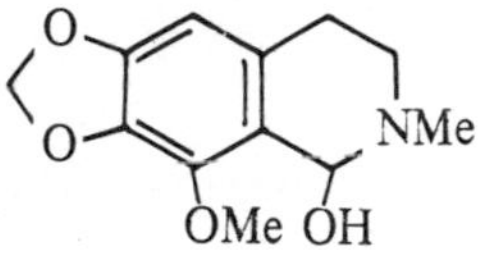

27

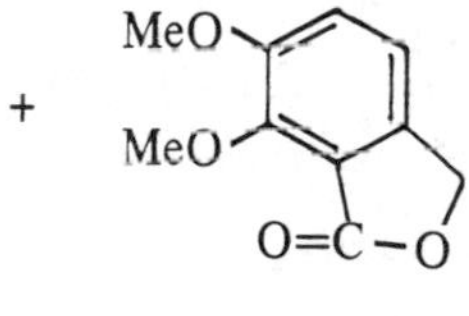

28

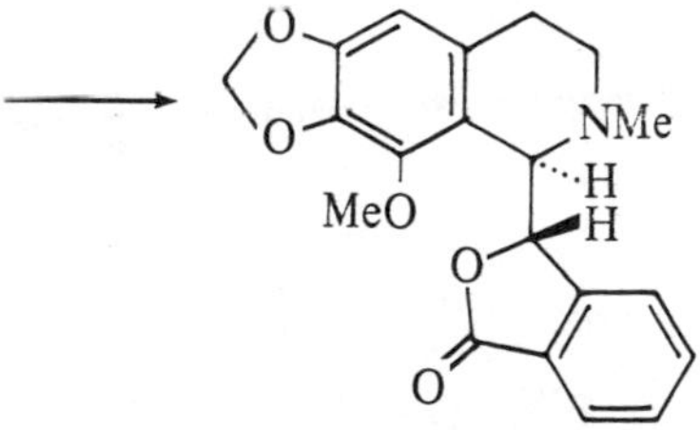

29 (±)-base
30 (−)-α-base

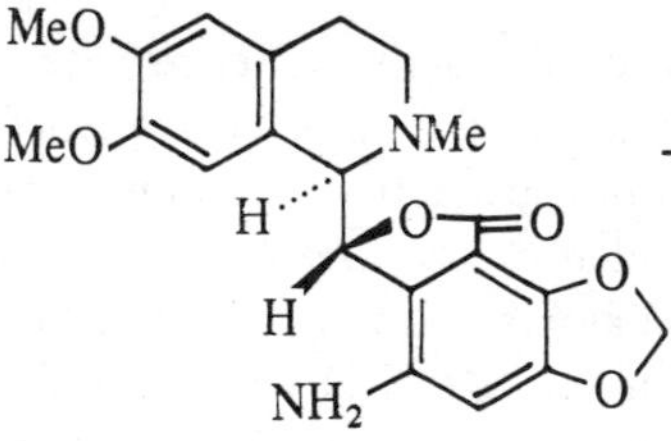

31

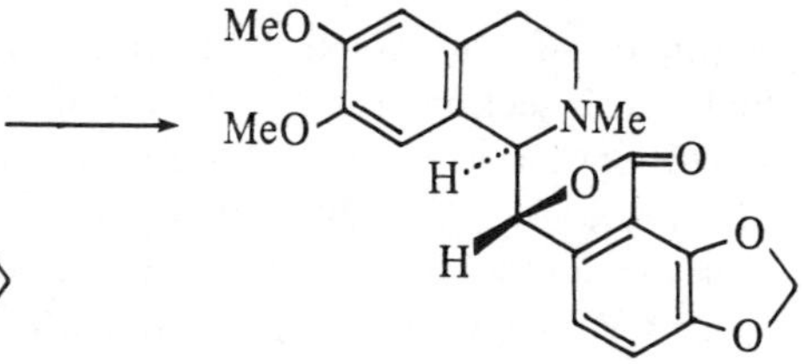

32 (±)-base
32a (−)-base
32b (+)-base

Optical resolution of some (±)-phthalideisoquinolines was carried out in the course of the total syntheses of these alkaloids. Bischler-Napieralski reaction of an amide **23**, followed by reduction of the cyclized product **24**, afforded (±)-hydrastine **25**, whose optical resolution with tartaric acid gave (−)-β-isomer **26**.[117] Optical resolution of (±)-narcotine **29**, which was obtained by condensation of isoquinoline **27** with the lactone **28**, resulted in the formation of (−)-α-narcotine **30**[118] as in case of **26**. The stereochemistry of these phthalide isoquinolines was examined by many investigators.[119,120,121] Sandmeyer reaction of aminoisoquinoline **31** gave (±)-corlumine **32**, whose resolution gave (−)-isomer **32a** and (+)-isomer **32b**[122] (Chart 2-6). As an another example, synthesis of (−)-tylophorine **4**[123] is cited in this category. The treatment of *N*-formyl derivative **33** with phosphoryl chloride, followed by the reduction of the quaternary salt with sodium borohydride, gave (±)-tylophorine, whose resolution with (+)-camphor-10-sulfonate yielded (−)-tylophorine **4** (Chart 2-7).

Chart 2-7.

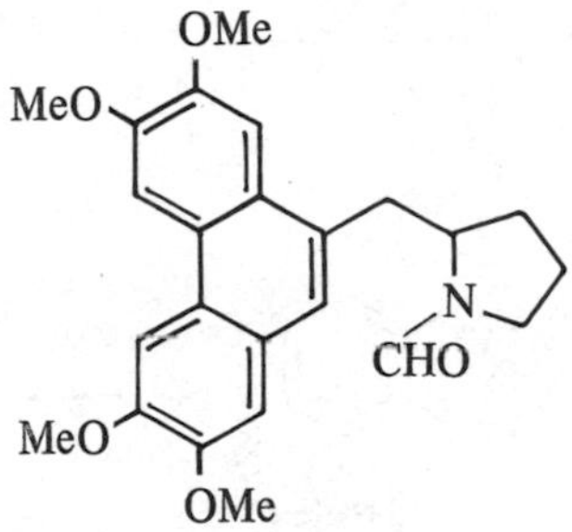

OMe
MeO
N
CHO
MeO
OMe

33

B. Total Syntheses Using Optically Active Intermediates

Simple 1-benzylisoquinolines

The syntheses of natural alkaloids using optically active intermediates are described in this section. Total synthesis of cularine **22**[124] was accomplished by the Eschweiler-Clarke reaction of synthetic, optically active cularimine **21**. As described previously, optically active intermediates have been used, in many cases, for the syntheses of the isoquinoline alkaloids. Although (±)-armepavine **1** was resolved to give 1(S)-armepavine, which was identical with natural one, optical resolution of (±)-*O*-benzoylarmepavine **10**,[125] followed by hydrolysis of (+)-isomer, also yielded armepavine **1** in good yield. In general, phenolic hydroxyl groups are protected by benzoylation, benzylation, acetylation, and ethoxycarbonylation in case of the synthesis of phenolic isoquinoline alkaloids, and, in many cases, optical resolution was achieved successfully before the removal of these protecting groups. Optical resolution of (±)-benzoyl derivatives, **34**[126] and **35**,[127] prepared in the usual manner, followed by the hydrolysis of the optical isomers, gave codamine **36** and 1(R)-laudanine **37**, respectively. The synthesis of the latter compound was also devised by the other workers. Acidic hydrolysis of 1(R)-laudanosine **38**[126] or diazotization of optically active amino derivatives, **39**[129] and **40**,[130] resulted in the formation of (+)-1(R)-laudanine **37** (Chart 2-8).

(+)-Reticuline **41**, an important precursor in the biosynthesis of morphine alkaloids,[131] was also obtained by the hydrolysis of optically active

Chart 2-8.

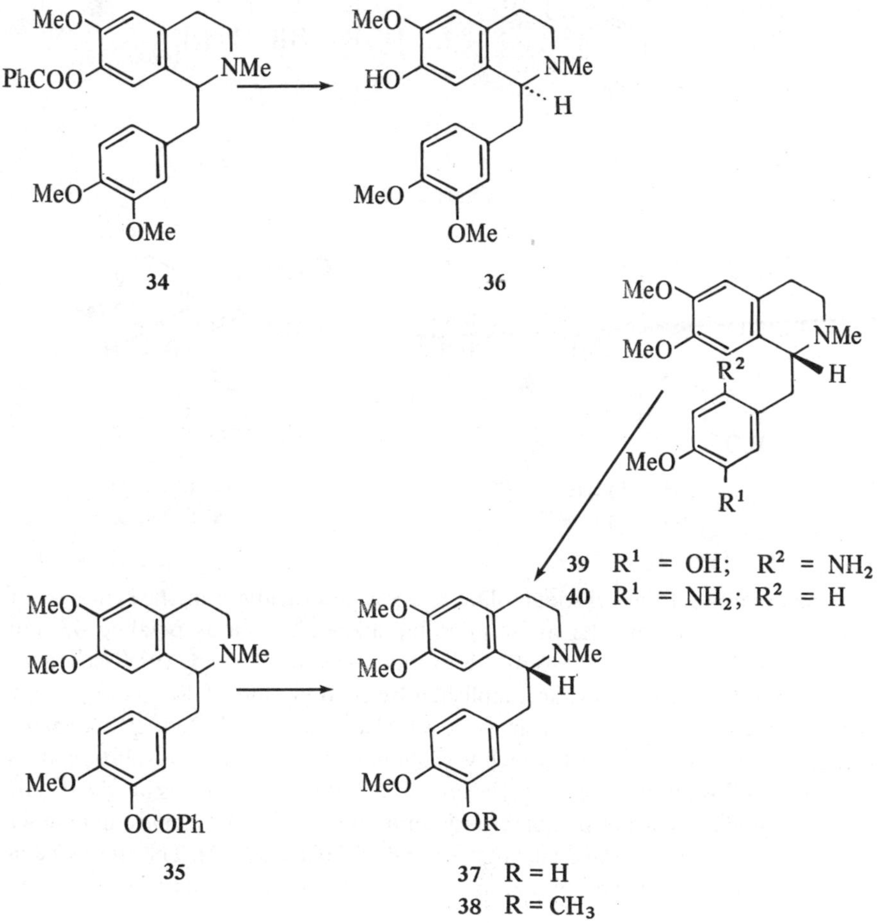

34 **36**

39 R^1 = OH; R^2 = NH_2
40 R^1 = NH_2; R^2 = H

35

37 R = H
38 R = CH_3

O,O-dibenzyl derivative **42**, which was formed by the optical resolution of its racemate **43** (Chart 2-9).

The synthesis of (±)-*N*-methylcoclaurine **44** was reported by many workers,[132,133] among which (−)-*N*-methylcoclaurine **45** involving the optical resolution of *O*-acetyl-*O*-benzylisoquinoline **46** was synthesized in good yield by Kametani.[134] Hydrolysis of **47** with sodium carbonate, followed by hydrochloric acid, yielded an optically active *N*-methylcoclaurine **45** (Chart 2-10).

Chart 2-9.

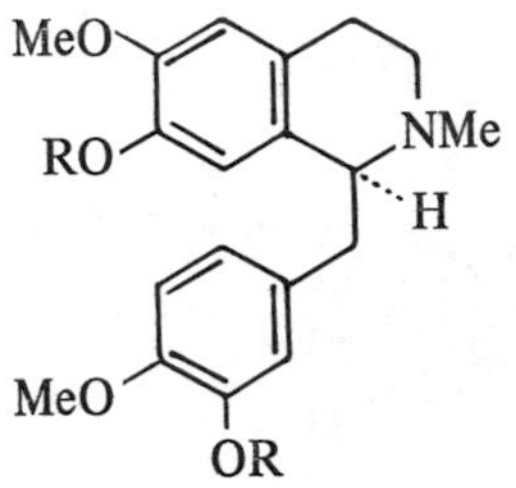

MeO, RO, NMe, H, MeO, OR

41 R = H
42 R = CH_2Ph
43 R = H[(±)-base]

Chart 2-10.

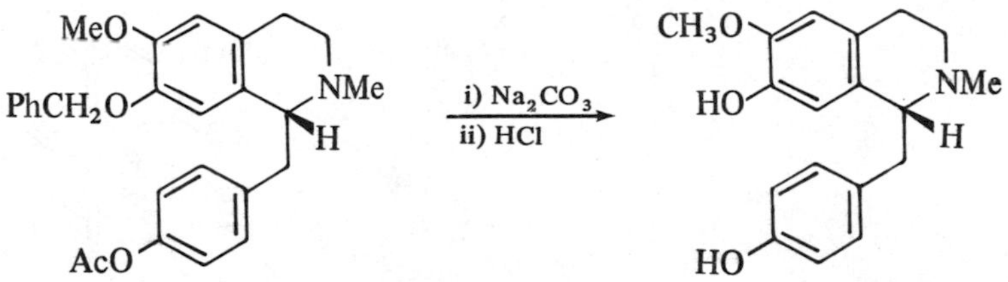

MeO, $PhCH_2O$, NMe, H, AcO

i) Na_2CO_3
ii) HCl

CH_3O, HO, NMe, H, HO

46 (±)-base
47 (−)-base

44 (±)-base
45 (−)-base

Optically active 1-benzylisoquinoline is a key intermediate in the synthesis of quaternary ammonium salts of isoquinoline alkaloids such as petaline **48** and lotusine **49**. The total synthesis of the former alkaloid **48** having the R-configuration at the C_1-position was accomplished by methylation of R-(−)-norpetaline **50**[135] with methyl iodide. Similar work was also reported.[136] Reductive debenzylation of the isoquinoline **51**, which was obtained by the optical resolution of its racemate, followed by the methylation of **52**, yielded (−)-lotusine **49** (Chart 2-11). Phenolic oxidation of quaternary ammonium salt **53** of (+)-reticuline **41** with ferric chloride afforded laurifoline iodide **54** (Chart 2-12). The stereochemistry of proaporphine, aporphine, and protoberberine alkaloids was correlated

Chart 2-11.

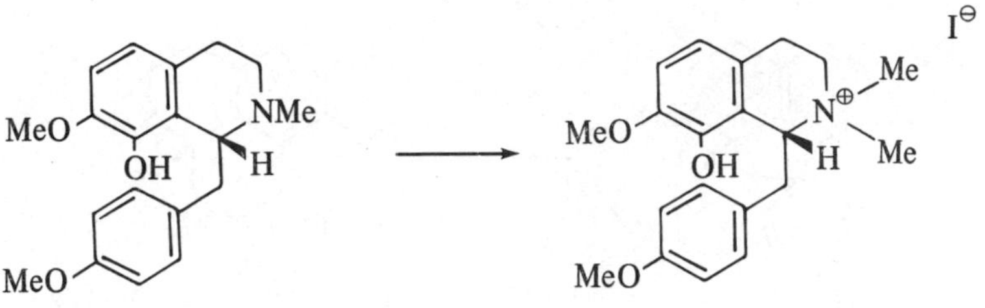

50

48

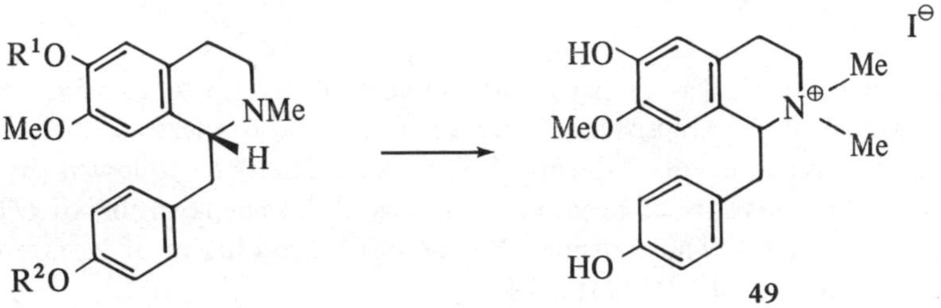

49

51 R^1 = Ac
R^2 = CH_2Ph
52 R^1 = R^2 = H

Chart 2-12.

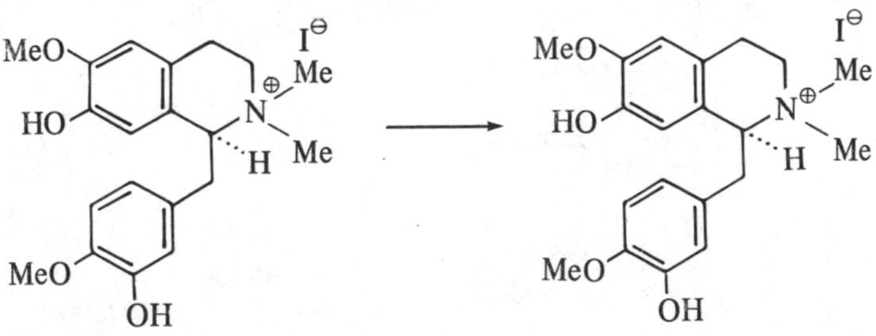

53

54

with that of 1-benzylisoquinoline whose stereochemistry had been known already.[137] Some of the proaporphine alkaloids were converted to aporphine alkaloids. For instance, tuduranine **55** was synthesized by acidic rearrangement of pronuciferine **56** (Chart 2-13).

Corrodi resolved norlaudanosine to give (−)-isomer **57**, whose configuration at the C-1 position was S-series. Mannich reaction of **57** afforded xylopinine **58**.[137] Scoulerine **59** and coreximine **60** were also synthesized by the similar method.[138]

Chart 2-13.

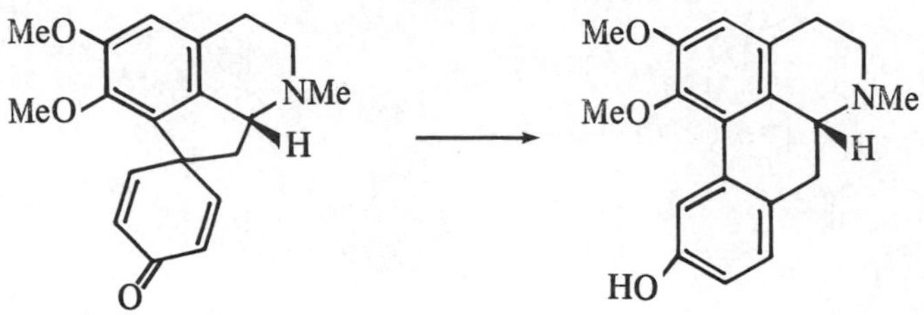

MeO MeO NMe H → MeO MeO NMe H HO

56 55

1(S)-Norreticuline **57** was cyclized with formalin at pH 6.3 to give **59** and **60**, both of which were synthesized in different ways. The former was obtained by the Mannich reaction of 2′-bromo-1(S)-norreticuline **61**, followed by the reductive debromination of bromoscoulerine **62**.[139] Debenzylation of *O,O*-dibenzylcoreximine **63**, which formed by the optical resolution of its racemate, afforded coreximine **60**[140] (Chart 2-14).

Chart 2-14.

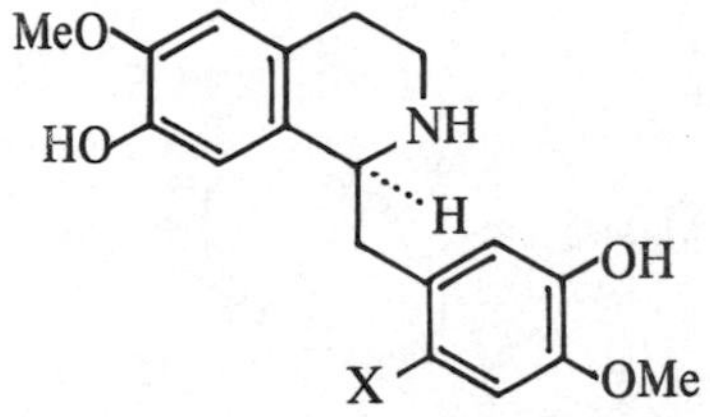

57 X = H
61 X = Br

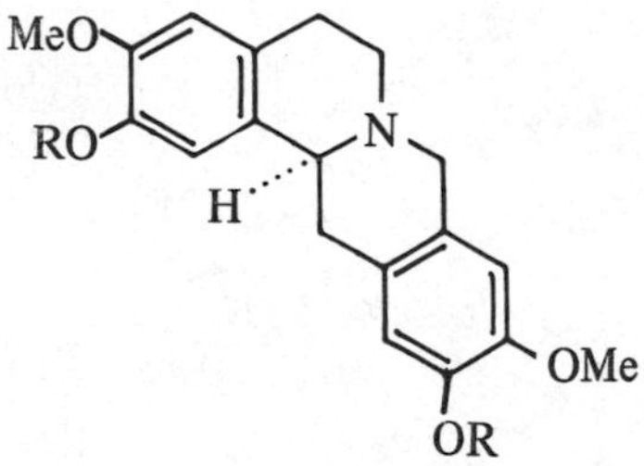

58 R = CH_3
60 R = H
63 R = CH_2Ph

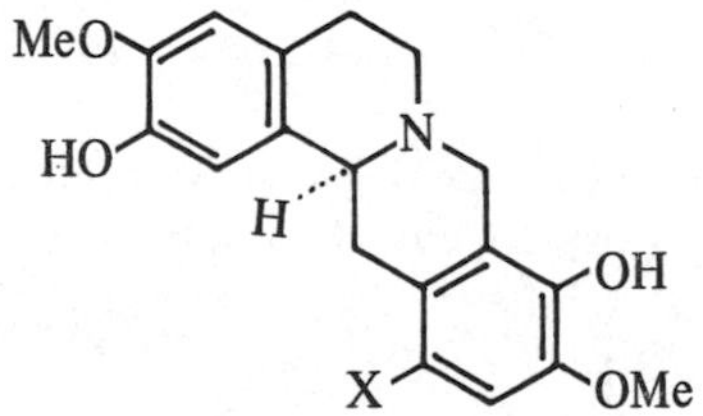

59 X = H
62 X = Br

Bisbenzylisoquinolines

Isoquinolines possessing two asymmetric centers such as bisbenzylisoquinoline involve another problem in the syntheses of racemate from the point of separation of two diastereoisomers. However, in many cases, the separation of a diastereoisomeric mixture is very difficult especially in case of bisbenzylisoquinolines having one diphenyl ether linkage. Since Ullmann reaction between a bromoisoquinoline and a phenolic isoquinoline is very useful for synthesizing bisbenzylisoquinoline alkaloids, the use of optically active isoquinolines would be necessary for the formation of natural products. In fact, the Ullmann reaction between (+)-bromoisoquinoline **64** and (+)-*O*-benzylcoclaurine **65** afforded *O,O,O*-tribenzylmagnoline **66**[141] and the similar reaction with the use of (−)-bromoisoquinoline **67** and **65** afforded *O,O,O*-tribenzylberbamunine **68**,[142] which was one of the diastereoisomers of **66**. Debenzylation of both compounds **66** and **68** afforded magnoline **69** and berbamunine **70**,[141,142] respectively (Chart 2-15). Furthermore, the Ullmann reaction between two suitable isoquinolines was applied to the total syntheses of liensinine **71**,[143] isoliensinine **72**,[144] neferine **73**,[145] thalicarpine **74**,[146] and tetrandrine **75**[147] via their corresponding *O*-benzyl derivatives. These results are shown in Table 2-1.

Chart 2-15.

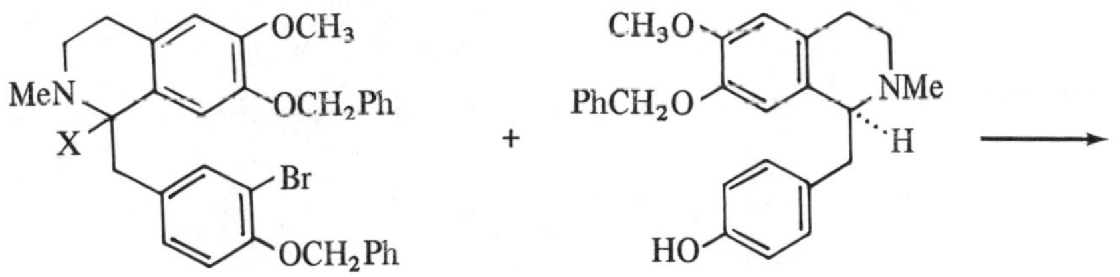

64 X = ◄H
67 X = ···H

65

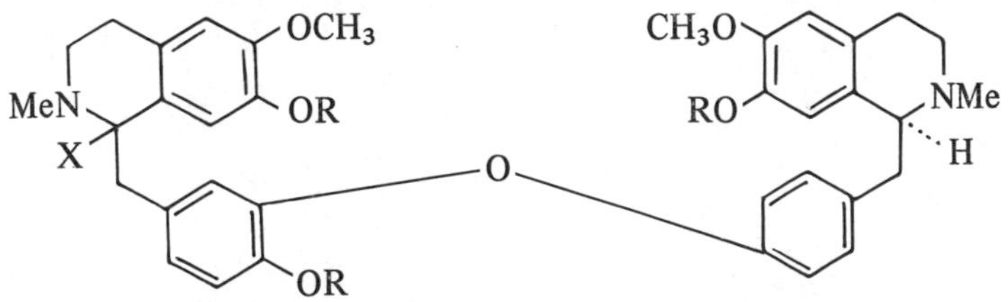

66 R = CH_2Ph; X = ◄H
68 R = CH_2Ph; X = ···H
69 R = H; X = ◄H
70 R = H; X = ···H

Table 2-1.

Starting material	Natural product
	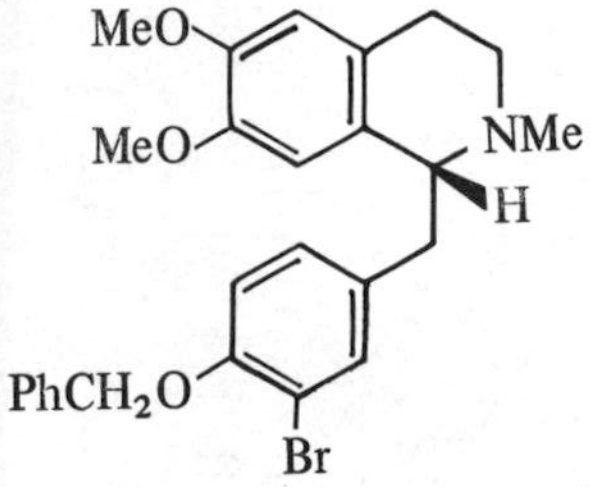 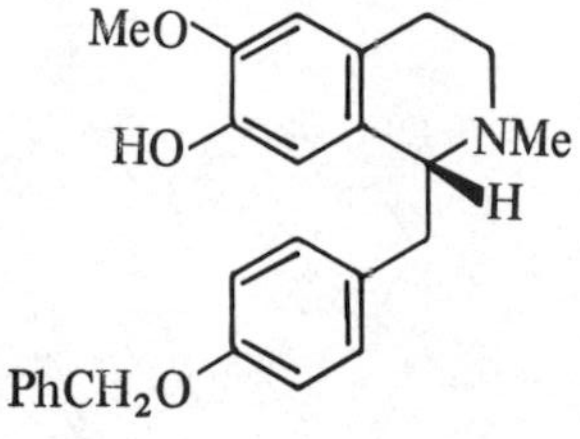 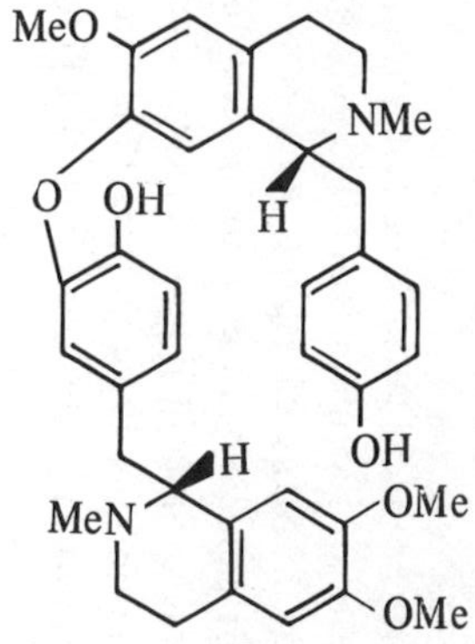 **71**
	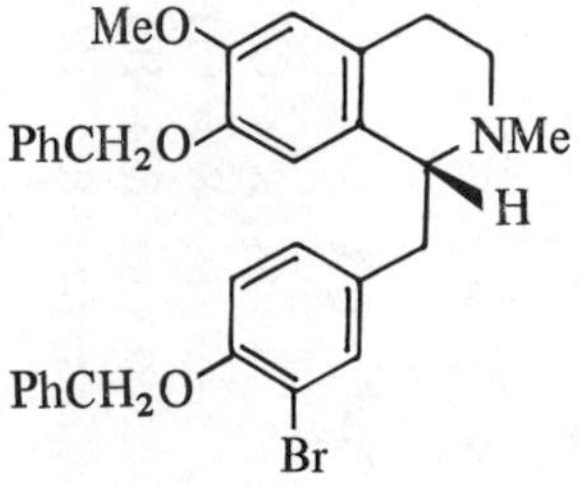 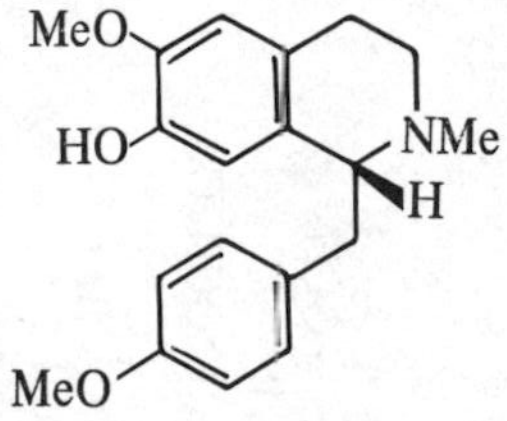 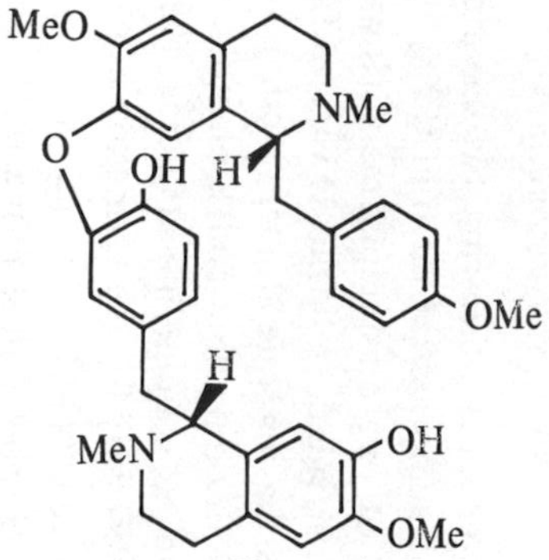 **72**

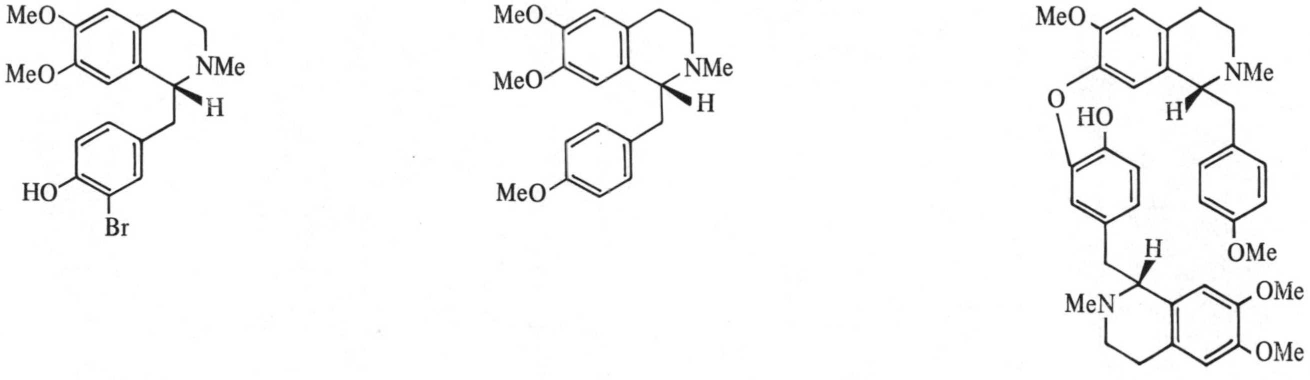

73

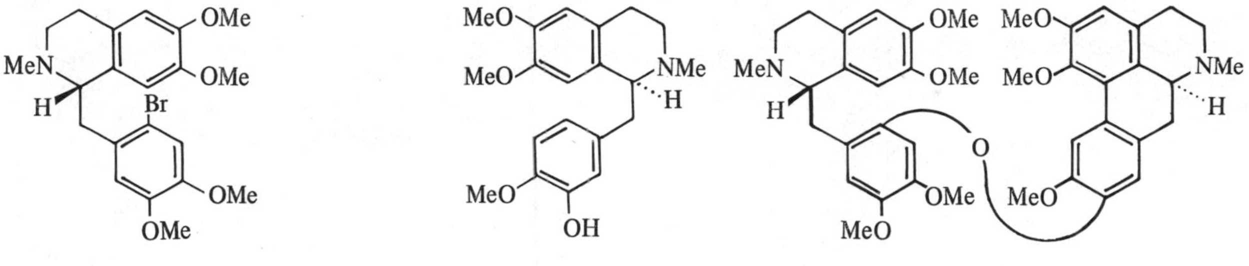

74

Table 2.1 (continued) .

Starting material	Natural product
	75

Recently, Inubushi reported an asymmetric synthesis of bisbenzylisoquinoline[148] having two biphenyl ether linkage such as tetrandrine **75**. Ullmann reaction between the phenolic isoquinoline **76** and methyl *p*-bromophenylacetate gave diphenoxyisoquinoline **77**, which was converted to dihydroisoquinoline derivative **79** through the amide **78**. Sodium borohydride reduction of **79**, followed by the Eschweiler-Clarke reaction, yielded a mixture of isotetrandrine **80** and pheanthine [(−)-tetrandrine] **81**, which were separated to give corresponding stereoisomers at the C_1-position (Chart 2-16).

The bisphenethylisoquinoline such as melanthioidine **82** was also synthesized by the application of the Ullmann reaction, as in the case of the bisbenzylisoquinoline alkaloids. The double Ullmann reaction of (−)-bromophenolic isoquinoline **83** afforded bisphenethylisoquinoline **84** possessing two biphenyl

Chart 2-16.

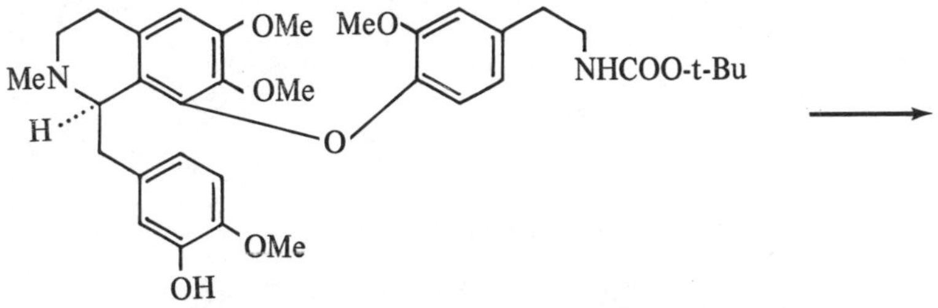

OMe MeO
NHCOO-t-Bu
MeN
OMe
H
O
OMe
OH

76

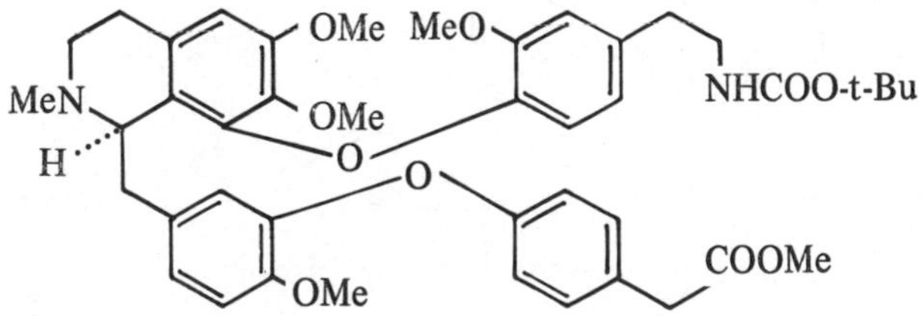

OMe MeO
NHCOO-t-Bu
MeN
OMe
H
O
O
COOMe
OMe

77

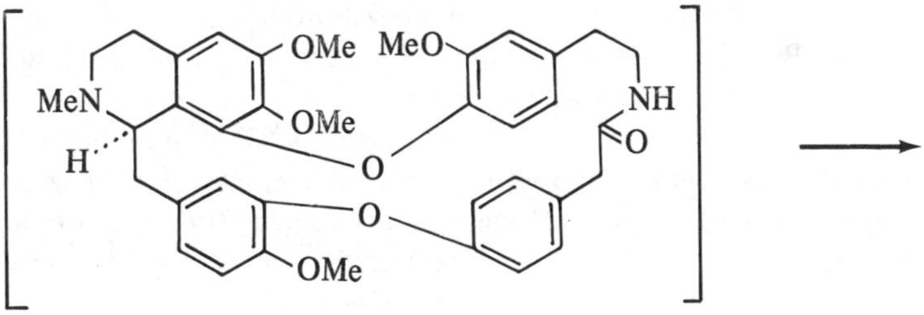

OMe MeO
MeN
NH
OMe
H
O
O
O
OMe

78

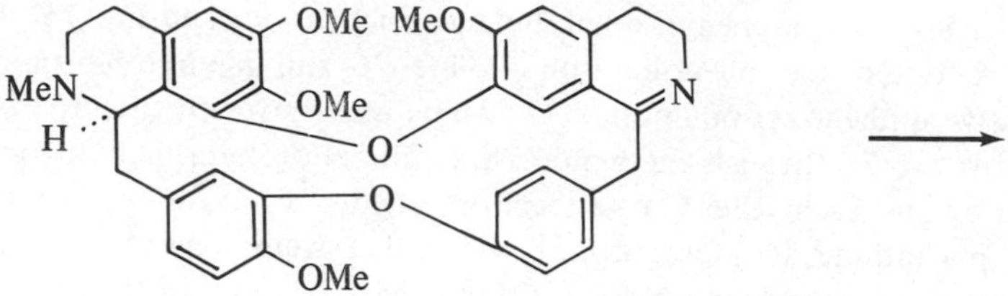

79

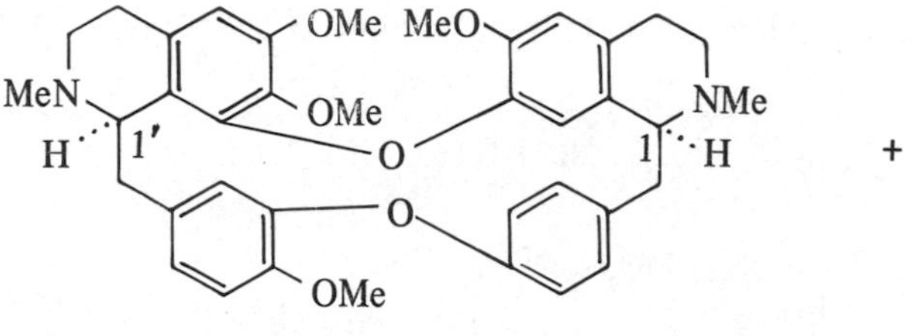

80

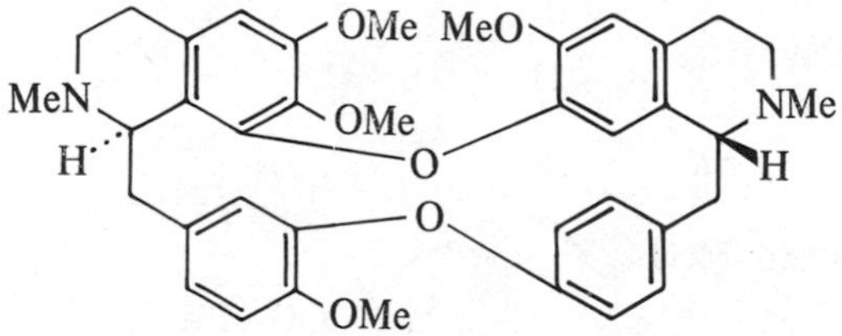

81

ether linkages. Reductive debenzylation of **84** gave melanthioidine **82**[149] (Chart 2-17).

Synthesis of some of homoaporphine alkaloids was established by Battersby,[150] Kametani,[151] and Brossi.[152] They reported the syntheses of multifloramine **88** and kreysigine **89** through phenolic oxidation of **86**, which was obtained by debenzylation of **85**. Dienone-phenol rearrangment of **87** gave multifloramine **88**, whose *O*-methylation with diazomethane yielded kreysigine **89** (Chart 2-18). Similarly, cryptostylines I, II, and III were also obtained from optically active norcryptostylines resolved by (−)-diacetone-2-keto-L-gluconic acid, and their absolute configurations were determined by X-ray analysis.[152a]

Chart 2-17.

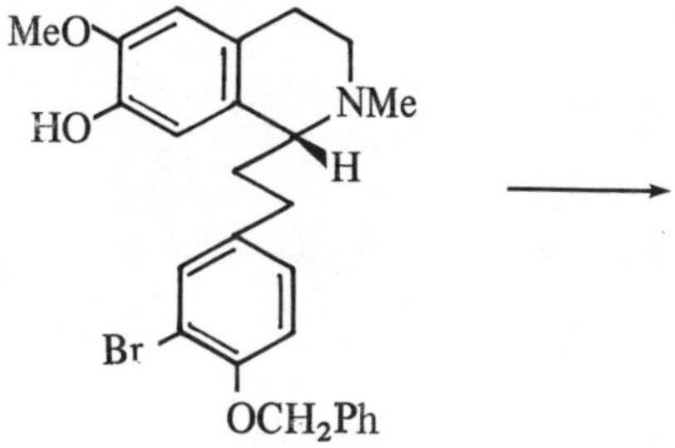

MeO
HO
NMe
H
Br
OCH_2Ph
83

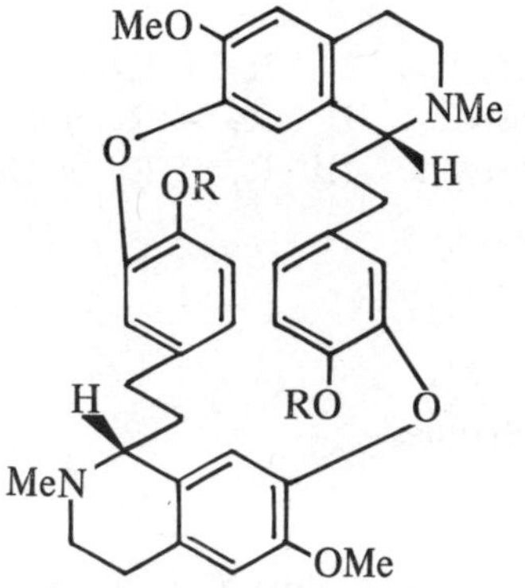

MeO
NMe
O
OR
H
H
RO
O
MeN
OMe

82 R = H
84 R = CH_2Ph

Chart 2-18.

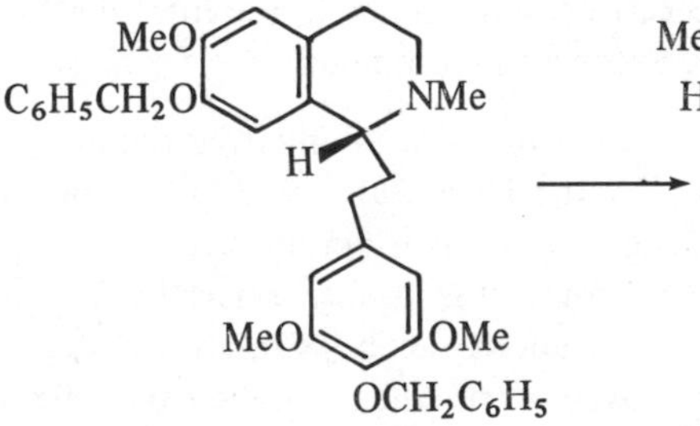

MeO
$C_6H_5CH_2O$
NMe
H
MeO
OMe
$OCH_2C_6H_5$
85

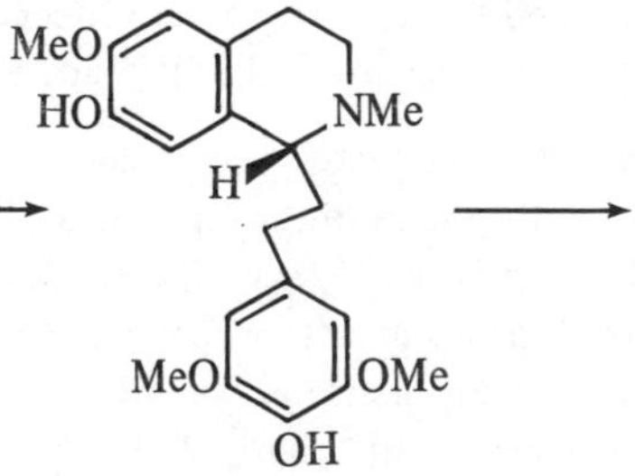

MeO
HO
NMe
H
MeO
OMe
OH
86

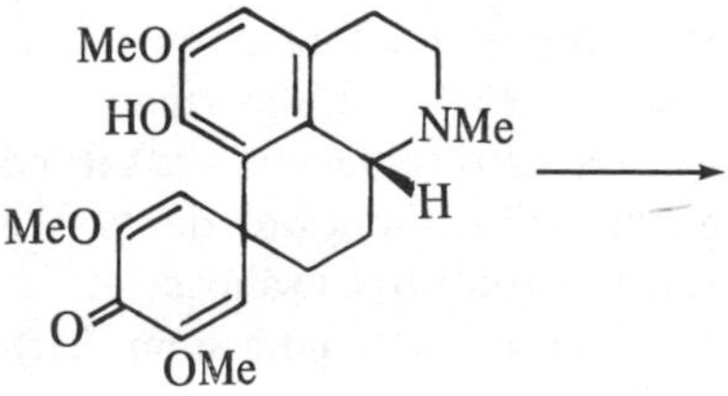

MeO
HO
NMe
H
MeO
O
OMe
87

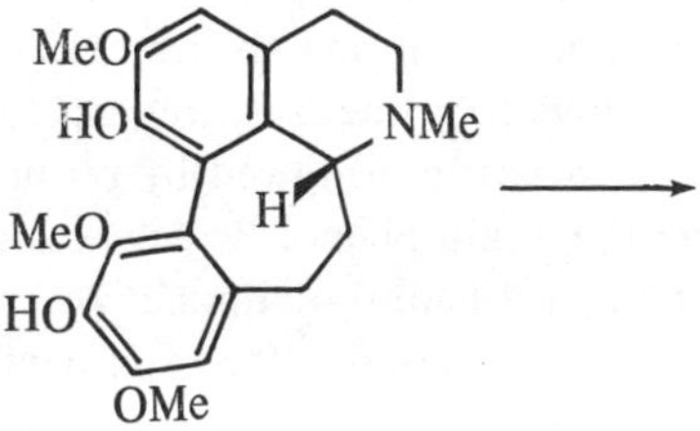

MeO
HO
NMe
H
MeO
HO
OMe
88

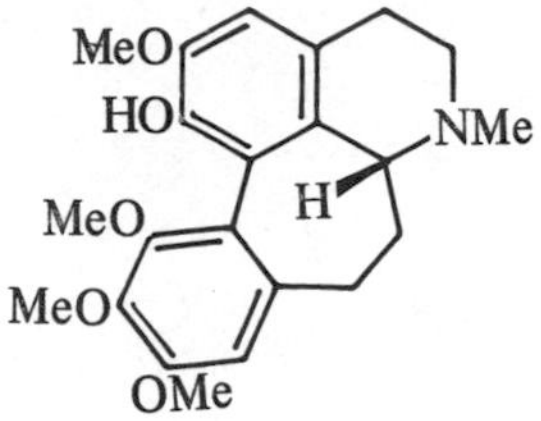

89

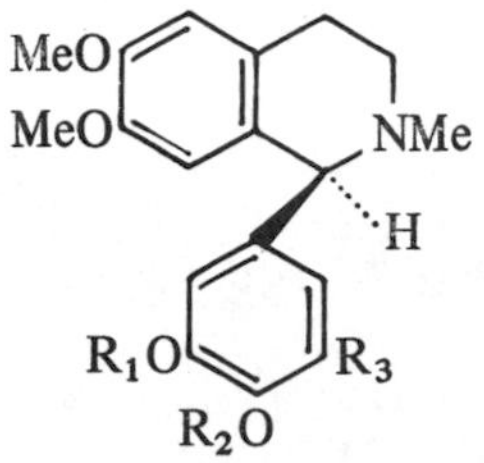

Cryptostyline I $R_1 + R_2 = -CH_2-$, $R_3 = H$
Cryptostyline II $R_1 = R_2 = Me$, $R_3 = H$
Cryptostyline III $R_1 = R_2 = Me$, $R_3 = OMe$

C. Stereospecific Total Syntheses

Morphine Series

The stereospecific syntheses of the isoquinoline alkaloids possessing the rather complex ring system such as morphine **90** (cateogry (C)) are described here. Many papyrochemical synthetic approaches to morphine have been investigated[153-157] and some of them have successfully culminated in the total synthesis of **90**, first by Gates and Tschudi[153] and recently by Kametani.[157]

Gates Method. The Gates' approach to the synthesis of **90** involved the preparation of a hydrophenanthrene precursor by the Diels-Alder reaction of bicyclic dienophilic component **98** with butadiene. The Diels-Alder adduct contained a cyanomethyl group at the proper position for further transformations. The key dienophilic starting material was obtained in about ten steps in an overall yield of 20% as shown in Chart 2-19. Nitrosation of the monobenzoate **92** of 2,6-dihydroxynaphthalene **91**, afforded the nitrosophenol **93**, which was catalytically reduced to aminophenol. Without separation, this was successively oxidized by ferric chloride to give 1,2-naphthoquinone **94**. Reduction to the diphenol, followed by methylation, gave the benzoate **95**, which was saponified. Analogous nitrosation, followed by reduction of the nitrosophenol, was carried out to give the aminophenol, which was again oxidized without isolation to the 5,6-dimethoxy-1,2-naphthoquinone **96**. Treatment with ethyl cyanoacetate in the presence of triethylamine, followed by oxidation with potassium ferricyanide gave the substituted cyanoacetic ester **97**, whose hydrolysis with alkali

Chart 2-19.

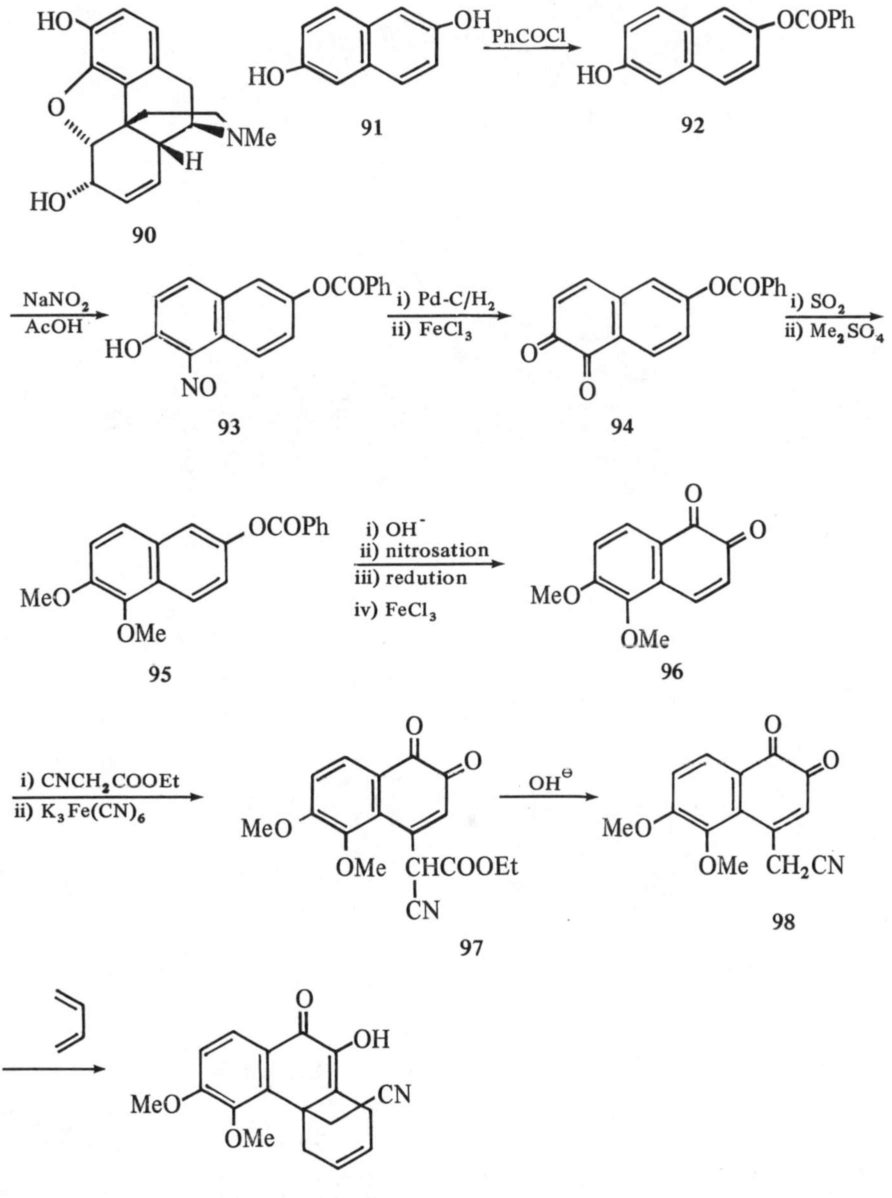

HO, O, HO, NMe, H

90

OH, HO

91

$PhCOCl$

OCOPh, HO

92

$NaNO_2$ / $AcOH$

OCOPh, HO, NO

93

i) $Pd\text{-}C/H_2$ ii) $FeCl_3$

OCOPh, O, O

94

i) SO_2 ii) Me_2SO_4

OCOPh, MeO, OMe

95

i) OH^- ii) nitrosation iii) redution iv) $FeCl_3$

O, O, MeO, OMe

96

i) $CNCH_2COOEt$ ii) $K_3Fe(CN)_6$

O, O, MeO, OMe, CHCOOEt, CN

97

$OH^{\ominus}$

O, O, MeO, OMe, CH_2CN

98

O, OH, CN, MeO, OMe

99

Chart 2-20.

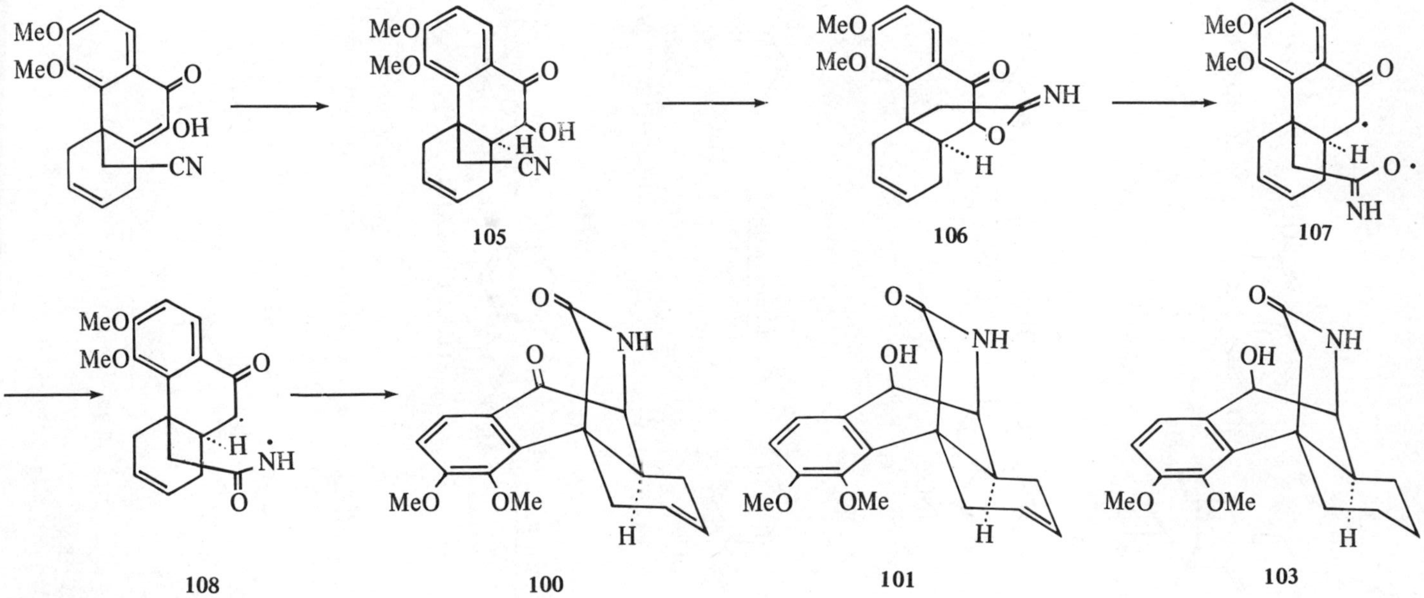

105 106 107

108 100 101 103

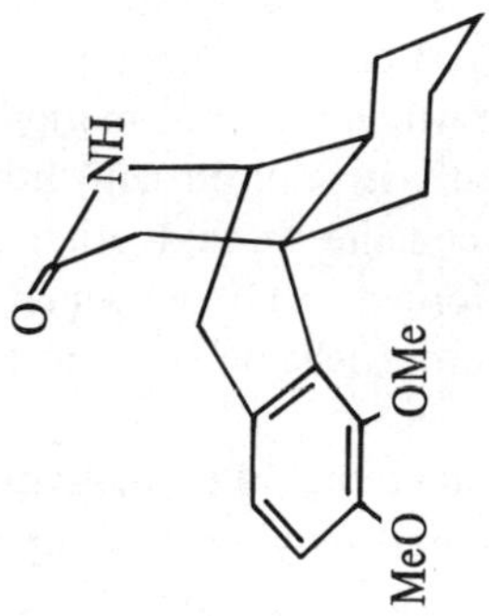

102

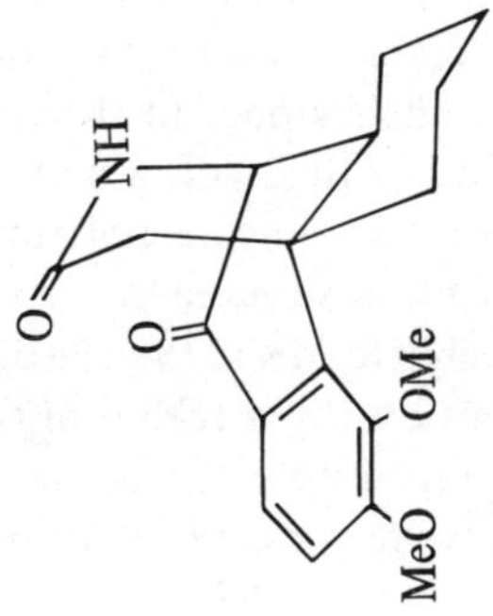

104

gave 4-cyanomethyl-5,6-dimethoxy-1,2-naphthoquinone **98** as the desired key starting material. Further condensation with butadiene afforded the substituted hydrophenanthrene derivative **99**, whose cyanomethyl group was situated in the angular position (Chart 2-19).

Reduction of **99** with copper chromite afforded five products, namely dihydro derivative **100**, two tetrahydro derivatives, **101** and **102**, hexahydro derivative **103**, and the desoxy derivative **104** as six-membered lactam. The mechanism of this cyclization would be explained as follows. Back-side attack of **99** would lead to the ketol **105**, which would be tautomerized to give an imino ether **106**. Homolytic cleavage of the C_9—O bond of **106** would give diradical **107**, which could be rewritten as **108**, and recyclization then gave the dihydro reduction product **100** (Chart 2-20).

The Wolff-Kishner reduction of **100** afforded **109**, which was then methylated with methyl iodide. Methyllactam **110** thus obtained was reduced with lithium aluminum hydride to give (±)-β-Δ^6-dihydrodesoxycodeine methyl ether **111**. Resolution of **111** with dibenzoyltartaric acid afforded (+)-base, which was identical with the sample obtained by conversion from thebaine through transformation of β-thebainone **112** (Chart 2-21).

The double bond in ring C of optically active compound **111** was hydrated with dilute sulfuric acid to give a mixture of the desired 6-hydroxy compound **113** and its isomer **114**. Partial demethylation of **113** was carried out by heating potassium hydroxide in diethylene glycol, and the resulting product was oxidized with potassium *tert*-butoxide and benzophenone to the corresponding ketone. β-Dihydrothebaine **115** thus obtained was brominated to afford the dibromo derivative **116**, which on treatment with 2,4-dinitrophenylhydrazine gave the 2,4-dinitrophenylhydrazone **117** of 1-bromothebainone. In the course of this reaction, the desired inversion of configuration of the hydrogen atom at the C_{14}-position took place and the reaction was found to occur polarimetrically. Since the hydrogen atom at the C_{14}-position in **117** is situated in the vinylogous α-position to the C=N bond at C_6, it was possible to invert the configuration at this point under alkaline or acidic conditions by the mechanism formulated as in the partial structure **117a** and **117b**, respectively. Thus the compound **116** with unnatural configuration in the tetracyclic system was converted to the more stable compound **117** (Chart 2-22).

Treatment of **117** with acetone and hydrochloric acid afforded 1-bromothebainone **118** as a free ketone, which was further hydrogenated to 1-bromodihydrothebainone **119** on platinum or to dihydrothebainone **120** on palladium by the removal of the bromine atom. Dibromination of the former compound **119** or tribromination of the latter **120**, followed in each case by treatment with 2,4-dinitrophenylhydrazine, gave 2,4-dinitrophenylhydrazone of 1-bromocodeinone **121**. Treatment of **121** with acetone and hydrochloric acid afforded a free ketone **122** in comparatively poor yield because of the ease with

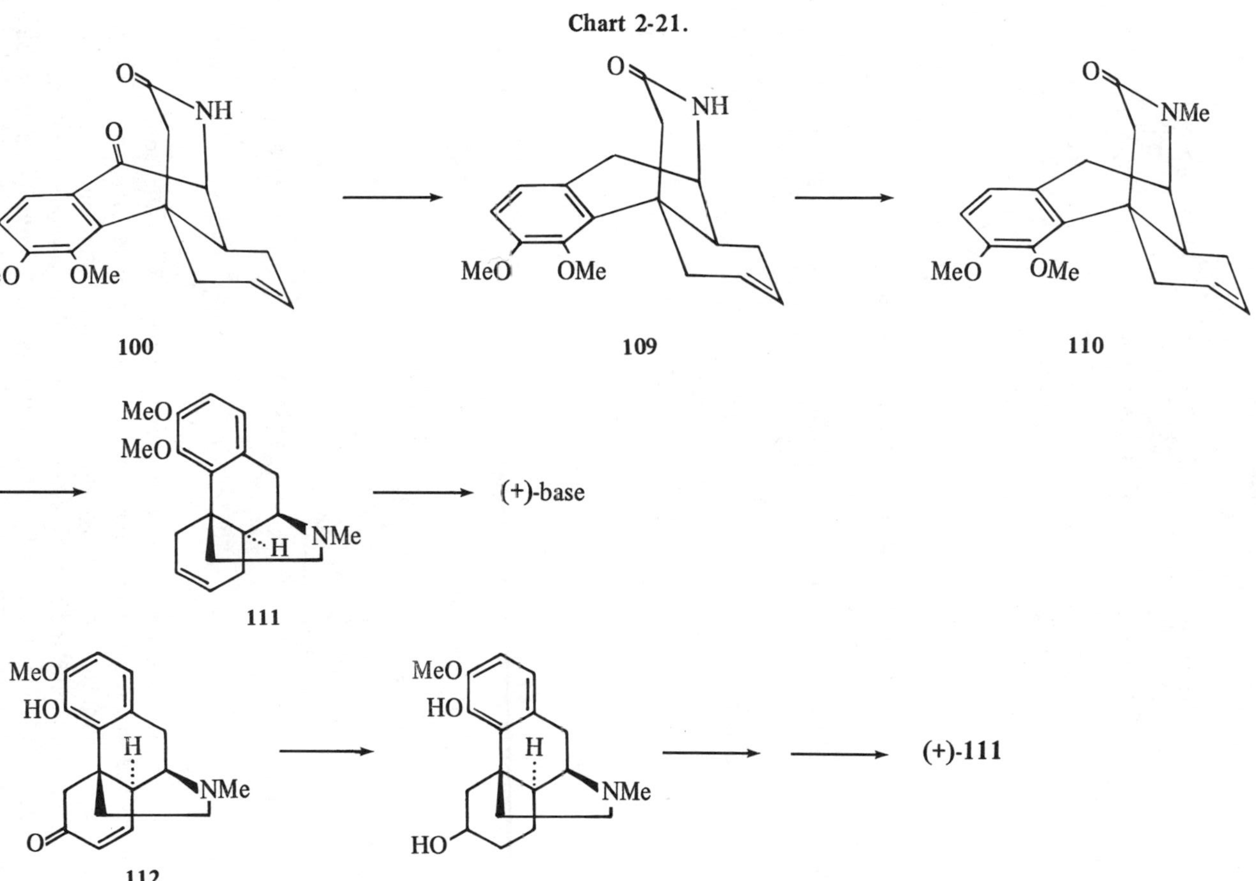
Chart 2-21.
O
NH
O
MeO
OMe
100
O
NH
MeO
OMe
109
O
NMe
MeO
OMe
110
MeO
MeO
H
NMe
111
(+)-base
MeO
HO
H
NMe
O
112
MeO
HO
H
NMe
HO
(+)-111

Chart 2-22.

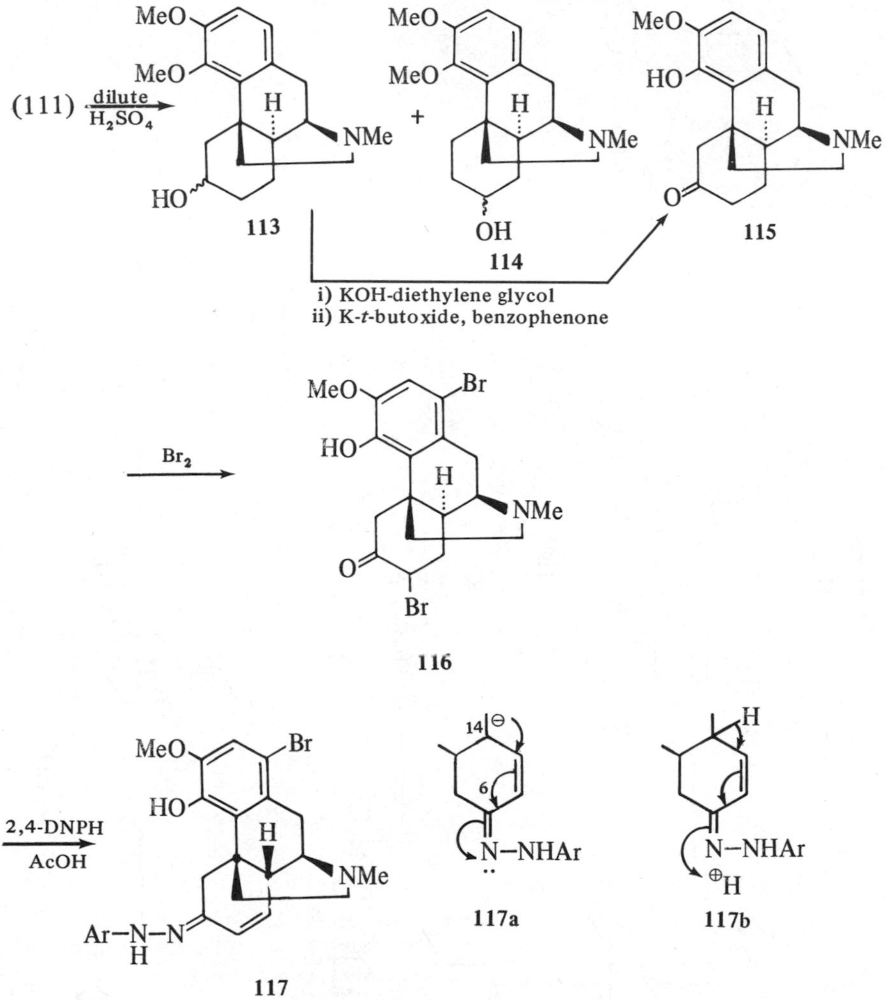

(111) $\xrightarrow[H_2SO_4]{dilute}$ 113 + 114 $\xrightarrow[\text{ii) K-}t\text{-butoxide, benzophenone}]{\text{i) KOH-diethylene glycol}}$ 115 $\xrightarrow{Br_2}$ 116 $\xrightarrow[AcOH]{2,4\text{-DNPH}}$ 117

117a 117b

which 1-bromocodeinone underwent rearrangement in acidic solution. The ketone **122** was identical with Oppenauer oxidation product of 1-bromocodeine. Energetic treatment of 1-bromocodeinone with lithium aluminum hydride leads, with loss of bromine and stereospecific reduction of the carbonyl group, directly to codeine **123**. Finally, demethylation of the synthetic optically active codeine thus obtained with pyridine hydrochloride led to totally synthetic morphine **90**. This cleavage of codeine to morphine has already been described by Rapoport (Chart 2-23).

Chart 2-23.

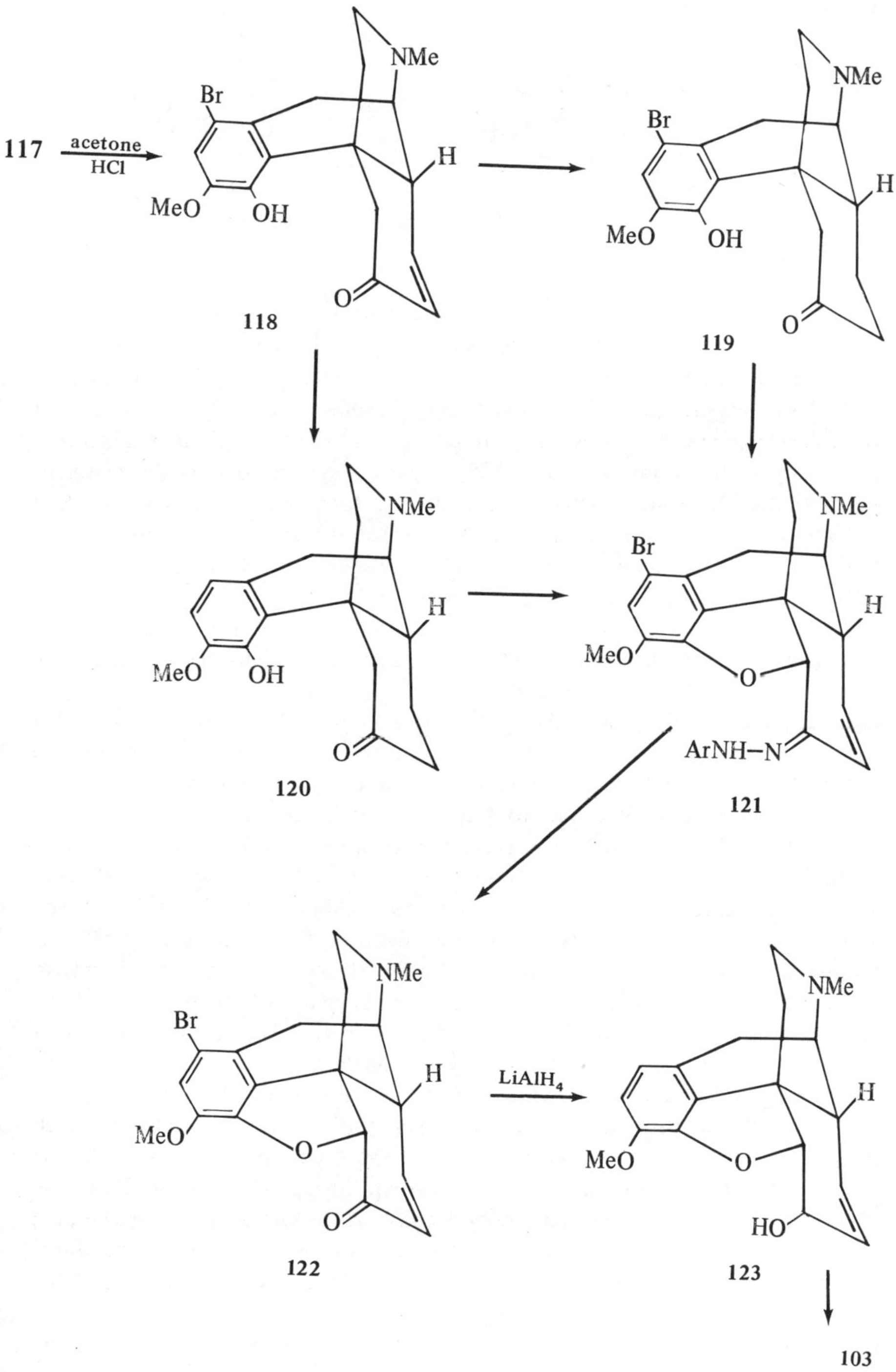

117 $\xrightarrow[\text{HCl}]{\text{acetone}}$

Br
NMe
H
MeO
OH
O

118

Br
NMe
H
MeO
OH
O

119

NMe
H
MeO
OH
O

120

Br
NMe
H
MeO
O
ArNH–N

121

Br
NMe
H
MeO
O
O

122

$\xrightarrow{\text{LiAlH}_4}$

NMe
H
MeO
O
HO

123

103

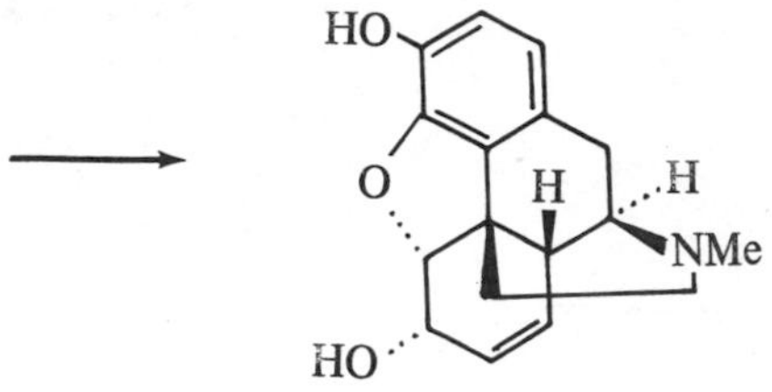

90

Ginsburg Method. The second synthesis of morphine was successfully carried out by Elad and Ginsburg.[154] Their papyrochemical approach was based on a different dissection of the morphine molecule, which was considered to be phenylcyclohexane derivative (heavy lines in **90a**). Thus the formation of phenylcyclohexane derivative containing ring A and C of morphine was involved as the key synthetic intermediates **125**, which was synthesized on the treatment of 3-lithioveratrole **124** with cyclohexanone, followed by dehydration and oxidation with selenium dioxide. The ketone **126** was also obtained by the acid hydrolysis of α,β-unsaturated oxime **127**, prepared from dehydration product **128** via nitroso-chloride **129** and **130** (Chart 2-24).

Michael condensation of the α,β-unsaturated ketone **126** with dibenzyl malonate **131** gave the adduct **132**, which, upon hydrogenolysis of the benzyl groups and thermal decarboxylation of the resulting dicarboxylic acid, gave the keto acid **133**. Cyclization of the keto acid **133** with hydrogen fluoride afforded the ketone **134**. There is a substantial difference in reactivity between the two carbonyl groups in **134** since one of these is an ordinary alicyclic ketone group and the other is adjacent to an aromatic ring. Ketalization of **134** with ethylene glycol afforded **135**, which was treated with amyl nitrite to give 9-oximino derivative **136**. Catalytic hydrogenation of **136** on palladised carbon in the presence of acid gave the thermodynamically more stable equatorial amino-ketone **137**, the ketal group being lost during the reduction. The amino-ketone **137** was treated with acetylglycolyl chloride to give a 10-acetoxyacetamido derivative **138**. When ketalization was attempted, cyclization of the nitrogen-containing ring occurred; furthermore, the methoxyl group at the C_4-position in the tetracyclic system was demethylated easily, in analogy to the behavior of this methoxyl group in various derivatives of morphine, to give **139**. The *cis*-BC ring fusion in **139** was verified through its further synthetic transformations, which ultimately led to a compound belonging to the "natural" rather than to the β-series. The blocking of the C_{10} carbonyl group through ketalization during the cyclization reaction was particularly convenient since further work necessitated use of the C_5 carbonyl group as a key for the introduction of an oxygen function at C_6. Treatment of **139** with amyl nitrite followed by hydrolysis of the

Chart 2-24.

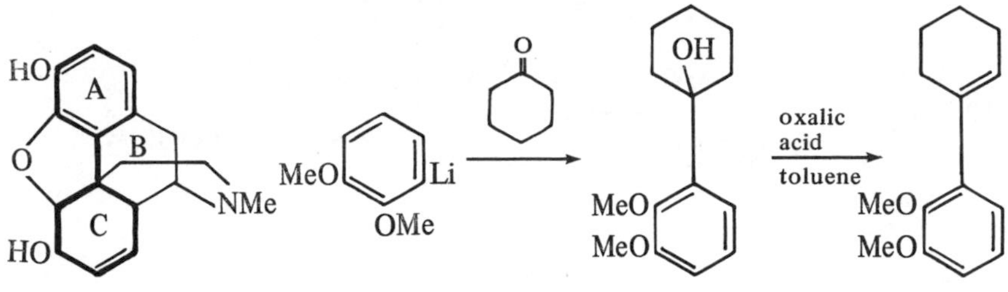

90a 124 125 128

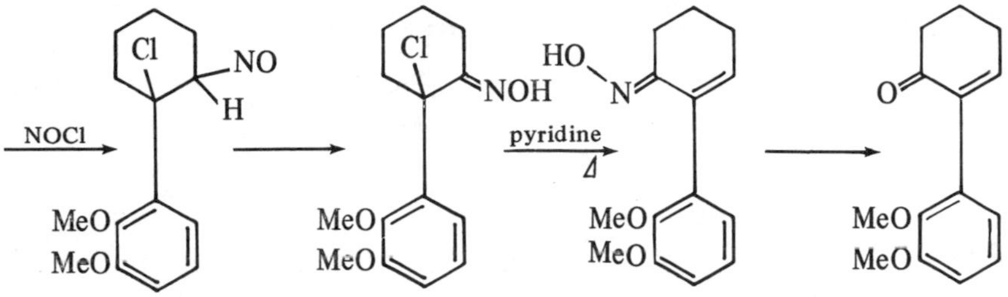

129 130 127 126

ketal gave the 6-oximino derivative **140**. The removal of the carbonyl groups at the C_5- and C_{10}-position by the Wolff-Kishner reduction under alkaline condition was carried out by heating with hydrazine in diethylene glycol to give **141**. Acidic hydrolysis of **141** afforded the ketone **142**. Treatment of **142** with lithium aluminum hydride caused reduction of the lactam ring and carbonyl group to give **143**. Methylation of the secondary amino group, followed by oxidation with potassium *tert*-butoxide and benzophenone, gave (±)-dihydrothebainone **144**, whose resolution with (+)-tartaric acid afforded (−)-isomer **145**, identical with the same product from natural sources. Since dihydrothebainone had previously been converted into morphine, its preparation constitutes a total synthesis of morpine (Chart 2-25).

Barton Method. It has been firmly established that the precursor of morphine is the phenolic isoquinoline **146** and [1-^{14}C] norlaudanosoline **147** is incorporated specifically into morphine **90**. Battersby et al. reported the correlation between stereochemical study of **90** and its precursor by the feeding experiment of

Chart 2-25.

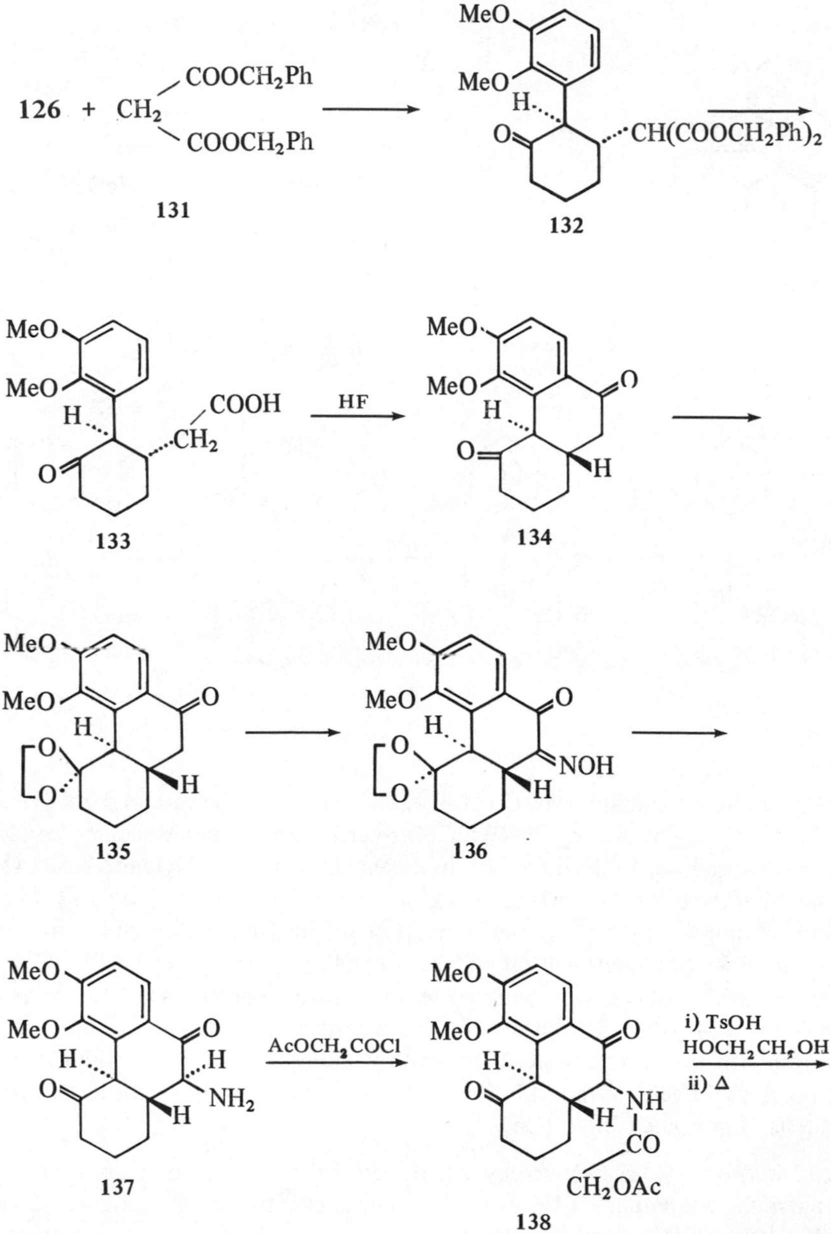

126 + $CH_2(COOCH_2Ph)_2$ → 131

MeO, MeO, H, O, $CH(COOCH_2Ph)_2$

132

MeO, MeO, H, O, COOH, CH_2

133

HF

MeO, MeO, H, O, O, H

134

MeO, MeO, H, O, O, O, H

135

MeO, MeO, H, O, O, O, NOH, H

136

MeO, MeO, H, O, O, H, NH_2, H

137

$AcOCH_2COCl$

MeO, MeO, H, O, O, NH, H, CO, CH_2OAc

138

i) TsOH
$HOCH_2CH_2OH$
ii) Δ

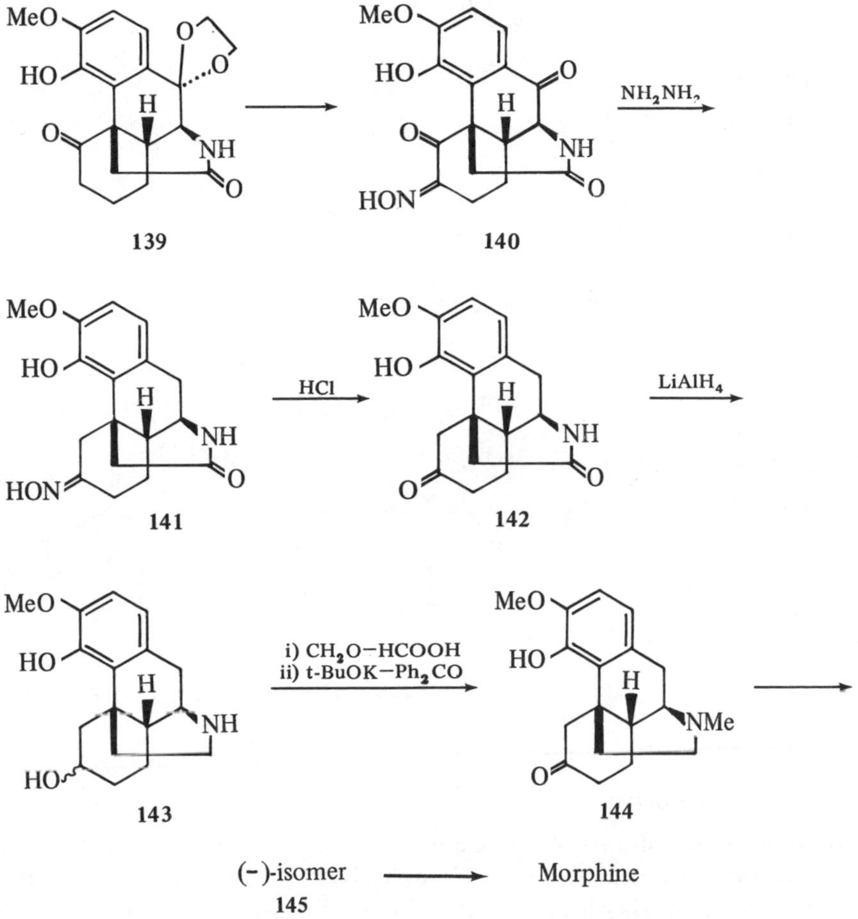

139 → 140 $\xrightarrow{NH_2NH_2}$

141 $\xrightarrow{HCl}$ 142 $\xrightarrow{LiAlH_4}$

143 $\xrightarrow[\text{ii) t-BuOK}-Ph_2CO]{\text{i) }CH_2O-HCOOH}$ 144 →

(–)-isomer 145 → Morphine

optically active base (**146** and enantiomer).[131,155] Among the morphinandienone-type alkaloids, such as flavinantine and sinoacutine, salutaridine is a key intermediate for the synthesis of morphine alkaloids from a chemical as well as from a biogenetical point of view. Barton and his co-workers synthesized radioactive salutaridine **148** along the biosynthetic lines by phenolic oxidation of radioactive reticuline, and converted salutaridine **148** into thebaine **149**.[156] Thebaine **149** has been chemically correlated with morphine **90** via dihydrothebainone **150** (Chart 2-26).

Kametani Method. Recently Kametani[157] synthesized salutaridine that was converted to thebaine, through a modified Pschorr reaction described later. This constitutes a formal total synthesis of morphine.

Chart 2-26.

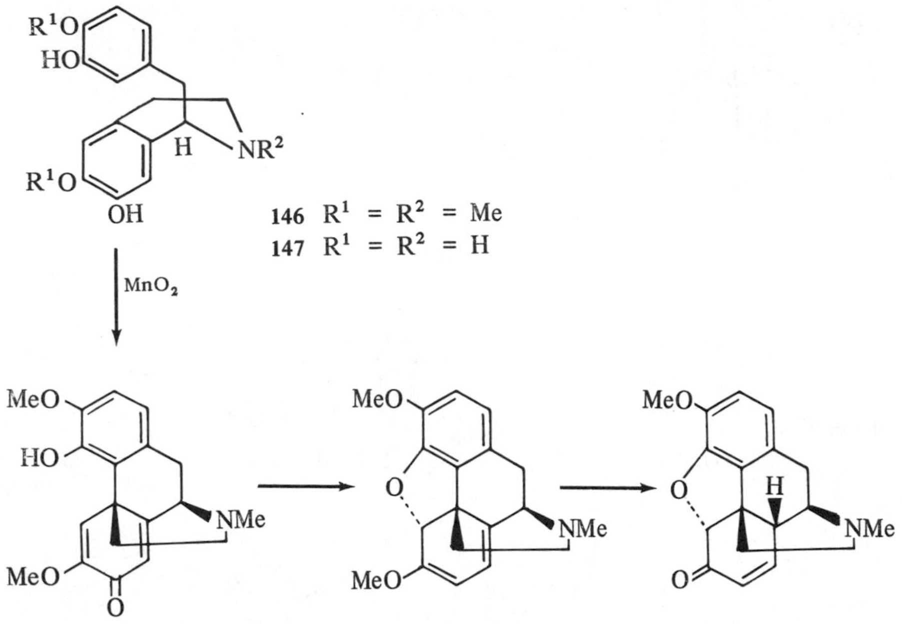

146 $R^1 = R^2 = Me$
147 $R^1 = R^2 = H$

MnO_2

148 149 150

Emetine Series

The asymmetric synthesis of emetine is discussed as one of the examples in the total syntheses of complex isoquinoline alkaloids possessing some asymmetric centers. The stereochemistry of emetine was fully investigated and assigned to the structure as shown in **151**.[158] The total synthesis of emetine has been devised by many workers, and first, Barash's approach[159] to the synthesis of emetine is considered.

Condensation of the pyridone **152** with 3,4-dimethoxyphenethyl iodide in *tert*-butanol gave *N*-substituted pyridone **153**; this was condensed with ethyl oxalate to give the pyruvic ester **154**, leading to the acetic acid **155**. Catalytic hydrogenation of **155** afforded two stereoisomeric 2-piperidone acids **156** and **157** (*cis* and *trans* isomer), which were separated into pure state. The *trans* isomer **156** was condensed with 3,4-dimethoxyphenethylamine to give the amide **158**, whose dehydrative cyclization with phosphorous pentoxide, followed by catalytic hydrogenation of **159**, gave an isomeric mixture of **160** and **161**. The cyclization of **160** with phosphoryl chloride, followed by catalytic reduction of the product **162**, afforded racemic emetine **151**; similarly, the isomer **161** was converted to iso-emetine **163**, which was epimeric with emetine at the C_1-position (Chart 2-27).

Since emetine contains four asymmetric centers, racemic emetine should be one of the eight possible racemic forms of the gross structure. The different routes used by Preobrazhensky and his co-workers[160] and by Barash and Osbond[159] gave a mixture of isomers, all eight being isolated by the former

Chart 2-27.

MeO MeO I

152 → 153 —$(CO_2Et)_2$→ 154 (COCOOEt) →

155 (COOH) → 156 *cis* / 156 *trans* (COOH) →

158 → 159 →

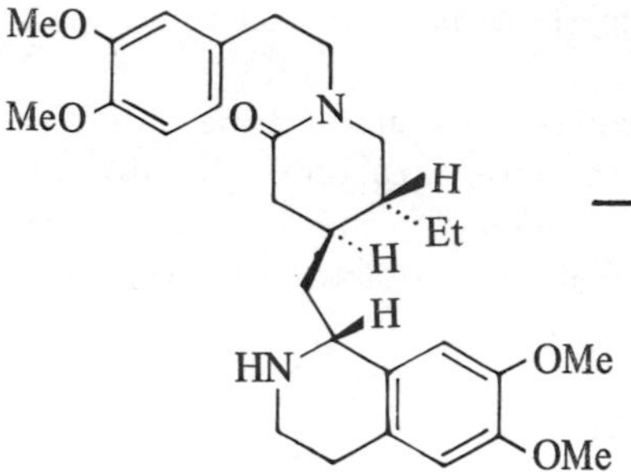

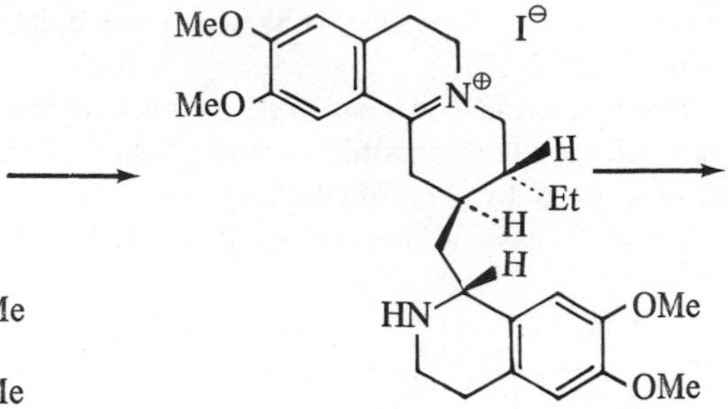

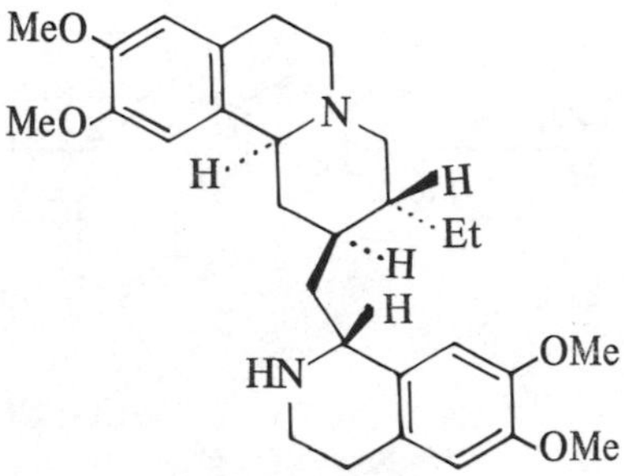

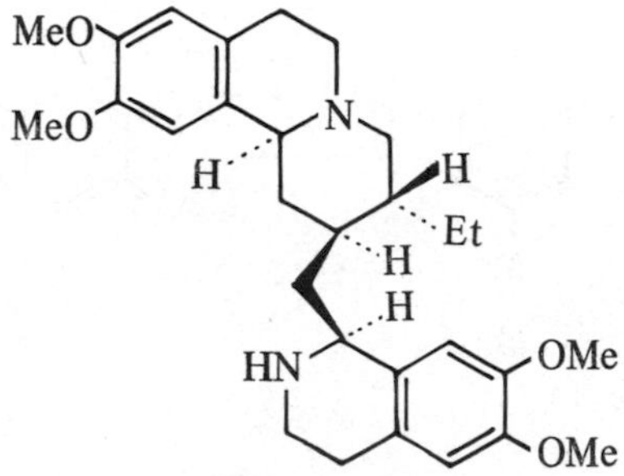

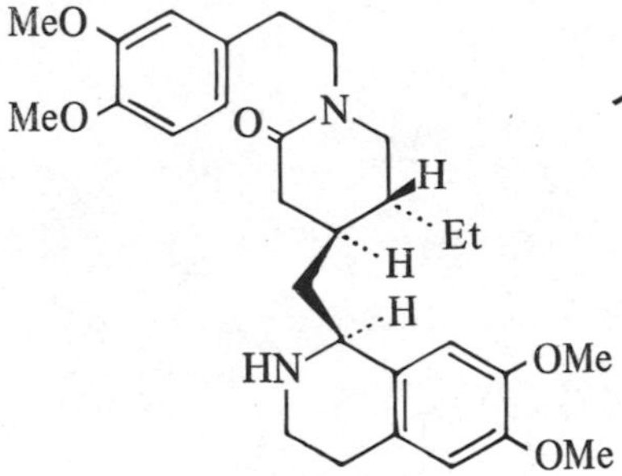

workers and six by the latter. Battersby[161] also succeeded in the stereospecific synthesis of (−)-emetine and (+)-*O*-methylpsychotrine **164**. Dieckmann cyclization of the diester **165**, followed by catalytic hydrogenation or reduction of **166** with sodium borohydride, afforded the epimeric alcohol **167**, which was converted to dihydropyridone **168** on the treatment with acetic anhydride. Michael addition of the malonate anion to **168** gave *trans*-diacid **169**, whose decarboxylation afforded *trans*-carboxylic acid **170**, leading to an important key intermediate **172** through the reduction of the cyclized product **171**. The condensation of the mixed anhydride of **173** with 3,4-dimethoxyphenethylamine gave the amide **174**, whose cyclization with phosphoryl chloride gave racemic *O*-methylpsychotrine **164**. The (±)-base was resolved as its dibenzoyltartrate and the (+)-isomer so obtained was hydrogenated to (−)-emetine. Battersby, furthermore, reported an alternative synthesis of emetine[162] (Chart 2-28).

O-Methylpsychotrine **164** and psychotrine **175** were synthesized through a similar method to that above by Tietel and Brossi.[163] Condensation of the ester **176** with 3-benzyloxy-4-methoxyphenethylamine gave the amide **177**. Bischler-Napieralski reaction of **177**, followed by debenzylation of **178**, afforded the compound **175**, whose *O*-methylation with diazomethane resulted in the formation of **164** (Chart 2-29). The amide **177** was obtained together with its isomer **183** on the catalytic reduction of the amide **182**,[164] which was formed by the dehydrative condensation of carboxylic acid **181** with 3,4-dimethoxyphenethylamine (Chart 2-30). Grussner[165] used the acid **184** in the synthesis of emetine as an intermediate. Dieckmann cyclization of diester **185** gave the keto ester **186**, which led to the acid **187** through four steps. Arndt-Eistert reaction of **187** gave **184**, which was condensed with 3,4-dimethoxyphenethylamine to afford the amide **188**. Dehydrative cyclization of **188**, followed by lithium aluminum hydride reduction of **189**, gave emetine **151** and iso-emetine **163** (Chart 2-31).

The ester **176** or its analog is an important intermediate in the synthesis of not only emetine but also deoxytubulosine **190**, tubulosine **191**, isotubulosine **192**, and emetamine **193**. Condensation of **176** with tryptamine, followed by dehydrative cyclization of **194**, yielded **195**, whose sodium borohydride reduction afforded deoxytubulosine **190** and its isomer **196**.[166] The same reaction was carried out by Openshow and Whittker[167] by using optically active **176**. Furthermore, they accomplished the total syntheses of tubulosine **191** and isotubulosine **192** by the similar reaction through four steps.[167] Pictet-Spengler reaction of protoemetine **197** with 5-hydroxytryptamine also yielded tubulosine **191**[168] (Chart 2-32). Thus the ester **176** is an important intermediate, which was synthesized by Openshow[169] and recently by van Tamelen et al.[170] The former used the ketone **198** as the starting material. Wittig reaction of **198**, followed by the catalytic hydrogenation, gave methyl ester of **176** together with its isomer **199** (Chart 2-33).

Mannich reaction of 3,4-dimethoxyphenethylamine with triester **200**, followed by spontaneous lactamization of the initially formed Mannich base **201**, afforded a diester **202**. Bischler-Napieralski reaction of **202**, followed by catalytic hydrogenation, afforded **203**, whose hydrolysis and decarboxylation of the *β*-keto ester function were carried out by heating in dilute hydrochloric acid. Conversion of acetyl to ethyl group was accomplished by preparation from keto ester **204** of the ethylene dithioketal ester, which was then desulfurized with Raney nickel (Chart 2-34).

Chart 2-28.

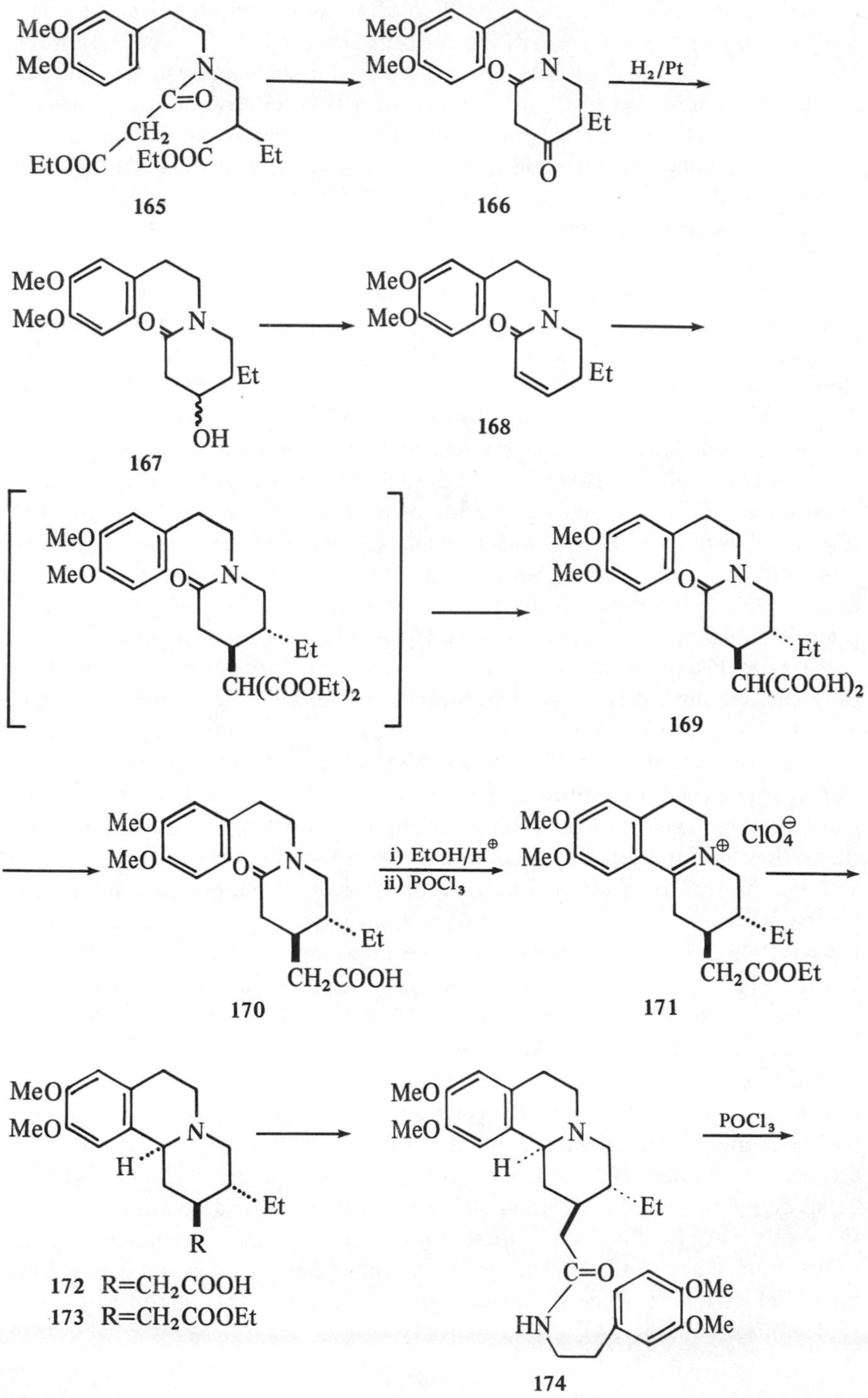

MeO
MeO
N
C=O
CH_2
EtOOC EtOOC Et

165

→

MeO
MeO
O N
Et
O

166

H_2/Pt →

MeO
MeO
O N
Et
OH

167

→

MeO
MeO
O N
Et

168

→

[MeO
MeO
O N
Et
$CH(COOEt)_2$]

→

MeO
MeO
O N
Et
$CH(COOH)_2$

169

→

MeO
MeO
O N
Et
CH_2COOH

170

i) EtOH/$H^{\oplus}$
ii) $POCl_3$ →

MeO
MeO
$N^{\oplus}$ $ClO_4^{\ominus}$
Et
CH_2COOEt

171

→

MeO
MeO
N
H
Et
R

172 R=CH_2COOH
173 R=CH_2COOEt

→

MeO
MeO
N
H
Et
C=O
HN
OMe
OMe

174

$POCl_3$ →

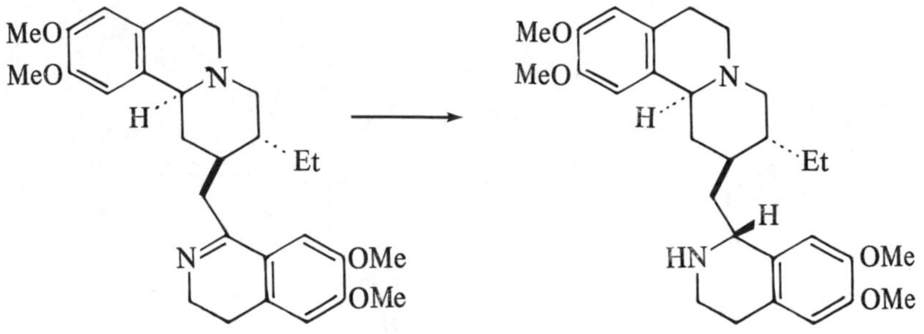

MeO MeO N H Et N OMe OMe

164

MeO MeO N H Et H HN OMe OMe

151 + 163

Chart 2-29.

MeO MeO N H Et COOEt

176

MeO MeO N H Et O HN OMe OCH_2Ph

177

MeO MeO N H Et N OMe OR

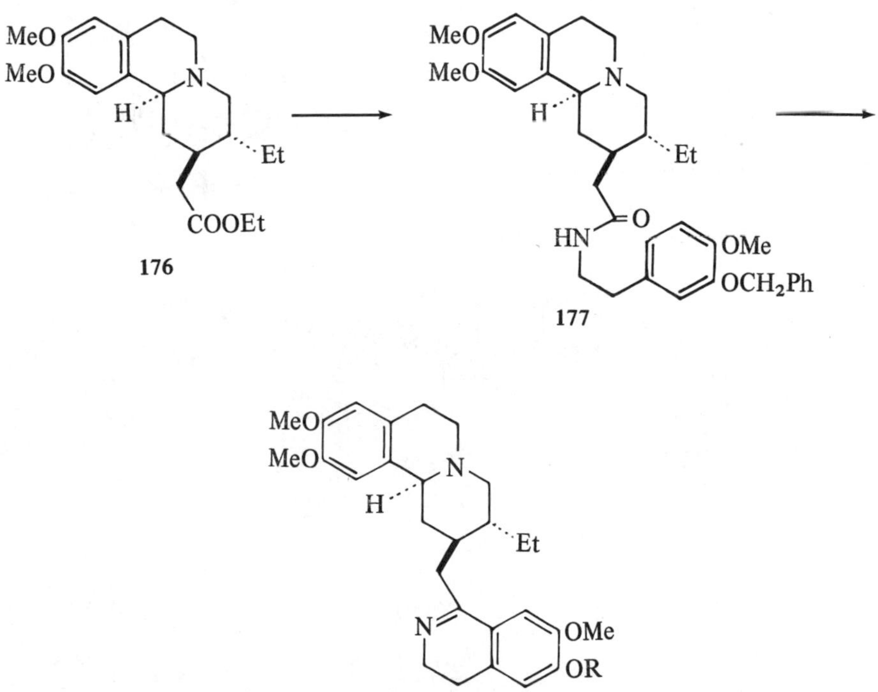

178 $R{=}CH_2Ph$
175 R=H

Chart 2-30.

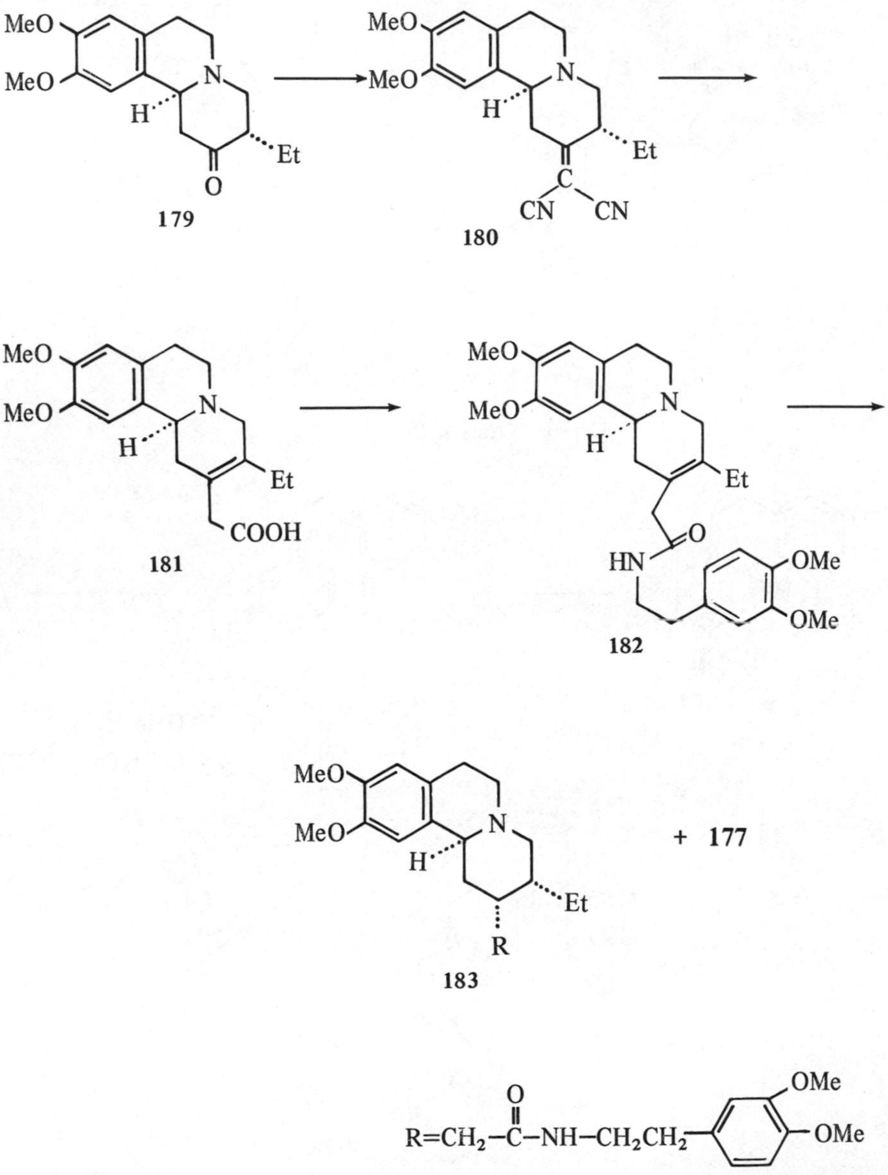

MeO
MeO
N
H
Et
O
179

MeO
MeO
N
H
Et
C
CN CN
180

MeO
MeO
N
H
Et
COOH
181

MeO
MeO
N
H
Et
O
HN
OMe
OMe
182

MeO
MeO
N
H
Et
R
183

+ 177

R=CH_2–C(=O)–NH–CH_2CH_2–(3,4-dimethoxyphenyl) OMe OMe

Chart 2-31.

MeO MeO N O COOEt Et COOEt 185

MeO MeO N O O Et COOEt 186

MeO MeO N O H Et COOH 187

MeO MeO N O H Et COOH 184

MeO MeO N O H Et O HN OMe OMe 188

MeO MeO N O H Et N OMe OMe 189

151 + 163

MeO MeO N H Et N OMe OMe 193

Chart 2-32.

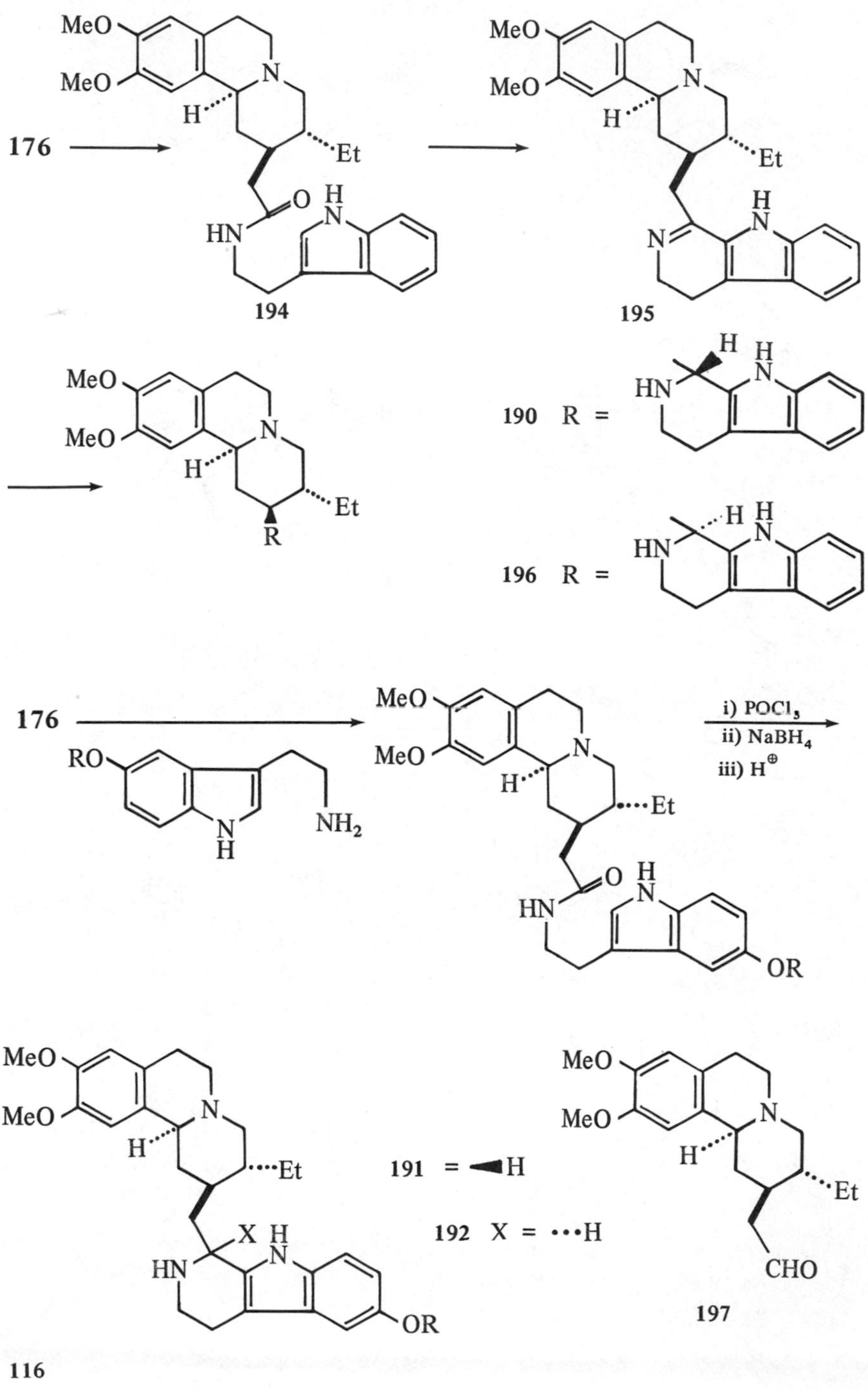

176 → 194 → 195 →

190 R =

196 R =

176 (RO-tryptamine) →

i) $POCl_3$
ii) $NaBH_4$
iii) $H^{\oplus}$

191 X = ◄H

192 X = •••H

197

Chart 2-33.

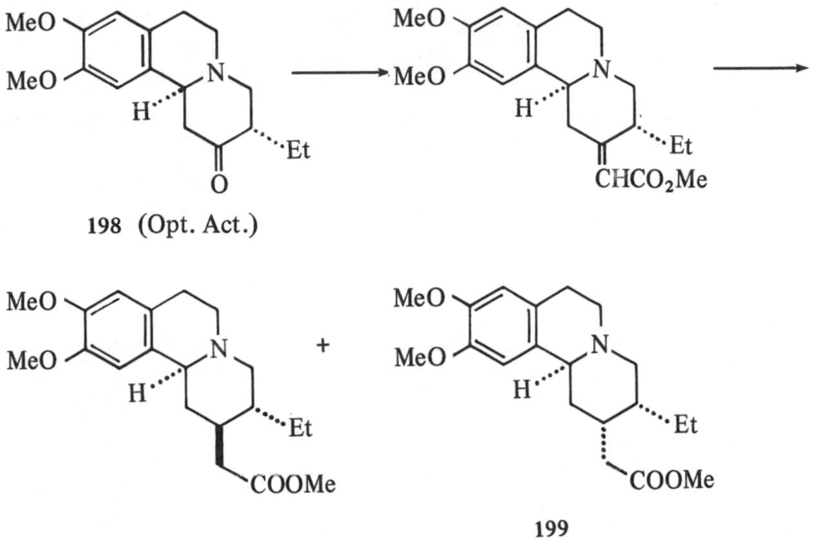

198 (Opt. Act.)

199

Rhoeadine Series

Rhoeadine, found in a variety of plants of the genus Papaver of the *Papaveraceae*, is an alkaloid of the group characterized by its 3-benzazepine structure along with the presence of a six-membered acetal ring in the molecule.

Brossi Method.[171] Based on model experiment for the preparation of benzazepines from the phthalide alkaloids, (−)-α-narcotine and (−)-β-narcotine, the phthalideisoquinoline (−)-bicuculline **205** has been converted by a new, straightforward synthesis into the benzapine alkaloids, (+)-rhoeadine **212** and its unnatural antipode. Since **205** was obtained from (−)-β-hydrastine, which has been previously synthesized, the following transformations by Brossi and his coworkers constitute the total synthesis of natural rhoeadine. Reaction of **205** with phenyl chloroformate and di-isopropylethylamine, followed by dehydrohalogenation with a mixture of dimethyl sulfoxide and di-isopropylethylamine, yielded the urethane **206**, which was treated with 2 N sodium hydroxide to afford the dihydrobenzazepine sodium salt **207** in 80% of the yield. Acidification of this product with acetic acid effected cyclization to the spirolactone **208**, which was oxidized by air in ethanolic solution to provide the keto-lactone **209** in 62% of the yield. Lithium borohydride reduction of **209** in tetrahydrofuran followed by acidification with acetic acid afforded, via the transient *cis*-hydroxy acid **210**, the *cis*-lactone, (±)-oxyrhoeagenine **211** in 75% of the yield.

Chart 2-34.

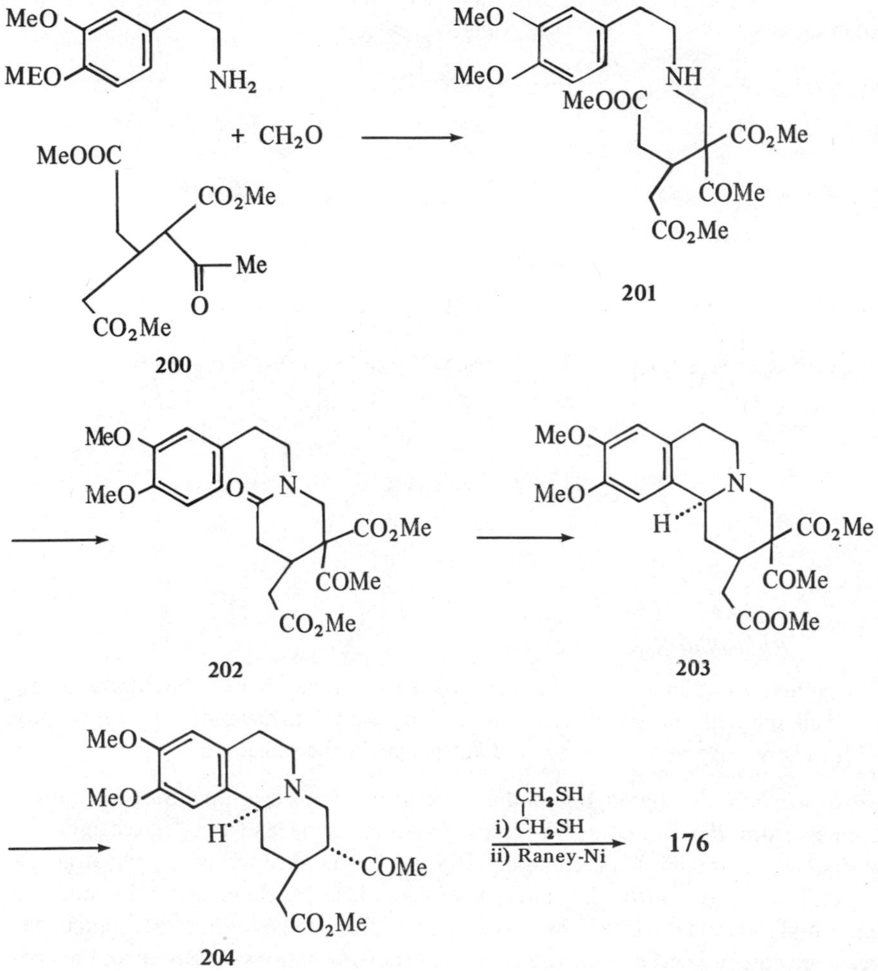

MeO MEO NH2 + CH_2O MeOOC CO_2Me Me O CO_2Me **200** → MeO MeO NH MeOOC CO_2Me COMe CO_2Me **201**

→ MeO MeO O N CO_2Me COMe CO_2Me **202** → MeO MeO N H CO_2Me COMe COOMe **203**

→ MeO MeO N H COMe CO_2Me **204** —i) CH_2SH–CH_2SH ii) Raney-Ni→ **176**

Resolution of (±)-**211** with (+)-10-camphorsulfonic acid in methanol and neutralization of the precipitated diastereoisomeric salt provided (−)-oxyrhoeagenine (the mirror image of **211**). Treatment of the mother liquors with (−)-10-camphorsulfonic acid, followed by neutralization, yielded (+)-**211**. Partial reduction of **211** in pyridine at −70° with sodium bis-(2-methoxyethoxy)aluminum hydride, followed by storage at −20°, yielded a mixture of anomeric lactols, which were etherified in methanol with trimethyl orthoformate catalyzed by mineral acid, to afford (+)-rhoeadine **112** (Chart 2-35).

Chart 2-35.

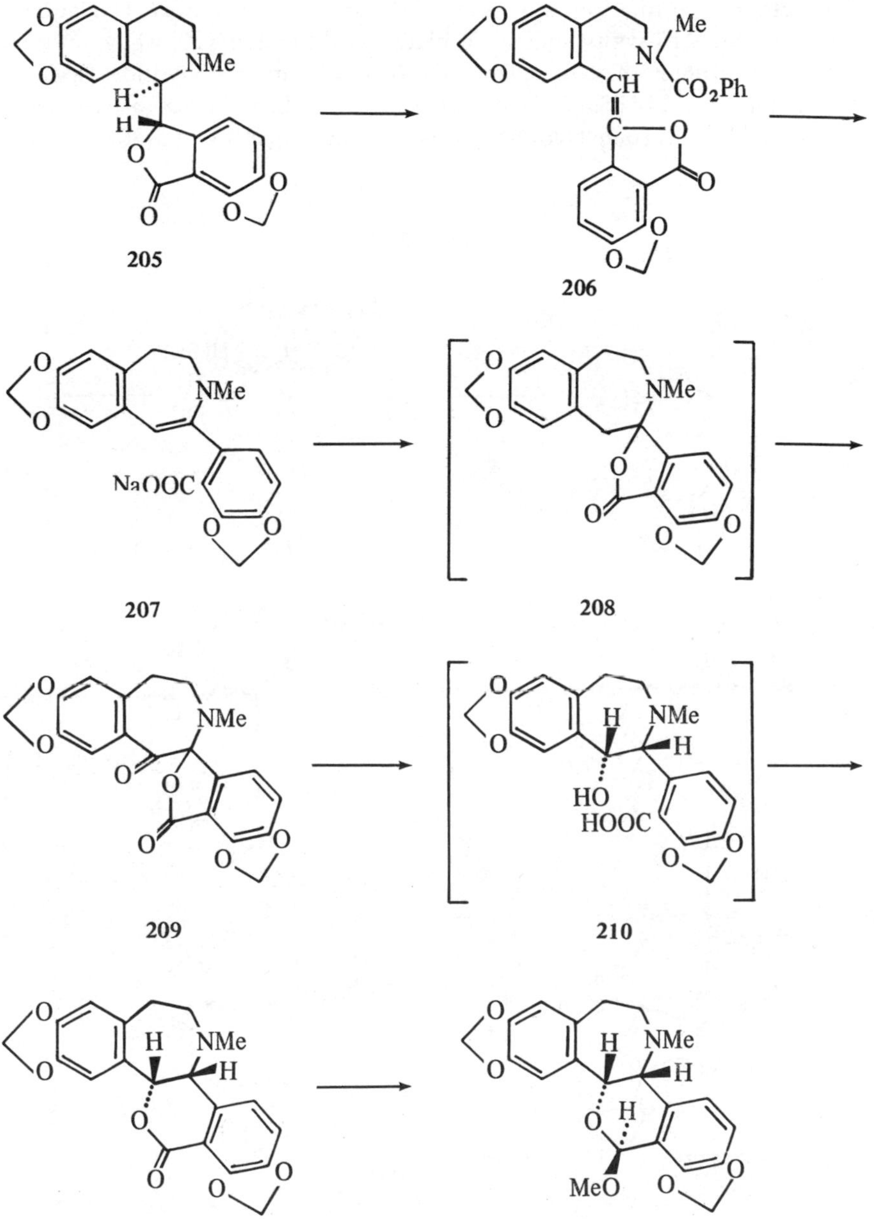

NMe
H
O
Me
N
CH
CO_2Ph
C
NaOOC
HO
HOOC
MeO

205 206 207 208 209 210 211 212

Irie Method.[172] The spiro-isoquinoline **213** was reduced with lithium aluminum hydride to give the *N*-methyl-alcohol **124**, which was treated with mesyl chloride and triethylamine to afford a mixture of the amines **215** and **216**. Oxidation of the allylamine **215** with osmium tetroxide, gave the diol **127**, which on treatment with sodium metaperiodate and then sodium borohydride gave (±)-rhoeagenine diol **218**. Since rhoeagenine diol had already been correlated with rhoeadine **212**, this completed the synthesis of rhoeadine (Chart 2-36).

Chart 2-36.

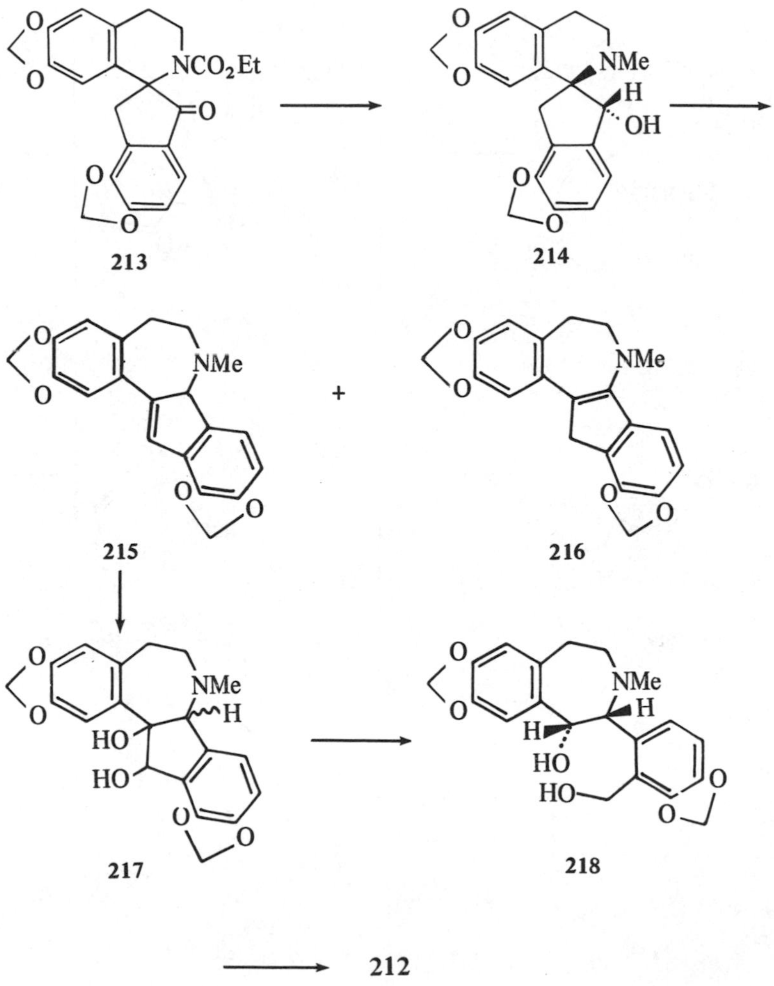

3. TOTAL SYNTHESIS BY PHENOL OXIDATION

Phenolic oxidation[173-182] has become one of the new research areas of synthetic organic methodology during the past twenty years, although it has long been recognized that a dimeric product can be derived from the oxidation of phenols with such reagents as ferric chloride and potassium ferricyanide.[183]

The concept that some isoquinoline alkaloids are built up in nature by oxidative coupling within the molecule of a benzylisoquinoline is not new; Gadamer[184] in 1911 drew attention to the relationship between laudanosoline and glaucine, and similar ideas were promulgated by Robinson[185] and Schöpf.[186] In 1957, Barton[174] proposed that the new C—C or C—O bond in isoquinoline alkaloids was formed by pairing of radicals from the substrate involved in the oxidative step. A very detailed knowledge of the biosynthetic pathway to the isoquinoline alkaloids has been gained from extensive tracer experiments,[186a] which showed that phenolic oxidation is an important step in the biogenesis of isoquinoline alkaloids.

Since 1960, the biogenetical applicability of phenolic oxidation in the synthesis of isoquinoline alkaloids has been repeatedly investigated, and the total syntheses of a number of isoquinoline alkaloids have been achieved using biogenetic-type reaction steps. We review recent advances in the synthetic aspects of the isoquinoline and related alkaloids considering biomimetic paths.

A. Simple Isoquinoline Alkaloids

Pilocereine **1**, which exists with lophocerine 2 in *Lophocereus schottii,* is the only example of a phenolic oxidation product in simple isoquinoline alkaloids. Tracer experiments showed that phenolic oxidation of 2 by enzymes, in the plant furnishes pilocereine **1**.[187,188] This coupling process has been reproduced in the laboratory by potassium ferricyanide oxidation of **2** at pH 6 to yield pilocereine **1** in 0.3% yield together with the dimer in 32% yield.[189] Moreover, pilocereine trimethiodide was formed by oxidation of lophocerine methiodide with potassium ferricyanide.[190] Phenolic oxidation of simple isoquinolines has been examined by several groups[191-197] (Chart 3-1).

B. 2-Benzylisoquinoline Series

Sendaverine **3** is found in *Corydalis aurea* together with several 1-benzylisoquinoline derivatives. Tracer investigations have not so far been carried out, but it is probable that coclaurine **4** is a precursor of sendaverine through an aziridine-type compound, and a valuable synthesis based upon this route has been developed.[196] Thus enzymic oxidation of *N*-norarmepavine **5** with peroxidase from horseradish in the presence of 3% hydrogen peroxide at pH 6.5, followed

Chart 3-1.

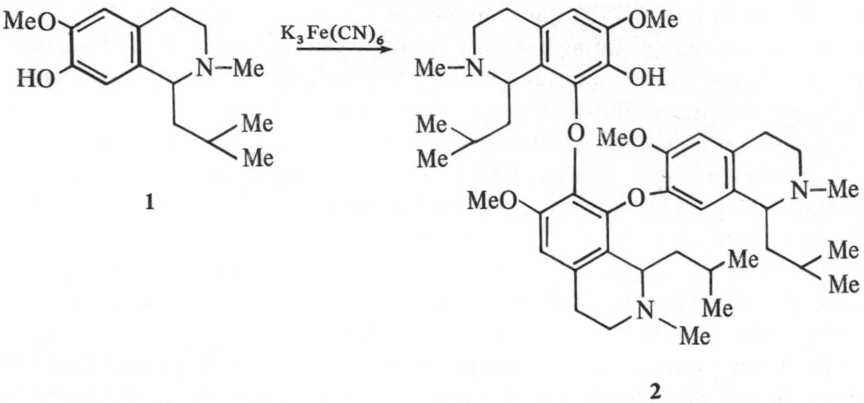

by sodium borohydride reduction, produces *N*-benzylisoquinoline **6**, which is a biogenetic equivalent of sendaverine **3** (Chart 3-2).

C. 1-Benzylisoquinoline Series

Tracer experiments[186a] have shown that 1-benzylisoquinoline **7**, having phenolic groups, is transformed into several types of isoquinoline alkaloids by phenolic oxidation with enzymes in the plant, as shown in Chart 3-3 (substituents are omitted).

Cularine and Related Alkaloids

Cularine **8** and related alkaloids may be biosynthesized by appropriate C–O coupling of diphenolic isoquinoline **9**, followed by *O*-methylation, to give either cularine **8** or compound **10**. Moreover, phenolic oxidation of a second diphenolic isoquinoline **11** gives the dienone **12**, which can be converted into cularine **8** by dienone-phenol rearrangement (Chart 3-4). In the laboratory, oxidation of diphenolic isoquinoline **9** with potassium ferricyanide in a two-phase system affords *O*-demethylcularine **13** in 2.5 to 8% yield, in addition to the ortho coupling product **14**;[199,200] the former, **13**, was then transformed into cularine **8**. Furthermore, the second isoquinoline **11** on oxidation with the same reagent gave dienone **12**.[200]

Bisbenzylisoquinoline Alkaloids

The synthesis of the aforementioned cularine group exemplifies intramolecular C–O bond formation to generate a biphenyl ether linkage. On the other hand,

Chart 3-2.

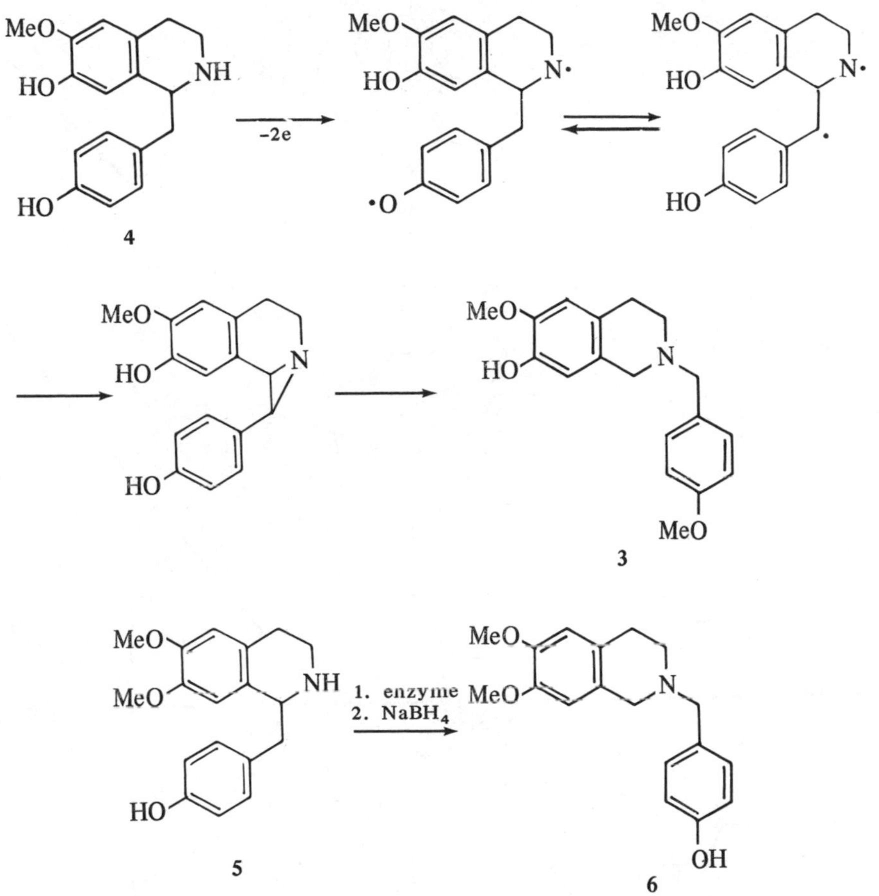

intermolecular C—O coupling of two *N*-methylcoclaurine units **15** by phenolic oxidation should afford bisbenzylisoquinoline alkaloids. This was confirmed by Barton, Kirby, and Wiechers[201] through feeding experiments showing that half of the epistephanine molecule **16** is derived exclusively from D-(−)-*N*-methylcoclaurine (Chart 3-5). Laboratory experiments have been reported for the conversion of *N*-methylcoclaurine units into bisbenzylisoquinoline analog, by phenolic oxidation.[196,201-207] Two experiments afforded coupled products having a biphenyl ether linkage; both have been found in nature. Thus enzymatic oxidation of *N*-methylcoclaurine **15** with homogenized potato peelings and hydrogen peroxide gave **17**, which is coupled in the "head-to-tail" mode.[206]

Chart 3-3.

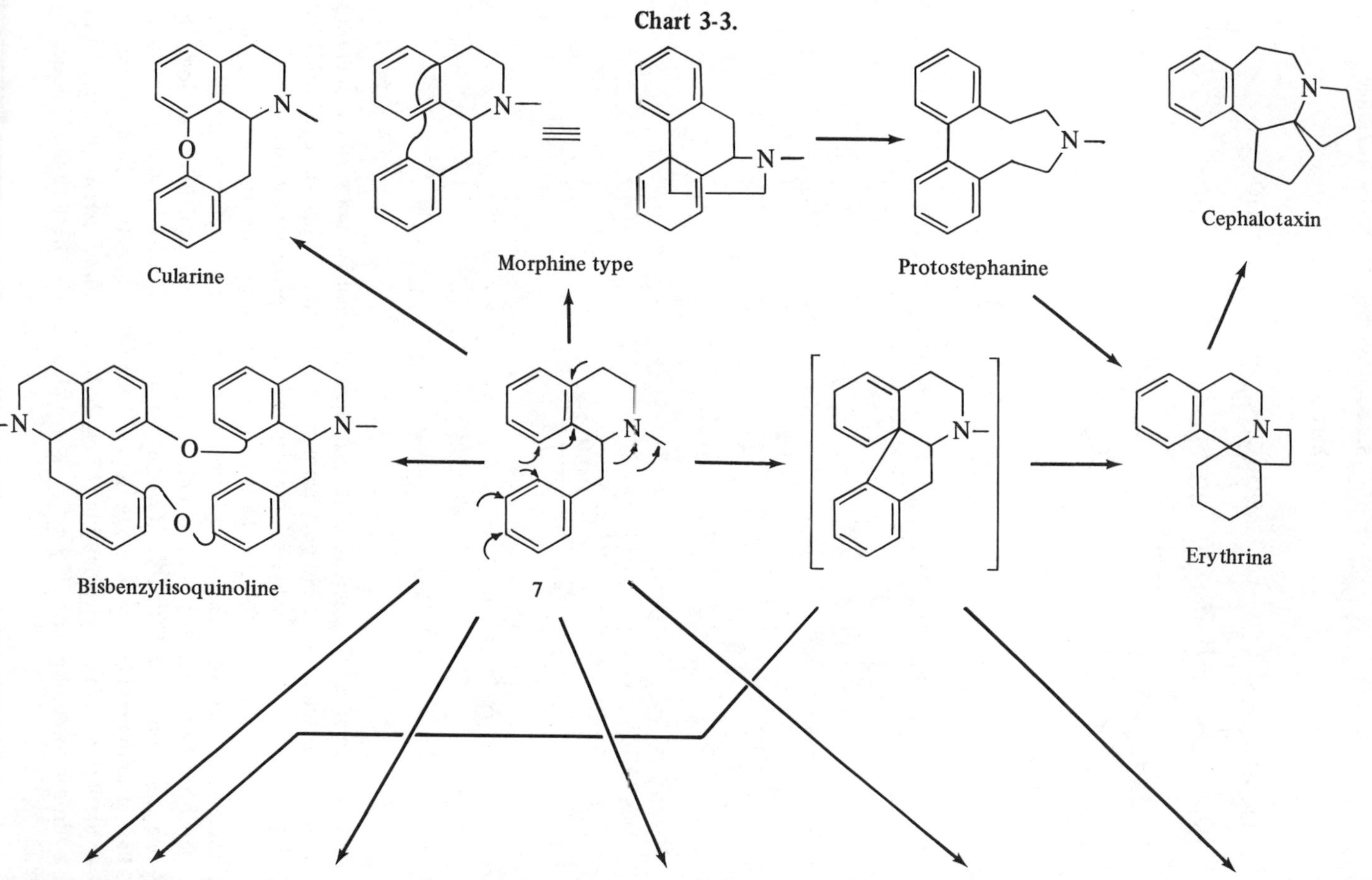
Cularine
Morphine type
Protostephanine
Cephalotaxin
Bisbenzylisoquinoline
7
Erythrina
N
O

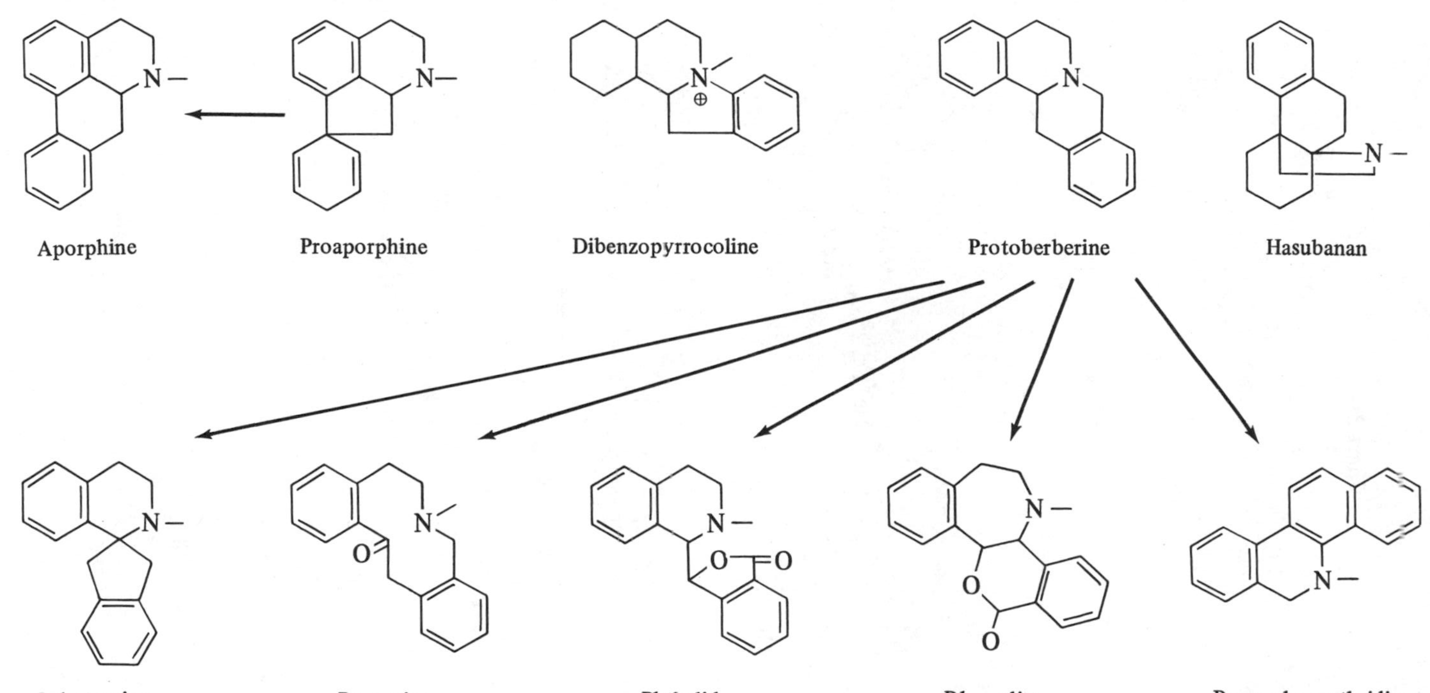
Aporphine
Proaporphine
Dibenzopyrrocoline
Protoberberine
Hasubanan
Ochotensine
Protopine
Phthalide
Rhoeadine
Benzophenanthridine

Chart 3-4.

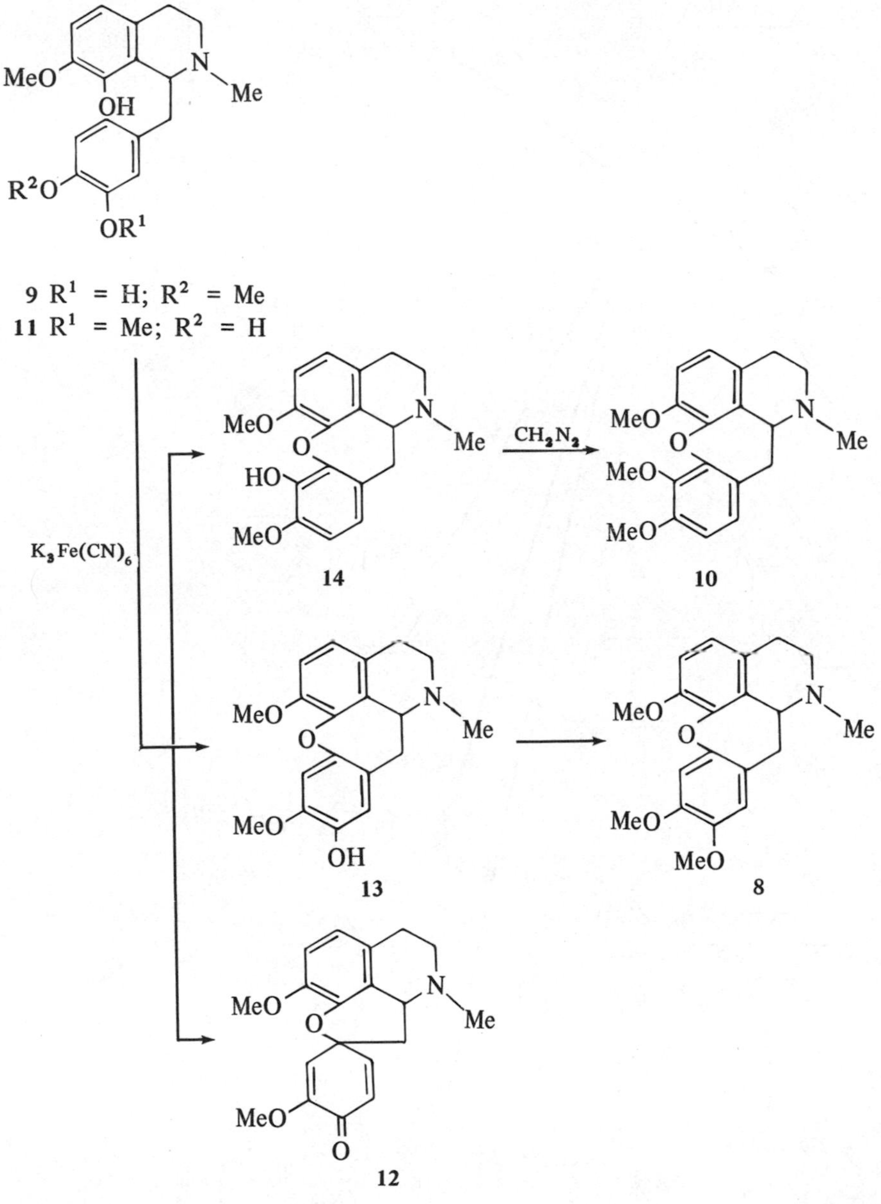

MeO OH N Me
R^{2}O OR^1

9 $R^1 = H; R^2 = Me$
11 $R^1 = Me; R^2 = H$

$K_3Fe(CN)_6$

MeO O HO MeO N Me

14

CH_2N_2

MeO O MeO MeO N Me

10

MeO O MeO OH N Me

13

MeO O MeO MeO N Me

8

MeO O MeO O N Me

12

Chart 3-5

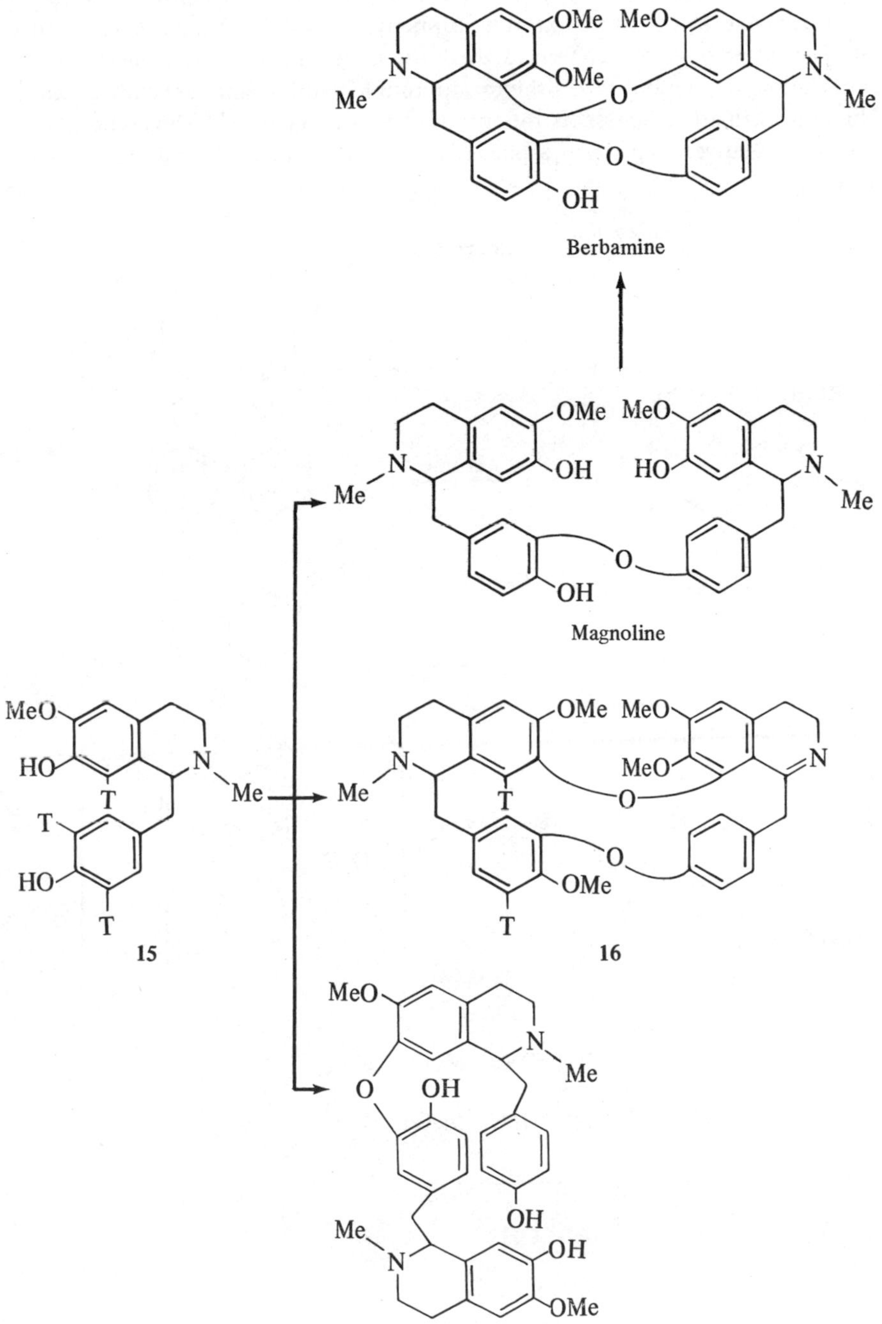

OMe MeO N Me OMe O Me N O OH

Berbamine

OMe MeO N Me OH HO N Me O OH

Magnoline

MeO HO N T Me T HO T

15

OMe MeO N Me MeO N T O O OMe T

16

MeO N Me O OH OH Me N OH OMe

However, the electrolytic oxidation of the sodium salt of *N*-ethoxycarbonyl-*N*-norarmepavine **18** led to a mixture of biphenyl derivative **19** and biphenyl ether **20**. The latter base was converted, without isolation, into dauricine **21** via *O*-benzylation, reduction with lithium aluminum hydride and debenzylation.[207] This transformation represents the first synthesis of a natural bisbenzylisoquinoline by oxidative coupling of a phenolic monomeric benzylisoquinoline (Chart 3-6).

Chart 3-6.

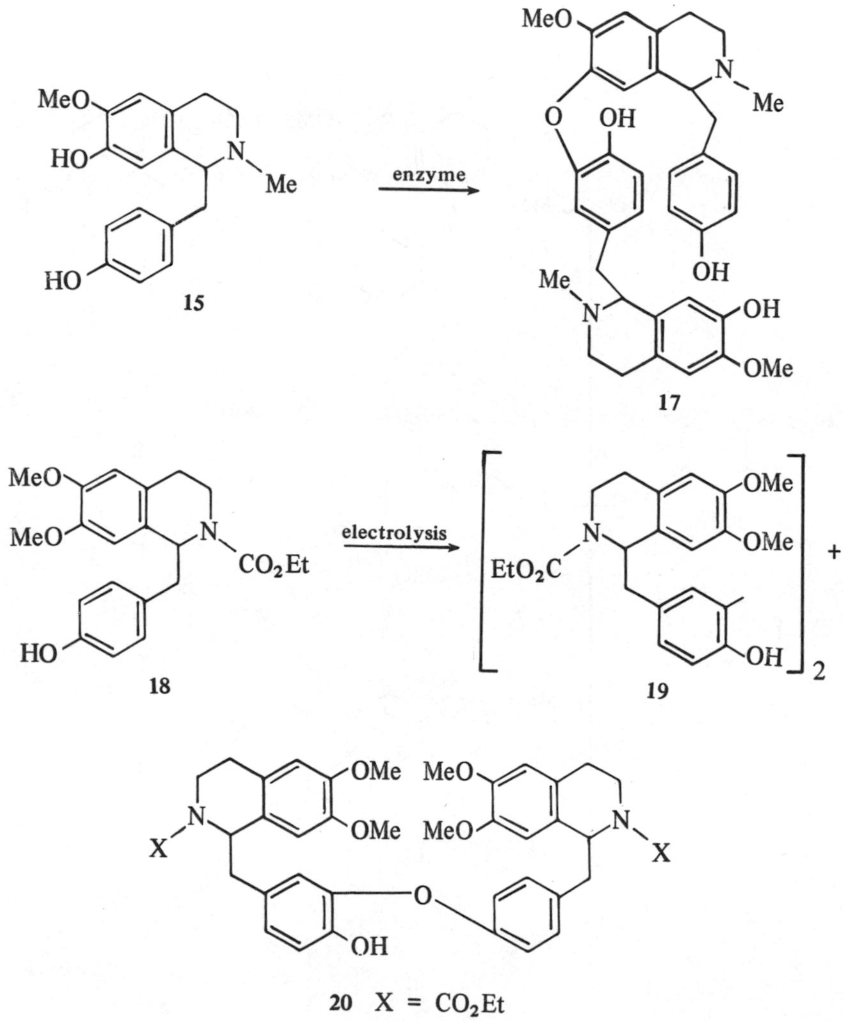

MeO HO N Me HO **15** —enzyme→ MeO N Me O OH OH Me N OH OMe **17**

MeO MeO N CO_2Et HO **18** —electrolysis→ [EtO_2C N OMe OMe OH]$_2$ **19** +

X N OMe OMe MeO MeO N X O OH

20 X = CO_2Et
21 X = Me

Aporphine and Proaporphine Alkaloids

The relationship between laudanosoline 22 and glaucine 23 was noted at the beginning of this century.[184,208] In 1957, Barton and Cohen[173] proposed the theory of phenol oxidation in the biogenesis of aporphine alkaloids; thus the majority of the aporphine bases can be regarded as being formed by phenolic coupling from phenolic base **24** (≡**25**). This compound type may be oxidized to corytuberine-type alkaloids **26** and glaucine-type alkaloids 27 (Chart 3-7). The hypothesis that phenolic oxidation generates the bond between the two aro-

Chart 3-7.

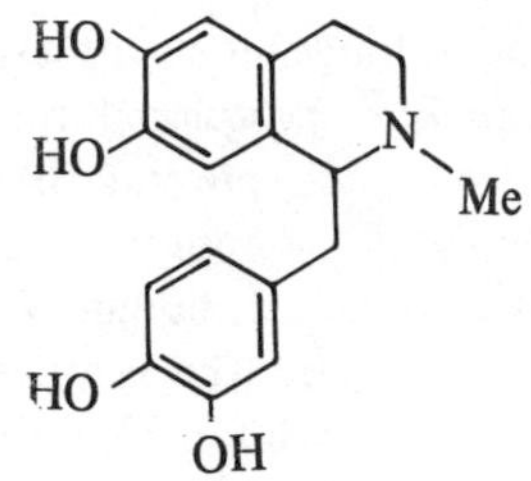

HO HO N Me HO OH **22**

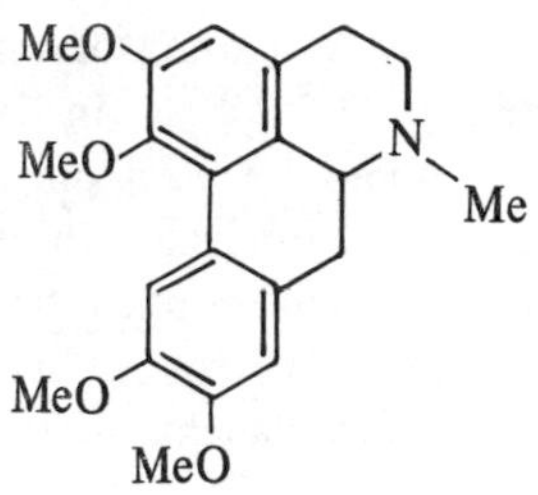

MeO MeO N Me MeO MeO **23**

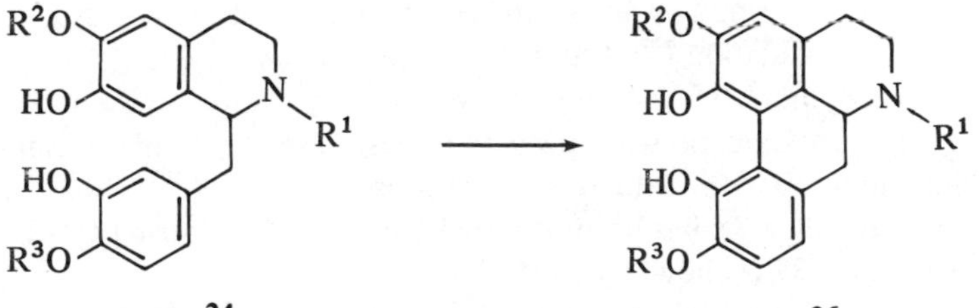

R^2O HO N R^1 HO R^3O **24** → R^2O HO N R^1 HO R^3O **26**

|||

R^2O HO N R^1 R^3O OH **25** → R^2O HO N R^1 R^3O OH **27**

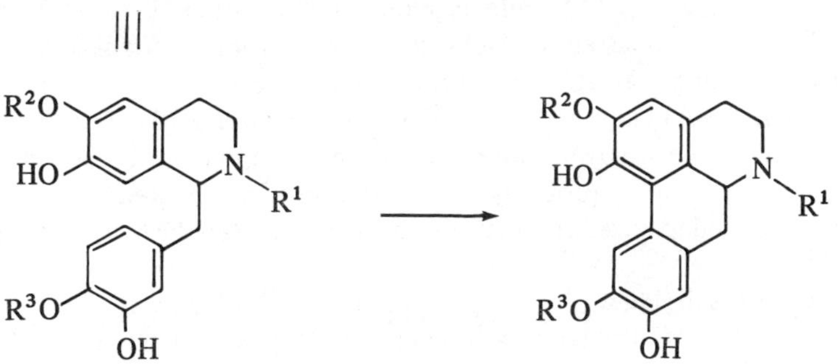

Chart 3-8.

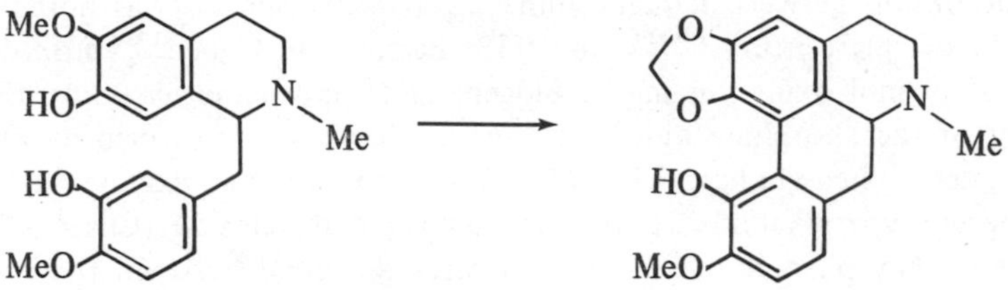

MeO HO N Me HO MeO 28 → O O N Me HO MeO 29

matic rings of the aporphines has been confirmed by conversion of a labeled reticuline **28** into a radioactive bulbocapnine **29**[209] (Chart 3-8).

Until recently, no laboratory analogy for this type of biogenetic synthesis had been realized.[185,186,210,211] In 1962, however, Franck[212] postulated that the presence of the lone electron pair on the nitrogen would prevent the C–C coupling, and accomplished the first in vitro synthesis of the aporphine derivative **30** in 60% yield via phenolic oxidation of laudanosoline methiodide **31** with ferric chloride at 20° for 20 hr. Similarly, but at a later date, the syntheses of laurifoline **32** and glaucine methopicrate **33** by oxidation with ferric chloride of the phenolic benzylisoquinolinium salts were reported.[213,214] Interestingly, treatment of pentahydric phenol **34** with ferric chloride gave, instead of the expected cularine or morphine analog, aporphine **30**, as is obtained from the tetrahydric series **31**.[214a] This reaction appears to involve a Lewis acid-catalyzed condensation rather than an oxidation.[177] (Chart 3-9).

N-Ethoxycarbonylnorreticuline **35** is converted into aporphine **36** in 5 to 7% yield by phenolic oxidation with potassium ferricyanide in dilute ammonia; lithium aluminum hydride reduction affords isoboldine **37**[215] (Chart 3-10). Furthermore, reticuline **38** has been oxidized with several oxidizing agents[216-221] to afford isoboldine **37** as shown in Table 3-1.

Isoboldine **37** has also been synthesized by potassium ferricyanide oxidation of 6′-bromoreticuline **39**,[215,216] and the mode of elimination of a bromine atom in the oxidation was discussed by Jackson.[216] Elimination of the halogen atom in oxidative coupling was also observed in a transformation of 2′-bromo-*N*-ethoxycarbonylnorreticuline into aporphine **36**[215] (Chart 3-11). The difficulty in oxidizing reticuline or 6′-bromoreticuline-type compounds to corytuberine-type aporphines (cf. **26**) is probably due to the influence of steric factors that prevent direct coupling of radicals ortho to both of the phenolic hydroxyl groups concerned.[216]

Methylation of isoboldine **37** gave glaucine **23**,[215-218] which was also obtained from the aporphine **41**. The latter compound was formed by phenolic oxidation of norlaudanosoline **40** with ferric chloride.[222] (Chart 3-12).

Chart 3-9.

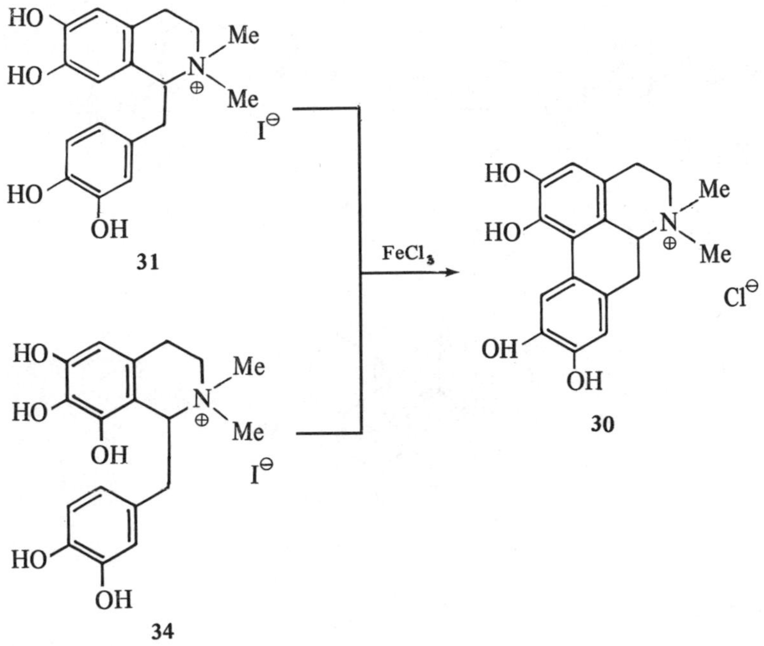

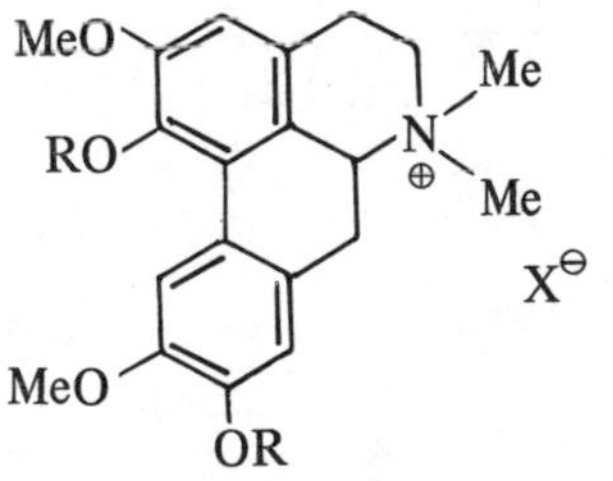

31

34

$FeCl_3$

30

32 R = H
33 R = Me

Franck has shown that the phenolic benzylisoquinolines may be converted to the aporphines in good yields not only through initial quaternization of the nitrogen atom, but also by repressing the formation of the *o*-quinone intermediates by complexation with ferric chloride. For example, laudanosoline is transformed into the corresponding aporphine by a concentrated solution of ferric chloride, which forms a complex with the phenolic oxygens of the substrate while simultaneously acting as the oxidizing agent[220, 222] (Chart 3-13).

Chart 3-10.

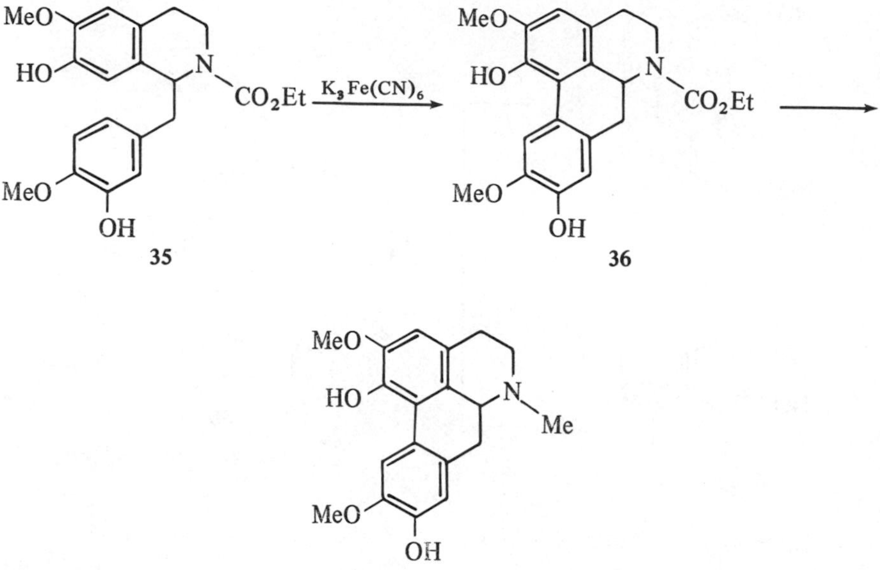

MeO HO N CO_2Et MeO OH **35** $\xrightarrow{K_3Fe(CN)_6}$ MeO HO N CO_2Et MeO OH **36** ⟶

MeO HO N Me MeO OH **37**

Chart 3-11.

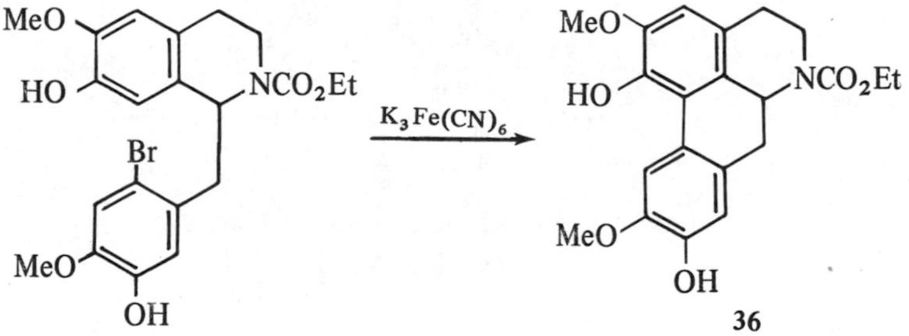

MeO HO NCO_2Et Br MeO OH $\xrightarrow{K_3Fe(CN)_6}$ MeO HO NCO_2Et MeO OH **36**

Recently, Battersby et al.[223] elucidated a surprising biogenetic route to glaucine **23** and corydine **42** by means of tracer experiments. Thus phenolic oxidation of a protosinomenine **43**-type isoquinoline could yield dienones **44** and **45**. Dienone-phenol rearrangement of dienone **45** could yield boldine **46**, affording glaucine **23** and the aporphine **47**, which in turn leads to corydine **42**, formed from the dienone **44**. Dienone **49** has been synthesized in 2% yield

Table 3-1. Conversion of Reticuline **38** into Isoboldine **37** Using Various Oxidizing Agents

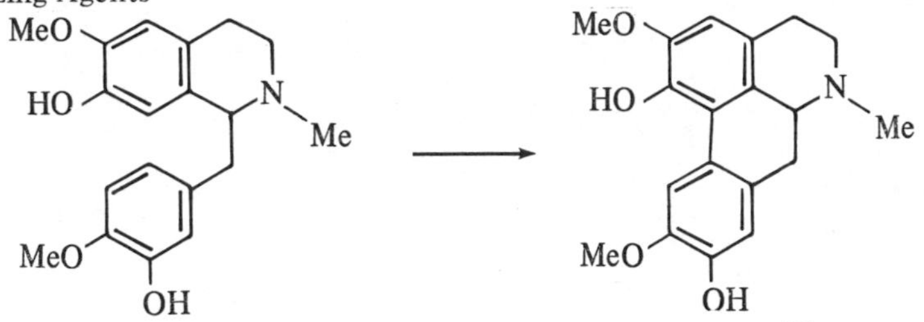

Reagents	Base	Reaction conditions			References
		Temperature	Time (hr)	Yield (%)	
$K_3Fe(CN)_6$	8% $AcONH_4$	$-12°$	14	0.55	216
$K_3Fe(CN)_6$	0.2 *N* KH_2PO_4 0.2 *N* NaOH	r.t.	3	5-6	217
$K_3Fe(CN)_6$	5% $NaHCO_3$	5-10°	0.5	0.4	218
	5% $NaHCO_3$	r.t.	1	5	219,221
Ag_2CO_3/celite		r.t.	1.5	3	219
$VOCl_3$		r.t.	2	+	219
MnO_2/SiO_2				6	220

in the laboratory by phenolic oxidation of *N*-ethoxycarbonyl-*N*-norprotosinomenine **48** with potassium ferricyanide in the presence of dilute ammonia and ammonium acetate, and the rearrangement of **49** was studied[224] (Chart 3-14).

Several aporphine alkaloids, exemplified by roemerine **50** and anonaine **51**, are constructed in such a way that their biogenesis by direct phenolic oxidation is not reasonable or involves unlikely precursors. Barton and Cohen[173] therefore proposed that a coclaurine analog would initially be oxidized to the dienone proaporphine **52**, which would be subjected to dienone-phenol rearrangement to generate anonaine **51**, and that reduction of dienone **52** to dienol **53**, followed by dienol-benzene rearrangement, would then furnish roemerine **50**. This hypothesis has been confirmed by feeding experiments showing that (+)-coclaurine **54** is incorporated into crotonosine **55**[225,226] and, moreover, the orientalinone **56** (biosynthesized from orientaline **57**) is incorporated into isothebaine **58** via orientalinol **59** in *Papaver somniferum*[227,228] (Chart 3-15).

This coupling process was first reproduced in the laboratory in the case of orientaline **57**, which underwent phenolic oxidation with potassium ferricyanide

Chart 3-12.

MeO HO N Me Br MeO OH

39

$K_3Fe(CN)_6$

MeO HO N Me MeO OH

37

HO HO NH HO OH

41

FeCl.

HO HO NH HO OH

40

MeO MeO N Me MeO MeO

23

Chart 3-13.

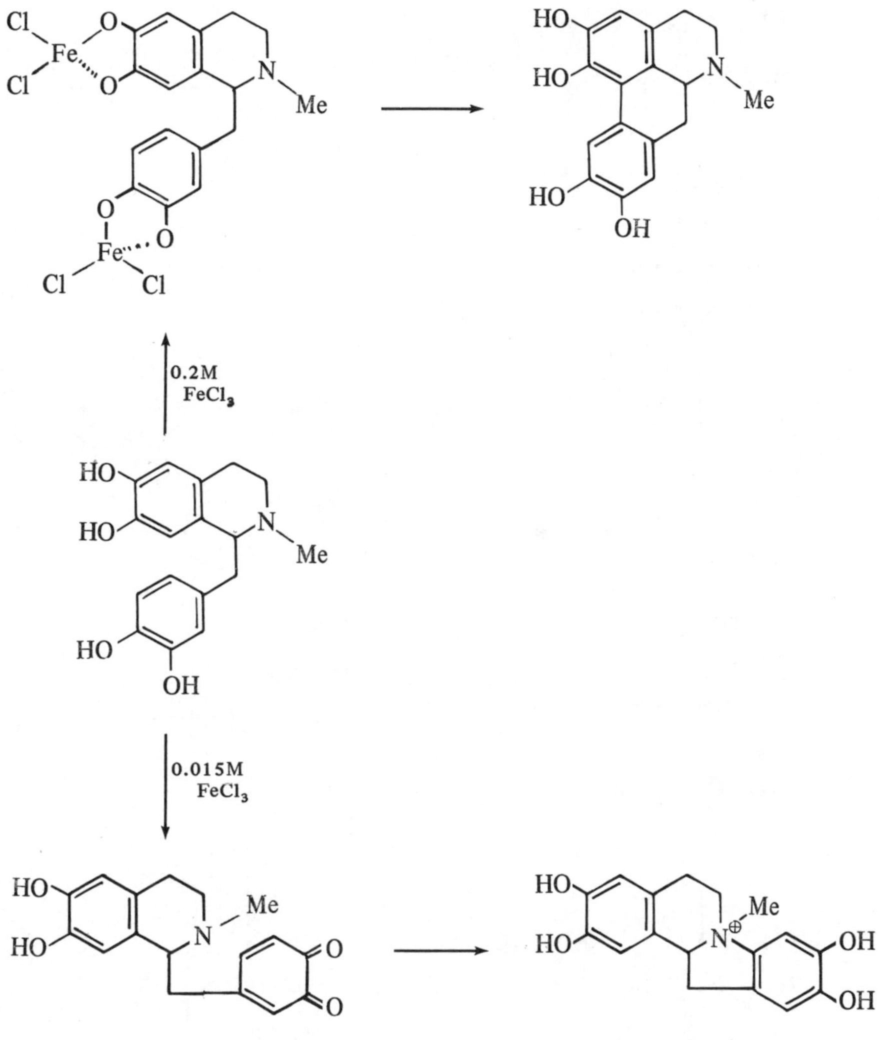

in 8% ammonium acetate solution to afford orientalinone **56**; and orientalinol **59**, obtained from sodium borohydride reduction of orientalinone, underwent dienol-benzene rearrangement to give isothebaine **58**.[228] Recently, reexamination of this sequence revealed that orientalinone **56** (in 20% yield) and its spiro-isomer (1% yield) are obtained in good yield in the oxidation step, and that the rearrangement of the oxidized products gave xylopine-type aporphine **60**

Chart 3-14.

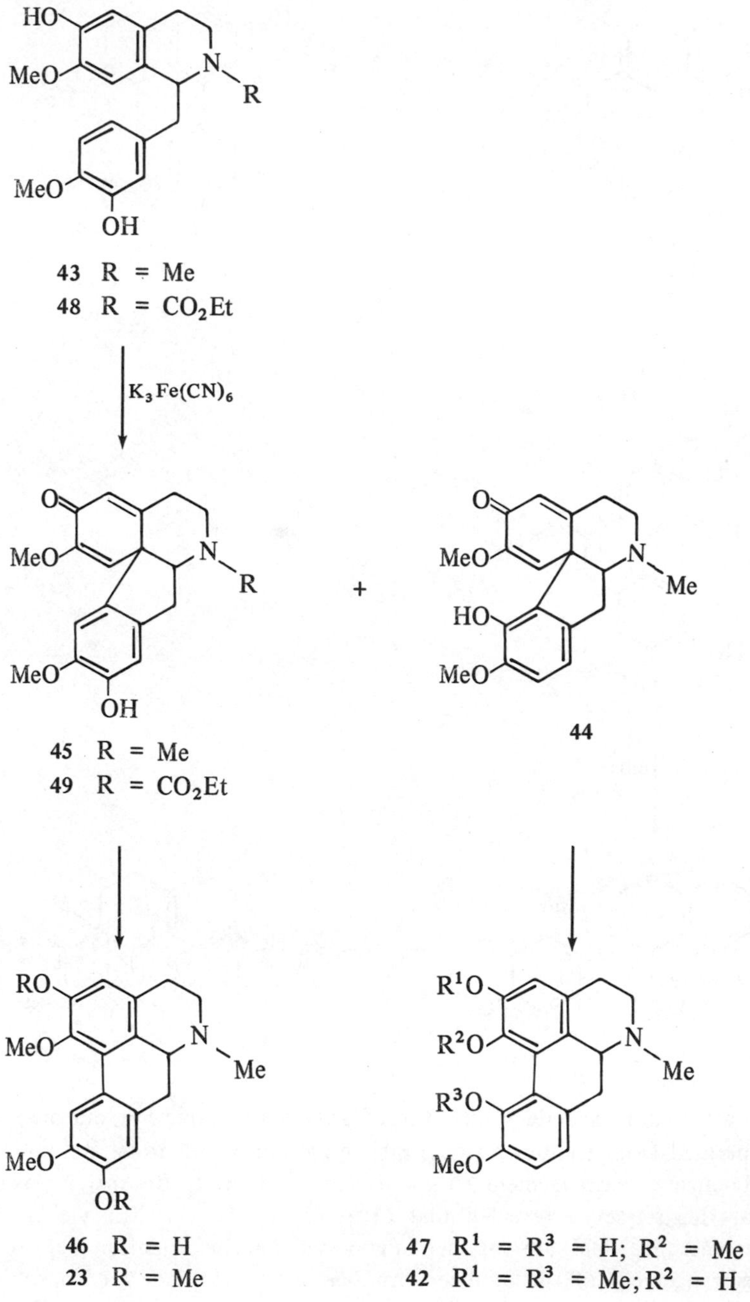

HO
MeO
N
R
MeO
OH

43 R = Me
48 R = CO_2Et

$K_3Fe(CN)_6$

O
MeO
N
R
MeO
OH

45 R = Me
49 R = CO_2Et

+

O
MeO
N
Me
HO
MeO

44

RO
MeO
N
Me
MeO
OR

46 R = H
23 R = Me

R^1O
R^2O
N
Me
R^3O
MeO

47 $R^1 = R^3 = H$; $R^2 = Me$
42 $R^1 = R^3 = Me$; $R^2 = H$

Chart 3-15.

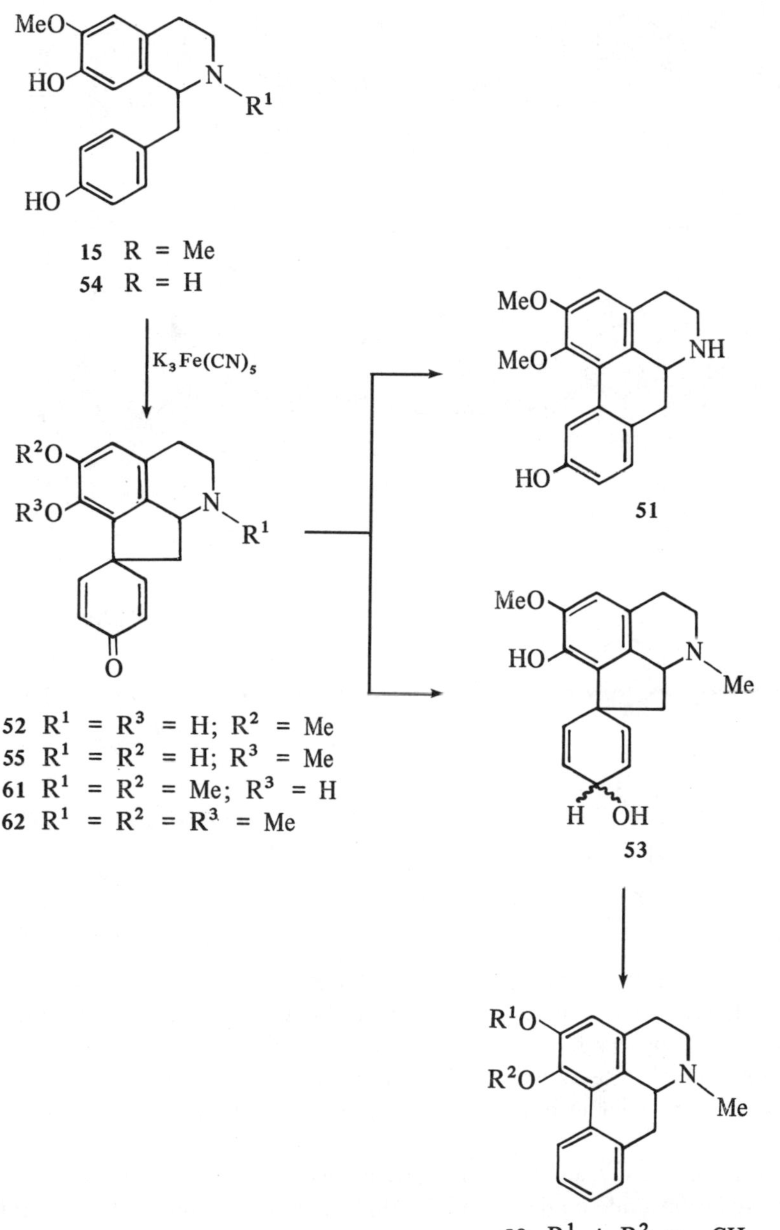

MeO
HO
N
R^1
HO

15 R = Me
54 R = H

$K_3Fe(CN)_5$

R^2O
R^3O
N
R^1
O

52 $R^1 = R^3 = H$; $R^2 = Me$
55 $R^1 = R^2 = H$; $R^3 = Me$
61 $R^1 = R^2 = Me$; $R^3 = H$
62 $R^1 = R^2 = R^3 = Me$

MeO
MeO
NH
HO
51

MeO
HO
N
Me
H OH
53

R^1O
R^2O
N
Me

50 $R^1 + R^2 = -CH_2-$
63 $R^1 = R^2 = Me$

cont.

continued

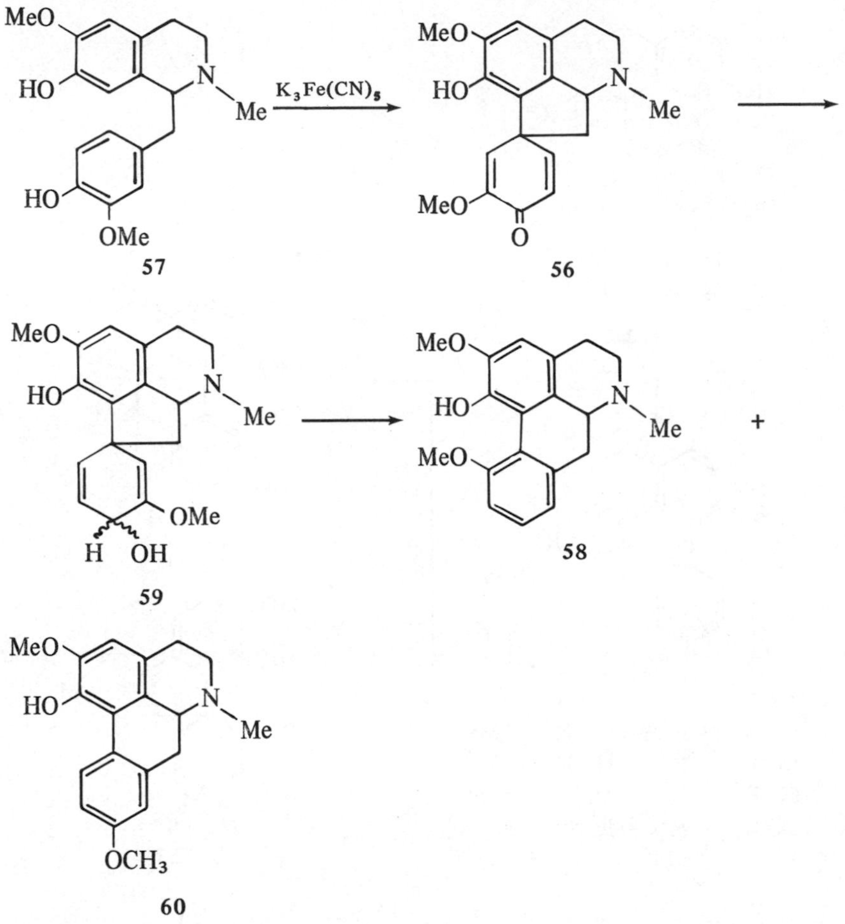

together with isothebaine.[227] Similarly, *N*-methylcoclaurine **15** on oxidation with potassium ferricyanide gave glaziovine **61**, which was converted into pronuciferine **62** by methylation.[229,230] This in turn could be transformed into nuciferine **63**.[231] Alternatively, glaziovine **61** was obtained in 1% of the yield by potassium ferricyanide oxidation of 8-bromo-*N*-methylcoclaurine[231a] (Chart 3-16).

Orientalinone **56**[232] and its *O*-methyl ether **64**[233] were prepared by a sequence involving ferricyanide oxidation of the diphenolic isoquinoline **65**, but not by a biogenetic route. The *ortho*-dienones **66** and **67**, obtained in 25-50% yield from the diphenolic isoquinolines, gave on reduction the corresponding

Chart 3-16.

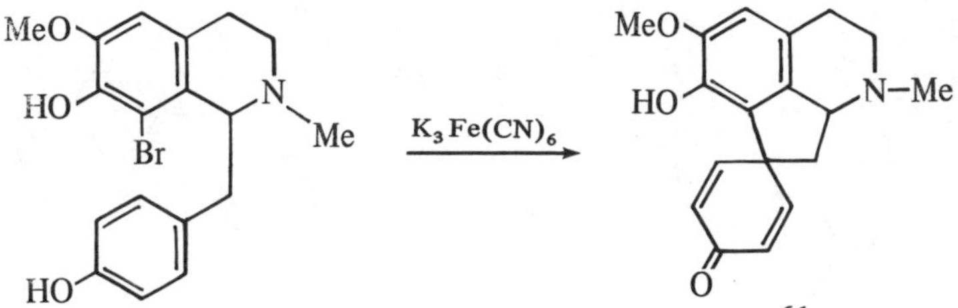

MeO HO Br N Me HO $K_3Fe(CN)_6$ → MeO HO N–Me O

61

dienols (**68** and **69**), which were converted into orientalinone **56** or *O*-methylorientalinone **64** with dilute methanolic hydrochloric acid. Moreover, orientalinone **56** was rearranged with dry hydrogen chloride to corydine **42**, which was also obtained from the dienol **68** with dry hydrogen chloride, but isocorytuberine **70** was obtained using aqueous acid.[232] Similarly, *O*-methylorientalinone gave pseudocorydine **71** in acidic aqueous solution, but on treatment with dry methanolic hydrogen chloride, a mixture of glaucine **23** and *O*-methylcorydine **72** was obtained (Chart 3-17).

Thalicsimidine **73** was synthesized by two biogenetic routes, one a direct coupling reaction and the other through proaporphine **74**.[234] Ferricyanide oxidation of the 5,3′-diphenolic benzylisoquinoline gave the aporphine that led to thalicsimidine **73** upon *O*-methylation with diazomethane. Alternatively, phenolic oxidative coupling of the 5,4′-diphenolic isoquinoline with ferricyanide afforded a mixture of the *spiro* isomers of dienones **74a** and **74b**, which on acid-catalyzed dienone-phenol rearrangement furnished the 1,2,3,10,11-*penta*-subsituted aporphine and the 1,2,3,9,10-*penta*-substituted isomer. The latter aporphine afforded thalicsimidine **73** by treatment with diazomethane (Chart 3-18).

An aporphine synthesis by rearrangement of the *ortho*-dienone derived from the diphenolic isoquinoline has been reported,[235,236] but this route was not biogenetic. For example, *N*-methylcaaverine has been prepared by the ferric chloride oxidation of 1,2,3,4-tetrahydro-7-hydroxy-1-(2-hydroxybenzyl)-6-methoxy-2-methylisoquinoline to the *ortho*-dienone followed by reduction to the dienol and acid-catalyzed dienol-benzene rearrangement[235] (Chart 3-19).

Hoshino, Tobinaga, and Umezawa[236a] reported a new route to the aporphine alkaloids from the 1-benzyl-7-hydroxyisoquinolines via the *para*-quinol acetates, although this route was not biogenetic. This synthesis involves treatment of codamine with lead tetraacetate to yield only *para*-quinol acetate, which without purification was treated with acetic anhydride and sulfuric acid to generate *O*-acetylthaliporphine in 18% of the yield. Domesticine was also synthesized by the same method from 1-(3,4-methylenedioxybenzyl)isoquinoline. The hydrolysis of this gives thaliporphine. Alternatively, treatment of *para*-quinol

Chart 3-17.

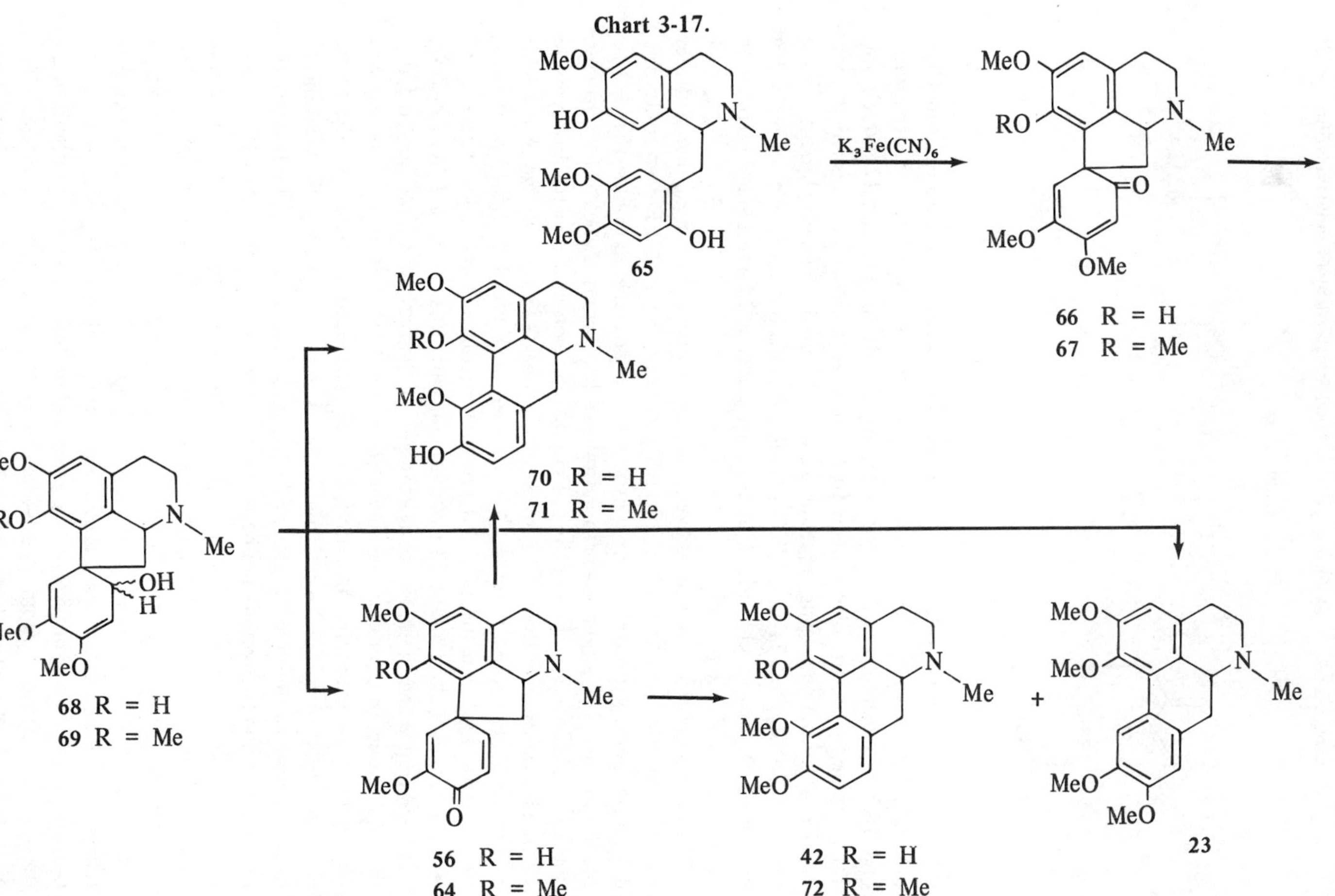

MeO
HO
N
Me
MeO
MeO
OH
65
K3Fe(CN)6
RO
O
OMe
66 R = H
67 R = Me
70 R = H
71 R = Me
OH
H
68 R = H
69 R = Me
56 R = H
64 R = Me
42 R = H
72 R = Me
+
23

Chart 3-18.

OH
MeO
MeO
N
Me
MeO
OH

$K_3Fe(CN)_6$

OH
MeO
MeO
N
Me
MeO
OH

OMe
MeO
MeO
N
Me
MeO
MeO

73

OH
MeO
MeO
N
H
Me
MeO
O

74a

OH
MeO
MeO
N
Me
HO
MeO

$K_3Fe(CN)_6$

+

OH
MeO
MeO
N
H
Me
OMe
O

74b

Chart 3-19.

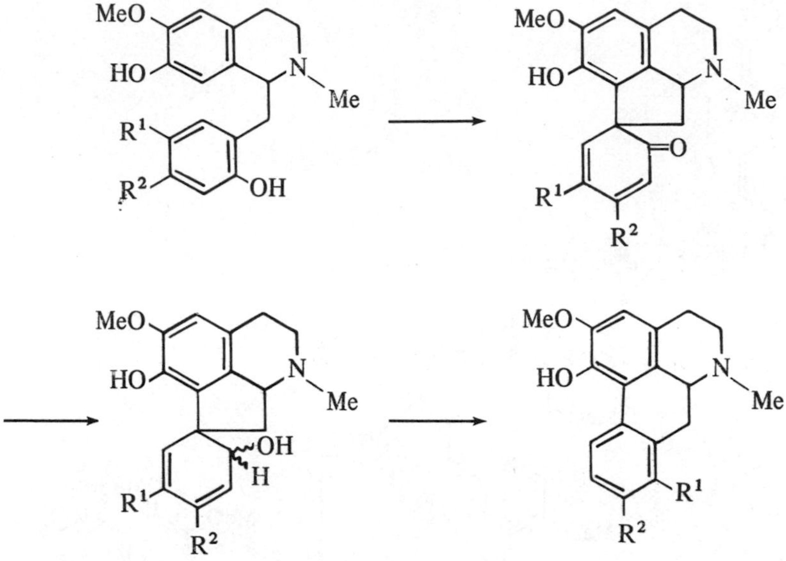

$R^1 = R^2 = H$
or $R^1 + R^2 = -O-CH_2-O-$

$R^1 = R^2 = H$ *N*-Methylcaaverine

Chart 3-20.

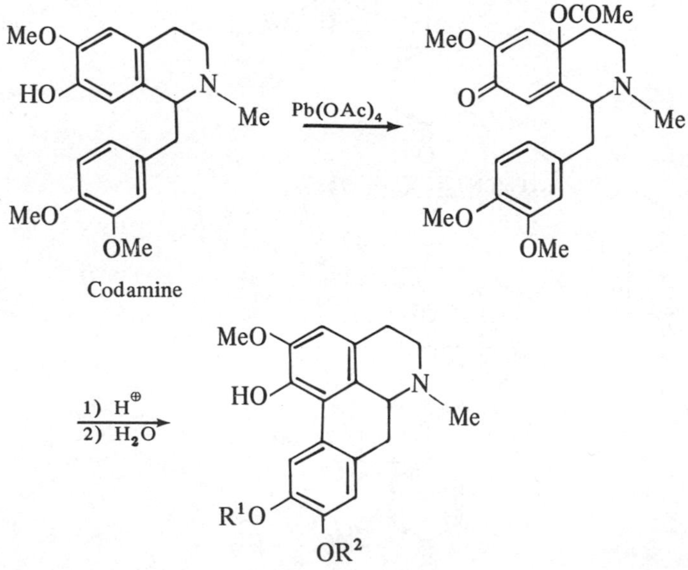

$R^1 = R^2 = Me$ Thaliporphine
$R^1 + R^2 = CH_2$ Domesticine

Chart 3-21.

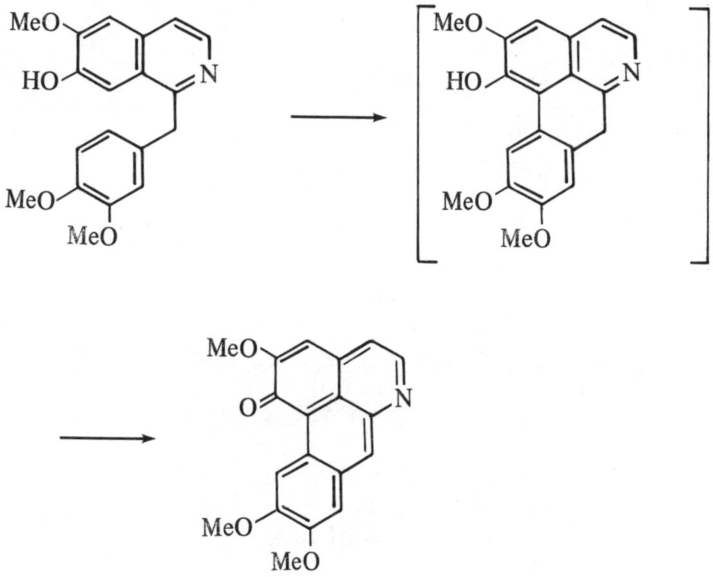

Oxidant	Medium	Temperature (°C)	Yield (%)
$CeSO_4$	10% H_2SO_4	0	25
$Co(OH)_3$	10% H_2SO_4	25	15
MnO_2	CF_3CO_2H	0	30
CrO_3	aq. $H_2SO_4-CH_3CO_2H$	0	25
$Tl(OCOCF_3)_3$	CF_3CO_2H	25	12
Pb_3O_4	CF_3CO_2H	0	22
VOF_3	CF_3CO_2H	0	59
$MoOCl_4$	$CF_3CO_2H-CHCl_3$	25	62

acetate with trifluoroacetic acid in methylene chloride gave directly domesticine in 60% of the yield (Chart 3-20). Recently, an efficient intramolecular oxidative coupling of the monophenolic benzylisoquinoline to oxoaporphine was reported by Kupchan and Liepa[236b] (Chart 3-21). This reaction may proceed by abstraction of hydrogen atoms from the activated diarylmethylene function and the phenolic hydroxyl group to yield an intermediate, and subsequently, further oxidation to oxoaprophine.

Dibenzo[b,g]pyrrocoline (Indolo[2,1-a]isoquinoline) Alkaloids

For many years, there has existed interest in synthesizing morphine- or aporphine-type alkaloids according to biogenetic routes.[85] In 1932, Robinson[210] and

Chart 3-22.

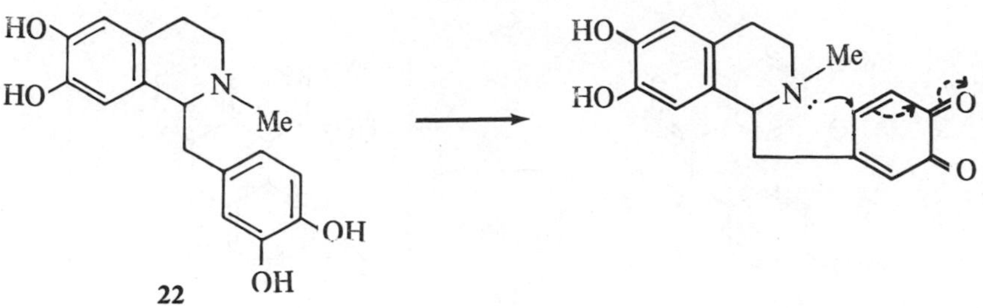

HO HO N Me 22 OH OH

HO HO N Me O O

R¹O HO N ⊕ Me I⊖ OR² OR³

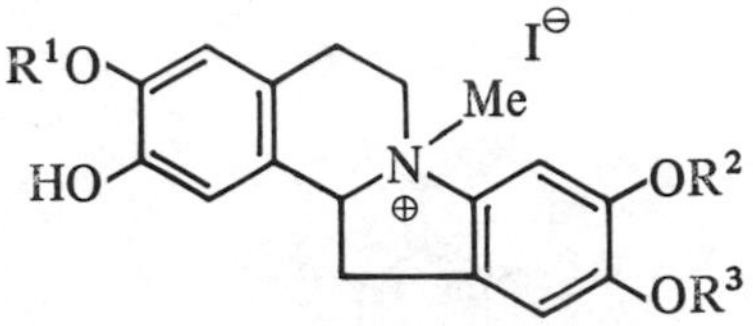

75 $R^1 = R^2 = R^3 = H$
76 $R^1 = Me; R^2+R^3 = -CH_2-$
77 $R^1 = R^2 = R^3 = Me$

Chart 3-23.

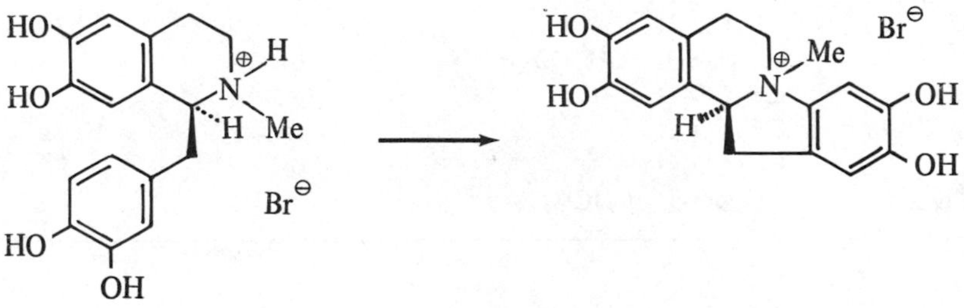

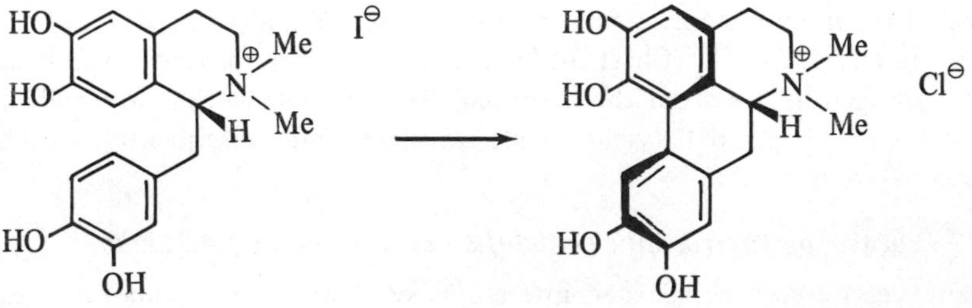

HO HO ⊕ H N H Me Br⊖ HO OH

HO HO ⊕ Me N H OH OH Br⊖

HO HO ⊕ Me N H Me I⊖ HO OH

HO HO ⊕ Me N H Me Cl⊖ HO OH

Schöpf[186] independently investigated the oxidation of laudanosoline 22 methiodide under mild conditions by chloranil[210] or tetrabromo-*o*-quinone.[186] They obtained no morphine or aporphine, but the dibenzo [b,g] pyrrocoline-type base 75. This type of alkaloid was isolated some 20 years later from *Cryptocrya bowiei* as the quaternary alkaloids, cryptowoline 76 and cryptaustoline 77 (Chart 3-22).

Similar reactions have been reported by two other groups.[222,237] Recently, the synthesis of aporphine and dibenzopyrrocoline has been achieved by an enzymic oxidative coupling of optically active laudanosoline and its methiodide.[237a] Thus oxidative coupling of (1S)-(+)-laudanosoline hydrobromide and (1R)-(−)-laudanosoline methiodide by use of the purified enzyme horseradish peroxidase under controlled reaction condition could be effected with great facility and in a preparative manner to afford the quaternary dibenzopyrrocoline (81% of the yield) and the quaternary aporphine (60% of the yield), respectively, with retention of absolute configuration (Chart 3-23).

Protoberberine Alkaloids

The protoberberine could be biosynthesized from reticuline, and the "berberine bridge" is derived from the *N*-methyl group of 1-benzylisoquinoline as shown in Chart 3-24.[186a] Kametani reported the conversion of reticuline into coreximine in rats. However, the phenolic isoquinolines that do not have a phenolic hydroxyl group at the 3′-position of the benzyl residue could not be transformed into protoberberines.[237b] Dimeric protoberberine alkaloid, bisjatrorrhizine, was synthesized by a catalytic oxidative coupling of jatrorrhizine chloride[237c] (Chart 3-25).

Morphinandienone Alkaloids

Tracer experiments have shown that morphine 78 is biosynthesized from salutaridine 79, formed via phenolic oxidation of reticuline 38, through thebaine 80 and codeine 81 as shown in Chart 3-26. This biogenetic scheme has been duplicated in the laboratory by oxidizing reticuline 38. The desired conversion was achieved by Barton et al.[238] using tritium-labeled reticuline; the product was isolated using unlabeled salutaridine as a carrier. Oxidation was carried out with manganese dioxide (0.024% yield), potassium ferricyanide (0.015% yield), ferric chloride (0.0007% yield), or Fremy's salt (0.0054% yield), and reduction of salutaridine 79 gave an epimeric mixture of salutaridinol-I and -II, which was dehydrated using an acidic catalyst to give thebaine 80. The conversion of thebaine into morphine 78 has been accomplished, so this work constitutes a total synthesis of morphine and relatives analogous to the biogenetic route.[238]

Another coupling mode of reticuline 38 would give a second morphinandienone pallidine 81, which could be converted in nature into amurine 82 or flavinantine 83 by transmethylation.[239] In 1967, Franck[220,240] oxidized reticuline 38

Chart 3-24.

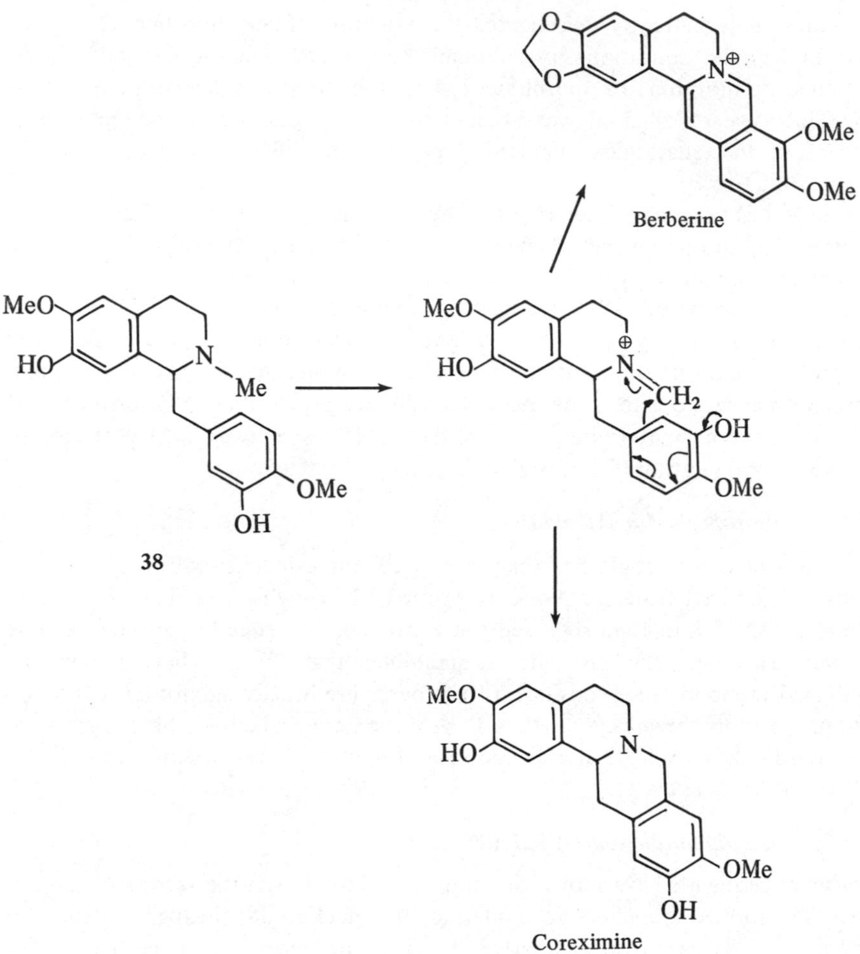

MeO, HO, N, Me, OMe, OH

38

MeO, HO, ⊕ N, CH_2, OH, OMe

O, O, N⊕, OMe, OMe

Berberine

MeO, HO, N, OMe, OH

Coreximine

Chart 3-25.

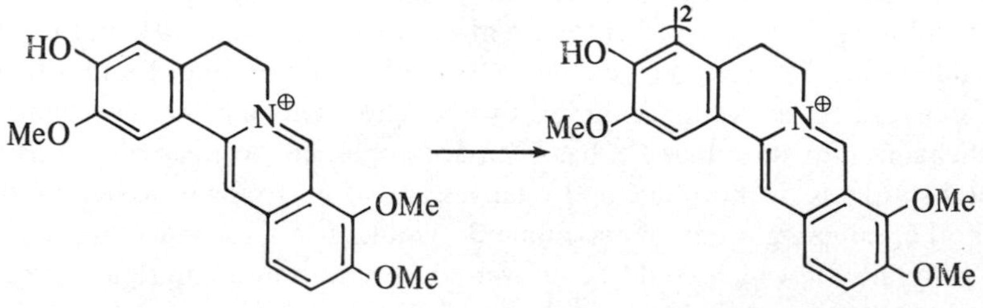

HO, MeO, N⊕, OMe, OMe

Jatrorrhizine

2, HO, MeO, N⊕, OMe, OMe

Bisjatrorrhizine

Chart 3-26.

MeO, HO, N, Me, MeO, OH, 38 ≡ MeO, HO, Me, N, MeO, OH

MnO_2 or $K_3Fe(CN)_6$ → MeO, HO, Me, N, MeO, O, 79 → MeO, HO, Me, N, MeO, H OH

→ MeO, O, Me, N, MeO, 80 → MeO, O, Me, N, O

→ MeO, O, Me, N, H, HO, 81 → HO, O, Me, N, H, HO, 78

Chart 3-27.

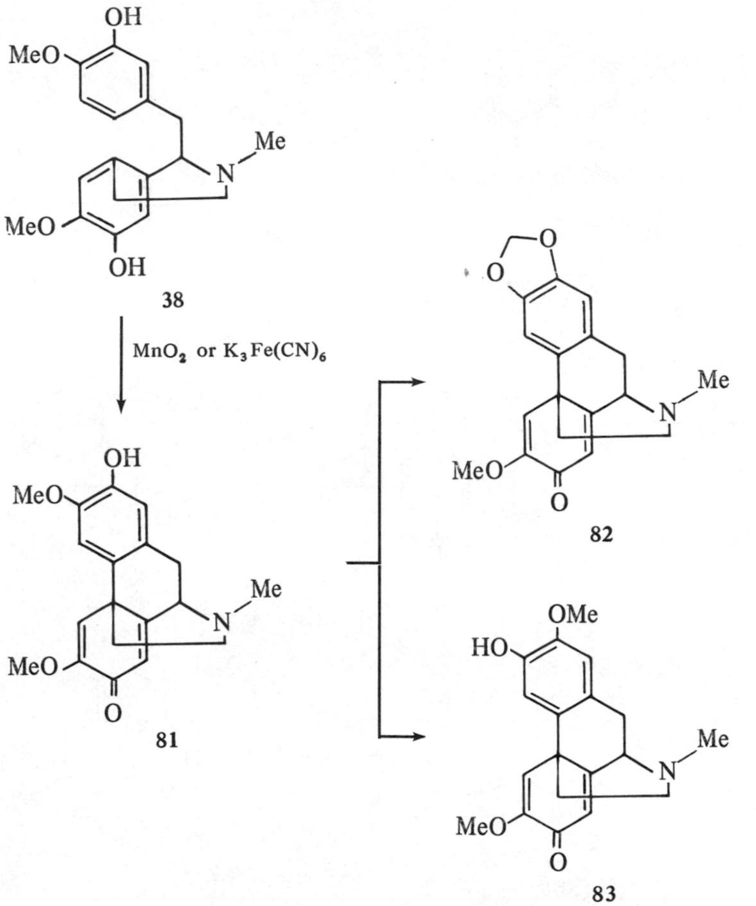

with manganese dioxide on silica gel to obtain pallidine **81** together with isoboldine **37**. In this reaction, salutaridine **79** could not be detected, due to steric preference for *para* coupling compared to *ortho* coupling. Moreover, since reticuline **38** was absorbed on the surface of silica gel during oxidation, the system represented a dilution method, in which the concentration of reticuline molecules on the surface of the oxidant was low and the distance between oxidant and substrate large, which hindered intermolecular reaction. Furthermore, Franck pointed out that phenolic oxidation of the compound absorbed on silica gel would be similar in process to an enzymic reaction in the plant cell (Chart 3-27). Pallidine **81** was also synthesized from reticuline **38** by phenolic oxidation with potassium ferricyanide in the presence of 5% sodium hydrogen carbonate,[218,219,221] silver carbonate on celite,[219] or vanadium oxychloride.[219]

A similar attempt to synthesize morphinandienone by oxidation of reticuline with homogenized *Papaver rhoeas* in the presence of hydrogen peroxide failed, but afforded β-hydroxyreticuline. This reaction may offer an initial clue to the biogenesis of the rhoeadine alkaloids[240a] (Chart 3-28). Miller, Stermitz, and Falck

Chart 3-28.

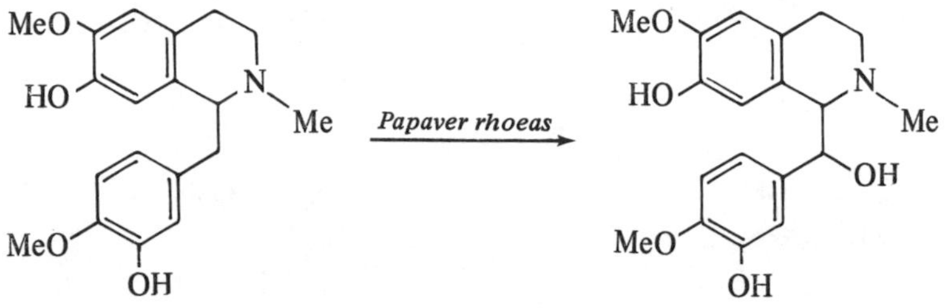

MeO HO N Me MeO OH *Papaver rhoeas* MeO HO N Me OH MeO OH

38 **β-Hydroxyreticuline**

have found a short and efficient synthesis of the morphinandienones by an electrooxidative coupling reaction of the nonphenolic base, laudanosine.[240b] Thus laudanosine was oxidized on platinum in a three-compartment cell at 1.1 V in acetonitrile at 0° in the presence of sodium carbonate. Either lithium perchlorate or tetraethylammonium borotetrafluoride was the background elecrolyte. *O*-Methylflavinantine was isolated from the anolyte in 52% yield. Similar oxidation of *O*-benzylcodamine, *O*-benzylpseudocodamine, *O*-benzyllaudanine, and *O*-benzylpseudolaudanine yielded *O*-methylflavinantine, *O*-benzylflavinantine, *O*-benzylpallidine, and 6-benzyloxy-2,3-dimethoxymorphinandienone, respectively. Oxidation of laudanosine in the presence of equimolar bis(acetonitrile)palladium (I) chloride enhanced the yield of *O*-methylflavinantine to 63% (Chart 3-29).

Chart 3-29.

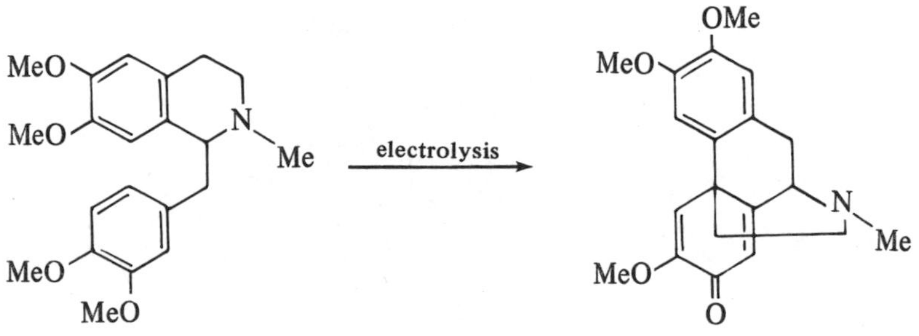

MeO MeO N Me MeO MeO electrolysis OMe MeO N Me MeO O

Chart 3-30.

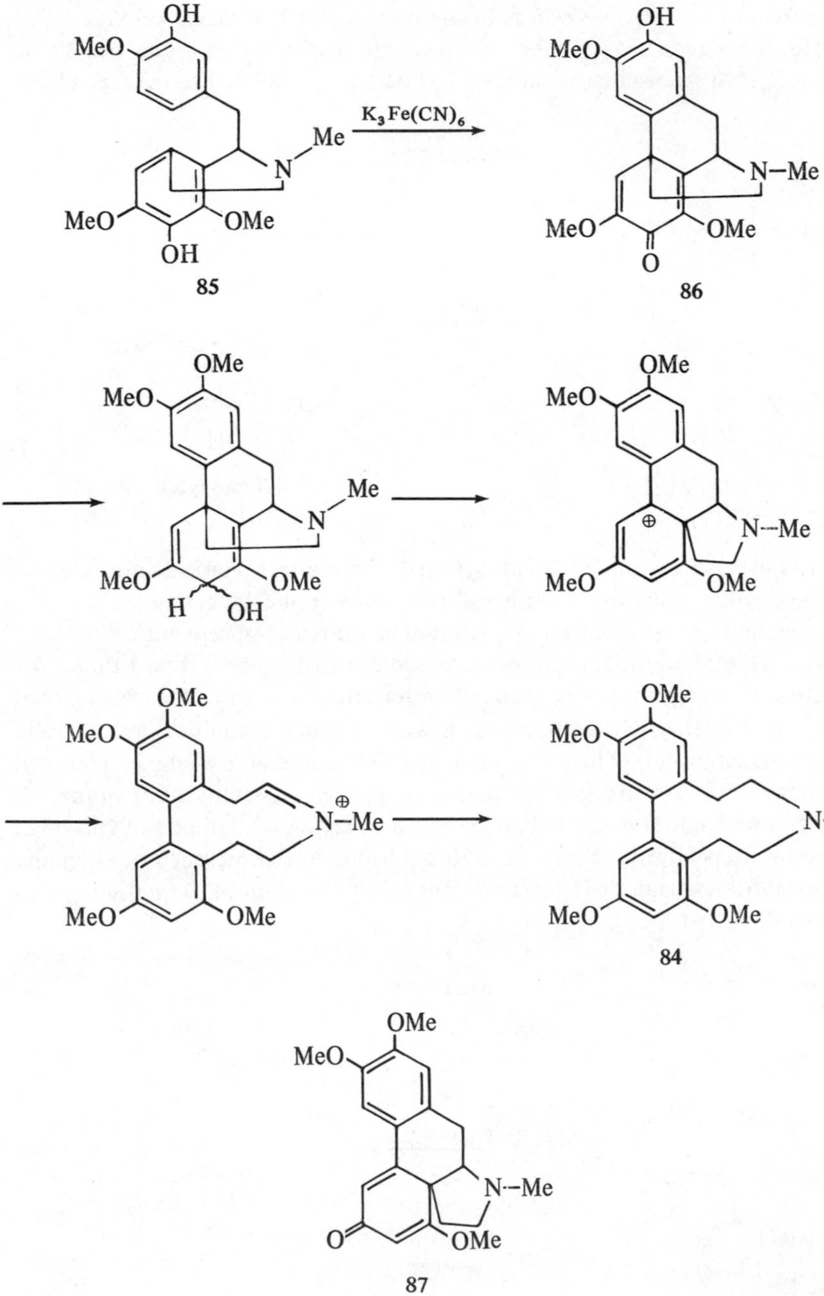

OH
MeO
Me
N
MeO
OMe
OH
85
K$_3$Fe(CN)$_6$
OH
MeO
N—Me
MeO
OMe
O
86
OMe
MeO
Me
N
MeO
OMe
H
OH
OMe
MeO
⊕
N—Me
MeO
OMe
OMe
MeO
⊕
N—Me
MeO
OMe
OMe
MeO
N—Me
MeO
OMe
84
OMe
MeO
N—Me
O
OMe
87

Tobinaga also synthesized flavinanthine, pallidine, and amurine in high yield by electrooxidative coupling of the nonphenolic 1-benzylisoquinolines followed by debenzylation.[240c] Protostephanine **84** is probably biosynthesized from the diphenolic isoquinoline **85** through a phenolic oxidation product, morphinandienone **86**.[241]

Battersby synthesized this alkaloid by a route related to that outlined in Chart 3-30. Ferricyanide oxidation of the diphenolic isoquinoline **85** afforded morphinandienone **86**. Reduction of the *O*-methyl ether of the dienone yielded the epimeric dienols, which upon treatment with sulfuric acid gave the dienone **87**. Heating of dienone **87** with magnesium iodide, followed by reduction and *O*-methylation, yielded protostephanine **84**.

Cepharamine **88**, a hasubanan-type alkaloid from *Menispermaceae*, is probably biosynthesized by rearrangement of dienone **44**, a precursor to the aporphine alkaloids formed from protosinomenine **43** by phenolic oxidation, although tracer experiments have not yet been carried out. In one laboratory, a dienone **49** having the same molecular skeleton as precursor **44** was synthesized by phenolic oxidation (cf. Scheme 1), but an acidic rearrangement of

Chart 3-31.

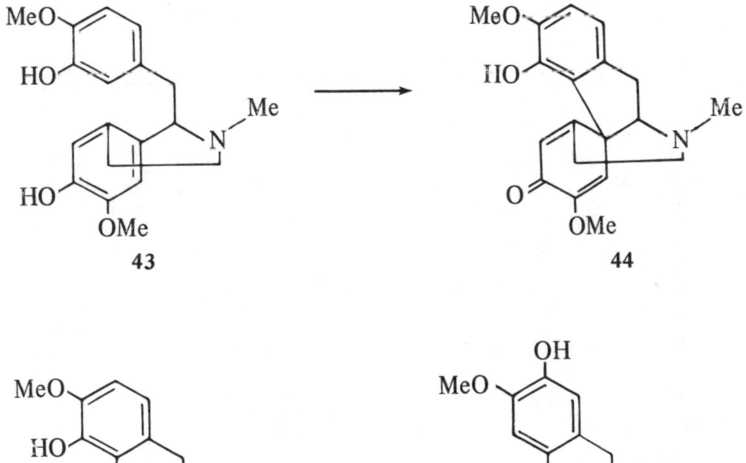

MeO HO HO OMe N Me **43** → MeO HO O OMe N Me **44**

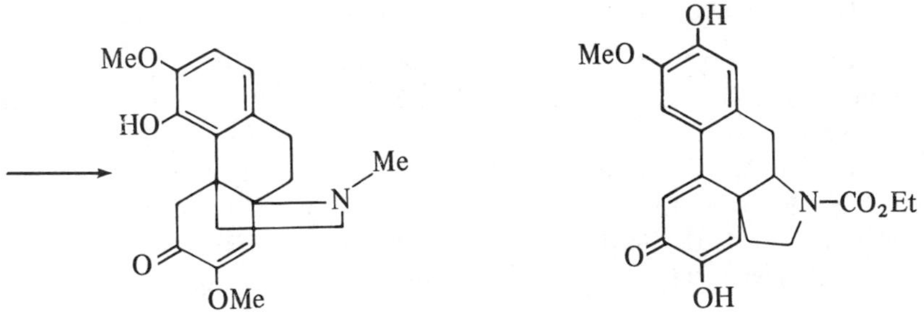

→ MeO HO O OMe N Me **88**

MeO OH O OH $N-CO_2Et$ **89**

dienone **49** gave dienone **89**, which was similar to the precursor **87** of protosinomenine **84**[224] (Chart 3-31).

On the other hand, the hasubanan molecular skeleton was synthesized by a nonbiogenetic route, in which phenolic oxidation was a key step. Vanadium oxychloride oxidation of the diphenolic compound **90**, followed by hydrolysis, gave a hasubanan-type enone **91**, which was converted into a cepharamine analog[242] (Chart 3-32).

Chart 3-32.

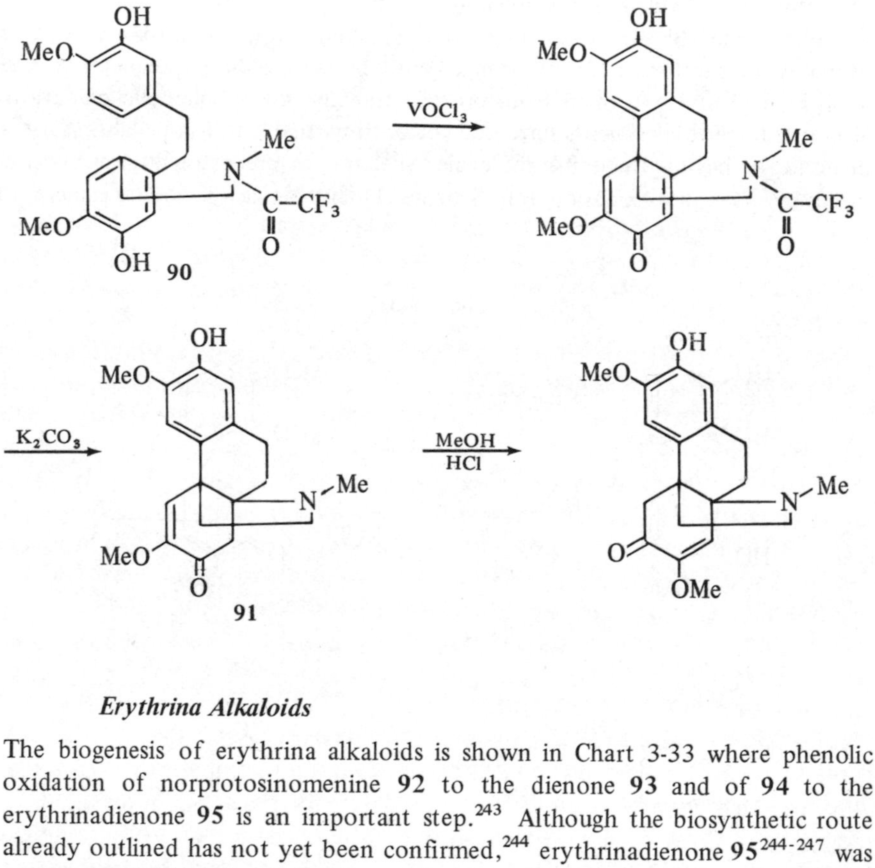

Erythrina Alkaloids

The biogenesis of erythrina alkaloids is shown in Chart 3-33 where phenolic oxidation of norprotosinomenine **92** to the dienone **93** and of **94** to the erythrinadienone **95** is an important step.[243] Although the biosynthetic route already outlined has not yet been confirmed,[244] erythrinadienone **95**[244-247] was obtained by phenolic oxidation of bisphenethylamine **96** and of biphenyl **94**[242,247] with potassium ferricyanide in the presence of 5% sodium hydrogen carbonate. Biphenyl **94** was converted into *O*-methylerybidine in the usual way[247a] (Chart 3-34).

Chart 3-33.

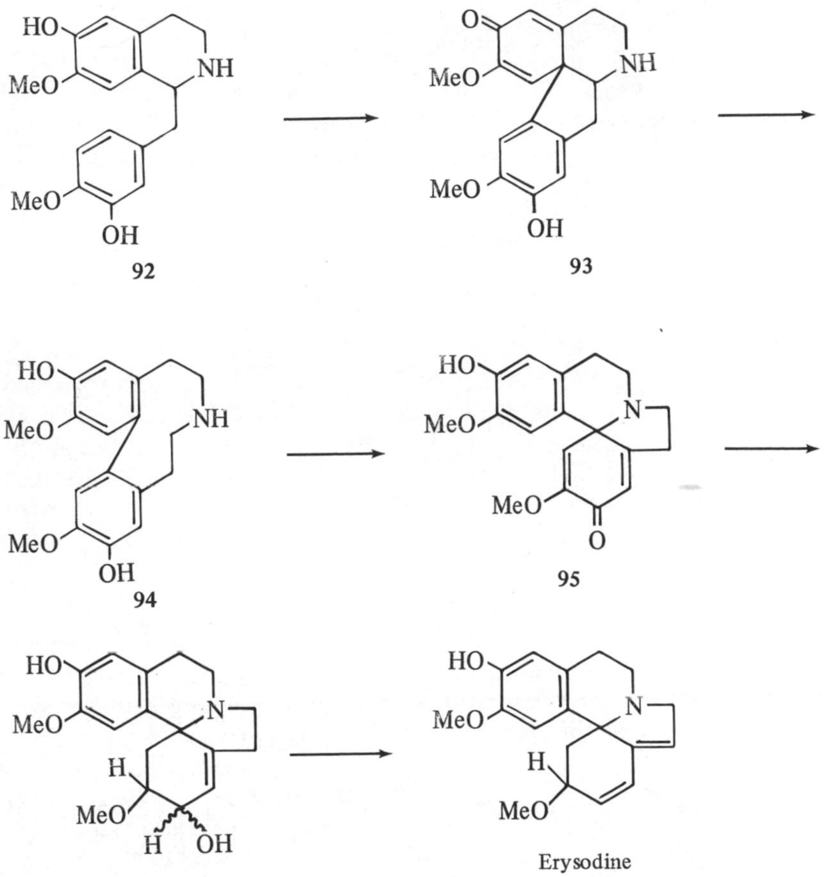

HO MeO NH MeO OH 92 → O MeO NH MeO OH 93 →

HO MeO NH MeO OH 94 → HO MeO N MeO O 95 →

HO MeO N H MeO H OH → HO MeO N H MeO Erysodine

Recently, Franck[248] succeeded in the conversion of the benzylisoquinoline **97** into erythrinadienone **99** through morphinandienone **98**, in which phenolic oxidations were employed in key steps. Thus oxidation of the diphenolic isoquinoline **97** with vanadium oxychloride gave the morphinandienone **98**, which was treated with boron trifluoride, followed by reduction and hydrolysis, to give biphenyl derivative **100**. Phenolic oxidation of this compound with potassium ferricyanide afforded erythrinadienone **99** in 61% yield (Chart 3-35).

Phenanthroquinolizidine Alkaloids

Cryptopleurine **101**, as alkaloid from *Cryptocarya pleurosperma,* is coexistent with pleurospermine **102**. Tracer experiments have not been carried out, but it is

Chart 3-34.

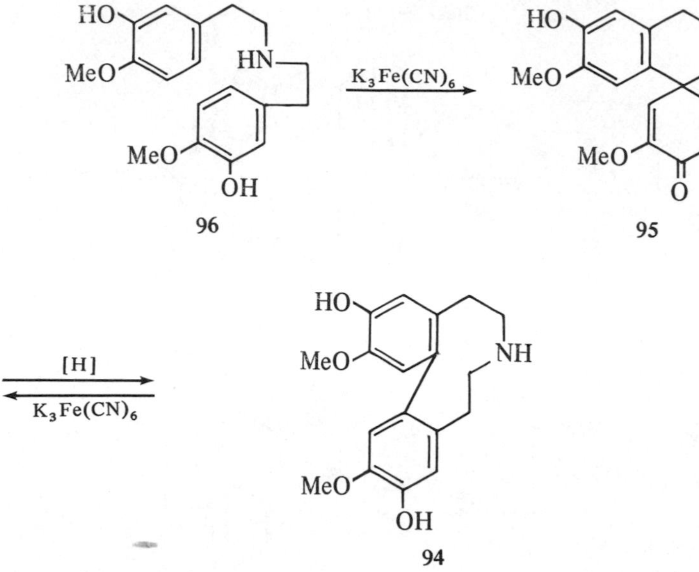

HO MeO HN MeO OH

96

$K_3Fe(CN)_6$

HO MeO N MeO O

95

[H]

$K_3Fe(CN)_6$

HO MeO NH MeO OH

94

Chart 3-35.

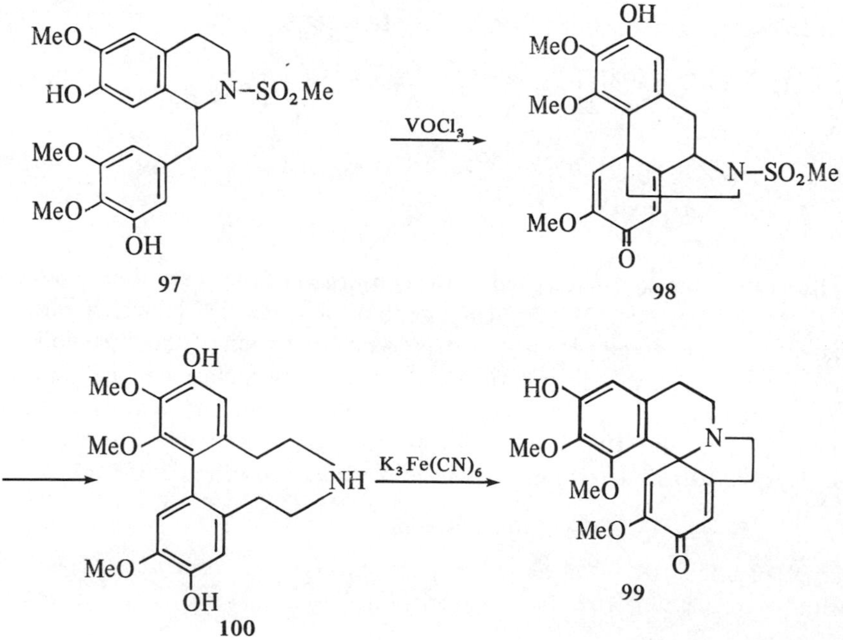

MeO HO N−SO_2Me MeO MeO OH

97

$VOCl_3$

OH MeO MeO N−SO_2Me MeO O

98

OH MeO MeO NH MeO OH

100

$K_3Fe(CN)_6$

HO MeO N MeO MeO O

99

Chart 3-36.

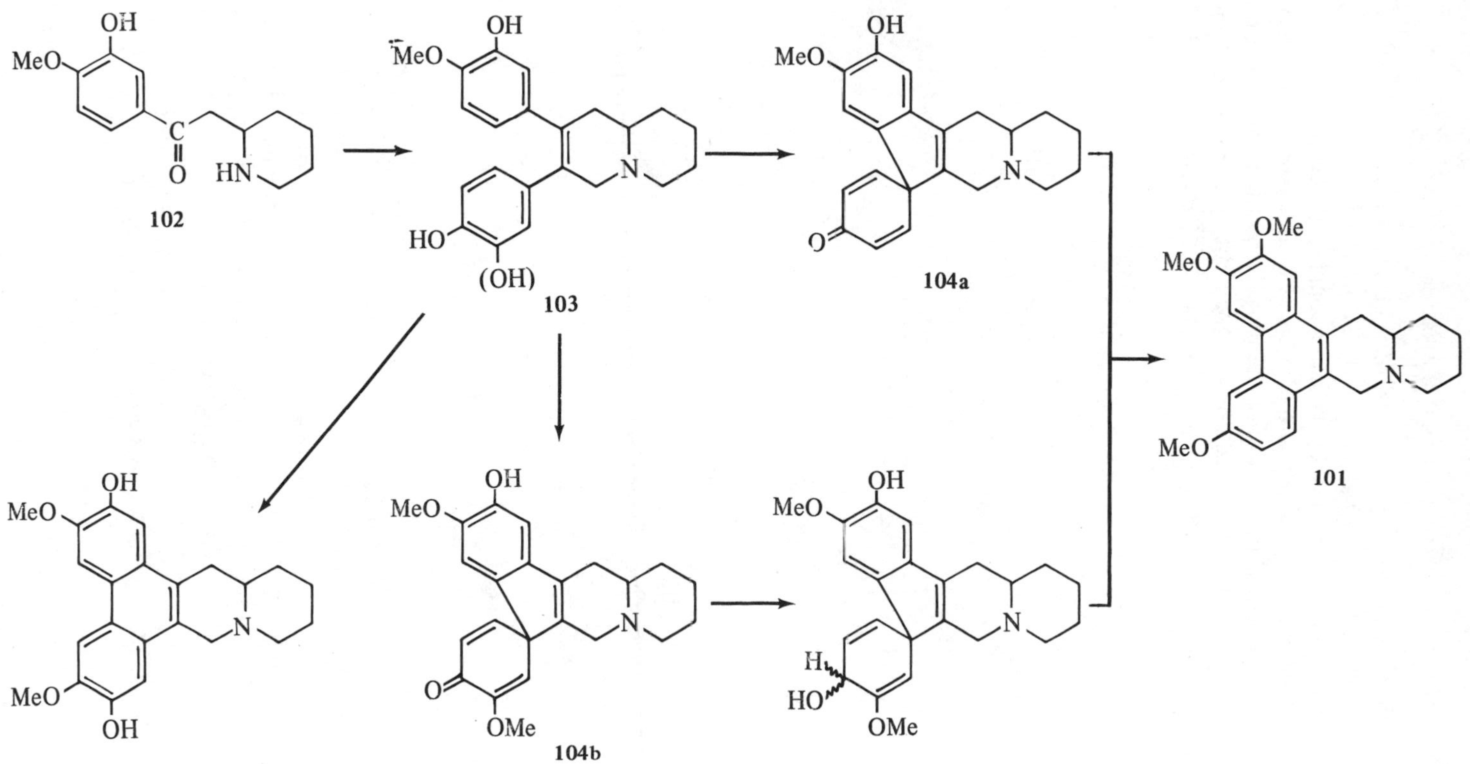

102

103

104a

104b

101

very probable that this type of alkaloid is biosynthesized by phenolic oxidation of the quinolizidine derivative (cf. **103**) by direct coupling or rearrangement of the dienone (cf. **104**) formed by oxidative coupling (Chart 3-36).

In the laboratory, a successful oxidation of quinolizidine **105** to dienone **106** has been brought about by manganese dioxide, and the conversion of dienone **106** into cryptopleurine **101** has been carried out by dienone-phenol rearrangement, followed by *O*-methylation and reduction[249] (Chart 3-37). Alternatively,

Chart 3-37.

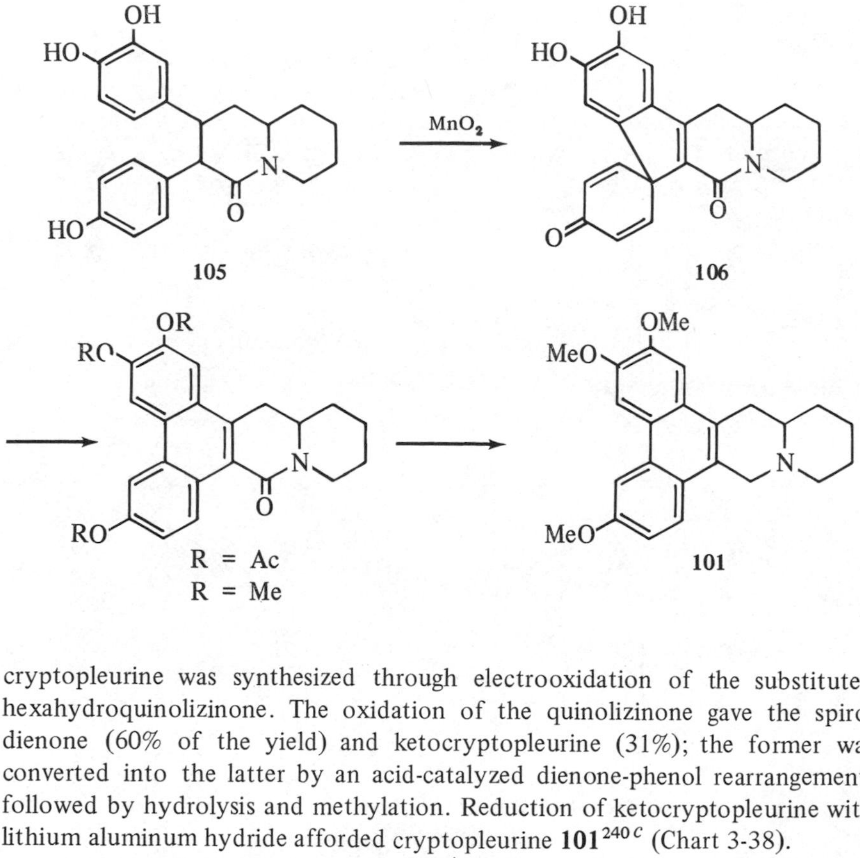

OH HO HO N O **105** $\xrightarrow{MnO_2}$ OH HO N O O **106**

OR RO N O RO R = Ac R = Me → OMe MeO N MeO **101**

cryptopleurine was synthesized through electrooxidation of the substituted hexahydroquinolizinone. The oxidation of the quinolizinone gave the spirodienone (60% of the yield) and ketocryptopleurine (31%); the former was converted into the latter by an acid-catalyzed dienone-phenol rearrangement, followed by hydrolysis and methylation. Reduction of ketocryptopleurine with lithium aluminum hydride afforded cryptopleurine **101**[240c] (Chart 3-38).

D. Amaryllidaceae Alkaloids

Barton and Cohen[173] suggested that a phenolic precursor such as amine **107**, now known as the natural product norbelladine, represents a single precursor for

Chart 3-38.

MeO
OMe
N
O
MeO

MeO
OMe
N
O
O

+

MeO
MeO
N
O
MeO

MeO
OMe
N
O
RO

R = Ac
R = H

MeO
OMe
N
MeO

101

Chart 3-39.

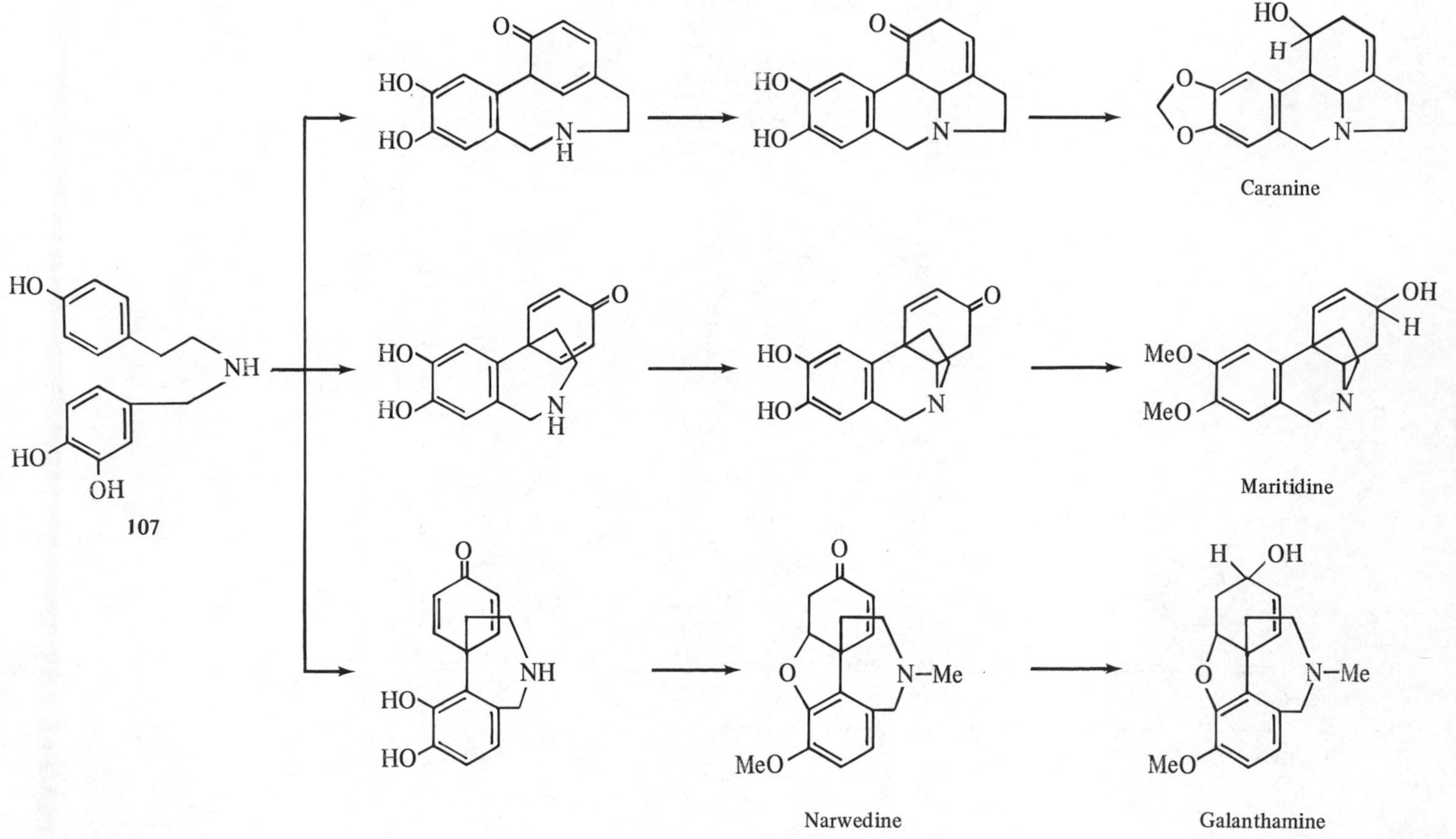

107

Caranine

Maritidine

Narwedine

Galanthamine

the main three types of the *Amaryllidaceae* alkaloids. The validity of this scheme has been conclusively demonstrated by independent investigators[250] (Chart 3-39).

Of the three possible alkaloid types obtained by phenolic oxidation of norbelladine, crinan- and galanthamine-type alkaloids have been obtained by a biogenetic-type synthesis. Schwartz[251] oxidized diphenol **108** with vanadium oxychloride, followed by hydrolysis to obtain the crinan-type compound **109**, which was converted into maritidine **110**. Tobinaga used the ferric chloride-DMF complex as an oxidant in this reaction[251a] (Chart 3-40).

Chart 3-40.

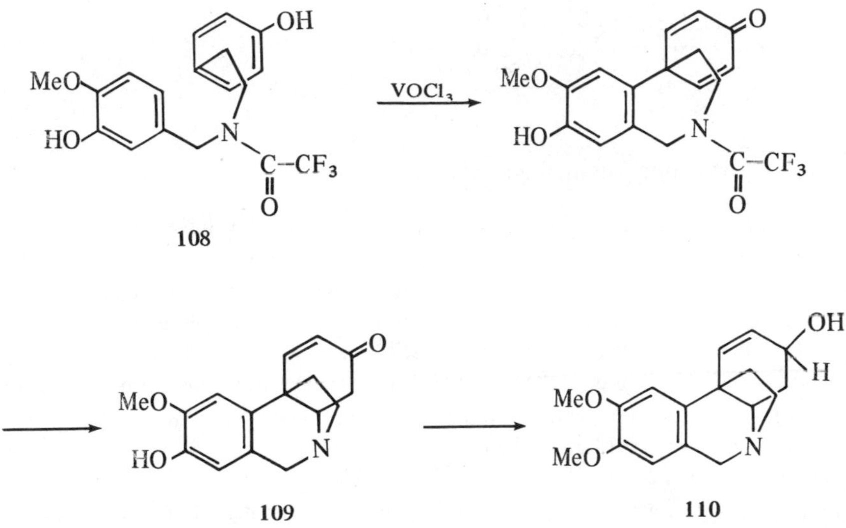

MeO HO OH N C–CF₃ O **108** $\xrightarrow{VOCl_3}$ MeO HO O N C–CF₃ O → MeO HO O N **109** → MeO MeO OH H N **110**

On the other hand, epicrinine was synthesized from the monophenolic base by intramolecular oxidative phenol coupling by two-electron oxidation with thallium trifluoroacetate[251b] (Chart 3-41). Narwedine **111**,[252-254] which was reduced to galanthamine **112**, was obtained in poor yield by manganese dioxide oxidation of belladine derivative **113**. Other oxidants, such as potassium ferricyanide or lead oxide, were studied, but the yields, determined by the radiochemical dilution method, were very low.[255]

Kametani reported improved methods[256-258] involving the use of the bromoamide as a substrate for phenolic oxidation. Amidation protects the basic nitrogen against oxidation, and the bromine atom prevents oxidative coupling at an undesired position. Potassium ferricyanide oxidation of the amides **114**, **115**, and **116** in the presence of sodium hydrogen carbonate yielded the corresponding enones (**117**, **118**, and **119**), which were reduced to galanthamine **112** by standard methods[256-258] (Chart 3-42).

Chart 3-41.

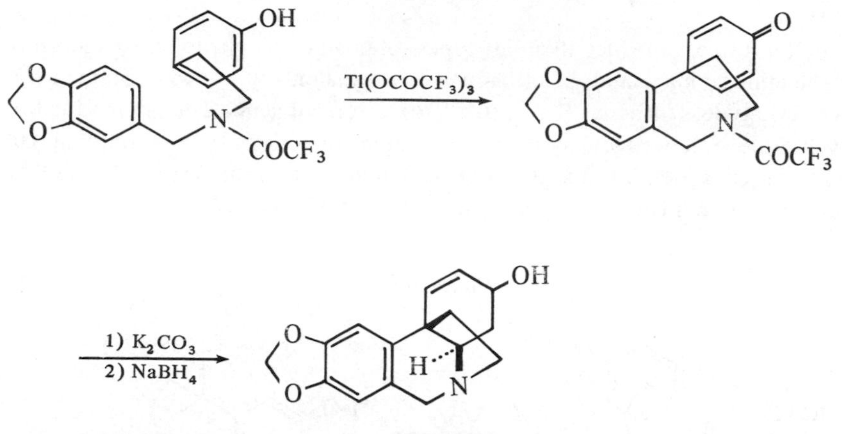

$Tl(OCOCF_3)_3$

$COCF_3$

1) K_2CO_3
2) $NaBH_4$

E. Phenethylisoquinoline Alkaloids

In the last decade, the phenethylisoquinoline alkaloids have been isolated, characterized, and divided into five groups: bisphenethylisoquinoline, homoproaporphine, homomorphinandienone, and homoerythrina alkaloids; corresponding to bisbenzylisoquinoline, proaporphine, aporphine, morphinandienone, and erythrina alkaloids in the benzylisoquinoline series. These alkaloids are probably biosynthesized from autumnaline or its biogenetic equivalents by phenolic oxidation, just as the benzylisoquinoline alkaloids could be formed from coclaurine or its analog (Chart 3-43).

Melanthioidine **120**, the only example of the bisphenethylisoquinolines, is probably biosynthesized from *N*-methylhomococlaurine **121** by oxidative coupling.[259] Several unsuccessful attempts have been carried out to biogenetically synthesize melanthioidine from phenethylisoquinoline **121**; chemical oxidation of **121** gave the intramolecular oxidation product, homoproaporphine **122**.[260,261] On the other hand, enzymatic phenolic oxidation with homogenized potato peelings and hydrogen peroxide or homogenized horseradish in the presence of hydrogen peroxide occurred through intermolecular oxidative coupling to afford "promelanthioidine" **123** or the head-to-head coupling product bisphenethylisoquinoline **124**, respectively, but no melanthioidine[262,263] (Chart 3-43A).

The homoproaporphine alkaloids are known only in the form of their representative kreysiginone **125**, which is probably biosynthesized from homoorientaline **126**, because kreysiginone corresponds to a homolog of orientalinone **56** biosynthesized from orientaline **57** by phenolic oxidation. Laboratory experiments

Chart 3-42.

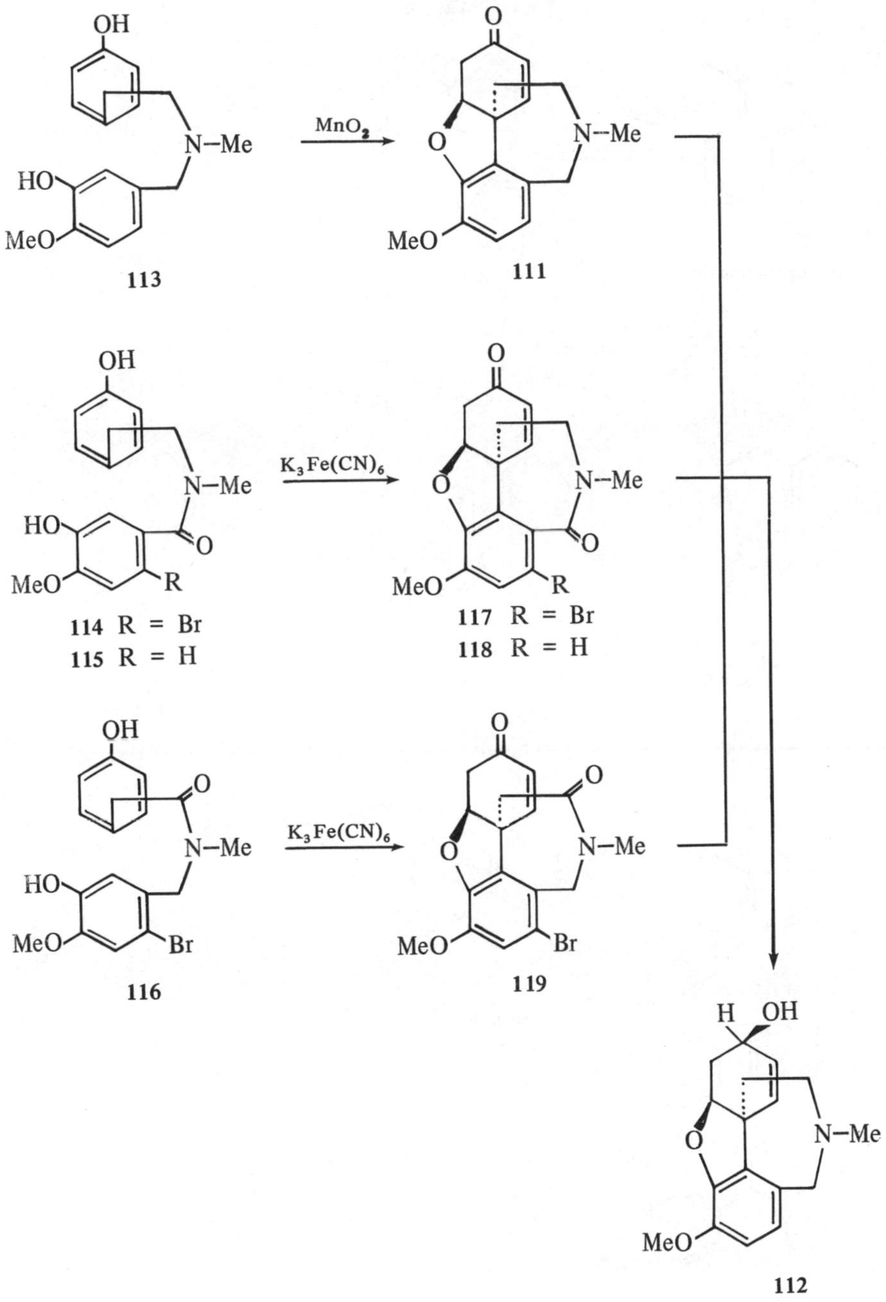

OH
N—Me
HO
MeO
113

MnO_2

O
O
N—Me
MeO
111

OH
N—Me
HO
O
MeO
R
114 R = Br
115 R = H

$K_3Fe(CN)_6$

O
O
N—Me
O
MeO
R
117 R = Br
118 R = H

OH
O
N—Me
HO
MeO
Br
116

$K_3Fe(CN)_6$

O
O
O
N—Me
MeO
Br
119

H OH
O
N—Me
MeO
112

Chart 3-43.

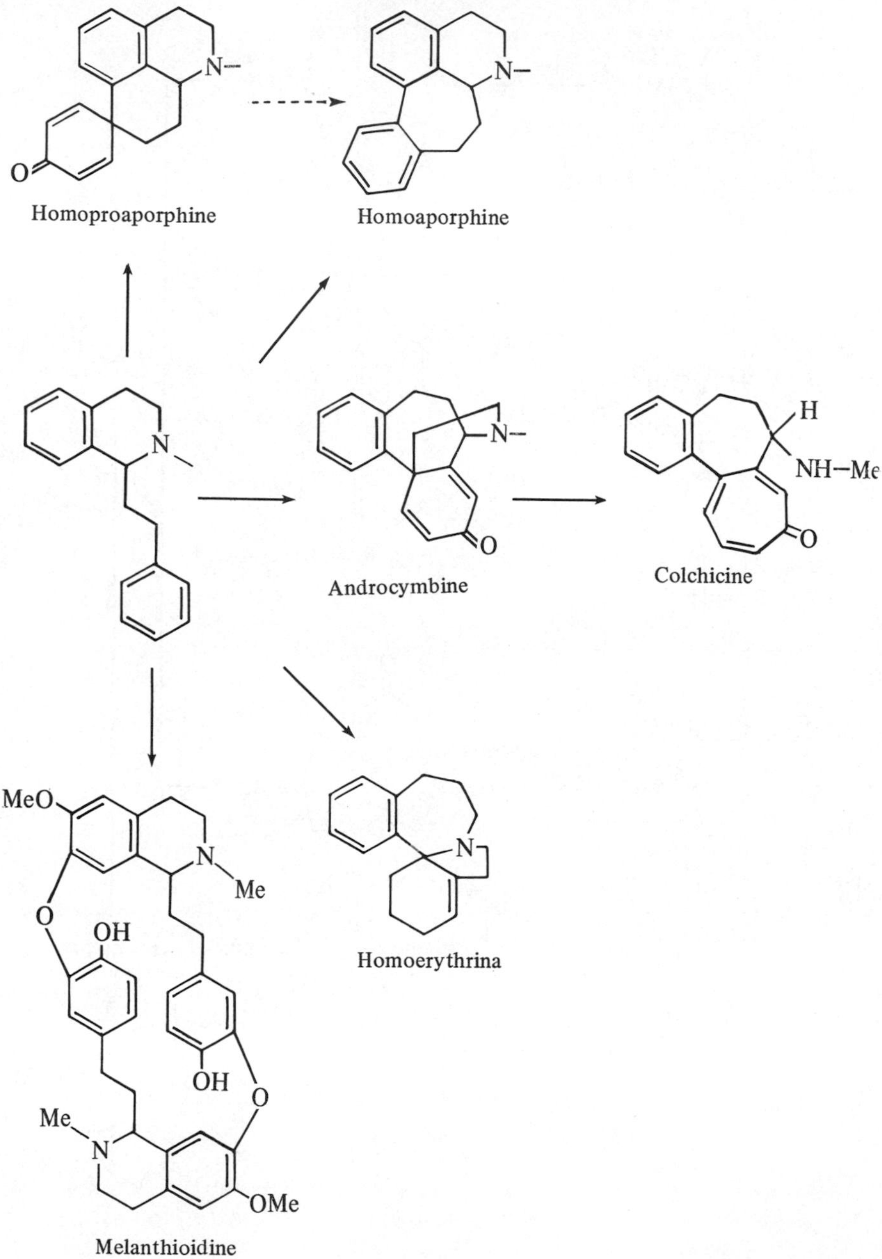

O N— N— Homoproaporphine Homoaporphine N H NH—Me O O Androcymbine Colchicine MeO N Me O OH OH O Me N OMe Homoerythrina Melanthioidine

Chart 3-43A.

MeO
HO
N
Me
OH
121

$FeCl_3$

enzymes

MeO
N
Me
O
OH
OH
O
Me
N
OMe
120

MeO
HO
N
Me
O
122

MeO
N
Me
O
OH
OH
Me
N
OH
OMe
123

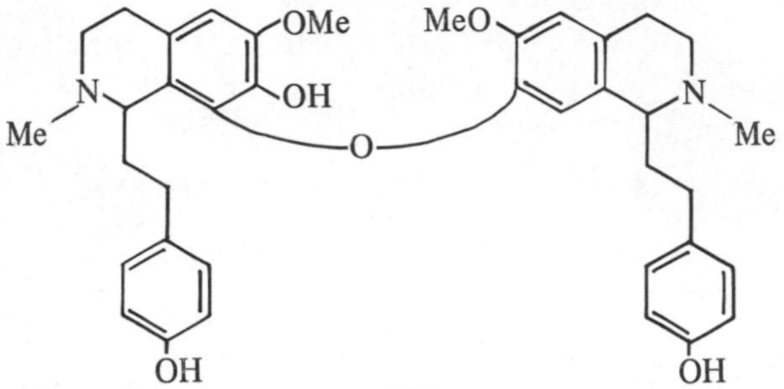

124

attempting conversion of homoorientaline **126** into kreysiginone have been carried out; phenolic oxidation of **126** with potassium ferricyanide[260,264,265] or ferric chloride[260,264,266,267] afforded a mixture of two dienones that differed in configuration at the *spiro* center,[268] one of which was identical with kreysiginone (Chart 3-44).

Chart 3-44.

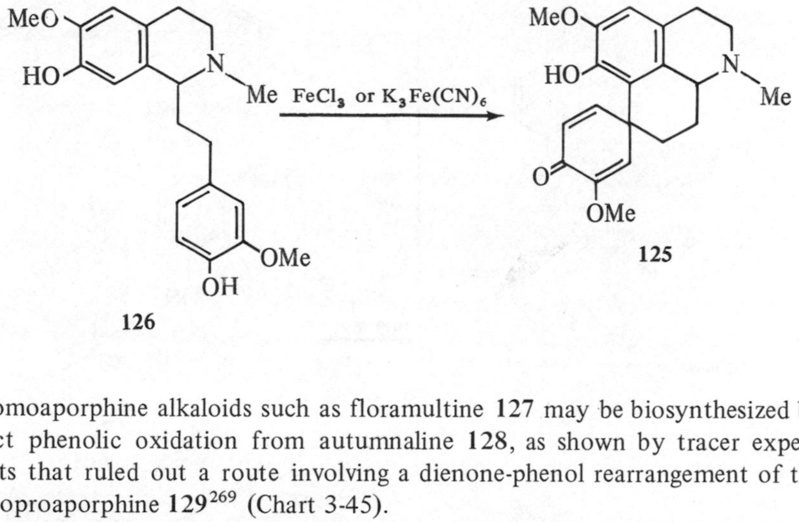

$FeCl_3$ or $K_3Fe(CN)_6$

126 **125**

Homoaporphine alkaloids such as floramultine **127** may be biosynthesized by direct phenolic oxidation from autumnaline **128**, as shown by tracer experiments that ruled out a route involving a dienone-phenol rearrangement of the homoproaporphine **129**[269] (Chart 3-45).

The laboratory analog of this biogenetic scheme was reported by Battersby;[265] thus ferricyanide oxidation of autumnaline **128** yielded 25% of homoaporphine **130**, a biogenetic equivalent of floramultine **127**, directly formed by *ortho-ortho* coupling. On the other hand, the synthesis of homoaporphine by a nonbiogenetic route through the homoproaporphine **129** has been carried out as

Chart 3-45.

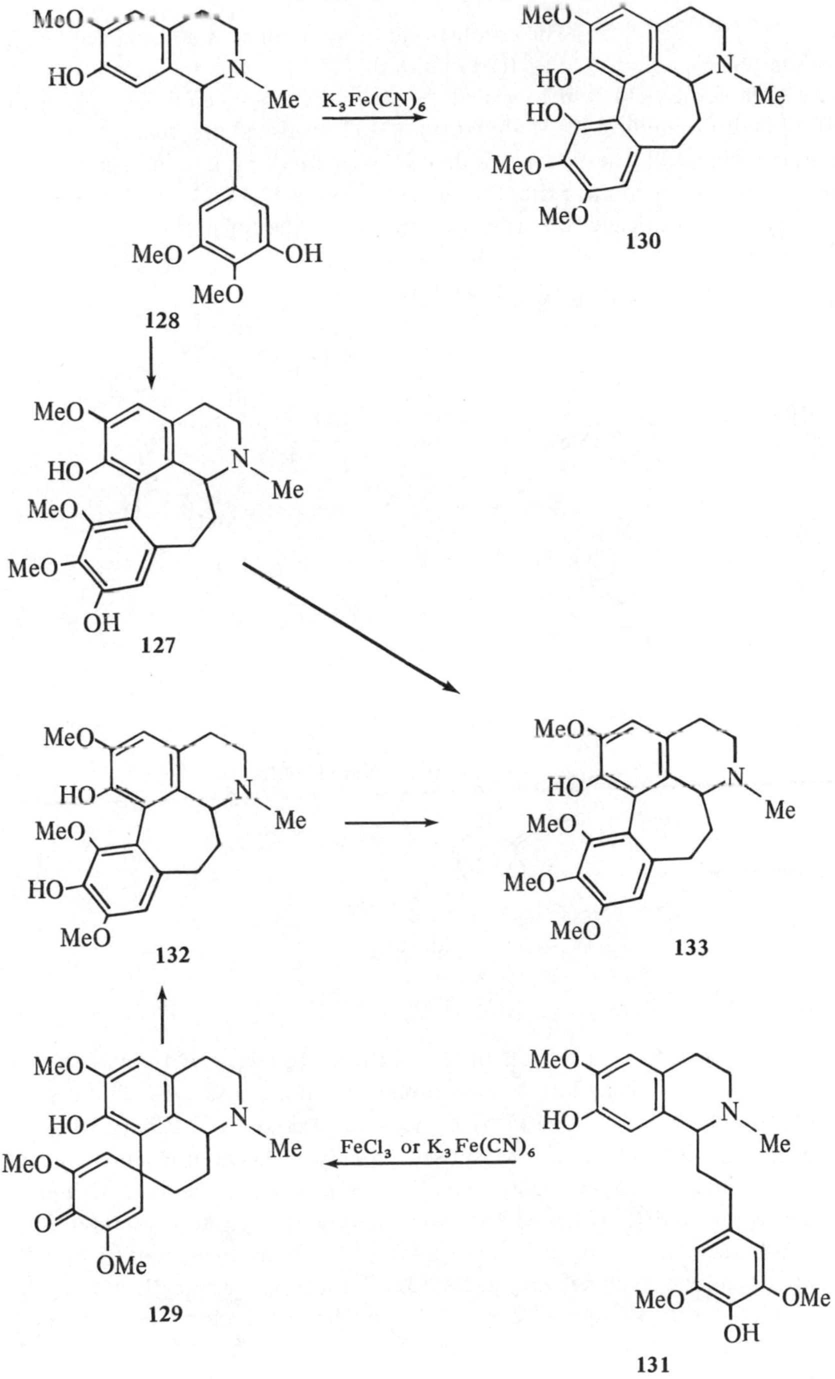

MeO
HO
N
Me
MeO
OH
MeO
128

$K_3Fe(CN)_6$

MeO
HO
N
Me
HO
MeO
MeO
130

MeO
HO
N
Me
MeO
MeO
OH
127

MeO
HO
N
Me
MeO
HO
MeO
132

MeO
HO
N
Me
MeO
MeO
MeO
133

MeO
HO
N
Me
MeO
O
OMe
129

$FeCl_3$ or $K_3Fe(CN)_6$

MeO
HO
N
Me
MeO
OMe
OH
131

follows.[260,264,266,268,270,271] The diphenolic isoquinoline was oxidized with potassium ferricyanide (49%) or ferric chloride (25%) to give the homoproaporphine **129** in good yield, which was subjected to dienone-phenol rearrangement to afford multifloramine **132**. Methylation of **132** furnished kreysigine **133**.

Androcymbine **134**, the first example of 1-phenethylisoquinoline alkaloids, is biosynthesized by phenolic oxidation of autumnaline **128** and transformed biogenetically into colchicine **135**, as shown by tracer experiments[272] (Chart 3-46).

Chart 3-46.

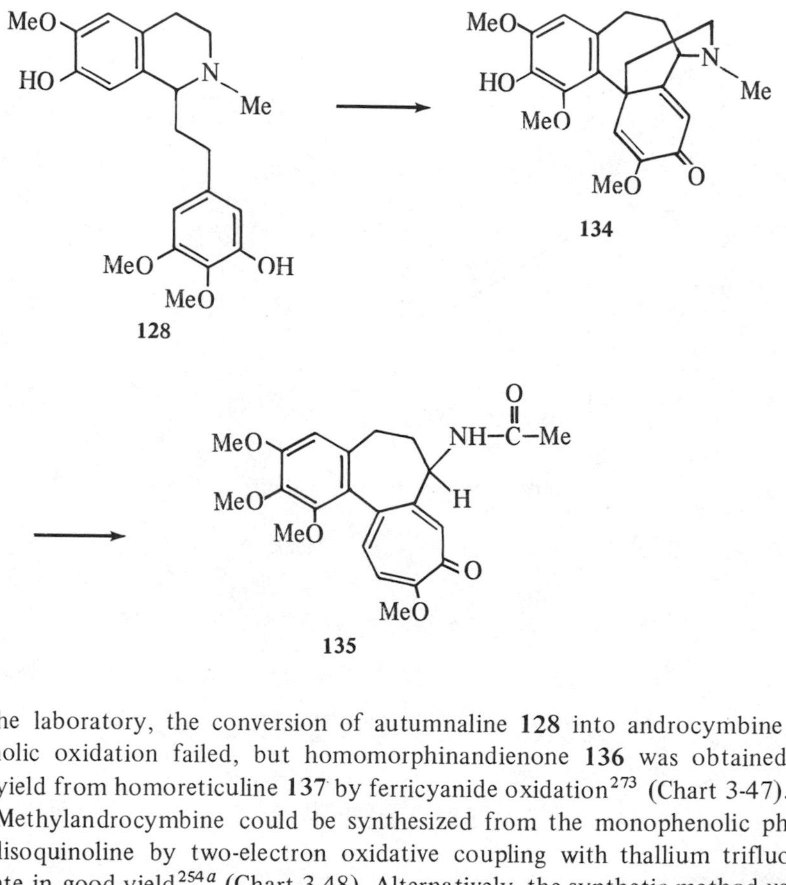

In the laboratory, the conversion of autumnaline **128** into androcymbine by phenolic oxidation failed, but homomorphinandienone **136** was obtained in low yield from homoreticuline **137** by ferricyanide oxidation[273] (Chart 3-47).

O-Methylandrocymbine could be synthesized from the monophenolic phenethylisoquinoline by two-electron oxidative coupling with thallium trifluoroacetate in good yield[254a] (Chart 3-48). Alternatively, the synthetic method using *para*-quinol acetates to prepare aporphines[236a] has been extended to the synthesis of homoaporphine, kreysigine **133**. Treatment with acetic anhydride-sulfuric acid of the *para*-quinol acetate derived from the monophenolic phenethylisoquinoline gave kreysigine in 18% yield after hydrolysis[273a] (Chart 3-49).

Chart 3-47.

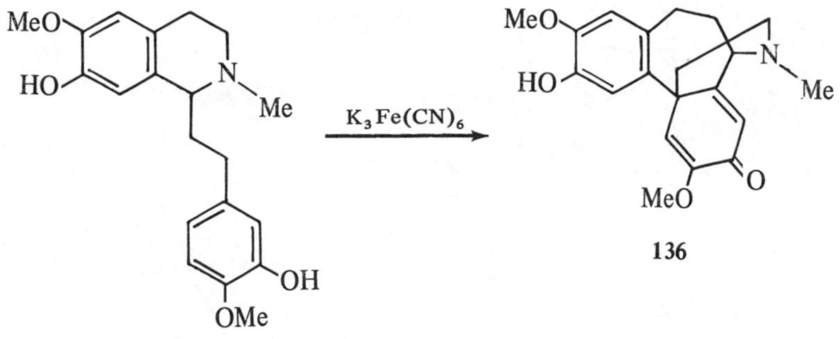

$K_3Fe(CN)_6$

137

136

Chart 3-48.

$Tl(OCOCF_3)_3$

Chart 3-49.

$Pb(OAc)_4$

1) H_2SO_4, Ac_2O
2) H_2O

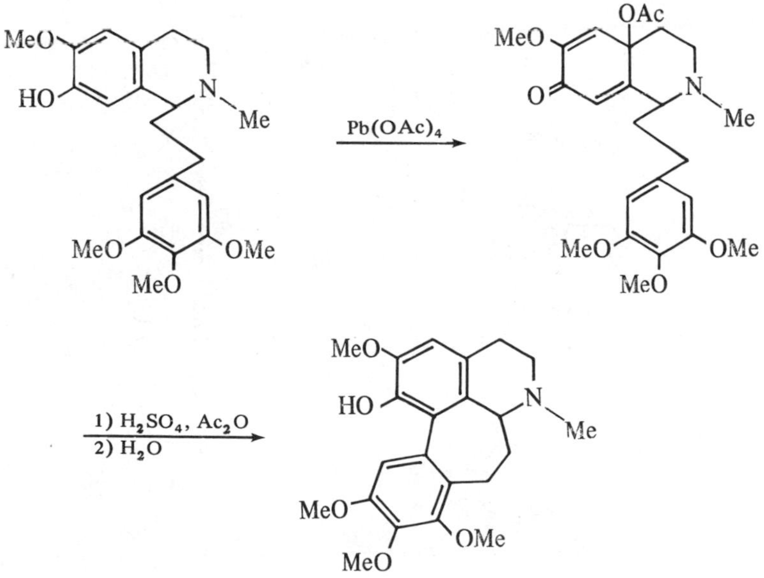

133

Chart 3-50.

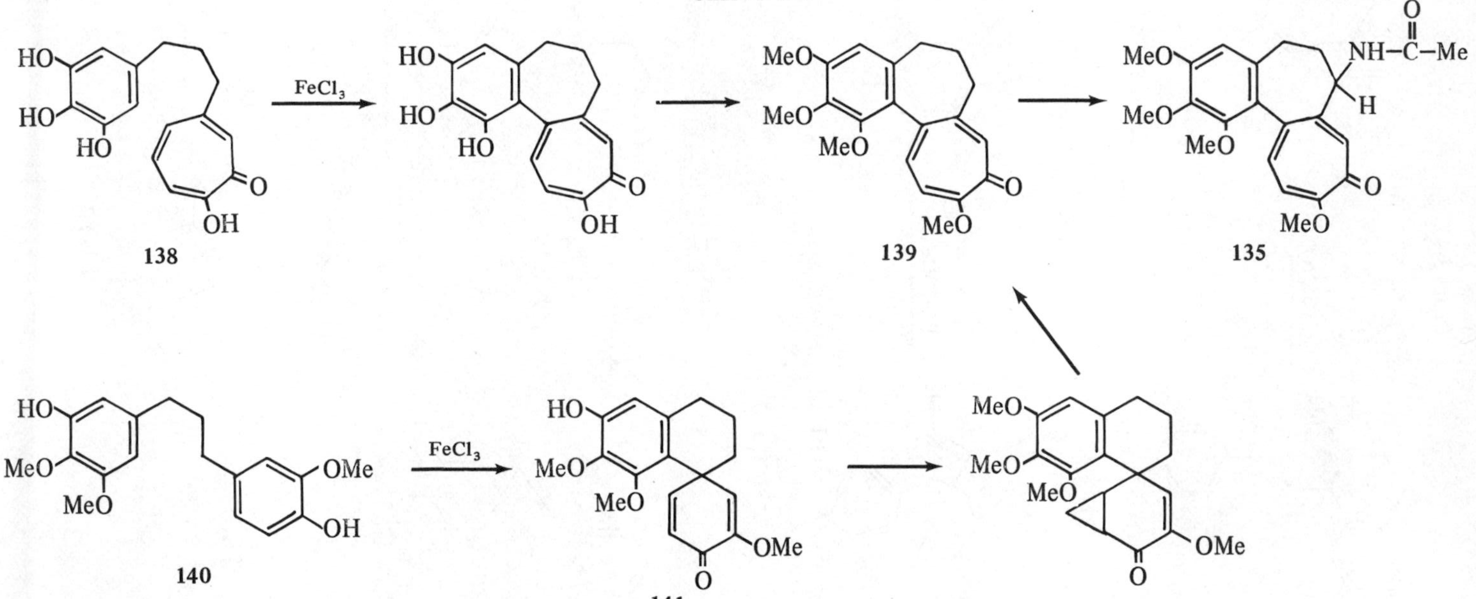

HO HO HO OH O

138

$FeCl_3$

HO HO HO OH O

MeO MeO MeO MeO O

139

MeO MeO MeO NH–C(=O)–Me H MeO O

135

HO MeO MeO OMe OH

140

$FeCl_3$

HO MeO MeO OMe O

141

MeO MeO MeO OMe O

Chart 3-51.

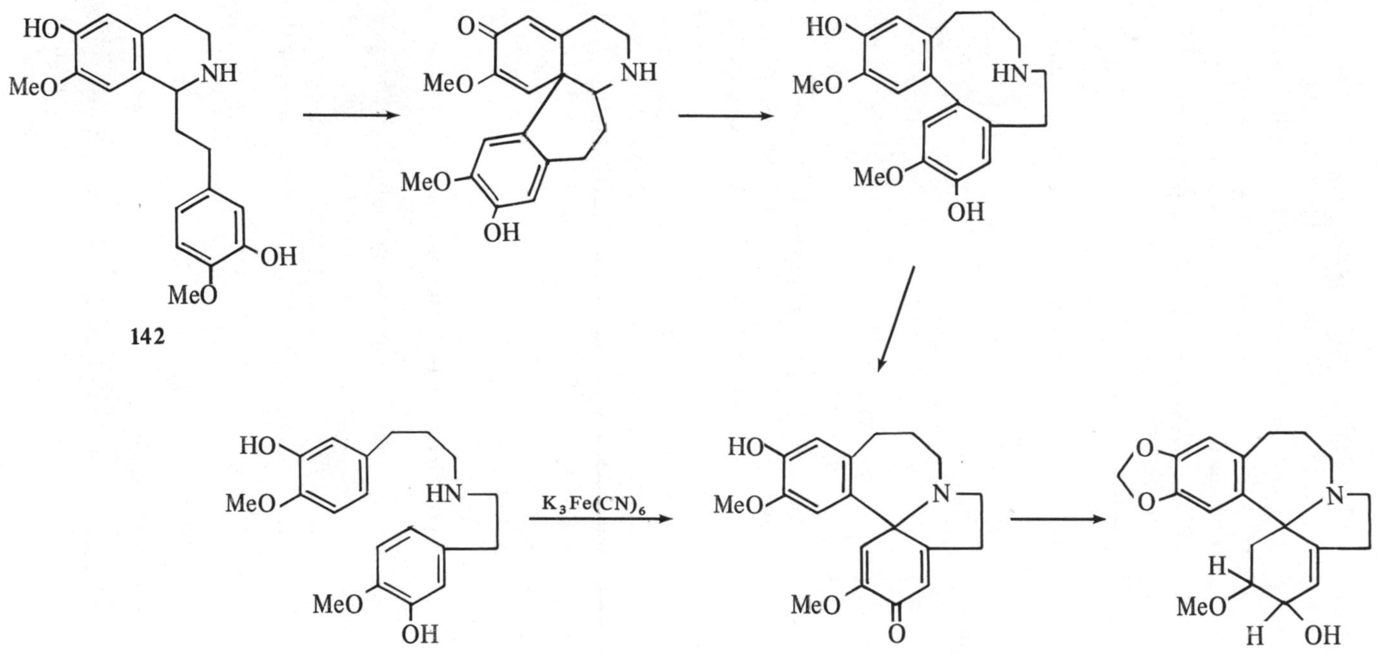

142

$K_3Fe(CN)_6$

144

143

On the other hand, colchicine **135** was synthesized through nonbiogenetic routes by phenolic oxidation. A two phase oxidizing medium consisting of ferric chloride, sulfuric acid, and chloroform converted pyrogallol derivative **138** into the tricyclic product **139**, which could be related to colchicine **135**.[274] Tobinaga and Kotani[275] also obtained the tricyclic product **139** by ferric chloride oxidation of diphenol **140** through dienone **141**, giving support to the Robinson-Anet theory[185] of colchicine biogenesis (Chart 3-50).

It is likely that the ring system of the homoerythrina alkaloids is derived from homoprotosinomenine **142** by a route analogous to that involved in the biogenesis of the erythrina alkaloids, for which a protosinomenine precursor has been established. A trial in the laboratory resulted in failure,[276] but homoerythrinadienone **143** has been obtained from amine **144** by potassium ferricyanide oxidation[277] (Chart 3-51).

4. PHOTOCHEMICAL SYNTHESIS

The photochemical reaction of organic compounds started to be investigated in the early years of this century,[278] but only in the last decade has photochemistry become a sophisticated field.[279,280] The improved methods for the isolation of products and for the determination of structure that have been developed since World War II have overcome the former reluctance of organic chemists to utilize photochemical methods of synthesis. Photochemical syntheses of strained or complicated molecules are widely employed. In particular, a number of alkaloids have been synthesized by photochemical reactions often from starting materials of rather simple structure. Such syntheses are the subject of this section.

A. Photolytic Electrocyclic Reaction[279]

Conjugated polyene systems often undergo photolytic electrocyclization. Thus *trans*-stilbene **1** undergoes a rapid *cis-trans* isomerization under the influence of ultraviolet light and *cis*-stilbene **2** then cyclizes to the *trans*-dihydrophenanthrene **3** upon further irradiation. Mild oxidation of the latter with air or iodine produces phenanthrene **4**.[281,282] This type of hexatriene-cyclohexadiene isomerization has been widely applied to the synthesis of several types of isoquinoline alkaloids (Chart 4-1).

Aporphine Alkaloids

1-Benzylidine-1,2,3,4-tetrahydroisoquinoline is a substituted stilbene and may undergo photolytic electrocyclization to give a dehydroaporphine system that is easily converted into the aporphine. Thus ultraviolet irradiation (Hanovia

Chart 4-1.

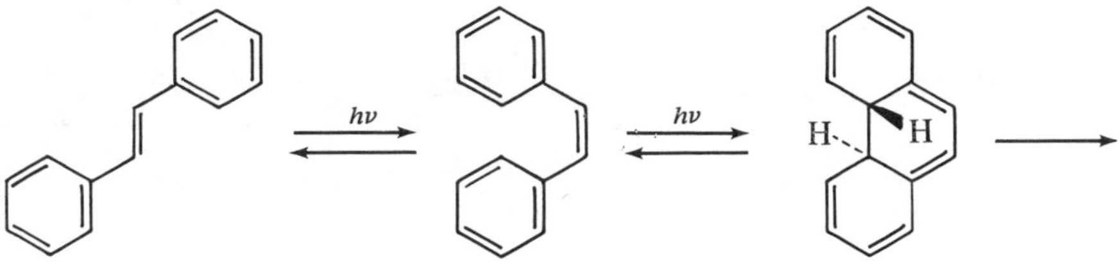

1 2 3

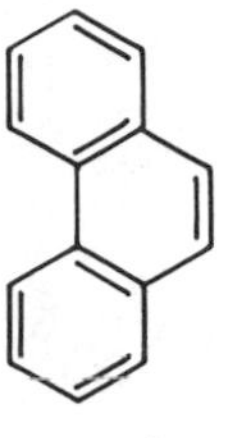

4

450-W mercury lamp housed in a water-cooled quartz insert) of 1-benzylidene-2-ethoxycarbonylisoquinoline **5** in the presence of iodine and cupric acetate in ethanol yielded 35% of the dehydronuciferine analog **6**,[283,284] which was converted by lithium aluminum hydride-aluminum chloride and then zinc-amalgam into nuciferine **7** (Chart 4-2). On the other hand, photolysis of 1-benzylidene-2-methylisoquinoline **8**, either in the presence or in the absence of oxidizing agents, afforded no cyclized product **9**. The same reaction on its *N*-ethoxycarbonyl analog **10** gave the expected aporphine **11**. This fact is interpreted as follows: The absorption spectrum of **8** revealed only a transition assigned to the styrylamine system that was absent in the stilbene chromophore. Therefore, the energy absorbed by the compound causes **8** to undergo rapid internal conversion to the low-lying excited state of the styrylamine system, which is then deactivated to the ground state, leaving the stilbene system unreacted. On the other hand, an interception of the conjugation between the nonbonding π-electron on the nitrogen and the styrene system by *N*-acylation reveals the stilbene chromophore in its untraviolet spectrum and leads to the aporphine system.[285] Moreover, the *N*-ethoxycarbonylstilbene derivative **12** gave the dehydroaporphine **13**,[284,285] which was converted to glaucine **14**.[284]

Although these cyclizations were carried out in the presence of an oxidizing agent, Cava et al. recently reported the nonoxidative photocyclization of halogenated stilbenes to the dehydroaporphine ring system.[284] Halogenated stilbene

Chart 4-2.

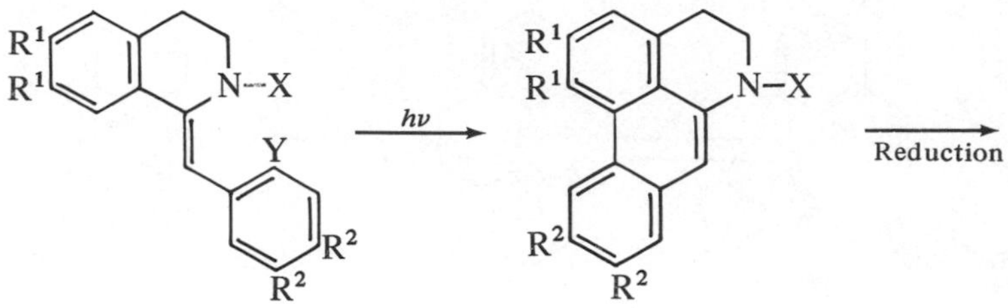

5 R^1 = OMe; R^2 = H; X = CO_2Et; Y = H
8 R^1 = R^2 = H; X = Me; Y = H
10 R^1 = R^2 = H; X = CO_2Et; Y = H
12 R^1 = R^2 = OMe; X = CO_2Et; Y = H
15 R^1 = OMe; R^2 = H; X = CO_2Et; Y = Cl
16 R^1 = R^2 = OMe; X = CO_2Et; Y = Br

6 R^1 = OMe; R^2 = H; X = CO_2Et
9 R^1 = R^2 = H; X = Me
11 R^1 = R^2 = H; X = CO_2Et
13 R^1 = R^2 = OMe; X = CO_2Et

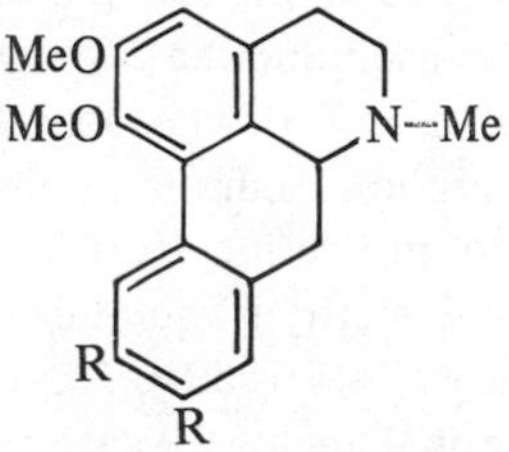

7 R = H Nuciferine
14 R = OMe Glaucine

derivatives **15** and **16** were photolyzed with Hanovia 450-W lamp fitted by Vycor filter in anhydrous ethanol in the presence of calcium carbonate as an acid scavenger, but in the absence of any added oxidant, to give dehydroaporphines **6** and **13**. The mechanism of this cyclization was suggested to be as follows: The primary photochemical conversion of the *trans*-stilbene **15** to its *cis* isomer **17** was followed by reversible isomerization of the latter to the dihydrophenanthrene intermediate **18**, and the irreversible loss of hydrogen halide from **18** yielded the dehydroaporphine **6**. However, the initial formation of the *cis* radical **19** from the *cis*-stilbene **17** by photochemical homolysis of the

Chart 4-3.

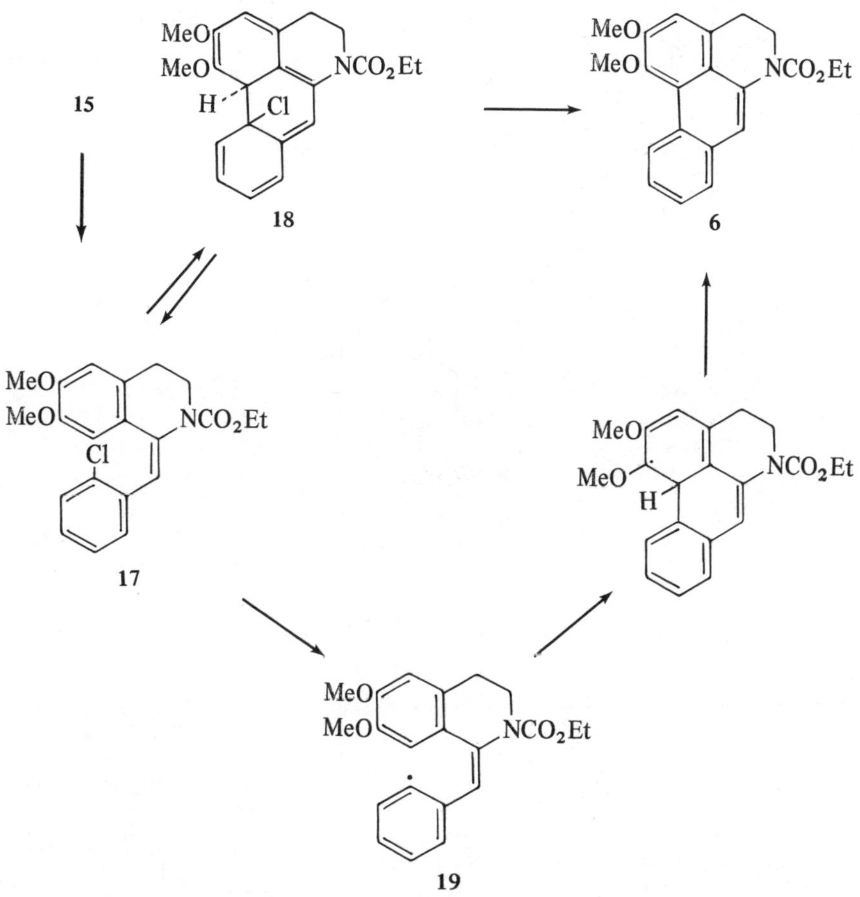

carbon-halogen bond, followed by ring closure and loss of a hydrogen atom to the aporphine ring, was also at least partially operative in this reaction (Chart 4-3). Recently, neolitsine **21** was synthesized from 2-ethoxycarbonyl-1,2,3,4-tetrahydro-1-(3,4-methylenedioxybenzylidene)-6,7-methylenedioxyisoquinoline **20** by a photolytic ring closure[286] (Chart 4-4).

The efficient nonoxidative photochemical route to the aporphines involves irradiation of the bromo-urethane **22** in the presence of potassium t-butoxide. The *N*-ethoxycarbonyldehydroaporphine **23** was thus obtained in 72% of the yield, and this was converted into cassameridine **24** and dicentrine **25**. This method is a distinct improvement over the original use of calcium carbonate in place of potassium t-butoxide as the acid scavenger[287] (Chart 4-5).

Chart 4-4.

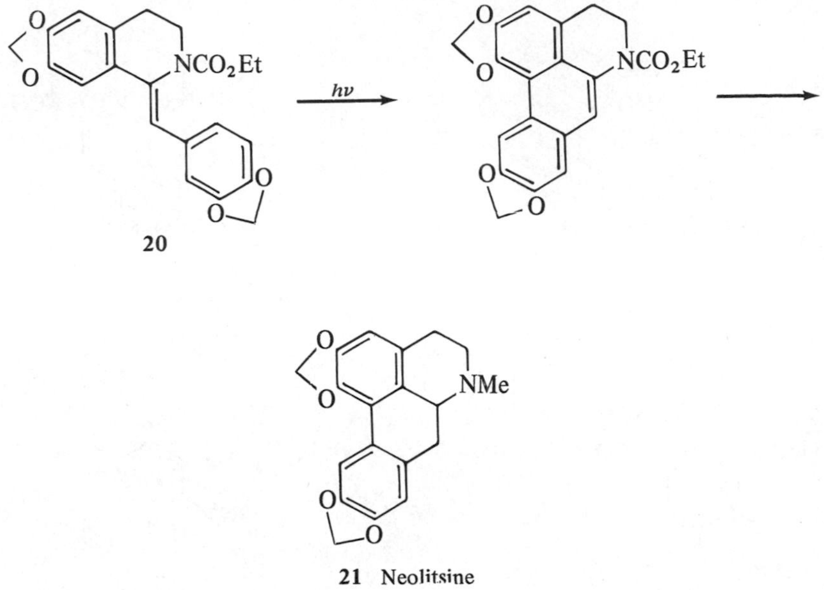

O O NCO$_2$Et

$h\nu$

O O NCO$_2$Et

O O

20

O O

O O NMe

O O

21 Neolitsine

Chart 4-5.

O O NMe O

O O

24 Cassameridine

RO RO NCO$_2$Et X OR RO

$h\nu$

RO RO NCO$_2$Et RO OR

22 X = Cl or Br

23

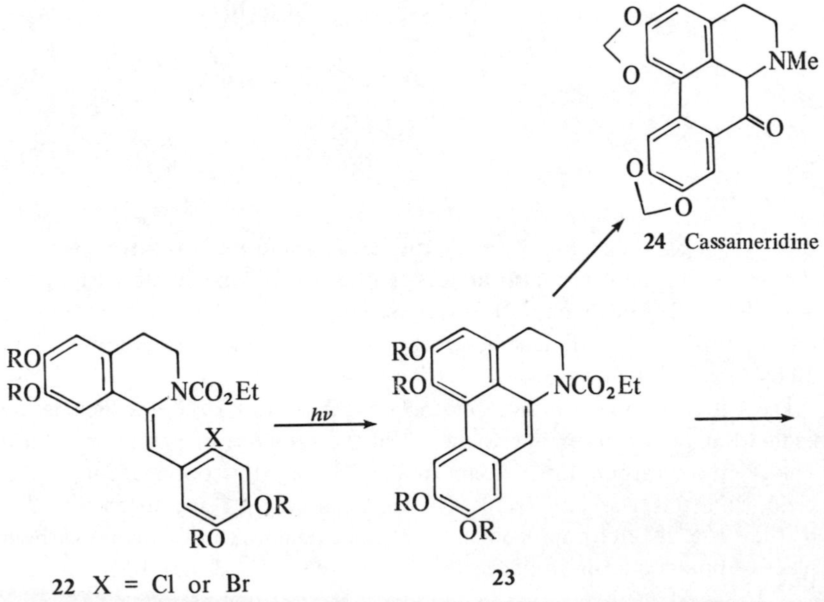

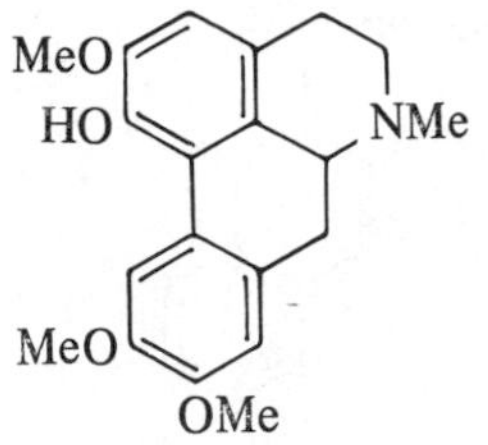

MeO
HO
NMe
MeO
OMe

25 Dicentrine

Protoberberine Alkaloids

Molecular orbital calculations on 1-benzylidene-2-ethoxycarbonylisoquinoline call for localization of electron density at the *ortho* position of stilbene in the excited state. The aromatic system is thus activated in the excited state, and intramolecular acylation occurs. In fact, the irradiation of *cis*-1-benzylidene-2-ethoxycarbonyl-1,2,3,4-tetrahydroisoquinoline **10** gave the dehydroprotoberberine **26** in 10 to 20% yield in addition to the dehydroaporphine **11** (Chart 4-6) (65% yield) depending upon reaction conditions.[288] A similar compound **27** was obtained in 50% yield of photoacylation of **26** with irradiation in the presence of iodine and cupric acetate. Interestingly, 2-benzoyl-1-benzylideneisoquinoline **28** also afforded **27**, but not the 8-phenyl analog **29**.[289] On

Chart 4-6.

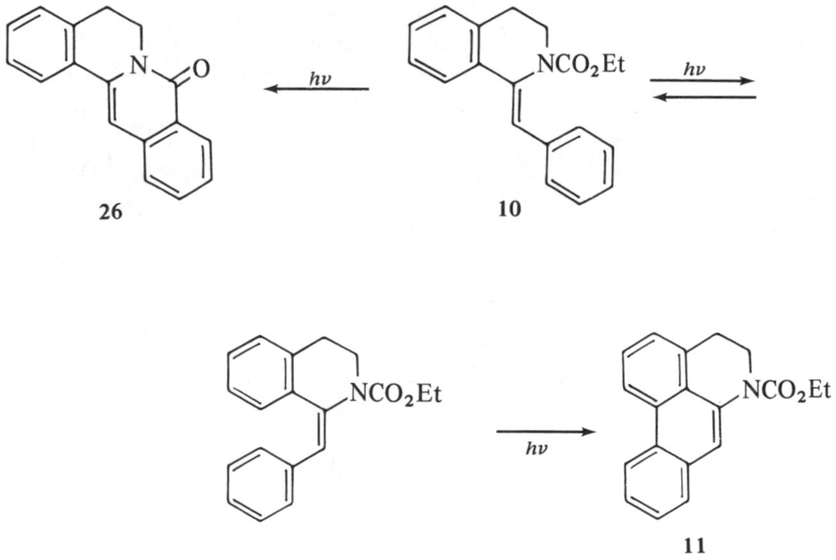

N, O — **26** ← $h\nu$ — NCO_2Et — **10** ⇄ $h\nu$

NCO_2Et —$h\nu$→ NCO_2Et — **11**

Chart 4-7.

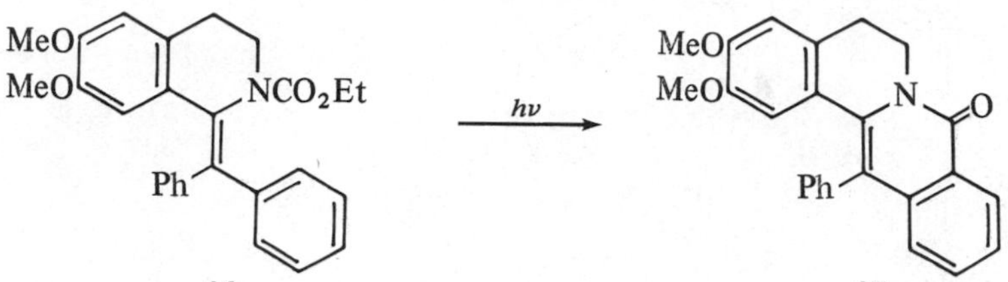

26

27

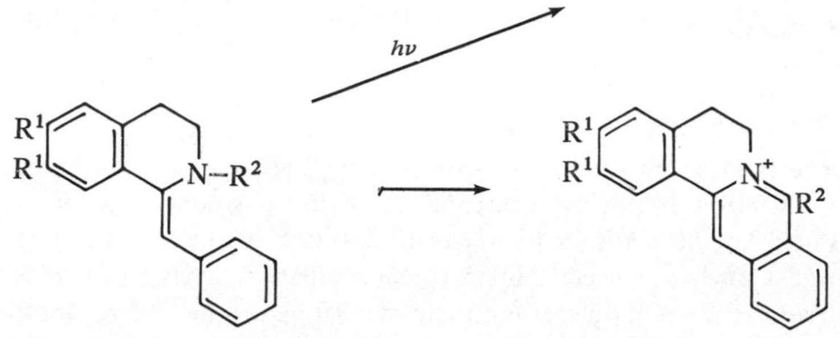

28 R^1 = OMe, R^2 = COPh
30 R^1 = H, R^2 = COMe

29 R^1 = OMe, R^2 = Ph
31 R^1 = H, R^2 = Me

Chart 4-8.

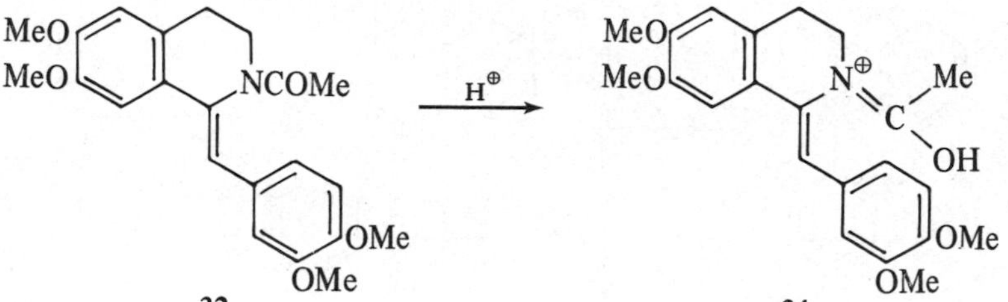

32

34

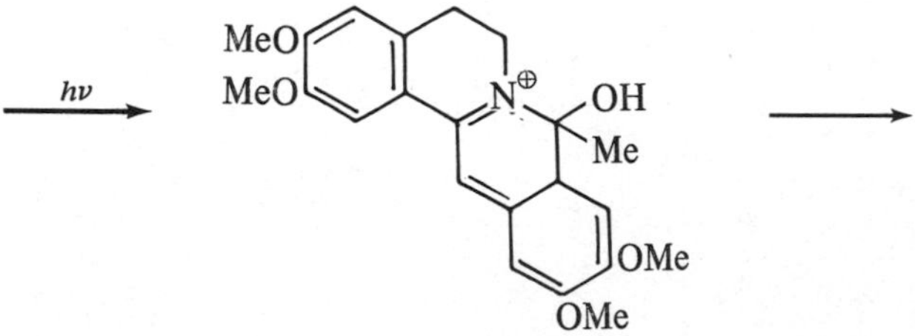

33 —reduction→ 35

the other hand, 2-acetyl-1-benzylideneisoquinoline **30** could be cyclized to **31** (Chart 4-7) by irradiation in the presence of iodine.[290] This reaction was extended to a tetramethoxy derivative **32**, which was irradiated in presence of iodine and hydriodic acid to give dehydro-β-coralydine **33** in a 75% yield. Presumably the acid protonated the amide group to give the immonium alcohol **34**, which increased the carbon to nitrogen double bond character, so that it reacted as a hexatriene system. Reduction of **33** with sodium borohydride afforded the protoberberine alkaloid, β-coralydine **35**[290] (Chart 4-8). Moreover, Ninomiya reports that benzoylation of the 1-methyl-3,4-dihydroisoquinoline **36**, followed by a photolytic cyclization of the resulting enamide **37**, provides a useful route to protoberberine alkaloid, xylopinine **38**[291] (Chart 4-9). Lenz[292] describes the elimination of *o*-substituent in the photocyclization of **39**.

Tylophorine

The total synthesis of tylophorine **43** by photocyclization of a stilbene derivative to a phenanthrene has been reported.[293] Irradiation of 3,4-dimethoxy-α-(3,4-dimethoxyphenyl)cinnamide **40** in the presence of iodine gave a phenanthrene derivative **41**, which was converted into tylophorine **43** by the usual method (Chart 4-10).

Benzophenanthridine Alkaloids

Benzophenanthridine, a basic skeleton of alkaloids such as chelerythrine **53a**, has been synthesized by photocyclization of the *N*-benzoylenamine of a cyclic ketone, which reacts as a hexatriene system (Chart 4-11). A methanolic solution of α-tetralone *N*-benzoylmethylenamine **44** was irradiated to give, in 55% yield, the *trans*-benzophenanthridone **45a**, which was reduced to benzophenanthridine **46a**. This stereoselective photocyclization to the *trans*-fused ring system could be considered to proceed through an intermediate **47**.[294] The analogous transformation of **48** to **49** was also carried out via **46b**.

Similarly, an irradiation of the enamide **50** gave an approximately one-to-one mixture of the octahydrophenanthridines (**51** and **52**)[295] (Chart 4-12). More-

Chart 4-9.

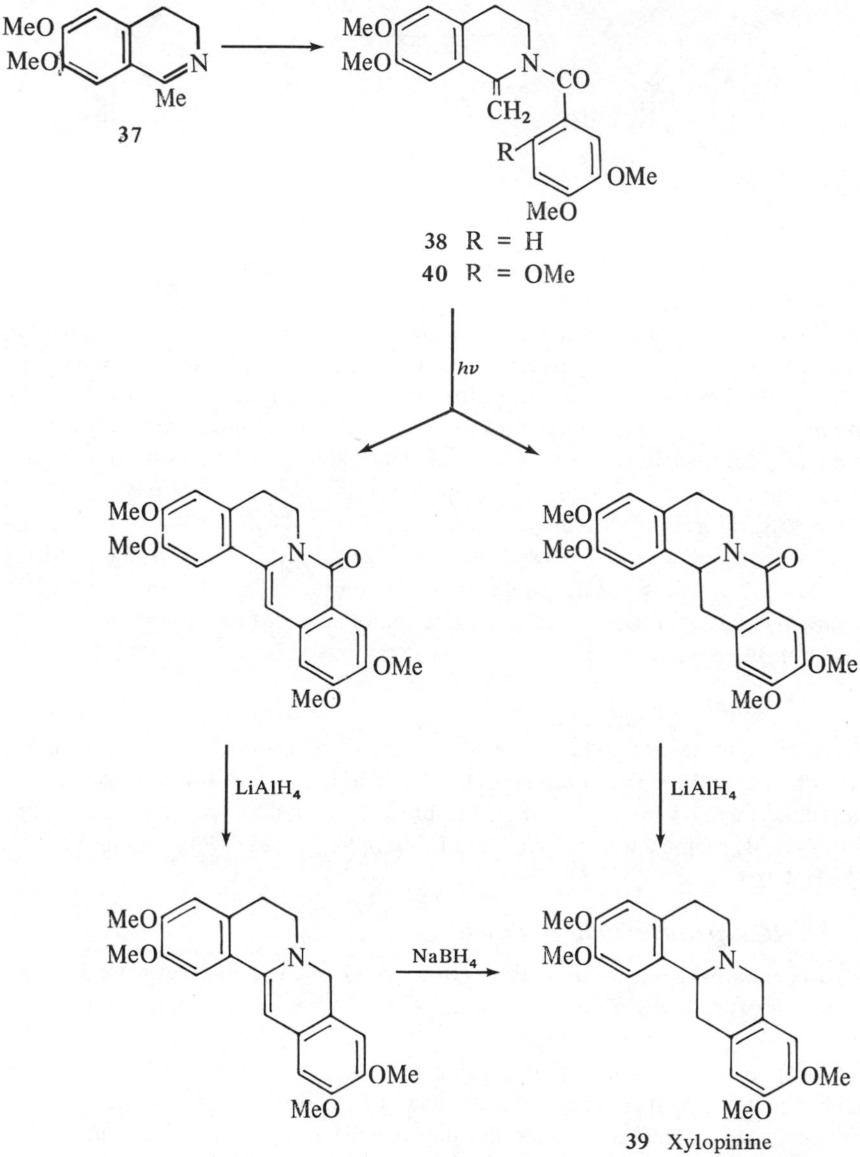

MeO MeO N Me 37 → MeO MeO N CO CH$_2$ R OMe MeO

38 R = H
40 R = OMe

$h\nu$

MeO MeO N O OMe MeO

MeO MeO N O OMe MeO

LiAlH$_4$

LiAlH$_4$

MeO MeO N OMe MeO

NaBH$_4$

MeO MeO N OMe MeO

39 Xylopinine

Chart 4-10.

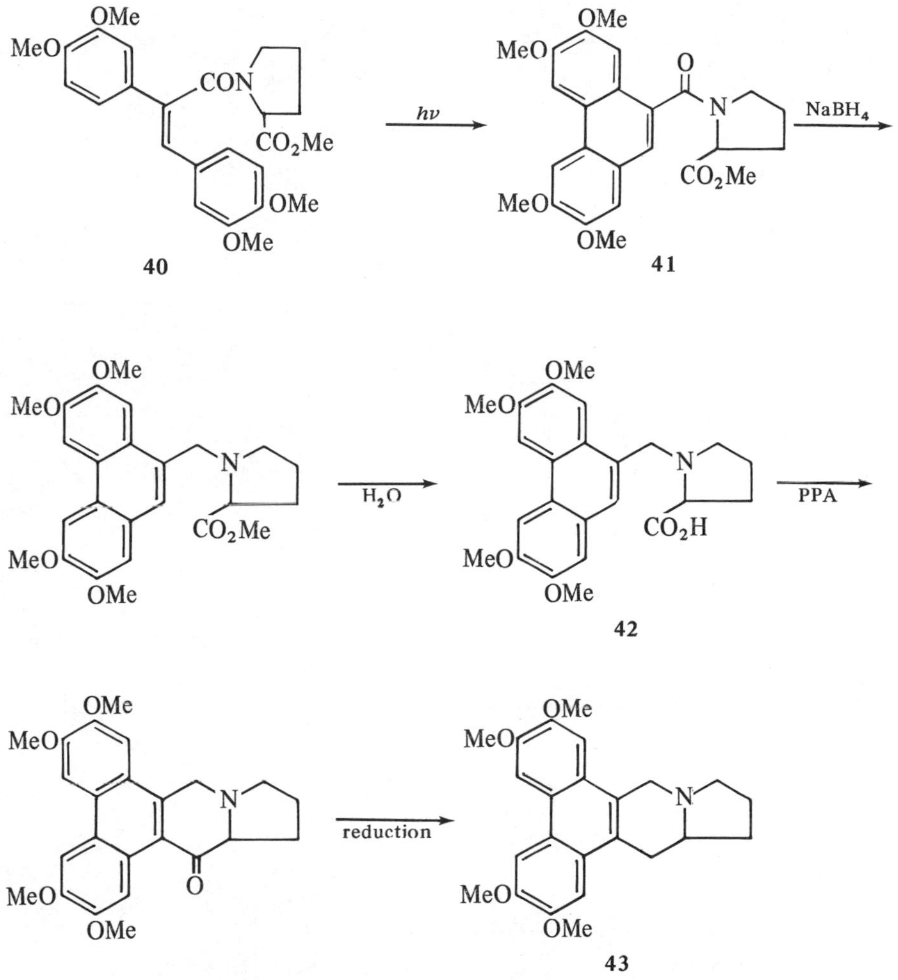

over, Onda synthesized chelerythrine **53a** and sanguinarine **53b** by a photolytic electrocyclic reaction from the methine bases **54** derived from the protoberberine **55** and protopine alkaloids **56**[296] (Chart 4-13). Irradiation of **54a** and **54b** gave the cyclized products **57a** and **57b**, respectively, which were converted into chelerythrine **53a** and sanguinarine **53b**.

The mechanism of the photocyclization from the methine base **54** to the benzophenanthidine system **57** was shown by a trapping experiment to proceed

Chart 4-11.

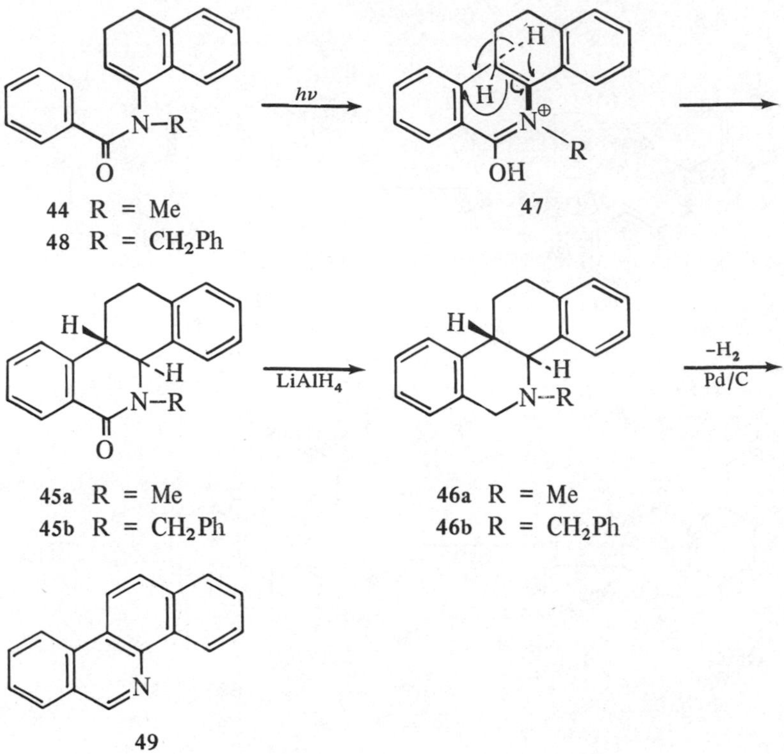

$h\nu$

N–R

O

44 R = Me

48 R = CH_2Ph

H

H

$N^{\oplus}$

R

OH

47

H

H

N–R

O

LiAlH$_4$

45a R = Me

45b R = CH_2Ph

H

H

N–R

46a R = Me

46b R = CH_2Ph

$-H_2$

Pd/C

N

49

Chart 4-12.

O

O

NCH_2Ph

O

$h\nu$

50

O

O

H

H

NCH_2Ph

O

51

+

O

O

H

H

NCH_2Ph

O

52

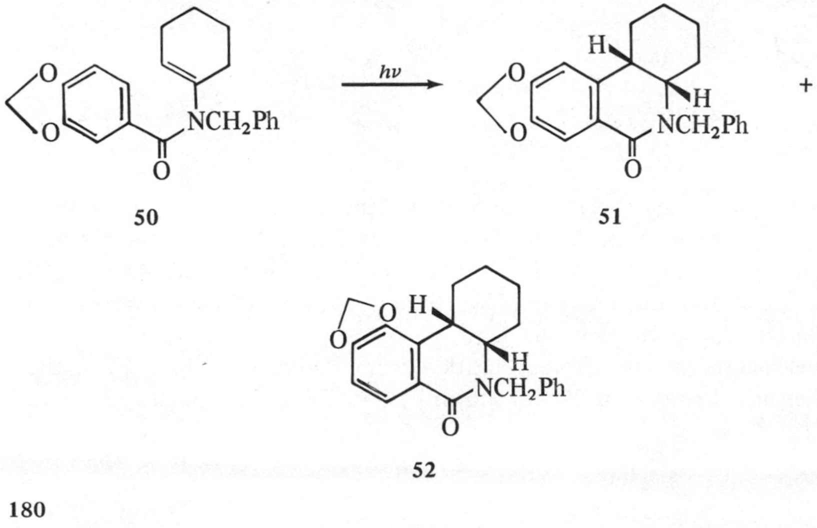

Chart 4-13.

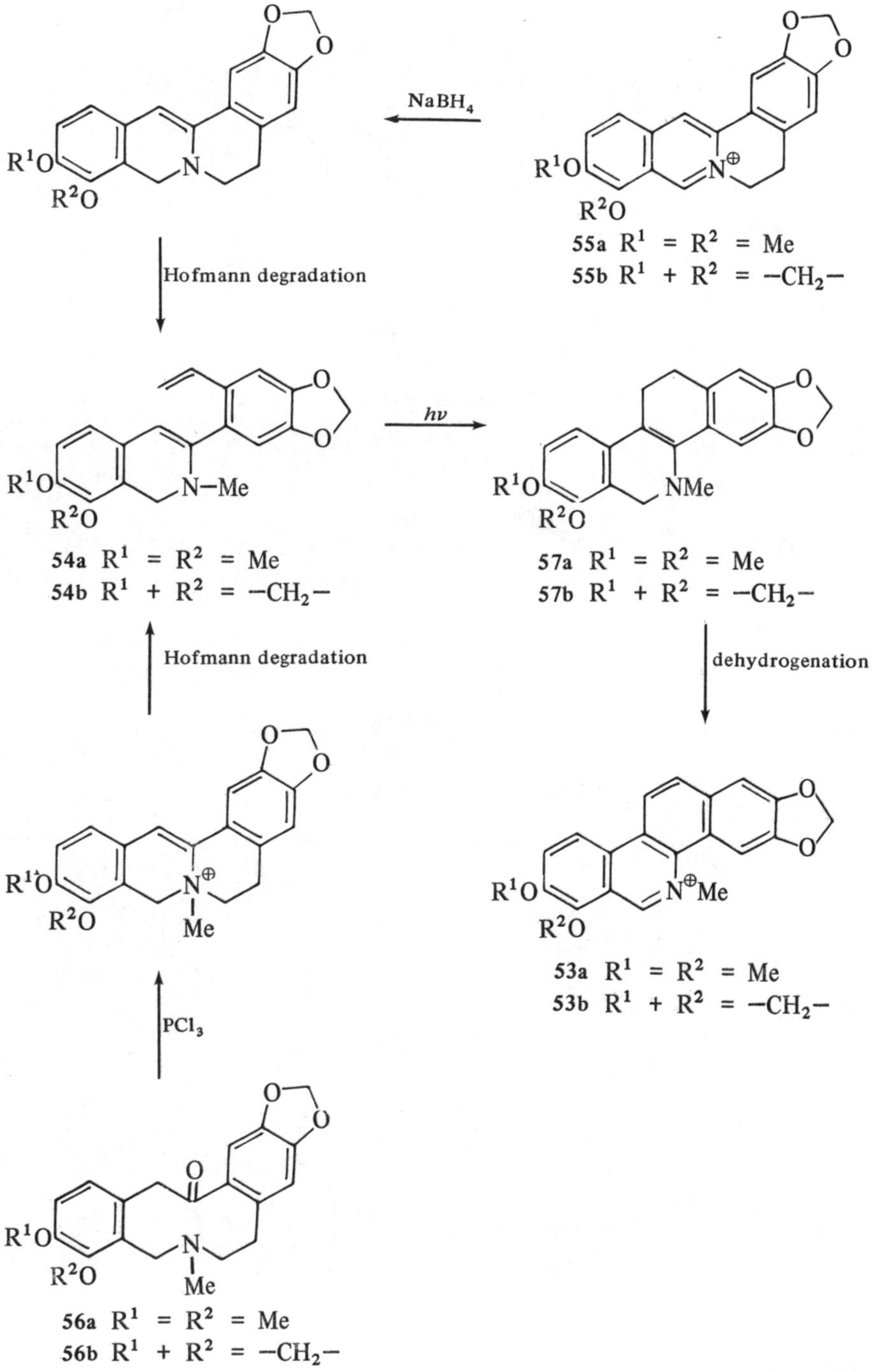

NaBH$_4$

R^1O, R^2O, N

R^1O, R^2O, $N^{\oplus}$

55a $R^1 = R^2 = Me$
55b $R^1 + R^2 = -CH_2-$

Hofmann degradation

R^1O, R^2O, N–Me

54a $R^1 = R^2 = Me$
54b $R^1 + R^2 = -CH_2-$

$h\nu$

R^1O, R^2O, NMe

57a $R^1 = R^2 = Me$
57b $R^1 + R^2 = -CH_2-$

Hofmann degradation

dehydrogenation

R^1O, R^2O, $N^{\oplus}$, Me

R^1O, R^2O, $N^{\oplus}Me$

53a $R^1 = R^2 = Me$
53b $R^1 + R^2 = -CH_2-$

PCl_3

R^1O, R^2O, O, N, Me

56a $R^1 = R^2 = Me$
56b $R^1 + R^2 = -CH_2-$

Chart 4-14.

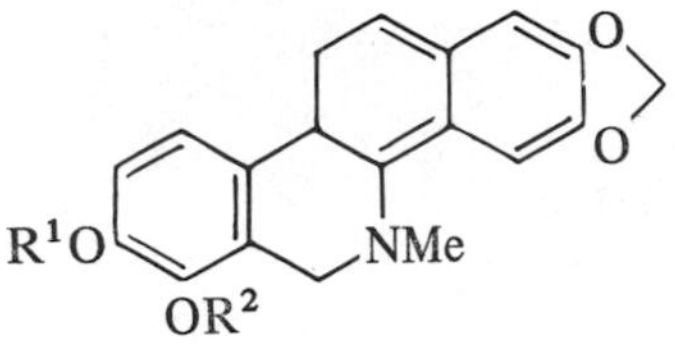

R^1O OR^2 NMe O O

58

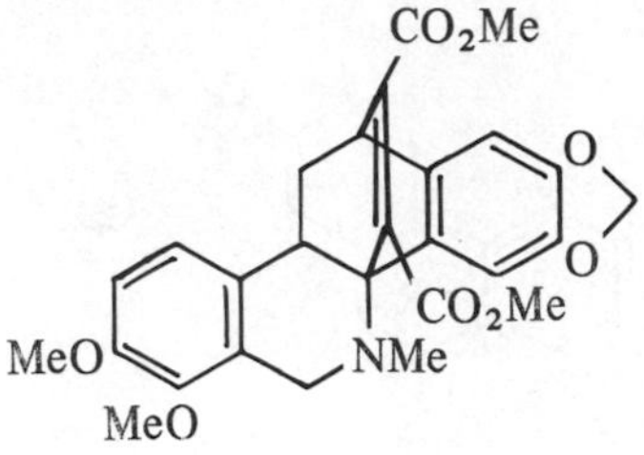

CO_2Me CO_2Me MeO MeO NMe O O

59

Chart 4-15.

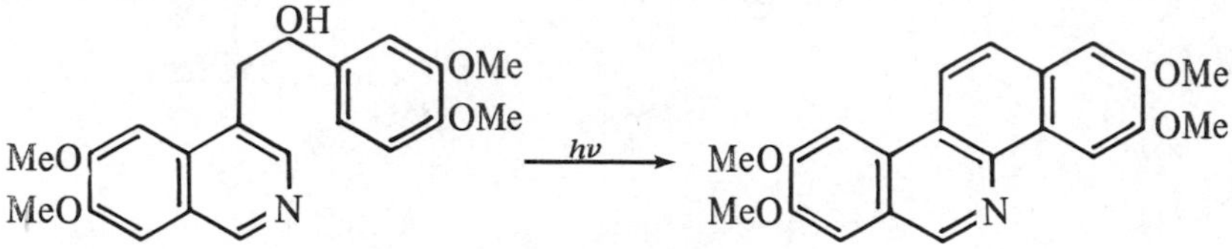

OH OMe OMe MeO MeO N $\xrightarrow{h\nu}$ MeO MeO N OMe OMe

Chart 4-16.

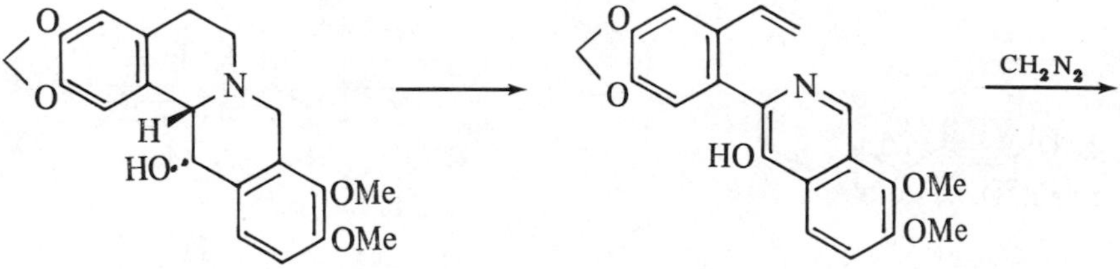

O O N H HO OMe OMe

61

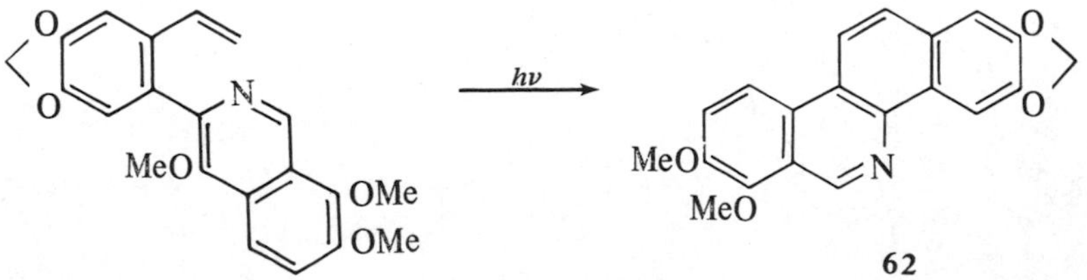

$\longrightarrow$ O O N HO OMe OMe $\xrightarrow{CH_2N_2}$

O O N MeO OMe OMe

60

$\xrightarrow{h\nu}$ MeO MeO N O O

62

via the intermediate **58**, formed by an electrocyclic reaction. When the irradiation was achieved in the presence of dimethyl acetylenedicarboxylate, the adduct **59** was obtained[297] (Chart 4-14). A similar type of cyclization was also carried out as seen in Chart 4-15.[298] Furthermore, a photolysis of the methyl ether of the isoquinoline **60**, obtained from ophiocarpine **61** by Oppenauer oxidation, also produced *N*-norchelerythrine **62**. This is the second known conversion of a protoberberine into a benzophenanthridine[299] (Chart 4-16).

Onda reported a conversion of the phthalideisoquinoline alkaloids into the benzophenanthridine derivative.[300] Thus 1-α-narcotine **63** was converted into the dihydroprotoberberine methiodide **64**, which on Hofmann degradation gave the methine base **65**. The photocyclization and dehydrogenation of this afforded the benzophenanthridine **65** (Chart 4-17). Recently, Onda applied this reaction to the synthesis of chelerythrine analog **67** as seen in Chart 4-18.[301]

Chart 4-17.

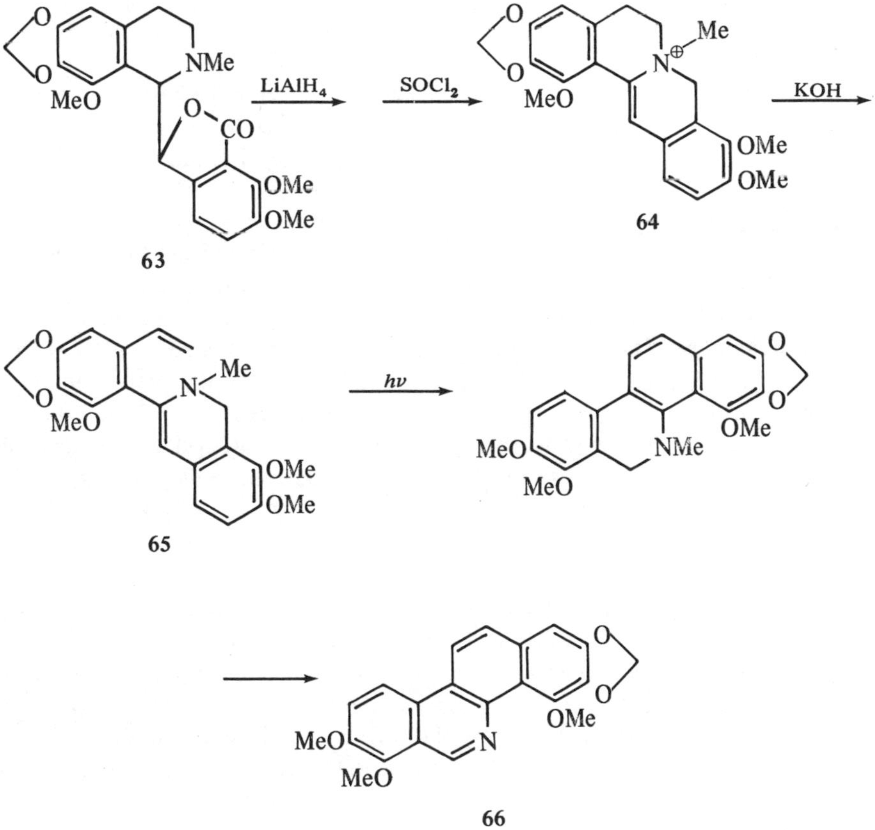

Chart 4-18.

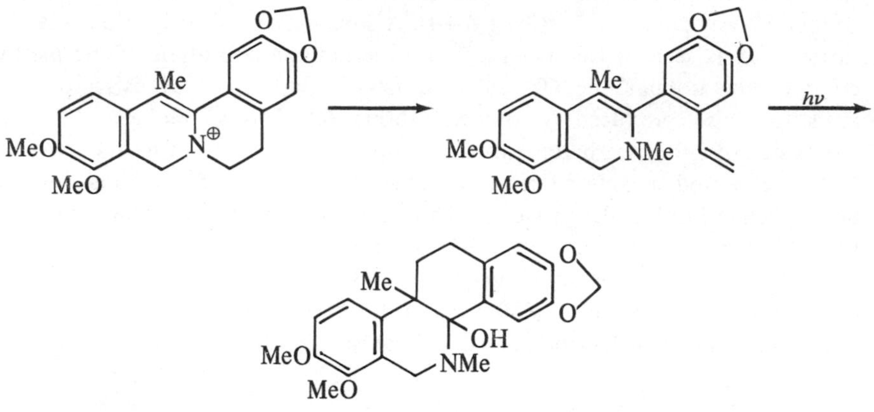

MeO MeO Me N⊕ O O → MeO MeO Me NMe O O $\xrightarrow{h\nu}$

MeO MeO Me OH NMe O O

67

Crinan Ring System

A new synthesis of (±)-crinan **68**, a compound possessing the basic ring skeleton of the alkaloid crinine, using a stereoselective photocyclization reaction has been devised.[302] The imine, readily prepared from 2-allylcyclohexanone and benzylamine, was benzoylated with piperonoyl chloride to give the *N*-acylenamine **69**. Photolysis of **69** in methanol solution gave compound **70** (15% yield) whose structure and stereochemistry were established by spectral means and by consideration of the proposed electrocyclic nature of the reaction. Successive ozonolysis and lithium aluminum hydride reduction provided the aminoalcohol **71**, which upon catalytic debenzylation followed by treatment with thionyl chloride gave (±)-crinan **68**. Alternatively, compound **71** was converted via its tosylate into the corresponding iodide, which upon hydrogenolysis also produced (±)-crinan. It may be envisaged that the key photocyclization step could also be used for the synthesis of functionalized crinan alkaloids (Chart 4-19).

B. Photochemical Transannular Reaction

The conversion of the protopine alkaloids into the berberine alkaloids has been accomplished by the photochemical transannular reaction of a carbonyl with a tertiary amine function in a ten-membered ring. Irradiation of cryptopine **72a** in 95% ethanol afforded epiberberine **73a** in moderate yield. The same reaction with protopine **72b** and α-allocryptopine **72c** gave coptisine **73b** and berberine **73c**, respectively[303] (Chart 4-20). Interestingly, the use of chloroform in place of ethanol as the irradiation solvent gave the same products, but more rapidly and

Chart 4-19.

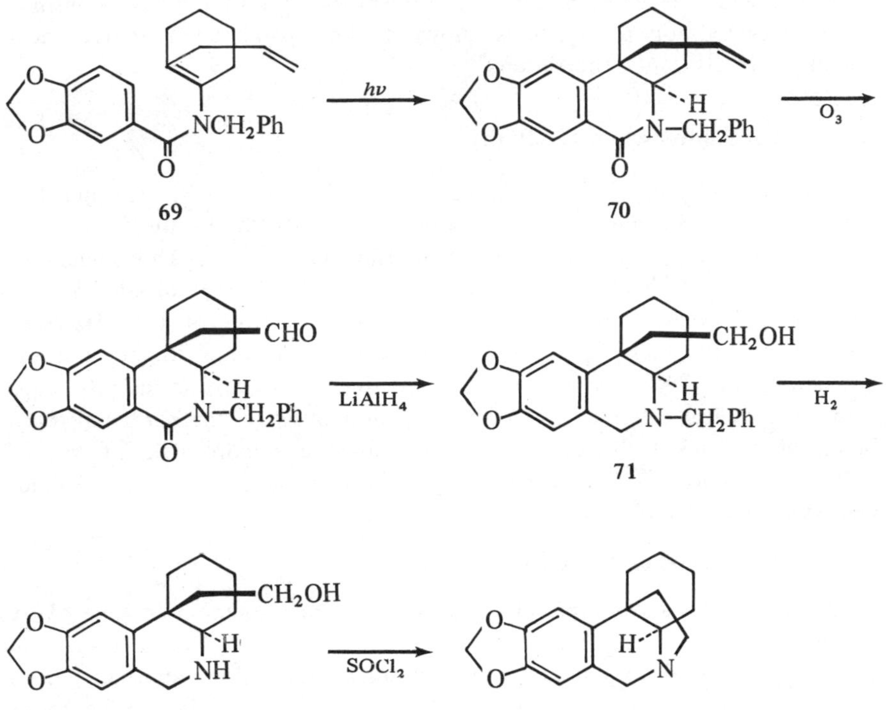

O, O, NCH$_2$Ph, O, 69, hν, O, O, H, N–CH$_2$Ph, O, 70, O_3

O, O, CHO, H, N–CH$_2$Ph, O, $LiAlH_4$, O, O, CH$_2$OH, H, N–CH$_2$Ph, 71, H_2

O, O, CH$_2$OH, H, NH, $SOCl_2$, O, O, H, N, 68

Chart 4-20.

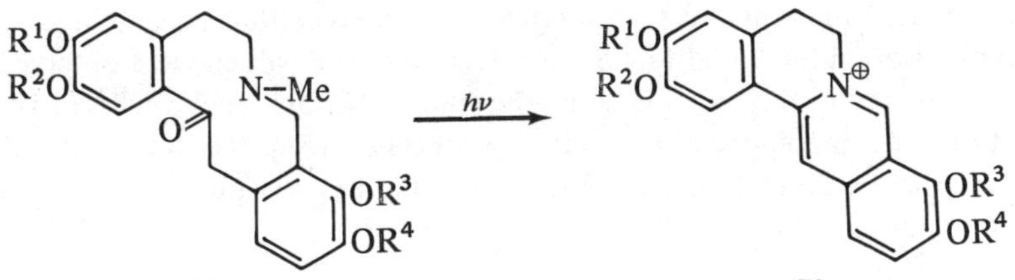

R^{1}O, R^{2}O, N–Me, O, OR^3, OR^4, 72, hν, R^{1}O, R^{2}O, N$^{\oplus}$, OR^3, OR^4, 73

a $R^1 = R^2 = Me$; $R^3 + R^4 = -CH_2-$
b $R^1 = R^2 = -CH_2-$; $R^3 + R^4 = -CH_2-$
c $R^1 + R^2 = -CH_2-$; $R^3 = R^4 = Me$

in higher yield. This could be due to the operation of a different mechanism in the halogenated solvent, which is known to be a good radical source under conditions of ultraviolet irradiation.[303]

C. Photo-Pschorr Reaction

Since its discovery in 1896, the Pschorr reaction[304] has been widely applied to the synthesis of aporphine-type compounds. Recently, Kametani[305] extended this reaction to a general synthesis of the morphinandienone-type alkaloids, such as salutaridine **74**, by thermal decomposition of the diazonium salt **75** from 1-(2-aminobenzyl)isoquinoline **76**.[306] The same reaction of the diazotized phenethylisoquinoline **77** gave the homomorphinandienone **78**, but not the homoaporphine **79**[307,308] (Chart 4-21). If this type of reaction proceeded through a radical intermediate formed by homolysis of the carbon-nitrogen bond, photolysis of the diazonium salt would be a more efficient way of effecting a homolysis. On the basis of this assumption, five types of alkaloids were synthesized as follows.

Benzylisoquinoline Series

Irradiation of the diazonium salt **80**, prepared from 6′-aminoorientaline **81** by diazotization, gave flavinantine **82** and bracteoline **83**.[309] The same reaction with the diazotized isoquinoline **84** (Chart 4-22) afforded the aporphine **85** in 17% of the yield; this was converted into *N,O*^{10}-dimethylhernovine **86** as shown[310,311] in Chart 4-23.

The usual Pschorr reaction of **84** gave the aporphine **85** in only 3% of the yield; thus the photo-Pschorr reaction is an improved method for the synthesis of the aporphine.[311] The same phenomenon was observed in the synthesis of *N*-methyllindecarpine **87** and isocorydine **88** from the diazotized isoquinoline **89**. The former alkaloid **87** was obtained directly by irradiation of the phenolic diazotized isoquinoline **90**.[312] Moreover, the side reactions, diazo coupling in the phenolic isoquinoline and loss of a protecting group, which were observed in the Pschorr reaction, did not occur in the photo-Pschorr reaction.[312] In the latter reaction, the presence of an *N*-ethoxycarbonyl group **91** did not lead to the protoberberine formation **92**, which was found in the photolytic electrocyclic reaction already described.[313]

Phenethylisoquinoline Series

The photo-Pschorr reaction was also applied to the phenethylisoquinoline series under the same conditions as in the benzylisoquinoline series (Chart 4-24). Photolysis of the diazonium salts **93a** and **93b** gave *O*-methylandrocymbine **94**,[314] but thermal decomposition of **93a** and **93b** afforded the abnormal compounds **95a** and **95b**, respectively.[314] Moreover, irradiation of **96** afforded *O*-

Chart 4-21.

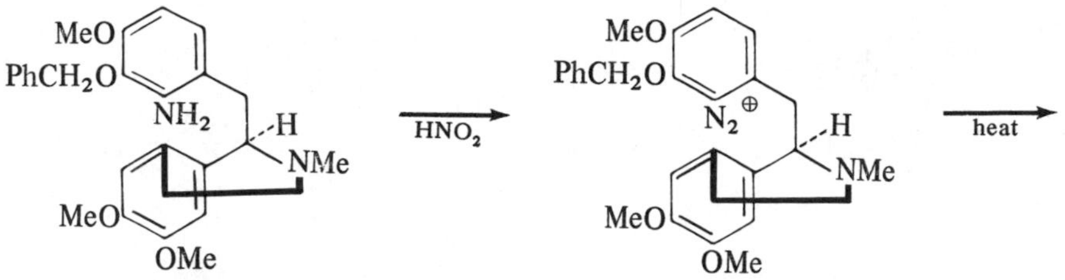

MeO
PhCH$_2$O
NH$_2$
H
NMe
MeO
OMe
76

HNO$_2$

MeO
PhCH$_2$O
N$_2^{\oplus}$
H
NMe
MeO
OMe
75

heat

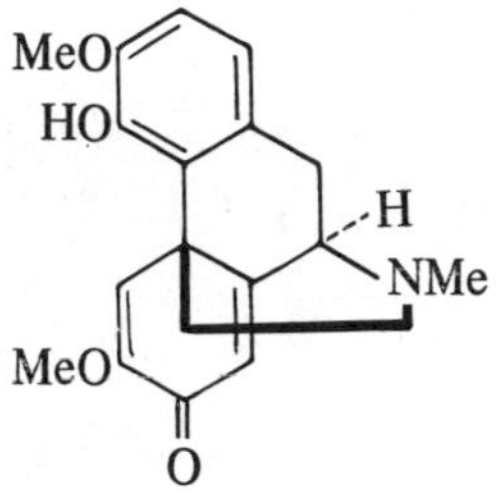

MeO
HO
H
NMe
MeO
O
74

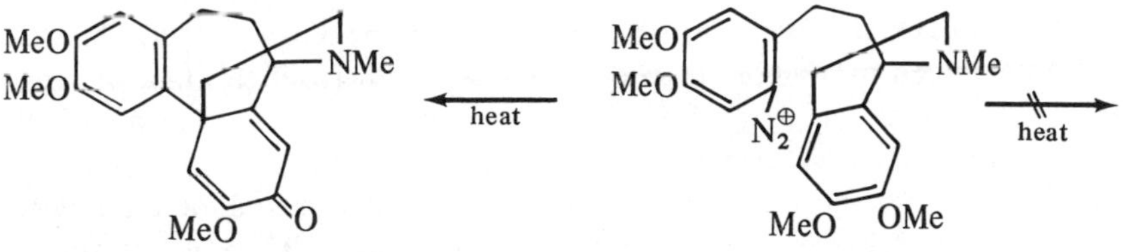

MeO
MeO
NMe
MeO
O
78

heat

MeO
MeO
NMe
N$_2^{\oplus}$
MeO
OMe
77

heat

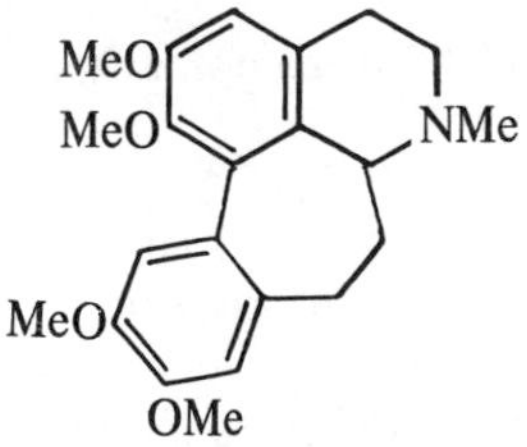

MeO
MeO
NMe
MeO
OMe
79

Chart 4-22.

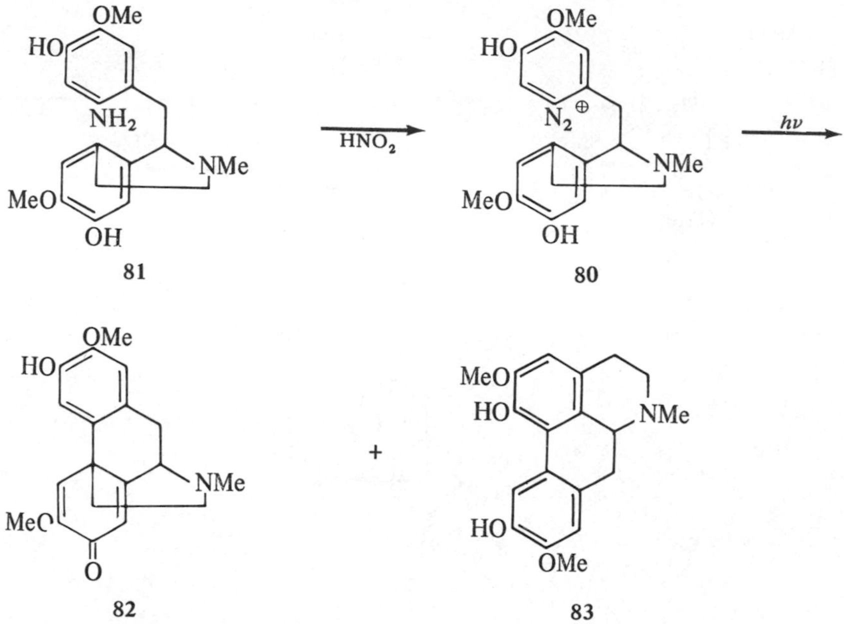

81 $\xrightarrow{HNO_2}$ 80 $\xrightarrow{h\nu}$ 82 + 83

methylandrocymbine **94** and kreysigine **97**.[311] Homoaporphine **99**, which could not be obtained by Pschorr reaction, was also synthesized from **98** by a photo-Pschorr reaction[315] (Chart 4-25). The total synthesis of androcymbine **100** has been accomplished by an application of the photo-Pschorr reaction as seen in Chart 4-26. However, direct conversion of **101** to androcymbine did not occur; instead this reaction produced the homoproaporphine **102**, probably via radical intermediates as illustrated[316] in Chart 4-27.

D. Photolytic Cyclodehydrohalogenation

It is well known that photolysis of aromatic halides in benzene results in the formation of biphenyl derivatives by the reaction of the aryl radicals produced by homolytic cleavage of the carbon-halogen bond.[317] Intramolecular reactions of this type also give the cyclization products. Intramolecular cyclization by photolysis is utilized in the total synthesis of the isoquinoline alkaloids (Chart 4-28).

Aporphine Alkaloids

Kupchan synthesized nornuciferine **103** and nuciferine **7** (30-67% yield) from

Chart 4-23.

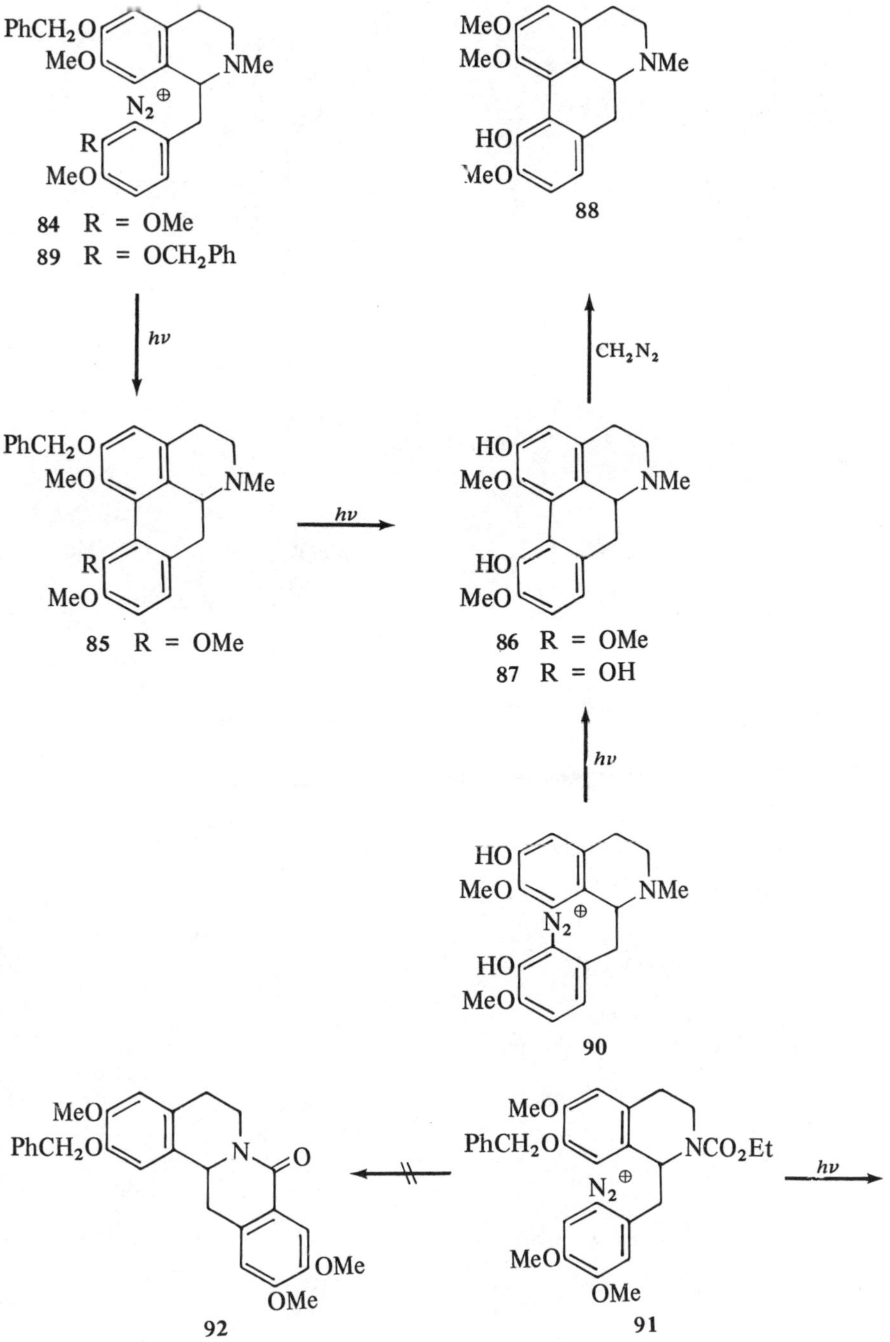

PhCH$_2$O
MeO
NMe
N$_2^{\oplus}$
R
MeO

84 R = OMe
89 R = OCH$_2$Ph

$h\nu$

PhCH$_2$O
MeO
NMe
R
MeO

85 R = OMe

$h\nu$

MeO
MeO
NMe
HO
MeO

88

CH$_2$N$_2$

HO
MeO
NMe
HO
MeO

86 R = OMe
87 R = OH

$h\nu$

HO
MeO
NMe
N$_2^{\oplus}$
HO
MeO

90

MeO
PhCH$_2$O
N
O
OMe
OMe

92

MeO
PhCH$_2$O
NCO$_2$Et
N$_2^{\oplus}$
MeO
OMe

91

$h\nu$

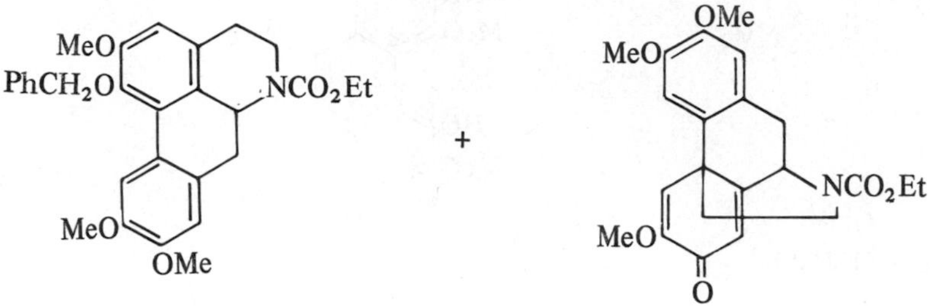

MeO
PhCH$_2$O
NCO$_2$Et
MeO
OMe

+

OMe
MeO
NCO$_2$Et
MeO
O

Chart 4-24.

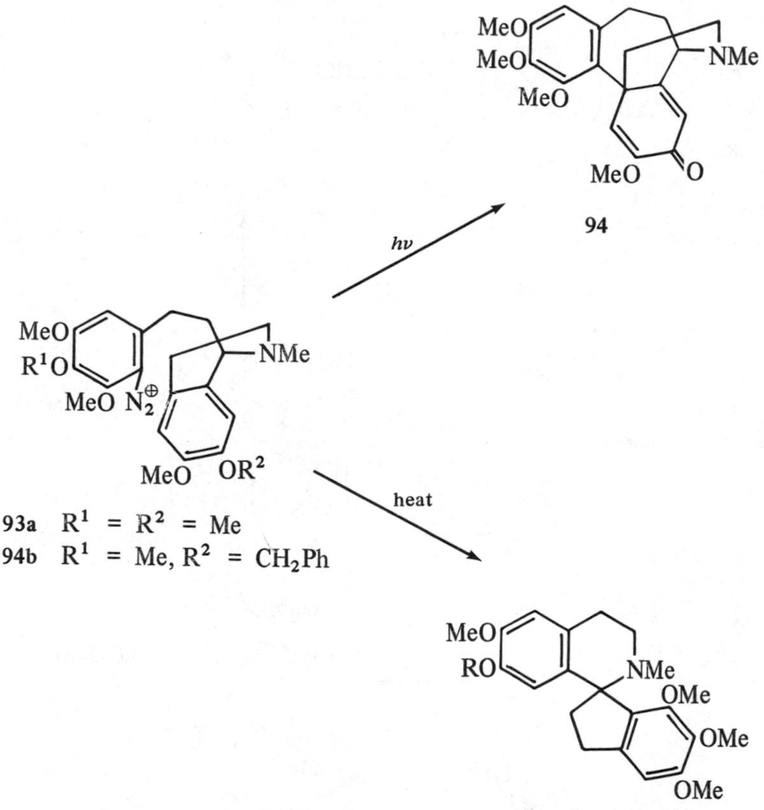

MeO
MeO
NMe
MeO
MeO
O

94

hν

MeO
R^{1}O
NMe
MeO $N_2^{\oplus}$
MeO OR^2

93a $R^1 = R^2 = Me$
94b $R^1 = Me, R^2 = CH_2Ph$

heat

MeO
RO
NMe
OMe
OMe
OMe

95a $R = Me$
95b $R = CH_2Ph$

Chart 4-25.

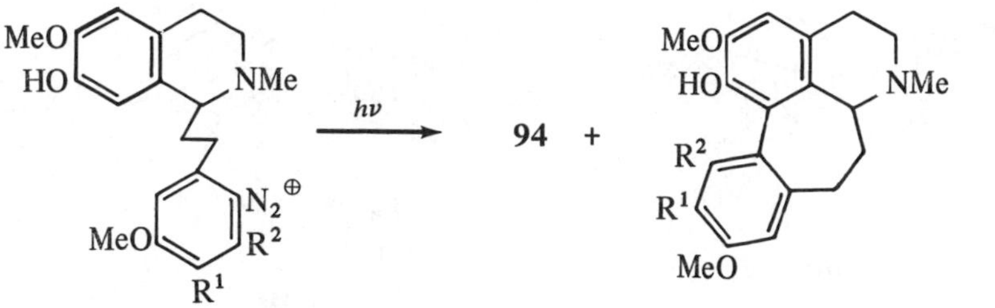

MeO HO NMe $N_2^{\oplus}$ MeO R^2 R^1 $\xrightarrow{h\nu}$ 94 + MeO HO NMe R^2 R^1 MeO

96 $R^1 = R^2 = OMe$
98 $R^1 = OH, R^2 = H$

97 $R^1 = R^2 = OMe$
99 $R^1 = OH, R_2 = H$

Chart 4-26.

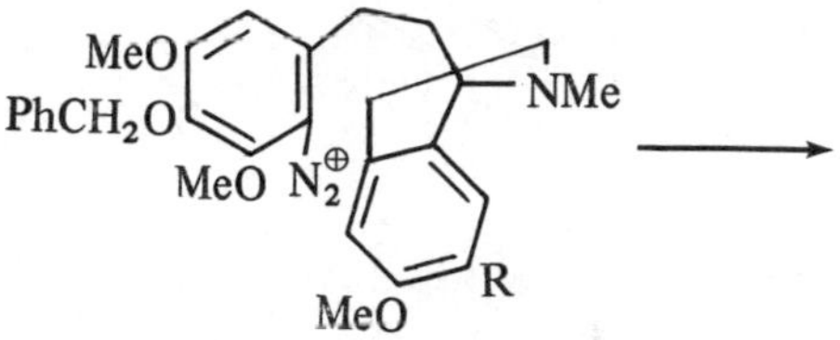

MeO $PhCH_2O$ MeO $N_2^{\oplus}$ NMe R MeO $\longrightarrow$

R = OMe and OCH_2Ph

MeO $PhCH_2O$ MeO NMe MeO O $\xrightarrow{H^{\oplus}}$ MeO HO MeO NMe MeO O

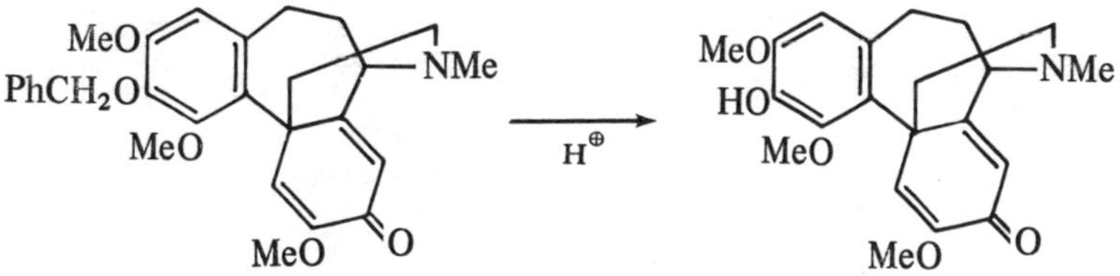

100

the corresponding 1,2,3,4-tetrahydro-1-(2-iodobenzyl)isoquinoline hydrochlorides by photolytic intramolecular cyclization. Although the same reaction of *N*-acyl derivatives (acetyl, benzoyl, ethoxycarbonyl and phenoxycarbonyl) in the presence of sodium thiosulfate produced the corresponding aporphines, the free base of 1-benzylisoquinolines gave no aporphines. Apparently, the presence of the free electron lone pair on nitrogen was detrimental to the desired photocyclization. *N*-Acyl derivatives of isoquinoline hydrochlorides,

Chart 4-27.

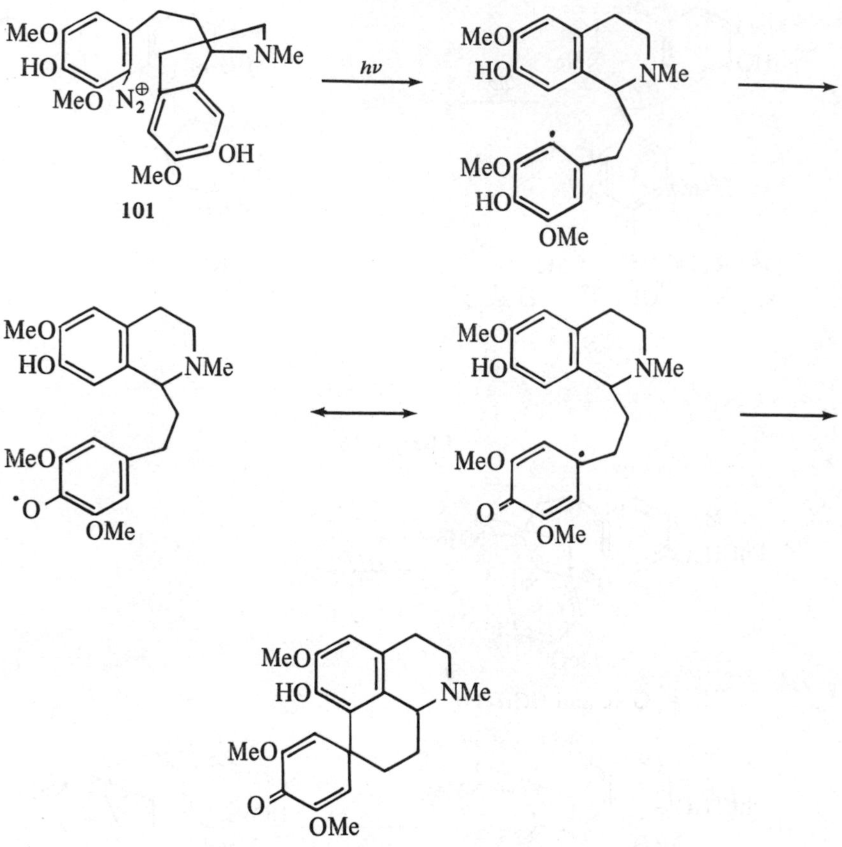

MeO HO MeO $N_2^{\oplus}$ NMe OH MeO

101

$h\nu$

MeO HO NMe MeO HO OMe

MeO HO NMe MeO •O OMe

MeO HO NMe MeO O OMe

MeO HO NMe MeO O OMe

102

Chart 4-28.

I NH

$h\nu$

NH

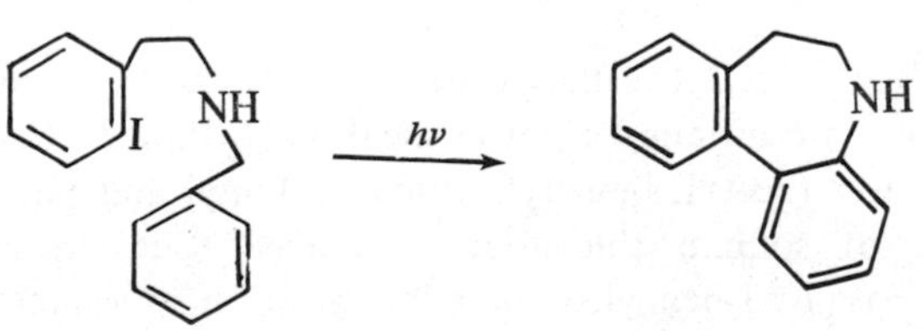

which circumvented the detrimental effect of the basic nitrogen, showed smooth cyclization[318] (Chart 4-29).

Moreover, synthesis of the aporphine alkaloids from the phenolic bromoisoquinolines has been reported by the present authors. In these reactions irradiation of 6′-bromoorientaline **104a**, 6′-bromoreticuline **104b**, and its methylenedioxy analog **104c** in the presence of base gave bracteoline **83**, isoboldine **105**, and domesticine **106**, respectively, in moderate yields, in addition to the morphinandienone alkaloids described later[319,320] (Chart 4-30). Similarly, the photolysis of the 1-(3-hydroxybenzyl)-8-bromoisoquinolines **107** gave pukateine **108a**, cassythicine **108b**, and *N*-methyllaurotetanine **108c**[321] (Chart 4-31). Recently, bracteoline **83** was obtained in 52% of the yield from the diphenolic bromoisoquinoline **104a** by irradiation with a 2537-Å light source.[322]

Chart 4-29.

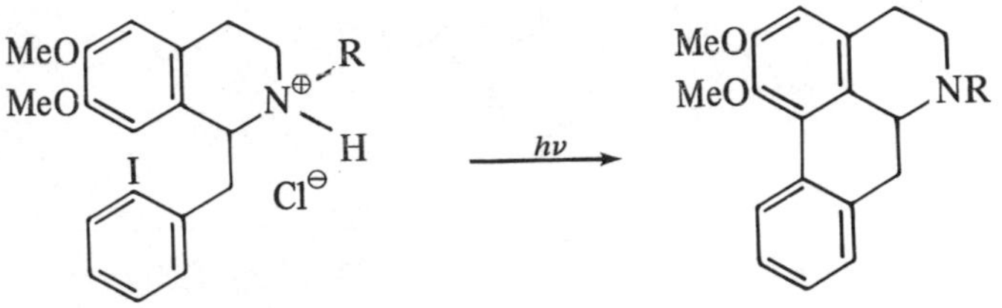

R = H and Me

103 R = H
7 R = Me

Chart 4-30.

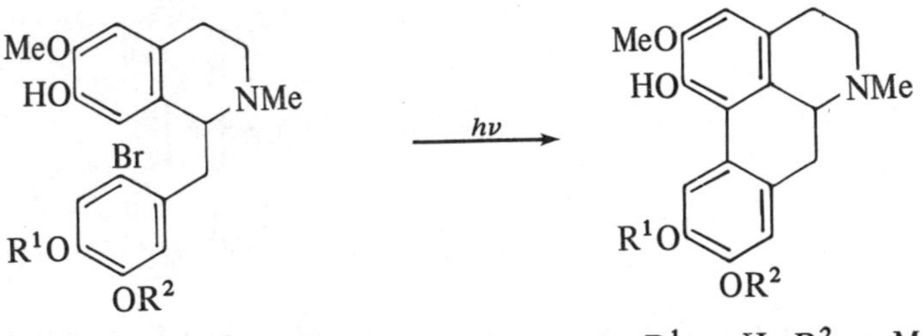

104a R^1 = H, R^2 = Me
104b R^1 = Me, R^2 = H
104c R^1 + R^2 = $-CH_2-$

83 R^1 = H, R^2 = Me
105 R^1 = Me, R^2 = H
106 R^1 + R^2 = $-CH_2-$

Chart 4-31.

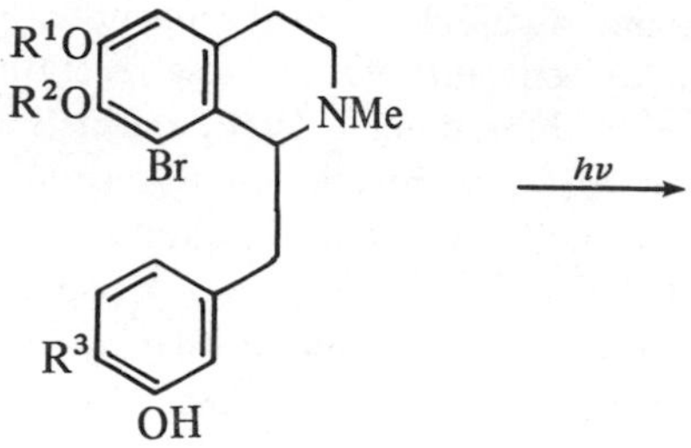

R^1O R^2O NMe Br R^3 OH $\xrightarrow{h\nu}$

107a $R^1 + R^2 = -CH_2-$, $R^3 = H$
107b $R^1 + R^2 = -CH_2-$, $R^3 = OMe$
107c $R^1 = R^2 = Me$, $R^3 = OMe$

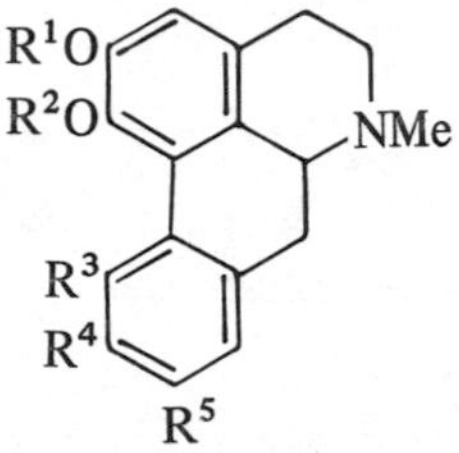

R^1O R^2O NMe R^3 R^4 R^5

108a $R^1 + R^2 = -CH_2-$, $R^3 = OH$, $R^4 = R^5 = H$
108b $R^1 + R^2 = -CH_2-$, $R^3 = H$, $R^4 = OMe$, $R^5 = OH$
108c $R^1 = R^2 = Me$, $R^3 = H$, $R^4 = OMe$, $R^5 = OH$

Chart 4-32.

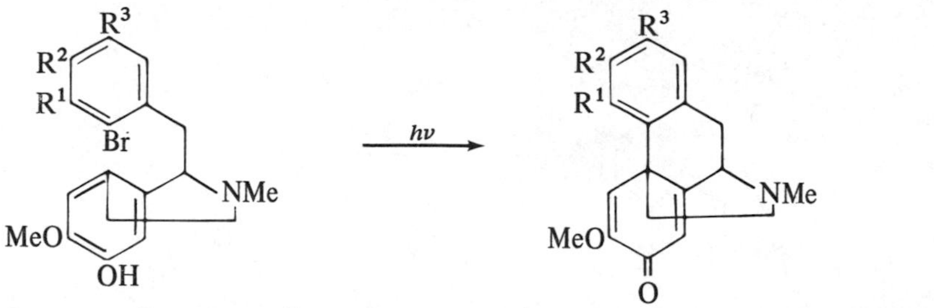

R^3 R^2 R^1 Br NMe MeO OH $\xrightarrow{h\nu}$ R^3 R^2 R^1 NMe MeO O

104a $R^1 = H$, $R^2 = OH$, $R^3 = OMe$ **82** $R^1 = H$, $R^2 = OH$, $R^3 = OMe$
104b $R^1 = H$, $R^2 = OMe$, $R^3 = OH$ **109** $R^1 = H$, $R^2 = OMe$, $R^3 = OH$
104c $R^1 = H$, $R^2 + R^3 = -OCH_2O-$ **110** $R^1 = H$, $R^2 + R^3 = -OCH_2O-$
104d $R^1 = OH$, $R^2 = OMe$, $R^3 = H$ **111** $R^1 = OH$, $R^2 = OMe$, $R^3 = H$

Chart 4-33.

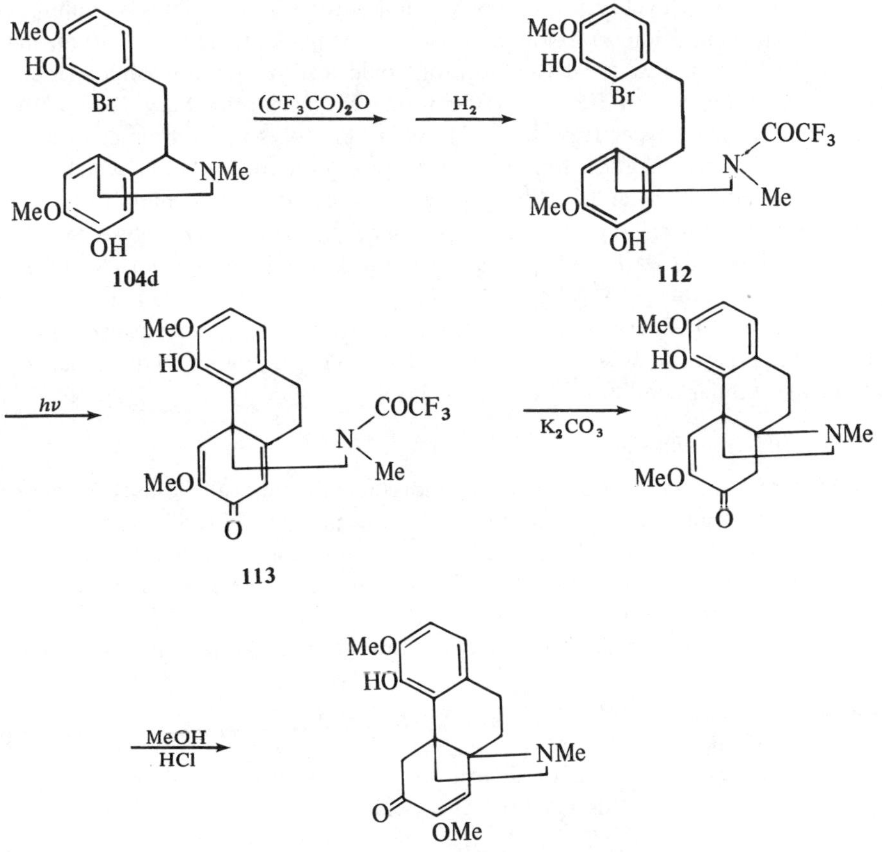

MeO
HO
Br
NMe
MeO
OH
104d

$(CF_3CO)_2O$ → H_2 →

MeO
HO
Br
$COCF_3$
N
Me
MeO
OH
112

hν →

MeO
HO
$COCF_3$
N
Me
MeO
O
113

→ K_2CO_3

MeO
HO
NMe
MeO
O

MeOH / HCl →

MeO
HO
NMe
O
OMe
114

Chart 4-34.

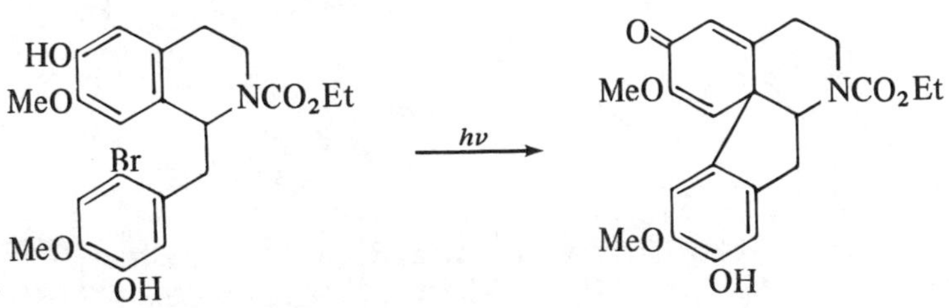

HO
MeO
NCO_2Et
Br
MeO
OH
116

hν →

O
MeO
NCO_2Et
MeO
OH
115

Morphinandienone Alkaloids

In the photolytic cyclization of the phenolic bromoisoquinolines, coupling is possible at the position *ortho* as well as *para* to the phenolic hydroxyl group, and a coupling reaction at the latter position could lead to the morphinandienone alkaloids. Irradiation of **104b** and **104c** produced flavinantine **82**, pallidine **109**, and amurine **110**, respectively.[319,320] Moreover, photolysis of 2′-bromoreticuline **104d** in the presence of sodium iodide gave salutaridine **111**, which was converted into morphine.[321,323] A new synthesis of cepharamine **114** by a photolysis has been reported in a preliminary form.[324] Photolysis of the dihydrostilbene derivative **112**, obtained from 2′-bromoreticuline **104d** in two steps, gave dienone **113**, whose Michael reaction and isomerization afforded cepharamine **114** (Charts 4-32 and 4-33). Furthermore, the proerythrinadienone-type compound **115** was synthesized in 12% yield from its 6-hydroxy analog **116** under similar conditions[325] (Chart 4-34).

Proaporphine Alkaloids

Because of the information that the morphinandienone alkaloids have been synthesized by photocyclization, the synthesis of proaporphine alkaloids from 8-bromo-1-(4-hydroxybenzyl)isoquinoline by photolysis was examined. The phenolic bromoisoquinolines (**117a**, **117b**, and **117c**) were irradiated to give mecambrine **118a**, pronuciferine **118b**, and glaziovine **118c**.[310,326,327,328] Alkaloid **118b** was also obtained by irradiation of **117b** in the presence of sodium borohydride and sodium hydroxide, followed by oxidation of the resulting dienol.[329] *O*-Methylorientalinone **119** was obtained from **117d** in addition to *O*-methylisoorientalinone, a *spiro* isomer of **119** (Chart 4-35). Orietalinone **120** was also obtained by this method from the bromophenol **89c**.[321] When copper powder was added in this reaction, the yield of glaziovine **118c** increased to 17%.[326,328]

Chart 4-35.

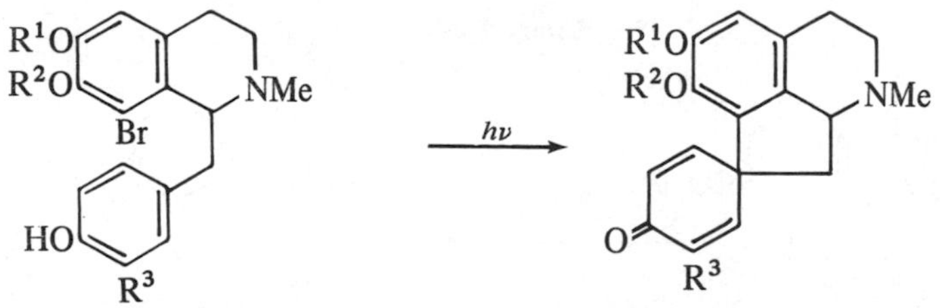

117a $R^1 + R^2 = -CH_2-$, $R^3 = H$ **118a** $R^1 + R^2 = -CH_2-$, $R^3 = H$
117b $R^1 = R^2 = Me$, $R^3 = H$ **118b** $R^1 = R^2 = Me$, $R^3 = H$
117c $R^1 = Me$, $R^2 = R^3 = H$ **118c** $R^1 = Me$, $R^2 = R^3 = H$
117d $R^1 = R^2 = Me$, $R^3 = OMe$ **119** $R^1 = R^2 = Me$, $R^3 = OMe$
89c $R^1 = Me$, $R^2 = H$, $R^3 = OMe$ **120** $R^1 = Me$, $R^2 = H$, $R^3 = OMe$

Phenethylisoquinoline Series

Photolytic cyclization of the phenolic bromobenzylisoquinolines to the aporphine, morphinandienone, and proaporphine alkaloids was extended, as follows, to the synthesis of the corresponding alkaloids in the phenethylisoquinoline group. 1-(2-Bromophenethyl)-7-hydroxyisoquinoline **121** was irradiated to afford *O*-methylandrocymbine **94** and kreysigine **97**.[330] Optically active bases were also obtained by this reaction.[331] Androcymbine **100** was obtained by the same reaction from **122**,[332] and 8-bromo-1-(4-hydroxyphenethyl)isoquinoline **123** (Chart 4-36) also gave *O*-methylkreysiginone **124**[326] (Chart 4-37). The

Chart 4-36.

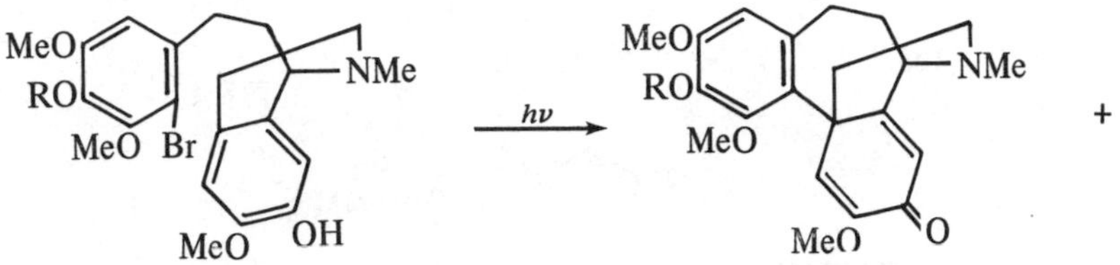

MeO, RO, MeO, Br, NMe, MeO, OH; $h\nu$; MeO, RO, MeO, NMe, MeO, O; +

121 R = Me
122 R = H

94 R = Me
100 R = H

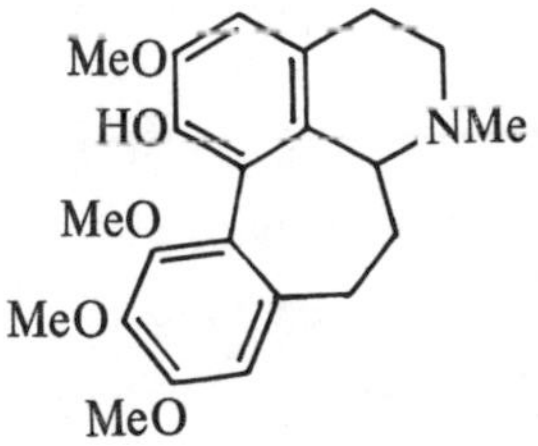

MeO, HO, NMe, MeO, MeO, MeO

97

Chart 4-37.

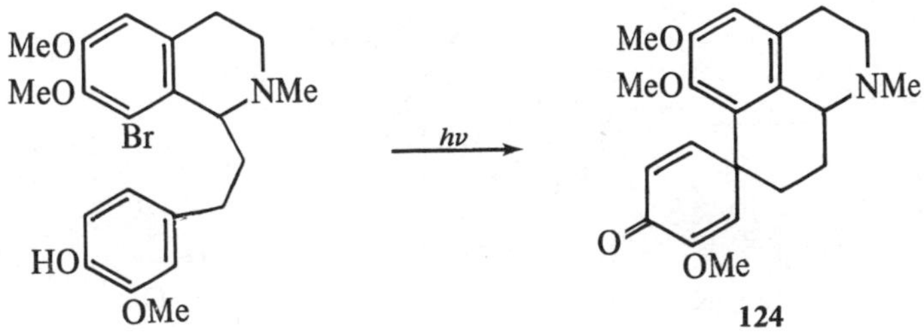

MeO, MeO, Br, NMe, HO, OMe; $h\nu$; MeO, MeO, NMe, O, OMe

123

124

kreysiginine-type compound **126** was synthesized by an irradiation of the diphenolic bromisoquinoline **125**, and in this reaction a phenolic oxygen added to the α,β-unsaturated ketone in a Michael reaction manner[333] (Chart 4-38).

Chart 4-38.

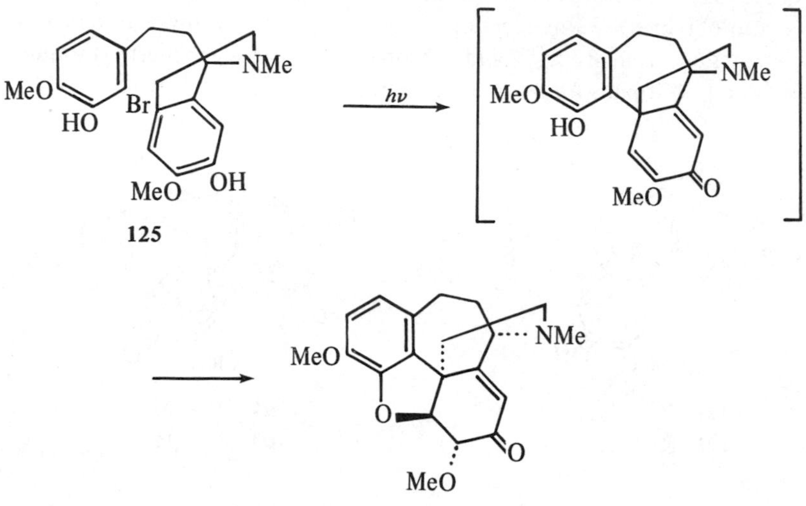

MeO HO Br NMe MeO OH **125** $\xrightarrow{h\nu}$ [MeO HO NMe MeO O] ⟶ MeO O NMe MeO O **126**

Chart 4-39.

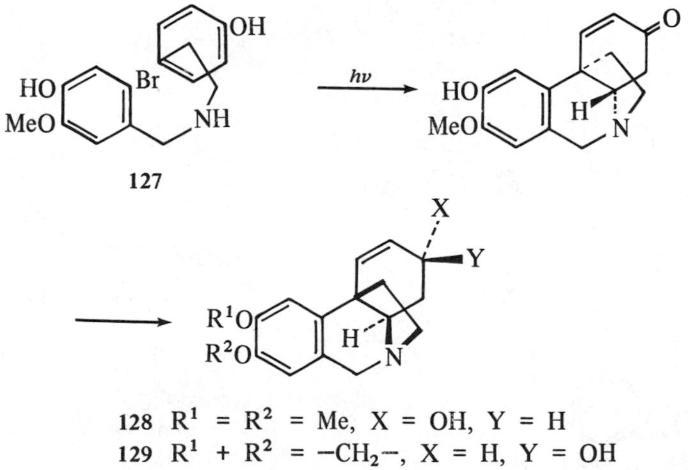

HO MeO Br OH NH **127** $\xrightarrow{h\nu}$ HO MeO H N O ⟶ R^1O R^2O H N X Y

128 $R^1 = R^2 = Me$, $X = OH$, $Y = H$

129 $R^1 + R^2 = -CH_2-$, $X = H$, $Y = OH$

Amaryllidaceae Alkaloids

The spirodienone synthesis by photolytic cyclization of the phenolic bromo compound was applied to the total synthesis of some *Amaryllidaceae* alkaloids. In this reaction, irradiation of the phenolic bromo-amine **127**, as usual, gave the enone, which has already been converted into maritidine **128**.[334] In a similar way, (±)-epicrinine **129** was synthesized by photolysis of the corresponding amine, followed by reduction.[335] Moreover, the phenolic bromo-amide **130** was cyclized to the narwedine-type compound **131**, which was a key intermediate to galanthamine **132**.[336] Mondon synthesized narciprimine **134** by a standard photochemical route from the bromo-amide **133**[337] (Charts 4-39, 4-40, and 4-41).

Chart 4-40.

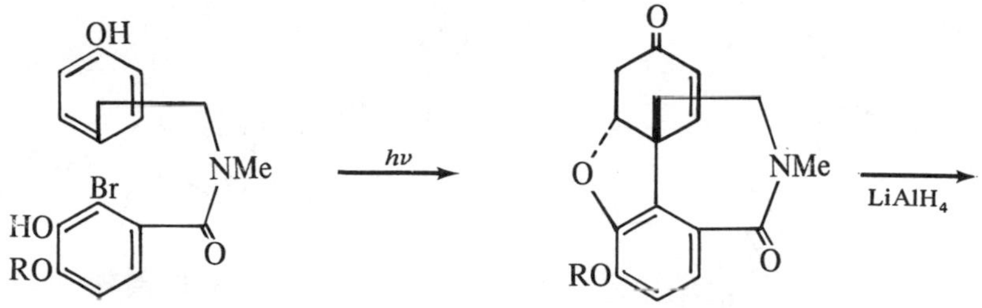

OH, NMe, Br, HO, RO, O — $h\nu$ → O, O, NMe, RO, O — $LiAlH_4$ →

130 R = Me and CH_2Ph **131** R = Me and CH_2Ph

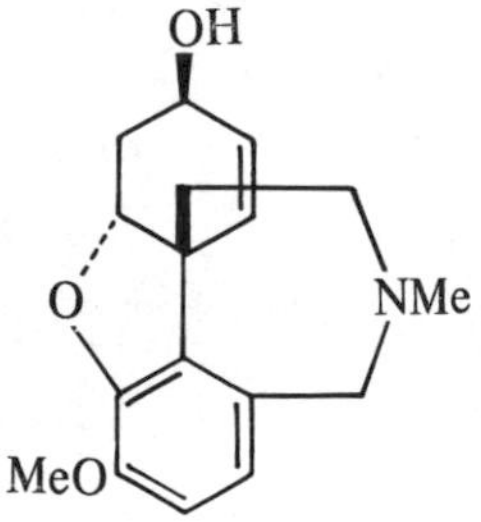

OH, O, NMe, MeO

132

E. Photolytic Cleavage

The irradiation of the 1-benzyl- or 1-phenethylisoquinolines with ultraviolet light in a deoxygenated state gives rise to oxidative cleavage to afford the

Chart 4-41.

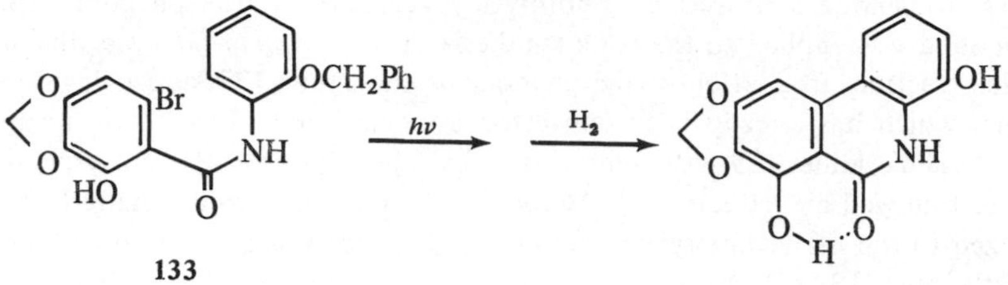

133

134

carbostyrils, and by this reaction, thalifoline **135** was synthesized from the diazotized isoquinoline **93**[314] and bromisoquinoline **104b**[320] (Chart 4-42). Similarly, *N*-methylcorydaline **136** was obtained from laudanosine.[338]

Chart 4-42.

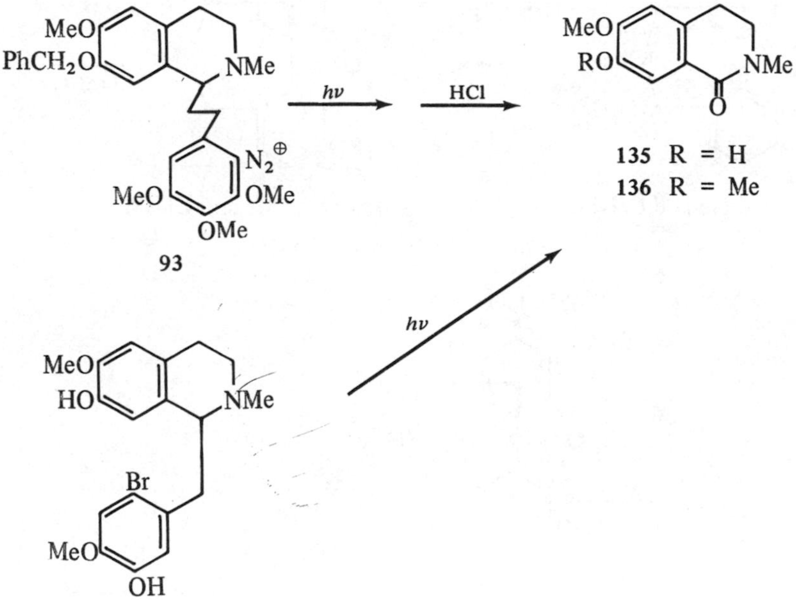

93

135 R = H
136 R = Me

104b

F. Azepine Synthesis

α-Haloketones form easily the methylene radicals by a homolytic fission between halogen and α-methylene carbon by ultraviolet irradiation. This reaction was applied for the synthesis of the azepine system **138** from α-haloamide **137**[339] and also for a synthesis of cephalotaxine **139**[340] (Chart 4-43).

Chart 4-43.

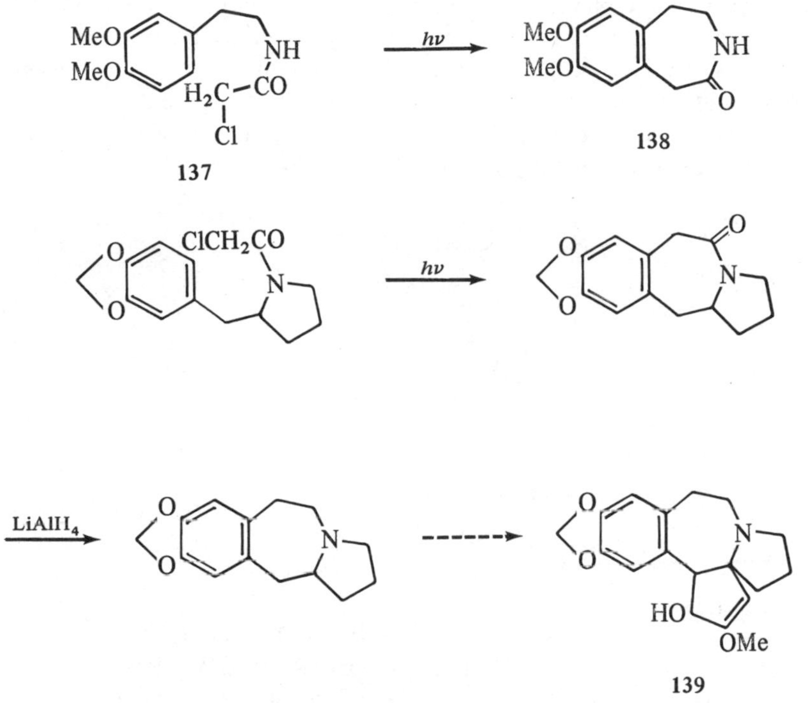

MeO MeO NH H_2C–CO Cl **137** —hν→ MeO MeO NH O **138**

$ClCH_2CO$ O O N —hν→ O O O N

—$LiAlH_4$→ O O N - - - - - → O O N HO OMe **139**

5. SPECIAL TOPICS

In this section, we describe the new synthetic methods developed in the latter half of the 1960's for the total syntheses of some groups in isoquinoline alkaloids. These methods are simple and elegant, as is phenol oxidation, and should be applied to the other fields in synthetic organic chemistry.

A. Pschorr Reaction

Morphinandienone-Type Alkaloids

Many morphinandienone-type alkaloids, represented by salutaridine, have recently been isolated, and the salutaridine, which has been synthesized by the phenol oxidation of reticuline by Barton and his co-workers, was converted chemically into thebaine. Synthesis of these morphinandienone-type alkaloids by phenol oxidation is appreciated, because it provides not only simple and elegant routes but also suggests the possible biogenetic pathway in nature. However, phenol oxidation also possesses the fatal defects that come from the nature of the reaction itself. As the coupling reaction proceeds without exception at the *ortho* or *para* position against the phenolic hydroxyl group, it could not be applied to the synthesis of the dienones, such as amurine, which has hydrogen at the *ortho* position and the methylenedioxy group at the *meta* and *para* coupling positions, and to flavinantine, flavinine and androcymbine which have *meta* hydroxyl groups at the coupling positions (Chart 5-1).

Chart 5-1.

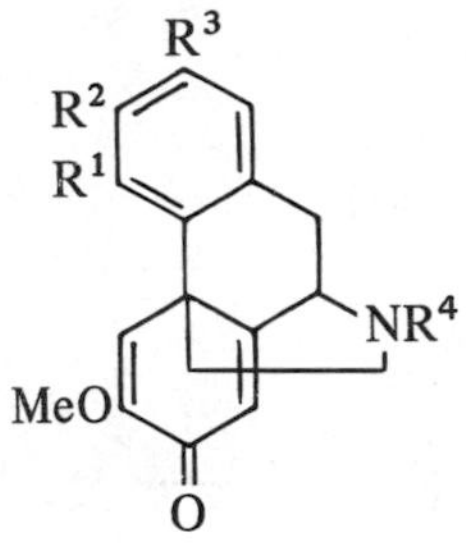

Salutaridine	R^1 = OH, R^2 = OMe, R^3 = H, R^4 = Me
Amurine	R^1 = H, R^2 + R^3 = OCH_2O, R^4 = Me
Flavinantine	R^1 = H, R^2 = OH, R^3 = OMe, R^4 = Me
Flavinine	R^1 = R^4 = H, R^2 = OH, R^3 = OMe
Pallidine	R^1 = H, R^2 = OMe, R^3 = OH, R^4 = Me
O-Methylflavinantine	R^1 = H, R^2 = R^3 = OMe, R^4 = Me
Norsinoacutine	R^1 = OH, R^2 = OMe, R^3 = R^4 = H

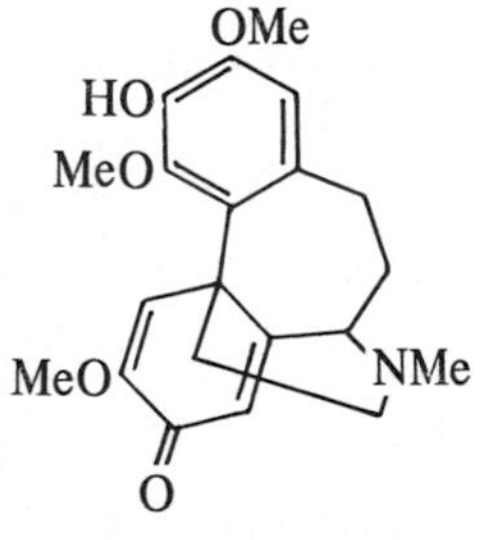

Androcymbine

Many chemists have searched for simple and general methods instead of using phenol oxidation; Kametani found the potentiality in Pschorr reaction.

Pschorr synthesized the phenanthrene derivatives **1** by a decomposition of diazotized α-aryl-*o*-aminocinnamic acids **2** in the presence of a copper catalyst at the end of the last century,[341] and he applied this reaction to the synthesis of glaucine **3** from 2′-aminolaudanosine **4** in 1904[342] (Chart 5-2). The reaction that

Chart 5-2.

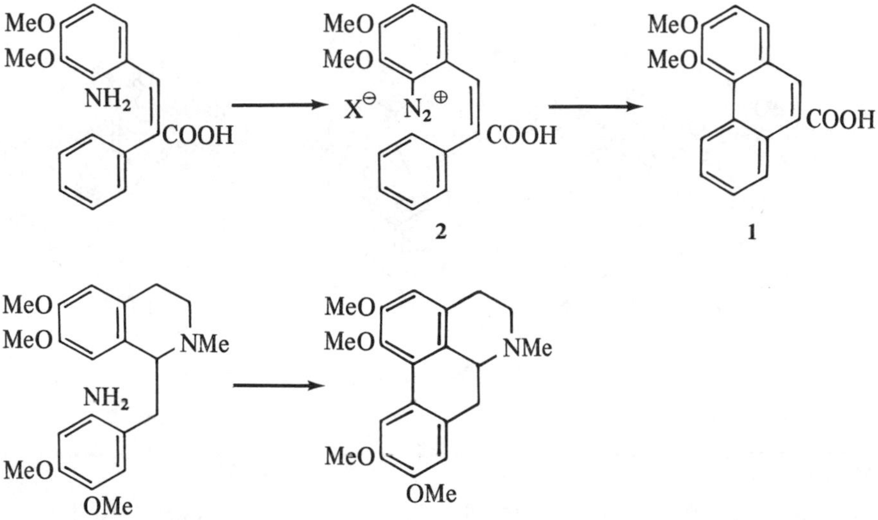

affords phenanthrene derivatives by a decomposition of the diazonium salts derived from *o*-aminostilbene on a metallic catalyst such as copper or zinc is called "Pschorr reaction," and this has been widely applied to a synthesis of aporphine alkaloids in isoquinoline alkaloids series.[343]

Pschorr aporphine synthesis is a nucleophilic reaction of the 8-position (arrow a) of the isoquinoline ring to the aromatic cation, which is derived from 1-(2-aminobenzyl)isoquinoline **4** via its diazonium salt **5**. However, it is possible that the nucleophilic attack from the carbon of **8a** (arrow b) and **4a** (arrow c) position to the aromatic cation would take place on consideration of the E-effect of alkoxy-groups on the isoquinoline ring, leading to the formation of dienone **6** or morphinandienone **7**, respectively. Moreover, an aromatic cation would react with the active methylene (arrow d) or basic nitrogen (arrow e) to form isopavine-type compound **8** or dibenzopyrrocoline **9**, respectively[344] (Chart 5-3).

Chart 5-3.

MeO MeO NMe NH$_2$ MeO OMe **4** $\xrightarrow{HNO_2}$ MeO MeO NMe $N_2^{\oplus}$ MeO OMe **5**

MeO MeO NMe MeO OMe **3**

MeO MeO NMe MeO OMe (a, b, c, d, e)

b → O MeO NMe MeO OMe ≡ OMe MeO NMe O OMe **6**

d → MeO MeO N Me OMe OMe **8**

e → MeO MeO $N^{\oplus}$ Me OMe OMe **9**

c → MeO Me—$O^{\oplus}$ NMe MeO OMe → MeO O NMe MeO OMe ≡ OMe MeO NMe MeO O **7**

Kametani diazotized 2′-aminolaudanosine **4** with a slight excess of sodium nitrite in *N*-sulfuric acid at 0 to 5°, and the resulting diazonium salt was decomposed thermally without a metallic catalyst at 70° to give the cyclohexadienone (A) together with laudanosine and glaucine **3**. Diazotization of the second aminoisoquinoline **10**, followed by thermal decomposition of the diazonium salt, gave the cyclohexadienone (B). If the structure of the cyclohexadienone (A) from the first aminoisoquinoline **4** were **6**, the product from the second one would be the compound **11**, which should be different from the product from the first one. However, both products (A and B) were proved to be identical by full spectroscopic data and mixed m.p. determination of their methiodide. These data proved the product to be the morphinandienone-type compound, (±)-*O*-methylflavinantine, and the infrared spectrum of the synthetic sample was superimposable on that of natural *O*-methylflavinantine, which was synthesized from flavinantine.[344,345] In this reaction, the compounds **6**, **8**, and **9** could not be obtained (Charts 5-3 and 5-3A).

Chart 5-3A.

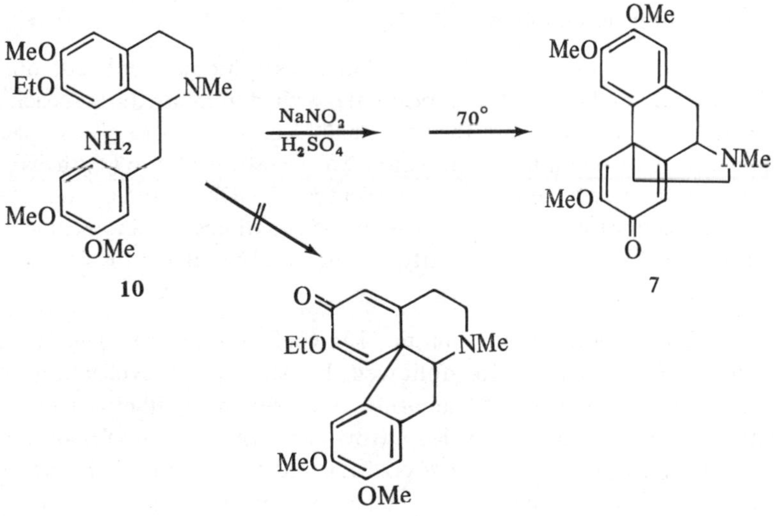

MeO EtO NMe NH$_2$ MeO OMe **10** $\xrightarrow[H_2SO_4]{NaNO_2}$ $\xrightarrow{70°}$ MeO OMe NMe MeO O **7**

O EtO NMe MeO OMe **11**

Thus a general synthetic method of the morphinandienone-type alkaloids has been established by a modified Pschorr reaction. Furthermore, this reaction displays a special character in the synthesis of the morphinandienones, which could not be synthesized by phenol oxidation; thus the coupling at the *ortho* position to a phenolic hydroxyl group is usually difficult in phenol oxidation; therefore, phenol oxidation of reticuline occurs at the preferable *para* position to the *ortho* position to afford pallidine described in Section 3. Moreover, the compounds, having a hydroxyl group at the position *meta* to the coupling site or having a methylenedioxy group at the *ortho* or *para* position to the site of coupling, could not be obtained by phenol oxidation. However, a coupling in Pschorr reaction occurs selectively at the position having an amino group, leading to synthesis of a morphinandienone, which has the substituents at the given positions from an appropriate starting material. Therefore, this reaction had been applied to the synthesis of salutaridine **12**, amurine **13**, and flavinantine **14**, which are described later.

Salutaridine, Thebaine, Codeine, and Morphine. Optical resolution of (±)-2′-aminobenzylisoquinoline was efficiently achieved by way of the salts with di-*p*-toluoyltartaric acid. The R-(−)-2′-aminobenzylisoquinoline **15** was diazotized with sodium nitrite and sulfuric acid, and the resulting diazonium salt was decomposed thermally without catalyst to give salutaridine **12**, which was reduced with sodium borohydride, followed by dehydration of the resulting epimeric salutaridinols **16** in the presence of *N*-hydrochloric acid, to furnish thebaine **17**. Thebaine has been converted into morphine **18** via codeine **19** as shown in Chart 5-4, so a formal total synthesis of morphine and its related alkaloids has been accomplished.[346]

Sinoacutine and Sinomenine. S-(+)-2′-Aminobenzylisoquinoline **20**, obtained by the optical resolution of the racemate with di-*p*-toluoyltartaric acid, was subjected to diazotization and then thermal decomposition to give sinoacutine **21**, an antipode of salutaridine **12**, which was transformed into (+)-thebaine **22**, an antipode of natural thebaine, via sinoacutinol **23**. Okabe and Goto succeeded in the conversion of natural thebaine into the enantiomer of sinomenine, so the synthesis of (+)-thebaine **22** constitutes the total synthesis of sinomenine[346] (Chart 5-5).

Amurine. The biogenesis of amurine **13** would involve *para-para* oxidative coupling of (−)-reticuline **25**, followed by subsequent cyclization of the *o*-methoxyphenol **26**. However, amurine could not be synthesized by phenol oxidation of the reticuline in the laboratory, since the structure for amurine has a methylenedioxy group at the *para* position and a hydrogen at the *ortho* position to the oxidative coupling site. Therefore, Pschorr reaction is a suitable method for the synthesis of amurine. Thus the 2′-aminobenzylisoquinolines **27** and **28** were independently diazotized as usual and decomposed thermally at 70° to give the same dienone, whose spectral data were superimposable on that natural amurine. Namely, the suggested structure **13** by Döpke was corroborated[347] (Chart 5-6).

Flavinantine. This alkaloid **14** is biosynthesized from reticuline **25** by the *para-para* oxidative coupling, but the reproduction of this reaction in the laboratory

Chart 5-4.

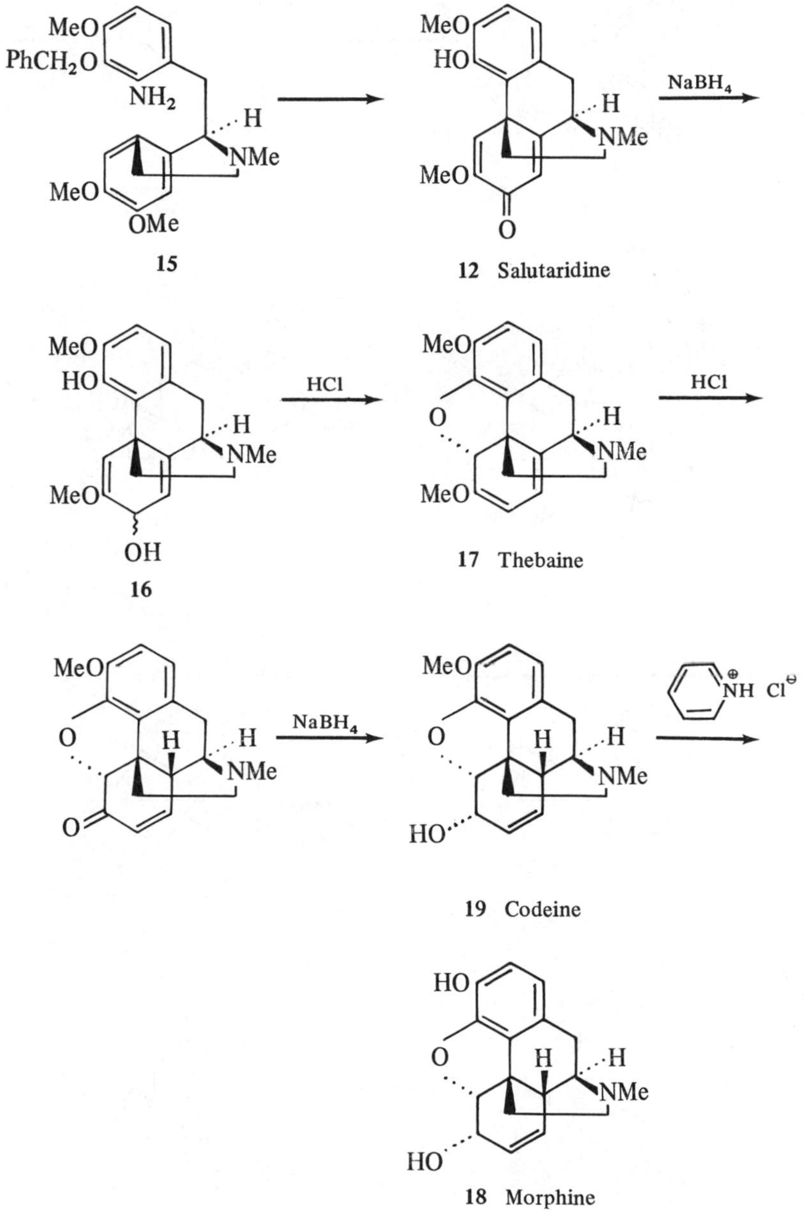

MeO
PhCH$_2$O
NH$_2$
H
NMe
MeO
OMe
15
MeO
HO
H
NMe
MeO
O
12 Salutaridine
NaBH$_4$
MeO
HO
H
NMe
MeO
OH
16
HCl
MeO
O
H
NMe
MeO
17 Thebaine
HCl
MeO
O
H
H
NMe
O
NaBH$_4$
MeO
O
H
H
NMe
HO
19 Codeine
NH Cl
HO
O
H
H
NMe
HO
18 Morphine

Chart 5-5.

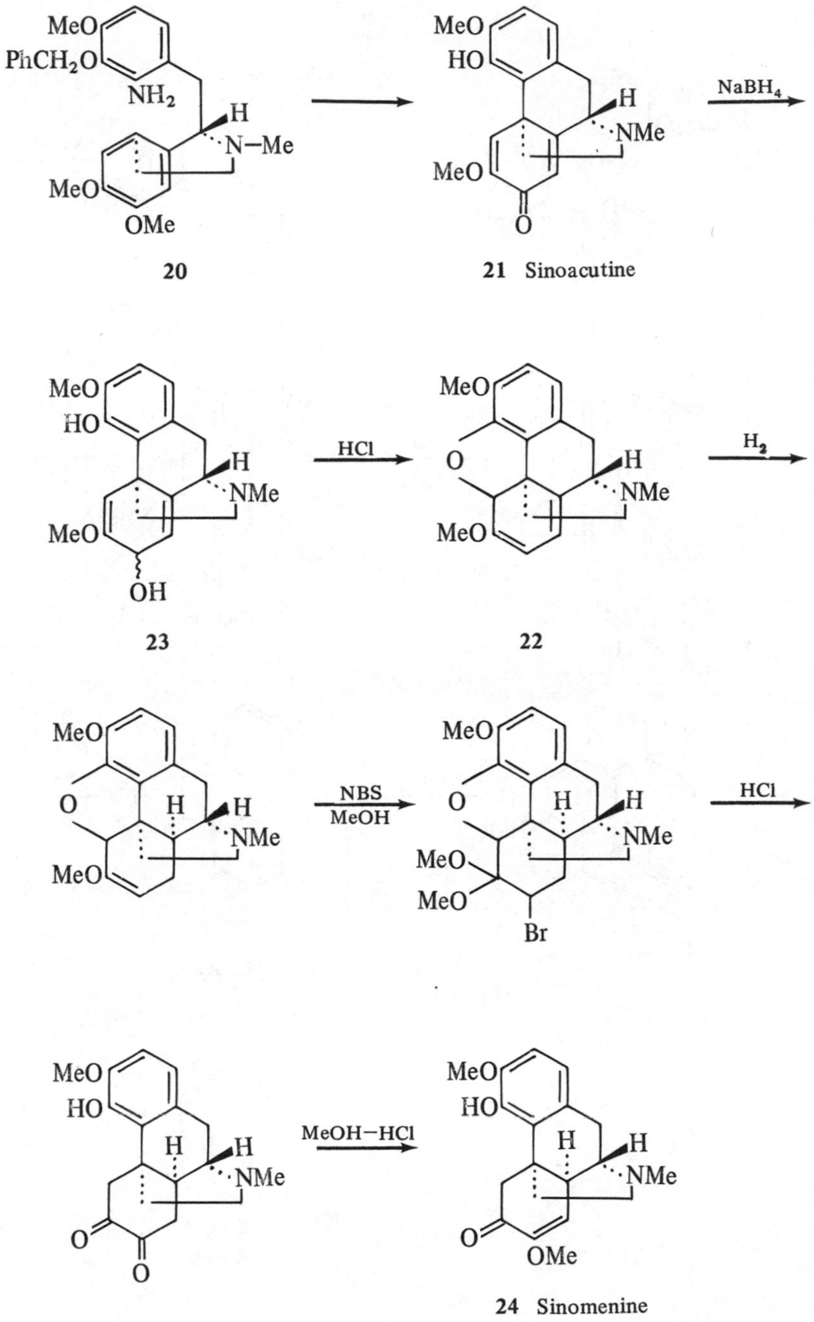

MeO
PhCH$_2$O
NH$_2$
H
N−Me
MeO
OMe

20

→

MeO
HO
H
NMe
MeO
O

21 Sinoacutine

$\xrightarrow{NaBH_4}$

MeO
HO
H
NMe
MeO
OH

23

$\xrightarrow{HCl}$

MeO
O
H
NMe
MeO

22

$\xrightarrow{H_2}$

MeO
O
H
H
NMe
MeO

$\xrightarrow[MeOH]{NBS}$

MeO
O
H
H
NMe
MeO
MeO
Br

$\xrightarrow{HCl}$

MeO
HO
H
H
NMe
O
O

$\xrightarrow{MeOH-HCl}$

MeO
HO
H
H
NMe
O
OMe

24 Sinomenine

Chart 5-6.

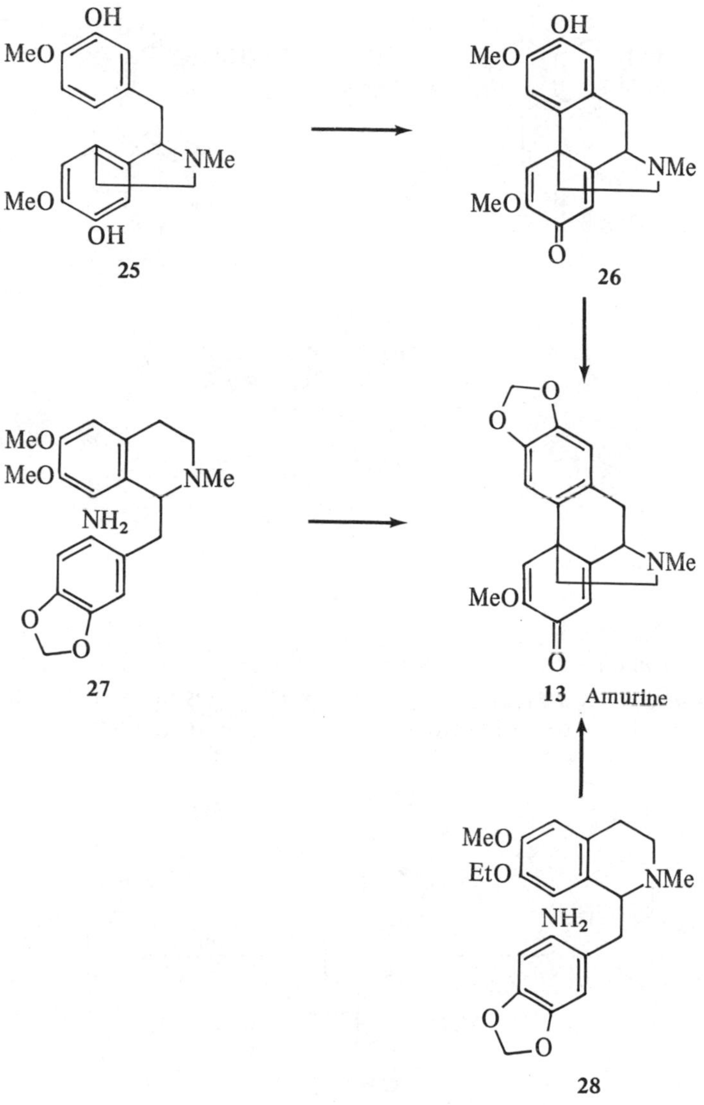

OH MeO MeO OH NMe 25 → 26 OH MeO MeO O NMe ↓ 13 Amurine

27 MeO MeO NH$_2$ NMe → 13; 28 MeO EtO NH$_2$ NMe ↑ 13 Amurine

would be difficult due to the fatal defects of phenol oxidation itself, so Pschorr reaction was applied to the synthesis of flavinantine by Kametani. Thus the 2′-aminobenzylisoquinoline was subjected to Pschorr reaction, and the resulting morphinandienone was debenzylated with 48% hydrobromic acid to afford flavinantine **14**[348] (Chart 5-7).

Chart 5-7.

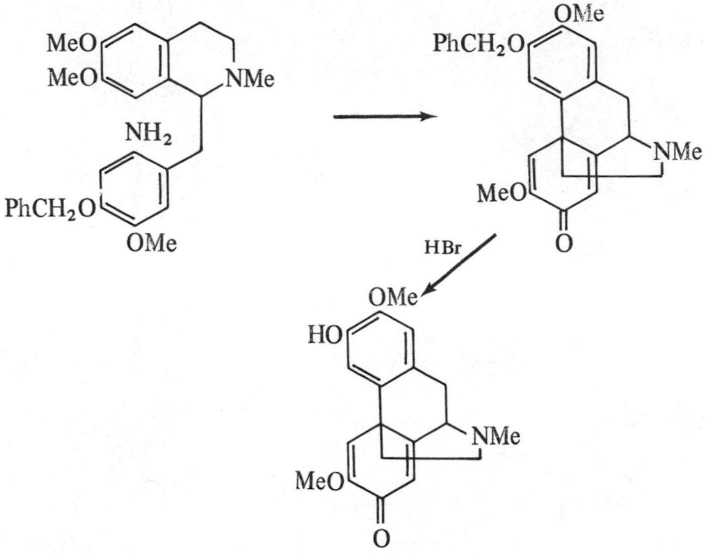

HBr

14 Flavinantine

Pallidine. Pallidine **26** was synthesized from 2′-aminobenzylisoquinoline **29** by the same way for the synthesis of flavinantine.[349] This method is better than a synthesis by phenol oxidation from the point of yield (Chart 5-8).

Chart 5-8.

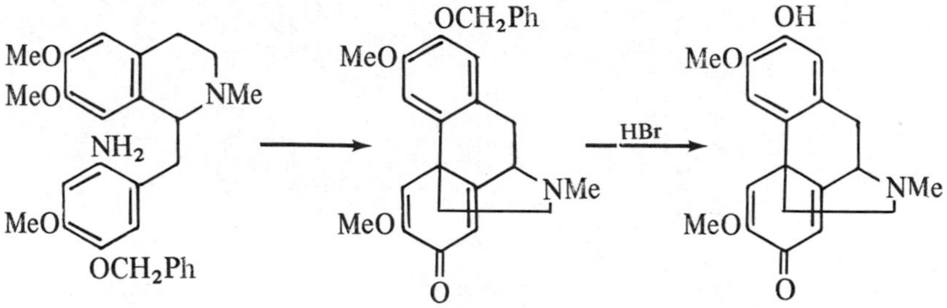

HBr

29

26 Pallidine

Protostephanine. The morphinandienone **30**, which was a key intermediate to protostephanine **31** in its biogenesis and chemical synthesis, was synthesized from the 2′-aminobenzylisoquinoline **32** by Pschorr reaction. The morphinandienone was converted into protostephanine as described in Section 3[350] (Chart 5-9).

Chart 5-9.

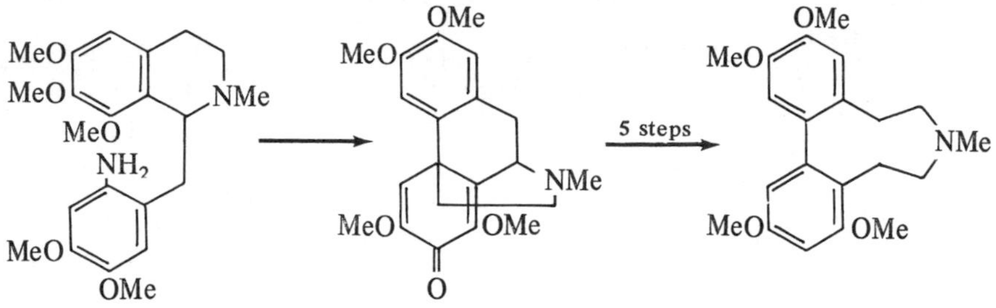

MeO MeO NMe MeO NH_2 MeO OMe → OMe MeO NMe MeO OMe O —5 steps→ OMe MeO NMe MeO OMe

32 **30** **31** Protostephanine

Proaporphine-Type Alkaloids

Pronuciferine. Ishiwata and co-workers[351-354] reported a new synthetic route to proaporphine alkaloids by a modification of the usual Pschorr reaction; thus 8-aminoisoquinoline derivative **33** was diazotized with a slight excess of sodium nitrite in 5% sulfuric acid at 5°, and the reduction of the resulting diazonium salt with hypophosphorous acid afforded pronuciferine **34** together with 1,2,10-trimethoxyaporphine **35** and a deamination product.[351] Moreover, a decomposition of the diazonium salt with an excess of sodium acetate at room temperature yielded 15% pronuciferine **34** in addition to aporphine **35**.[353] In a similar way, decomposition of the diazonium salt of phenolic 8-aminoisoquinoline **36** with 10% sodium hydroxide provided pronuciferine **34** in a 20% yield and 10-hydroxy-1,2-dimethoxyaporphine **37**.[353] They suggested that the formation of proaporphine by the reduction of a diazonium salt should proceed via radical intermediates as shown in Chart 5-10.[353]

Orientalinone. *O*-Methylorientalinone **38** and *O*-methylisoorientalinone, a *spiro* isomer of **38**, were synthesized by a Pschorr reaction of 8-aminoisoquinoline **39**.[354] Thus a general synthesis of a proaporphine system has been established (Chart 5-11).

Homolinearisine. Proaporphine-type alkaloids are synthesized by phenol oxidation: its fatal defect is that coupling occurs only at the position *para* or *ortho* to the hydroxyl group. Therefore, homolinearisine **40**, having a hydroxyl group at the position *meta* to the coupling site, cannot be synthesized by phenol oxidation. However, a starting material in Pschorr reaction is 8-aminoisoquinoline, having the appropriate substituents at the given positions; therefore, this reaction would be suitable for the synthesis of the compound that could not be obtained by phenol oxidation. Thus Ishiwata and Itakura[352] applied Pschorr reaction for the total synthesis of homolinearisine **40** as follows: Phenolic 8-aminoisoquinoline **41** was diazotized, then decomposition of the diazonium

Chart 5-10.

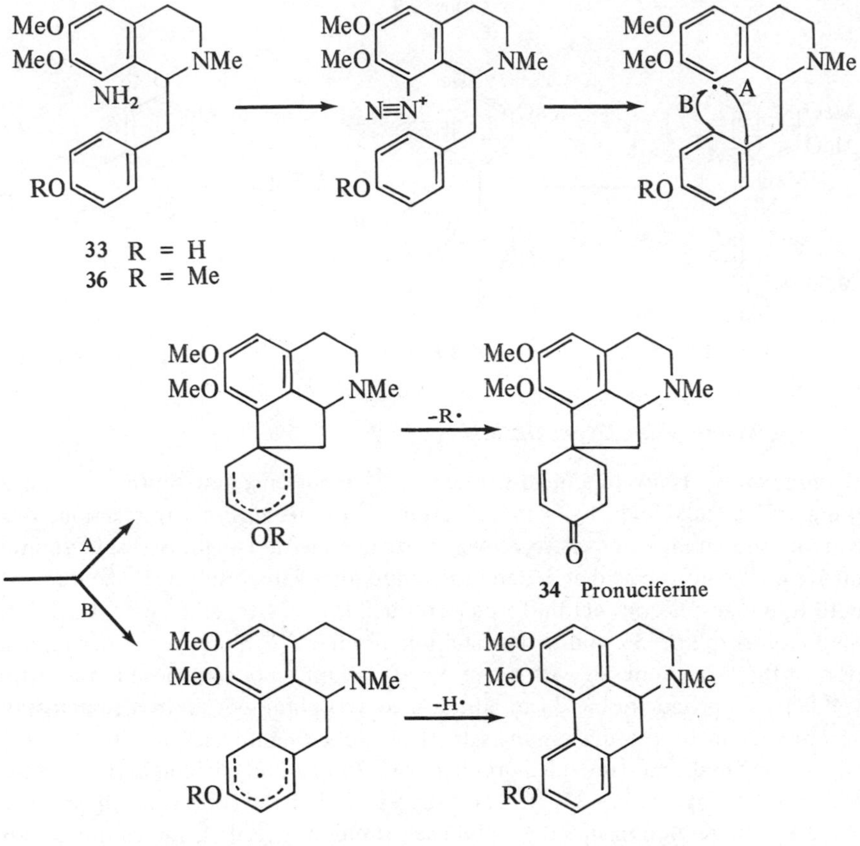

MeO
MeO
NMe
NH_2
RO
MeO
MeO
NMe
$N{\equiv}N^+$
RO
MeO
MeO
NMe
B
A
RO

33 R = H
36 R = Me

MeO
MeO
NMe
–R•
OR
MeO
MeO
NMe
O
A
B
34 Pronuciferine
MeO
MeO
NMe
–H•
RO
MeO
MeO
NMe
RO

35 R = Me
37 R = H

Chart 5-11.

MeO
MeO
NMe
NH_2
MeO
OMe
MeO
MeO
NMe
OMe
O

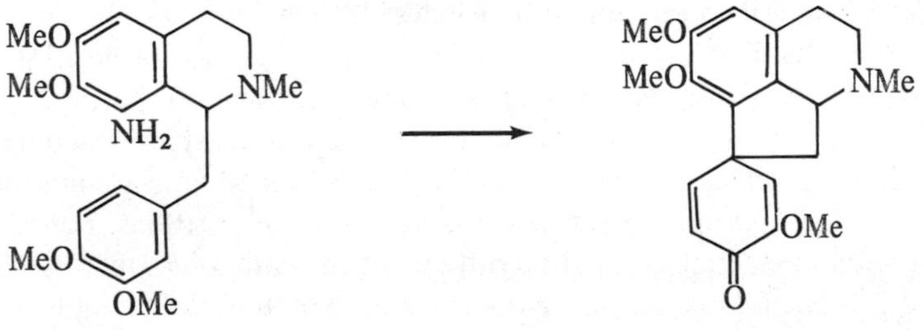

39

38

Chart 5-12.

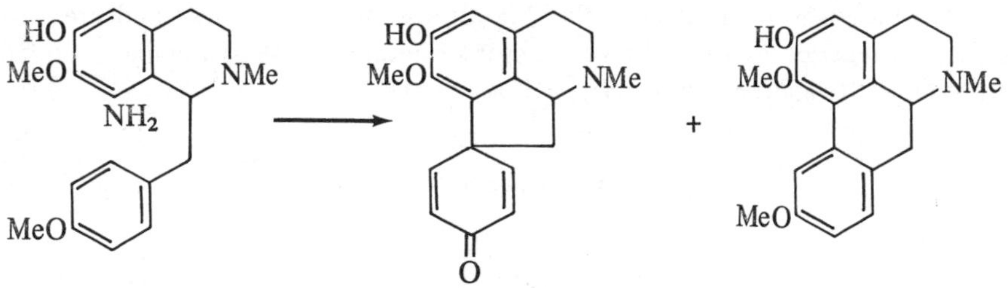

41 **40** Homolinearisine **42**

salt in the presence of sodium acetate furnished homolinearisine **40** in a 10% yield together with a corresponding aporphine **42**. However, the application of this method to the synthesis of homoproaporphine series failed[355] (Chart 5-12).

Androcymbine

The homomorphinandienone system **43** was synthesized by the Pschorr reaction from the 2′-aminophenethylisoquinoline **44** in the same method with the synthesis of morphinandienone. The structural assignment was carried out by

Chart 5-13.

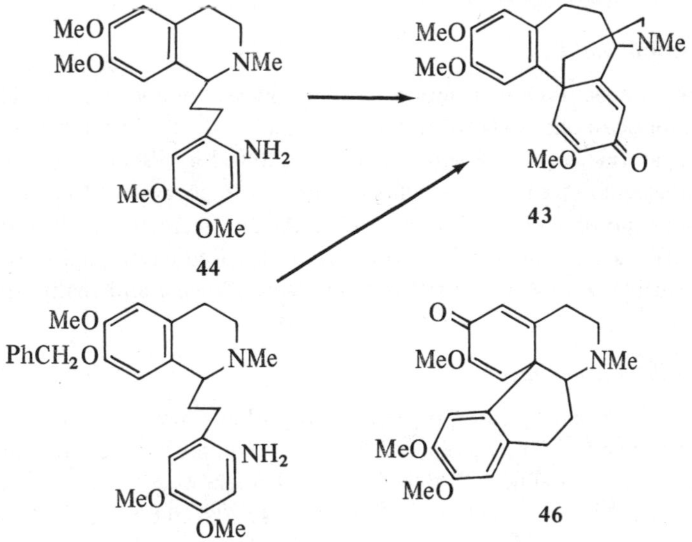

44 **43** **45** **46**

the spectral speculation and by an alternative synthesis from the second 2′-aminoisoquinoline **45**. This ruled out the other coupling product **46** [356] (Chart 5-13). On the basis of these facts, the 2′-aminoisoquinoline **47** was subjected to the Pschorr reaction as usual to give the spiro-isoquinoline **48** as a main product, but not *O*-methylandrocymbine **49**[357] (Chart 5-14).

Chart 5-14.

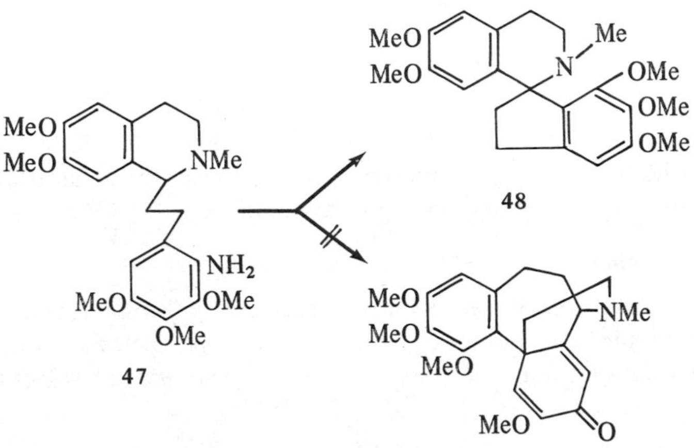

MeO MeO NMe NH$_2$ MeO OMe OMe **47**

MeO MeO N Me OMe OMe OMe **48**

MeO MeO NMe MeO MeO O **49**

Indenoisoquinoline

Cava developed a novel Pschorr reaction for the synthesis of aporphine alkaloids from the aminoisoquinolines via the indenoisoquinolines.[357a] Thus diazotization of 2′-aminopapaverine in a 46% sulfuric acid, followed by Pschorr cyclization gave the indenoisoquinoline in a 30% yield, in addition to papaverine (4.3% of the yield) and 1-oxoaporphine (2.4% of the yield). Acidic treatment of the indeno-isoquinoline afforded 1-oxoaporphine quantitatively. Reduction of 1-oxoaporphine, followed by *O*- or *N*-methylation, gave *N*-norglaucine and thaliporphine (Chart 5-14A).

Aporphine Alkaloids

The classic Pschorr reaction was applied to the synthesis of aporphine alkaloids such as atheroline,[357b] bracteoline,[357c] cassameridine,[357d] corunnine,[357e] domesticine,[357f] hernandaline,[357g] imenine,[357h] lanuginosine,[357i] laureline,[357j] *N*-methylovigerine,[357k] michelalbine,[357l] nantenine,[357m] nuciferine,[357n] predicentrine,[357o] oconovine,[357p] thalicsimidine,[357o, 357q] and xylopine[357i] (Chart 5-14B). This reaction was also used for the synthesis of the phenylisoquinoline

Chart 5-14A.

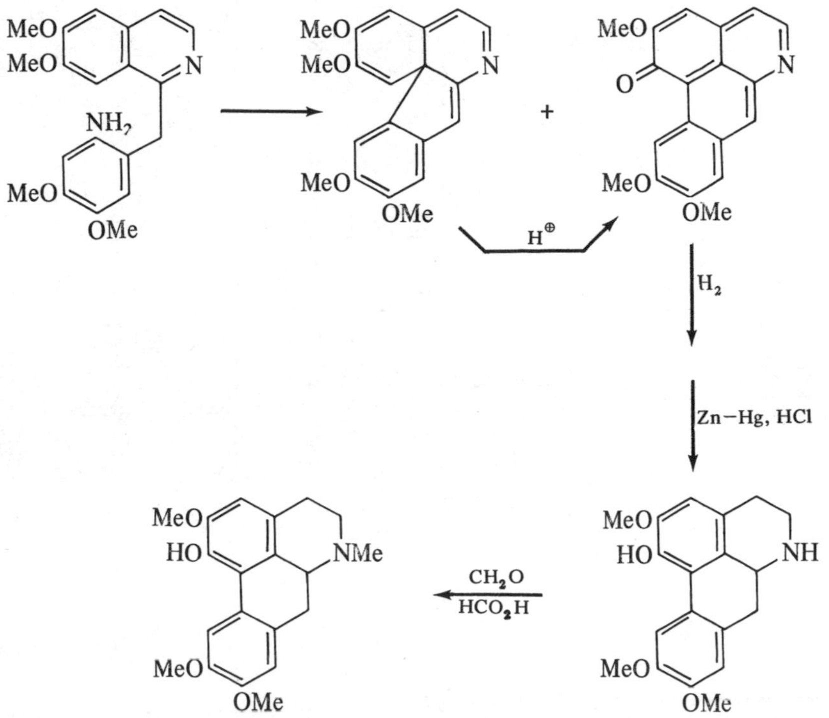

Thaliporpine

alkaloids, rufescine and imeluteine[357r] (Chart 5-14C). Dalton and Abraham reported the steric effects in the Pschorr cyclization at the *N*-position,[357s] and Kupchan described an improved Pschorr reaction of general synthetic utility[357t] (Chart 5-14D). Ishiwata described a new synthesis of aporphine alkaloids from 8-amino-1-benzylisoquinolines by Pschorr reaction[354, 357u] (Chart 5-14E). Moreover, the Pschorr reaction was used for the synthesis of the phenanthridine[357v] and *Amaryllidaceae* alkaloids[357w] (Chart 5-14F).

B. Benzyne Reaction

The benzyne reaction,[358] developed by Wittig, has played an important role in synthetic organic chemistry. In this field, an intramolecular addition reaction of the hetero atom to benzyne system has provided a novel synthetic method of the heterocyclic compounds (Chart 5-15). This intramolecular addition reaction was applied to a synthesis of tetrahydrodibenzo[b,g]pyrrocoline alkaloids by two groups.

Chart 5-14B.

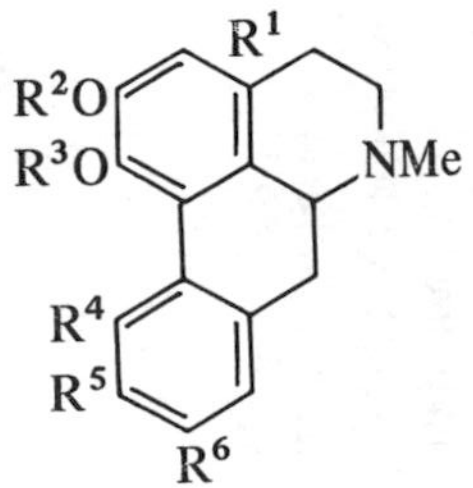

Bracteoline	$R^1 = R^3 = R^4 = H, R^2 = Me$ $R^5 = OH, R^6 = OMe$
Domesticine	$R^1 = R^3 = R^4 = H, R^2 = Me$ $R^5 + R^6 = OCH_2O$
Hernandaline	$R^1 = R^4 = H, R^2 = R^3 = Me, R^5 = OMe$ $R^6 = OC_6H_2(OMe)_2(CHO)$-4,5,2
Laureline	$R^1 = R^4 = R^6 = H, R^2 + R^3 = CH_2$ $R^5 = OMe$
N-Methylovigerine	$R^1 = R^6 = H, R^2 + R^3 = CH_2$ $R^4 + R^5 = OCH_2O$
Nantenine	$R^1 = R^4 = H, R^2 = R^3 = Me$ $R^5 + R^6 = OCH_2O$
Nuciferine	$R^1 = R^4 = R^5 = R^6 = H, R^2 = R^3 = Me$
Predicentrine	$R^1 = R^2 = R^4 = H, R^3 = Me,$ $R^5 = R^6 = OMe$
Oconovine	$R^1 = R^5 = OMe, R^2 = R^3 = OMe$ $R^4 = OH, R^6 = H$
Thalicsimidine	$R^1 = R^5 = R^6 = OMe$ $R^2 = R^3 = Me, R^4 = H$

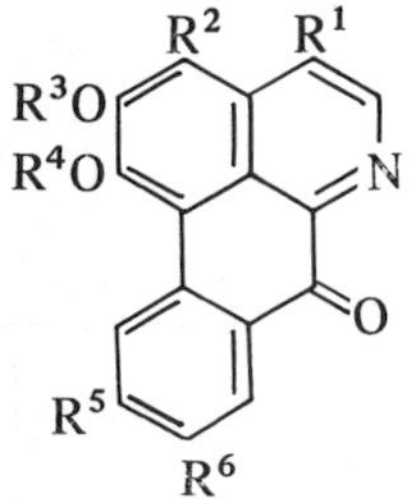

Atheroline	$R^1 = R^2 = H, R^3 = R^4 = Me$ $R^5 = OMe, R^6 = OH$
Cassameridine	$R^1 = R^2 = H, R^3 + R^4 = CH_2$ $R^5 + R^6 = OCH_2O$
Corunnine	$R^5 = R^6 = OMe$, methiodide $R^1 = R^2 = R^4 = H, R^3 = OMe$
Imenine	$R^1 = R^2 = OMe, R^3 = R^4 = Me$ $R^5 = R^6 = H$
Lanuginosine	$R^1 = R^2 = R^5 = H, R^3 + R^4 = CH_2$ $R^6 = OMe$

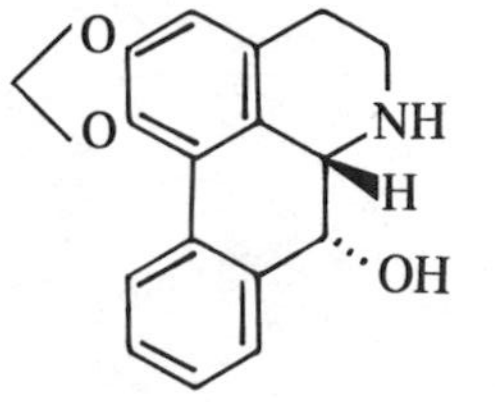

Michelalbine

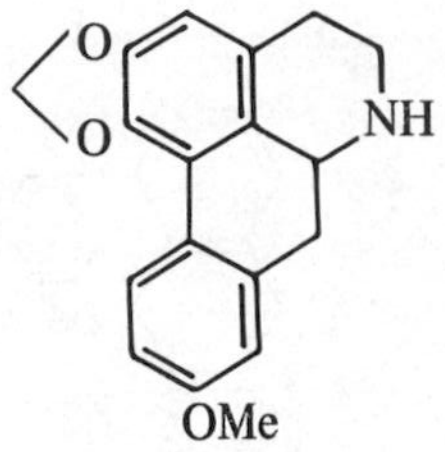

Xylopine

Chart 5-14C.

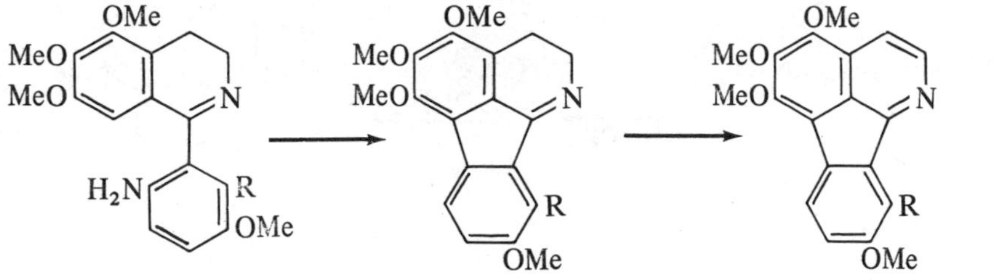

OMe MeO MeO N H$_2$N R OMe → OMe MeO MeO N R OMe → OMe MeO MeO N R OMe

Rufescine R = H
Imeluteine R = OMe

Chart 5-14D.

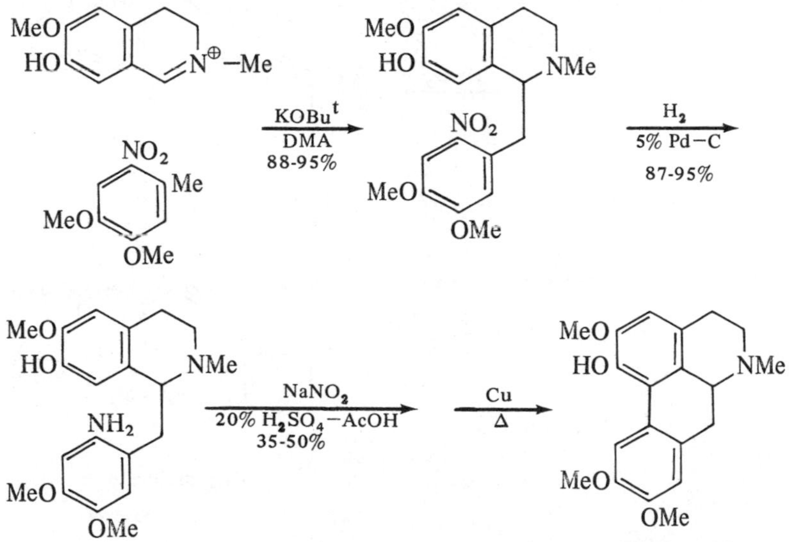

MeO HO N$^{\oplus}$–Me + NO$_2$ Me MeO OMe —KOBu^t, DMA, 88-95%→ MeO HO NMe NO$_2$ MeO OMe —H$_2$, 5% Pd–C, 87-95%→

MeO HO NMe NH$_2$ MeO OMe —NaNO$_2$, 20% H$_2$SO$_4$–AcOH, 35-50%→ —Cu, Δ→ MeO HO NMe MeO OMe

Thaliporphine

2′-Chloro-1,2,3,4-tetrahydroisoquinoline **50** was subjected to the benzyne reaction with sodium amide in liquid ammonia and tetrahydrofuran to provide a tetrahydrodibenzopyrrocoline **52**,[359] which was also obtained from the 2′-bromo analog **51** of **50** by the treatment of potassium amide in liquid ammonia.[360] The methiodide **53** was treated with hydrochloric acid to give cryptowoline **54**.[359] In the same way, 2′-bromoisoquinoline **55** afforded cryptaustoline **56**.[359] The proof that the reaction proceeded via a benzyne intermediate in the formation of the tetrahydrodibenzopyrrocolines from the bromoisoquinolines was provided

Chart 5-14E.

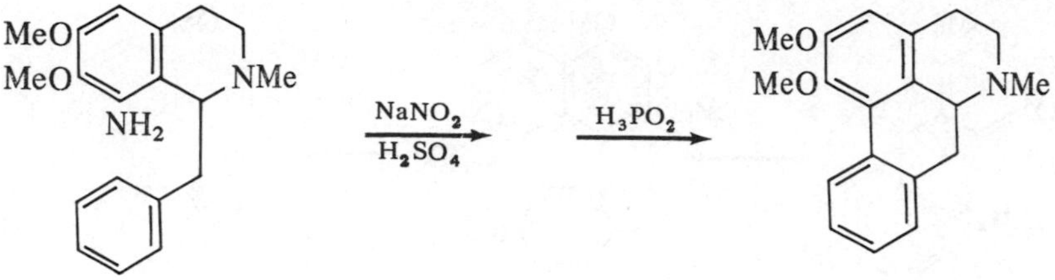

MeO
MeO
NMe
NH_2
$NaNO_2$
H_2SO_4
H_3PO_2
MeO
MeO
NMe

Nuciferine

Chart 5-14F.

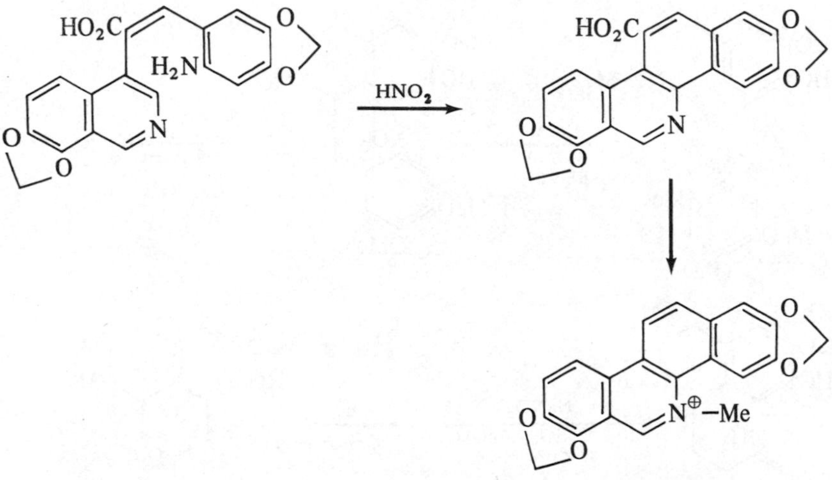

HO_2C
H_2N
O
O
N
O
O
HNO_2
HO_2C
O
O
N
O
O

O
O
$N^{\oplus}$—Me
O
O

Sanguinarine

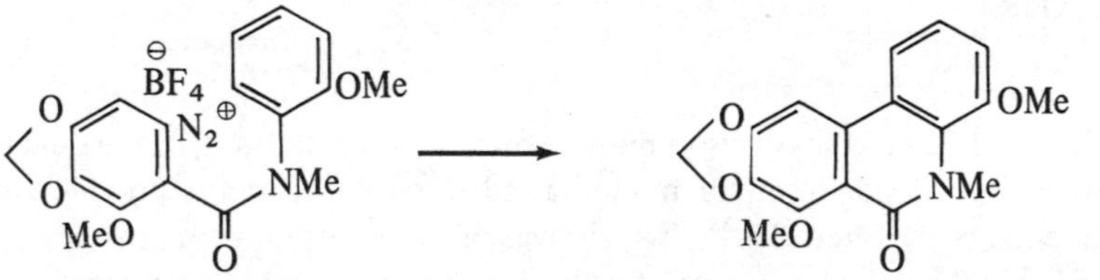

$\overset{\ominus}{BF_4}$
$N_2^{\oplus}$
OMe
O
O
NMe
MeO
O

O
O
OMe
NMe
MeO
O

Permethylnarciprimine

Chart 5-15.

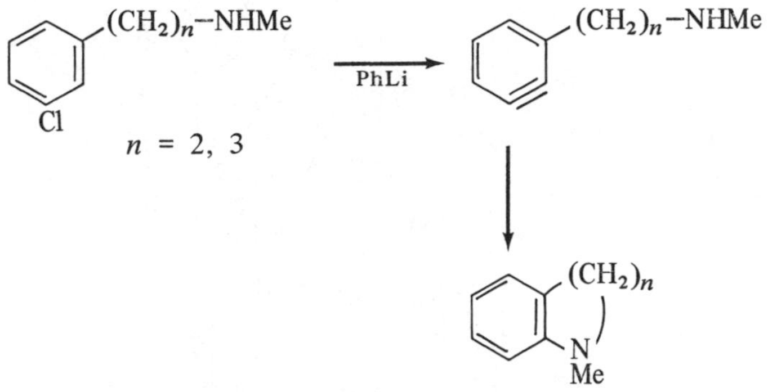

$(CH_2)_n$–NHMe PhLi $(CH_2)_n$–NHMe
Cl
$n = 2, 3$
$(CH_2)_n$
N
Me

by the *cine* substitution observed in the conversion of the 3′-bromoisoquinoline **57** into the dibenzopyrrocoline **58** with sodium amide in liquid ammonia and tetrahydrofuran[359] (Chart 5-16). The second application of the benzyne reaction for the isoquinoline alkaloids was carried out for the total synthesis of the aporphine[361,362] and morphinandienone alkaloids[362] by C–C coupling. Thus the phenolic 6′-bromobenzylisoquinoline **59** was treated with sodium amide in liquid ammonia to afford domesticine **60** and amurine **13** (Chart 5-17). Moreover, the phenolic 5′-bromobenzylisoquinoline **61** also gave the aporphine **62**, the fact of which proved this reaction to be a benzyne reaction.[363] Surprisingly, **61** also yielded the dibenzopyrrocolinium salt **63** in addition to aporphine **62**, and a one-step synthesis of cryptaustoline **56** was achieved by benzyne reaction of **64**.[364]

Recently, Kessar, Randhawa, and Gandhi[364a] and Kametani et al.[364b] independently investigated this type of benzyne reaction and synthesized many alkaloids of the aporphine, morphinandienone, and dibenzopyrrocoline groups.

On the other hand, an aromatic primary amine, which could be converted into an aporphine alkaloid by a Pschorr reaction, was synthesized from a halobenzylisoquinoline by treatment with potassium amide in liquid ammonia via a benzyne intermediate[357j] (Chart 5-17A). Moreover, a benzophenanthridine ring system was synthesized by treatment of an *o*-chlorobenzalamine with potassium amide in liquid ammonia.[364c] This reaction probably proceeds through a benzyne intermediate (Chart 5-17B). Recently, cephalotaxine was also synthesized by reaction of the benzyne system with active methylene[364d] (Chart 5-17C). Similarly, α-lycorane was obtained from the keto-bromide through a benzyne intermediate.[364c] (Chart 5-17D).

Chart 5-16.

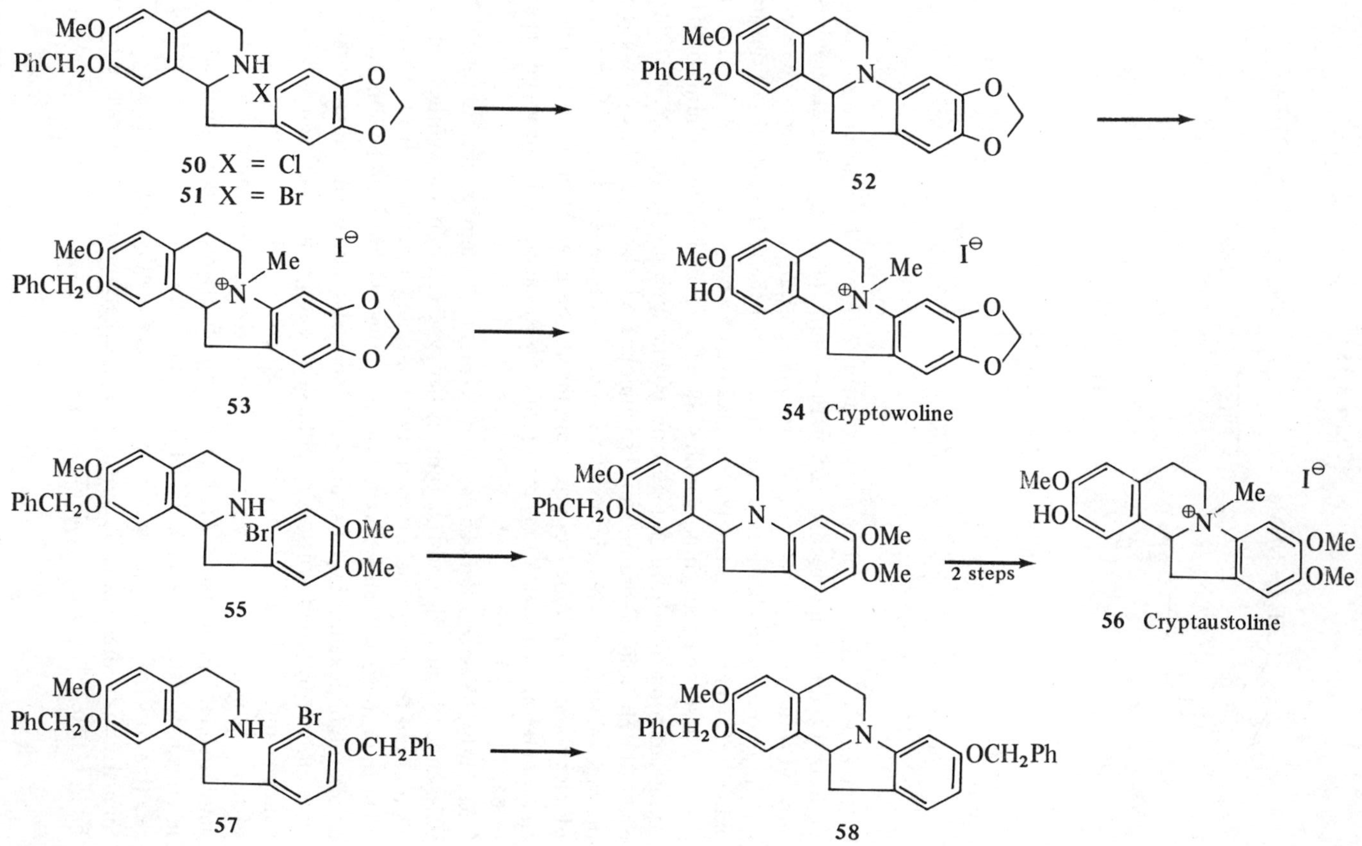

MeO
$PhCH_2O$
NH
X
50 X = Cl
51 X = Br

→

MeO
$PhCH_2O$
N
52

→

MeO
$PhCH_2O$
$\oplus$N
Me
$I^{\ominus}$
53

→

MeO
HO
$\oplus$N
Me
$I^{\ominus}$
54 Cryptowoline

MeO
$PhCH_2O$
NH
Br
OMe
OMe
55

→

MeO
$PhCH_2O$
N
OMe
OMe

2 steps →

MeO
HO
$\oplus$N
Me
$I^{\ominus}$
OMe
OMe
56 Cryptaustoline

MeO
$PhCH_2O$
NH
Br
OCH_2Ph
57

→

MeO
$PhCH_2O$
N
OCH_2Ph
58

Chart 5-17.

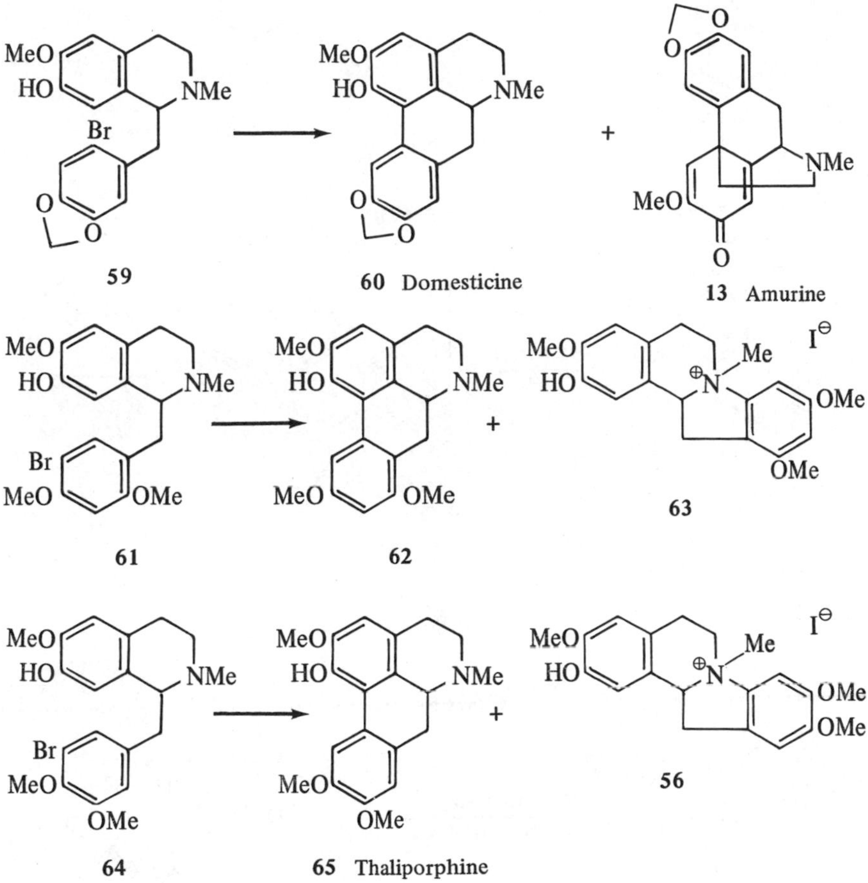

MeO HO NMe Br O O

59

MeO HO NMe O O

60 Domesticine

+

O O NMe MeO O

13 Amurine

MeO HO NMe Br MeO OMe

61

MeO HO NMe MeO OMe

62

+

MeO HO Me ⊕N OMe OMe I⊖

63

MeO HO NMe Br MeO OMe

64

MeO HO NMe MeO OMe

65 Thaliporphine

+

MeO HO Me ⊕N OMe OMe I⊖

56

Chart 5-17A.

O O NMe Br MeO

$\xrightarrow[NH_3]{KNH_2}$

[O O NMe MeO]

⟶

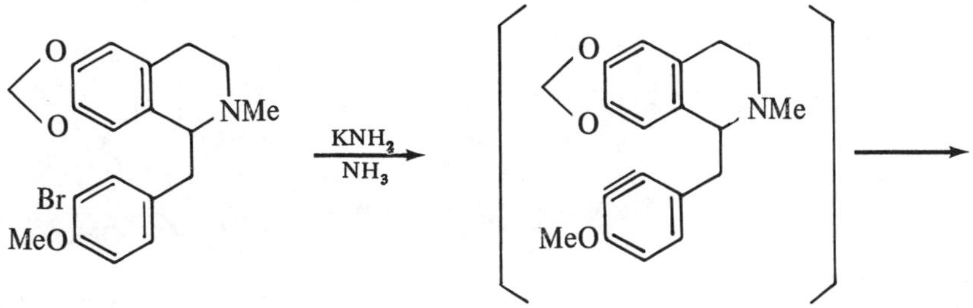

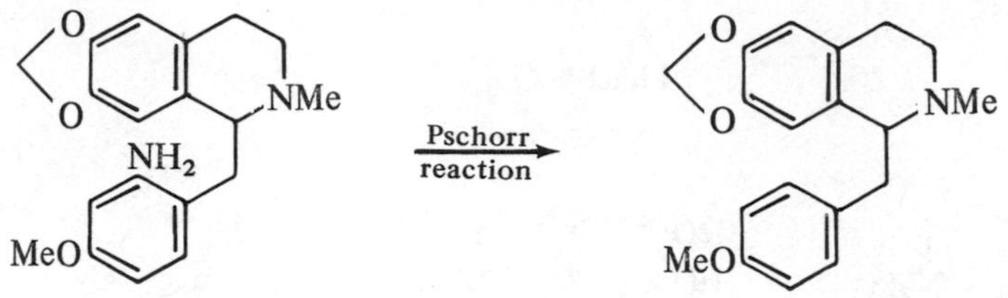

Pschorr reaction

Laureline

Chart 5-17B.

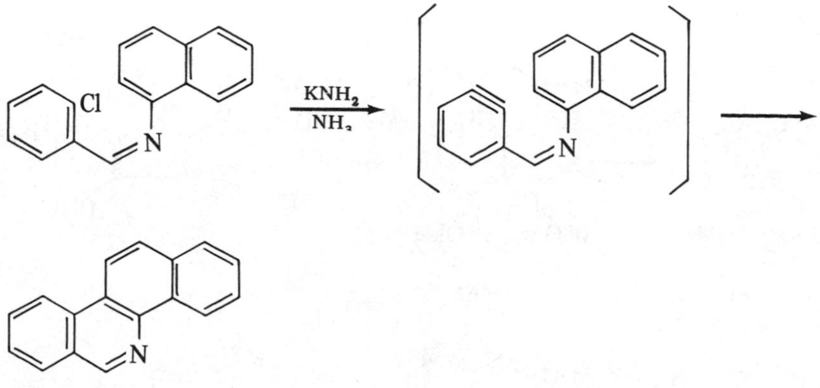

KNH_2 / NH_3

Chart 5-17C.

$KOCPh_3$ / $MeO(CH_2)_2OMe$

$^iBu_2AlH$

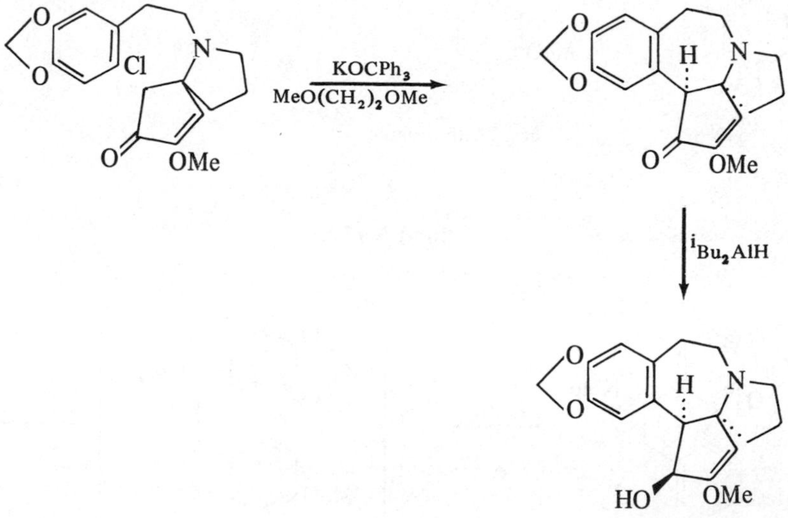

Cephalotaxine

Chart 5-17D.

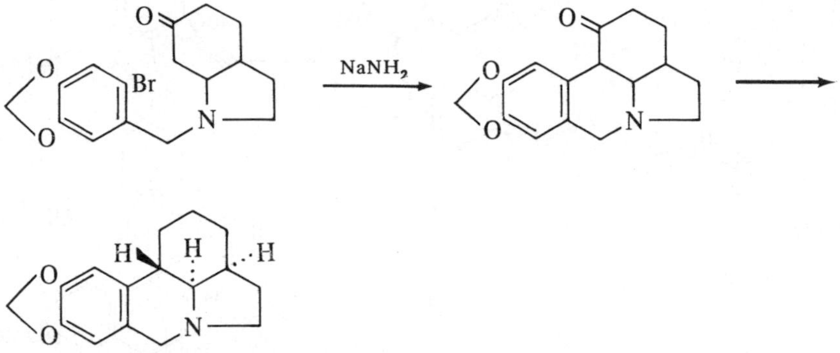

NaNH$_2$

C. Ullmann Reaction

The Ullmann reaction affords biphenyls from two kinds of the aromatic halides, or biphenyl ethers from aryl halides and phenols in the presence of copper. The biphenyl system is found in the aporphine, homoaporphine, benzophenanthridine, protostephanine, and tylophorine alkaloids; the biphenyl ether groups are present in the bisbenzylisoquinoline, bisphenethylisoquinoline, and cularine alkaloids. In the classic synthesis of these alkaloids, the biphenyl system is

Chart 5-18.

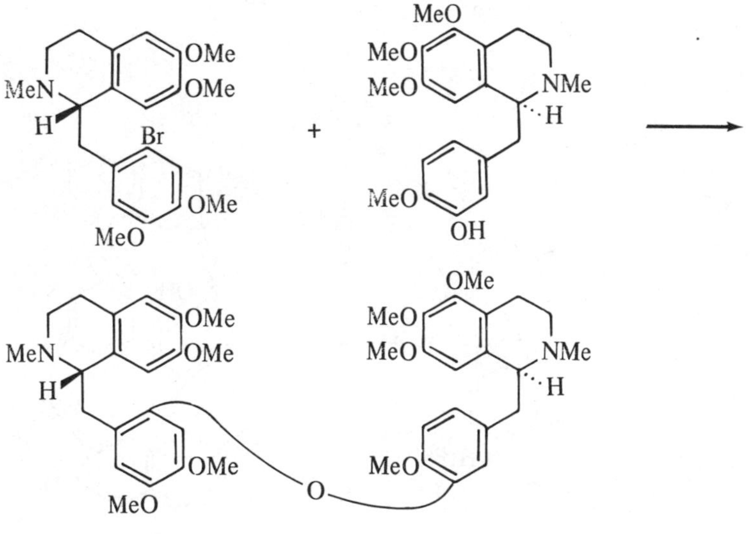

66

Chart 5-19.

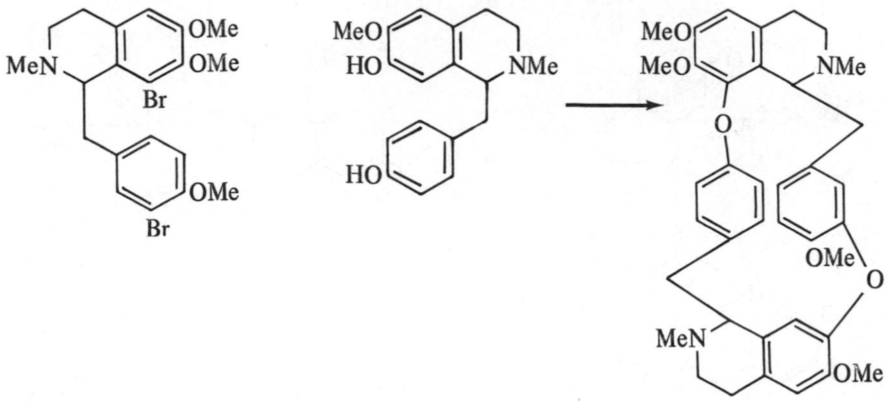

67

Chart 5-20.

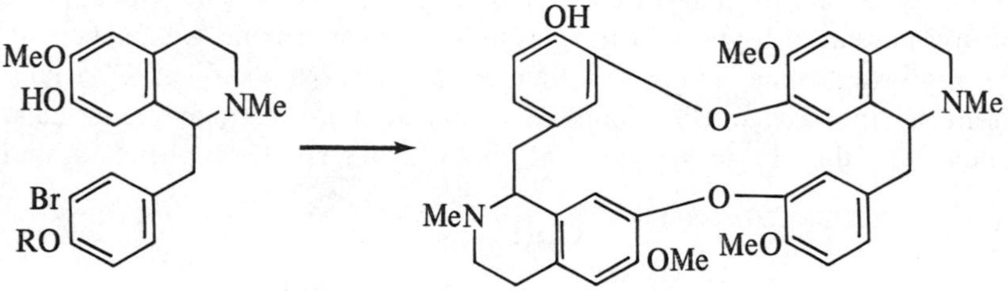

68

Chart 5-21.

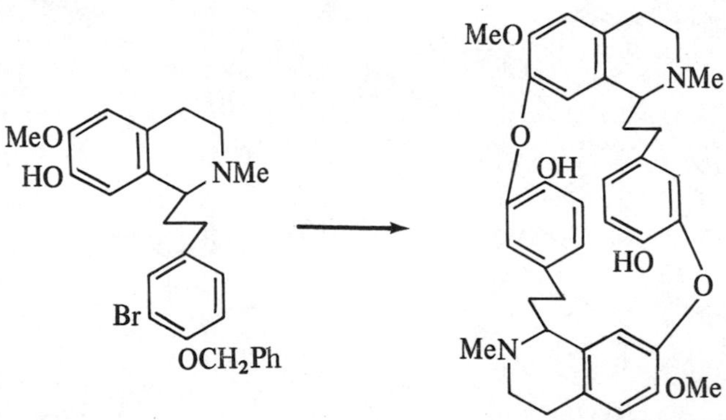

69

usually prepared by the Pschorr reaction, and the biphenyl ether function is synthesized by the Ullmann reaction. In this section, we describe the total synthesis of the bisbenzylisoquinoline, bisphenethylisoquinoline, and cularine alkaloids by an Ullmann reaction from the reports appeared since 1970.

A typical synthetic example of a bisbenzylisoquinoline alkaloid from the Ullmann reaction is that of adiantifoline **66** as shown in Chart 5-18.[365] Thalicarpine[366] and magnoline[367] were also synthesized by this method: *O,O*-Dimethylcurine **67** is obtained by an Ullmann reaction of the diphenolic isoquinoline and dibromoisoquinoline in the presence of cuprous chloride and potassium carbonate in pyridine[368] (Chart 5-19). The intermolecular Ullmann reaction of the phenolic bromoisoquinoline gave the symmetric bisbenzylisoquinoline, the so-called hayatine **68**[369] (Chart 5-20). Similarly, a double Ullmann reaction in the presence of copper powder in dry pyridine, followed by a debenzylation gave melanthioidine **69**[370] (Chart 5-21). The intramolecular Ullmann reaction of 1-(2-bromobenzyl)-8-hydroxyisoquinolines was found to give cularine-type compounds,[371] and this reaction was applied to a total synthesis of cularicine **70**[372] and cularidine **71**[373] (Chart 5-22). Moreover, intramolecular Ullmann reaction of 8-bromo-1-(2-hydroxybenzyl)isoquinoline afforded cularine **72**[374] (Chart 5-23).

Chart 5-22.

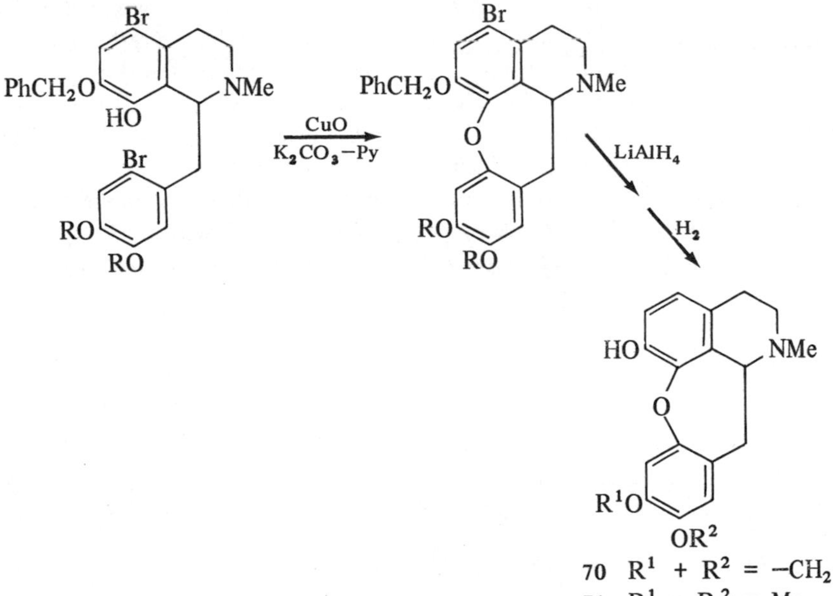

70 $R^1 + R^2 = -CH_2-$
71 $R^1 = R^2 = Me$

Chart 5-23.

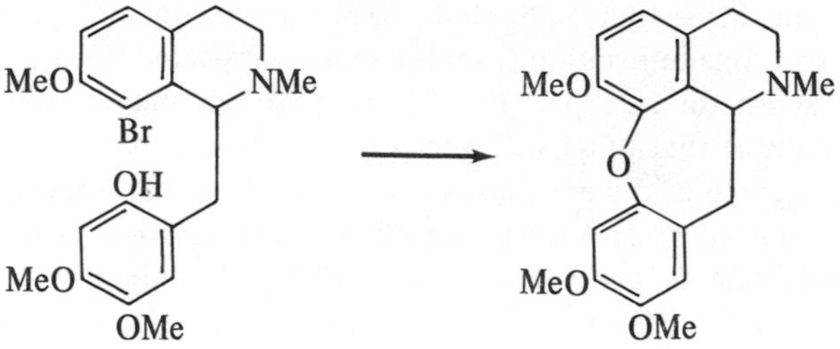

MeO NMe Br OH MeO OMe → MeO NMe O MeO OMe 72

D. Rearrangement

Thermal Reaction

It is well known that benzocyclobutenes **73**[375] upon heating afford reactive *o*-quinodimethanes **74**, which easily react intra- and intermolecularly with olefins **75** to give the tetralin system **76** in a regioselective and/or stereoselective manner[376,377] (Chart 5-24). This type of cycloaddition or electrocyclic reaction could be applied to the total syntheses of the benzophenanthridine-, protoberberine-, and ochotensine-type alkaloids.

The first example, the total synthesis of chelidonine **77**, was reported by Oppolzer and Keller.[378] Thus the pyrolysis of the acetylenic urethane **78** gave the benzophenanthridine system **79**, which on hydroboration furnished two diastereoisomeric alcohols. The *cis* B/C-fused isomer **80** upon oxidation with chromic acid, reduction with sodium borohydride, removal of the urethane group, and *N*-methylation gave chelidonine **77** (Chart 5-25). The second example was the synthesis of the tetrahydroprotoberberine alkaloids by an electrocyclic reaction. Xylopinine **81a** has been synthesized from the 1-benzocyclobutenylisoquinoline hydrochloride **82** in good yield by thermolysis in *o*-dichlorobenzene or

Chart 5-24.

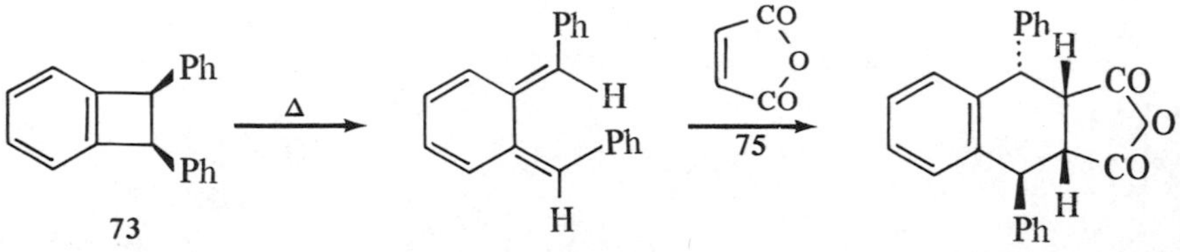

Ph Ph 73 →Δ Ph H Ph H 74 → CO O CO 75 Ph H CO O CO H Ph 76

Chart 5-25.

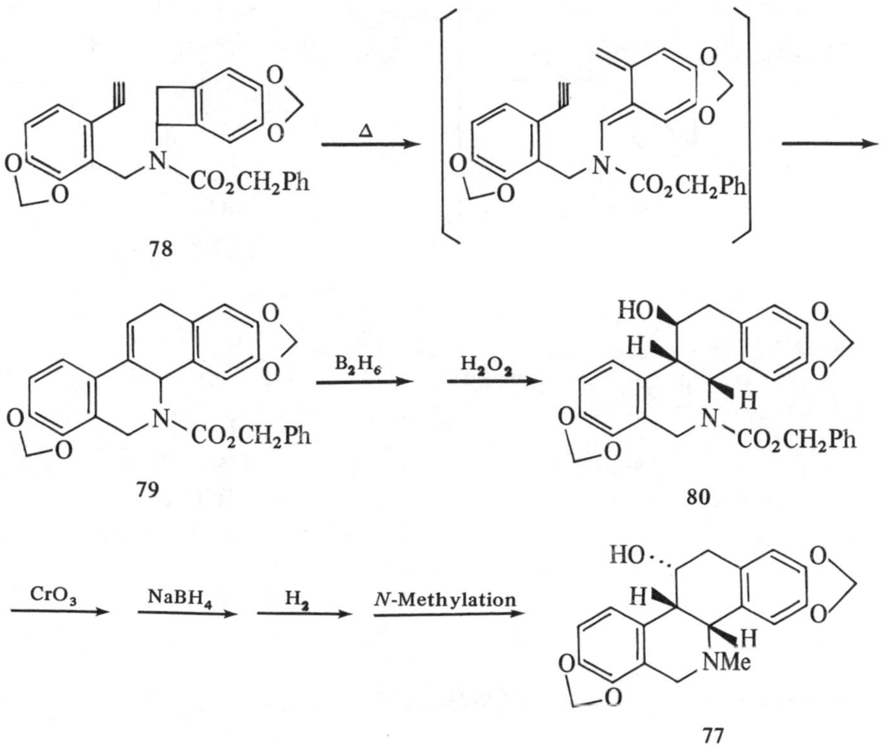

78

Δ

CO_2CH_2Ph

79

B_2H_6 H_2O_2

80

CrO_3 $NaBH_4$ H_2 *N*-Methylation

77

bromobenzene followed by catalytic reduction.[379] Similarly, discretine **81b**[380] and coreximine **81c**[381] were obtained from the corresponding benzocyclobutenylisoquinolines (Chart 5-26). However, the thermolysis of the free base of 1-benzocyclobutenylisoquinoline **82** did not afford the expected protoberberine. The different behavior of the free base and its hydrochloride on thermolysis might be due to the direction of the benzocyclobutene ring opening. During the thermolytic ring cleavage, a more stabilized *o*-quinodimethane **83** having the Z-system formed at first, and then only the hydrochloride, which possessed an activated group $>C=\overset{+}{N}H-$, tautomerized to a required *o*-quinodimethane **84**, having the E form to give the protoberberinium salt[379] (Charts 5-27A and 5-27B).

Alternatively, xylopinine **81a** was obtained in a regioselective manner by thermolysis of the benzocyclobutenol **85** in the presence of 3,4-dihydroisoquinoline **86**[382] (Chart 5-28). Similarly, 1-bromobenzocyclobutene[383] and 1-cyanobenzocyclobutene[384] gave protoberberines in good yields (Chart 5-29).

Chart 5-26.

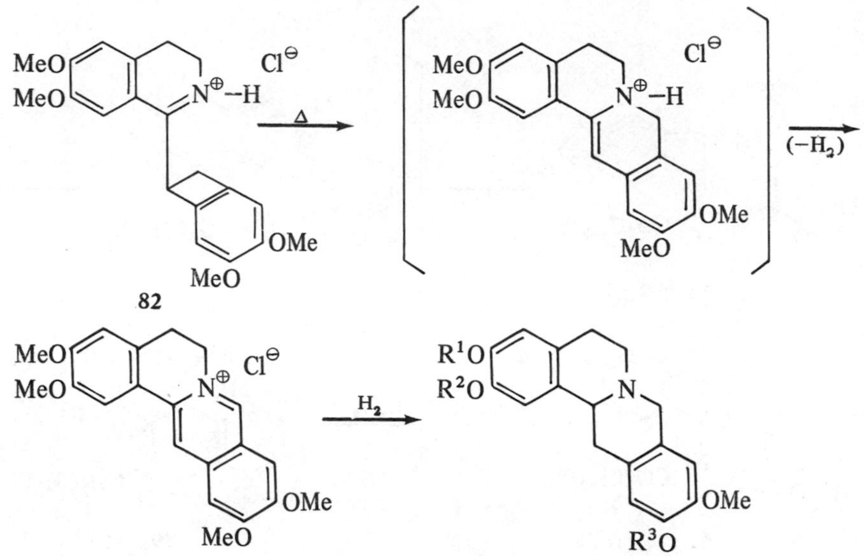

MeO MeO $N^{\oplus}$–H $Cl^{\ominus}$ OMe MeO **82** $\xrightarrow{\Delta}$ [MeO MeO $N^{\oplus}$–H $Cl^{\ominus}$ OMe MeO] $\xrightarrow{(-H_2)}$

MeO MeO $N^{\oplus}$ $Cl^{\ominus}$ OMe MeO $\xrightarrow{H_2}$ R^1O R^2O N OMe R^3O

81a $R^1 = R^2 = R^3 = Me$
81b $R^1 = H, R^2 = R^3 = Me$
81c $R^1 = Me, R^2 = R^3 = H$

Chart 5-27A.

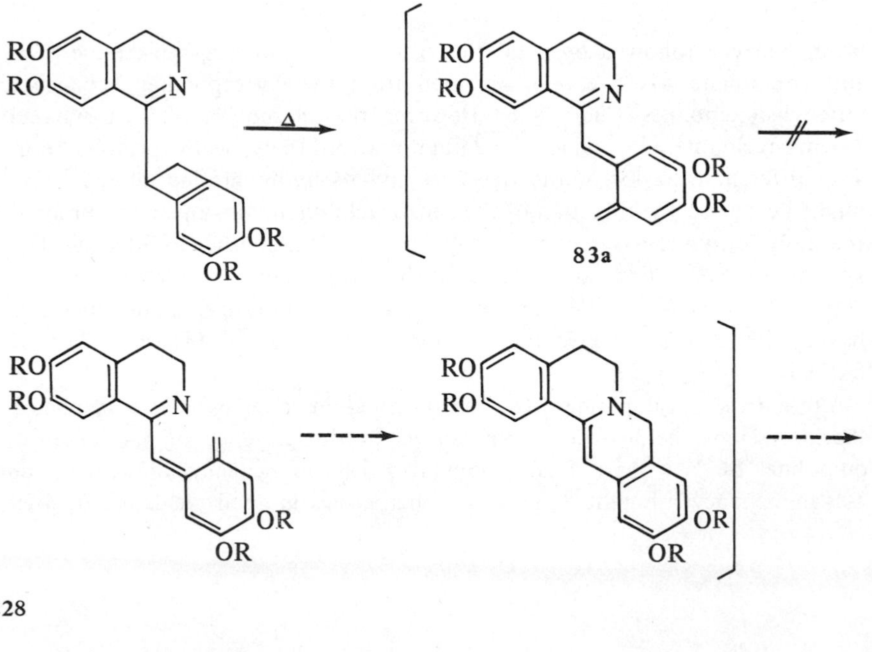

RO RO N OR OR $\xrightarrow{\Delta}$ [RO RO N OR OR **83a** $\not\longrightarrow$

RO RO N OR OR - - - → RO RO N OR OR] - - - →

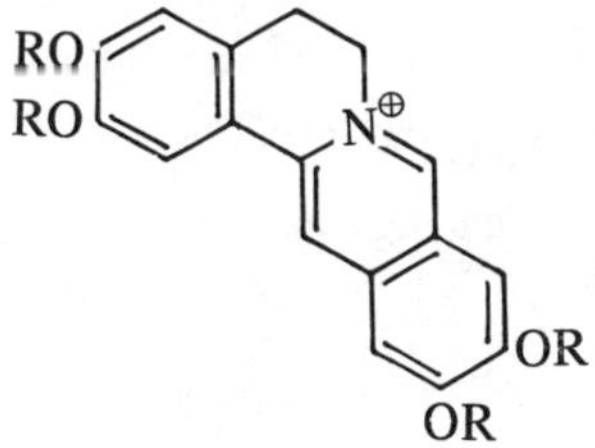

Chart 5-27B.

RO
RO
N⊕
OR
OR
82
Δ
MeO
MeO
N⊕–H
OMe
OMe
83b
MeO
MeO
NH
O⊕–Me
OMe
MeO
MeO
NH
OMe
OMe
⊕
MeO
MeO
N⊕–H
OMe
MeO
12
MeO
MeO
N
OMe
MeO
MeO
MeO
N⊕
OMe
MeO

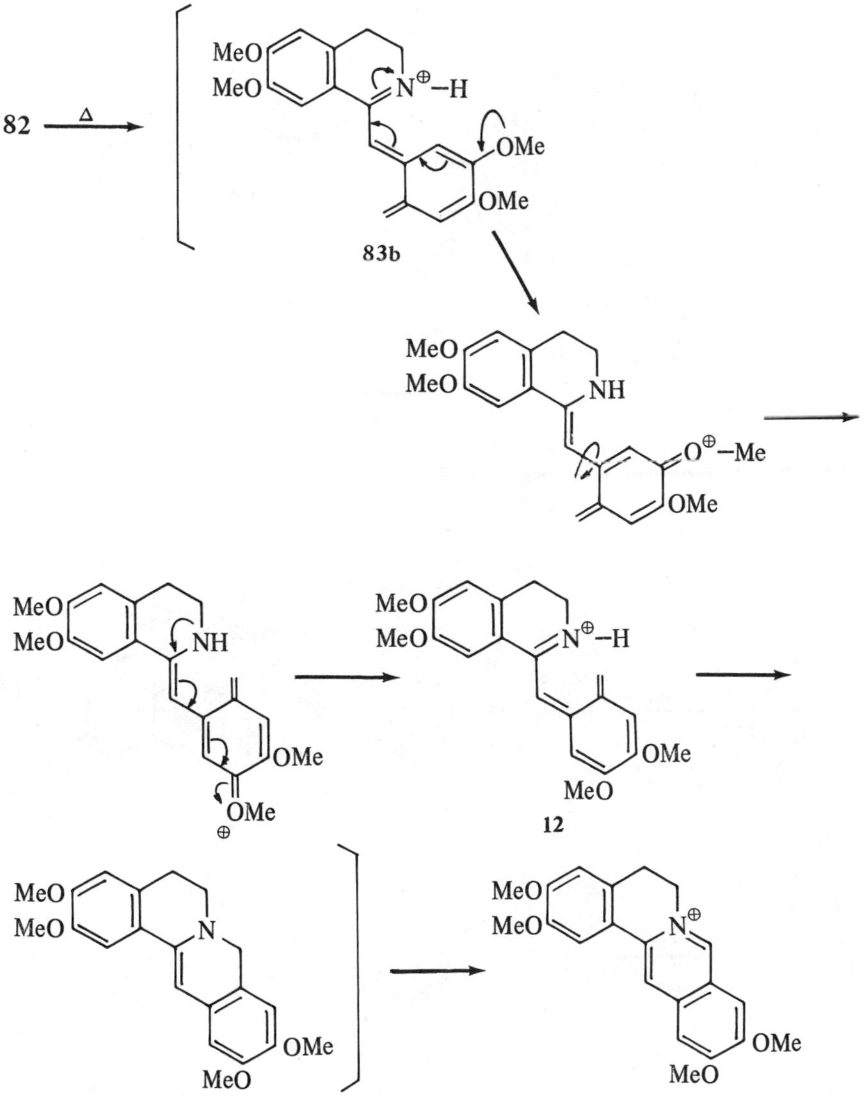

Chart 5-28.

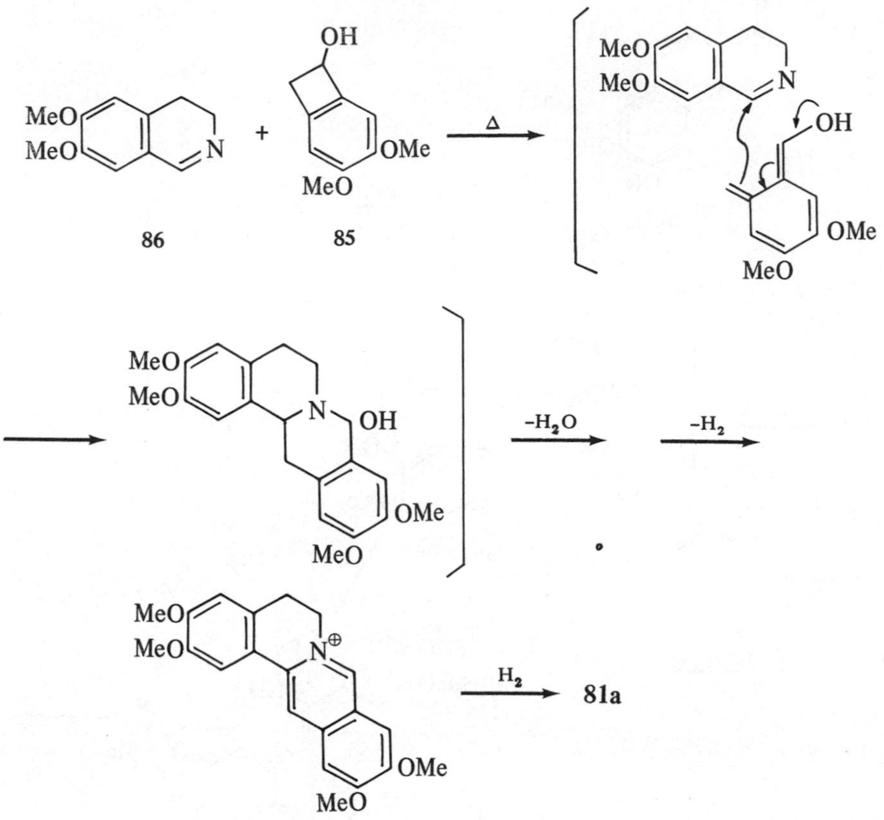

MeO MeO N + OH OMe MeO Δ → [MeO MeO N OH OMe MeO]

86 85

→ MeO MeO N OH OMe MeO] $-H_2O$ → $-H_2$ →

MeO MeO $N^{\oplus}$ OMe MeO $\xrightarrow{H_2}$ **81a**

Chart 5-29.

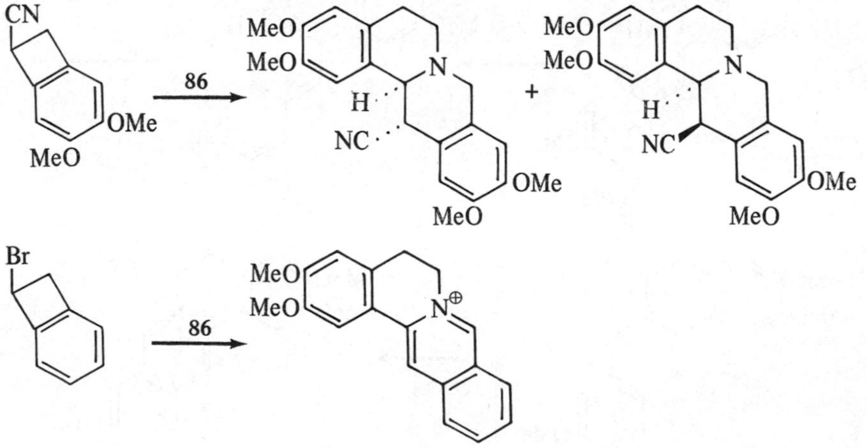

CN OMe MeO $\xrightarrow{86}$ MeO MeO N H NC OMe MeO + MeO MeO N H NC OMe MeO

Br $\xrightarrow{86}$ MeO MeO $N^{\oplus}$

Moreover, thermolysis of a benzocyclobutene in the presence of an imine gave, regioselectively, an isoquinoline, which would be an intermediate for a benzophenanthridine alkaloid[385] (Chart 5-30). The ochotensine analog **87**[386] was synthesized by thermal rearrangement of a benzocyclobutene derivative, follow-

Chart 5-30.

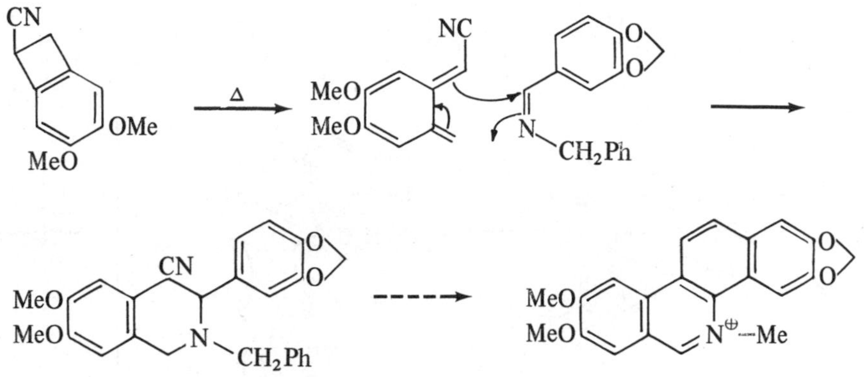

ing the biogenetic hypothesis by Shamma described later.[387] Thus a Bischler-Napieralski reaction of the benzocyclobutenecarboxylic amide **88** did not give the expected 3,4-dihydroisoquinolinium chloride **89** but gave an ochotensine analog **87**. This reaction could be explained as follows: A normal product **89** formed by cyclization is unstable to heat and would immediately give *o*-quinodimethane **90**, which would cyclize to the ochotensine analog **87** via the spirane **91** (Chart 5-31).

Stevens Rearrangement

This reaction is a rearrangement of a quaternary ammonium salt to tertiary base under the influence of a strong base. The pavine ring system **93** was obtained by the electrophilic molecular rearrangement of a tetracyclic dibenzopyrrocoline *N*-metho salt **92**[381] (Chart 5-32). An interesting synthesis of petaline **94**, which takes advantage of a Stevens rearrangement, has been reported by two groups[389,390] as shown in Chart 5-33. Moreover, a quaternary protoberberine **95** rearranged to an ochotensine-type base **96** by treatment with strong base[391] (Chart 5-34). This reaction is also a part of the protoberberine → ochotensine transformation described later.

Dienone-Phenol and Dienol-Benzene Rearrangements[392-395]

An acidic treatment of dienones and dienols gives phenols and benzene

Chart 5-31.

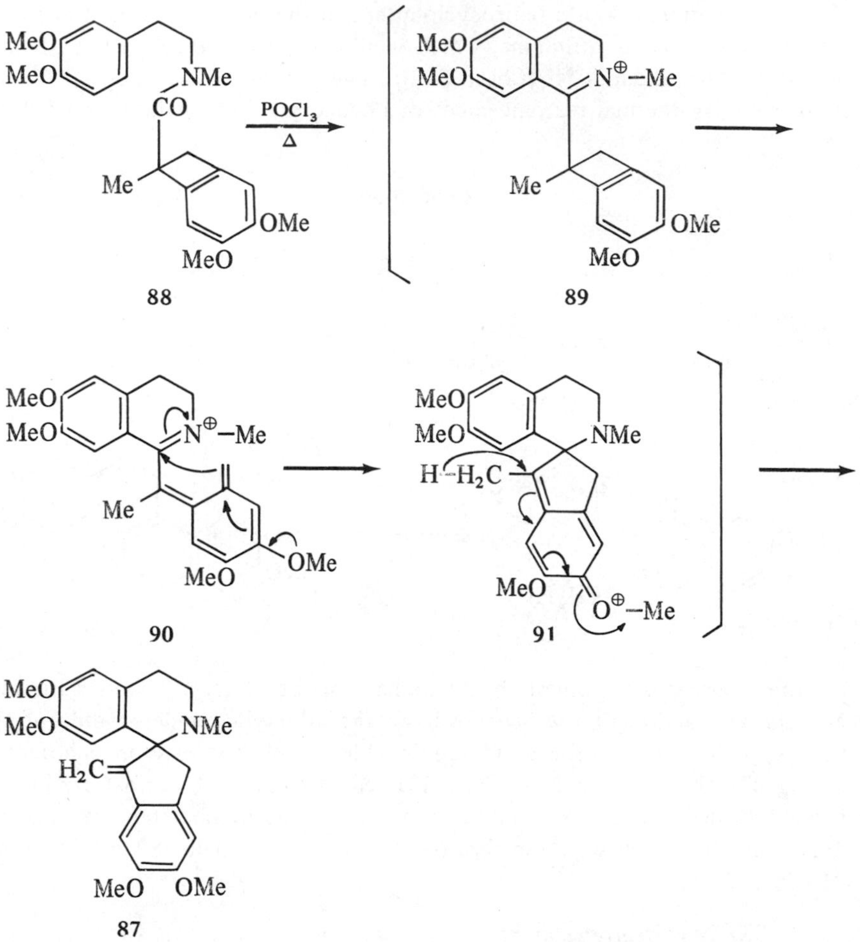

MeO MeO NMe CO Me OMe MeO POCl$_3$ Δ

88

MeO MeO N$^{\oplus}$—Me Me OMe MeO

89

MeO MeO N$^{\oplus}$—Me Me MeO OMe

90

MeO MeO NMe H—H$_2$C MeO O$^{\oplus}$—Me

91

MeO MeO NMe H$_2$C= MeO OMe

87

Chart 5-32.

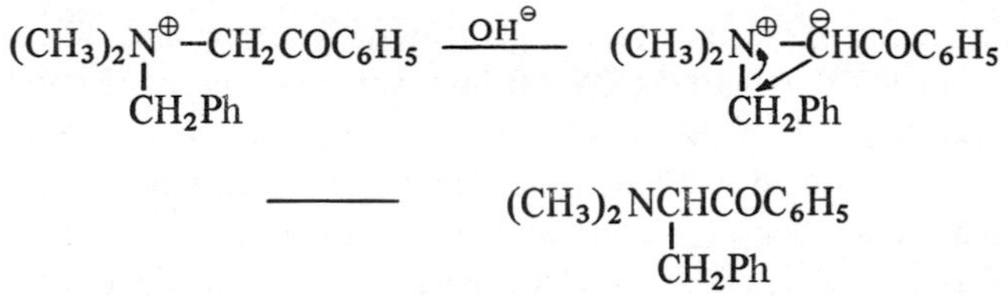

$(CH_3)_2N^{\oplus}$—$CH_2COC_6H_5$ (CH$_2$Ph) $\xrightarrow{OH^{\ominus}}$ $(CH_3)_2N^{\oplus}$—$\overset{\ominus}{C}HCOC_6H_5$ (CH$_2$Ph)

⟶ $(CH_3)_2NCHCOC_6H_5$ (CH$_2$Ph)

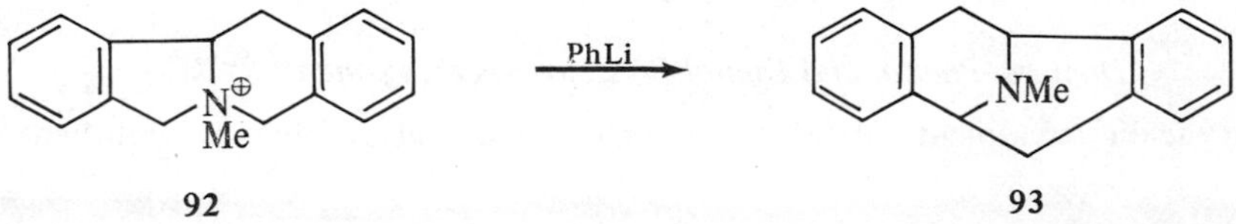

N$^{\oplus}$ Me $\xrightarrow{PhLi}$ NMe

92 93

Chart 5-33.

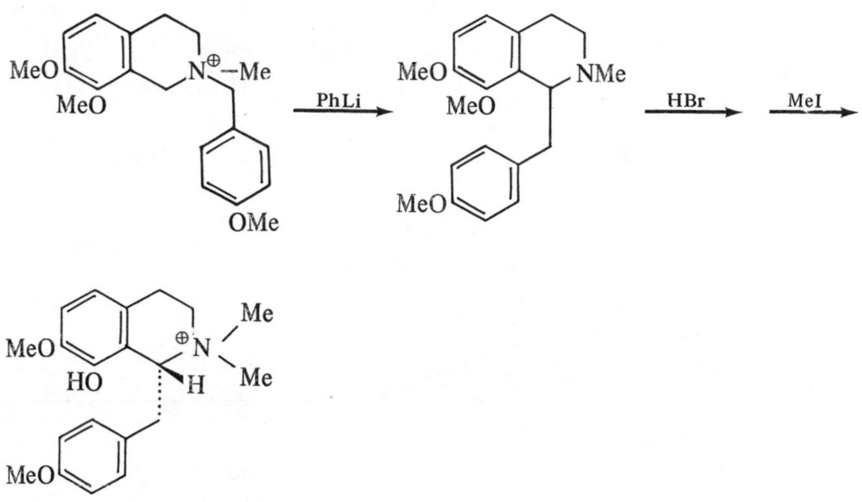

PhLi → HBr → MeI →

94

Chart 5-34.

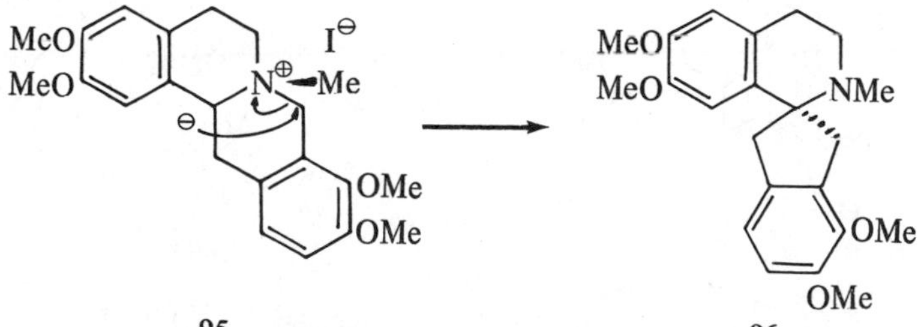

95 → **96**

derivatives, respectively. This type of rearrangement could be applied to the synthesis of aporphine and homoaporphine alkaloids, described in Section 3. Several examples are listed in Charts 5-35A and 5-35B.

Schmidt Rearrangement

Uyeo and his collaborators prepared many types of the *Amaryllidaceae* alkaloids by the utilization of a Schmidt rearrangement in the synthetic sequence. The indanone ester **97** was converted to the corresponding alcohol **98**, which was subjected to the Schmidt rearrangement with sodium azide to give the desired isocarbostyril **99**. Homologation of the side chain in the usual way afforded the carboxylic acid, which was cyclized to the iodolactone **100**. Upon treatment of

Chart 5-35A.

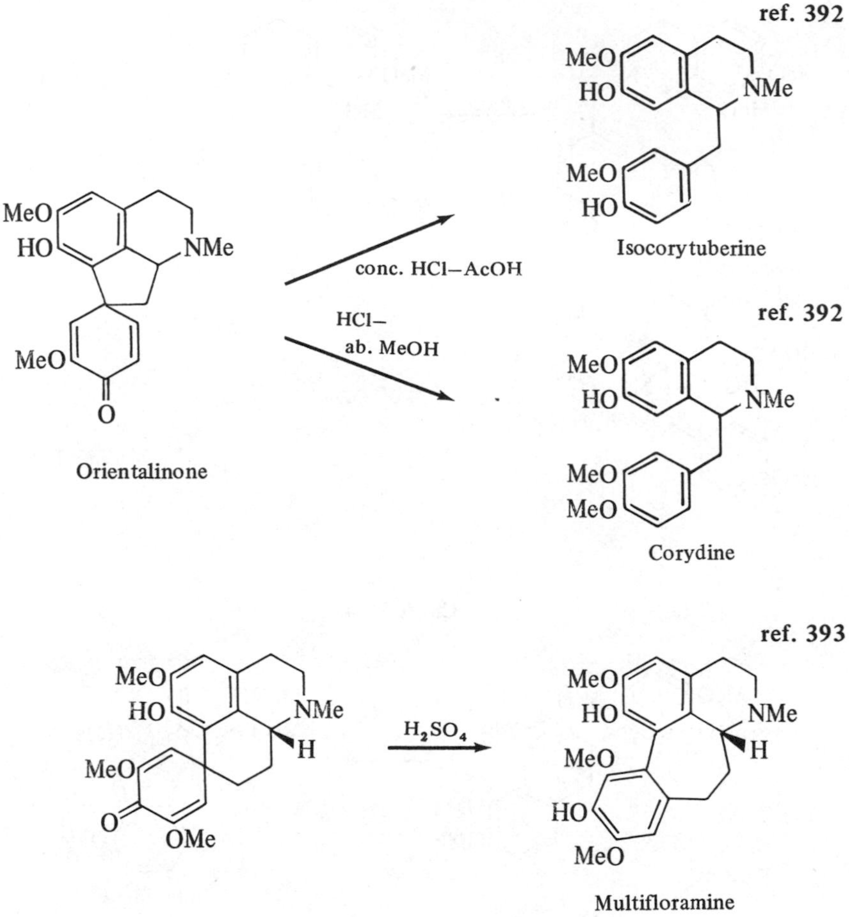

ref. 392

MeO
HO
NMe
MeO
HO
Isocorytuberine

MeO
HO
NMe
MeO
O
Orientalinone

conc. HCl–AcOH

HCl–
ab. MeOH

ref. 392

MeO
HO
NMe
MeO
MeO
Corydine

ref. 393

MeO
HO
NMe
H
MeO
O
OMe

H_2SO_4

MeO
HO
NMe
H
MeO
HO
MeO
Multifloramine

100 with acetic acid and acetic anhydride, it rearranged to the imide **101**, which was transformed into the olefin **102** upon treatment with lithium chloride. Epoxidation of **102**, followed by reduction with lithium aluminum hydride and zinc chloride, gave dihydrolycorine **103**[396] (Chart 5-36). Lycoramine **104** synthesis was also based upon the use of the Schmidt rearrangement for the introduction of the nitrogen atom[397] (Chart 5-37). Similarly, the introduction of the nitrogen atom by the Schmidt rearrangement was applied to the synthesis of dihydrocrinine.[398]

Chart 5-35B.

ref. 394

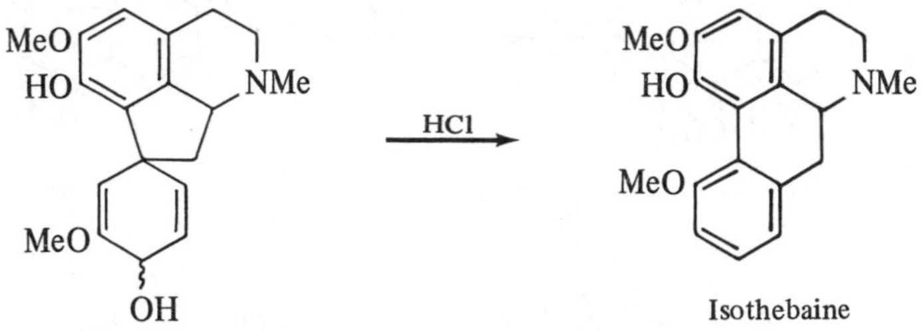

MeO HO NMe MeO OH Orientalinol $\xrightarrow{HCl}$ MeO HO NMe MeO Isothebaine

ref. 395

MeO HO MeO NCO_2Et MeO OMe $\xrightarrow{FSO_2OMe}$ MeO MeO NCO_2Et MeO OMe

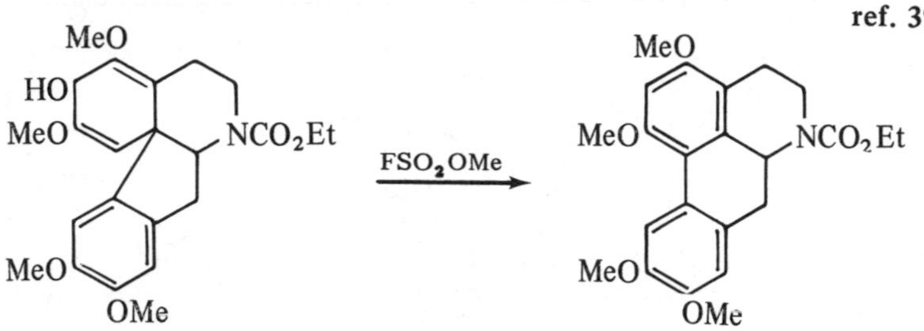

Transformation of Protoberberine to Ochotensine

Knowledge that the protoberberine-type alkaloids co-occur with the spirobenzylisoquinoline alkaloids, such as ochotensine, suggested that the latter type of alkaloids may be derived in vitro from the former by the prototype rearrangement as shown in the following scheme; the presence of the C_{13}-methyl and C_{13}-C_{13a} double bond are required[399] (Chart 5-38). Along with this hypothesis, ochotensine-type compounds were synthesized by Shamma and coworkers.[399-402] Treatment of the diphenolic base **105** with boiling aqueous ethanolic sodium hydroxide produced the quinone methide **106**, which was dissolved in dimethyl sulfoxide to give the expected diphenolic spirobenzylisoquinoline **107**[399] (Chart 5-39). An alternative biogenetic model in which the phenolic groups are in different rings gave the spirobenzylisoquinoline **108** directly by prolonged heating with a strong base[400,402] (Chart 5-40).

In a similar manner, base-catalyzed rearrangement of the monophenolic **109** leads to the spirobenzylisoquinolines **110**[401,402] (Chart 5-41). Stevens

Chart 5-36.

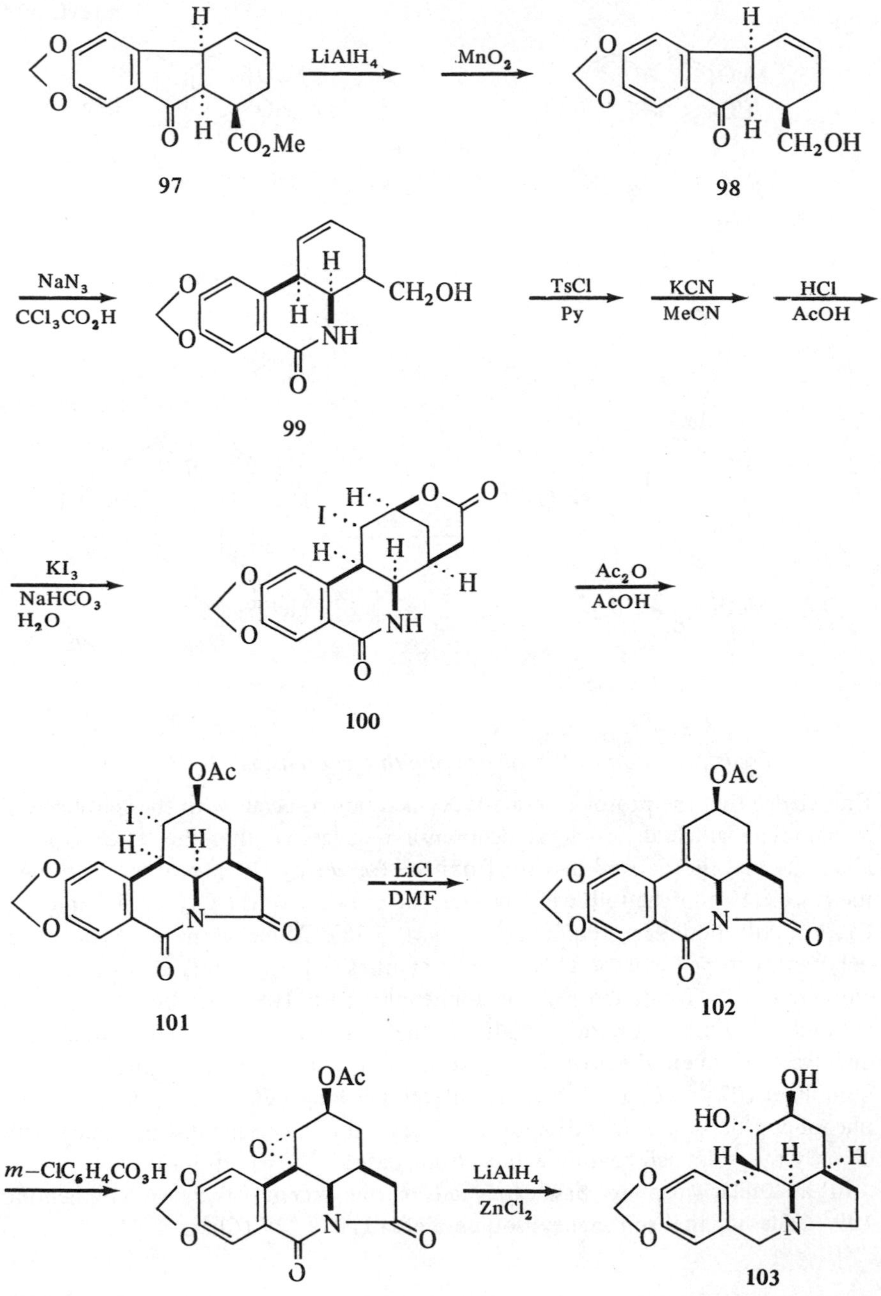
LiAlH$_4$
MnO$_2$
97
98
NaN$_3$
CCl$_3$CO$_2$H
99
TsCl
Py
KCN
MeCN
HCl
AcOH
KI$_3$
NaHCO$_3$
H$_2$O
100
Ac$_2$O
AcOH
LiCl
DMF
101
102
m–ClC$_6$H$_4$CO$_3$H
LiAlH$_4$
ZnCl$_2$
103

Chart 5-37.

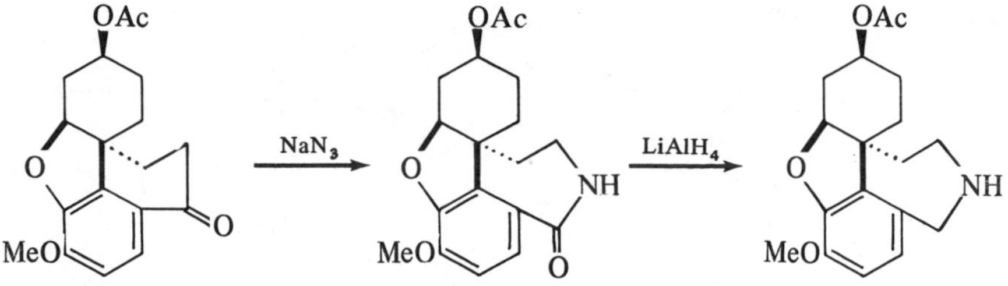

NaN$_3$ LiAlH$_4$

104

Chart 5-38.

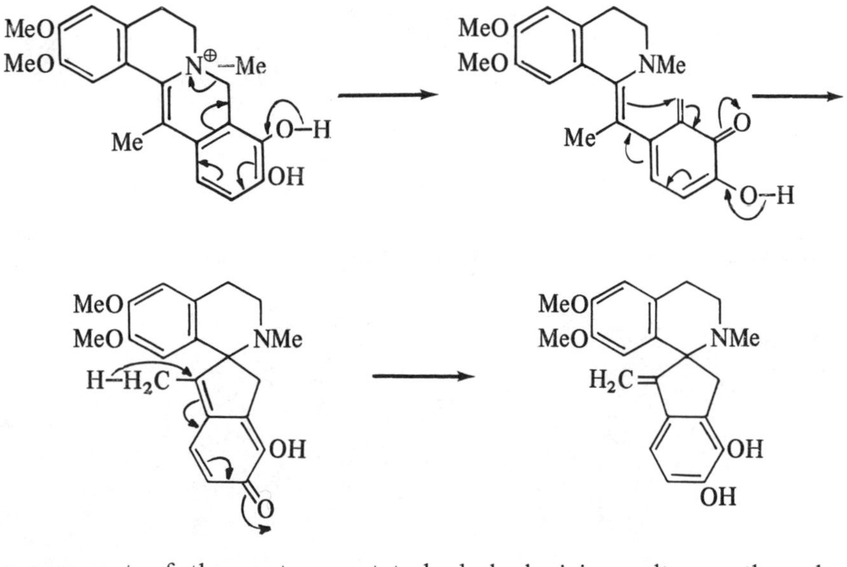

rearrangement of the quaternary tetrahydroberberinium salt gave the ochotensine system as described previously[391] (Chart 5-42). Alternatively, photolysis of 13-keto-7,8,13,13a-tetrahydroprotoberberine methiodide **111** caused rearrangement to the spirobenzylisoquinoline **112** by a mechanism mentioned later[403] (Chart 5-43).

Transformation of Protoberberine to Protopine

The protoberberine-type compounds were transformed into the protopine-type bases by two methods. Treatment of methine base **114**, derived from

Chart 5-39.

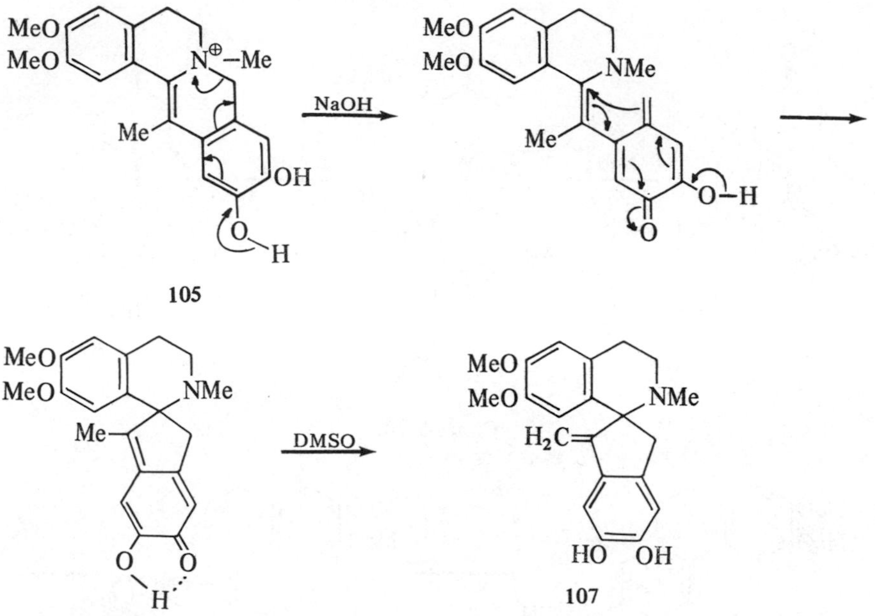

MeO MeO N⊕–Me Me OH O H NaOH MeO MeO NMe Me O O–H

105

MeO MeO NMe Me O O H DMSO MeO MeO NMe H_2C HO OH

106

107

Chart 5-40.

H–O MeO Br⊖ N⊕–Me Me OH MeO NaOH O MeO NMe Me MeO O–H

O⊖ MeO NMe H–CH_2 MeO O work-up HO MeO NMe H_2C MeO OH

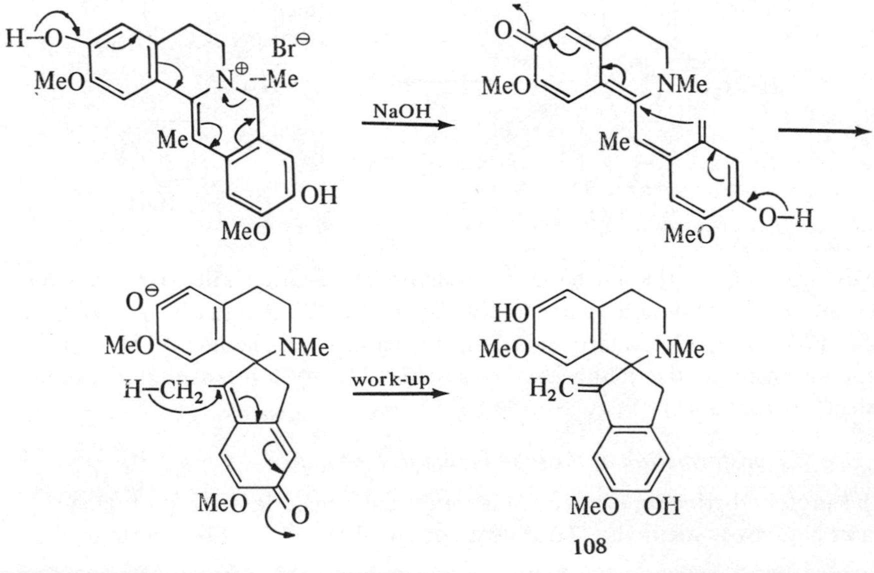

108

Chart 5-41.

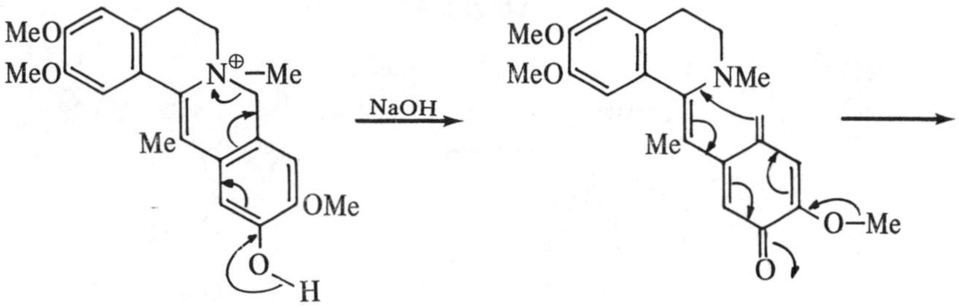

NaOH

109a

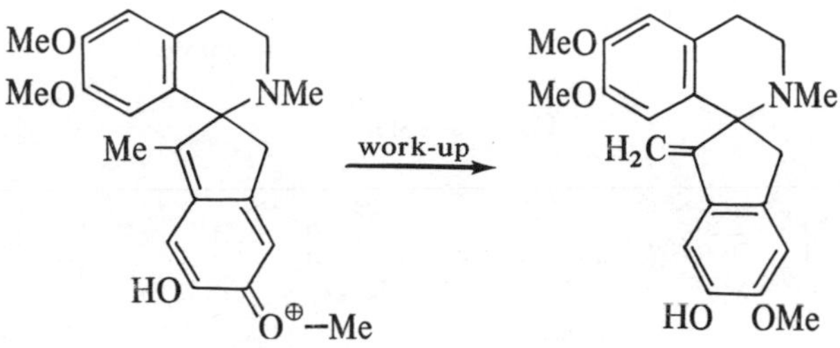

work-up

110a

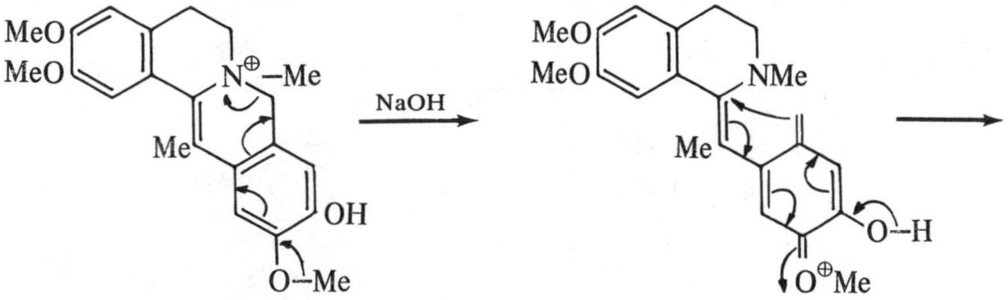

NaOH

109b

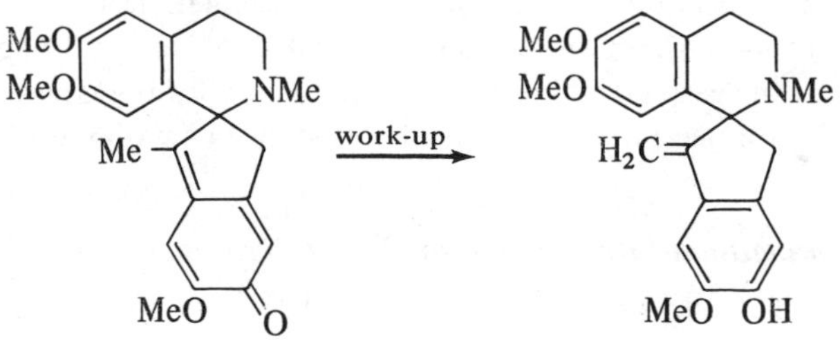

work-up

110b

Chart 5-42.

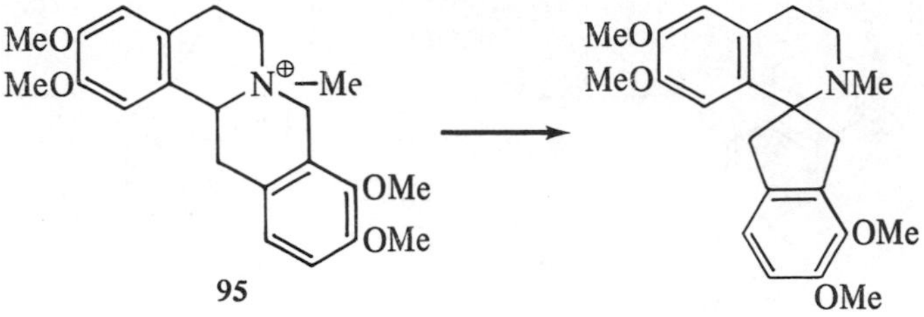

Chart 5-43.

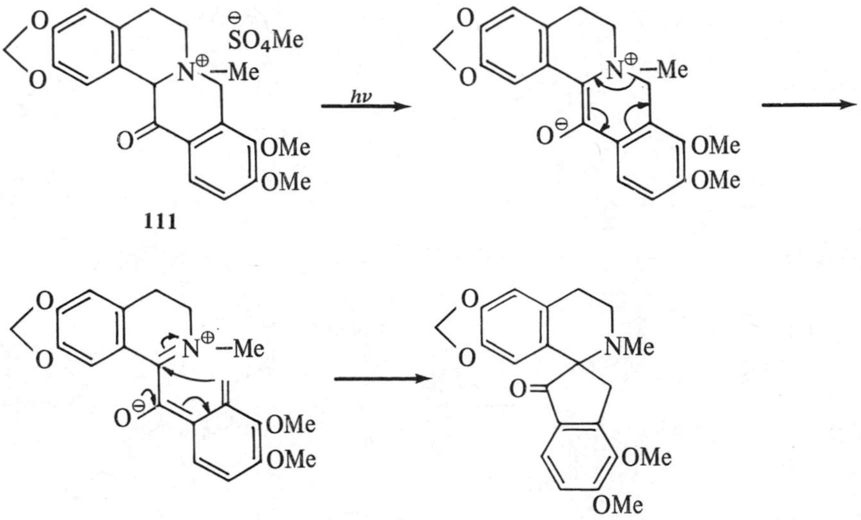

tetrahydroepiberberine **113** by Hofmann degradation with perbenzoic acid gives the *N*-oxide **115** which on acid-catalyzed rearrangement leads to cryptopine **116**[404,405] (Chart 5-44). Alternatively, the *N*-oxide **117** of tetrahydroepiberberine was converted to the carbinolamine **118** by treatment with potassium chromate. This base was transformed into cryptopine methiodide **119** with methyl iodide[406] (Chart 5-45).

Transformation of Rhoeadine-type Base to Protoberberine

Rhoeageninediol **121**, derived from rhoeadine **120**, was converted to the dihydroprotoberberine by treatment with boiling thionyl chloride[407] (Chart 5-46).

Chart 5-44.

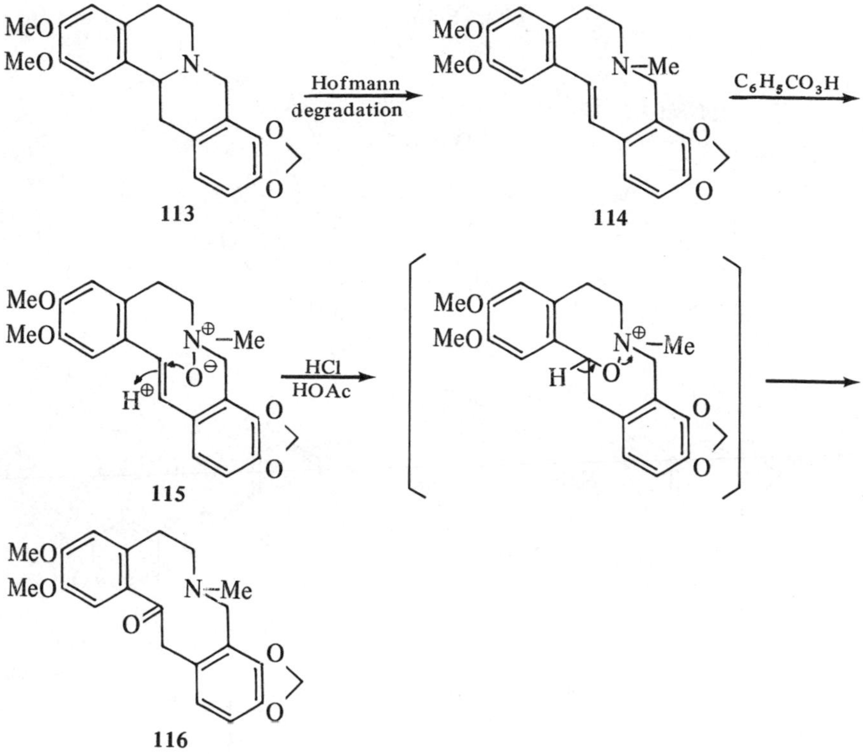

MeO
MeO
N
O
O
113

Hofmann degradation →

MeO
MeO
N—Me
O
O
114

$C_6H_5CO_3H$ →

MeO
MeO
N$^{\oplus}$—Me
O$^{\ominus}$
H$^{\oplus}$
O
O
115

HCl / HOAc →

[MeO
MeO
N$^{\oplus}$—Me
H
O
O
O] →

MeO
MeO
N—Me
O
O
O
116

Chart 5-45.

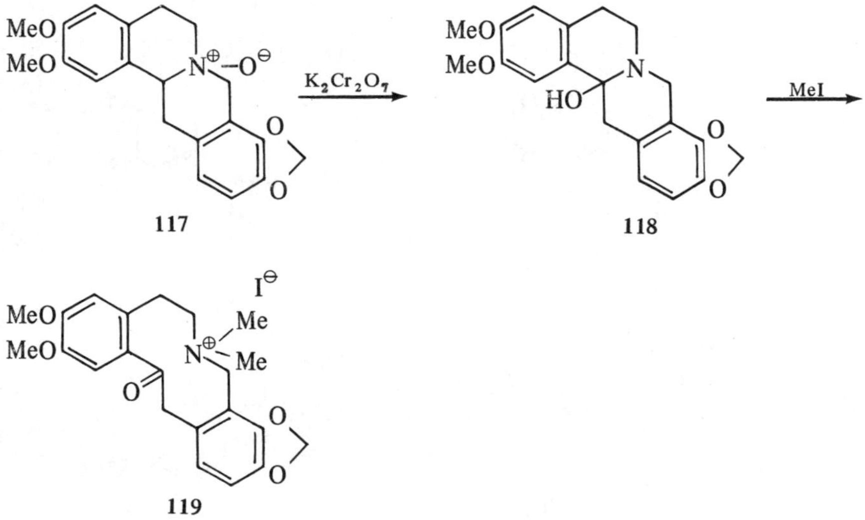

MeO
MeO
N$^{\oplus}$—O$^{\ominus}$
O
O
117

$K_2Cr_2O_7$ →

MeO
MeO
N
HO
O
O
118

MeI →

I$^{\ominus}$
MeO
MeO
N$^{\oplus}$
Me
Me
O
O
O
119

Chart 5-46.

120

121

SOCl$_2$

Transformation of Ochotensine-type Base to Rhoeadine

Rhoeadine **120** and alpinigenine have been synthesized by a Wagner-Meerwein rearrangement of the indanol **122**, followed by ring opening by a Criegee-type reaction.[408] Thus Wagner-Meerwein rearrangement of **122** with mesyl chloride gave the benzazepine **123**. Oxidation of **123** with osmium tetroxide gave the diol **124**, which was converted into rhoeagenine diol **121** with sodium periodate followed by sodium borohydride. Since **121** had been previously converted into rhoeadine **120**, the total synthesis was formally completed. Similarly, alpinigenine was synthesized by this route [408] (Chart 5-47).

Chart 5-47.

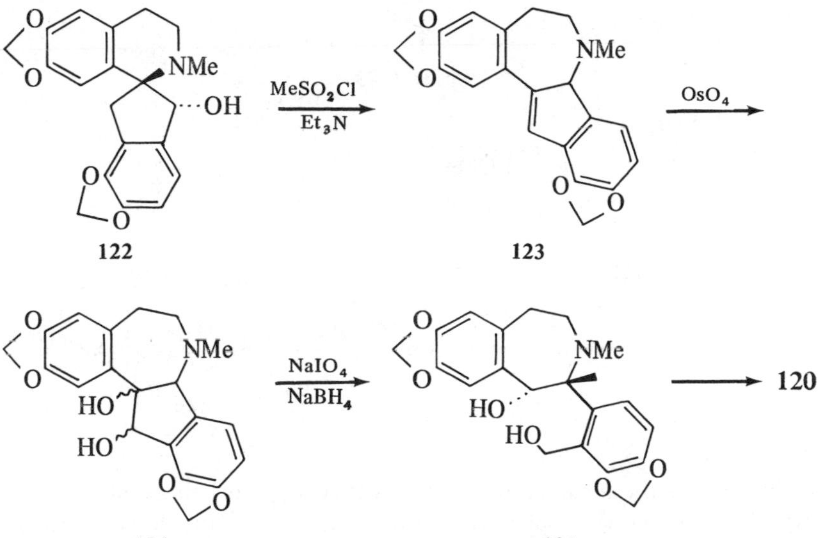

NMe
OH
$MeSO_2Cl$
Et_3N
NMe
OsO_4
122
123
NMe
HO
HO
$NaIO_4$
$NaBH_4$
NMe
HO
HO
120
124
121

E. Azepine Alkaloids

There are many azepine alkaloids that are derived biogenetically from benzylisoquinoline alkaloids. These azepine alkaloids are subdivided into three types: rhoeadine, cephalotaxine, and isopavine.

In this section, we mention the general synthesis of the azepine system and the total synthesis of some alkaloids that belong to these three base types.

The classic synthesis of the azepine ring system is the dehydrative cyclization of *N*-phenethylaminoethanol[409] (Chart 5-48). This reaction can be used for the synthesis of the isopavine alkaloids. Thus an acidic treatment of 4-hydroxy-

Chart 5-48.

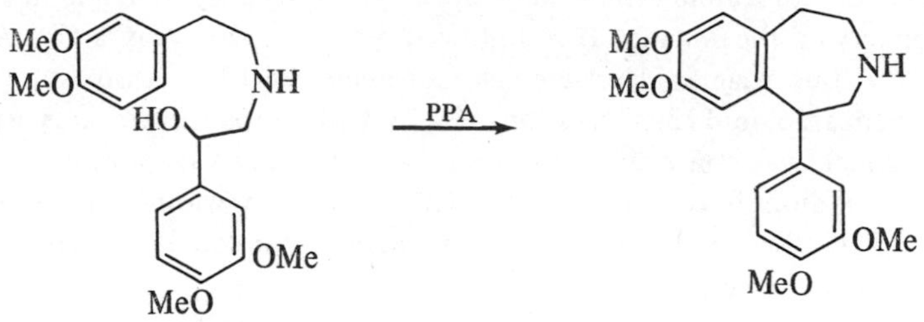

MeO MeO HO NH OMe MeO PPA MeO MeO NH OMe MeO

1,2,3,4-tetrahydroisoquinoline **125** gave *O*-methylthalisopavine **126**.[410] Reframidine and reframine were also synthesized by this method[410] (Chart 5-49).

Chart 5-49.

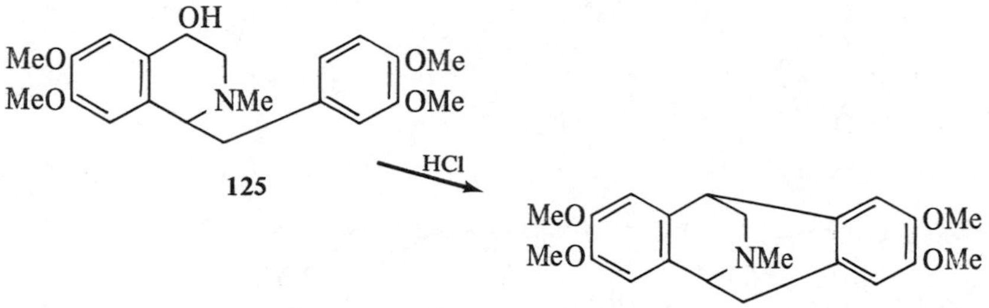

OH MeO MeO NMe OMe OMe **125** HCl MeO MeO NMe OMe OMe **126**

Cyclization of the aminoacetal by a Lewis acid yields benzazepine[411] (Chart 5-50). This reaction was applied to the total synthesis of the isopavine alkaloids as follows. The aminoacetal **127** was treated with sulfuric acid, followed by a reductive *N*-methylation to give thalisopavine **126**.[412] In this reaction, 4-hydroxyisoquinoline **128** is presumed an intermediate (Chart 5-51). Similarly, the amino-aldehyde **129** was cyclized by a Lewis acid to the benzazepine **130**;

Chart 5-50.

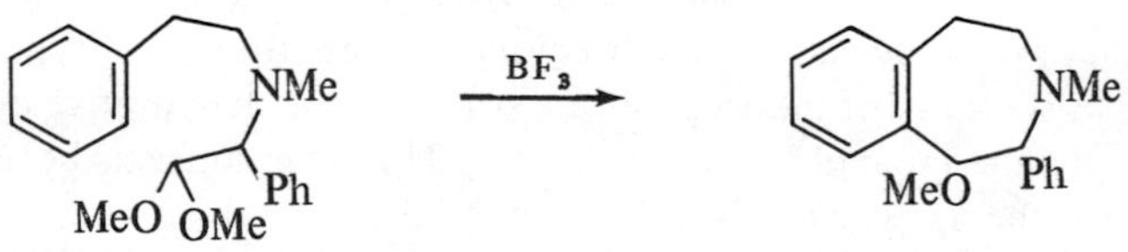

NMe MeO OMe Ph BF_3 NMe MeO Ph

Chart 5-51.

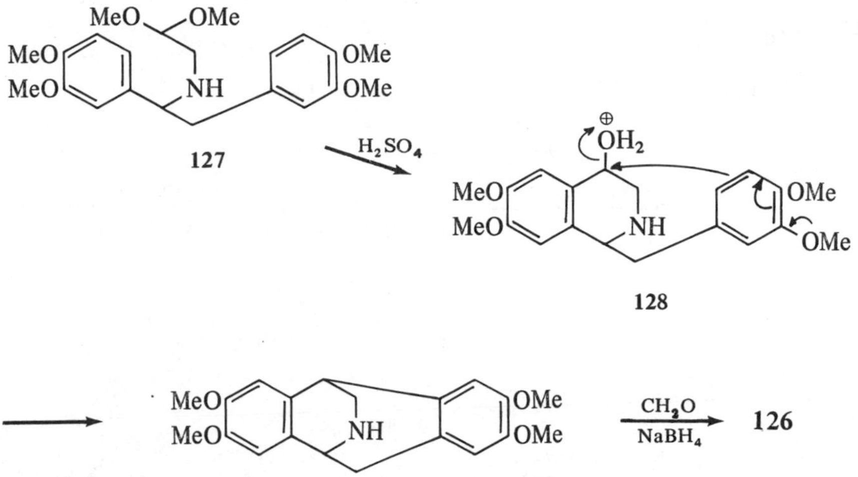

127 $\xrightarrow{H_2SO_4}$ 128 $\longrightarrow$ $\xrightarrow[NaBH_4]{CH_2O}$ 126

this reaction was used for the first total synthesis of cephalotaxine **131**[413] as shown in Chart 5-52.

N-Phenethylchloroacetamide on ultraviolet irradiation gave the benzazepine[414-416] (Chart 5-53). An interesting and novel synthesis, reported by Semmelhack,[364d] utilizes the benzyne reaction as a key step in the total synthesis of cephalotaxine **131** (Chart 5-54). Intramolecular amidation of the amino ester **132** gave the benzazepine **133** from which the *trans* and *cis* isomers of 7,8-dimethoxy-*N*-methyl-2-phenyl-1,2,4,5-tetrahydro-3H-3-benzazepin-1-ol were synthesized by Inubushi[417] (Chart 5-55).

Dehydrative cyclization of the amino-ketone also afforded the azepine, and this reaction was used by Brossi for the total synthesis of rhoeadine **120** from the phthalideisoquinoline **134**.[418,419] An amino-ketone **135**, obtained from (−)-bicuculline **134**, was heated in the presence of the base to afford the benzazepine **136**, which cyclized in dilute acetic acid to the unstable spirolactone **137**. After standing in ethanol, lactone **137** was oxidized to the keto-lactone **138**. Lithium borohydride reduction of **138**, followed by acidification with acetic acid afforded, via the transient *cis*-hydroxy acid **139**, the *cis*-lactone **140**, which was subjected to optical resolution with 10-camphorsulfonic acid (Chart 5-56). Partial reduction of (−)-base **140** with sodium bis-(2-methoxyethoxy)aluminum hydride, followed by *O*-methylation with trimethyl orthoformate, gave rhoeadine **120**.

The conversion of isoquinolines into benzazepines by ring expansion is reported. Thus treatment of the 3,4-dihydroisoquinoline metho-salt **141** with diazoalkane led to the aziridinium salt **142**, which underwent a reaction with

Chart 5-52.

129

$BF_3 \cdot Et_2O$

130

$LiAlH_4$

$CH_3COO\overset{CH_3}{C}HCOCl$

K_2CO_3

PbO_2

O

Me

OAc

$Mg(OMe)_2$

MeOH

TsOH

HO

O

Me

O

H

$NaBH_4$

OMe

HO

131

Chart 5-53.

NH

$h\nu$

NH

Cl

O

Cl

$h\nu$

131

Chart 5-54.

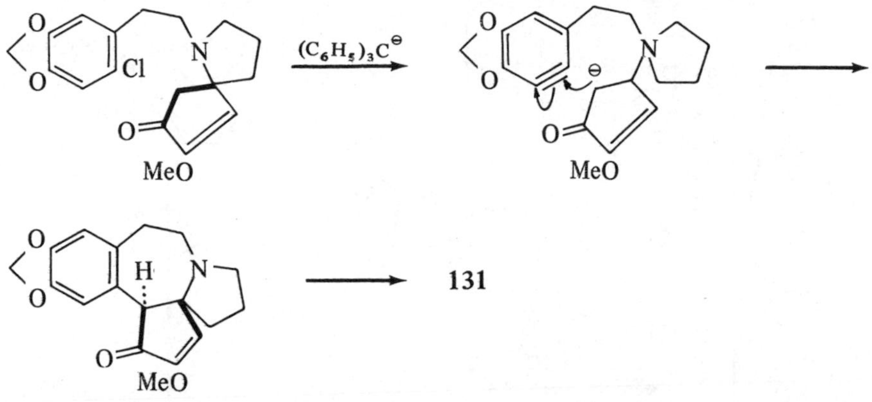

O O Cl N O MeO $(C_6H_5)_3C^{\ominus}$ O O N O MeO

O O H N O MeO 131

Chart 5-55.

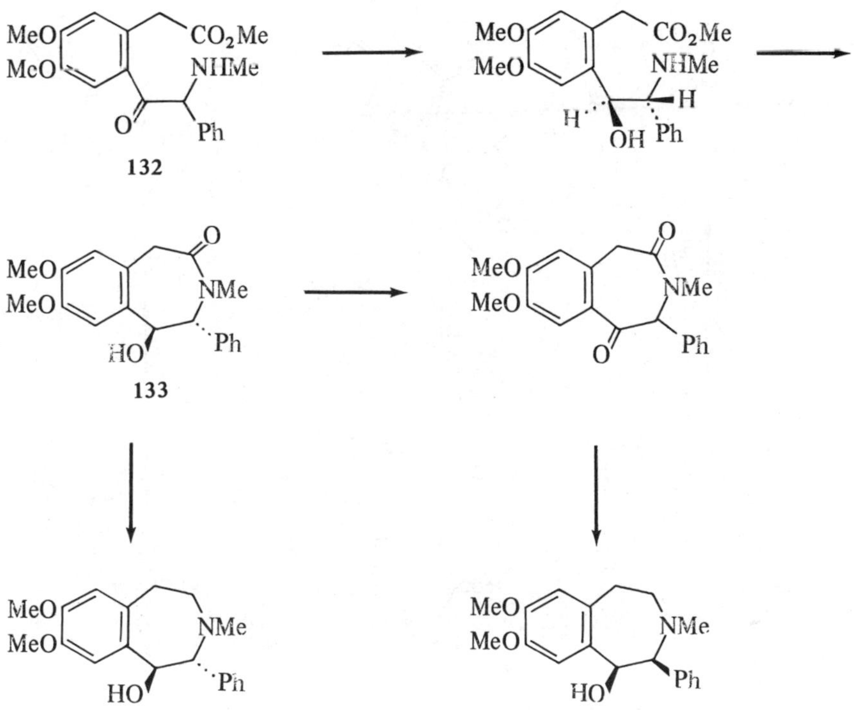

MeO MeO CO_2Me NHMe O Ph 132 MeO MeO CO_2Me NHMe H H OH Ph

MeO MeO O NMe HO Ph 133 MeO MeO O NMe O Ph

MeO MeO NMe HO Ph MeO MeO NMe HO Ph

Chart 5-56.

134 $\xrightarrow[(^iPr)_2NH_2]{ClCO_2Ph}$ $\xrightarrow[(^iPr)_2NH]{DMSO}$ $\longrightarrow$ 135 $\xrightarrow{NaOH}$

136 $\xrightarrow{AcOH}$ 137 $\xrightarrow{air}$ 138 $\xrightarrow{LiBH_4}$

139 $\longrightarrow$ 140 $\xrightarrow{Na(MeOCH_2CH_2O)_2AlH_2}$ $\xrightarrow[H^{\oplus}]{HC(OMe)_3}$ 120

Chart 5-57.

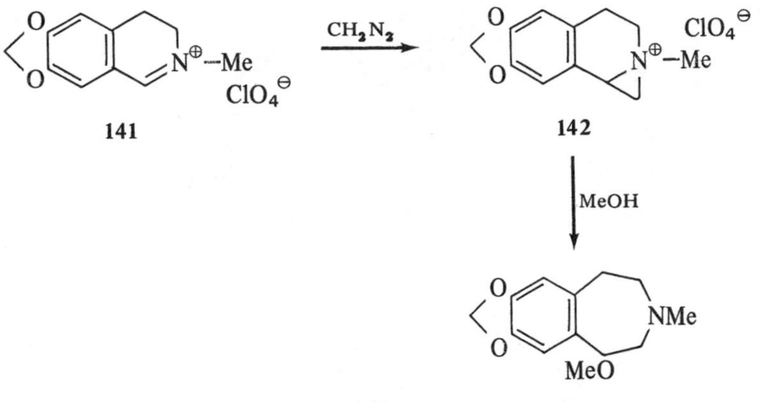

141

142

143

methanol to produce the benzazepine **143**[420,421] (Chart 5-57). A new synthetic method for the isopavine alkaloid, reframidine **144**, was carried out by a one-step ring opening and ring closure of an aziridinium intermediate[422] (Chart 5-58). Alternatively, the phenylbenzazepines are obtained by an acid-catalyzed rearrangement of 1-benzylisoquinolines. When either 2-methyl-1,2,3,4-tetrahydropapaveraldine **145** or 3,4-dihydropapaveraldine metho-salt **146** was treated with zinc and propionic acid, 4-phenylbenzazepine **147** was isolated[423] (Chart 5-59). On the other hand, β-hydroxylaudanosine **148**, which is the reduction product of **145**, was treated with tosyl chloride in the presence of triethylamine to afford the 5-phenylbenzazepine **149**[409] (Chart 5-60). The first total synthesis of rhoeadine **120** by Irie was also accomplished by rearrangement of the spiro-

Chart 5-58.

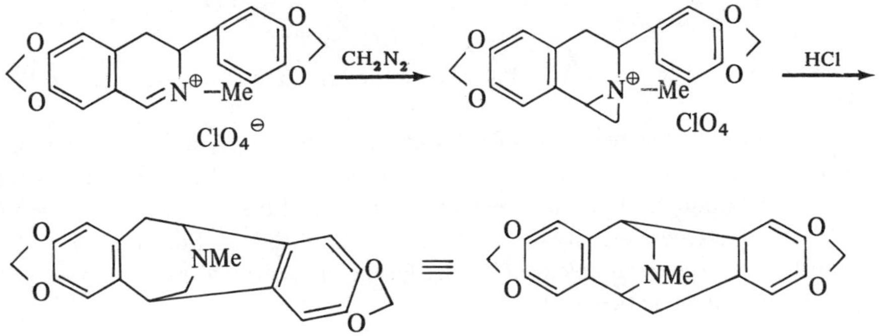

144

Chart 5-59.

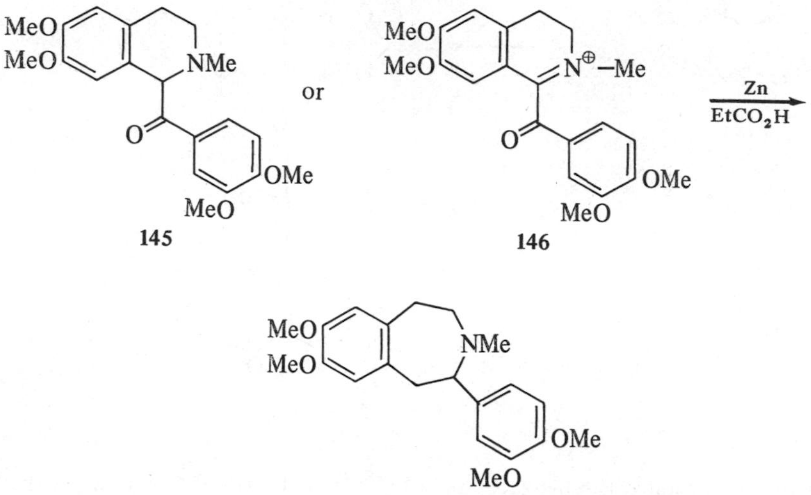

MeO MeO NMe O OMe MeO **145** or MeO MeO N⊕–Me O OMe MeO **146** $\xrightarrow[EtCO_2H]{Zn}$

MeO MeO NMe OMe MeO **147**

Chart 5-60.

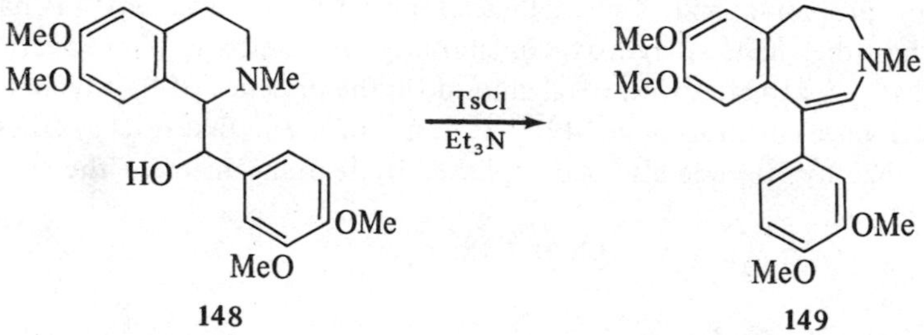

MeO MeO NMe HO OMe MeO **148** $\xrightarrow[Et_3N]{TsCl}$ MeO MeO NMe OMe MeO **149**

isoquinoline **122** to the benzazepine **123** as described in rearrangement (Section 5D).

This survey has shown a simple synthesis of several types of isoquinoline alkaloids, but these reactions provide novel methods for the total synthesis of the natural products. The improvement of yields and the possession of a stereospecificity or stereoselectivity by the modification of the reaction conditions and reagents will become simpler, and more beautiful synthetic methods than the classic ways can be anticipated.

REFERENCES

1. G. Goldschmidt, *Monatsh.*, **2**, 349, 762, 778 (1888).
2. T. Kametani, *The Chemistry of the Isoquinoline Alkaloids,* Hirokawa, Tokyo, and Elsevier, Amsterdam, 1969.
3. C. A. Coulson and H. C. Longuet-Higgins, *Rev. Sci. Paris,* **85**, 929 (1947); C. A. Coulson, P. Daudel, and R. Daudel, *Bull. Soc. Chim. France,* **15**, 1181 (1948).
4. W. M. Whaley and T. R. Govindachari, *Org. Reactions,* **6**, 74 (1951).
5. R. H. F. Manske, *The Alkaloids,* Vol. I - XI, Academic, New York.
6. T. Kametani and M. Ihara, *J. Pharm. Soc. Japan,* **87**, 174 (1967).
7. L. E. Craig and C. S. Tarbell, *J. Amer. Chem. Soc.,* **70**, 2738 (1948).
8. J. Kunitomo, *J. Pharm. Soc. Japan,* **81**, 1253 (1961).

8*a*. E. Gellert and R. E. Summons, *Tetrahedron Lett.,* 5055 (1969).

9. E. Spáth and A. Burger, *Berichte,* **60**, 704 (1927).
10. A. Pictet and A. Gams, *Berichte,* **42**, 2943 (1909); **43**, 2384 (1910).
11. C. Mannich and O. Walter, *Arch. Pharm.,* **265**, 1 (1927); K. W. Rosenmund, M. Nothnagel, and H. Riesenfeldt, *Berichte,* **60**, 392 (1927); C. Mannich and M. Falber, *Arch. Pharm.,* **267**, 601 (1929).
12. T. Nakada and K. Nishihara, *J. Pharm. Soc. Japan,* **64**, 74 (1944).
13. A. Kaufmann and R. Radoseveić, *Berichte,* **49**, 675 (1916).
14. *Ger. Pat.,* 576,532; 579, 277; *Frdl.,* **20**, 719, 722 (1933).
15. H. J. Barber, S. J. Holt, and W. R. Wragg, *Brit. Pat.,* 631,651. [*C. A.,* **44**, 5401 (1950)] ; J. Cymerman and W. F. Short, *J. Chem. Soc.,* **703** (1949).
16. Boots Pure Drug Co. Ltd., *Brit. Pat.,* 642,286 [*C. A.,* **45**, 7155 (1951)].
17. E. Spáth and A. Dobrowsky, *Berichte,* **58**, 1274 (1925).
18. A. Brossi, F. Schenker, and W. Leimgruber, *Helv. Chim. Acta,* **47**, 2089 (1964).

18*a*. J. B. Hendrickson, T. L. Bogard, and M. E. Fisch, *J. Amer. Chem. Soc.,* **92**, 5538 (1970).

19. O. Schales, *Berichte,* **68**, 1579 (1935).
20. K. Kindler and W. Peschke, *Arch. Pharm.,* **271**, 431 (1933).
21. S. Sugasawa, S. Toda, and H. Tomisawa, *J. Pharm. Soc. Japan,* **72**, 252 (1952).
22. S. Sugasawa and H. Shigehara, *J. Pharm. Soc. Japan,* **65**, 370 (1945).
23. T. Kametani and H. Iida, *J. Pharm. Soc. Japan,* **70**, 260 (1950).
24. T. Kametani, K. Fukumoto, H. Iida, and T. Kikuchi, *J. Chem. Soc. (C),* 1178 (1968).

24*a*. A. R. Battersby and J. C. Turner, *J. Chem. Soc.,* 717 (1960).

25. T. Kametani and K. Fukumoto, *J. Chem. Soc.,* 614 (1964).

25*a*. S. M. Kupchan and H. W. Altland, *J. Med. Chem.,* **16**, 913 (1973).

25*b*. M. Tomita, K. Fujitani, and Y. Aoyagi, *Tetrahedron Lett.,* **1967**, 1201.

26. T. Kametani, T. Terui, H. Agui, and K. Fukumoto, *J. Heterocyclic Chem.,* **5**, 753 (1968).
27. J. J. Ritter and F. X. Murphy, *J. Amer. Chem. Soc.,* **74**, 763 (1952).

27*a*. S. Sugasawa and H. Shigehara, *Berichte,* **74**, 459 (1941).

27*b*. T. Kametani, K. Ogasawara, and T. Takahashi, *Tetrahedron,* **29**, 73 (1973); T. Kametani, Y. Hirai, F. Satoh, K. Ogasawara, and K. Fukumoto, *Chem. Pharm. Bull.,* **21**, 907 (1973).

28. T. Kametani, T. Kikuchi, and K. Fukumoto, *Chem. Pharm. Bull.,* **17**, 709 (1969).

29. T. Kametani, O. Kusama, and K. Fukumoto, *Chem. Comm.,* 1212 (1967); *J. Chem. Soc. (C),* 1798 (1968).

29*a*. T. Kametani, K. Wakisaka, and K. Kigasawa, *Chem. Comm.,* 277 (1970); *J. Heterocyclic Chem.,* 7, 509 (1970).

30. R. D. Haworth and W. H. Perkin, Jr., *J. Chem. Soc.,* **127**, 1448 (1925).

30*a*. T. Kametani, T. Kobari, K. Fukumoto, and M. Fujihara, *J. Chem. Soc. (C),* 1796 (1971); cf. Ref. 71.

30*b*. C. Tani, S. Takao, H. Endo, and E. Oda, *J. Pharm. Soc. Japan,* **93**, 268 (1973).

30*c*. T. Kametani, T. Nakano, K. Shishido, and K. Fukumoto, *J. Chem. Soc. (C),* 3350 (1971); H. Iida, H.-C. Hsu, H. Miyano, and T. Kikuchi, *J. Pharm. Soc. Japan,* **91**, 795 (1971).

30*d*. S. Ishiwata, T. Fujii, N. Miyagi, Y. Satoh, and K. Itakura, *Chem. Pharm. Bull.,* **18**, 1850 (1970).

31. M. P. Cava, S. C. Havlicek, A. Lindert, and R. J. Spangler, *Tetrahedron Lett.,* 2937 (1966).

31*a*. I. Ninomiya and T. Naito, *Chem. Comm.,* 137 (1973).

32. A. McCourbery and D. W. Mathieson, *J. Chem. Soc.,* 696 (1949).

33. B. B. Dey and T. R. Govindachari, *Proc. Nat. Inst. Sci. India,* **6**, 219 (1940).

34. B. B. Dey and V. S. Ramanathan, *Proc. Nat. Inst. Sci. India,* **9**, 193 (1943).

35. S. Ishiwata and K. Itakura, *Chem. Pharm. Bull.,* **17**, 2256 (1969).

35*a*. T. Kametani, K. Fukumoto, and M. Fujihara, *J. Chem. Soc. Perkin I,* 394 (1972).

35*b*. Y. Kanaoka, E. Sato, O. Yonemitsu, and Y. Ban, *Tetrahedron Lett.,* 2419 (1964); *Chem. Pharm. Bull.,* **12**, 793 (1964).

35*c*. S. Teitel and A. Brossi, *J. Heterocyclic Chem.,* **5**, 825 (1968).

35*d*. A. Brossi and S. Teitel, *J. Org. Chem.,* **35**, 3559 (1970).

35*e*. A. Brossi and S. Teitel, *Helv. Chim. Acta,* **54**, 1564 (1971).

35*f*. E. E. van Tamelen, C. Placeway, G. P. Schiemenz, and I. G. Wright, *J. Amer. Chem. Soc.,* **91**, 7359 (1969).

35*g*. Y. Inubushi, Y. Masaki, S. Matsumoto, and F. Takami, *J. Chem. Soc. (C),* 1547 (1969).

35*h*. E. Fujita, A. Sumi, and Y. Yoshimura, *Chem. Pharm. Bull.,* **20**, 368 (1972).

35*i*. V. G. Voronin, O. N. Tolkachev, A. B. Prokhorov, V. P. Chernova, and N. A. Preobrazhenskii, *Khim. Geterosikl. Soed.,* 4, 606 (1969).

35*j*. B. Anjaneyule, T. R. Govindachari, and N. Viswanathan, *Tetrahedron,* **27**, 439 (1970).

35*k*. A. R. Battersby, R. Southgate, J. Staunton, and M. Hirst, *J. Chem. Soc. (C),* 1052 (1966).

35*l*. N. T. LeQuang Thuan and J. Gardent, *Bull. Soc. Chim. France,* 2401 (1966).

35*m*. R. D. Haworth, W. H. Perkin, Jr., and H. S. Pink, *J. Chem. Soc.,* **127**, 1709 (1925).

35*n*. W. H. Perkin, Jr., J. N. Ray, and R. Robinson, *J. Chem. Soc.,* **127**, 740 (1925).

35*o*. P. Kerekes and R. Bognar, *J. Prakt. Chem.,* **313**, 923 (1971); *Magy. Kem. Foly.,* 77, 655 (1971).

35*p*. K. Y. Zeecheng and C. C. Cheng, *J. Heterocyclic Chem.*, **10**, 85 (1973).

35*q*. Y. Tsuda, A. Ukai, and K. Isobe, *Tetrahedron Lett.*, 3153 (1972).

36. W. M. Whaley and T. R. Govindachari, *Org. Reactions*, **6**, 151 (1951).

36*a*. E. Späth and H. Röder, *Monatsh.*, **43**, 96, 108, 111 (1922).

37. F. F. Blicke, *Org. Reactions*, **1**, 303 (1942).

38. A. Pictet and T. Spengler, *Berichte*, **44**, 2030 (1911).

39. H. Decker and P. Becker, *Annalen*, **395**, 342 (1913).

40. A. Pictet and T. Q. Chou, *Berichte*, **49**, 370 (1916).

40*a*. T. Kametani, M. Takeshita, and S. Takano, *J. Chem. Soc. Perkin I*, 2834 (1972).

41. C. Schöpf and W. Salzer, *Annalen*, **544**, 1 (1940).

42. G. Hahn, L. Bärwarld, O. Schales, and H. Werner, *Annalen*, **520**, 107 (1935).

43. G. Hahn and H. Werner, *Annalen*, **520**, 123 (1935).

43*a*. G. E. Krejcark, B. W. Dominy, and R. G. Lawton, *Chem. Comm.*, 1450 (1968).

43*b*. S. Yamada, M. Konda, and T. Shioiri, *Tetrahedron Lett.*, 2215 (1972).

44. L. Helfer, *Helv. Chim. Acta*, **7**, 945 (1924).

45. T. Kametani, H. Iida, and T. Kikuchi, *J. Pharm. Soc. Japan*, **88**, 1185 (1968).

45*a*. M. Tomita, M. Kozuka, H. Ohyabu, and K. Fujitani, *J. Pharm. Soc. Japan*, **90**, 82 (1970).

45*b*. S. Ishiwata and K. Itakura, *Chem. Pharm. Bull.*, **18**, 896 (1970).

46. A. Pictet and A. Gams, *Berichte*, **44**, 2480 (1911).

47. A. Pictet and A. Gams, *Compt. Rend.*, **153**, 386 (1911).

48. R. D. Haworth, W. H. Perkin, Jr., and J. Rankin, *J. Chem. Soc.*, **125**, 1686 (1924).

49. E. Späth and E. Kruta, *Monatsh.*, **50**, 341 (1928).

50. A. R. Battersby, R. Southgate, J. Staunton, and M. Hirst, *J. Chem. Soc. (C)*, 1052 (1966).

50*a*. T. Kametani, K. Fukumoto, T. Terui, K. Yamaki, and E. Taguchi, *J. Chem. Soc. (C)*, 2709 (1971).

50*b*. T. Kametani, T. Honda, and M. Ihara, *J. Chem. Soc. (C)*, 3318 (1971).

50*c*. T. Kametani, T. Honda, and M. Ihara, *J. Chem. Soc. (C)*, 2396 (1971).

51. C. Schöpf, *Angew. Chem.*, **50**, 797 (1938).

52. T. Kametani and M. Ihara, *J. Chem. Soc. (C)*, 530 (1967); 1305 (1968).

52*a*. T. Kametani, H. Iida, T. Kikuchi, T. Honda, and M. Ihara, *J. Heterocyclic Chem.*, **7**, 491 (1970).

53. H. Kondo and E. Ochiai, *J. Pharm. Soc. Japan*, **495**, 313 (1923).

54. H. Kondo and S. Tanaka, *J. Pharm. Soc. Japan*, **50**, 923 (1930).

55. S. Frankel and K. Zeimer, *Biochem. Z.*, **110**, 234 (1920).

56. J. Wellisch, *Biochem. Z.*, **49**, 173 (1913).

57. W. S. Ide and J. S. Buck, *J. Amer. Chem. Soc.*, **59**, 726 (1937).

58. S. N. Chakravarti, N. A. Vaidynathan, and A. Venkatasubban, *J. Indian Chem. Soc.*, **9**, 573 (1932).

59. R. Campbell, R. D. Haworth, and W. H. Perkin, Jr., *J. Chem. Soc.*, 32 (1926).

60. C. Schöpf and H. Bayerler, *Annalen*, **513**, 190 (1934).

61. G. Hahn and S. Schales, *Berichte*, **58**, 26 (1925).

62. E. Späth, F. Kuffer, and F. Kesztler, *Berichte*, **69**, 378 (1936).

63. T. Kametani, K. Fukumoto, H. Agui, H. Yagi, K. Kigasawa, H. Sugahara, M. Hiiragi, T. Hayasaka, and H. Ishimaru, *J. Chem. Soc. (C)*, 174 (1968).
64. T. Kametani, S. Shibuya, and M. Satoh, *Chem. Pharm. Bull.*, **16**, 953 (1968).
65. G. Hahn and H. J. Schuls, *Berichte*, **71**, 2135 (1938).
66. W. O. Kermack and R. H. Slatter, *J. Chem. Soc.*, 32 (1928).
67. R. D. Haworth, W. H. Perkin, Jr., and J. Rankin, *J. Chem. Soc.*, **125**, 1686 (1924).
68. E. Späth and E. Kruta, *Monatsh.*, **50**, 341 (1928).
69. E. Späth and F. Berger, *Berichte*, **63**, 3098 (1930).
70. T. Kametani, K. Kigasawa, M. Hiiragi, and H. Ishimaru, *J. Heterocyclic Chem.*, **7**, 51 (1970).
71. T. Kametani, K. Fukumoto, K. Kigasawa, M. Hiiragi, H. Ishimaru, and K. Wakisaka, *J. Chem. Soc. (C)*, 1805 (1971).
72. T. Kametani, T. Terui, A. Ogino, and K. Fukumoto, *J. Chem. Soc. (C)*, 874 (1969).
72*a*. M. R. Falco, J. X. Dervis, and G. Mann, *Z. Chem.*, **13**, 56 (1973).
72*b*. S. McLean, M.-S. Lin, and J. Whelan, *Tetrahedron Lett.*, 2425 (1968); *Can. J. Chem.*, **48**, 948 (1970).
72*c*. T. Kishimoto and S. Uyeo, *J. Chem. Soc. (C)*, 2600 (1969).
72*d*. T. Kametani, S. Hibino, and S. Takano, *Chem. Comm.*, 925 (1971); *J. Chem. Soc. Perkin I*, 391 (1972).
72*e*. B. Nalliah, Q. A. Ahmed, R. H. F. Manske, and R. D. Rodrigo, *Can. J. Chem.*, **50**, 1819 (1972).
72*f*. S. MaLean and J. Whelan, *Can. J. Chem.*, **51**, 2457 (1973).
72*g*. T. Kishimoto and S. Uyeo, *J. Chem. Soc. (C)*, 1644 (1971).
72*h*. T. Kametani, K. Kigasawa, M. Hiiragi, and O. Kusama, *J. Heterocyclic Chem.*, **10**, 31 (1973).
72*i*. W. Wiegrebe, L. Faber, and H. Budzikiewicz, *Annalen*, **733**, 125 (1970).
72*j*. R. V. Stevens and L. E. DuPree, Jr., *Chem. Comm.*, 1585 (1970); R. V. Stevens, L. E. DuPree, Jr., and P. L. Loewenstein, *J. Org. Chem.*, **37**, 977 (1972).
72*k*. Y. Tsuda and K. Isobe, *Chem. Comm.*, 1555 (1971).
72*l*. M. Lora-Tamayo, R. Madroñero, G. G. Muñoz, J. M. Marzal, and M. Stud, *Berichte*, **94**, 199 (1961).
72*m*. P. A. Bather, J. L. Lindsay Smith, and R. O. C. Norman, *J. Chem. Soc. (C)*, 3060 (1971).
72*n*. A. R. Battersby, R. Binks, and P. S. Uzzell, *Chem. Ind.*, 1039 (1955).
72*o*. T. O. Soine, C.-H. Chen, and K. H. Lee, *J. Pharm. Sci.*, **59**, 1529 (1970); C.-H. Chen, T. O. Soine, and K.-H. Lee, *J. Pharm. Sci.*, **60**, 1634 (1971); C.-H. Chen and T. O. Soine, *J. Pharm. Soc.*, **61**, 451 (1972).
72*p*. S. Natarajan and B. R. Pai, *Indian J. Chem.*, **10**, 451 (1972).
72*q*. F. R. Stermitz and D. K. Williams, *J. Org. Chem.*, **38**, 1761 (1973).
73. S. Gabriel, *Berichte*, **18**, 1251, 2433, 3470 (1885).
74. S. Gabriel, *Berichte*, **19**, 830, 1653 (1886).
74*a*. A. Mondon and K. Krohn, *Berichte*, **103**, 2729 (1970).
75. T. Kametani, S. Shibuya, S. Seino, and K. Fukumoto, *J. Chem. Soc.*, 4146 (1964).

75*a*. I. W. Elliot, *J. Heterocyclic Chem.*, 7, 1229 (1970).

75*b*. D. G. Farber and A. Giacomazzi, *Anales Asoc. Quim. Argentina*, 58, 133 (1970) [*C. A.*, 73, 131, 173p (1970)].

75*c*. I. W. Elliot, *J. Heterocyclic Chem.*, **9**, 853 (1972).

75*d*. W. E. Kreighbaum, W. F. Kavanaugh, and W. T. Comer, *J. Heterocyclic Chem.*, **10**, 317 (1973).

75*e*. T. Kametani, H. Iida, and C. Kibayashi, *J. Heterocyclic Chem.*, 7, 339 (1970).

75*f*. A. S. Bailey and C. R. Worthing, *J. Chem. Soc.*, 4535 (1956).

76. T. Zincke, *Berichte*, **25**, 405, 1493 (1892).

77. E. Bamberger and M. Kischelt, *Berichte*, **25**, 1141 (1892).

78. E. Bamberger and W. Frew, *Berichte*, **27**, 198 (1894).

79. M. LeBlanc, *Berichte*, **21**, 2299 (1888).

80. S. Gabriel, *Berichte*, **20**, 2199 (1887).

81. R. Grewe and A. Mondon, *Berichte*, **81**, 279 (1948).

82. J. von Brown and F. Zobel, *Berichte*, **56**, 2151 (1923).

83. L. Helfer, *Helv. Chim. Acta*, **6**, 789 (1923).

84. D. Bain, W. H. Perkin, Jr., and R. Robinson, *J. Chem. Soc.*, **105**, 2392 (1914).

85. S. Gabriel, *Berichte*, **49**, 1608 (1916).

86. G. M. Badger, J. W. Cook, and G. M. S. Donald, *J. Chem. Soc.*, 1394 (1951).

87. M. Gates and G. Tschudi, *J. Amer. Chem. Soc.*, **74**, 1109 (1952); 78, 1380 (1956).

87*a*. W. H. Perkin, Jr., *J. Chem. Soc.*, **111**, 492 (1918).

87*b*. J. Comin and V. Deulofeu, *Tetrahedron*, **6**, 63 (1959).

87*c*. N. S. Narasimhan and B. H. Bhide, *Chem. Ind.*, 621 (1969).

88. S. Gabriel and J. Colman, *Berichte*, **33**, 980 (1900).

88*a*. J. Finkelstein and A. Brossi, *J. Heterocyclic Chem.*, **4**, 315 (1967).

88*b*. G. Rosen and F. D. Popp, *Can. J. Chem.*, **47**, 864 (1969).

89. S. Gabriel and J. Colman, *Berichte*, **33**, 2630 (1900).

89*a*. I. G. Hinton and F. G. Mann, *J. Chem. Soc.*, 599 (1959).

90. P. Staub, *Helv. Chim. Acta*, **5**, 888 (1922).

91. W. J. Gensler, *Org. Reactions*, **6**, 191 (1951).

92. P. Fritsch, *Berichte*, **26**, 419 (1893).

93. P. Fritsch, *Annalen*, **286**, 1 (1895).

94. E. Schlittler and J. Müller, *Helv. Chim. Acta*, **31**, 914 (1948).

95. C. K. Bradsher, *Chem. Rev.*, **38**, 447 (1946).

96. T. Kametani, S. Shibuya, and I. Noguchi, *J. Pharm. Soc. Japan*, **85**, 667 (1965).

97. R. B. Woodward and W. E. Doering, *J. Amer. Chem. Soc.*, **67**, 860 (1945).

98. E. Fischer, *Berichte*, **26**, 764 (1893); **27**, 165 (1894).

99. R. Forsyth, C. I. Kelly, and F. L. Pyman, *J. Chem. Soc.*, **127**, 1659 (1925).

100. B. B. Dey and T. R. Govindachari, *Arch. Pharm.*, **275**, 383 (1937).

101. J. M. Bobbitt, J. M. Kiely, K. L. Khanna, and R. Ebermann, *J. Org. Chem.*, **30**, 2247 (1965).

102. G. Grethe, M. Uskokovic and A. Brossi, *J. Org. Chem.*, **33**, 2500 (1968).

103. J. M. Bobbitt, D. N. Roy, A. Marchand, and C. W. Allen, *J. Org. Chem.*, **32**, 2225 (1967).

103*a*. M. Takido, K. L. Klanna, and A. G. Paul, *J. Pharm. Sci.*, **59**, 271 (1970).

103*b*. G. J. Kapagia, M. B. E. Fayez, M. L. Sethin, and G. S. Rao, *Chem. Comm.*, 856 (1970).

103*c*. A. J. Birch, A. H. Jackson, P. V. R. Shannon, and P. S. P. Varma, *Tetrahedron Lett.*, 4789 (1972).

103*d*. N. E. Cundasaway and D. B. MacLean, *Can. J. Chem.*, **50**, 3028 (1972).

103*e*. S. F. Dyke and E. P. Tiley, *Tetrahedron Lett.*, 5175 (1972).

103*f*. S. F. Dyke and M. Sainsbury, *Tetrahedron,* **23**, 3161 (1967).

104. G. Grethe, H. L. Lee, M. Uskokovic, and A. Brossi, *J. Org. Chem.*, **33**, 491 (1968).

105. G. Grethe, M. Uskokovic, T. Williams, and A. Brossi, *Helv. Chim. Acta,* **50**, 2397 (1967).

105*a*. A. Brossi, G. Grethe, S. Teitel, W. C. Wildman, and D. T. Bailey, *J. Org. Chem.*, **35**, 1100 (1970).

105*b*. R. B. Herbert and C. J. Moody, *Chem. Comm.*, 121 (1970).

105*c*. B. Chauncy and E. Gellert, *Austral. J. Chem.*, **23**, 2503 (1970).

106. A. Pictet and S. Popovici, *Berichte,* **25**, 733 (1892).

107. M. A. Schwartz and S. W. Scott, *J. Org. Chem.*, **36**, 1827 (1971).

108. J. Knabe and P. Horn, *Arch. Pharm.*, **300**, 547 (1967).

109. A. Pictet and M. Finkelstein, *Berichte,* **42**, 1979 (1909); L. E. Craig and D. S. Tarbell, *J. Amer. Chem. Soc.*, **70**, 2378 (1948).

110. F. Faltis and E. Alder, *Arch. Pharm.*, **284**, 281 (1951).

111. L. Marion and V. Grassie, *J. Amer. Chem. Soc.*, **73**, 1290 (1951).

112. I. Kikkawa, *J. Pharm. Soc. Japan,* **77**, 1244 (1957).

113. M. Shamma and W. A. Slusarchyk, *Chem. Rev.*, **64**, 73 (1964).

114. T. Kametani, B. Shibuya, S. Seino, and K. Fukumoto, *J. Chem. Soc.*, 4146 (1964).

115. T. Kametani and S. Shibuya, *J. Chem. Soc.*, 5565 (1965).

116. T. Kametani, T. Honda, H. Shimanouchi, and Y. Sasada, *Chem. Comm.*, 1072 (1972).

117. R. D. Haworth and A. R. Pinder, *J. Chem. Soc.*, 1776 (1950).

118. W. H. Perkin, Jr. and R. Robinson, *J. Chem. Soc.*, **99**, 775 (1911).

119. A. R. Battersby and H. Spencer, *J. Chem. Soc.*, 1087 (1965).

120. M. Ohta, T. Tani, S. Morozumi, S. Kodaira, and K. Kuriyama, *Tetrahedron Lett.*, 1857 (1963).

121. M. Ohta, H. Tani, S. Morozumi, S. Kodaira, and K. Kuriyama, *Tetrahedron Lett.*, 693 (1964).

122. W. M. Whaley and M. Meadow, *J. Chem. Soc.*, 1067 (1953).

123. T. R. Govindachari, M. V. Lakshmikam, and S. Rajadurai, *Chem. Ind.*, 664 (1960); *Tetrahedron,* **14**, 284 (1964).

124. T. Kametani and S. Shibuya, *Tetrahedron Lett.*, 1897 (1965).

125. D. G. Farber and A. Giacomazi, *Chem. Ind.*, 57 (1968).

126. B. K. Cassels and V. Deulofeu, *Tetrahedron, Suppl.* 8, Part II, 485.

127. B. Frydman, R. Bendisch, and V. Deulofeu, *Tetrahedron,* **4**, 342 (1958).

128. E. Spāth and A. Burger, *Monatsh.*, **47**, 733 (1926).

129. T. Kametani, M. Ihara, K. Fukumoto, and H. Yagi, *J. Chem. Soc. (C)*, 2030 (1969).

130. C. Ferrari and V. Deulofeu, *Tetrahedron*, **18**, 419 (1962).

131. A. R. Battersby, R. Binks, D. M. Foulkes, R. J. Francis, and D. J. McCaldin, *Proc. Chem. Soc.*, 203 (1963).

132. M. Tomita and F. Kusuda, *J. Pharm. Soc. Japan*, **72**, 280, 793 (1952).

133. T. Kametani, S. Takano, K. Masuko, and S. Kurihara, *J. Pharm. Soc. Japan*, **85**, 166 (1965).

134. T. Kametani and H. Yagi, *Chem. Pharm. Bull.*, **15**, 1283 (1967).

135. G. Grethe, M. Uskokovic, and A. Brossi, *J. Org. Chem.*, **33**, 2500 (1968).

136. T. Kametani, S. Takano, F. Sasaki, and K. Yamaki, *Chem. Pharm. Bull.*, **16**, 20 (1968).

137. H. Corrodi and E. Hardegger, *Helv. Chim. Acta*, **39**, 889 (1956).

138. A. R. Battersby, R. Southgate, J. Staunton, and M. Hirst, *J. Chem. Soc. (C)*, 1052 (1966).

139. T. Kametani and M. Ihara, *J. Chem. Soc. (C)*, 530 (1967); 1305 (1968).

140. T. Kametani and M. Ihara, *J. Chem. Soc. (C)*, 2010 (1966).

141. T. Kametani, H. Iida, and K. Sakurai, *Chem. Pharm. Bull.*, **16**, 1623 (1968); *J. Chem. Soc. (C)*, 500 (1969).

142. T. Kametani, K. Sakurai, and H. Iida, *J. Pharm. Soc. Japan*, 88, 1163 (1968).

143. Y.-Y. Hsieh, P.-C. Pan W.-C. Chen, and I.-S. Kao, *Sci. Sinica*, **12**, 2020 (1964) [*C. A.*, **62**, 9184 (1965)].

144. T. Kametani, S. Takano, H. Iida, and M. Shinbo, *J. Chem. Soc. (C)*, 298 (1969).

145. H. Furukawa, *J. Pharm. Soc. Japan*, **85**, 335 (1965).

146. M. Tomita, H. Furukawa, S.-T. Lu, and S. M. Kupchan, *Tetrahedron Lett.*, 4309 (1965); *Chem. Pharm. Bull.*, **15**, 959 (1967).

147. M. Tomita, K. Fujitani, and T. Kishimoto, *J. Pharm. Soc. Japan*, **82**, 1148 (1962).

148. Y. Inubushi, Y. Masaki, S. Matsumoto, and F. Takami, *J. Chem. Soc. (C)*, 1547 (1969).

149. A. R. Battersby, R. B. Herbert, L. Mo, and F. Šantavý, *J. Chem. Soc. (C)*, 1739 (1967).

150. A. R. Battersby, R. B. Bradbury, R. B. Herbert, M. H. G. Munro, and R. Ramage, *Chem. Comm.*, 450 (1967).

151. T. Kametani, H. Yagi, K. Fukumoto, and F. Satoh, *Chem. Pharm. Bull.*, **16**, 3297 (1968); T. Kametani, F. Satoh, H. Yagi, and K. Fukumoto, *J. Org. Chem.*, **33**, 690 (1968).

152. A. Brossi, J. O'Brien, and S. Teitel, *Helv. Chim. Acta*, **52**, 678 (1969).

152*a*. A. Brossi and S. Teitel, *Helv. Chim. Acta*, **54**, 1564 (1971).

153. M. Gates and G. Tshudi, *J. Amer. Chem. Soc.*, **74**, 1109 (1952); 78, 1380 (1956).

154. D. Elad and D. Ginsburg, *J. Amer. Chem. Soc.*, **76**, 312 (1954); *J. Chem. Soc.*, 3052 (1954).

155. A. R. Battersby, D. M. Foulkes, and R. Binks, *J. Chem. Soc.*, 3323 (1965).

156. D. H. R. Barton, G. W. Kirby, W. Steglich, and G. M. Thomas, *Proc. Chem. Soc.*, 203 (1963).

157. T. Kametani, M. Ihara, K. Fukumoto, and H. Yagi, *J. Chem. Soc. (C)*, 2030 (1969).

158. A. Brossi, A. Cohen, J. M. Osbond, Pl. A. Plattner, O. Schnider, and J. C. Wickens, *Chem. Ind.*, 491 (1958).

159. M. Barash and J. M. Osbond, *Chem. Ind.*, 490 (1958).
160. R. P. Evstignceva, R. S. Livshits, L. I. Zakharkin, M. S. Bainova, and N. A. Preobrazhensky, *Dokl. Akad, Nauk SSSR.*, **75**, 539 (1950).
161. A. R. Battersby and J. C. Turner, *Chem. Ind.*, 1324 (1958).
162. A. R. Battersby and J. C. Turner, *J. Chem. Soc.*, 717 (1960); *Brit. Pat.* 895,910 [*C. A.*, 58, 6878 (1963)].
163. S. Teitel and A. Brossi, *J. Amer. Chem. Soc.*, **88**, 4068 (1966).
164. A. Brossi, M. Baumann, and O. Schnider, *Helv. Chim. Acta*, **42**, 1515 (1959).
165. A. Grüssner, E. Jaeyer, J. Hellerbach, and O. Schnider, *Helv. Chim. Acta*, **42**, 2431 (1959).
166. A. R. Battersby, J. R. Merchant, E. A. Ruveda, and S. S. Salgar, *Chem. Comm.*, 315 (1965).
167. H. T. Openshow and N. Whittaker, *J. Chem. Soc. (C)*, 91 (1969).
168. Cs. Szántay and By. Kalaus, *Acta Chem. Scand. Sci. Hung.*, **49**, 427 (1966) [*C. A.*, **67**, 22073m (1967)].
169. H. T. Openshow and N. Whittaker, *J. Chem. Soc.*, 1461 (1963).
170. E. E. van Tamelen, C. Placeway, G. P. Schiemenz, and I. G. Wright, *J. Amer. Chem. Soc.*, **91**, 7759 (1969).
171. W. Klötzer, S. Teitel, and A. Brossi, *Helv. Chim. Acta*, **54**, 2057 (1971).
172. H. Irie, S. Tani, and H. Yamane, *Chem. Comm.*, 1713 (1970).
173. D. H. R. Barton and T. Cohen, *Festschrift A. Stoll*, Birkhauser, Basel, 1967, p. 117.
174. D. H. R. Barton, *Proc. Chem. Soc.*, 239 (1963).
175. T. Kametani and K. Fukumoto, *Jap. J. Pharm. Chem.*, **23**, 426 (1963).
176. B. Franck, G. Blaschke, and G. Schlingloff, *Angew. Chem.*, **75**, 957 (1963).
177. H. Musso, *Angew. Chem.*, **75**, 965 (1963).
178. A. I. Scott, *Quart. Rev.*, **19**, 1 (1965).
179. A. R. Battersby, *Oxidative Coupling of Phenols* (W. I. Taylor and A. R. Battersby, Eds.), Dekker, New York, 1967.
180. T. Kametani and K. Fukumoto, *Farumashia*, **4**, 144 (1968).
181. T. Kametani and K. Fukumoto, *Phenolic Oxidation*, Gihodo, Tokyo, 1970, p. 121.
182. T. Kametani and K. Fukumoto, *J. Heterocyclic Chem.*, **8**, 341 (1971).
183. R. Pummerer and E. Frankfurter, *Berichte*, **47**, 1472 (1914).
184. J. Gadamer, *Arch. Pharm.*, **249**, 498, 680 (1911).
185. R. Robinson, *The Structural Relationships of Natural Products*, Clarendon Press, Oxford, 1955.
186. C. Schöpe and K. Thierfelder, *Annalen*, **497**, 22 (1932).
186*a*. A recent review of biosynthesis of isoquinoline alkaloids is found in H. R. Schütte, *Biosynthese der Alkaloide* (K. Mothes and H. R. Schütte, Eds.), VEB Deutscher Verlag der Wissenschaften, Berlin, 1969, p. 367.
187. D. G. Odonovan and H. Noran, *J. Chem. Soc. (C)*, 1737 (1969).
188. D. G. Odonovan, E. Barry, and H. Horan, *J. Chem. Soc. (C)*, 2398 (1971).
189. J. M. Bobbitt, R. Ebermann, and M. Schubert, *Tetrahedron Lett.*, 575 (1963).
190. B. Franck and G. Blaschke, *Tetrahedron Lett.*, 569 (1963); B. Franck, G. Blaschke, and K. Lewejohann, *Annalen*, **685**, 207 (1965).

191. J. M. Bobbitt, J. T. Stock, A. Marchand, and K. H. Weisgraber, *Chem. Ind.*, 2117 (1966).

192. M. Tomita, K. Fujitani, Y. Masaki, and K. H. Lee, *Chem. Pharm. Bull.*, **16**, 251 (1968).

193. M. Tomita, Y. Masaki, and K. Fujitani, *Chem. Pharm. Bull.*, **16**, 257 (1968).

194. B. Umezawa, O. Hoshino, H. Hara, and J. Sakakibara, *Chem. Pharm. Bull.*, **16**, 381, 566 (1968).

195. O. Hoshino, H. Hara, M. Wada, and B. Umezawa, *Chem. Pharm. Bull.*, **18**, 637 (1970).

196. Y. Inubushi, Y. Aoyagi, and M. Matsuo, *Tetrahedron Lett.*, 2363 (1969).

197. J. M. Bobbitt, I. Noguchi, H. Yagi, and K. H. Weisgraber, *J. Amer. Chem. Soc.*, **93**, 3551 (1971).

198. T. Kametani, T. Kikuchi, and K. Fukumoto, *Chem. Comm.*, 546 (1967); *Chem. Pharm. Bull.*, **16**, 1003 (1968).

199. A. H. Jackson and G. W. Stewart, *Chem. Comm.*, 149 (1971).

200. T. Kametani, K. Fukumoto, and M. Fujihara, *Chem. Comm.*, 352 (1971).

201. D. H. R. Barton, G. W. Kirby, and A. Wiechers, *J. Chem. Soc. (C)*, 2312 (1966).

202. B. Franck, G. Blaschke, and G. Schlingloff, *Tetrahedron Lett.*, 439 (1962).

203. M. Tomita, Y. Masaki, K. Fujitani, and Y. Sakatani, *Chem. Pharm. Bull.*, **16**, 688 (1968).

204. B. Franck and G. Blaschke, *Annalen*, **668**, 145 (1963).

205. A. M. Choudhury, J. G. C. Coutts, A. K. Durbin, K. Schofield, and D. J. Humphreys, *J. Chem. Soc. (C)*, 2070 (1969).

206. T. Kametani, H. Nemoto, T. Kobari, and S. Takano, *J. Heterocyclic Chem.*, 7, 181 (1970).

207. J. M. Bobbitt, and R. C. Hallcher, *Chem. Comm.*, 543 (1971).

208. F. L. Pyman, *J. Chem. Soc.*, **95**, 1266 (1909).

209. G. Blaschke, *Arch. Pharm.*, **301**, 432 (1968); **303**, 358 (1970).

210. R. Robinson and S. Sugasawa, *J. Chem. Soc.*, 789 (1932).

211. E. E. van Tamelen, *Fortschr. Chem. Org. Naturstoffe*, **19**, 242 (1961).

212. B. Franck and G. Blaschke, *Annalen*, **659**, 132 (1962); B. Franck and G. Schlingloff, *Annalen*, **695**, 142 (1966).

213. S. M. Albonico, A. M. Kuck, and V. Deulofeu, *Chem. Ind.*, 1580 (1964); *Annalen*, **685**, 200 (1965).

214. T. Kametani and I. Noguchi, *J. Chem. Soc. (C)*, 1440 (1967).

214*a*. B. Franck, G. Blaschke, and G. Schlingloff, *Angew. Chem. Int. Ed.*, **3**, 192 (1964).

215. T. Kametani, T. Sugahara, and K. Fukumoto, *Tetrahedron*, **25**, 3667 (1969).

216. A. H. Jackson and J. A. Martin, *Chem. Comm.*, 420 (1965); *J. Chem. Soc. (C)*, 2061 (1969).

217. W. W. C. Chen and P. Maitland, *J. Chem. Soc. (C)*, 753 (1966).

218. T. Kametani, K. Fukumoto, A. Kozuka, and H. Yagi, *J. Chem. Soc. (C)*, 2034 (1969).

219. T. Kametani, A. Kozuka, and K. Fukumoto, *J. Chem. Soc. (C)*, 1021 (1971).

220. B. Franck, G. Dunkelmann, and H. J. Lubs, *Angew. Chem.*, **79**, 1066 (1967); *Angew. Chem. Int. Ed.*, **6**, 1075 (1967).

221. T. Kametani, K. Fukumoto, K. Kigasawa, and K. Wakisaka, *Chem. Pharm. Bull.*, **19**, 714 (1971).

222. B. Franck and L. F. Tietze, *Angew. Chem.*, **79**, 817 (1967); *Angew. Chem. Int. Ed.*, **6**, 799 (1967).

223. A. R. Battersby, J. L. McHugh, J. Staunton, and M. Todd, *Chem. Comm.*, 985 (1971).

224. T. Kametani, R. Charubala, M. Ihara, M. Koizumi, and K. Fukumoto, *Chem. Comm.*, 289 (1971); T. Kametani, R. Charubala, M. Ihara, M. Koizumi, K. Takahashi, and K. Fukumoto, *J. Chem. Soc. (C)*, 3315 (1971).

225. D. H. R. Barton, D. S. Bhakuni, G. M. Chapman, and G. W. Kirby, *J. Chem. Soc. (C)*, 2134 (1967).

226. D. H. R. Barton, D. S. Bhakuni, G. M. Chapman, G. W. Kirby, L. J. Haynes, and K. L. Stuart, *J. Chem. Soc. (C)*, 1295 (1967).

227. A. R. Battersby, T. J. Brockson, and R. Ramage, *Chem. Comm.*, 464 (1969) and references cited therein.

228. A. R. Battersby, T. H. Brown, and J. H. Clements, *Proc. Chem. Soc.*, 85 (1964); *J. Chem. Soc.*, 4550 (1965).

229. T. Kametani, and H. Yagi, *Chem. Comm.*, 366 (1967); *J. Chem. Soc. (C)*, 2182 (1967).

229*a*. D. S. Bhakuni, S. Satish, and M. M. Dlar, *Tetrahedron*, **28**, 4579 (1972).

230. D. H. R. Barton, *Chemistry in Britain*, 330 (1968).

231. S. Pfeifer and L. Kuhn, *Pharmazie*, **20**, 394 (1965).

231. T. Kametani, S. Shibuaya, T. Nakano, and K. Fukumoto, *J. Chem. Soc. (C)*, 3818 (1971).

232. A. H. Jackson and J. A. Martin, *Chem. Comm.*, 142 (1965); *J. Chem. Soc. (C)*, 2222 (1966).

233. M. Shamma and W. A. Slusarchyk, *Chem. Comm.*, 528 (1965).

234. T. Kametani, K. Takahashi, and K. Fukumoto, *J. Chem. Soc. (C)*, 3617 (1971).

235. T. Kametani and I. Noguchi, *J. Chem. Soc. (C)*, 502 (1969).

236. T. Kametani and I. Noguchi, *Chem. Pharm. Bull.*, **16**, 2451 (1968).

236a. O. Hoshino, T. Tobinaga, and B. Umezawa, *Chem. Comm.*, 1533 (1971).

236b. S. M. Kupchan and A. J. Liepa, *J. Amer. Chem. Soc.*, **95**, 4062 (1973).

237. J. Harley-Mason, *J. Chem. Soc.*, 1465 (1953).

237*a*. A. Brossi, A. Ramel, J. O'Brien, and S. Teitel, *Chem. Pharm. Bull.*, **21**, 1839 (1973).

237*b*. T. Kametani, M. Ihara, and K. Takahashi, *Chem. Pharm. Bull.*, **20**, 1587 (1972).

237*c*. M. L. Carvalhas, *J. Chem. Soc. Perkin I*, 327 (1972).

238. D. H. R. Barton, G. W. Kirby, W. Steglich, and G. H. Thomas, *Proc. Chem. Soc.*, 203 (1963); D. H. R. Barton, D. S. Bhakuni, R. James, and G. W. Kirby, *J. Chem. Soc. (C)*, 128 (1967).

239. K. L. Stuart, V. Teetz, and B. Franck, *Chem. Comm.*, 333 (1969).

240. B. Franck, J. Lubs, and G. Dunkelmann, *Angew. Chem.*, **79**, 989 (1967); *Angew. Chem. Int. Ed.*, **6**, 969 (1967).

240*a*. T. Kametani, S. Takano, and T. Kobari, *J. Chem. Soc. (C)*, 1030 (1971).

240*b*. L. L. Miller, R. F. Stermitz, and J. R. Falck, *J. Amer. Chem. Soc.*, **93**, 5941 (1971); **95**, 2651 (1973).

240*c*. E. Kotani, M. Kitazawa, and S. Tobinaga, *Abstracts of 17th Symposium on the Chemistry of Natural Products*, Tokyo, Japan, 1973, p. 298.

241. A. R. Battersby, A. K. Bhatnagar, P. Hackett, C. W. Thornber, and J. Staunton, *Chem. Comm.*, 1214 (1968).

242. T. Kametani, T. Kobari, and K. Fukumoto, *Chem. Comm.*, 288 (1972).

243. D. H. R. Barton, R. B. Boar, and D. A. Widdowson, *J. Chem. Soc. (C)*, 1213 (1970).

244. J. E. Gervay, F. McCapra, T. Money, G. M. Sharma, and A. I. Scott, *Chem. Comm.*, 142 (1966).

245. A. Mondon and M. Ehrhardt, *Tetrahedron Lett.*, 2557 (1966).

246. D. H. R. Barton, R. James, G. W. Kirby, R. W. Turner, and D. A. Widdowson, *Chem. Comm.*, 266 (1967); *J. Chem. Soc. (C)*, 1529 (1968).

247. D. H. R. Barton, R. B. Boar, and D. A. Widdowson, *J. Chem. Soc. (C)*, 1208 (1970).

247a. K. Ito, H. Furukawa, and H. Tanaka, *Chem. Pharm. Bull.*, **19**, 1509 (1971).

248. B. Franck and V. Teetz, *Angew. Chem.*, **83**, 409 (1971); *Angew. Chem. Int. Ed.*, **10**, 411 (1971).

249. J. M. Paton, P. L. Pauson, and T. S. Stevens, *J. Chem. Soc. (C)*, 1309 (1969).

250. H. R. Schütte, *Biosynthese der Alkaloide* (K. Mothes and H. R. Schütte, Eds.), VEB Deutscher Verlag der Wissenschaften, Berlin, 1969, p. 420.

251. M. A. Schwartz and R. A. Holton, *J. Amer. Chem. Soc.*, **92**, 1090 (1970).

251*a*. E. Kotani, N. Takeuchi, and S. Tobinaga, *Tetrahedron Lett.*, 2735 (1973).

251*b*. M. A. Schwartz, B. F. Rose, and B. Vishnuvajjala, *J. Amer. Chem. Soc.*, **95**, 612 (1973).

252. R. A. Abramovitch and S. Takahashi, *Chem. Ind.*, 1039 (1963).

253. B. Franck and J. Lubs, *Annalen*, **720**, 131 (1968).

254. B. Franck and J. Lubs, *Angew. Chem.*, **80**, 238 (1968); *Angew. Chem. Int. Ed.*, 7, 223 (1968).

255. D. H. R. Barton and G. W. Kirby, *J. Chem. Soc.*, 806 (1962).

256. T. Kametani, K. Yamaki, H. Yagi, and K. Fukumoto, *Chem. Comm.*, 425 (1969); *J. Chem. Soc. (C)*, 2602 (1969).

257. T. Kametani, C. Seino, K. Yamaki, S. Shibuya, K. Fukumoto, K. Kigasawa, F. Satoh, M. Hiiragi, and T. Hayasaka, *J. Chem. Soc. (C)*, 1043 (1971).

258. T. Kametani, K. Shishido, E. Hayashi, C. Seino, T. Kohno, S. Shibuya, and K. Fukumoto, *J. Org. Chem.*, **36**, 1295 (1971).

259. A. R. Battersby, R. B. Herbert, L. Mo, and F. Šantavý, *J. Chem. Soc. (C)*, 1739 (1967).

260. T. Kametani, K. Fukumoto, H. Yagi, and F. Satoh, *Chem. Comm.*, 878 (1967).

261. T. Kametani, H. Yagi, F. Satoh, and K. Fukumoto, *J. Chem. Soc. (C)*, 271 (1968).

262. T. Kametani, S. Takano, and T. Kobari, *Tetrahedron Lett.*, 4565 (1968); *J. Chem. Soc. (C)*, 9 (1969).

263. T. Kametani, S. Takano, and T. Kobari, *J. Chem. Soc. (C)*, 2770 (1969).

264. T. Kametani, F. Satoh, H. Yagi, and K. Fukumoto, *J. Org. Chem.*, **33**, 690 (1968).

265. A. R. Battersby, M. McDonald, M. H. G. Munro, and R. Ramage, *Chem. Comm.*, 934 (1967).

266. A. Brossi, J. O'Brien, and S. Teitel, *Helv. Chim. Acta*, **52**, 678 (1969).

267. R. E. Harmon and B. L. Jensen, *J. Heterocyclic Chem.*, 7, 1077 (1970).

268. T. Kametani, F. Satoh, H. Yagi, and K. Fukumoto, *Chem. Comm.*, 1103 (1967); *J. Chem. Soc. (C)*, 1003 (1968); 382 (1970).

269. A. R. Battersby, P. Bohler, M. H. G. Munro, and R. Ramage, *Chem. Comm.*, 1066 (1969).

270. A. R. Battersby, R. B. Bradbury, R. B. Herbert, and M. H. G. Munro, *Chem. Comm.*, 450 (1967).

271. T. Kametani, H. Yagi, K. Fukumoto, and F. Satoh, *Chem. Pharm. Bull.*, **17**, 2297

(1969).

272. A. C. Baker, A. R. Battersby, M. McDonald, R. Ramage, and J. H. Clements, *Chem. Comm.*, 390 (1967) and references cited therein.

273. T. Kametani, K. Fukumoto, M. Koizumi, and A. Kozuka, *Chem. Comm.*, 1605 (1968); *J. Chem. Soc. (C)*, 1295 (1969).

273*a*. O. Hoshino, T. Toshioka, and B. Umezawa, *Chem. Comm.*, 740 (1970).

274. A. I. Scott, F. McCapra, R. L. Buchanan, A. C. Day, I. G. Wright, and D. W. Young, *Tetrahedron,* **21**, 3605 (1965).

275. S. Tobinaga and E. Kotani, *Abstracts of 12th Symposium on the Chemistry of Natural Products,* Sendai, Japan, 1968, p. 259.

276. T. Kametani, K. Fukumoto, M. Kawazu, and M. Fujihara, *J. Chem. Soc. (C),* 922 (1970).

277. T. Kametani and K. Fukumoto, *Chem. Comm.*, 26 (1968); *J. Chem. Soc. (C),* 2156 (1968).

278. P. deMayo, *Adv. Org. Chem.*, **1**, 357 (1960).

279. R. B. Woodward and R. Hoffmann, *The Conservation of Orbital Symmetry,* Academic, New York, 1970.

280. P. G. Sammes, *Quart. Rev.*, **24**, 37 (1970).

281. G. B. Gill, *Quart. Rev.*, **22**, 338 (1968).

282. R. F. Stermitz, *Organic Photochemistry,* Vol. I, (O. L. Chapman, Ed.), Dekker, New York, 1967, p. 247.

283. M. P. Cava, S. C. Havlicek, A. Lindert, and R. J. Spangler, *Tetrahedron Lett.*, 2937 (1966).

284. M. P. Cava, M. J. Mitchell, S. C. Havlicek, A. Lindert, and R. J. Spanger, *J. Org. Chem.*, **35**, 175 (1970).

285. N. C. Yang, G. R. Lenz, and A. Shani, *Tetrahedron Lett.*, 2941 (1966).

286. G. Y. Moltrasio, M. Sotelo, and D. Giacopello, *J. Chem. Soc. Perkin I,* 349 (1973).

287. M. P. Cava, P. Stern, and K. Wakisaka, *Tetrahedron,* **29**, 2245 (1973).

288. N. C. Yang, A. Shani, and G. R. Lenz, *J. Amer. Chem. Soc.*, **88**, 5369 (1966).

289. M. P. Cava and S. C. Havlicek, *Tetrahedron Lett.*, 2625 (1967).

290. G. R. Lenz and N. C. Yang, *Chem. Comm.*, 1136 (1967).

291. I. Ninomiya and T. Naito, *Chem. Comm.*, 137 (1973).

292. G. R. Lenz, *Tetrahedron Lett.*, 1763 (1973).

293. R. B. Herbert and C. J. Moody, *Chem. Comm.*, 121 (1970).

294. I. Ninomiya, T. Naito, and T. Mori, *Tetrahedron Lett.*, 3643 (1969).

295. I. Ninomiya, T. Naito, and T. Kiguchi, *Tetrahedron Lett.*, 4451 (1970).

296. M. Onda, K. Yonezawa, and K. Abe, *Chem. Pharm. Bull.*, **17**, 404 (1969); **19**, 31 (1971).

297. M. Onda, K. Yonezawa, K. Abe, H. Toyama, and T. Suzuki, *Chem. Pharm. Bull.*, **19**, 317 (1971).

298. S. F. Dyke and M. Sainsbury, *Tetrahedron,* **23**, 3161 (1967).

299. V. Šmula, R. H. F. Manske, and R. Rodrigo, *Can. J. Chem.*, **50**, 1544 (1972).

300. M. Onda and K. Kawakami, *Chem. Pharm. Bull.*, **20**, 1484 (1972).

301. M. Onda, K. Yuasa, J. Okada, A. Kataoka, and K. Abe, *Chem. Pharm. Bull.*, **21**, 1333 (1973).

302. I. Ninomiya, T. Naito, and T. Kiguchi, *Chem. Comm.*, 1969, (1970).

303. X. A. Dominguez, J. G. Delgado, W. P. Reeves, and P. D. Gardner, *Tetrahedron Lett.*, 2493 (1967).

304. D. F. DeTar, *Org. Reactions,* **9**, 409 (1957).

305. T. Kametani, K. Fukumoto, F. Satoh, and H. Yagi, *Chem. Comm.*, 1398 (1968); *J. Chem. Soc. (C)*, 520 (1969).

306. T. Kametani, M. Ihara, K. Fukumoto, and H. Yagi, *J. Chem. Soc.*, 2030 (1969).

307. T. Kametani, K. Fukumoto, F. Satoh, and H. Yagi, *Chem. Comm.*, 1001 (1968); *J. Chem. Soc. (C)*, 3084 (1968).

308. A review of the morphinandienone and homomorphinandienone synthesis by the Pschorr reaction is found in T. Kametani and K. Fukumoto, *J. Heterocyclic Chem.*, 8, 341 (1971).

309. T. Kametani, H. Sugi, S. Shibuya, and K. Fukumoto, *Chem. Pharm. Bull.*, **19**, 1513 (1971); *Tetrahedron,* **27**, 5375 (1971).

310. T. Kametani, K. Fukumoto, and K. Shishido, *Chem. Ind.*, 1506 (1970).

311. T. Kametani, M. Koizumi, K. Shishido, and K. Fukumoto, *J. Chem. Soc. (C)*, 1923 (1971).

312. T. Kametani, T. Sugahara, and K. Fukumoto, *Tetrahedron,* **27**, 5367 (1971).

313. T. Kametani, R. Charubala, M. Ihara, M. Koizumi, K. Takahashi, and K. Fukumoto, *Chem. Comm.*, 289 (1971).

314. T. Kametani, M. Koizumi, and K. Fukumoto, *Chem. Comm.*, 1157 (1970); *J. Chem. Soc. (C)*, 1792 (1971).

315. T. Kametani, T. Nakano, C. Seino, S. Shibuya, K. Fukumoto, T. R. Govindachari, N. Nagarajan, B. R. Pai, and P. S. Subramaniani, *Chem. Pharm. Bull.*, **20**, 1507 (1972).

316. T. Kametani, M. Koizumi, and K. Fukumoto, *J. Org. Chem.*, **36**, 3729 (1971).

317. R. K. Sharma and K. Kharasch, *Angew. Chem.*, **80**, 69 (1968).

318. S. M. Kupchan and R. M. Kanojia, *Tetrahedron Lett.*, 5353 (1966); S. M. Kupchan, J. L. Maniot, R. M. Kanojia, and J. B. O'Brien, *J. Org. Chem.*, **36**, 2413 (1971).

319. T. Kametani, S. Shibuya, H. Sugi, O. Kusama, and K. Fukumoto, *J. Chem. Soc. (C)*, 2446 (1971).

320. T. Kametani, H. Sugi, S. Shibuya, and K. Fukumoto, *Chem. Ind.*, 818 (1971); *Tetrahedron,* **27**, 5379 (1971).

321. T. Kametani, K. Fukumoto, S. Shibuya, H. Nemoto, S. Nakano, T. Sugahara, T. Takahashi, Y. Aizawa, and M. Toriyama, *J. Chem. Soc. Perkin I,* 1435 (1972).

322. R. J. Spangler and D. C. Boop, *Tetrahedron Lett.*, 4851 (1971).

323. T. Kametani, H. Nemoto, T. Nakano, S. Shibuya, and K. Fukumoto, *Chem. Ind.*, 788 (1971).

324. T. Kametani, H. Nemoto, T. Kobari, K. Shishido, and K. Fukumoto, *Chem. Ind.*, 538 (1972).

325. T. Kametani, T. Honda, M. Ihara, and K. Fukumoto, *Chem. Ind.*, 119 (1972); T. Kametani, K. Takahashi, T. Honda, M. Ihara, and K. Fukumoto, *Chem. Pharm. Bull.*, **20**, 1793 (1972).

326. T. Kametani, T. Sugahara, H. Sugi, S. Shibuya, and K. Fukumoto, *Chem. Comm.*, 724 (1971).

327. T. Kametani, T. Sugahara, H. Sugi, S. Shibuya, and K. Fukumoto, *Tetrahedron,* **27**,

5993 (1971).

328. T. Kametani, S. Shibuya, T. Nakano, and K. Fukumoto, *J. Chem. Soc. (C)*, 3818 (1971).
329. Z. Horii, Y. Nakashita, and C. Iwata, *Tetrahedron Lett.*, 1167 (1971).
330. T. Kametani, Y. Satoh, S. Shibuya, M. Koizumi, and K. Fukumoto, *J. Org. Chem.*, **36**, 3733 (1971).
331. T. Kametani, Y. Satoh, and K. Fukumoto, *J. Chem. Soc. Perkin I*, 2160 (1972).
332. T. Kametani and M. Koizumi, *J. Chem. Soc. (C)*, 3976 (1971).
333. T. Kametani, T. Kohno, R. Charubala, and K. Fukumoto, *Tetrahedron*, **28**, 3227 (1972).
334. T. Kametani, T. Kohno, S. Shibuya, and K. Fukumoto, *Chem. Comm.*, 774 (1971); *Tetrahedron*, **27**, 5441 (1971).
335. T. Kametani and T. Kohno, *Tetrahedron Lett.*, 3155 (1971); T. Kametani, T. Kohno, R. Charubala, S. Shibuya, and K. Fukumoto, *Chem. Pharm. Bull.*, **20**, 1488 (1972).
336. T. Kametani, K. Yamaki, T. Terui, S. Shibuya, and K. Fukumoto, *J. Chem. Soc. Perkin I*, 1513 (1972).
337. A. Mondon and K. Krohn, *Tetrahedron Lett.*, 2123 (1970).
338. I. R. C. Bick, I. B. Bremner, and J. Wiriyachitra, *Tetrahedron Lett.*, 4795 (1971).
339. O. Yonemitsu, H. Nakai, Y. Kanaoka, I. L. Karle, and B. Witkop, *J. Amer. Chem. Soc.*, **92**, 5691 (1970); H. B. Bernhard and V. Snieckus, *Tetrahedron*, **27**, 2091 (1971).
340. L. J. Dolby, S. J. Nelson, and D. Senkovich, *J. Org. Chem.*, **37**, 3691 (1972).
341. R. Pschorr, *Berichte*, **29**, 496 (1896).
342. R. Pschorr *Berichte*, **37**, 1926 (1904).
343. D. F. DeTar, *Org. Reactions*, **9**, 409 (1957); P. H. Teake, *Chem. Rev.*, **56**, 27 (1956); L. F. Fieser and M. Fieser, *Natural Products Related to Phenanthrene*, New York, 1949, p. 8; T. Kametani and K. Fukumoto, *Farumashia*, **7**, 457 (1970).
344. T. Kametani, K. Fukumoto, F. Satoh, and H. Yagi, *J. Chem. Soc. (C)*, 520 (1969).
345. T. Kametani, K. Fukumoto, F. Satoh, and H. Yagi, *Chem. Comm.*, 1398 (1968).
346. T. Kametani, M. Ihara, K. Fukumoto, and H. Yagi, *J. Chem. Soc. (C)*, 2030 (1969).
347. T. Kametani, K. Fukumoto, and T. Sugahara, *Tetrahedron Lett.*, 5459 (1968); *J. Chem. Soc. (C)*, 801 (1969).
348. T. Kametani, T. Sugahara, H. Yagi, and K. Fukumoto, *J. Chem. Soc. (C)*, 1063 (1969).
349. T. Kametani, M. Koizumi, and K. Fukumoto, *Chem. Pharm. Bull.*, **17**, 2245 (1969).
350. A. R. Battersby, A. K. Bhatnagar, P. Hackett, C. W. Thornber, and J. Staunton, *Chem. Comm.*, 1214 (1968).
351. S. Ishiwata and K. Itakura, *Chem. Pharm. Bull.* **17**, 1299 (1969).
352. S. Ishiwata and K. Itakura, *Chem. Pharm. Bull.*, **18**, 416 (1970).
353. S. Ishiwata, K. Itakura, and K. Misawa, *Chem. Pharm. Bull.*, **18**, 1219 (1970).
354. S. Ishiwata and K. Itakura, *Chem. Pharm. Bull.*, **18**, 1224 (1970).
355. T. Kametani, K. Fukumoto, M. Kawazu, and M. Fujihara, *J. Chem. Soc. (C)*, 2213 (1970).
356. T. Kametani, K. Fukumoto, F. Satoh, and H. Yagi, *Chem. Comm.*, 1001 (1968); *J. Chem. Soc. (C)*, 3084 (1968).

357. T. Kametani, K. Takahashi, T. Sugahara, M. Koizumi, and K. Fukumoto, *J. Chem. Soc. (C)*, 1032 (1971).

357*a*. M. P. Cava, I. Noguchi, and K. T. Buck, *J. Org. Chem.*, **38**, 2394 (1973).

357*b*. M. P. Cava and I. Noguchi, *J. Org. Chem.*, **37**, 2936 (1972).

357*c*. P. Kerekes, K. Delenk-Heydenreich, and S. Pfeifer, *Tetrahedron Lett.*, 2483 (1970); *Berichte*, **105**, 609 (1972); *Magy. Kem. Foly.*, **78**, 410 (1972).

357*d*. F. N. Lahey and K. F. Mark, *Tetrahedron Lett.*, 4511 (1970).

357*e*. I. Ribas, J. Saa, and L. Castedo, *Tetrahedron Lett.*, 3617 (1973).

357*f*. T. R. Govindachari, N. Viswanathan, R. Charubala, and B. R. Pai, *Indian J. Chem.*, **7**, 841 (1969).

357*g*. S. M. Kupchan and A. J. Liepa, *Chem. Comm.*, 599 (1971); S. M. Kupchan, A. J. Liepa, V. Kameswaran, and K. Sempuku, *J. Amer. Chem. Soc.*, **95**, 2975 (1973).

357*h*. M. P. Cava and I. Noguchi, *J. Org. Chem.*, **28**, 60 (1973).

357*i*. C. Casagrande and G. Merrotti, *Farm. Ed. Sci.*, **21**, 799 (1970); T. R. Govindachari, N. Viswanathan, S. Narayanaswami, and B. R. Pai, *Indian J. Chem.*, **8**, 475 (1970).

357*j*. M. S. Gibson, G. W. Prenton, and J. M. Walthew, *J. Chem. Soc. (C)*, 2234 (1970).

357*k*. M. P. Cava and M. Srinivasan, *Tetrahedron*, **26**, 4649 (1970).

357*l*. J. Kunitomo, M. Miyoshi, E. Yuge, T.-H. Yang, and C.-M. Chen, *Chem. Pharm. Bull.*, **19**, 1502 (1971).

357*m*. J. R. Merchant and H. K. Desai, *Indian J. Chem.*, **11**, 342 (1973).

357*n*. J. Kunitomo, M. Miyoshi, M. Toyoko, and E. Yuge, *J. Pharm. Soc. Japan*, **91**, 896 (1971).

357*o*. T. Kametani, K. Takahashi, T. Sugahara, M. Koizumi, and K. Fukumoto, *J. Chem. Soc. (C)*, 1032 (1971).

357*p*. M. P. Cava and M. V. Lakshmikantham, *J. Org. Chem.*, **35**, 1867 (1970).

357*q*. R. W. Doskotch, J. D. Phillipson, A. B. Ray, and J. L. Beal, *J. Org. Chem.*, **36**, 2409 (1971).

357*r*. M. P. Cava, K. T. Buck, and A. I. DaRocha, *J. Amer. Chem. Soc.*, **94**, 5931 (1972).

357*s*. D. R. Dalton and A. A. Abraham, *Syn. Comm.*, **2**, 303 (1972).

357*t*. S. M. Kupchan, V. Kameswaran, and J. W. A. Findlary, *J. Org. Chem.*, **38**, 405 (1973).

357*u*. S. Ishiwata and K. Itakura, *Chem. Pharm. Bull.*, **17**, 2261 (1969).

357*v*. S. F. Dyke, B. J. Moon, and M. Sainsbury, *Tetrahedron Lett.*, 3933 (1968); *J. Chem. Soc. (C)*, 1797 (1970).

357*w*. G. Savona and F. Piozzi, *J. Heterocyclic Chem.*, **8**, 681 (1971).

358. G. Wittig, *Angew. Chem. Int. Ed.*, **4**, 731 (1965).

359. T. Kametani and K. Ogasawara, *J. Chem. Soc. (C)*, 2208 (1967).

360. F. Benington and R. D. Morin, *J. Org. Chem.*, **32**, 1050 (1967).

361. S. V. Kessar, S. Batra, and S. S. Gandhi, *Indian J. Chem.*, **8**, 468 (1970).

362. T. Kametani, S. Shibuya, K. Kigasawa, M. Hiiragi, and O. Kusama, *J. Chem. Soc. (C)*, 2712 (1971).

363. T. Kametani, K. Fukumoto, and T. Nakano, *Tetrahedron*, **28**, 4667 (1972).

364. T. Kametani, K. Fukumoto, and T. Nakano, *J. Heterocyclic Chem.*, **9**, 1363 (1972).

364*a*. S. V. Kessar, R. Randhawa , and S. S. Gandhi, *Tetrahedron Lett.*, 2923 (1973).

364*b*. T. Kametani, A. Ujiie, K. Takahashi, T. Nakano, T. Suzuki, and K. Fukumoto, *Chem. Pharm. Bull.*, **21**, 766 (1973).

364*c*. S. V. Kessar and M. Singh, *Tetrahedron Lett.*, 1155 (1969).

364*d*. M. F. Semmelhack, B. P. Chung, and L. D. Jones, *J. Amer. Chem. Soc.*, **84**, 8629 (1972).

364*e*. N. Ueda, T. Tokuyama, and T. Sakan, *Bull. Soc. Chem. Japan*, **39**, 2012 (1966).

365. R. W. Doskotch, J. D. Phillipson, A. B. Ray, and J. L. Berl, *J. Org. Chem.*, **36**, 2409 (1971).

366. M. Tomita, H. Furukawa, S. T. Lu, and S. M. Kupchan, *Chem. Pharm. Bull.*, **15**, 959 (1967).

367. T. Kametani, H. Iida, and K. Sakurai, *J. Chem. Soc. (C)*, 500 (1969).

368. T. Kametani, H. Iida, and K. Sakurai, *J. Chem. Soc. (C)*, 1024 (1971).

369. T. Kametani, H. Iida, and S. Tanaka, *J. Pharm. Soc. Japan*, **90**, 209 (1970).

370. A. R. Battersby, R. B. Herbert, L. Mo, and F. Šantavý, *J. Chem. Soc. (C)*, 1739 (1967).

371. T. Kametani, H. Iida, T. Kikuchi, M. Mizushima, and K. Fukumoto, *Chem. Pharm. Bull.*, **17**, 709 (1969).

372. H. Iida, H.-C. Hsu, T. Kikuchi, and K. Kawano, *Chem. Pharm. Bull.*, **20**, 1242 (1972).

373. H. Iida, H.-C. Hsu, and T. Kikuchi, *Chem. Pharm. Bull.*, **21**, 1001 (1973).

374. S. Ishiwata, T. Fujii, N. Miyaji, Y. Satoh, and K. Itakura, *Chem. Pharm. Bull.*, **18**, 1850 (1970).

375. I. L. Klundt, *Chem. Rev.*, **70**, 471 (1970).

376. R. Huisgen and H. Seidel, *Tetrahedron Lett.*, 3381 (1964).

377. R. B. Woodward and R. Hoffmann, *The Conservation of Orbital Symmetry*, Academic, New York, 1970.

378. W. Oppolzer and K. Keller, *J. Amer. Chem. Soc.*, **93**, 3836 (1971).

379. T. Kametani, K. Ogasawara, and T. Takahashi, *Chem. Comm.*, 675 (1972); *Tetrahedron*, **29**, 73 (1973).

380. T. Kametani, Y. Hirai, F. Satoh, K. Ogasawara, and K. Fukumoto, *Chem. Pharm. Bull.*, **21**, 907 (1973).

381. T. Kametani, M. Takemura, K. Ogasawara, and K. Fukumoto, *J. Heterocyclic Chem.*, **11**, 179 (1974).

382. T. Kametani, Y. Kato, and K. Fukumoto, *J. Chem. Soc. Perkin I*, 1712 (1974).

383. T. Kametani, Y. Kato, and K. Fukumoto, *Tetrahedron*, **30**, 1043 (1974).

384. T. Kametani, T. Takahashi, K. Ogasawara, T. Honda, and K. Fukumoto, *J. Org. Chem.*, **39**, 447 (1974).

385. T. Kametani, T. Takahashi, K. Ogasawara, and K. Fukumoto, *Tetrahedron*, **30**, 1047 (1974).

386. T. Kametani, T. Takahashi, and K. Ogasawara, *Tetrahedron Lett.*, 4847 (1972).

387. M. Shamma and C. D. Jones, *J. Amer. Chem. Soc.*, **91**, 4009 (1969); **92**, 4943 (1970).

388. G. Wittig, H. Tenhaeff, W. Schoch, and G. König, *Annalen*, **572**, 1 (1951).

389. G. Grethe, H. L. Lee, M. R. Uskokovic, and A. Brossi, *Helv. Chim. Acta*, **53**, 874 (1970).

390. T. Kametani, T. Kobari, K. Fukumoto, and M. Fujihara, *J. Chem. Soc. (C)*, 1796 (1971).

391. T. Kametani, S.-P. Huang, A. Ujiie, M. Ihara, and K. Fukumoto, *Heterocycles*, **4**, 1223 (1976).

392. A. H. Jackson and J. A. Martin, *J. Chem. Soc. (C)*, 2222 (1966).

393. A. Brossi, J. O'Brien, and S. Teitel, *Helv. Chim. Acta*, **52**, 678 (1969).

394. A. R. Battersby and T. H. Brown, *Proc. Chem. Soc.*, 85 (1964).

395. T. Kametani, K. Takahashi, K. Ogasawara, and K. Fukumoto, *Chem. Pharm. Bull.*, **21**, 662 (1973).

396. S. Uyeo, H. Irie, Y. Nishitani, and M. Sugita, *Chem. Comm.*, 1313 (1970); H. Irie, Y. Nishitani, M. Sugita, K. Tamoto, and S. Uyeo, *J. Chem. Soc. Perkin I*, 588 (1972).

397. Y. Masaki, T. Mizutani, M. Sekido, and S. Uyeo, *J. Chem. Soc. (C)*, 2954 (1968).

398. H. Irie, S. Uyeo, and A. Yoshitake, *J. Chem. Soc. (C)*, 1802 (1968).

399. M. Shamma and C. D. Jones, *J. Amer. Chem. Soc.*, **91**, 4009 (1969); **92**, 4943 (1970).

400. M. Shamma and J. F. Nugent, *Tetrahedron Lett.*, 2624 (1970).

401. M. Shamma and J. F. Nugent, *Chem. Comm.*, 1265 (1973).

402. M. Shamma and J. F. Nugent, *Tetrahedron*, **29**, 1265 (1973).

403. B. Malliah, R. H. F. Manske, R. Rodrigo, and D. B. MacLean, *Tetrahedron Lett.*, 2795 (1973).

404. R. D. Haworth and W. H. Perkin, Jr., *J. Chem. Soc.*, 1769 (1926).

405. P. B. Russell, *J. Amer. Chem. Soc.*, **78**, 3115 (1956).

406. K. W. Bentley and A. W. Murray, *J. Chem. Soc.*, 2491 (1963).

407. A. Klásek, V. Šimánek, and F. Šantavý, *Tetrahedron Lett.*, 4549 (1968).

408. H. Irie, S. Tani, and H. Yamane, *Chem. Comm.*, 1713 (1970); *J. Chem. Soc. Perkin I*, 2986 (1972).

409. T. Kametani, S. Hirata, S. Shibuya, and K. Fukumoto, *J. Chem. Soc. (C)*, 1927 (1971).

410. S. F. Dyke and A. C. Ellis, *Tetrahedron*, **27**, 3803 (1971).

411. J. Likforman and J. Gardent, *C. R. Acad. Sci. Ser. C*, **268**, 2340 (1969).

412. S. M. Kupchan and A. Yoshitake, *J. Org. Chem.*, **34**, 1062 (1969).

413. J. Auerbach and S. M. Weinreb, *J. Amer. Chem. Soc.*, **94**, 7172 (1972).

414. H. O. Bernhard and V. Snieckus, *Tetrahedron*, **27**, 2091 (1971).

415. S. Naruto and O. Yonemitsu, *Chem. Pharm. Bull.*, **21**, 629 (1973).

416. L. J. Dolby, S. J. Nelson, and D. Senkovich, *J. Org. Chem.*, **37**, 3691 (1972).

417. Y. Inubushi, T. Harayama, and K. Takeshima, *Chem. Pharm. Bull.*, **20**, 689 (1972).

418. W. Klötzer, S. Teitel, J. F. Blount, and A. Brossi, *J. Amer. Chem. Soc.*, **93**, 4321 (1971).

419. W. Klötzer, S. Teitel, and A. Brossi, *Helv. Chim. Acta*, **54**, 2057 (1971); **55**, 2228 (1972).

420. B. Göbar and G. Engelhardt, *Pharmazie*, **24**, 423 (1969).

421. H. O. Bernhard and V. Snieckus, *Tetrahedron*, **27**, 2091 (1971).

422. T. Kametani, S. Hirata, and K. Ogasawara, *J. Chem. Soc. Perkin I*, 1466 (1973).

Chapter 1 describes work published up to the middle of 1973. The following list of references, divided by subject heading, is intended to update the work in this area of synthesis to the end of 1975.

I. Syntheses via Bischler-Napieralski Reaction

423. J. Gal, R. J. Wienkam, and N. Castagnoli, Jr., *J. Org. Chem.*, **39**, 418 (1974).
424. T. Okawara and T. Kametani, *Heterocycles*, **2**, 571 (1974).
425. T. Fujii and S. Yoshifuji, *Tetrahedron Lett.*, 731 (1975).
426. T. Fujii, S. Yoshifuji, and K. Yamada, *Tetrahedron Lett.*, 1527 (1975).
427. E. Wenkert, H. P. S. Chawla, and F. M. Schell, *Syn. Comm.*, **3**, 381 (1973).
428. K. Torsell, *Tetrahedron Lett.*, 623 (1974).
429. J. B. Hendrickson, T. L. Bogard, M. E. Fisch, S. Grossert, and N. Yoshimura, *J. Amer. Chem. Soc.*, **96**, 7781 (1974).
430. Y. Tsuda, A. Ukai, and K. Isobe, *Tetrahedron Lett.*, 3153 (1972).
431. Y. Tsuda, T. Sano, J. Taga, K. Isobe, J. Toda, H. Irie, H. Tanaka, S. Takagi, M. Yamaki, and M. Maruta, *Chem. Comm.*, 933 (1975).

II. Syntheses via Pictet-Spengler Reaction

432. G. J. Kapadia, G. S. Rao, M. H. Hussain, and B. K. Chowdhury, *J. Heterocyclic Chem.*, **10**, 135 (1973).
433. M. Konda, T. Shioiri, and S. Yamada, *Chem. Pharm. Bull.*, **23**, 1063 (1975).
434. M. R. Falco, J. X. DeVries, and G. Mann, *Z. Chem.*, **13**, 56 (1973).
435. A. Brossi, A. Focella, and S. Teitel, *Helv. Chim. Acta*, **55**, 15 (1972).
436. S. McLean and J. Whelan, *Can. J. Chem.*, **51**, 2457 (1973).
437. A. Shoeb, K. Raj, P. S. Kapil, and S. P. Popli, *J. Chem. Soc. Perkin I*, 1245 (1975).
438. T. Kametani, M. Takeshita, and S. Takano, *J. Chem. Soc. Perkin I*, 2834 (1972).
439. T. Kametani, A. Ujiie, and K. Fukumoto, *Heterocycles*, **2**, 55 (1974); *J. Chem. Soc. Perkin I*, 1959 (1974).
440. T. Kametani, A. Ujiie, M. Ihara, and K. Fukumoto, *Heterocycles*, **3**, 143 (1975); *J. Chem. Soc. Perkin I*, 1822 (1975).
441. H. Muxfeldt, J. P. Bell, J. A. Baker, and U. Cuntze, *Tetrahedron Lett.*, 4587 (1973).
442. R. M. Coomes, J. R. Falck, D. K. Williams, and F. R. Stermitz, *J. Org. Chem.*, **38**, 3701 (1973).
443. D. S. Walsh and R. E. Lyle, *Tetrahedron Lett.*, 3849 (1973).

III. Syntheses by Type 5 Synthesis

444. I. Noguchi and D. B. MacLean, *Can. J. Chem.*, **53**, 125 (1975).
445. A. R. Battersby, J. Staunton, H. R. Wiltshine, R. J. Bircher, and C. Fuganti, *J. Chem. Soc. Perkin I*, 1162 (1975).

446. S. F. Dyke and E. P. Tiley, *Tetrahedron,* **31**, 561 (1975).
447. N. E. Cundasawmy and D. B. MacLean, *Can. J. Chem.,* **50**, 3028 (1972).
448. A. J. Birch, A. H. Jackson, and P. V. R. Shannon, *J. Chem. Soc. Perkin I,* 2190 (1974).
449. S. F. Dyke, A. C. Ellis, R. F. Kinsman, and A. W. C. White, *Tetrahedron,* **30**, 1193 (1974).
450. T. Kametani, K. Takahashi, and C. V. Loc, *Tetrahedron,* **31**, 235 (1975).

IV. Racemization of Isoquinoline Alkaloids

451. T. Kametani and M. Ihara, *J. Chem. Soc. (C),* 191 (1968).
452. T. Kametani, M. Ihara, and K. Shima, *J. Chem. Soc. (C),* 1619 (1968).

V. Syntheses by Phenol Oxidation

453. T. Kametani, K. Fukumoto, and M. Fujihara, *Bioorg. Chem.,* **1**, 40 (1973).
454. A. H. Jackson, G. W. Stewart, G. A. Charnock, and J. A. Martin, *J. Chem. Soc. Perkin I,* 1911 (1974).
455. A. R. Battersby, R. B. Bradbury, R. B. Herbert, M. H. G. Munro, and R. Ramage, *J. Chem. Soc. Perkin I,* 1394 (1974).
456. T. Kametani, T. Kobari, K. Shishido, and K. Fukumoto, *Tetrahedron,* **30**, 1059 (1974).
457. M. A. Schwartz, *Syn. Comm.,* **3**, 33 (1973).

VI. Oxidative Coupling by Chemical Oxidation

458. S. M. Kupchan and A. J. Liepa, *J. Amer. Chem. Soc.,* **95**, 4062 (1973).
459. S. M. Kupchan, A. J. Liepa, J. Andris, V. Kameswaran, and R. F. Bryan, *J. Amer. Chem. Soc.,* **95**, 6861 (1973).
460. T. Kametani, K. Takahashi, K. Ogasawara, and K. Fukumoto, *Tetrahedron Lett.,* 4219 (1973); *Coll. Czech. Chem. Comm.,* **40**, 712 (1975).
461. B. Umezawa and O. Hoshino, *Heterocycles,* **3**, 1005 (1975).
462. O. Hoshino, T. Toshioka, and B. Umezawa, *Chem. Pharm. Bull.,* **22**, 1302 (1974).
463. O. Hoshino, H. Hara, N. Serizawa, and B. Umezawa, *Chem Pharm. Bull.,* **23**, 2048 (1975).
464. O. Hoshino, T. Toshioka, K. Ohyama, and B. Umezawa, *Chem. Pharm. Bull.,* **22**, 1307 (1974).
465. O. Hoshino, H. Hara, M. Ogawa, and B. Umezawa, *Chem. Comm.,* 306 (1975); *Chem. Pharm. Bull.,* **23**, 2578 (1975).
466. O. Hoshino, M. Taga, and B. Umezawa, *Heterocycles,* **1**, 223 (1973).

VII. Electrooxidation

467. S. M. Kupchan and C.-K. Kim, *J. Amer. Chem. Soc.,* **97**, 5623 (1975).
468. E. Kotani, M. Kitazawa, and S. Tobinaga, *Tetrahedron,* **30**, 3027 (1974).
469. J. M. Bobbitt, I. Noguchi, R. S. Ware, K. N. Chong, and S. J. Huang, *J. Org. Chem.,* **40**, 2924 (1975).

VIII. Photochemical Syntheses

470. I. Ninomiya, *Heterocycles,* **2**, 105 (1974).
471. I. Ninomiya, H. Takasugi, and T. Naito, *Heterocycles,* **1**, 17 (1973).
472. I. Ninomiya, T. Naito, and H. Takasugi, *J. Chem. Soc. Perkin I,* 1720 (1975).
473. I. Ninomiya, T. Naito, and T. Mori, *J. Chem. Soc. Perkin I,* 505 (1973); I. Ninomiya, T. Naito, T. Kiguchi, and T. Mori, *J. Chem. Soc. Perkin I,* 1696 (1973).
474. I. Ninomiya, T. Naito, H. Ishii, T. Ishida, M. Ueda, and K. Harada, *J. Chem. Soc. Perkin I,* 762 (1975).
475. I. Ninomiya, T. Naito, and T. Kiguchi, *J. Chem. Soc. Perkin I,* 2261 (1973).
476. L. Faber and W. Wiegrebe, *Helv. Chim. Acta,* **56**, 2882 (1973).
477. T. Kametani, T. Nakano, C. Seino, S. Shibuya, K. Fukumoto, T. R. Govindachari, K. Nagarajan, B. R. Pai, and P. S. Subramanian, *Chem. Pharm. Bull.*, **20**, 1507 (1972).
478. M. P. Cava, P. Stern, and K. Wakisaka, *Tetrahedron,* **29**, 2245 (1973).
479. M. P. Cava and S. S. Libsch, *J. Org. Chem.*, **39**, 577 (1974).
480. S. M. Kupchan and F. P. O'Brien, *Chem. Comm.*, 915 (1973).
481. Z. Horii, S. Uchida, Y. Nakatsuka, E. Tsuchida, and C. Iwata, *Chem. Pharm. Bull.*, **22**, 583 (1974).
482. T. Kametani, K. Fukumoto, M. Ihara, M. Takemura, H. Matsumoto, B. R. Pai, K. Nagarajan, M. S. Premila, and H. Suguna, *Heterocycles,* **3**, 811 (1975).
483. T. Kametani, Y. Satoh, and K. Fukumoto, *Tetrahedron,* **29**, 2027 (1973).
484. T. Kametani, R. Nitadori, H. Terasawa, K. Takahashi, and M. Ihara, *Heterocycles,* **3**, 821 (1975).
485. M. Shamma and D.-Y. Hwang, *Heterocycles,* **1**, 31 (1973).
486. K. Ito and H. Tanaka, *Chem. Pharm. Bull.*, **22**, 2108 (1974).
487. A. Mondon and K. Krohn, *Berichte,* **105**, 3726 (1972).
488. H. Hara, O. Hoshino, and B. Umezawa, *Tetrahedron Lett.*, 5031 (1972).
489. H. Irie, K. Akagi, S. Tani, K. Yabusaki, and H. Yamane, *Chem. Pharm. Bull.*, **21**, 855 (1973).

IX. Syntheses by Pschorr Reaction

490. S. M. Kupchan, V. Kameswaran, and J. W. A. Findlay, *J. Org. Chem.*, **38**, 406 (1973).
491. D. R. Dalton and A. A. Abraham, *Syn. Comm.*, **2**, 303 (1972).
492. I. Ahmad and M. S. Gibson, *Can. J. Chem.*, **53**, 3660 (1975).
493. T. Kametani, S. Shibuya, R. Charubala, M. S. Premila, and B. R. Pai, *Heterocycles,* **3**, 439 (1975).
494. J. Kunitomo, M. Miyoshi, E. Yuge, T.-S. Yang, and C.-M. Chen, *Chem. Pharm. Bull.*, **19**, 1502 (1971).
495. M. P. Cava, I. Noguchi, and K. T. Buck, *J. Org. Chem.*, **38**, 2394 (1973).
496. M. Shamma, *The Alkaloids,* Vol. 4, The Chemical Society, London, 1974, p. 227.
497. M. P. Cava and I. Noguchi, *J. Org. Chem.*, **37**, 2936 (1972).
498. I. Ribas, J. Saa, and L. Castedo, *Tetrahedron Lett.*, 3617 (1973).
499. M. P. Cava and I. Noguchi, *J. Org. Chem.*, **38**, 60 (1973).

500. M. P. Cava, K. T. Buck, I. Noguchi, M. Srinivasan, M. G. Rao, and A. I. D. DaRocha, *Tetrahedron,* **31**, 1667 (1975).

X. Syntheses via Benzyne Intermediate

501. J. P. Gillespie, L. G. Amoros, and F. R. Stermitz, *J. Org. Chem.,* **39**, 3239 (1974); F. R. Stermitz, J. P. Gillespie, L. G. Amoros, R. Romero, and T. A. Stermitz, *J. Med. Chem.,* **18**, 708 (1975).

XI. Syntheses via Ullmann Reaction

502. H.-C. Hsu, T. Kikuchi, S. Aoyagi, and H. Iida, *J. Pharm. Soc. Japan,* **92**, 1030 (1972).
503. H. Iida, H.-C. Hsu, T. Kikuchi, and K. Kawano, *J. Pharm. Soc. Japan,* **92**, 1242 (1972).
504. T. Kametani, K. Fukumoto, and M. Fujihara, *Chem. Pharm. Bull.,* **20**, 1800 (1972).

XII. Syntheses by Thermolysis

505. T. Kametani, K. Takahashi, C. V. Loc, and M. Hirata, *Heterocycles,* **1**, 247 (1973).
506. T. Kametani, Y. Hirai, H. Nemoto, and K. Fukumoto, *J. Heterocyclic Chem.,* **12**, 185 (1975).
507. T. Kametani, M. Kajiwara, T. Takahashi, and K. Fukumoto, *J. Chem. Soc. Perkin I,* 737 (1975).

XIII. Syntheses via Rearrangement

508. J. P. Marino and J. M. Samanen, *Tetrahedron Lett.,* 4553 (1973).
509. T. Kametani, T. Kohno, and K. Fukumoto, *Chem. Pharm. Bull.,* **20**, 1678 (1972).
510. V. Smula, N. E. Cundasawmy, H. L. Holland, and D. B. MacLean, *Can. J. Chem.,* **51**, 3287 (1973).
511. T. Kametani, T. Honda, H. Inoue, and K. Fukumoto, *Heterocycles,* **3**, 1091 (1975).
512. S. O. Desilova, K. Orito, R. H. F. Manske, and R. Rodrigo, *Tetrahedron Lett.,* 3243 (1974).

XIV. Azepine Alkaloids

513. T. Kametani and K. Fukumoto, *Heterocycles,* **3**, 931 (1975).
514. S. M. Weinreb and J. Auerbach, *J. Amer. Chem. Soc.,* **97**, 2503 (1975).
515. M. F. Semmelhack, B. P. Chong, R. D. Stauffer, T. D. Rogerson, A. Chong, and L. D. Jones, *J. Amer. Chem. Soc.,* **97**, 2507 (1975).
516. H. Irie, S. Tani, and H. Yamane, *Chem. Comm.,* 1713 (1970); *J. Chem. Soc. Perkin I,* 2986 (1972).
517. K. Orito, R. H. Manske, and R. Rodrigo, *J. Amer. Chem. Soc.,* **96**, 1944 (1974).
518. S. O. de Silva, K. Orito, R. H. Manske, and R. Rodrigo, *Tetrahedron Lett.,* 3243

(1974).

519. B. Nalliah, R. H. Manske, and R. Rodrigo, *Tetrahedron Lett.*, 2853 (1974).

520. T. Kametani, S. Hirata, F. Satoh, and K. Fukumoto, *J. Chem. Soc. Perkin I*, 2509 (1974).

521. T. Kametani, S. Hirata, H. Nemoto, M. Ihara, S. Hibino, and K. Fukumoto, *J. Chem. Soc. Perkin I*, 2028 (1975).

522. T. Kametani, S. Hirata, M. Ihara, and K. Fukumoto, *Heterocycles*, **3**, 405 (1975).

523. T. Kametani, M. S. Premila, S. Hirata, H. Seto, H. Nemoto, and K. Fukumoto, *Can. J. Chem.*, **53**, 3824 (1975).

524. M. Shamma and L. Töke, *Chem. Comm.*, 740 (1973).

The Synthesis of Indole Alkaloids

J. P. KUTNEY

Dept. of Chemistry
University of British Columbia
Vancoover, Canada

1. INTRODUCTION

The indole alkaloids are a very large family of natural products, and the synthetic approaches to these compounds represent a massive effort by a large number of chemists. It is clearly beyond the scope of this chapter to provide a complete discussion of all the advances that have been made in this area. It is hoped that the present organization and selection of the synthetic endeavours will provide some appreciation of the various factors that are normally considered in planning the framework of the synthetic pathway.

The impressive array of structures inherent in this family has provided the opportunity for considerable diversification in the synthetic approaches to these molecules. Some of the syntheses are impressive in the stereospecificity of the crucial constructive steps, others are characterized by their generality and versatility, whereas still others are patterned along routes that are reasonable approximations of biosynthetic pathways.

The architectural plans of all of the syntheses under discussion may be divided into two main categories:

1. The route provides a synthesis of the indole ring as one of the operations after the other factors such as stereochemistry, and functionality in the synthetic templates have been successfully attained.
2. The starting material is indolic, and elaboration of the intermediates normally represents selective utilization of the highly reactive indole ring in the key synthetic operations.

To set forth an introduction that may serve as a guide for those unacquainted with the field, it is appropriate to make a few brief comments about the synthetic characteristics that normally prevail in these categories.

A. Synthesis of the Indole Ring

The synthesis of an indole system from nonindolic starting materials has received considerable attention, and numerous approaches are now well developed.[1-3] The most versatile and widely applied reaction particularly in the syntheses of the natural alkaloids is the Fischer indole synthesis. Its synthetic utility is well documented in several recent review articles,[3-5] and only brief mention is required here. In general, the reaction involves the cylization of arylhydrazones **1** under a variety of acidic conditions. The resultant indole **2** may possess appropriate substitution in either the aromatic or heterocyclic rings. As a result of thorough investigations into the nature of the cyclization process, its mechanistic consequences, and so forth, it is possible to predict rather precisely the course of the reaction and thereby its appropriate utilization. Thus the Fischer synthesis

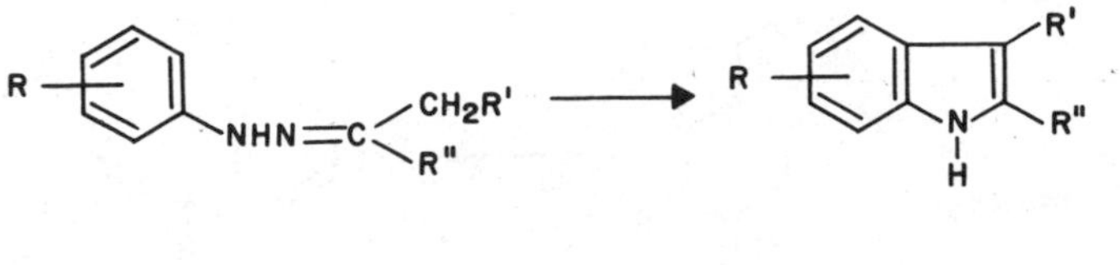

(1) (2)

finds application in the early stages of a synthetic sequence, as for example, the three-step preparation of substituted tryptamines **7** developed by Abramovitch and Shapiro[6,7] or the later stages of a sequence as in the synthesis of the

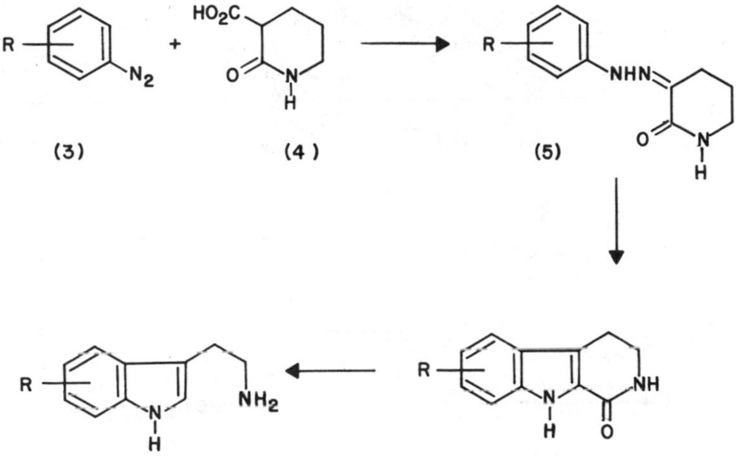

(3) (4) (5)

(7) (6)

yohimbine ring system[8] or the more recent work of Stork and Dolfini[9] and Ban et al.[10] in the *Aspidosperma* series **13**.

One of the most common substituents in the benzenoid ring of the indole alkaloids is the methoxyl group, and obviously the syntheses of such substituted tryptamine analogs are required as synthetic templates. Often these compounds are prepared from the benzenoid analogs containing the methoxyl substituent **15** in a reductive cyclization process. The resultant indole **16** is then converted to the tryptamine system **18** by appropriate electrophilic substitution reactions. In this way, 6-methoxytryptamine **18** was synthesized and subsequently employed in the synthesis of reserpine[11] and more recently in the *Aspidosperma* series.[12]

Another monomolecular cyclization process originally studied by von Braun, Heider, and Muller[13] effectively involving an internal Friedel-Crafts reaction, has found application in the synthesis of ring-substituted indole derivatives. The initially formed dihydroindole **20** or oxindole **24** can be readily converted to the

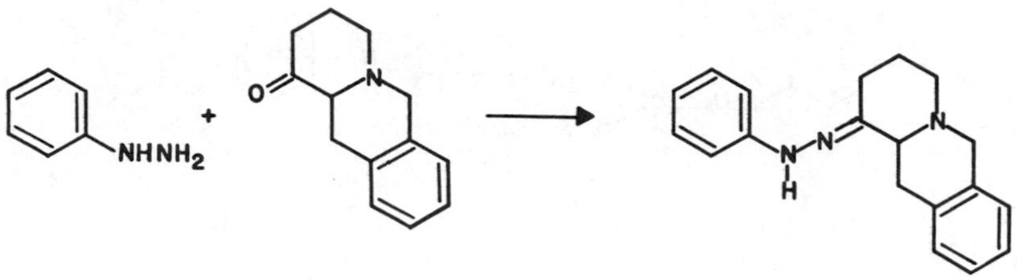

$NHNH_2$ + O N → N N N H

(8) (9) (10)

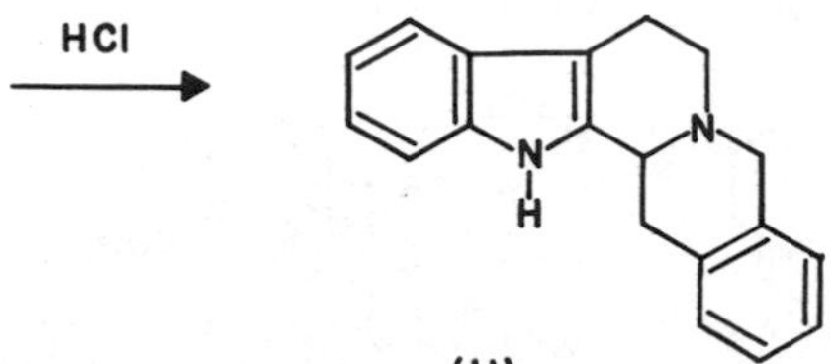

HCl → N H N

(11)

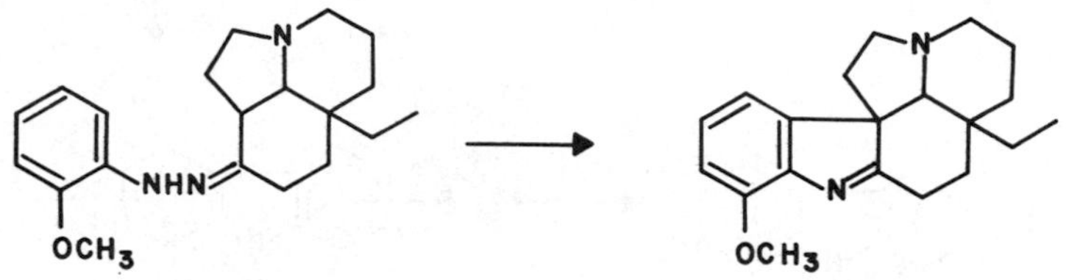

NHN OCH_3 N → N N OCH_3

(12) (13)

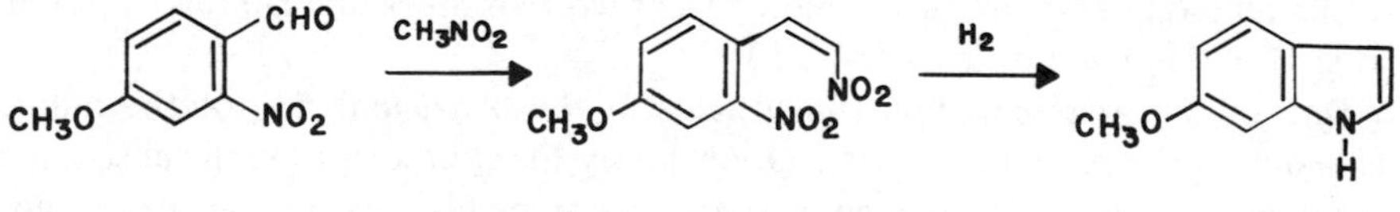

CH_3O CHO NO_2 $\xrightarrow{CH_3NO_2}$ CH_3O NO_2 NO_2 $\xrightarrow{H_2}$ CH_3O N H

(14) (15) (16)

$\xrightarrow{(COCl)_2}$ CH_3O N H O O Cl $\xrightarrow[2)\ LiAlH_4]{1)\ NH_4OH}$ CH_3O N H NH_2

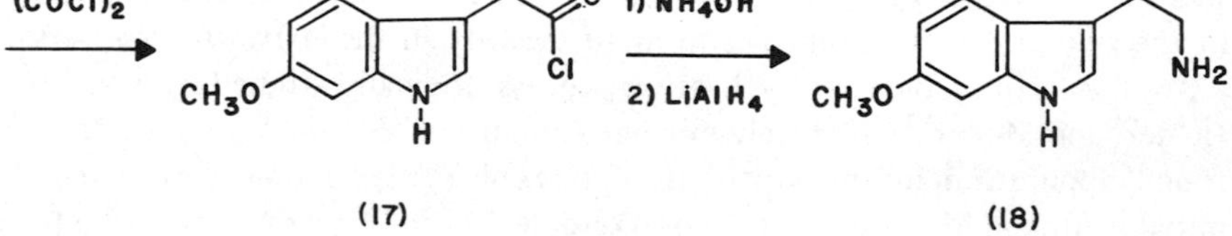

(17) (18)

$AlCl_3$; $-H_2$

(19) (20) (21)

$Cl-C(=O)-CH_2Cl$

(22) (23) (24)

$LiAlH_4$

(25)

(26) + (27) → [ ] $AlCl_3$

$ClCH_2CN$, NaOEt

(28) (29)

1) H_2, Pd-C
2) PhCHO, CH_3I
(Decker method)

1) Na, EtOH
2) $AlCl_3$

(30) (31)

indole system **21**. An example of the application of this approach is seen in the synthesis of the physostigmine[14-16] system **31**.

B. Elaboration of the Indole Ring

The well-known reactivity of indoles toward electrophilic substitution[17] provides a vast array of possibilities toward the synthesis of substituted indole derivatives. Since the β-carboline nucleus **1** forms a portion of the skeleton of

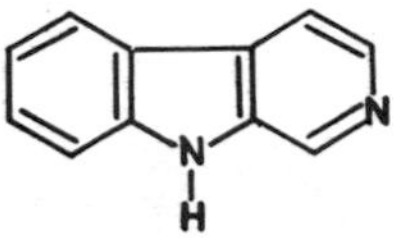

(1) 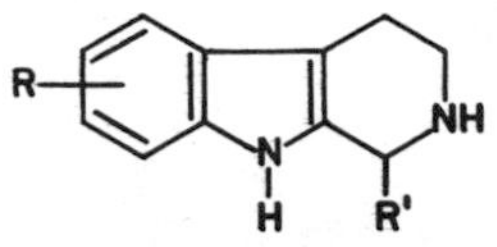 (2)

many indole alkaloids, a great deal of effort has been expended into synthetic developments in this area. Many of these natural products contain the pyridine ring in a reduced form, that is, tetrahydro-β-carboline **2** and the obvious emphasis on the latter system is essential.

The synthetic approaches of the requisite tetrahydro-β-carbolines have usually followed two pathways depending on whether substitution is required in the aromatic ring of this heterocyclic system. These are:

1. Synthetic elaboration of the parent indole ring through electrophilic substitution.
2. Conversion of tryptamine (or its appropriate derivative) to the essential β-carboline unit.

The first approach has been normally employed in the instances where appropriate functionality (usually hydroxyl or methoxyl) is required in the benzene ring. A few examples will suffice to illustrate this objective.

Synthesis of Tryptamines

An important route to tryptamines from indoles involves reaction of the latter with oxalyl chloride and conversion of the resultant glyoxylyl chlorides **4** to the final products **5**. The bufotenines **8**, which are hallucinatory substances obtained from certain shrubs as well as skin secretions of the toad, have been prepared by this route.[18]

The Mannich condensation represents a versatile approach to the synthesis of 3-substituted indole analogs. It provides, for example, a direct route to the barley alkaloid, gramine **10**.[19,20] The further utility of this approach is seen in

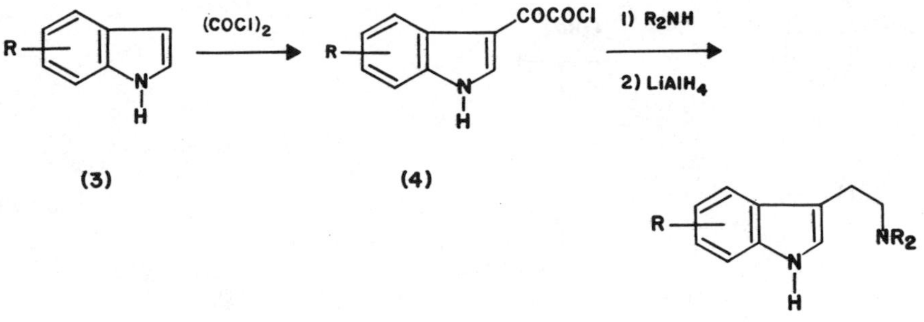

(3) (4) (5)

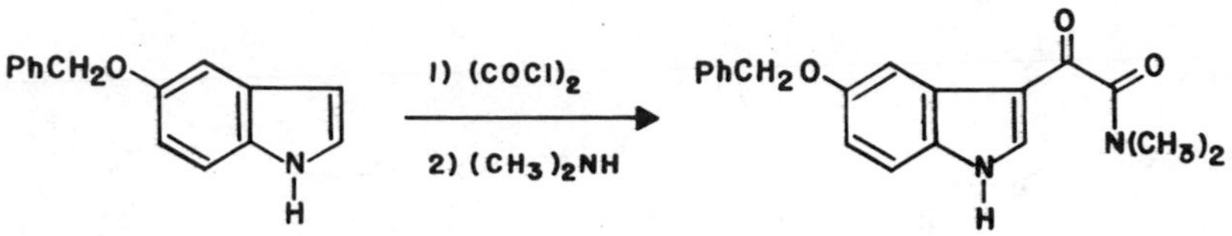

(6) (7)

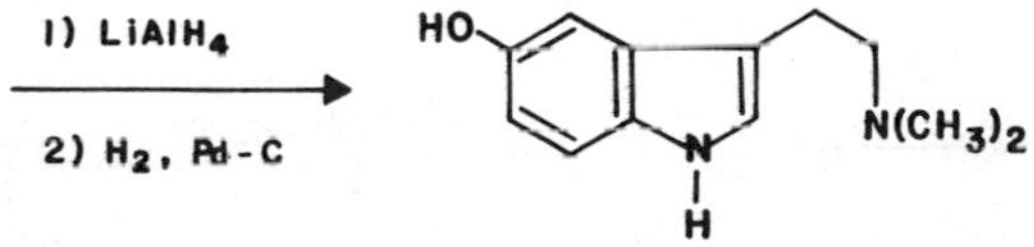

(8)

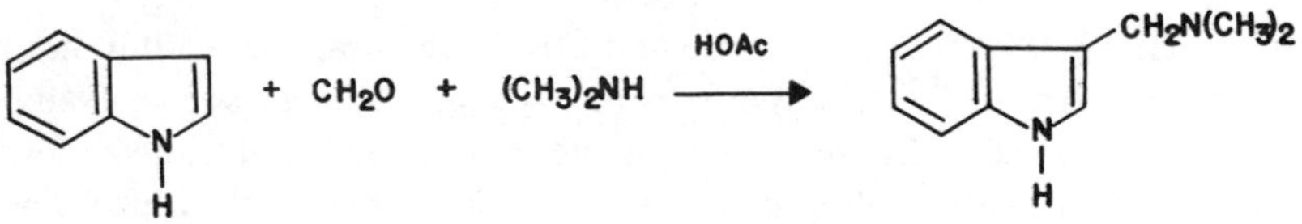

(9) (10)

the synthesis of indole-3-acetic acid or heteroauxin **13**, a plant growth hormone.[21-23] In this synthesis, the methiodide derivative of gramine **11** provides a very good system for nucleophilic displacement and thereby an entry into a number of valuable indole intermediates **12** becomes available.

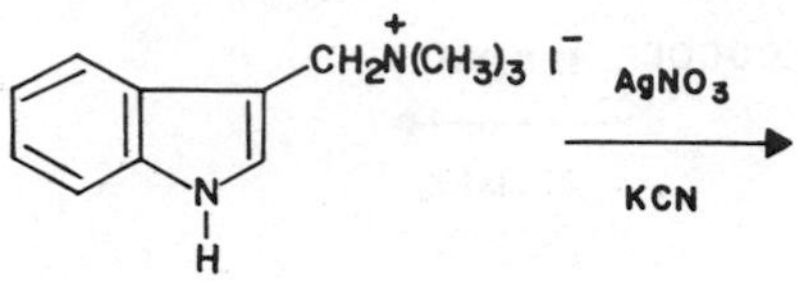

$CH_2\overset{+}{N}(CH_3)_3\ I^-$ $AgNO_3$ / KCN

(11)

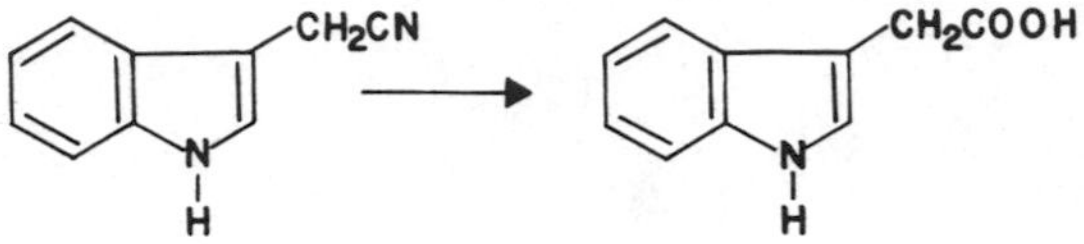

CH_2CN → CH_2COOH

(12) (13)

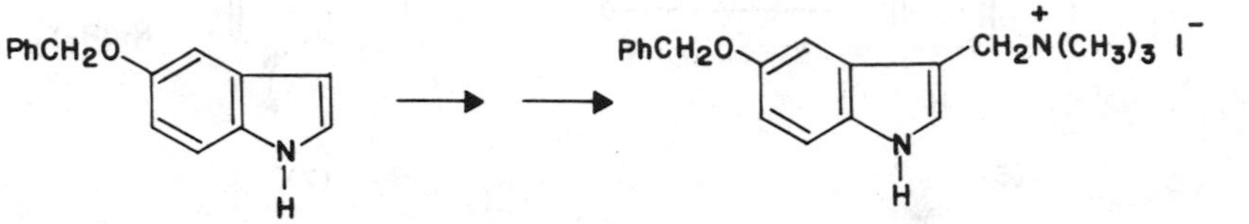

$PhCH_2O$ → → $PhCH_2O$ $CH_2\overset{+}{N}(CH_3)_3\ I^-$

(14) (15)

1) KCN 2) $LiAlH_4$ $PhCH_2O$ NH_2 H_2, Pd-C HO NH_2

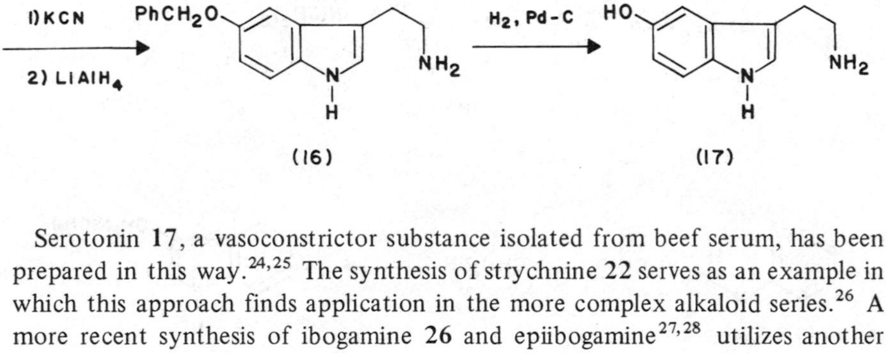

(16) (17)

Serotonin **17**, a vasoconstrictor substance isolated from beef serum, has been prepared in this way.[24,25] The synthesis of strychnine **22** serves as an example in which this approach finds application in the more complex alkaloid series.[26] A more recent synthesis of ibogamine **26** and epiibogamine[27,28] utilizes another acid derivative, indole-3-acetyl chloride **24** as a crucial intermediate.

Formylation and acylation of indoles have been accomplished using acid-catalyzed reactions of hydrogen cyanide (Gatterman synthesis) and nitriles (Houben-Hoesch synthesis), although the more recent Vilsmeier-Haack reaction employing *N,N*-dialkylamides and phosphorus oxychloride also provides a convenient route for the synthesis of 3-formyl derivatives **27**. These latter substances then provide satisfactory starting materials for further conversion to the tryptamine series.

(18) $\xrightarrow{HCHO-HNMe_2}$ (19) $\xrightarrow[2)\ NaCN]{1)\ MeI}$

(20) $\xrightarrow{LiAlH_4}$ (21) → → Strychnine (22)

(23) + (24) → (25)

→ → Ibogamine (26)

$\xrightarrow[HCl]{HCN}$ (27) $\xrightarrow{CH_3NO_2}$

(28) $\xrightarrow{LiAlH_4}$ (29)

C. Tetrahydro-β-carbolines from Tryptamines

As already mentioned, the tetrahydro-β-carboline nucleus, or some variation of it, is present in a very large number of indole alkaloids. Consequently, the large majority of indole alkaloid syntheses incorporate at some crucial stage the synthesis of this type of skeleton. In effect, the various methods available for this purpose basically involve overall substitution at the 2-position of the indole ring. Until recently, the generally accepted mechanism for this reaction has invoked direct electrophilic attack at C_2 of the indole system. However, investigations by Jackson and Smith[29-32] have suggested that in 3-substituted indole derivatives, initial attack occurs at the 3-position, and the resultant 3,3-spirocyclic indolenine **5** undergoes rearrangement to the 2,3-disubstituted system **6**. As is illustrated later, this consideration has interesting implications in indole alkaloid synthesis and biosynthesis.

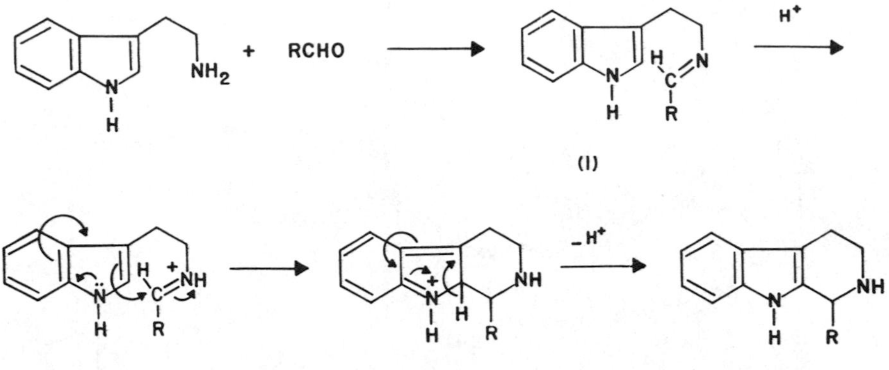

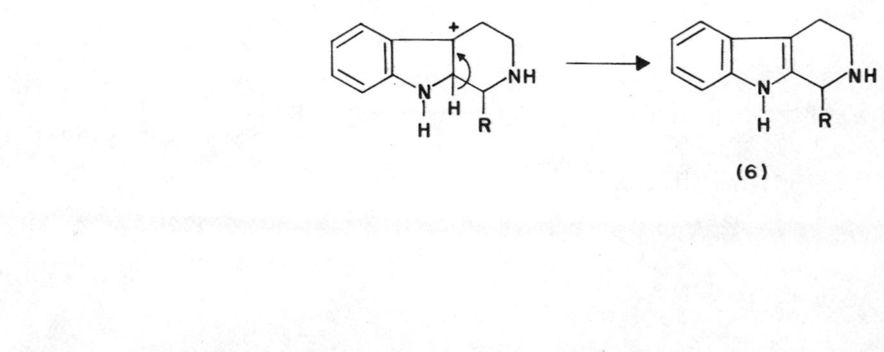

The most frequently employed reactions for the construction of tetrahydro-β-carbolines and related systems are conveniently divided into two main classes.

1. Cyclizations involving the application of the well-known Pictet-Spengler reaction.
2. Cyclizations involving the application of the Bischler-Napieralski reaction.

Application of the Pictet-Spengler Reaction

The original study by Pictet and Spengler[33] involved the formation of tetrahydroisoquinoline derivatives **9** by condensation of β-arylethylamines **7** with

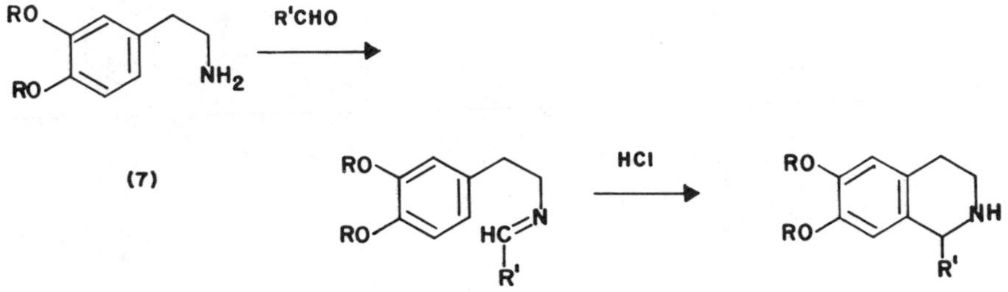

(7) (8) (9)

carbonyl compounds. The obvious extension of this reaction to the synthesis of tetrahydro-β-carbolines has provided one of the most versatile and general approaches to this system. Whaley and Govindachari[34] have reviewed the literature up to about 1950. More recent applications to alkaloid syntheses continue to appear in the literature.

The syntheses of 3-methylaspidospermidine **15** and eburnamine **13** represent two more recent examples of the Pictet-Spengler approach involving an aldehydo ester **10** as the condensing unit. The higher reactivity of the aldehydic carbonyl function allows preferential reaction at this site.[35] Kuehne[36] similarly employed this type of cyclization in the synthesis of the Vinca alkaloid, vincamine **18**. A general study on the utility of tetrahydropyridine derivatives (**19**, **22**, and **25**) in indole alkaloid syntheses has been made by Wenkert.[37] These investigations, which are discussed more fully in a later section, are mentioned here, since the tetrahydropyridine intermediates provide an interesting variant of the Pictet-Spengler cyclization. The syntheses of d,l-dihydrogambirtannine **23** and d,l-epialloyohimbane **26**[38,39] illustrate this approach.

It is often possible to effect Pictet-Spengler cyclizations to the appropriate tetrahydro-β-carboline systems by generating the requisite carbonyl intermediates **29** *in situ*. van Tamelen and his co-workers have made effective use of such reactions in some of their indole alkaloid syntheses. An illustration from

NH_2 + CHO CO_2Me → (10) (11)

BF_3

1) OsO_4

2) $NaIO_4$

(14)

$LiAlH_4$

(12)

$LiAlH_4$

(15)

(13)

NH_2 + CHO CO_2Et MeO_2C → MeO_2C

(16)

(17)

→ →

HO CO_2Me

(18)

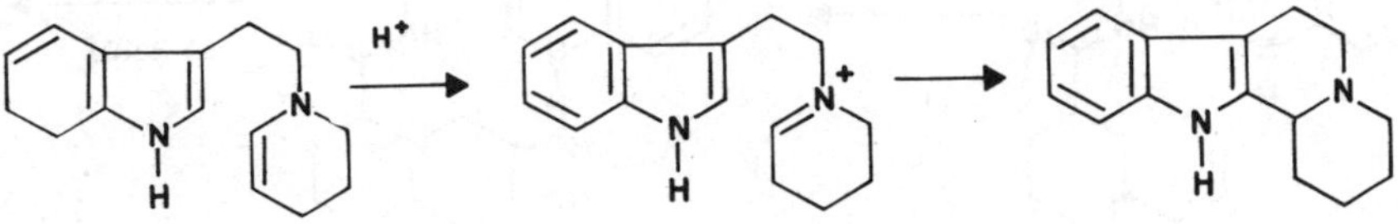

their studies on the synthesis of yohimbine portrays the effectiveness of this approach.[40,41] Additional examples of the Pictet-Spengler reaction are cited during the various syntheses of the alkaloid classes elaborated in the later sections.

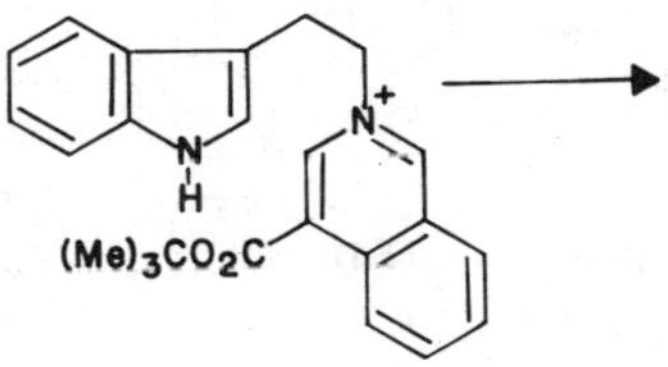

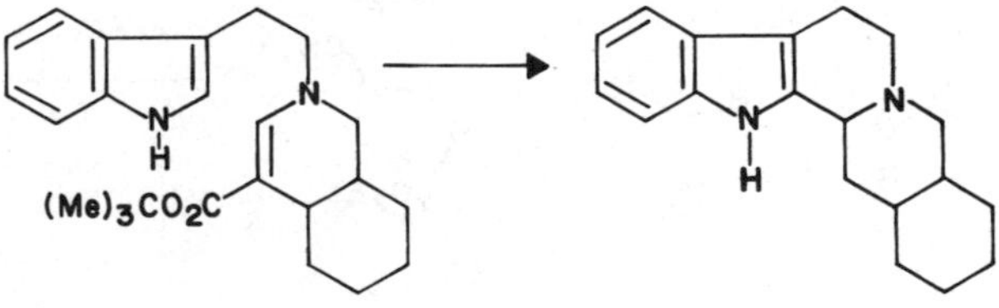

(27)

OsO_4

(28)

1) H_2/Pt

2) $NaIO_4$

(29)

H_3PO_4

(30)

(31)

P_2O_5

$\triangle$

(32)

H_2

(33)

+

(34)

H_2/Ni

(35)

$POCl_3$

(36)

H_2/Pt

(37)

Application of the Bischler-Napieralski Reaction

The Bischler-Napieralski synthesis had its initial development in the tetrahydroisoquinoline family.[42] The reaction normally employs a suitable amide derivative **31**, and cyclization is usually effected by phosphorus oxychloride or phosphorus pentoxide.

Direct extension of this reaction involving conversion of the amides of tryptamines to the tetrahydro-β-carbolines has seen enormous application in indole alkaloid syntheses. Several reviews[43,44] on this reaction tabulate and discuss a number of successful cyclizations, but a few examples are included here for illustrative purposes. The recently described synthesis of the corynantheine series[45,46] reveals an effective use of the Bischler-Napieralski reaction. In a

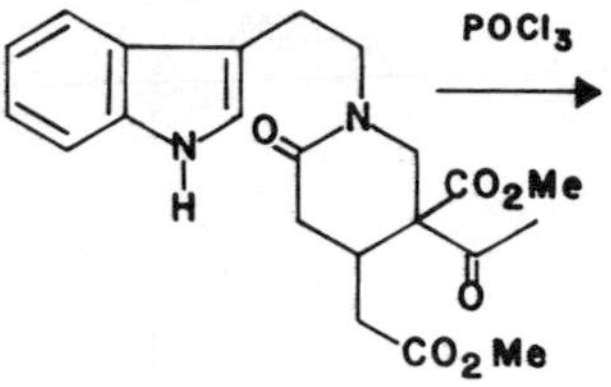

$POCl_3$

(38)

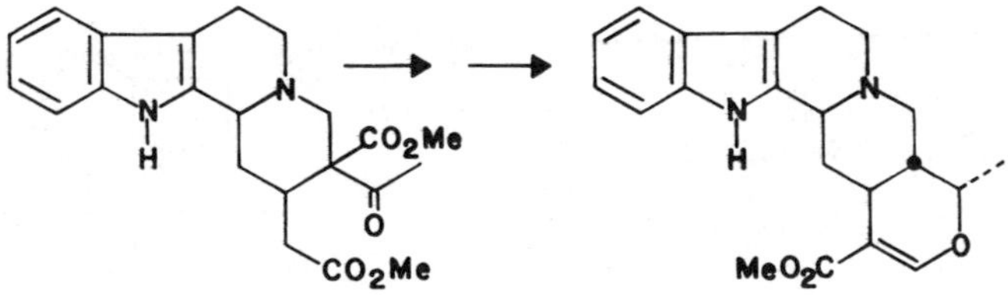

(39)

(40)

similar manner, van Tamelen employed the Bischler-Napieralski synthesis as a crucial step in the total synthesis of ajmalicine **40**.[47] The synthesis of reserpine **46**[11] involved this cyclization after the critical substitution and stereochemisty had been established. In the eburnamine series, a recent synthesis of eburnamonine **50**[48] embodies a combination of Pictet-Spengler and Bischler-Napieralski cyclizations. Additional applications of this reaction will be seen from the later discussions of the individual alkaloid classes.

(41) + (42) → (43)

$NaBH_4$ → (44) → $POCl_3$ → (45) → →

(46)

(47) → $POCl_3$ → (48) → 1) OH^- 2) ICH_2CO_2Et →

(49) → 1) H_2 2) OH^- → (50)

2. TOTAL SYNTHESIS OF ALKALOID FAMILIES

A. The Simple Indole Bases

Some of the simpler indole alkaloids contain one substituent only, and this is usually located in the 3-position of the indole ring. Examples of such bases include gramine **1** and its mono- and dimethyl derivatives, abrine **2**, hypaphorine

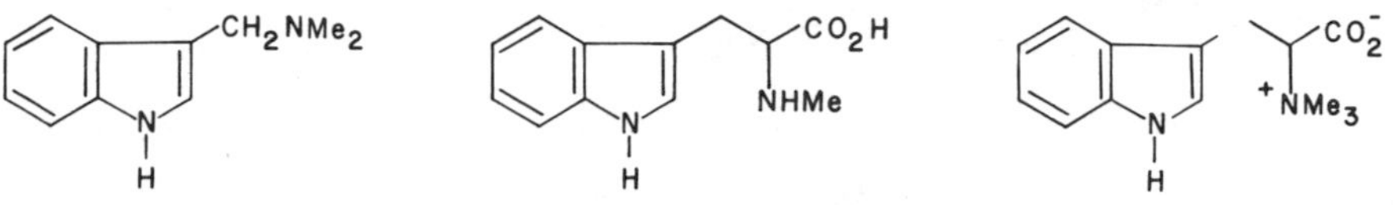

(1) gramine

(2) abrine

(3) hypaphorine

3 (the methylbetaine of methyl abrine) and tryptamine and its derivatives. Gramine is certainly one of the simplest alkaloids known, and its original isolation from chlorophyll-deficient barley mutants led to the belief that the presence of this alkaloid was genetically related to the chlorophyll deficiency. Its subsequent isolation from normal sprouting barley made this belief untenable.[49-52]

The synthesis of gramine by the application of the Mannich reaction has already been mentioned. Since the yield in this reaction is almost quantitative, gramine becomes readily available as a valuable intermediate in other alkaloid syntheses. As already noted, the methiodide of this alkaloid **4** has seen extensive application due to its facile displacement by a variety of nucleophiles. For

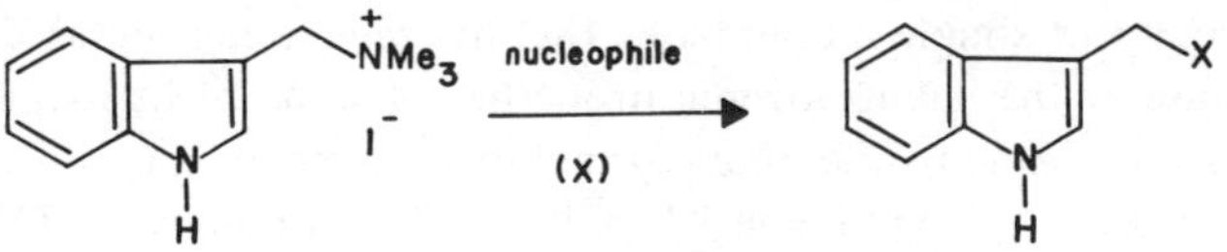

(4) —nucleophile (X)→ (5)

example, facile conversion of gramine to the tryptophan series, of which abrine and hypaphorine are members, can be accomplished in this way.[53-56] In a similar manner, anions derived from α-nitrocarboxylate esters **9** can be employed.[57,58]

NMe_2 + $^{-}C(CO_2Et)_2$ / NH / $COCH_3$ → $(CO_2Et)_2$ / NH / $COCH_3$

(6) (7)

hydrolysis ($-CO_2$) → CO_2H / NH_2

(8)

NMe_2 + $^{-}C(CO_2Et)_2$ / NO_2 → $(CO_2Et)_2$ / NO_2

(9) (10)

→ → CO_2Et / NH_2

(11)

Another family of simple indole bases that has seen a considerable revival of interest because of the hallucinogenic properties of some of its members is the tryptamine group. Among the most prominent members are bufotenine (or 5-hydroxy-*N,N*-dimethyltryptamine **12**, psilocin **13**, and psilocybin **14**).

Bufotenine has been isolated from the leguminous shrubs, *Piptadenia peregrina* Benth[59,60] and *P. macrocarpa* Benth. It is interesting to note that the seeds of these plants have been used by Indian tribes of Latin America for ceremonial purposes. The narcotic snuff called "cohoba" is apparently inhaled through a bifurcated tube and, in small doses, produced hallucinatory effects. Several syntheses of bufotenine are available and the routes employed are rather different. One of these, reported by Speeter and Anthony, exemplifies the utility of glyoxyl chlorides and has already been presented in the brief discussion on syntheses of tryptamine.

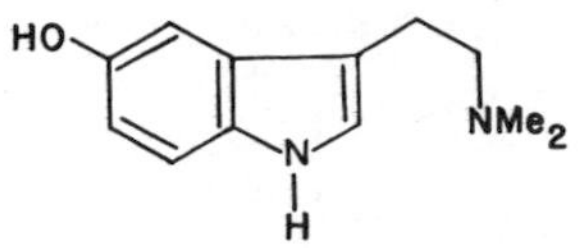

(12)
bufotenine

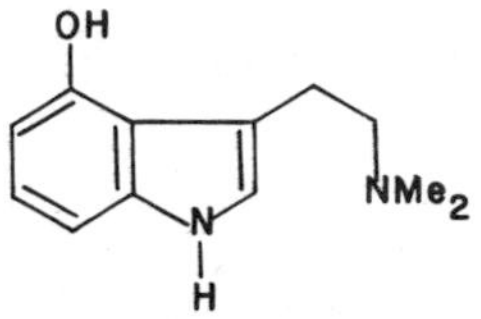

(13)
psilocin

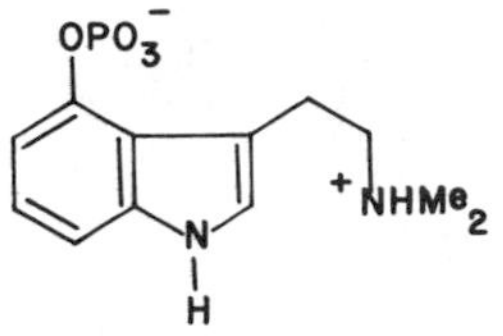

(14)
psilocybin

Another synthesis of bufotenine[61] involves the application of a novel route to the indole system. In this sequence, the reactive methylene group of 2,5-dimethoxybenzyl cyanide **15** is alkylated with dimethylaminoethyl chloride and the resultant intermediate **16** is elaborated to the final product. The most interesting step in this sequence represents the formation of the indole ring via the quinone **18** obtained in the potassium ferricyanide oxidation step. Stoll,

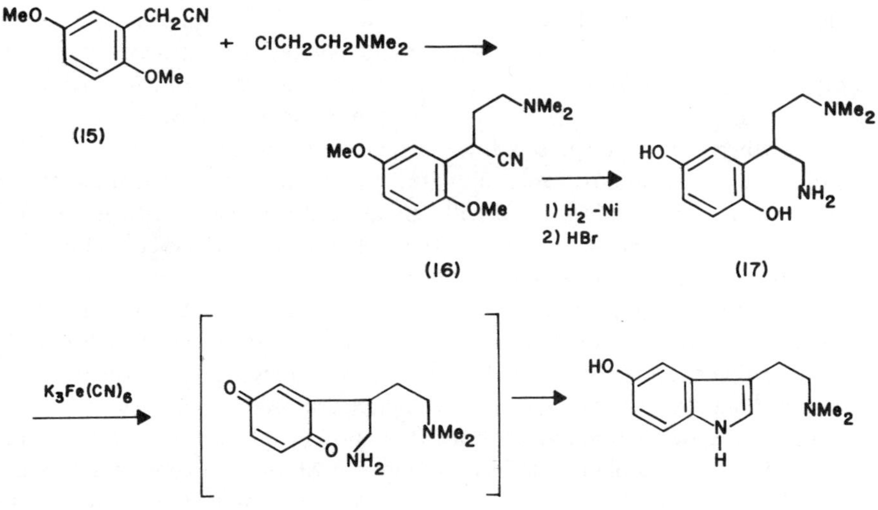

(15) + $ClCH_2CH_2NMe_2$ →

(16) → 1) H_2-Ni 2) HBr → (17)

$K_3Fe(CN)_6$ → (18) → (19)

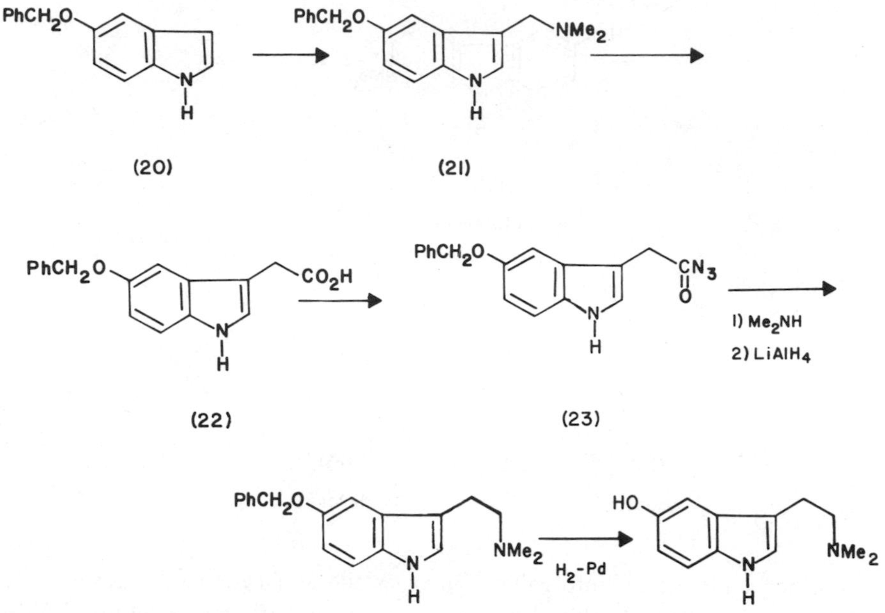

(20) → (21)

(22) → (23) → 1) Me_2NH 2) $LiAlH_4$

(24) → H_2-Pd

Hofmann and their co-workers[62] have used the gramine route **21** in achieving still another synthesis of this alkaloid.

Various fungi contain the very interesting derivatives of 4-hydroxy indole, psilocybin, and psilocin. Investigations by Heim and his group in France and Hofmann and his co-workers in Switzerland portray a particularly intriguing story. Psilocybin was first isolated in 1958 from *Psilocybe mexicana* Heim,[63] a Mexican fungus. Other fungi belonging to the *Psilocybe* and *Stropharia* genera have now been shown to possess these two alkaloids. It is pertinent to note that all of these investigations were stimulated by the realization that the Mexican Indians in the use of narcotic and hallucinogenic drugs in their rituals prepared their drugs from such fungi. Qualitatively, the effects of these compounds are similar to mescaline and lysergic acid diethylamide.

The synthesis of psilocin and psilocybin, which provided the necessary substantiation for the proposed formulations of these interesting phosphorylated indole alkaloids, utilizes the oxalyl chloride route to tryptamine derivatives.[64-66]

Another simple base, 5-hydroxytryptamine (serotonin, enteramine, thrombocytin), has received repeated attention. Its important physiological activity and its function as a neurohormone has undoubtedly stimulated considerable interest.[67,68] For example, it has been implicated as the active irritant in such plants as *Mucuna pruriens* D.C. (cowhage) and *Urtica dioica* L. (stinging nettle). In

OCH_2Ph — 1) $(COCl)_2$ 2) $HNMe_2$ → OCH_2Ph, O, O, NMe_2 — 1) $LiAlH_4$ 2) H_2-Pd → OH, NMe_2

(25) (26) (27)

$(PhCH_2O)_2POCl$ → $OPO(OCH_2Ph)_2$, NMe_2 — H_2-Pd → OPO_3^-, $H\overset{+}{N}Me_2$

(28) (29)

mammals, this alkaloid is found in the tissues of the stomach, lungs, and intestines, as well as in the brain and in the blood. It appears to play an important role in the central nervous system. Interestingly, its presence has also been reported in such edible fruits as the banana, plum, pineapple, avocado, and tomato.

Numerous convenient syntheses of this alkaloid have been described and most established routes to tryptamine derivatives have been employed. The first

$PhCH_2O$ — $HCONMe_2$, $POCl_3$ → $PhCH_2O$, CHO — CH_3NO_2 →

(30) (31)

$PhCH_2O$, NO_2 — 1) $LiAlH_4$ 2) H_2-Pd → HO, NH_2

(32)

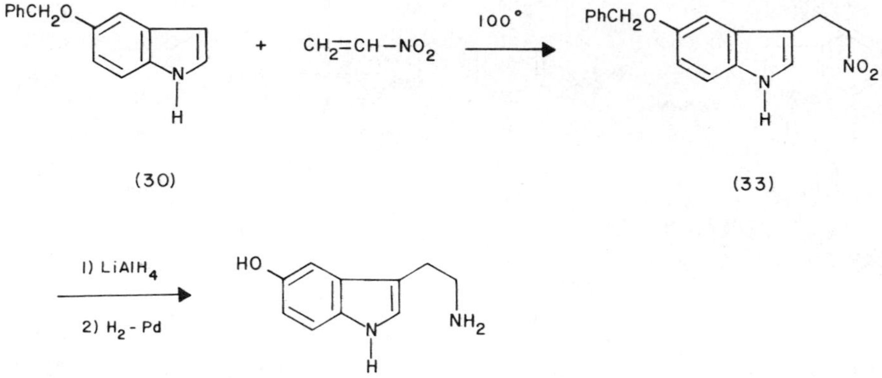

PhCH$_2$O (indole, N–H) + $CH_2{=}CH{-}NO_2$ $\xrightarrow{100°}$ PhCH$_2$O (indole, N–H) $CH_2CH_2NO_2$

(30) (33)

$\xrightarrow[\text{2) } H_2\text{ - Pd}]{\text{1) LiAlH}_4}$ HO (indole, N–H) $CH_2CH_2NH_2$

PhCH$_2$O (indole, N–Mg Br) + CH_2–CH_2 (N–H aziridine) $\longrightarrow$ PhCH$_2$O (indole, N–H) $CH_2CH_2NH_2$

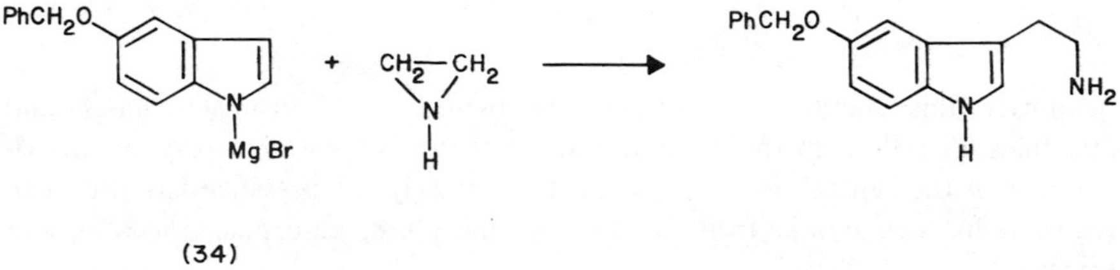

(34)

MeO (C_6H_4) $\overset{+}{N_2}$ Cl^- + HO_2C (piperidin-2-one, N–H) $\longrightarrow$ MeO (C_6H_4) N(H)–N= (piperidin-2-one, NH, O)

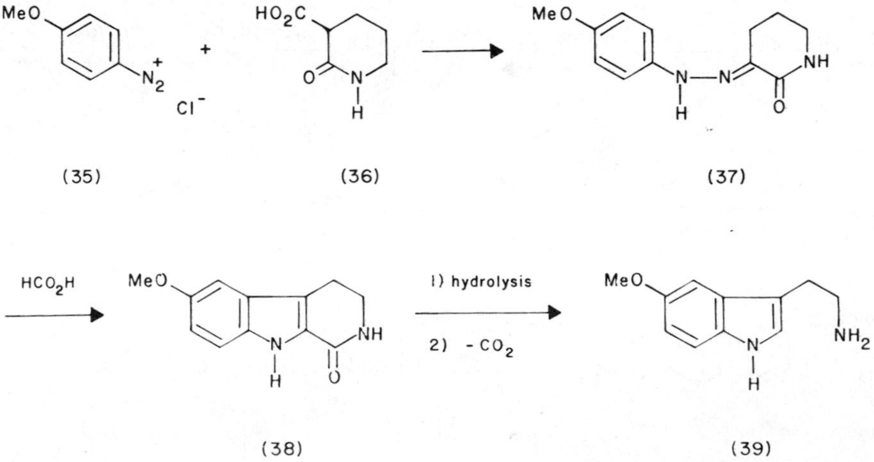

(35) (36) (37)

$\xrightarrow{HCO_2H}$ MeO (N–H, NH, O) $\xrightarrow[\text{2) } -CO_2]{\text{1) hydrolysis}}$ MeO (indole, N–H) $CH_2CH_2NH_2$

(38) (39)

synthesis utilized the gramine route,[25] whereas a second synthesis started with 5-benzyloxyindole acetonitrile, prepared by the reaction of 5-benzyloxyindole magnesium iodide with chloroacetonitrile.[24] The procedure of Speeter and Anthony via 5-benzyloxyindoleglyoxylyl chloride represents a high yielding preparation.[18] A few other routes, which have received little mention in the previous discussion, are now illustrated.

Utilization of indole-3-aldehydes **31** is seen in the following sequence of reactions.[69,70] An example of substitution with nitro-olefins in the indole series is available from the following synthesis of 5-hydroxytryptamine.[71] Perhaps the most direct and shortest route to this alkaloid has been reported by a Japanese group.[72] A number of synthetic routes obviate the preparation of 5-benzyloxyindole **30**. In these instances, the indole ring is formed after appropriate provision for the ethanamine side chain has been made.

The interesting approach developed by Abramovitch and Shapiro[7,73] has already been mentioned as a versatile route to substituted tryptamines. The initial Fischer cyclization reaction furnishes a suitable tetrahydro-β-carboline intermediate **38** for further elaboration to the desired product. A different approach adopted by an Italian group[74] employs the Japp-Klingemann reaction in the initial hydrazone formation with a cyclopentanone carboxylic ester **41**. The resultant intermediate **42** is then transformed via Fischer cyclization to the substituted indole.

There are a number of simple indole alkaloids that possess the carbazole system. Two of these, murrayanine **44** and glycozoline **45**, have been studied recently and may be cited as examples. A synthesis of murrayanine has been achieved according to the following sequence of reactions.[75] A similar approach

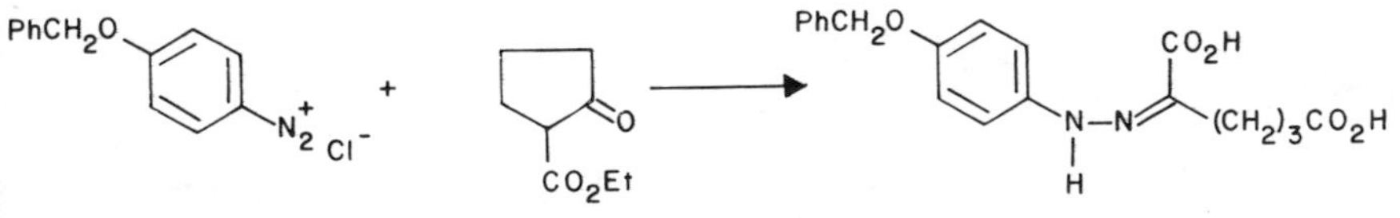

$PhCH_2O$ … $\overset{+}{N}_2$ Cl^- + … CO_2Et → $PhCH_2O$ … CO_2H … N—N= … $(CH_2)_3CO_2H$ … H

(40) (41) (42)

→ $PhCH_2O$ … CO_2H … N … H —HN_3, etc.→ $PhCH_2O$ … NH_2 … N … H

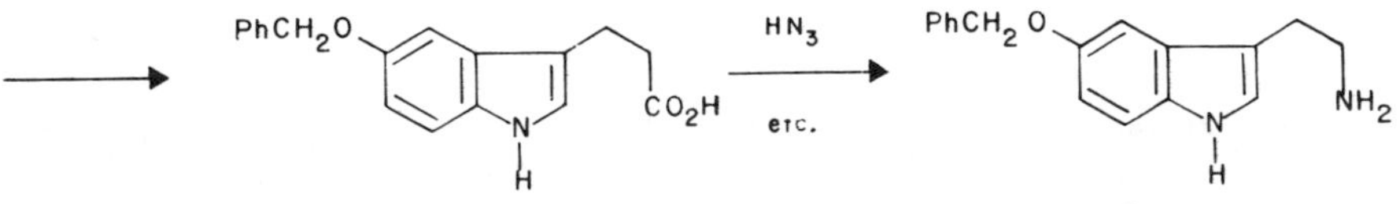

(43)

murrayanine

(44)

glycozoline

(45)

(46) + cyclohexanone → (47) —HOAc→ (48)

—chloranil (−H_2)→ (49) —1) n-BuLi 2) Ph–N(CH$_3$)–CHO→ (44)

(50) + N_2^+ Cl^- →

(51) → (52)

—Wolff-Kishner reduction→ (53) —chloranil (−H_2)→ (45)

(54) $\xrightarrow{\text{nitration}}$ (55) $\xrightarrow[\text{quaternary methosulphate}]{\text{via}}$

(56) $\xrightarrow{\text{HCl / EtOH}}$ (57) $\xrightarrow{\text{1) sodium dithionite reduction; 2) OH}^-}$

(58) $\xrightarrow{\text{dicyclohexyl carbodiimide}}$ (59) $\xrightarrow{B_2H_6}$

(60) $\xrightarrow{\text{MeI}}$ (61)

$\xrightarrow{H_2/Pd\text{-}C/MeOH}$ (62) $\xrightarrow{\text{HCl / MeOH}}$ (63)

Dehydrobufotenine

employing the Japp-Klingemann reaction and the Fischer indole synthesis provides glycozolidine.[76] Dehydrobufotenine **63** is the principal constituent from the parotid glands of the South American toad *Bufo marinus,* and a synthesis of this interesting heterocyclic system has been described.[77]

B. The Carboline Alkaloids

The carboline alkaloids, often referred to as the harmala alkaloids since they were first found in *Peganum harmala* L.[78], contain the fundamental carboline system in which an indole nucleus is fused with a pyridine ring. Various reduced forms of the latter are known, and the prominent members are shown.

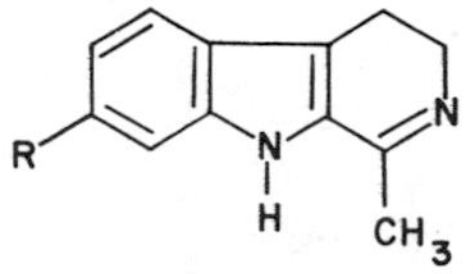

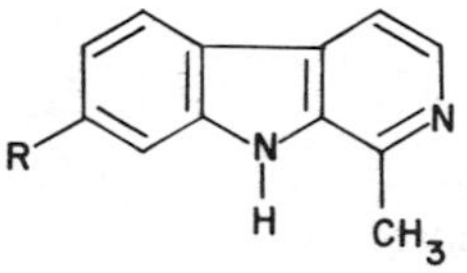

R N N H CH_3

R N N H CH_3

(1) harmaline (R = MeO)

(2) harmalan (R = H)

(3) harmine (R = MeO)

(4) harman (R = H)

The syntheses in the family involve elaboration of the indole nucleus to appropriately disubstituted tryptamine derivatives and cyclization to form the pyridine ring in the final stages of the sequence. For example, the substituted hydrazone derivative **8** under Fischer conditions leads to **9**, which is then cyclized and oxidized to yield harmine **3**.[79] Spath and Lederer[80] employed a Bischler-Napieralski cyclization in another synthesis of harmine **3** and harmaline **1**. In this synthesis, one must bear in mind that the preparation of 6-methoxytryptamine **12** is complicated by the fact that the 4-methoxy isomer is also obtained, since the alternative cyclization in the zinc chloride reaction is also observed. This method was also used for the synthesis of harman **4** and harmalan **2**. Phenylhydrazine was used as starting material instead of 3-methoxyphenylhydrazine **10**. A more recent synthesis of Spenser[81] ,provides good overall yields of these alkaloids. The last step of the synthesis involves a dehydration of the 1-hydroxymethyl-1,2,3,4-tetrahydro-β-carboline **14** and rearrangement of the resultant exocyclic double bond into the endocyclic position.

C. The Ergot Alkaloids

The tremendous interest that has surrounded this family of alkaloids is undoubtedly because of the remarkable pharmacological properties associated with these

NaOH

(5)

MeO–C₆H₄–N_2^+ Cl^-

(6)

Ac_2O

(7)

(8)

HCl / n-BuOH

NH_2NH_2 (alc·)

HCl

(9)

chromic acid

(1) → (3)

CH_3O–C6H4–NHNH2 (10) + (11) $\xrightarrow{ZnCl_2}$ (12)

$\xrightarrow{Ac_2O}$ (13) $\xrightarrow{P_2O_5,\ xylene}$ harmaline (1)

$\xrightarrow{Pd}$ harmine (3)

(12) $\xrightarrow{CH_2OHCHO}$ (14)

$\xrightarrow{H_3PO_4}$ (1)

molecules.[82] Ergot, the dark brown tuberous growth protruding from the ears of rye, is a fungus, *Claviceps purpurea,* which produces one of the most remarkable drugs known to man. The best known member of this family, lysergic acid, in the form of its diethylamide (LSD) receives almost daily attention because of its use as a hallucinatory agent. The obvious interest associated with structure-activity relationships in this area has demanded a considerable synthetic effort of not only the natural products by synthetic analogues as well.

The fundamental tetracyclic system of this family has been named ergoline **15**, and it is convenient to subdivide the various members into two main groups, one group belonging to the lysergic acid series and the other to the so-called clavine family.

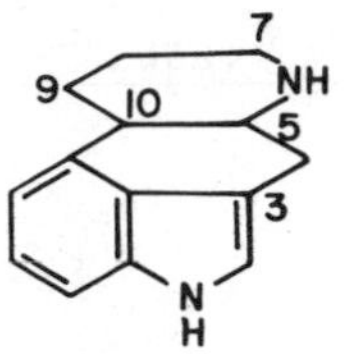

(15) ergoline

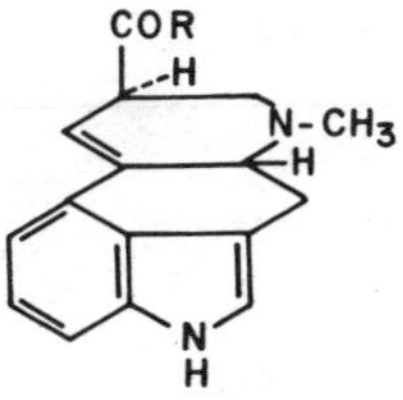

(16) R = OH, lysergic acid

(17) R = NH_2, ergine

(18) R = $NHCH(CH_3)CH_2OH$, ergotametrine

(19) R = [cyclic peptide, $CH_2C_6H_5$], ergotamine

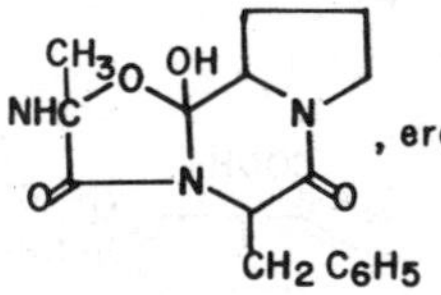

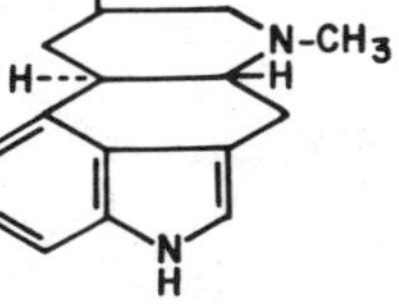

(20) R = H, agroclavine

(21) R = OH, elymoclavine

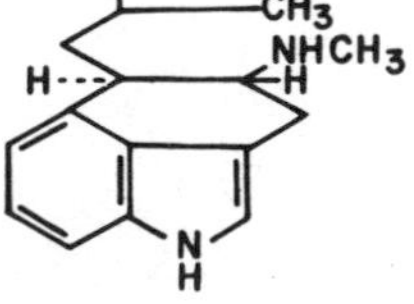

(22) chanoclavine

The first synthesis of racemic dihydrolysergic acid was accomplished by Uhle and Jacobs in 1945.[83] The starting material for this synthesis was 4-aminonaphthostyril **26** in which three rings of the ergoline system are already present. It is interesting to note that during this work a new synthesis of 3-substituted quinolines was developed. Thus 3-nitroquinolines and 3-quinoline carboxylic acids **36** may be formed directly in this sequence.

Stoll and his group made substantial contributions to syntheses in this area. In 1950, they accomplished the synthesis of optically active dihydrolysergic acids.[84] In this improved synthesis, the initially obtained racemic dihydronorly-

$(C_2H_5O)_2CH-CH_2Br \xrightarrow{KCN} (C_2H_5O)_2CH-CH_2CN \xrightarrow[Na]{HCOOC_2H_5} Na^{\oplus}\,{}^{\ominus}C(CHO)_2-CN$

(23) (24) (25)

(26) $\xrightarrow{HCl}$ (27) $\xrightarrow{1)\ ZnCl_2\ \ 2)\ HCl}$ (28)

$\xrightarrow{1)\ CH_3I\ \ 2)\ AgCl}$ (29) $\xrightarrow{Pt/H_2/HCl}$ (30) $\xrightarrow{Na/C_4H_9OH}$ (31)

rac-dihydrolysergic acid

CHO
CH—CN
CHO
NH2
(32)
H+
CHO CN
N
(33)

ZnCl2 CN (34) + CONH2 (35) HCl COOH (36)

sergic acid was converted to the homogeneous racemates and, in turn, to the optically active acids.[85]

The racemic dihydronorlysergic acid **41** thus obtained is a mixture of three stereoisomers that, after conversion to the methyl ester, may be separated by chromatographic means to give racemic dihydronorlysergic acid-I methyl ester, racemic dihydronorisolysergic acid-I methyl ester, and racemic dihydronoriso-lysergic acid-II methyl ester.

NH2 (26)
EtOCH=C(COOEt)2
EtOOC CH—CH=N EtOOC (37)
COOEt NH (38)

1.) Zn/HCl
2.) MeOH
COOMe NH (39) or COOMe NH (40)
Na/butanol
COOH NH (41)
rac-dihydronorlysergic acid

The dihydronorlysergic acids were converted to the corresponding racemic dihydrolysergic acids by migration of the methyl group in the esters. This migration was effected by heating the methyl esters of the nor series or by reduction in the presence of formaldehyde.[84,86,87] Finally, resolution of the resulting dihydrolysergic acid-I racemates in the form of the L-norephedride completed the synthesis of d-(−)-dihydrolysergic acid-I, the basic constituent of all the reduced natural ergot alkaloids.

Another approach to the 8-ketoergoline system as devised by Stoll and his group is also illustrated.[88] In this sequence, resolution was performed on racemic 6-methyl-8-hydroxyisoergoline **44** via its L-menthoxyacetyl ester. The resultant optically active amino alcohol was then oxidized under Oppenauer conditions to 6-methyl-ergolin-8-one **45**.

(26) → $ClCH_2CH(OH)CH_2Cl$ → (42)

(42) → 1) CH_3Br / NaOAc 2) cyclohexanol / $C_6H_5NEt_2$ → (43) → Na / butanol →

(44) → $Al(OPh)_3$/cyclohexanone → (45)

6-methyl-8-hydroxyisoergoline (44)

6-Methylergolin-8-one (45)

In still another appraoch to the dihydronorlysergic acid series, Stoll and his group utilized 4-methoxynaphthostyril **46** as the starting material and converted this substance to the benz[c,d]indoline system **47** by reduction with lithium aluminum hydride. The amino group is introduced at the appropriate step under the conditions of the Bucherer reaction. Subsequent elaboration of this intermediate **49** follows the chemical methods previously developed.[89,90]

(46) → 1) $LiAlH_4$ 2) acetylation → (47)

1) HBr 2) Ac_2O / pyridine → (48) → Bucherer → (49)

1) $(CHO)_2CHCN$ 2) $ZnCl_2$ / C_6H_5CN 3) H_2SO_4 EtOH → (50) → 1) Ac_2O 2) CH_3I 3) Pt/H_2 →

(51) → 1) Na/butanol 2) esterification → (52)

Finally, a different synthesis to dihydrolysergic acid was developed by a British group,[91] In this sequence, 4-hydroxynaphthostyril **53** is converted directly to 4-methylaminonaphthostyril, and the latter is elaborated to the benzindoline analog **56** by standard reactions. Condensation with ethyl bis-(chloromethyl)malonate, cyclization, and hydrolysis provides an interesting route to these compounds **57**.

Although these syntheses made effective use of the naphthostyril or benz[c,d]-indoline system, such approaches proved fruitless in the synthesis of lysergic acid

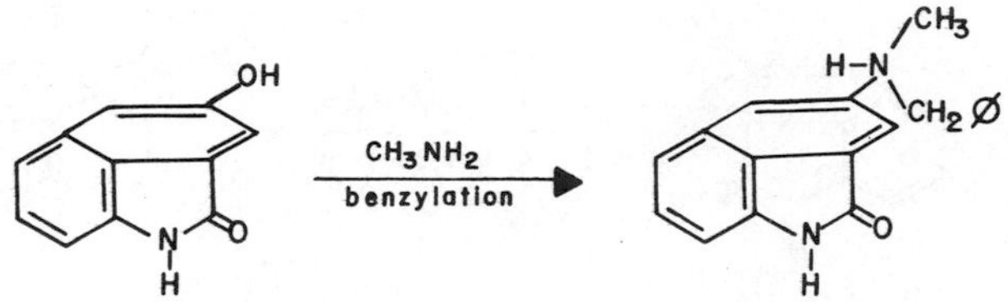
OH
CH3NH2
benzylation
CH3
H-N
CH2Ø
N
H
O

(53) (54)

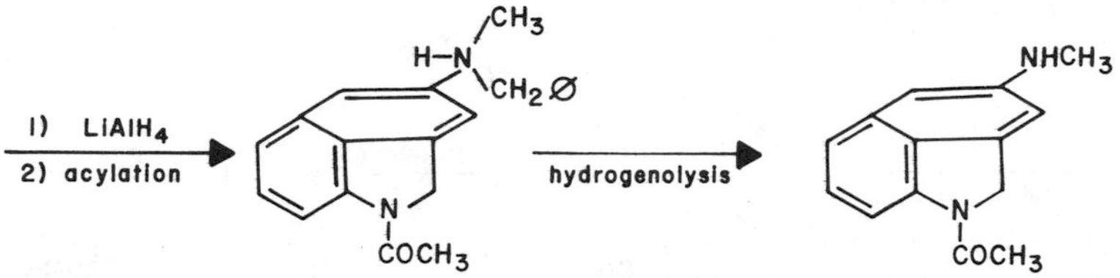
1) LiAlH4
2) acylation
CH3
H-N
CH2Ø
N
COCH3
hydrogenolysis
NHCH3

(55) (56)

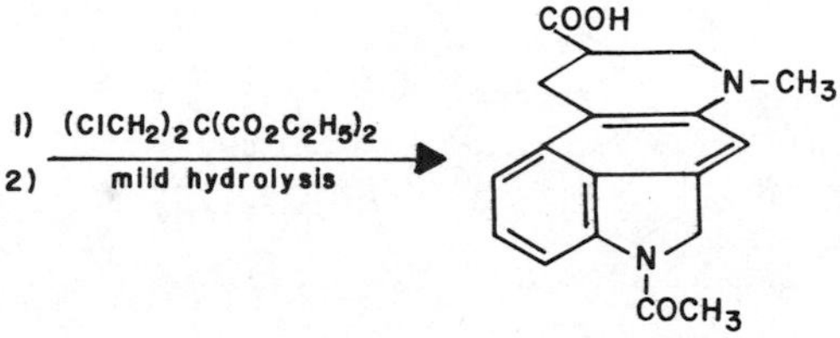
1) (ClCH2)2C(CO2C2H5)2
2) mild hydrolysis
COOH
N-CH3
N
COCH3

(57)

and its derivatives. These attempts fail because of the inability of introducing the necessary double bond in the 9,10-position. The ergolene system of lysergic acid **58** undergoes an irreversible rearrangement under the influence of strong acidic reagents, since the resultant benzindoline system **59** appears to possess a greater resonance energy.[89] On this basis the production of lysergic acid compounds via the readily available benzindoline derivatives was not possible.

The first total synthesis of lysergic acid was accomplished by the Lilly research groups of Kornfeld.[92,93] In this approach, the starting material was a *N*-benzoyl-

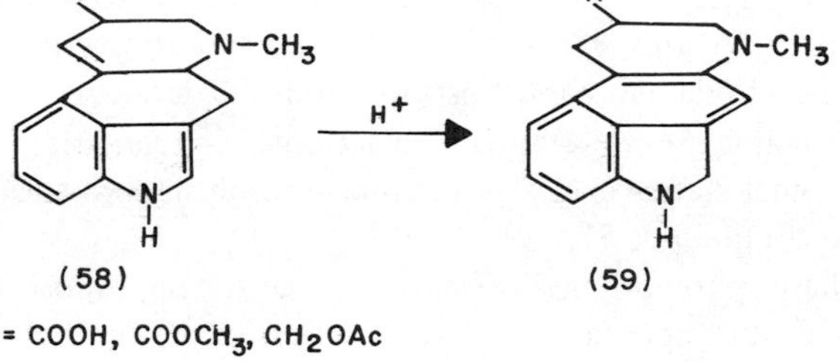
R
N-CH3
N
H
H+

(58) (59)

R = COOH, $COOCH_3$, CH_2OAc

2,3-dihydroindole derivative **60**, which was stable to the various, rather classical, methods employed to elaborate rings C and D. Once the tetracyclic molecule **65** was intact, dehydrogenation of the 2,3-dihydroindole system to the required indole system was accomplished in the last stage of the sequence. In this manner, formation of the benz[c,d]-indoline system was prevented.

D. The Physostigmine-Geneserine Group

The fruit of the African vine *Physostigma venenosum* Balf., more commonly known as Calabar bean, has been used by the West African natives as a poison. It

COOH
N
OC-C_6H_5
(60)
1) $SOCl_2$
2) $AlCl_3$
N
OC-C_6H_5
(61)

pyridine hydrobromide
perbromide/acetic acid
Br
N
OC-C_6H_5
(62)
$CH_3NHCH_2C(OCH_2)_2CH_3$

CH_3
$(CH_2O)_2$-C-CH_2
N-CH_3
N
OC-C_6H_5
(63)
1) HCl
2) NaOMe
3) Ac_2O
N-CH_3
N
OC-CH_3
(64)

1) $NaBH_4$
2) $SOCl_2$ / SO_2
3) NaCN
4) methanolysis
$COOCH_3$
N-CH_3
N
H
(65)
1) OH^-
2) deact. Raney Ni
COOH
N-CH_3
N
H
(16)
Lysergic acid

was administered to criminals during trials for various offenses, including witchcraft. The principal alkaloid, physostigmine **1** (also known as eserine) is extremely poisonous, and its main action on the parasympathetic nervous system is produced by inhibition of the enzyme acetylcholinesterase. Other related members in this family are geneserine **2** (also known as eseridine), eseramine **3**, and physovenine **4**.

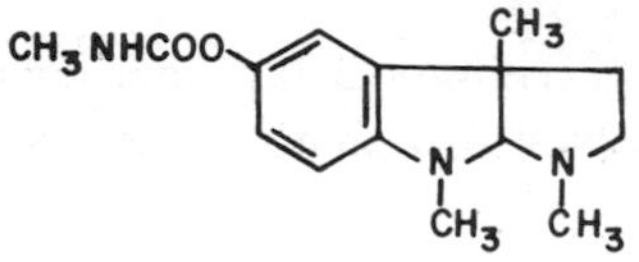

(1) 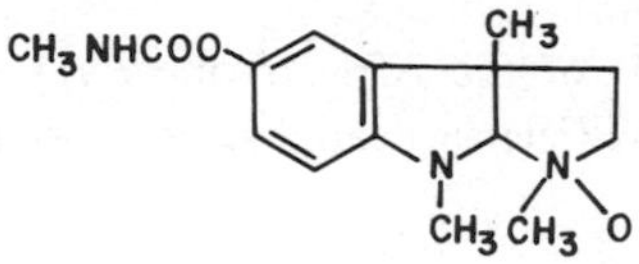 (2)

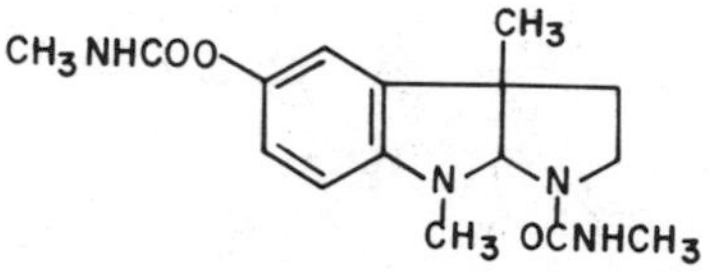

(3) 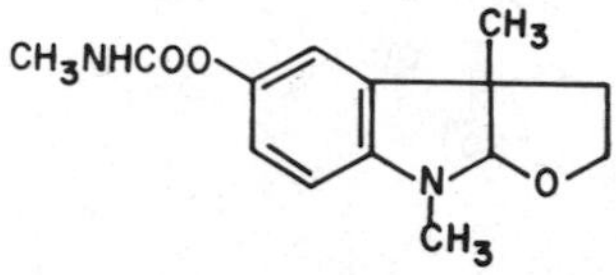 (4)

The first complete synthesis of physostigmine was carried out by Julian and Pill[14-16] in 1935. This approach, employing the reductive ring closure of an appropriately substituted oxindole **5**, has already been discussed. Its application to other members of this family has been reported by several groups.[94-96] A completely different approach to the synthesis of physostigmine was described by Harley-Mason and Jackson.[97] The crucial step is based on an earlier finding[61]

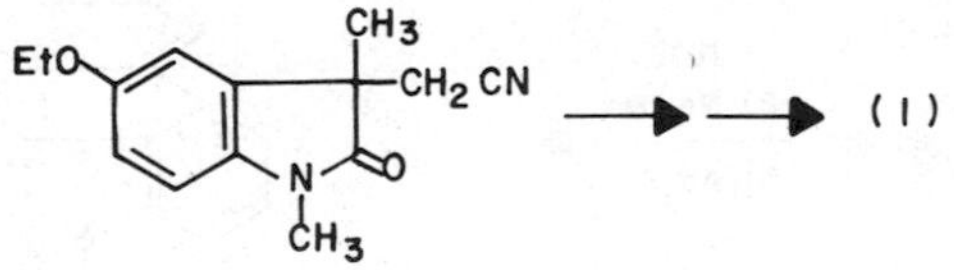

(5) → → (1)

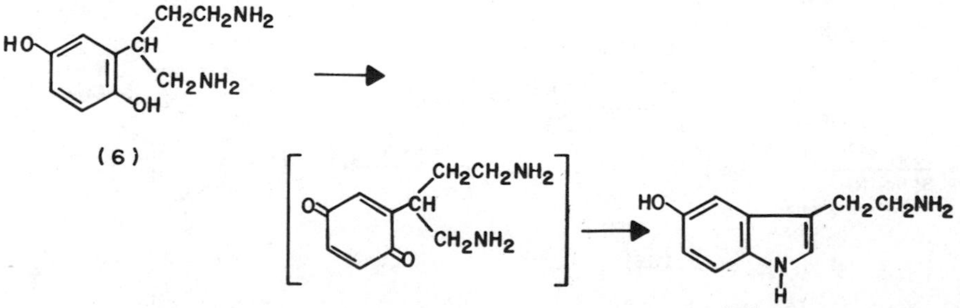

(6) → [ ] → (7)

MeO, COCH$_3$, OMe (8) + CNCH$_2$COOEt $\xrightarrow[\text{HOAc}]{\text{NH}_4\text{OAc}}$ MeO, CH$_3$, CN, C=C, CO$_2$Et, OMe (9) $\xrightarrow[\text{EtOH}]{\text{KCN}}$

MeO, CH$_3$, C, CH$_2$CN, CN, OMe (10) $\xrightarrow{\text{1) H}_2\text{/Pt; 2) C}_6\text{H}_5\text{CHO; 3) CH}_3\text{I; 4) HBr}}$ MeO, CH$_3$, C, CH$_2$CH$_2$NHCH$_3$, CH$_2$, NHCH$_3$, OH (11)

(11) $\xrightarrow{\text{K}_3\text{F(CN)}_6}$ HO, CH$_3$, N, N, CH$_3$, CH$_3$ (12)

CH$_3$CONHCH(CO$_2$Et)$_2$ (13) + CH$_2$=C(CH$_3$)–CHO $\xrightarrow{\text{NaOEt}}$ OHC–CH(CH$_3$)–CH$_2$–C(CO$_2$Et)$_2$, NHCOCH$_3$ (14)

$\xrightarrow{\text{C}_6\text{H}_5\text{NHNH}_2}$ N–N=CH, H, HC, CH$_3$, CH$_2$, C(CO$_2$Et)$_2$, NHCOCH$_3$ (15) $\xrightarrow{\text{HOAc}}$ [CH$_3$, N, NHCOCH$_3$, CO$_2$Et, CO$_2$Et]

$\longrightarrow$ CH$_3$, N–H, N, COCH$_3$, CO$_2$Et, CO$_2$Et (16) $\xrightarrow{\text{1) KOH; 2) Ba(OH)}_2}$ CH$_3$, N–H, N–H, COOH (17) mentioned previously in connection with a synthesis of bufotenin, that the ferricyanide oxidation of β-aminoethylhydroquinones **6** led directly to the corresponding 5-hydroxyindoles **7**. Adaptation of this observation led to the successful synthesis of eseroline **12**. Consideration of the biosynthetic route to physostigmine led Withop and Hill[98] to develop a new method of synthesis of this ring system. A more recent approach[99] allows elaboration to d,l-desoxyeseroline **22**. Appropriate modifications in the oxindole approach already described have provided a synthesis of d,l-physovenine **4**.[100]

H^+ ; 1) KOH 2) H^+

(18) (19)

CH_3NH_2

(20)

$LiAlH_4$

(21) (22)

CH_2—CH_2 (O) / NaEtO

1) Na, ETOH 2) H^+ 3) OH^-

(23) (24)

$AlCl_3$ / CH_3NCO

(25) (4)

E. Indole Alkaloids of the *Calycanthaceae* Family

Indole and quinoline alkaloids occur together in various species of *Calycanthus* and *Chimonanthus*. Among the most thoroughly investigated members of the family may be mentioned the alkaloids chimonanthine **1**, folicanthine **2**, and calycanthine **3**.

(2)

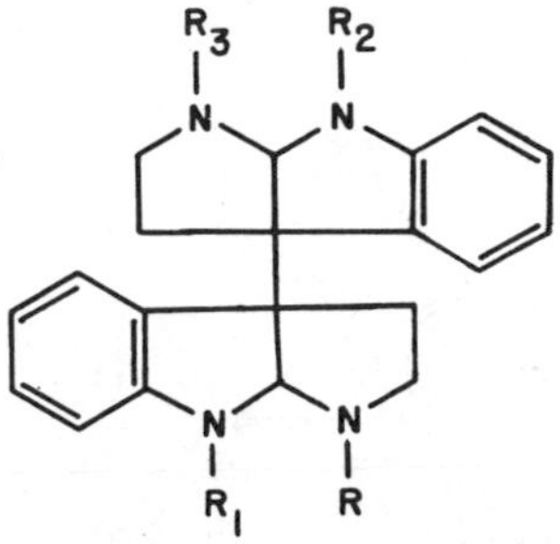

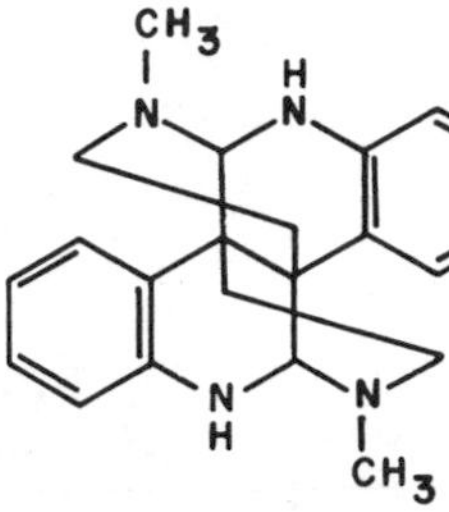

(3)

(1, $R = R_3 = CH_3$; $R_1 = R_2 = H$)

(2, $R = R_1 = R_2 = R_3 = CH_3$)

The synthetic approaches employed in this series provide an excellent example of how biogenetic speculations often allow the development of elegant synthetic pathways. The suggested[101,102] biosynthesis of the calycanthaceous alkaloids involved an oxidative coupling of *N*-methyltryptamine **4** via its resonance stabilized radical intermediate **5**. The resultant indolenine **6**, via hydrolysis, may be considered as equivalent to the tetraamino-dialdehyde **7**, which may, under well-established processes, convert to the five structural isomers (**1** and **3**). Such postulates provided the stimulus for several research groups to design successful laboratory syntheses of these alkaloids.

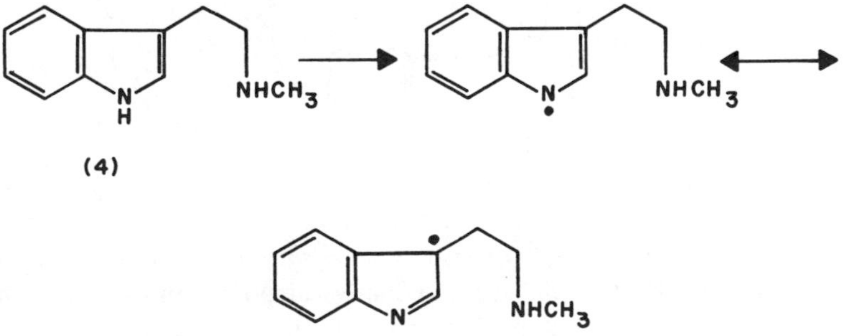

(4)

(5)

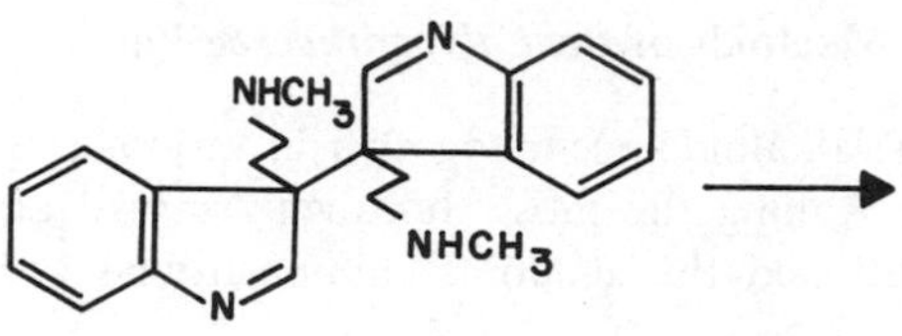
NHCH3
N
NHCH3
N

(6)

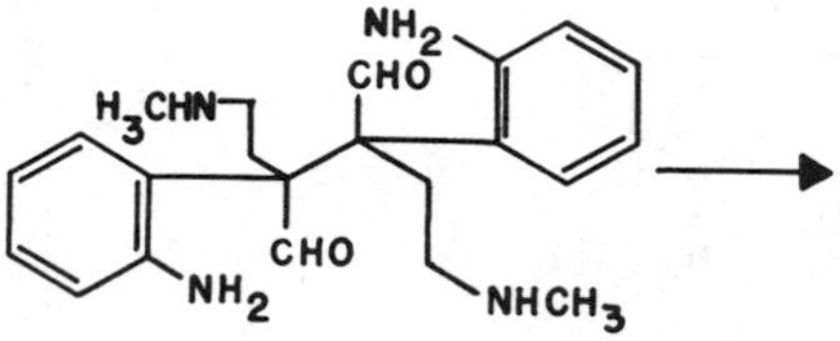
NH2
CHO
H3CHN
CHO
NH2
NHCH3

(7)

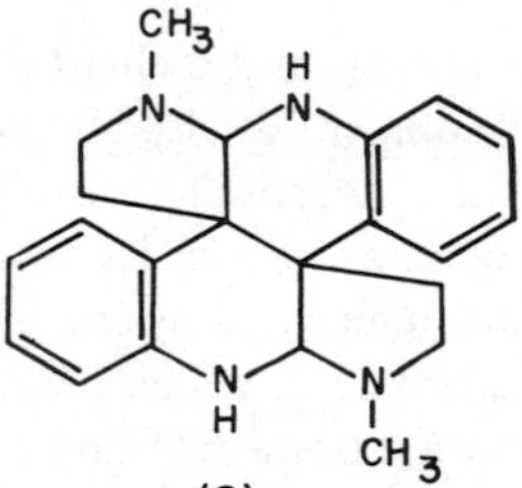
CH3
N
H
N
N
H
N
CH3

(8)

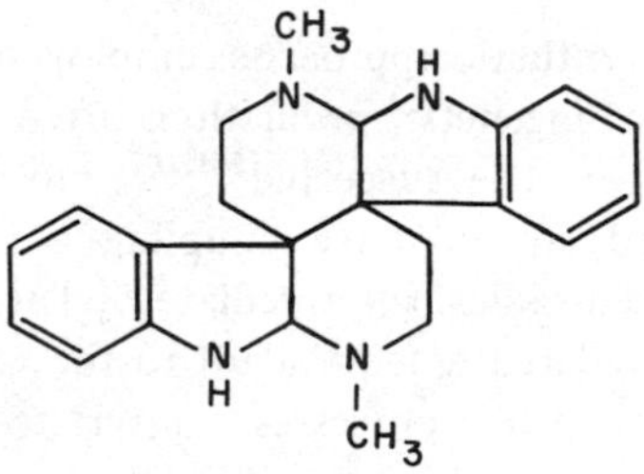
CH3
N
H
N
N
H
N
CH3

(9)

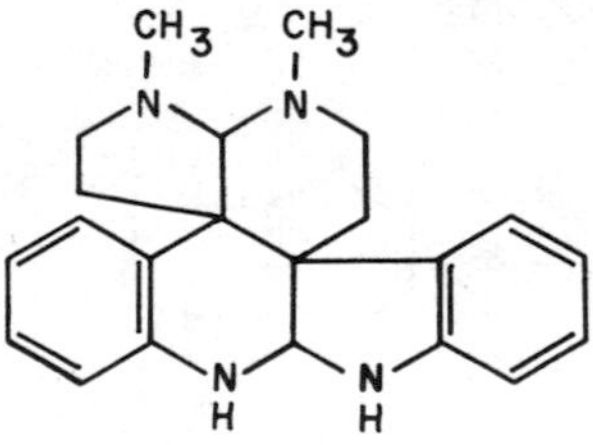
CH3
CH3
N
N
N
H
N
H

(10)

Hendrickson and his group[103] describe a total synthesis of d,l-calycanthine 3 and d,l-chimonanthine 1 in which the oxidative dimerization of *N*-carbethoxytryptamine 12 is a crucial step. The oxindole analog was chosen since it is expected to react exclusively at the β-position of the indole ring. It should be noted that the oxidative dimerization of the sodium salt of the urethane **12** with iodine in tetrahydrofuran proceeds in low yield to two diastereoisomeric dimers. Reduction of one of these isomers yielded an isomer of calycanthine, whereas similar reduction of the other isomer afforded racemic chimonanthine **1**.

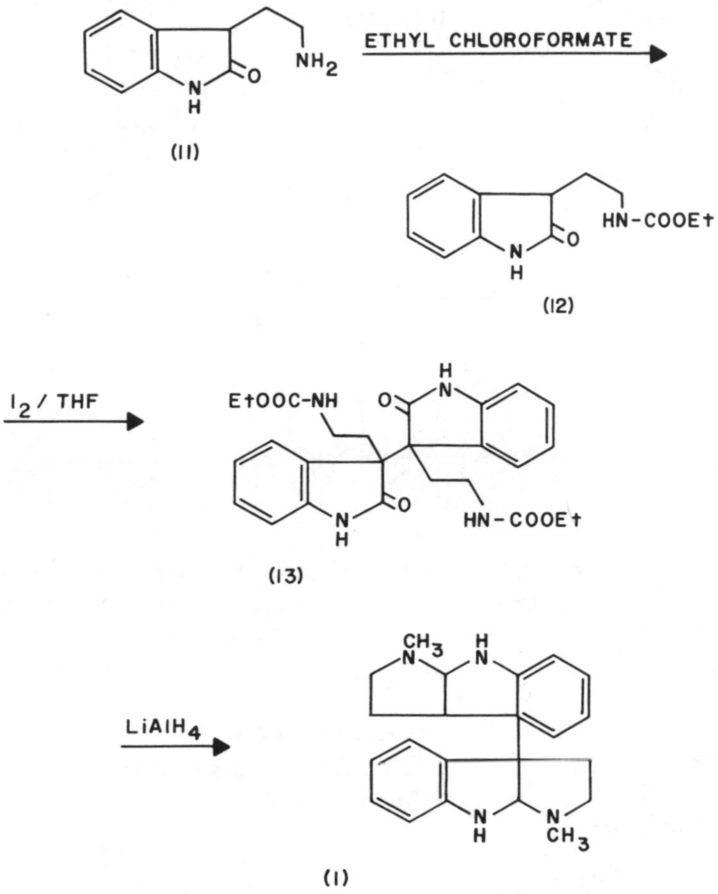

The next successful synthesis[104] involved oxidative dimerization of *N*-methyltryptamine in the form of its magnesium iodide derivative (Grignard reagent). In this approach, use was made of the ionic character of indolyl magnesium halides[105] and the ability of anhydrous ferric chloride to convert aromatic Grignard reagent to their respective coupling products.[106] The resultant product mixture obtained in this sequence provided, after chromatographic separation, racemic and mesochimonanthine **1**.

The total synthesis of racemic folicanthine **2** has been reported by Hino and Yamada.[107] In this case, dimeric oxindole derivatives available from previous

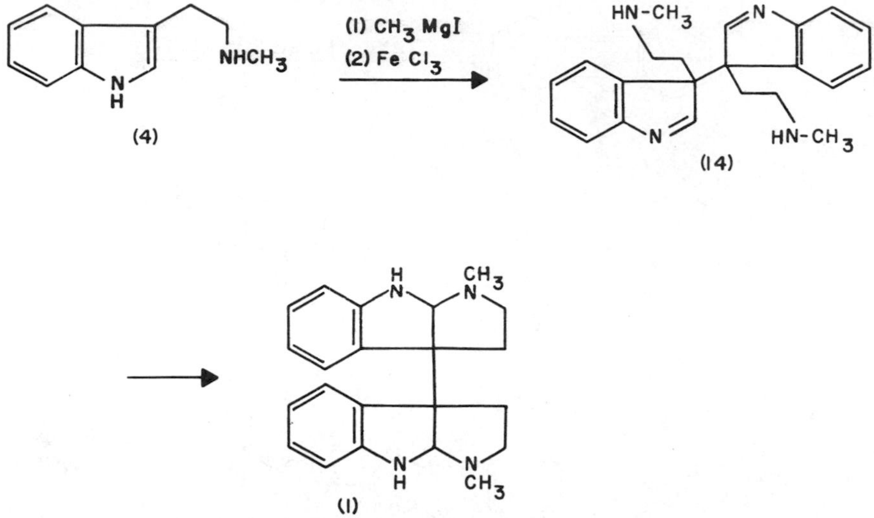

(4)

(1) CH_3MgI
(2) $FeCl_3$

(14)

(1)

investigations[108] were employed as the effective reactants. The required starting material **16** is obtained in good yield through the condensation of 1-methylisatin and 1-methyl oxindole **15**, followed by reduction of the resultant product wih zinc in acidic conditions. After appropriate functionalization of the oxin-

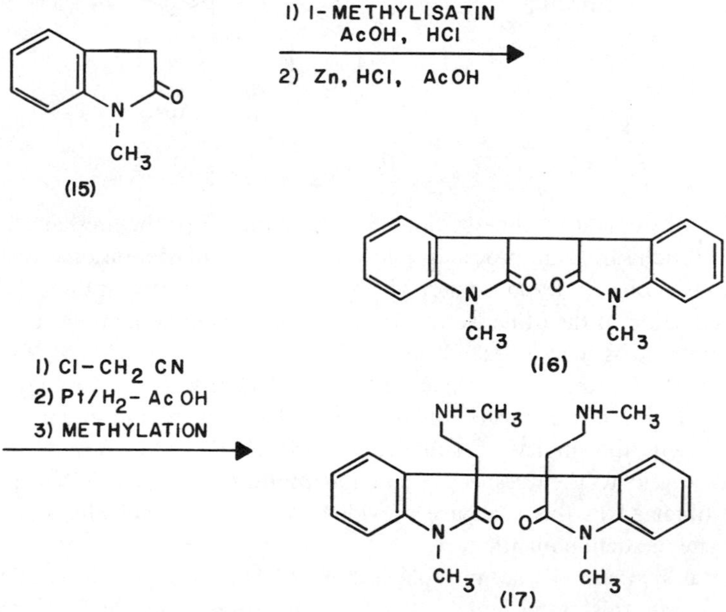

(15)

1) 1-METHYLISATIN
AcOH, HCl
2) Zn, HCl, AcOH

(16)

1) $Cl-CH_2CN$
2) Pt/H_2-AcOH
3) METHYLATION

(17)

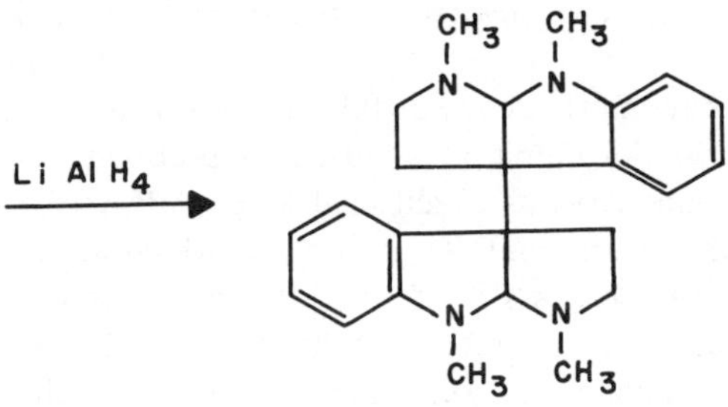

Li Al H_4

(2)

dole derivative to **17**, the latter is subjected to an interesting reductive cyclization as the final step leading to the racemic alkaloid.

F. The Corynantheine Family

The corynantheine series of alkaloids, for which corynantheine **1** dihydrocorynantheine **2**, and corynantheidine **3** may be cited as members, has been known for many years.[109,110] Initial interest, particularly from a structural point of view, dates back to the early 1900's when the presence of these substances as well as the various yohimbine isomers was noted in yohimbehe bark (*Pausinystalia yohimba* Pierre, syn., *Corynanthe yohimbe* K. Schum.) and related genera. More

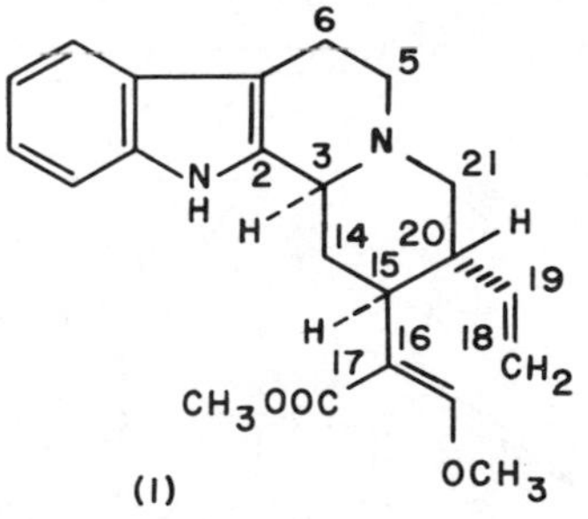

(1)

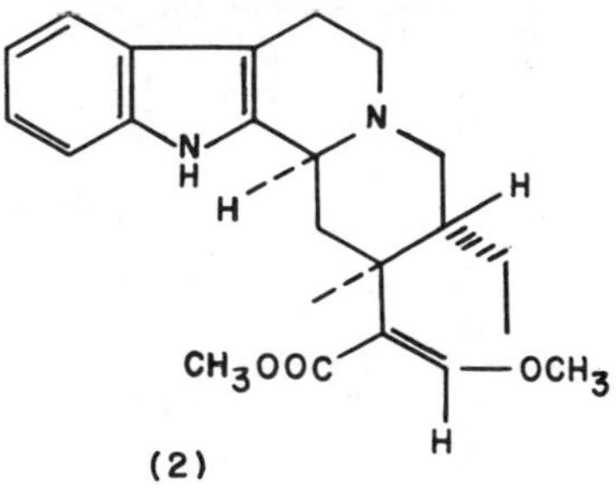

(2)

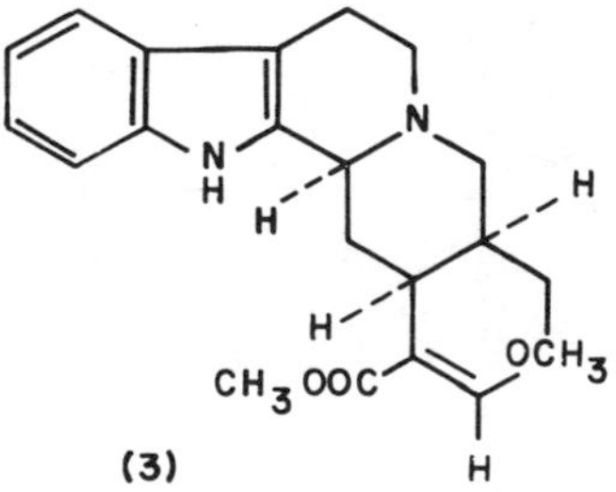

(3)

recent attention has been devoted to the synthesis and biosynthesis of these compounds.

Although it is beyond the scope of this discussion to elaborate in any detail on the biosynthesis of the indole alkaloids, it is pertinent to note that the tetracyclic corynantheine series is established as the biosynthetic precursor of the *Strychnos, Aspidosperma,* and *Iboga* families, which are discussed later. Reference to several recent reviews[111-115] provides an excellent survey of the various biosynthetic investigations. In brief, it has been established that the precursor for the indole template is tryptophan or a closely related biointermediate, whereas the C_9–C_{10} "nontryptophan" segment normally seen in these structures is of isoprenoid origin. Although additional research is essential to place some of the biosynthetic details on a firm basis, the structural features and interrelationships among the various classes may be summarized as shown. Secologanin **4**, the secoiridoid unit prevalent in various monoterpenoids, undergoes condensation with the indole segment to provide one of the crucial biointermediates vincoside **5**.[116,117]

Subsequent cyclization of the latter allows the genesis of geissoschizine **6**, the corynanthenoid alkaloid that bears the necessary functionality for elaboration to the various alkaloids in this class (see later). As the scheme reveals, the corynantheine series occupies a central role since it appears to be the biological precursor to the *Strychnos* family (for example, stemmadenine **7**), and the latter, via a

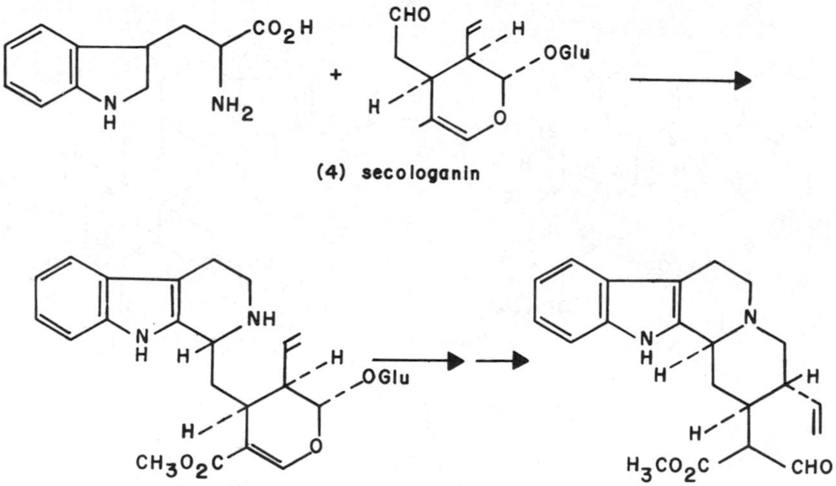

(4) secologanin

(5) vincoside

(6) geissoschizine (corynanthe)

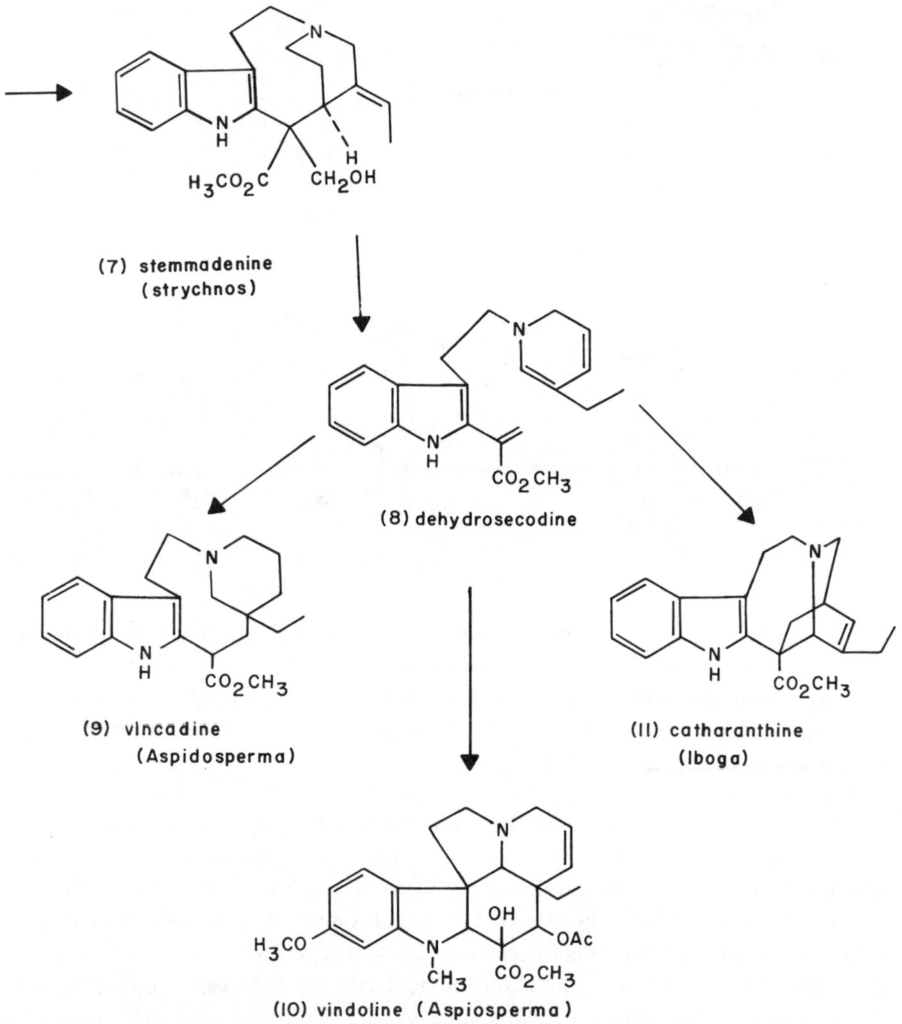

H_3CO_2C CH_2OH

(7) stemmadenine
(strychnos)

CO_2CH_3
(8) dehydrosecodine

CO_2CH_3
(9) vincadine
(Aspidosperma)

CO_2CH_3
(11) catharanthine
(Iboga)

OH
H_3CO OAc
CH_3 CO_2CH_3
(10) vindoline (Aspiosperma)

cleaved intermediate tentatively portrayed here as dehydrosecodine **8**, allows biosynthetic conversion to the *Aspidosperma* (vincadine, **9** and vindoline **10**) and *Iboga* (catharanthine **11**) families.[114]

Although the necessary biosynthetic experiments are still incomplete, it is tempting to interrelate the tetracyclic corynantheine skeleton with the well-known indole systems such as ajmalicine **12**, yohimbine **13**, polyneuridine **14**, and ajmaline **15**. Such biogenetic relationships have often been taken into serious consideration in the design of the synthetic pathways leading to these various alkaloids.

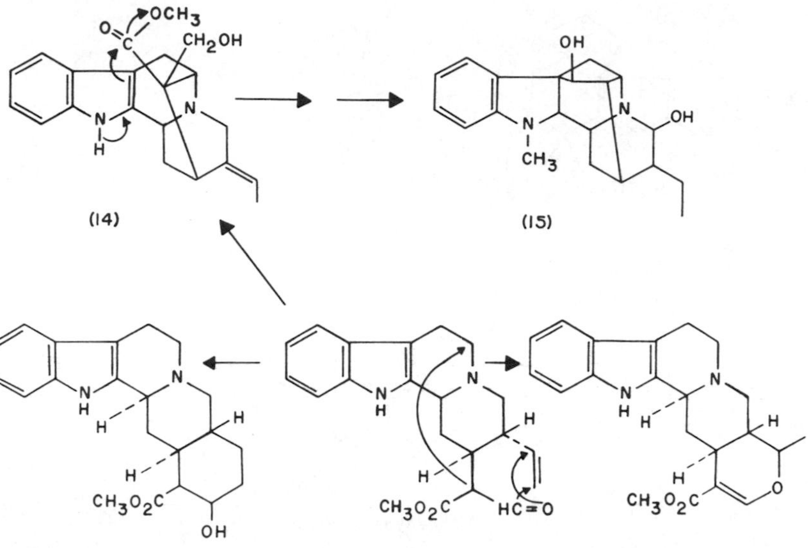

The order in which the syntheses of the various alkaloids is presented in the following discussion takes into account the presently accepted biosynthetic sequence. The first total synthesis of a representative of the corynantheine group was reported by van Tamelen in 1958.[45] A more recent publication provides a detailed account of this synthesis.[46]

The synthesis is based on a reductive alkylation of tryptamine with a suitable acyclic compound **21** to yield the basic tetracyclic ring system. The acyclic condensing unit was obtained through a method described by Preobrazhenskii.[118] A Michael condensation of diethylglutaconate **16** with ethyl cyanoacetate **17** yielded the triester **18**, which on alkylation with ethyl iodide allows formation of the cyanotriester **19**. Selective saponification followed by pyrolytic decomposition of the resulting acid **20** gave the desired cyano diester **21**. Precautions are required in this step, since **19** undergoes sodium ethoxide-induced elimination of the ethyl ethylcyanoacetate, regenerating **16** which in turn undergoes self-condensation to 22.

The cyano diester **21** was condensed with tryptamine **23** under reductive conditions. Unfortunately, the desired lactam **24** was obtained in low yield (13%), since the major product resulted from direct reduction of **21** and subsequent cyclization to the lactam **25**. Also contained in the reaction mixture was compound **26** formed by reductive alkylation of tryptamine with **18**, which had been carried through the sequence as an impurity. Saponification of the resultant mixture (**24** and **26**) and separation of the corresponding acids yielded, after esterification of the latter, the pure esters **24** and **26**. The structure and stereochemistry of the required lactam **24** was established by conversion to a known compound **27**, employing a reliable and proved degradative sequence.[119]

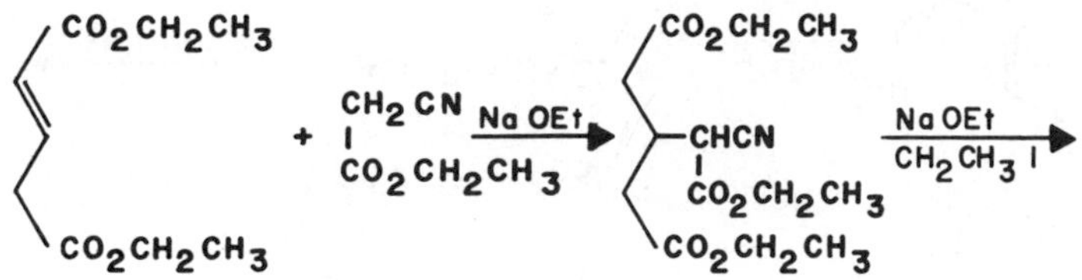

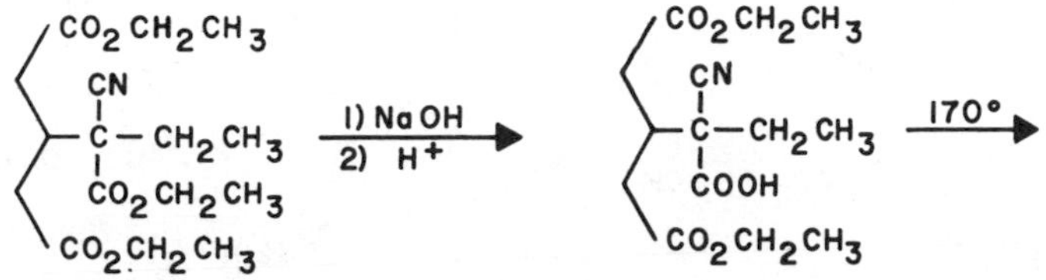

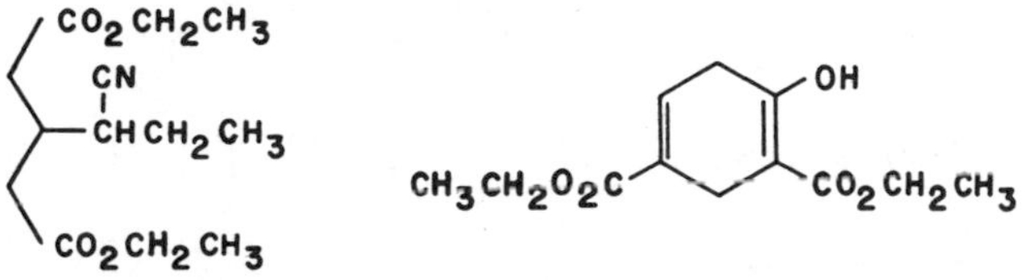

Although several plausible alternatives may be proposed for the reductive alkylation step, it appears that the following pathway is most attractive (**27a** → **27c**).

Formation of the tetracyclic quaternary iminium chloride **28** is achieved via Bischler-Napieralski conditions, and the necessary stereochemistry at C_3 results from catalytic reduction of this product. *Trans* esterification of the ethyl ester **29** to the methyl ester, formylation to the hydroxymethylene derivative **30**, and reaction of the latter with diazomethane afforded d,l-dihydrocorynantheine **2**.

In a complementary series of investigations, van Tamelen describes the total synthesis of d,l-corynantheine **1**.[45,120] In this sequence, a key tetracyclic intermediate previously employed in the ajmalicine synthesis (see later) was utilized. Thus a Mannich-type condensation of tryptamine **23**, formaldehyde, and the keto triester **31** led to the lactam **32**. Cyclization of the latter under Bischler-

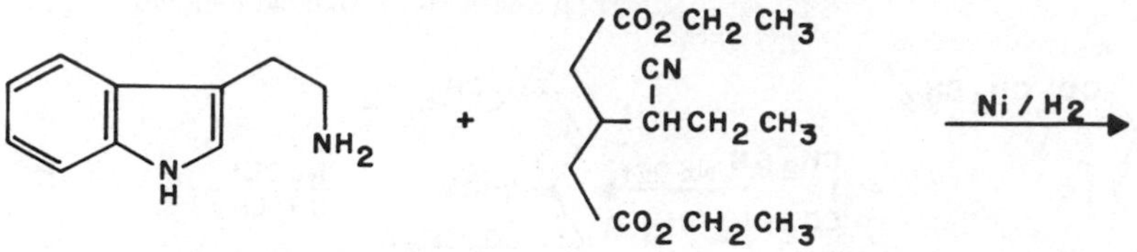

$CO_2CH_2CH_3$
CN
$CHCH_2CH_3$
$CO_2CH_2CH_3$
NH_2
N H
Ni / H_2
(23) (21)

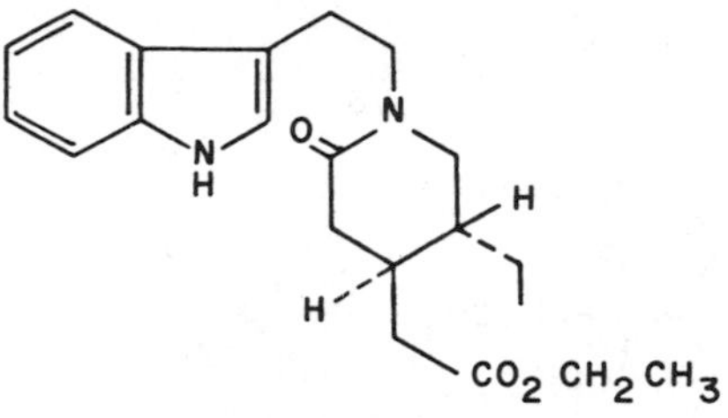

N H O N H H $CO_2CH_2CH_3$
(24)

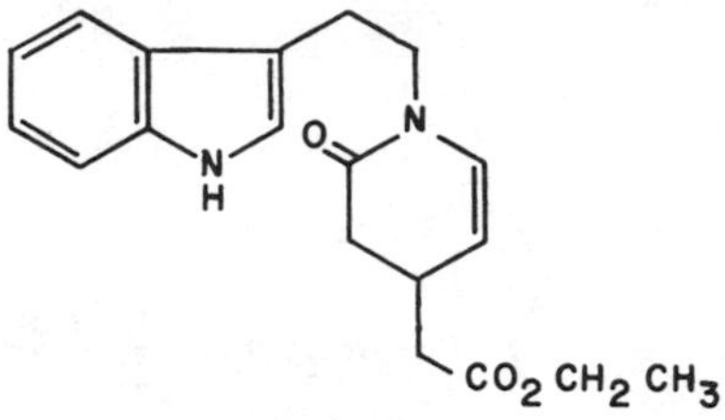

N H O N $CO_2CH_2CH_3$
(26)

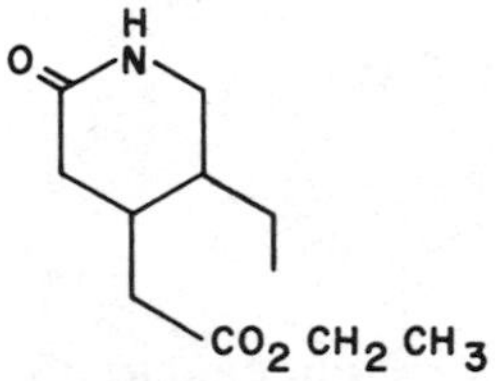

H O N $CO_2CH_2CH_3$
(25)

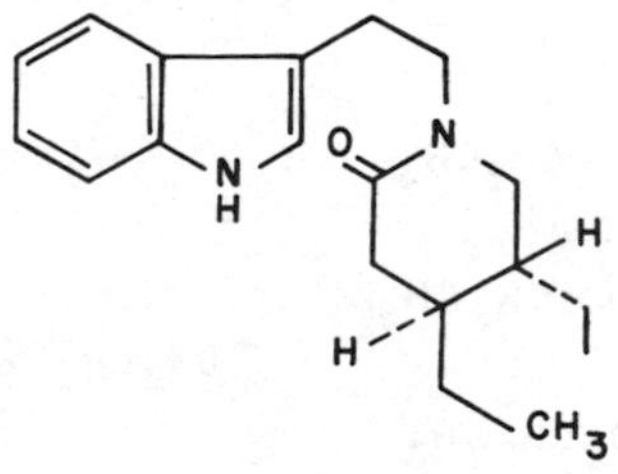

N H O N H H CH_3
(27)

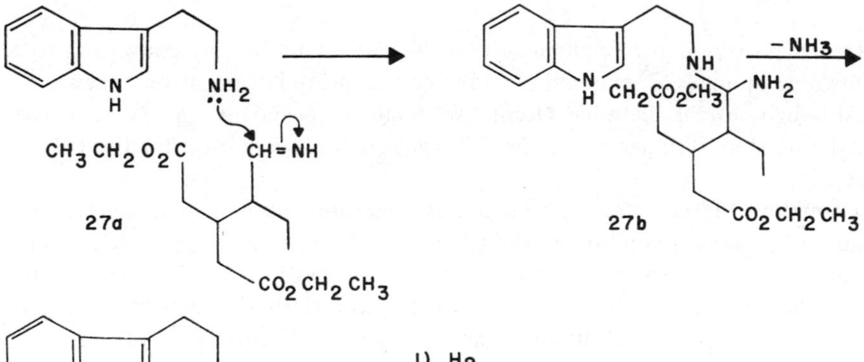

N H NH_2 $CH_3CH_2O_2C$ CH = NH $CO_2CH_2CH_3$
27a

N H NH $CH_2CO_2CH_3$ NH_2 $CO_2CH_2CH_3$
27b
$-NH_3$

N H N $CH_2CO_2CH_3$ $CO_2CH_2CH_3$
27c
1) H_2
2) CYCLIZATION
24

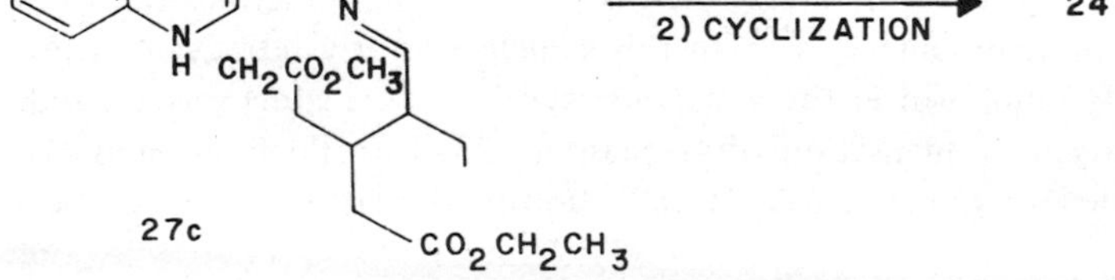

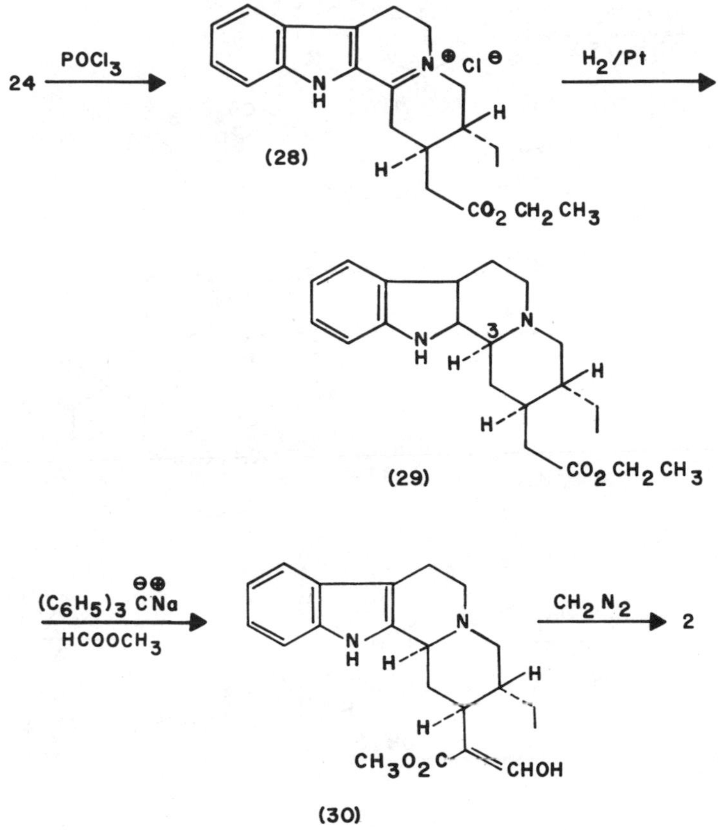

24 $\xrightarrow{POCl_3}$ (28) $\xrightarrow{H_2/Pt}$ (29)

$\xrightarrow[HCOOCH_3]{(C_6H_5)_3\overset{\ominus\oplus}{C}Na}$ (30) $\xrightarrow{CH_2N_2}$ 2

Napieralski conditions, followed by catalytic reduction, yielded the tetracyclic compound 33, which upon acid treatment afforded 34 as a mixture of *cis* and *trans* isomers.

The final stages of the sequence reveal the generation of the required vinyl side chain via thermal decomposition of sulfonylhydrazone salts, a method known to

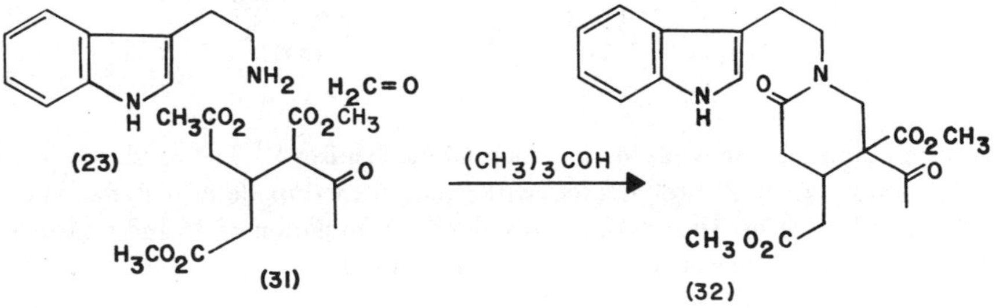

(23) + $H_2C{=}O$ + (31) $\xrightarrow{(CH_3)_3COH}$ (32)

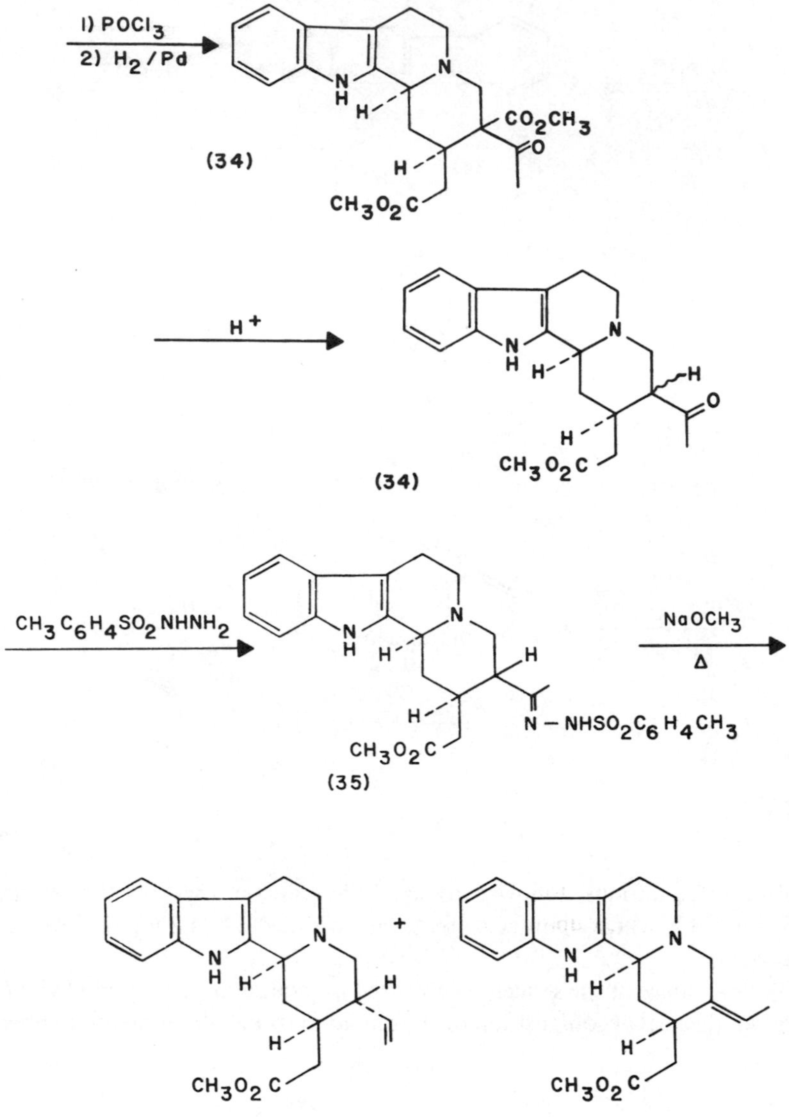

1) $POCl_3$
2) H_2/Pd

(34)

H^+

(34)

$CH_3C_6H_4SO_2NHNH_2$

$NaOCH_3$
Δ

(35)

(36) + (37)

produce olefinic compounds via a carbenoid mechanism.[121] The tosylhydrazone of the *trans* isomer **35** was treated with sodium methoxide in diglyme, and a mixture of the olefins **36** and **37** was obtained. Formylation of **36** and treatment of the resultant product **37a** gave d,l-corynantheine **1**.

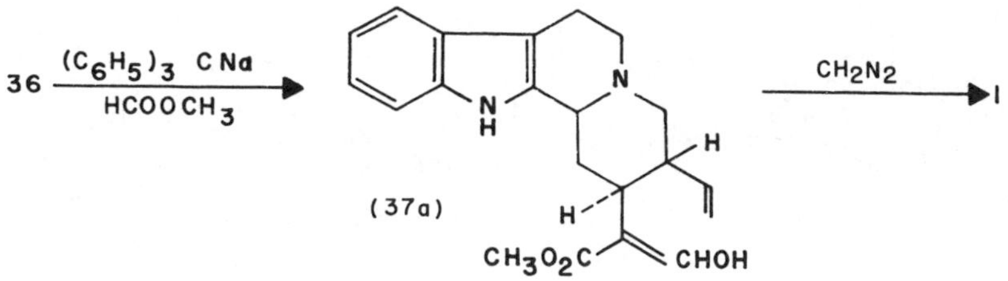

A rather different approach to the synthesis of corynantheine has been reported by Autrey and Scullard.[122] In this sequence, yohimbone **38**, a commercially available substance and one that has also been totally synthesized and resolved,[123] was selected as the starting material. This compound was converted to 18-formyl-yohimban-17-one **39** by previously published procedures, and this latter substance was then employed in an interesting series of reactions that result in an overall fragmentation of the yohimbine skeleton to the required corynantheine ring system.

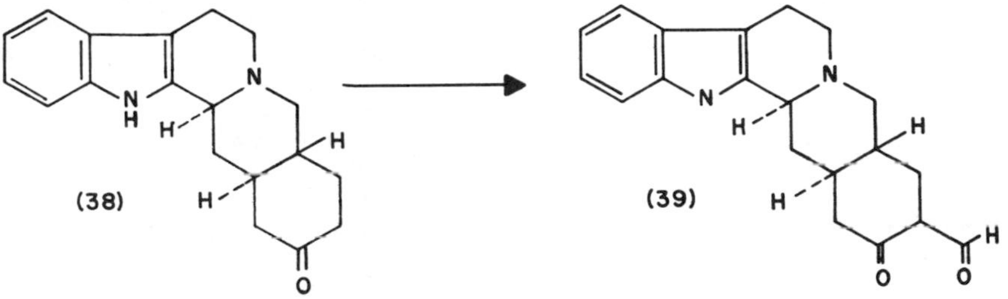

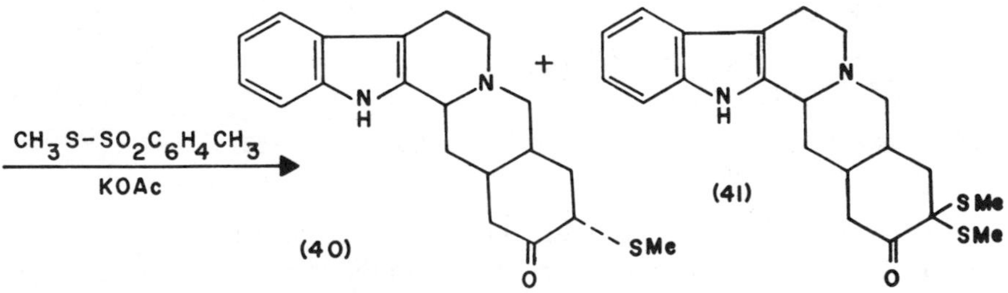

Introduction of the thiomethyl group was accomplished by reaction with methyl thiotosylate and subsequent deformylation of the resultant product by the base. The undesired disubstituted product **41**, presumably formed by further reaction of the intermediate anion **42** with methyl thiotosylate, could be suppressed so that the overall yield of the monomethioxy compound was in the range of 57 to 63%. Oximation of **40** and Beckmann fragmentation of the oxime **43** provided the enol thioether **44**, which now possesses the required tetracyclic skeleton and the essential stereochemistry. Desulfurization of the thioether proceeded smoothly with Raney nickel to corynanthenitrile **45**, and the latter was readily converted to methyl corynantheate **46** under Fischer esterification conditions.

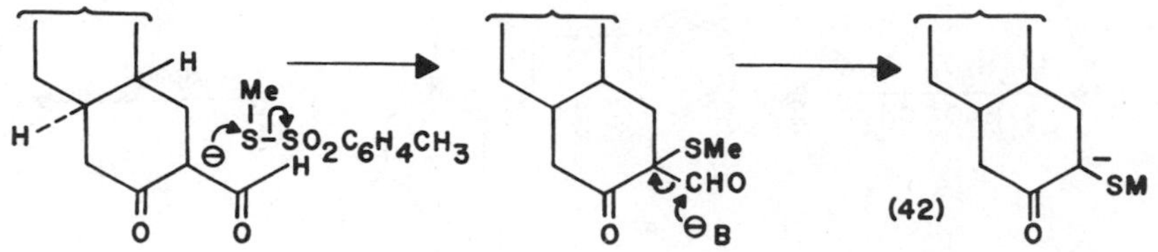

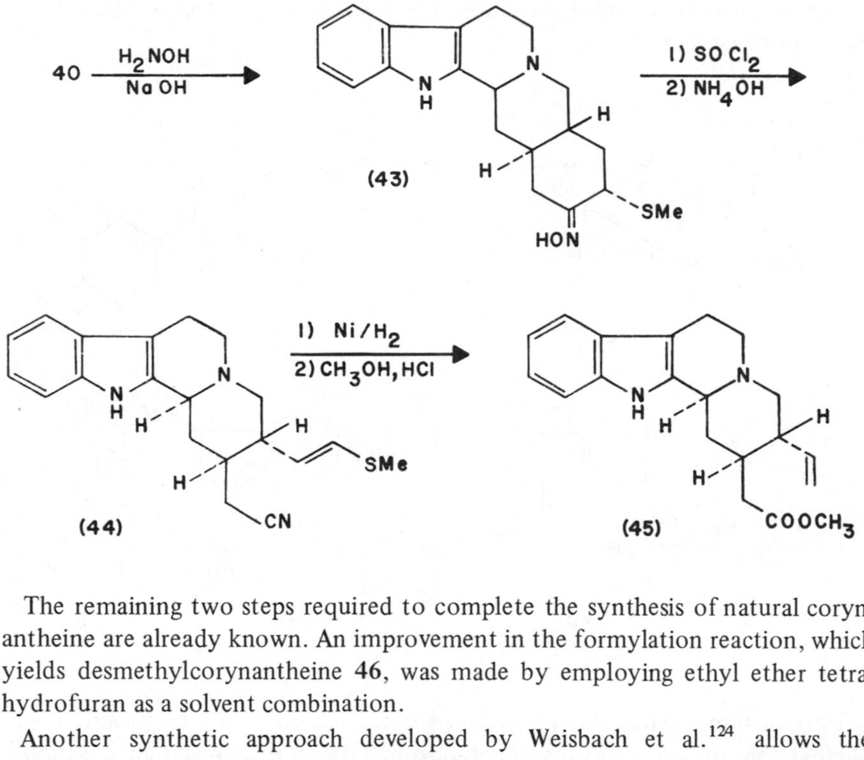

The remaining two steps required to complete the synthesis of natural corynantheine are already known. An improvement in the formylation reaction, which yields desmethylcorynantheine **46**, was made by employing ethyl ether tetrahydrofuran as a solvent combination.

Another synthetic approach developed by Weisbach et al.[124] allows the synthesis of both dihydrocorynantheine **2** and its isomer, corynantheidine **3**, from a common intermediate. The starting material **47**, available from tryptamine, was converted to the esters **48** and **49** by reaction with trimethyl and triethylphosphonoacetate, respectively. Catalytic reduction of these substances

45 $\xrightarrow[\text{HCOOCH}_3,\ \text{ether - THF}]{(C_6H_5)_3C^{\ominus}Na^{\oplus}}$ (46) $\xrightarrow[\text{HCl}]{\text{CH}_3\text{OH}}$ I

(46)

yielded the tetracyclic esters **50** and **51** possessing the desired stereochemistry at C_{15}. Formylation of **50** and methylation of the latter according to procedures already discussed previously afforded d,l-dihydrocorynantheine **2**.

The required stereochemistry prevalent in corynantheidine was achieved by a series of oxidation-reduction procedures. Thus the 3,4-unsaturated ethyl ester **52**, available by alkoxide treatment of **49**, was reacted with iodine-sodium acetate, the resultant salt **53** was obtained as a mixture of perchlorate and iodide was reduced directly to **54**. Conversion of the latter to methyl ester **55**, followed by formylation and methylation, completed the synthesis.

The alternative and more direct sequence (**49** → **53**), employing mercuric acetate as an oxidant, is not apparently advantageous in terms of yield.

The first total synthesis of the optically active, natural isomer of corynantheidine was reported by Szantay and Barczai-Beke.[125,126] The starting material was similar to that utilized in the previous synthesis. Condensation of **56** with methyl cyanoacetate provided **57** in 66% of the yield. Borohydride reduction of the latter and methanolysis of the resultant nitrile **58** allowed isolation of the diester **59**. An interesting controlled reduction converts the diester to the α-hydroxymethylene derivative **60**, which, in the form of its sodium salt, is methylated in essentially quantitative yield to racemic corynantheidine. Resolution of this racemate with dibenzoyl-d-tartaric acid completed the synthesis of (−)-corynantheidine.

In a series of investigations, Wenkert has developed a new procedure for the construction of the indoloquinolizidine skeleton so common to a large family of indole alkaloids. It is based on a partial reduction of 1-[β-(3-indolyl)ethyl]-3-acyl pyridinium salts and the subsequent acid-catalyzed cyclization of the resultant 2-piperideines. This approach has led to a number of syntheses among which is included the synthesis of d,l-corynantheidine.[127]

Alkylation of methyl 4-carbomethoxy-methyl-5-ethylnicotinate **61** with tryptophyl bromide **62** yielded the pyridinium salt, which was isolated and characterized as the perchlorate **63**. Hydrogenation of the latter produced the tetrahydropyridine **64**. Two procedures for the cyclization and decarboalkoxylation of such tetrahydropyridines had been previously developed and employed in the present case. Alkaline hydrolysis of **64** followed by re-esterification with

N
H
O
(47)

N
H
N
CH
CO_2R
(48) R = CH_3
(49) R = CH_2CH_3

H_2/Pd

N
H
N
H
H
15
H
CO_2R
(50) R = CH_3
(51) R = CH_2CH_3

1) $HCOOCH_3$, $(C_6H_5)_3C^{\ominus}Na^{\oplus}$
2) CH_3OH, HCl
→ 2

Then:

49 $\xrightarrow{^{\ominus}OR}$

N
H
N
(52)
$CO_2CH_2CH_3$

49 $\xrightarrow{Hg(OAc)_2}$

I_2, NaOAc

N
H
$N^{\oplus}X^{\ominus}$
(53)
$CO_2CH_2CH_3$

H_2/Pd
$NaOCH_2CH_3$

N
H
N
H
H
H
(54)
$CO_2CH_2CH_3$

$NaOCH_3$
CH_3OH

N
H
N
H
H
H
(55)
CO_2CH_3

→ → 3

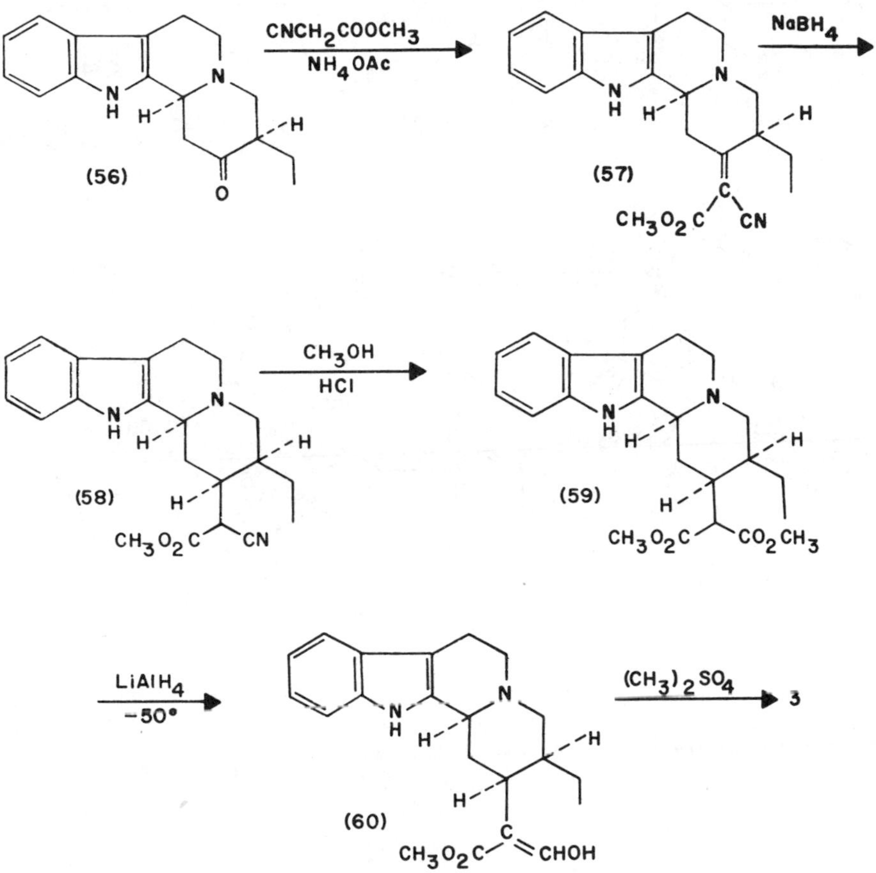

methanolic acid led to the tetracyclic ester **65**. Palladium dehydrogenation converted the latter to the tetradehydro compound **66**, which could be characterized in the form of its perchlorate salt. Sodium borohydride reduction of **66** afforded **67**. The alternate route to **66** involved acid-induced cyclization of **64**, acid hydrolysis of the resultant tetracyclic ester **67**, dehydrogenation, and re-esterification.

The two remaining reactions necessary to complete the synthesis of d,l-corynantheidine, namely formylation and methylation, were already in hand from the aforementioned syntheses.

Ziegler has reported a synthesis of dihydrocorynantheol **84** and 3-epidihydrocorynantheol **85** employing Claisen rearrangement to elaborate the necessary functionality at C_{15} and C_{20} of the tetracyclic skeleton.[128] In this sequence, 6-chloronicotinic acid **68** is converted to the required 2-chloro-5-(α-hydroxyethyl)pyridine **72** by standard methods, and this substance is then condensed with tryptophyl bromide to yield the pyridinium salt **73** in 27% yield. Reduction of **73** with sodium borohydride provided a diastereoisomeric mixture of allylic alcohols **74**.

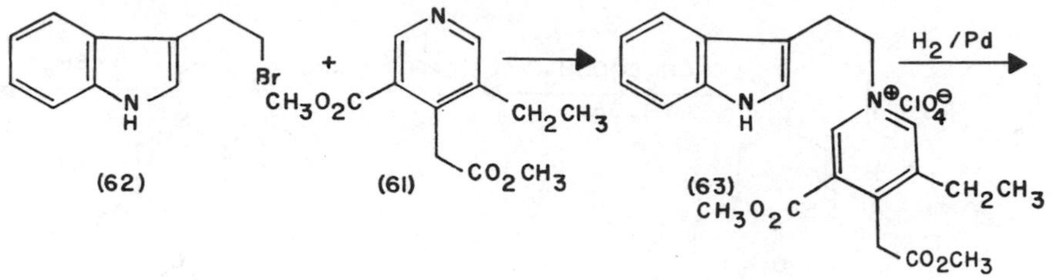

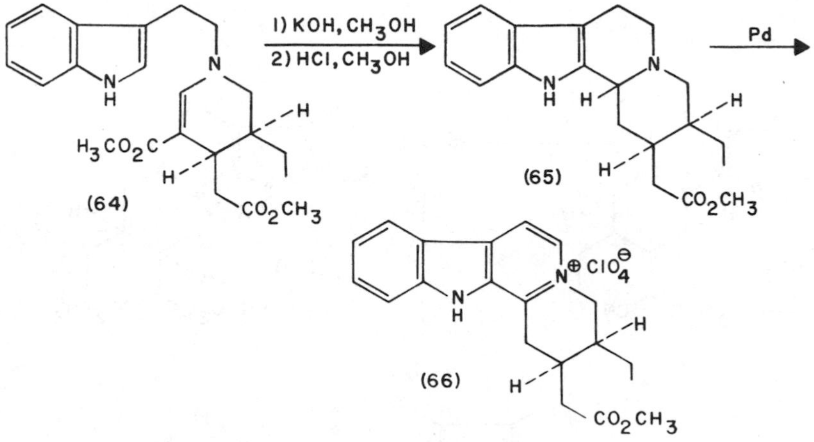

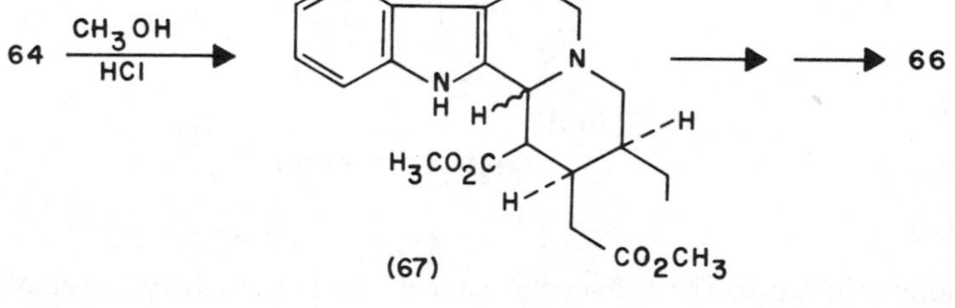

Further functionalization of **74** involved the application of a Claisen rearrangement previously studied by Eschenmoser,[129] in which the reaction between *N,N*-dimethylacetamide dimethyl acetal **75** and allylic as well as benzylic alcohols to form *N,N*-dimethylamides of γ,δ-unsaturated acids is presented. For example, when the allylic alcohol **76** is reacted with a mixture of **75** and its methanol

PCl5, POCl3

NaBH4

MnO2

(68) (69) (70)

CH3MgBr

(71) (72)

(73)

NaBH4

(74)

$(CH_3)_2N-C(OCH_3)(CH_3)-OCH_3$ (75) / $(CH_3)_2N-C(=CH_2)-OCH_3$ (77)

140°, XYLENE, 14 HRS.

(76)

(78)

(79) (80)

elimination product, 1-methoxy-1-dimethylaminoethylene **77**, the resulting amide **78** is obtained in 70% of the yield. The rearrangement is believed to proceed according to the intermediates of the structural type **79**, which convert readily to **80**.

Reaction of the mixture of allylic alcohols **74** with dimethylacetamide dimethylacetal gave rise to a diastereoisomeric mixture of amides (**81** and **82**). Saponification of this mixture followed by esterification afforded **83** as a mixture of double bond isomers. Finally, hydride and catalytic reduction provided a mixture of dihydrocorynantheol **84** and 3-epi-dihydrocorynantheol **85**.

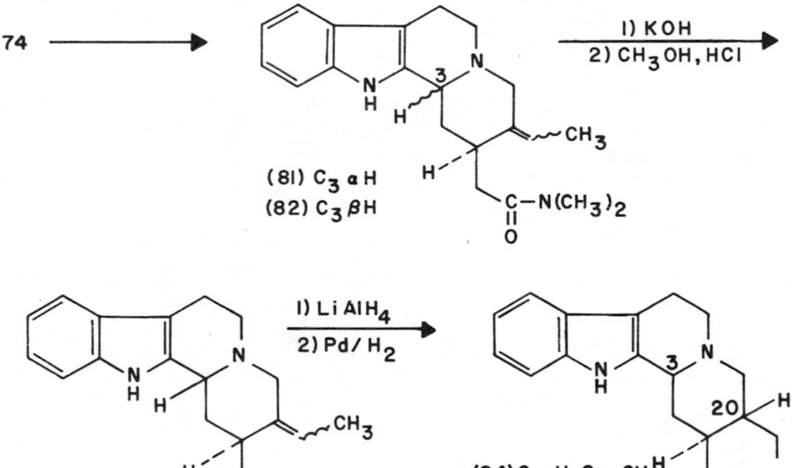

74 → (81) $C_3\,\alpha H$ (82) $C_3\,\beta H$ — 1) KOH 2) CH_3OH, HCl →

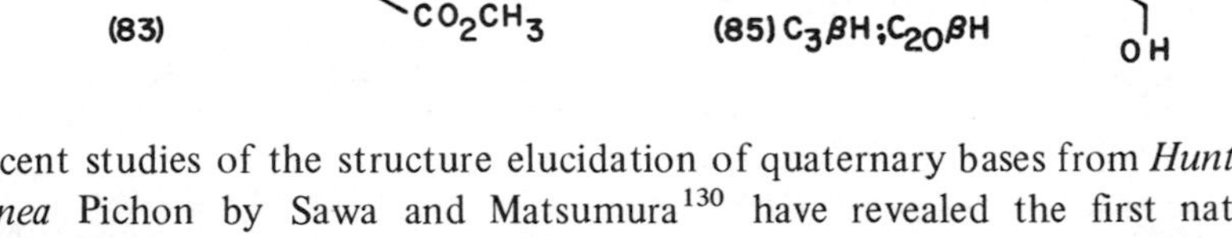

(83) — 1) $LiAlH_4$ 2) Pd/H_2 → (84) $C_3\alpha H; C_{20}\beta H$ (85) $C_3\beta H; C_{20}\beta H$

Recent studies of the structure elucidation of quaternary bases from *Hunteria eburnea* Pichon by Sawa and Matsumura[130] have revealed the first natural examples of corynantheinoid variants in which both possible isomers at N_b are evident. During these investigations in which the structures, including the absolute configuration, of hunterburnine α- and β-methochlorides (**86** and **87**) were being evaluated, interesting syntheses of several corynantheine derivatives

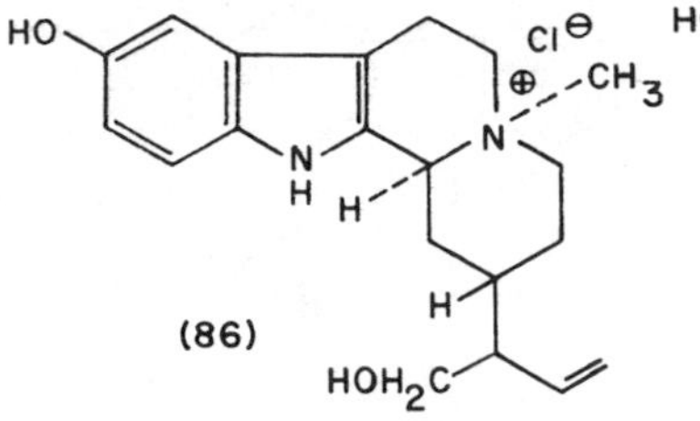

(86)

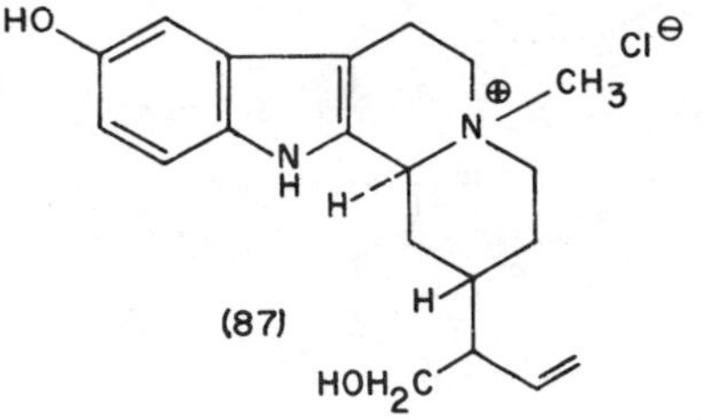

(87)

were also achieved. For example, 10-methoxydihydrocorynantheol **88**, ochrosandwine **89**, and dihydrocorynantheol **84** were synthesized from quinine **90**.

The rather large class of ring-E heterocyclic indole alkaloids is obviously closely related in a biogenetic sense to the corynantheinoid members already discussed. For example, vincoside **5** upon hydrolysis of the glycosidic linkage yields **91**, which after unexceptional cyclization would provide ajmalicine **12**, one of the well-known members in this series. It is therefore appropriate to discuss the syntheses of this alkaloid at this time.

The total synthesis of ajmalicine **12** in racemic form represents the first synthesis of a member of the ring-E heterocyclic indole alkaloids.[47,131] In this study the acyclic component **31**, corresponding to the nontryptophan portion of the alkaloid and already employed in a previously discussed synthesis of d,l-cory-

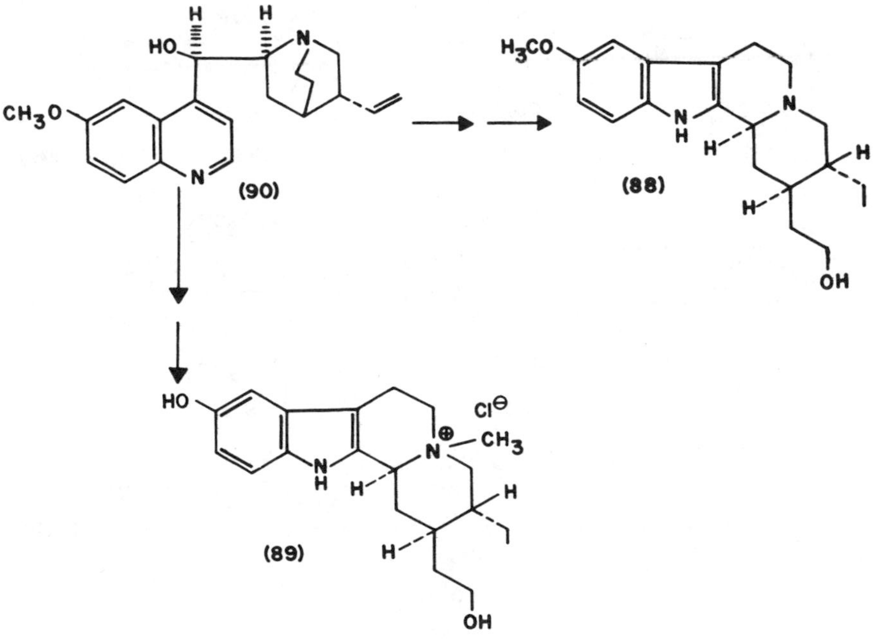

(90)

(88)

(89)

(5) → (91) →

(92) → (12)

(93) $\xrightarrow[Ni]{H_2}$ (94) $\xrightarrow{Ac_2O}$ (95) $\xrightarrow{\Delta}$ (96)

$\xrightarrow{CH_3COCH^{\ominus}CO_2CH_3}$ (31) + (97)

94 $\xrightarrow{PCl_5}$ (98) $\xrightarrow{CH_3COCH^{\ominus}CO_2CH_3}$ 31

nantheine, was prepared according to the alternative pathways (**93** → **31** and **94** → **98** → **31**). The sequence involving displacement of halogen in dimethyl β-chloroglutarate **98** provided a more convenient and better yielding route to **31**.

A simple Mannich reaction of tryptamine, formaldehyde, and the keto triester **31** yielded, in the manner already described, the lactam **32**, which was converted as before to the keto ester **34**. This compound was then elaborated via the δ-lactone **99** to the ajmalicine system. Low temperature reduction of **34** provided a nearly quantitative yield of the lactone **99**.

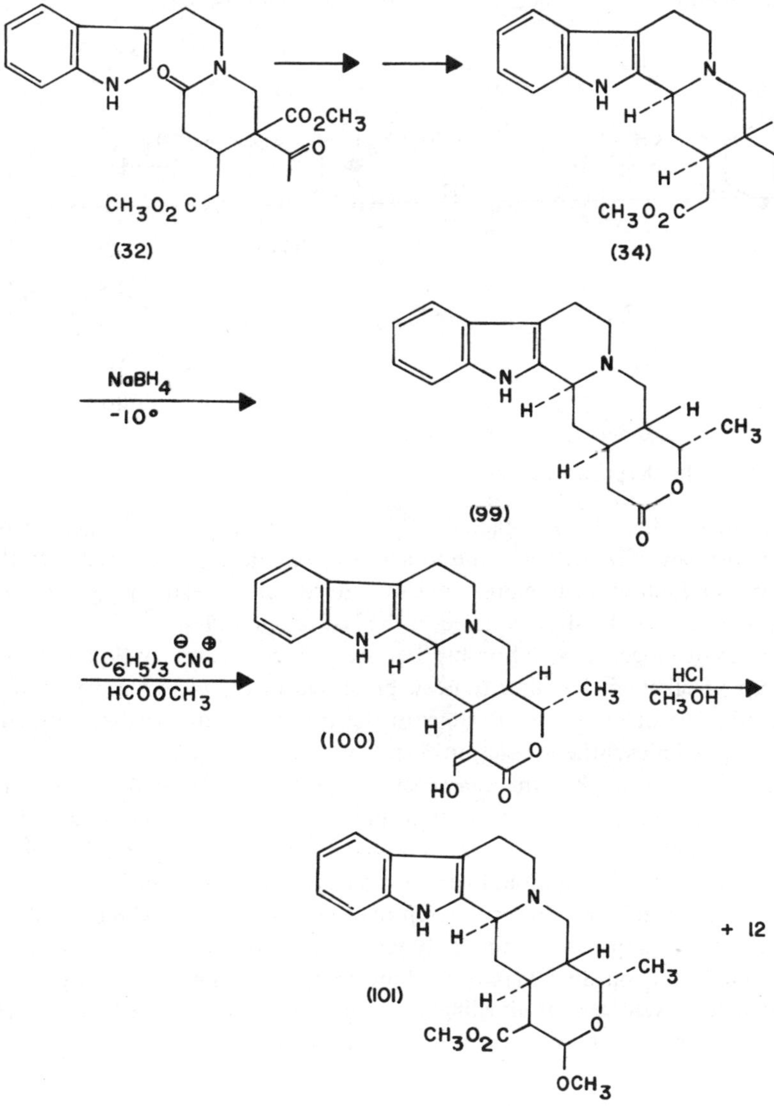

(32) → → (34)

$\xrightarrow[-10°]{NaBH_4}$ (99)

$\xrightarrow[HCOOCH_3]{(C_6H_5)_3\overset{\ominus}{C}\overset{\oplus}{Na}}$ (100) $\xrightarrow[CH_3OH]{HCl}$

(101) + 12

Although the isomeric composition of the lactone product mixture was not determined, it was expected that the *trans* isomer predominated. The α-hydroxymethylene lactone **100** was obtained by a routine formylation procedure, and this substance was then subjected to an acyl lactone rearrangement process previously studied by Korte and Machleidt.[132] In this study, Korte had demonstrated that an α-acyl-δ-lactone, for example **102**, undergoes a facile rearrangement under acidic conditions to the corresponding dihydropyrancarboxylic ester **104**. The reaction is visualized as proceeding through the cyclic acetal **103**, which upon elimination of methanol provides the final product.

CH_3 OH H_3C O O (102) → [CH_3 CO_2CH_3 CHO H_3C OH] → CH_3 CO_2CH_3 H_3C O OCH_3 (103) → CH_3 CO_2CH_3 H_3C O (104)

G. The Sarpagine-Ajmaline Family

The ajmaline-sarpagine family of alkaloids has been extensively investigated from a structural point of view,[133,134] and much of the chemistry associated with this group was undoubtedly developed because of its close relationship with the Rauwolfia bases so extensively studied in the last two decades.

The apparent biogenetic relationship between the corynantheinoid family and this group of alkaloids has already been presented in the introductory section, and indeed some of the synthetic steps in the ajmaline syntheses discussed later take the implied biosynthetic reactions into consideration.

The sarpagine group has the characteristic bridged skeleton portrayed in 1 (numbering system corresponds to that previously shown in corynantheine), whereas the additional bridging to the β-position of the indole system in 1 allows the elaboration of the dihydroindole skeleton typical of the ajmaline family 2. Specific members of the sarpagine group may be exemplified by vellosimine 3, normacusin B **4**, sarpagine **5**, and polyneuridine **6**, whereas tetraphyllicine **7**, vincamedine **8**, and ajmaline **9** are typical members of the ajmaline family.

The first total synthesis of ajmaline was reported by Masamune et al.[135] The sequence employed in this investigation is shown here.

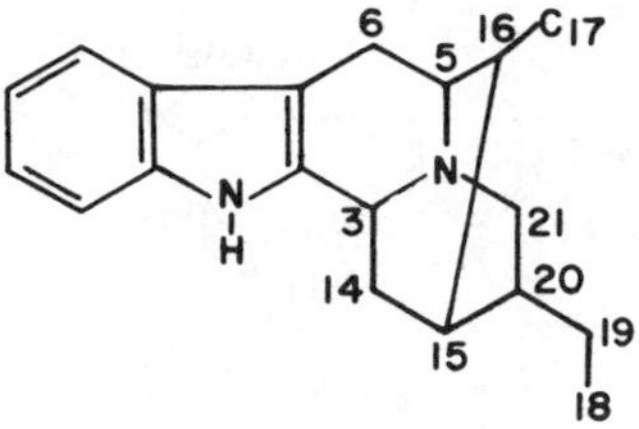

(1)

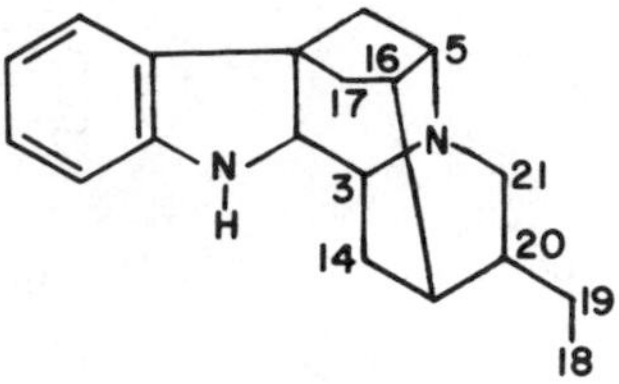

(2)

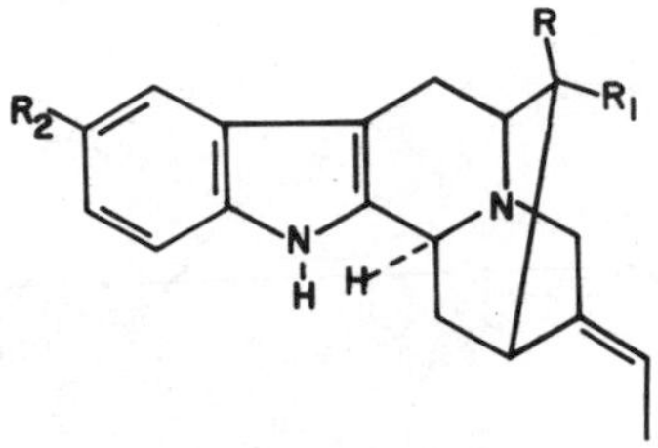

(3) $R = R_2 = H$; $R_1 = CHO$

(4) $R = R_2 = H$; $R_1 = CH_2OH$

(5) $R = H$; $R_1 = CH_2OH$; $R_2 = OH$

(6) $R = CH_2OH$; $R_1 = COOCH_3$; $R_2 = H$

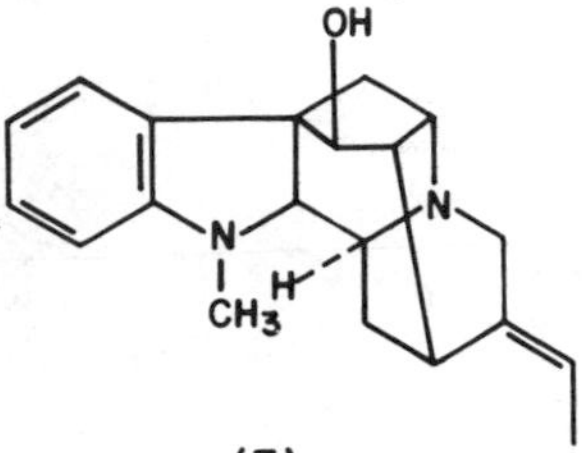

(7)

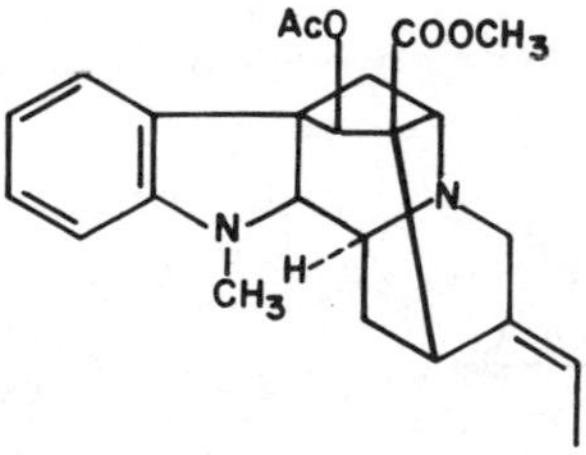

(8)

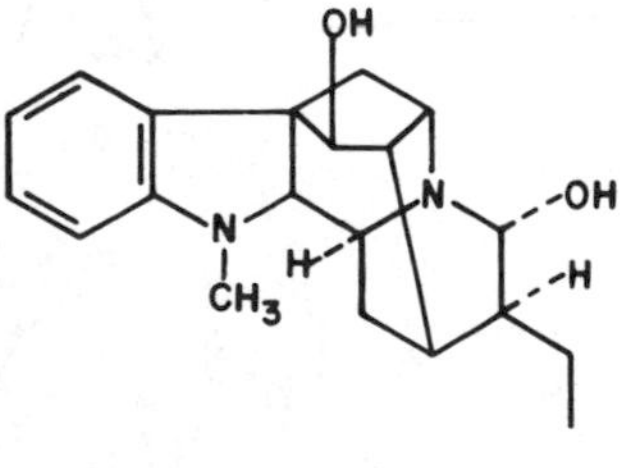

(9)

Condensation of *N*-methyl-3-indoleacetyl chloride **10** with the magnesium chelate of ethyl hydrogen Δ^3-cyclopentenylmalonate **11** afforded the keto ester **12** in 80% of the yield. Reaction of the latter with methoxyamine followed by reduction provided a 2:1 mixture of the epimeric amino alcohols **13**, which could be readily separated and characterized as their diacetyl and dibenzoyl derivatives. Since both epimeric series are useful and they are interconvertible at a later stage of the synthesis, no differentiation of these stereoisomers is made in this discussion. The dibenzoyl derivative **14** on reaction with osmium tetroxide

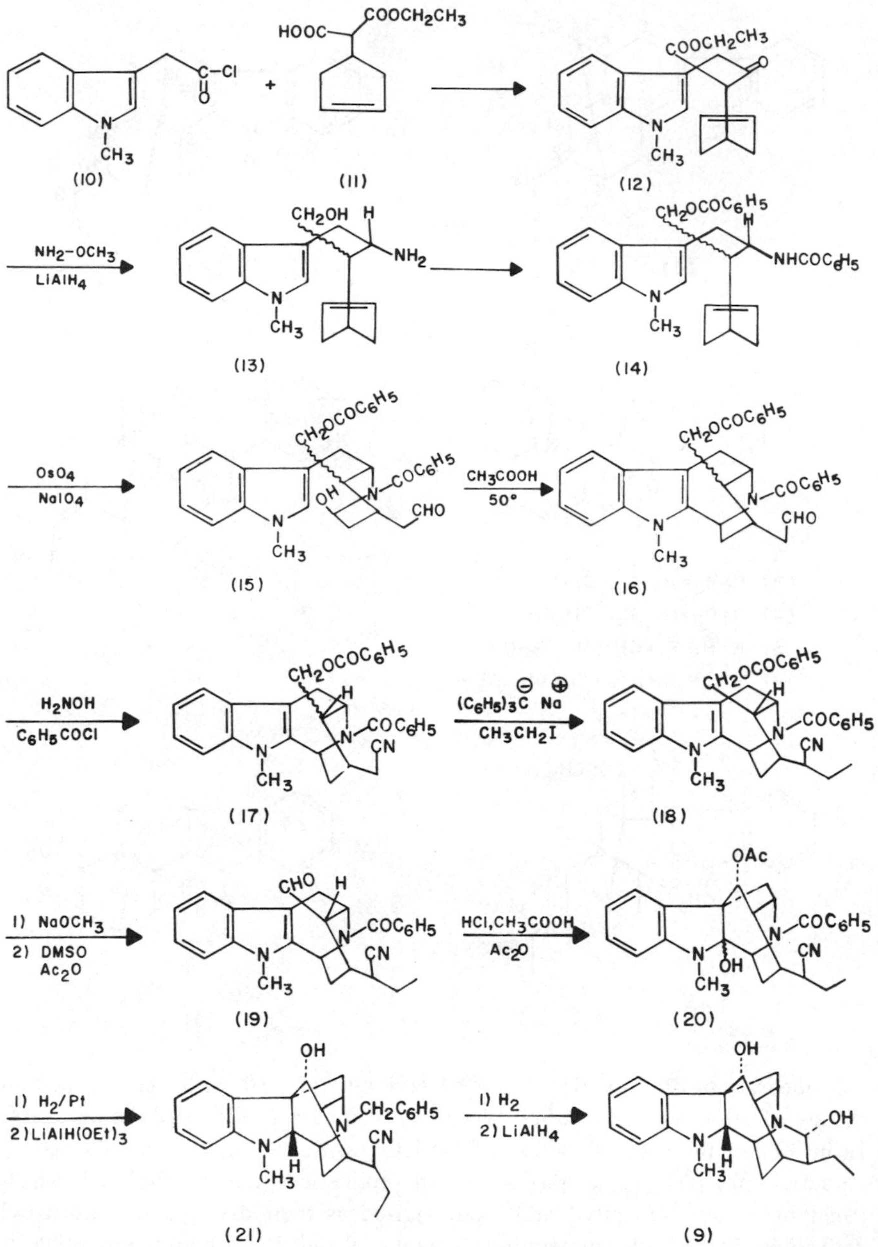

and sodium metaperiodate afforded the aldehyde mixture **15** in essentially quantitative yield. Acid-induced cyclization provided **16**, which is converted to the nitrile **17** by dehydration of the oxime derivative. Base-catalyzed alkylation allowed introduction of the ethyl group at the carbon atom bearing the nitrile function. At this point, the structures and stereochemistry of several synthetic intermediates **18**, and the alcohols obtained by hydrolysis of **18** were compared with the corresponding degradation products of ajmaline. The epimeric aldehydes **19** obtained from Moffatt oxidation were interconvertible by means of alumina; therefore, both epimeric series could be utilized.

A number of previous investigations had already revealed that the desired β,β-disubstituted skeleton characteristic of the ajmaline molecule could be achieved through the reaction of the electron-rich indole ring and an appropriate electron-deficient center (22 → 23). Application of this principle allows conversion of **19** to **20**. Catalytic hydrogenolysis and hydride reduction of the amide function in **20** provided a good overall yield of **21**. Removal of the *N*-benzyl group leads to the secondary amine. Since the latter has been converted to ajmaline by use of lithium aluminum hydride,[136] the synthesis of this alkaloid is now complete.

(22) (23)

A second synthesis of ajmaline has been reported by van Tamelen.[137] This approach attempts to utilize chemical operations that are likely to prevail in some of the biosynthetic steps leading to this natural product. The indolic material, *N*-methyl tryptophan **29**, is exposed to reaction with an appropriate C_9 unit **28**, which after the required carbonyl-amine condensations leads to the tetracyclic indole skeleton **32**. This latter substance then serves as the fundamental template for the remaining elaboration to ajmaline.

The synthesis of the essential C_9 unit starts with d,l-α(Δ^3-cyclopentenyl)-butyric acid **24** and proceeds according to the sequence (**24** → **28**).

Reductive alkylation of *N*-methyltryptophan with the aldehyde **28** provided the expected amino acid, which without further purification was saponified to the required diol **30**. This substance was now suitable for generating the intermediate dialdehyde **31**, which undergoes spontaneous Pictet-Spengler cyclization to the tetracyclic indole derivative **32**. This approach involving dialdehyde formation by cycloalkane-1,2-diol cleavage followed by cyclization with an appropriate tryptamine system was first applied by van Tamelen et al. in a yohimbine synthesis.[40,41] It has since seen application in other syntheses including the Masamune synthesis discussed previously.

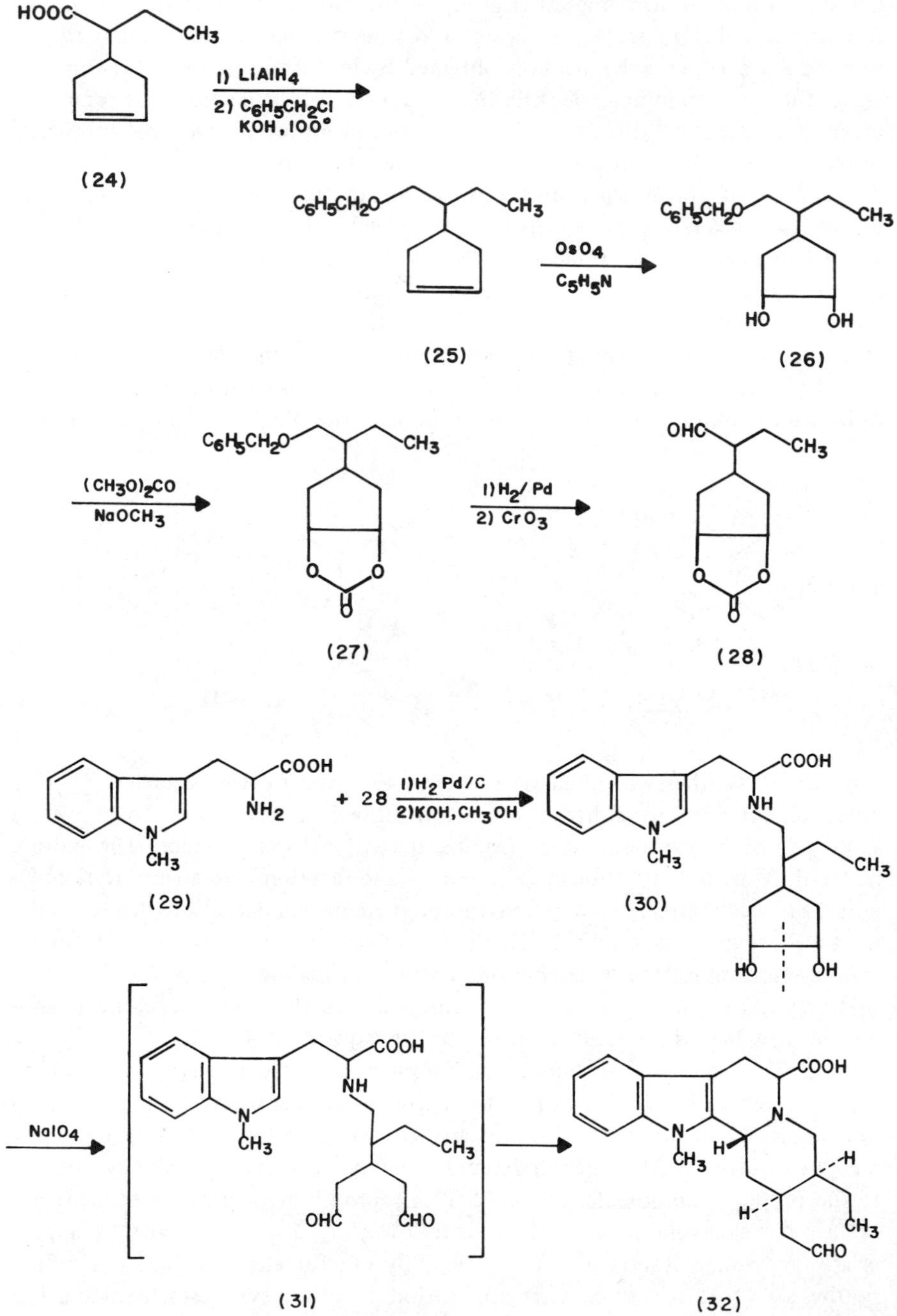

HOOC CH3

1) LiAlH4
2) C6H5CH2Cl KOH, 100°

(24)

C6H5CH2O CH3

OsO4
C5H5N

C6H5CH2O CH3 HO OH

(25)

(26)

(CH3O)2CO
NaOCH3

C6H5CH2O CH3 O O O

1) H2/Pd
2) CrO3

OHC CH3 O O O

(27)

(28)

COOH NH2 N CH3

+ 28

1) H2 Pd/C
2) KOH, CH3OH

COOH NH N CH3 CH3 HO OH

(29)

(30)

NaIO4

COOH NH N CH3 CH3 OHC CHO

COOH N N CH3 H H H CH3 CHO

(31)

(32)

It had been postulated previously that a plausible biosynthetic route linking the tetracyclic corynantheine skeleton as portrayed in 33, for example, with the bridged system characteristic of the ajmaline series could involve a cyclization reaction of an activated site in the side chain with an electrophilic center as shown in the conversion (33 → 34). It is clear that generation of the iminium

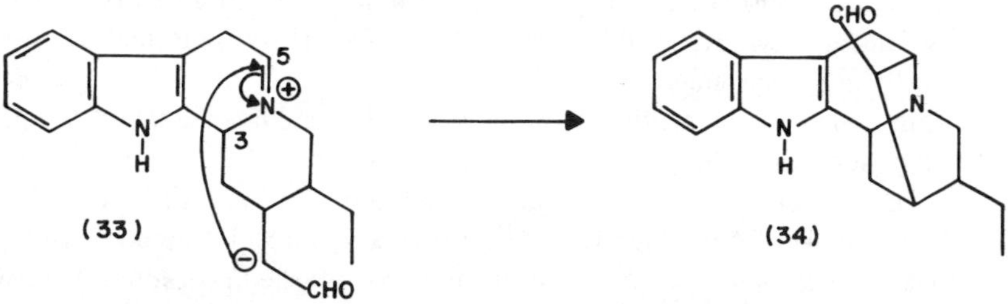

(33) (34)

(35) (36)

system as revealed in 33 required specific reaction at C_5; therefore, direct dehydrogenation of a tetrahydrocarboline derivative is not plausible, since the conjugated and thermodynamically more stable Δ^3-dehydro-β-carboline would

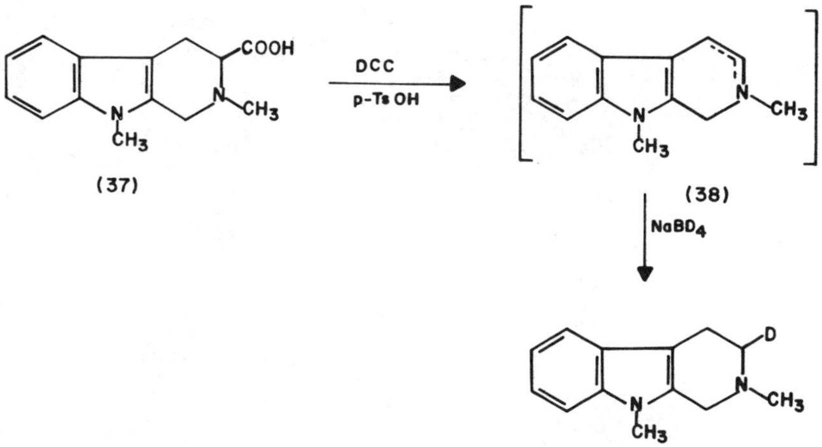

DCC, p-Ts OH; NaBD$_4$

(37) (38) (39)

result (see **35** → **36**). For this purpose these workers studied, initially in a model system, a decarbonylation reaction that places the double bond at the required site. The conversion **37** → **38** was evaluated by reducing the resultant product **38** with sodium borodeuteride to the 3-monodeuterio tetrahydrocarboline derivative **39**.

Under these conditions, the amino acid **32** was converted to the iminium salt **40**, which without isolation was cyclized spontaneously to d,l-deoxyajmalal-B **41**. This substance was resolved by use of D-camphor-10-sulfonic acid to provide the optically active specimen.

The completion of the synthetic route was performed by utilization of appropriate relay substances. Either the isomeric deoxyajmalal-A **42** or **41** on reaction with acetic acid-sodium acetate provides an equilibrium mixture (about 15% of **42** and 85% of **41**) from which the authentic deoxyajmalal-A could be isolated. Reductive cyclization of the latter according to a published procedure[138] provided deoxyajmaline **43**. Functionalization of the latter at C_{21} was accomplished by

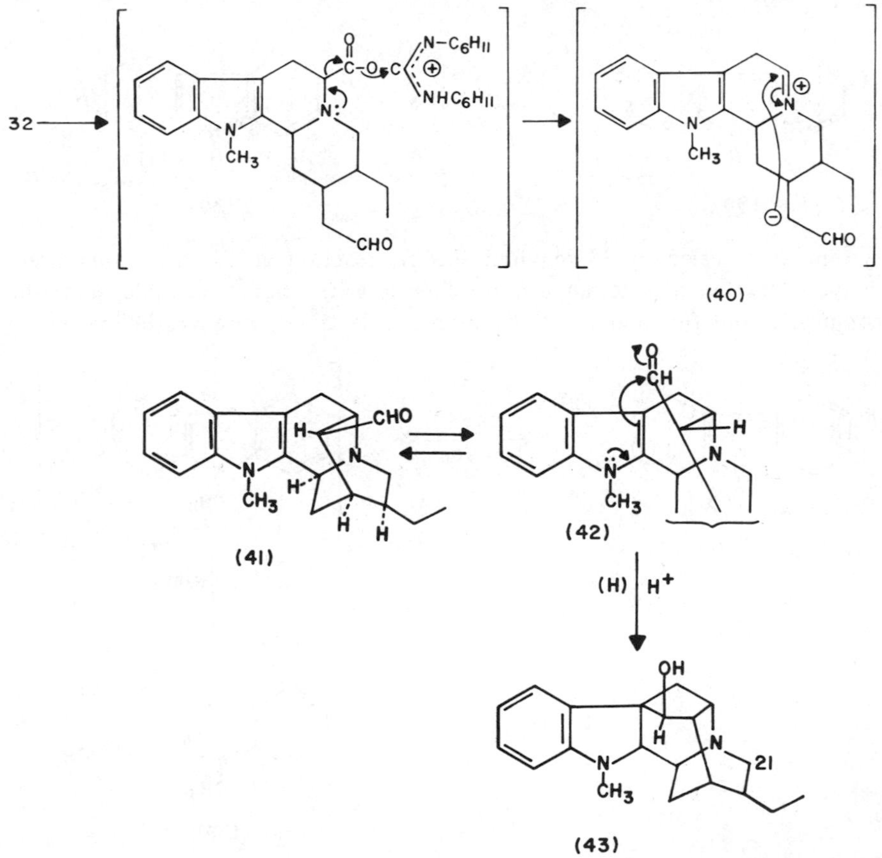

(40)

(41)

(42)

(43)

the phenyl chloroformate ring opening-oxidative ring closure method reported by Hobson and McCluskey[139]; whereupon ajmaline 9 was finally obtained.

A Japanese group[140] recently complete the total synthesis of isoajmaline 57, an alkaloid that is isomeric with ajmaline at C_{20} and C_{21}. The successful sequence is outlined here.

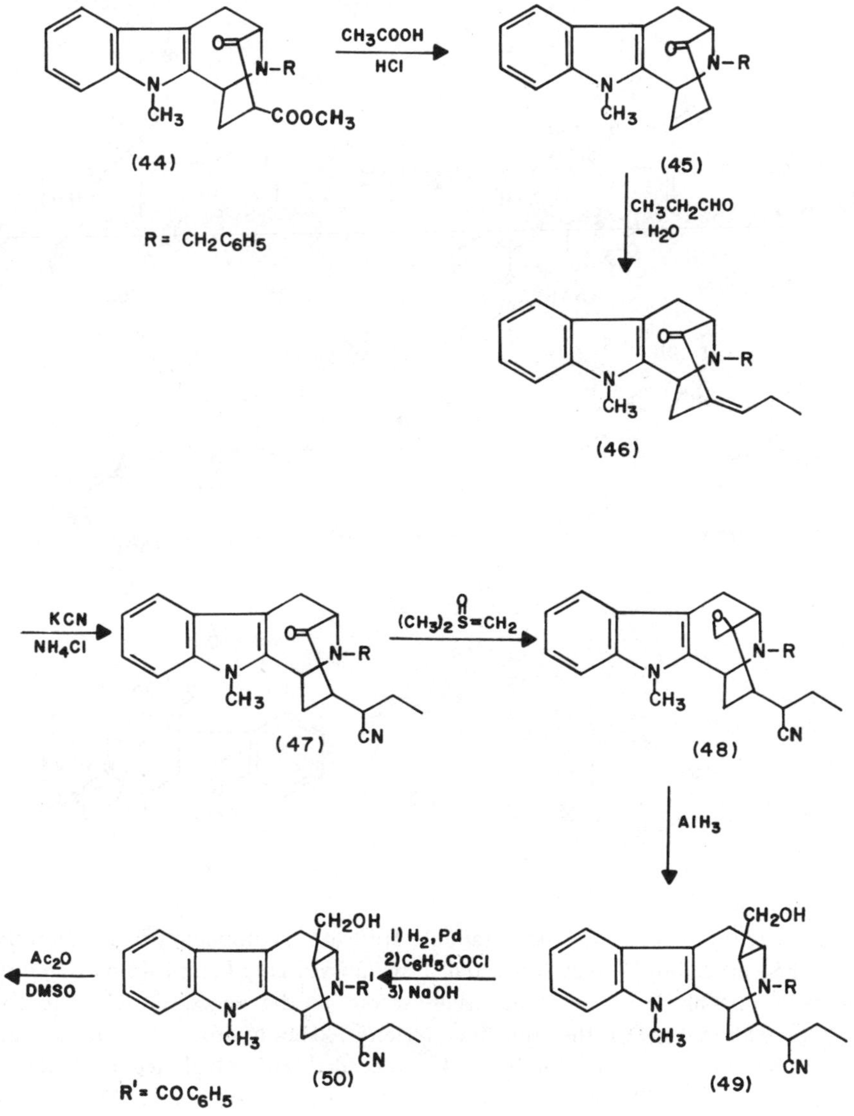

CH_3COOH, HCl
(44) → (45)
$R = CH_2C_6H_5$

CH_3CH_2CHO, $-H_2O$
(45) → (46)

KCN, NH_4Cl
→ (47)

$(CH_3)_2\overset{O}{S}{=}CH_2$
(47) → (48)

AlH_3
(48) → (49)

1) H_2, Pd 2) C_6H_5COCl 3) NaOH
(49) → (50)
$R' = COC_6H_5$

Ac_2O, DMSO
(50) →

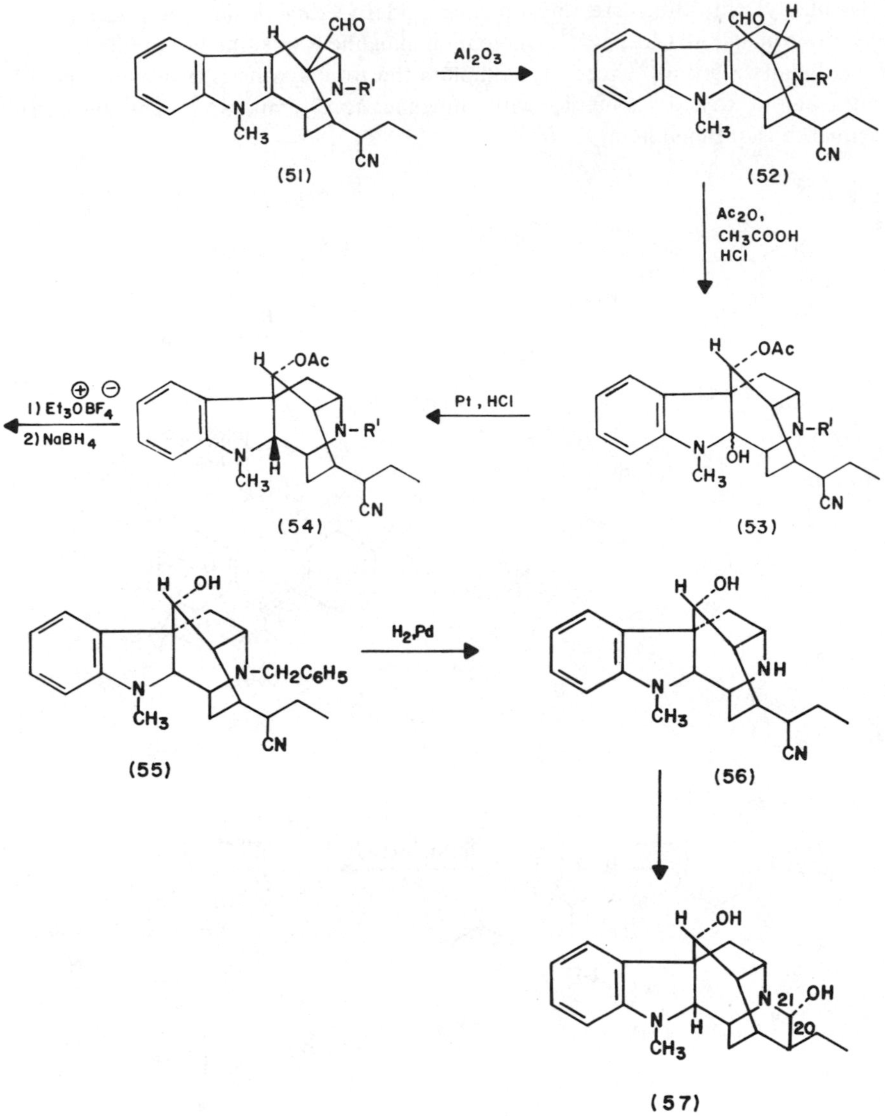

The starting β-keto ester **44**, available from d,l-tryptophan via a previously published procedure[141] was converted to the unsaturated ketone **46** by standard procedures. This latter substance then served as the necessary template for further elaboration to the ajmaline skeleton. The hydrocyanation reaction associated with the conversion **46** → **47** proceeded with a high degree of stereoselectivity, and a 70% yield of the ketonitrile **47** was obtained. Oxirane ring formation **48** according to the Corey method was achieved in a 50% yield, whereas the ring cleavage to the primary alcohol **49** was accomplished in a 54% yield with aluminum hydride. The mixed hydride $HAlCl_2$, successful in a model series allowed conversion in only a 20% yield. It is interesting to note that the oxidative step (**50** → **51**), normally difficult due to the facile oxidation of the indole ring, could be accomplished in high yield (80%) by means of DMSO-acetic anhydride. The isomeric aldehydes obtained by alumina treatment of **51** could not be separated, but the resultant alcohols (**50** and its isomer) prepared by $NaBH_4$ reduction of **51** and **52** could be isolated in pure form.

The structures of these alcohols were ascertained by comparison of the appropriate alcohols obtained by the known degradation of natural ajmaline. Spectroscopic comparison revealed that the synthetic materials belong to the isoajmaline series, that is, the stereochemistry of the nitrile group appears to be isomeric to that of the naturally derived materials. The total synthesis of isoajmaline then proceeded from this point employing the aldehyde **52**, obtained from natural sources, as a relay substance.

Acid catalyzed ring closure of **52** in a manner similar to that employed in the previous syntheses afforded, after reduction, the acetate **54**. Removal of the acetate and *N*-benzoyl groups was achieved by first treating **54** with the Meerwein reagent, followed by borohydride reduction, and finally catalytic debenzylation. During these reactions, the nitrile group remained intact and the nitrile-alcohol **56** was obtained. Since conversion of **56** into isoajmaline had already been described, a formal synthesis of this alkaloid was complete.

An extension of the foregoing route also allows a synthesis of ajmaline.[142] The ketone **45** via its pyrrolidine enamine was treated with chloroacetonitrile and the resulting product **58** was converted to the oxirane **59** in the manner previously described. Elaboration of the latter substance, essentially by the reaction sequence mentioned previously, leads to a compound **17**, which proved identical with the intermediate previously obtained by Masamune.

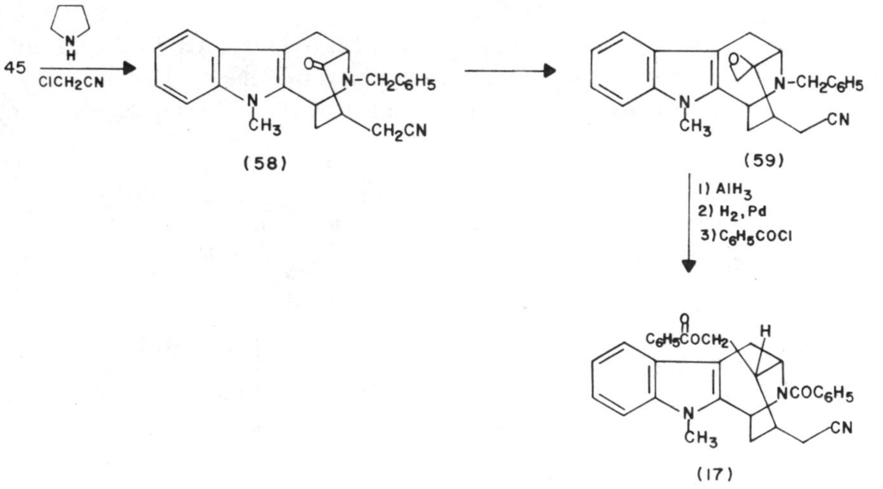

45 → (pyrrolidine, $ClCH_2CN$) → (58) → (59)

(59) → 1) AlH_3; 2) H_2, Pd; 3) C_6H_5COCl → (17)

H. The 2-Acyl Indole Family

The 2-acyl indole alkaloids now represent a large family, although the recognition of this structural type and its abundance in a variety of apocynaceous plants is largely the result of a concentrated effort within the last decade.[143] Biogenetically, it is rather clear that these alkaloids are related to the corynantheine and sarpagine families, although direct experimental evidence is lacking. Thus if one considers the transannular cyclization of a corynantheinoid system, as for example the alkaloid geissoschizine **1** in the manner presented during discussion of the sarpagine-ajmaline group and then ring fusion of the latter occurs, the 2-acyl indole skeleton is readily discernible (see **1** → **2** → **3**).

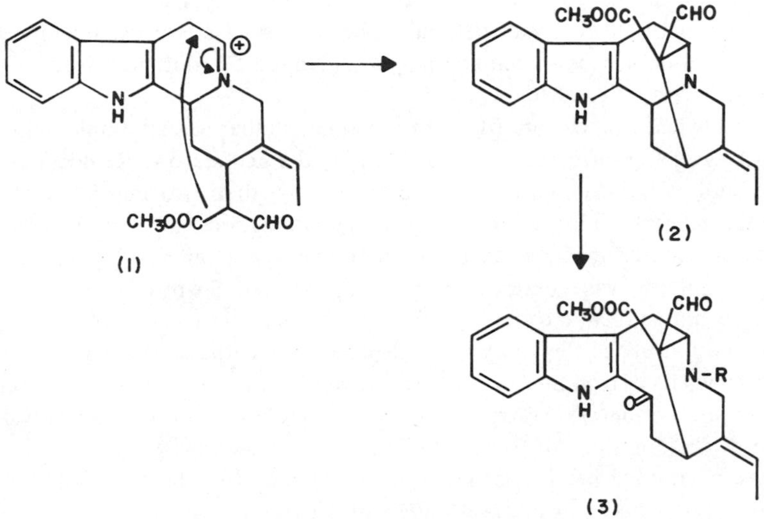

(1) (2) (3)

Thus far, the only synthetic work in this area has been reported by Shioiri and Yamada.[144] In this investigation, the first synthesis of 1-methyl-16-demethoxycarbonyl-20-desethylidenevobasine **14**, according to the following sequence, is described.

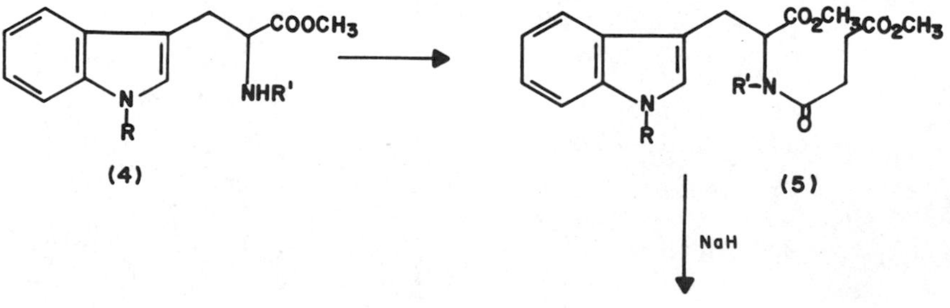

(4) (5)

NaH

1) $C_6H_5CH_2OH$
2) $BrCH_2CO_2CH_2C_6H_5$
K_2CO_3

(6)

R = H or CH_3
R' = $CH_2C_6H_5$

(7)

H_2/Pd
$-CO_2$

(8)

H_2N-NH_2
KOH

(9)

polyphosphoric acid

(10)

(11)

1) $LiAlH_4$
2) (O)

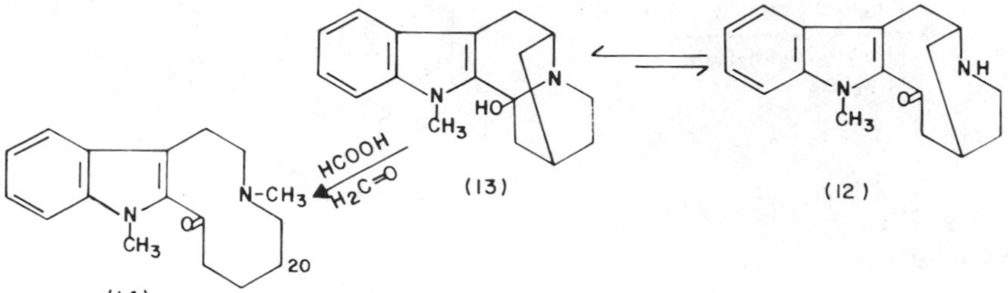

The starting tryptophan derivative **4**, in the parent series or possessing the *N*-methyl group, was transformed to the keto ester **5** by reaction with 3-chloroformylpropionate under Schotten-Baumann conditions. Dieckmann cyclization of the resulting esters was studied under a variety of conditions, but NaH gave the optimum yields. Transesterification and alkylation with benzyl bromoacetate furnished a mixture of diastereoisomers **7**, which upon debenzylation and decarbonylation allowed conversion to the required keto acids **8**. For the purpose of this discussion, only the 4,6-*cis* isomer is utilized for the remaining elaboration to the final product. Huang-Minlon reduction of **8** and standard debenzylation conditions provided **10**, which upon cyclization with polyphosphoric acid allowed construction of the requisite 2-acyl indole skeleton. Removal of the amide carbonyl and mild oxidation of the expected amino alcohols with chromium trioxide in pyridine furnished the corresponding ketone **12**, which exhibited a behavior characteristic of such systems. The facile transannular cyclization of this amino ketone destroys the carbonyl chromophore, as evidenced by infrared and ultraviolet spectroscopy. Clearly, the carbinolamine **13** is predominant species in solution. However, reaction of **13** under Clarke-Eschweiler conditions forces ring opening, and the final product **14** is obtained.

I. The Yohimbine Family

The yohimbine alkaloids were extensively investigated from structural and synthetic points of view during the 1950's when the pharmacological potential of reserpine **2** and related analogs were clearly evident. Such investigations brought forth a number of synthetic routes to these substances. The parent alkaloid yohimbine **1** was synthesized by van Tamelen in 1958,[40,41] two years after Woodward announced his stereospecific synthesis of reserpine **2**.[11]

These alkaloids, which differ considerably in the different arrangement of asymmetric centers, portray the requirement for different strategy in the synthetic plan. The necessary synthetic development of the D/E *trans* fusion of the yohimbine skeleton cannot be easily modified to accommodate the D/E *cis*-fused system present in reserpine. In fact, this stereochemical variation was approached in rather different ways in the synthetic routes.

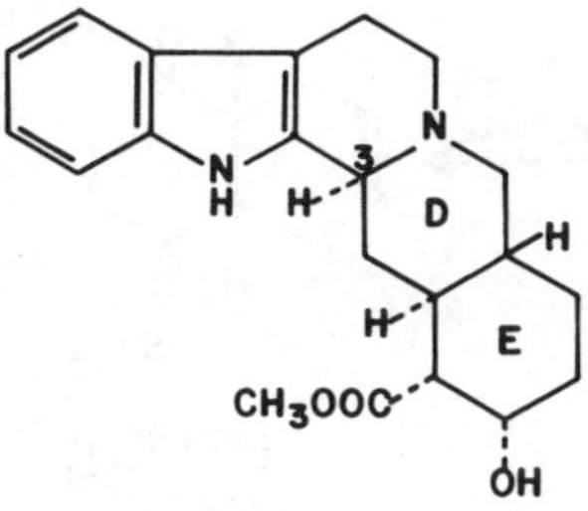

(1) 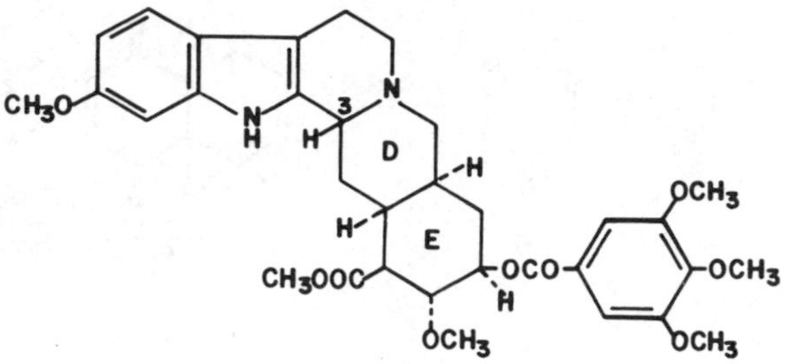 (2)

The large majority of the alkaloids in this family possess the more stable D/E *trans* system; therefore, careful stereochemical control in the early phases is not an essential requirement in these instances.

In the yohimbine synthesis,[40,41] the general plan involved the initial construction of an appropriate nonindolic unit **3**, which would constitute rings D and E of the alkaloids. This unit, upon subsequent condensation with tryptamine **4**, would provide an intermediate that could then be converted to the natural system:

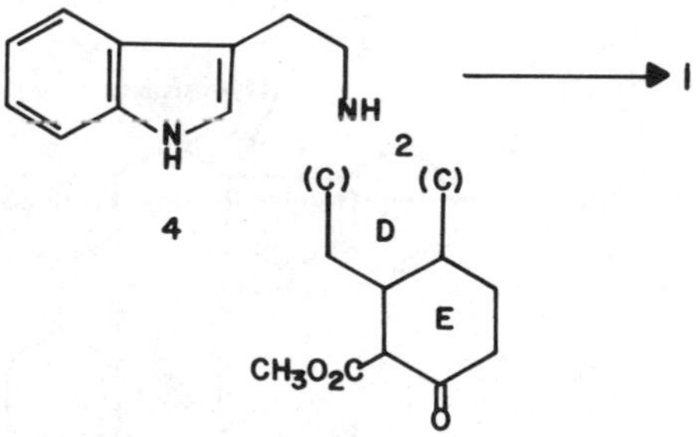

4 3

The initial phase led to the desired keto acid **12**, according to the scheme shown (**7** → **8** → **9** → **10** → **11** → **12**), in which a Diels-Alder reaction is employed to construct the requisite bicyclic skeleton **7**. The most convenient method for incorporating the tryptamine unit was by means of amide bonding (**13** → **14**). The latter intermediate was then elaborated to the desired natural product in the indicated manner. It is of interest to note that the crucial step, leading to the pentacyclic skeleton demanded by the natural system, proceeds by spontaneous cyclization of a dialdehyde formed in situ in the reaction medium (**16** → **17**). A series of extensive investigations by the Woodward group culminated in a brilliant stereospecific synthesis of reserpine.[11] The bicyclic starting material (**22**), which eventually constitutes the D/E rings of the alkaloid was prepared in a Diels-Alder reaction involving *p*-benzoquinone and vinylacrylic acid. One the assumption that the favored conformation of 22 is the one in which the carboxyl group is quasi-equatorial and that the approach of reagents would occur from the less hindered side (the "convex" face) of the molecule, the stereochemistry in the reduction and epoxidation products is a portrayed in 23 and 24, respectively. The hydroxy acid 23 does not spontaneously form a lactone, indicating that, for lactonization to occur, the molecule must first be converted into the less stable conformation, in which the carboxyl and hydroxyl groups are axially disposed. Further, lactone formation is possible *only* if the hydroxy acid possesses the detailed stereochemistry at C_1, C_5, C_9, and C_{10} implied in 23.

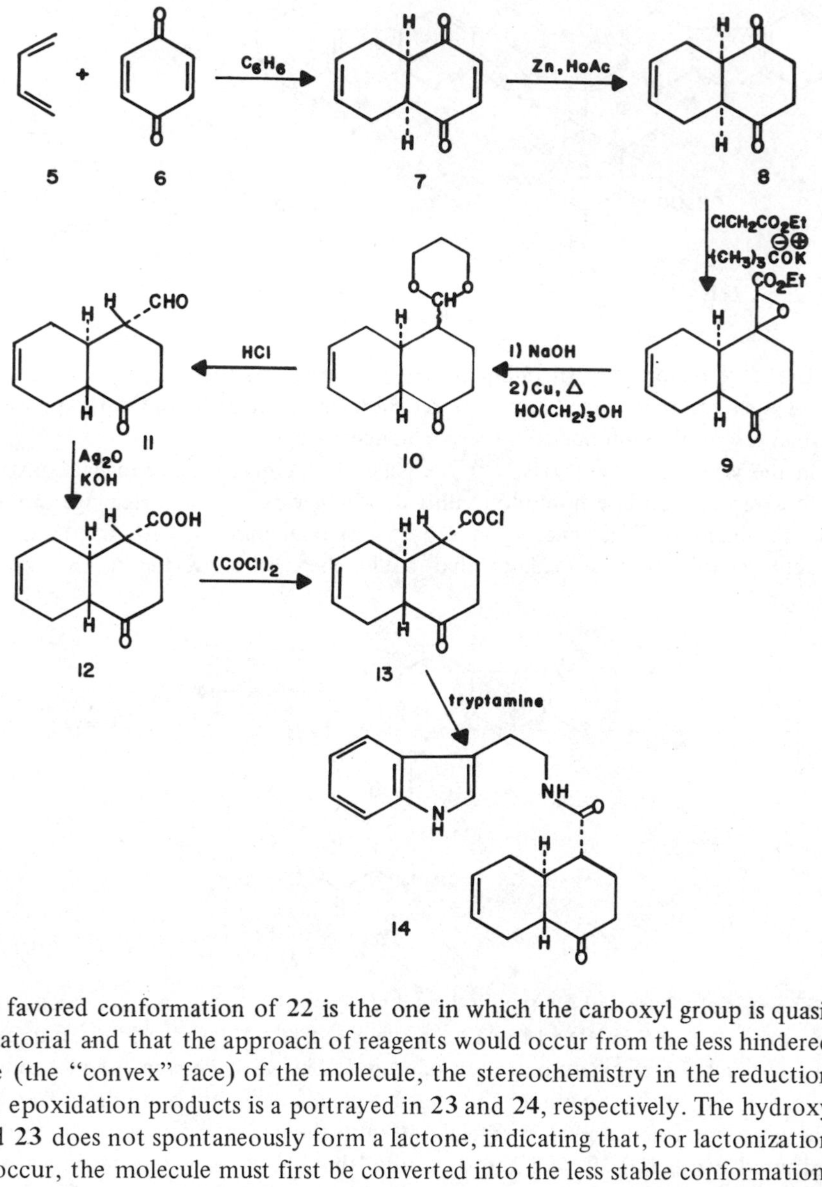

The opening of the oxide ring in 24 to give an intermediate containing the correct stereochemistry required for ring E of reserpine presents a further stereo-

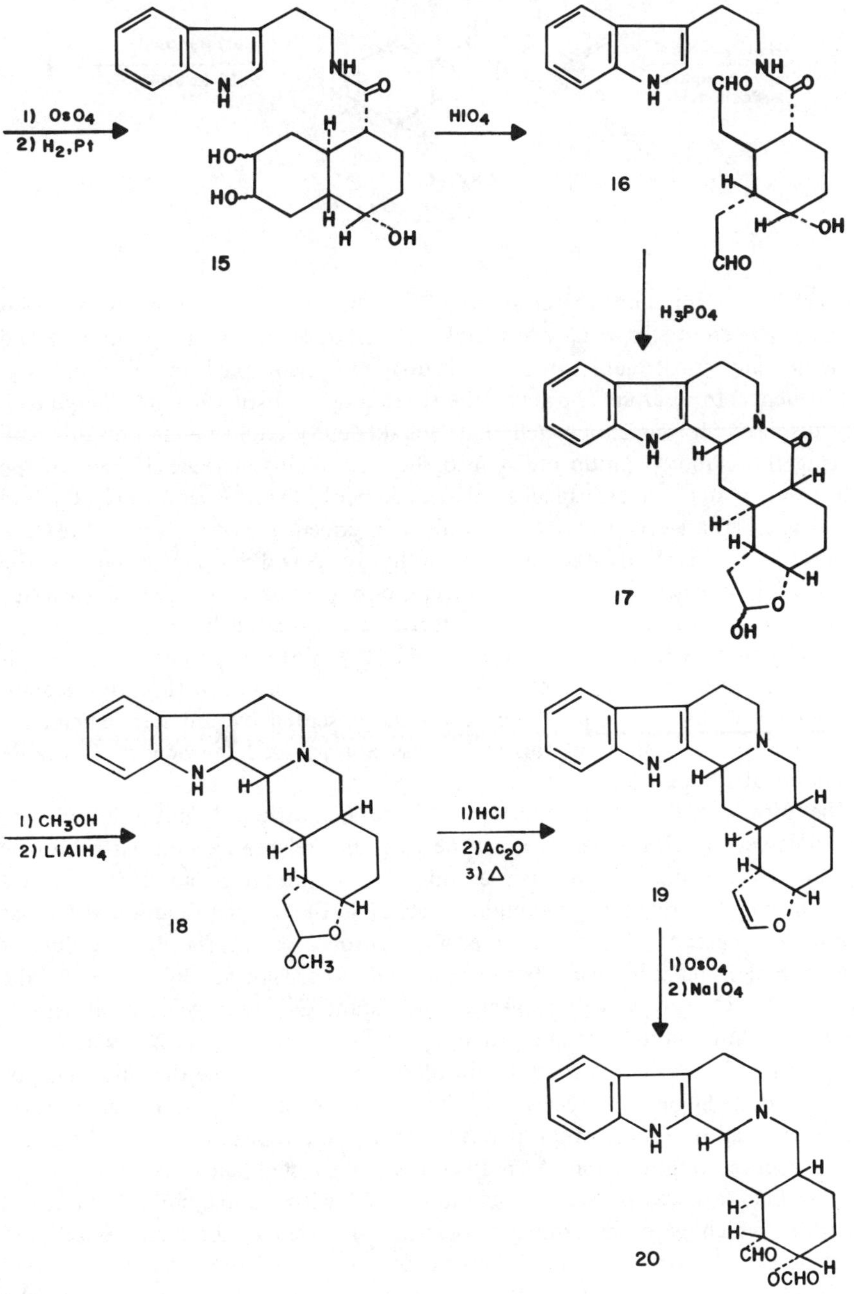

1) OsO_4
2) H_2, Pt

15

HIO_4

16

H_3PO_4

17

1) CH_3OH
2) $LiAlH_4$

18

1) HCl
2) Ac_2O
3) Δ

19

1) OsO_4
2) $NaIO_4$

20

1) CrO_3, CH_3OH
2) l-camphor sulfonic acid

1) $Hg(OAc)_2$
2) $NaBH_4$
→ I

21

chemical problem, since the cleavage of cyclohexane oxides normally proceeds to give, preferentially, axial *trans* diols. In the case of **24**, the product would contain one equatorial (carboxyl group) and two axial (at C_2 and C_3) substituents. In reserpine, however, the three ring E substituents are all equatorially disposed. It was anticipated that this difficulty could be circumvented by forcing the carboxyl group of 24 into the axial conformation; cleavage of the oxide ring would then give an all-axial arrangement of substituents at C_1, C_2, and C_3. The alternative conformation would consequently possess the desired all-equatorial arrangement. Accordingly, the hydroxy acid was converted into the lactone **25**, in which the lactone carbonyl group must be axial oriented. Cleavage of the oxide ring in **25** by boiling acetic acid gave the hydroxy acetate **41** (R = H), from which the ether acetate **41** (R = Me) was prepared by methylation with silver oxide and methyl iodide. The preparation of this intermediate provided the first solution of the problems presented by the stereochemistry of ring E of reserpine, but this series was not pursued, in view of favorable developments in parallel investigations.

The Meerwein-Ponndorf reduction of the C_8 carbonyl group in **25** was accompanied by cleavage of the lactone ring, and relactonization with the new C_8 hydroxyl group, to give a γ-lactone. The hydroxyl group released at C_5 severed the oxide ring by nucleophilic attack, and the oxygen function at C_2 was eliminated together with the activated C_1 hydrogen atom. The final product of this complex reaction was the unsaturated ether-lactone **26** in which the hydrogen at C_8 was *cis* with respect to the C_9 and C_{10} hydrogens. Treatment of **26** with sodium methoxide in methanol gave the methoxy ether **27** by attack of the methoxide ion at the convex side of the molecule. Since the final stage of the reaction involved discharge of an anion under conditions sufficiently vigorous to allow equilibration, it was necessary to establish the stereochemistry of the lactone ring junction. A *cis* fusion was suspected, since the product was very stable; this was proved by reaction of **27** with stannic chloride in acetyl chloride, which gave the aromatic δ-lactone **42**. Hence, the lactone carbonyl must be *cis* with respect to the ether oxygen at C_3, and *trans*, to the methoxyl group at C_2. The stereochemistry of **27** at C_1, C_2, C_3, C_9, and C_{10} is thus the same as that of ring E in reserpine.

An even more remarkable stereospecific synthesis of **27** was developed later from the primary adduct **22** (R = Me) by way of **31**, its Meerwein-Ponndorf reduction product. When **31** was treated with bromine, preferably in an inert solvent but even in methanol solution, attack by bromine at C_2 was followed by attack at C_3 by the C_5 hydroxyl group, with loss of a proton and formation of the bromolactone **43**. As anticipated, reaction of this product with

22 → ($NaBH_4$) → 23 → (C_6H_5COOOH) → 24 → (NaOAc, Ac_2O) → 25

22 → (Al(i-oPr)$_3$, i-PrOH) → 31

25 → (Al(i-oPr)$_3$, i-PrOH) → 26

31 → (1) Br_2 + MeOH, 2) Na OMe) → 27

26 → (Na OMe, Me OH) → 27

27 → ($NBS/H_2O/H_2SO_4$) → 28 → (CrO_3, AcOH) → 29 → (Zn + AcOH) → 30

30 → (1) CH_2N_2, 2) Ac_2O Pyridine) → 32 → (OsO_4) → 33 → (1) HIO_4, 2) CH_2N_2) → 34

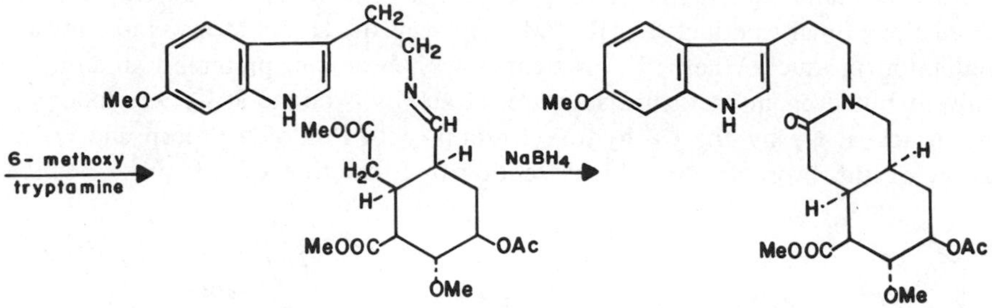

6- methoxy tryptamine

NaBH$_4$

35

36

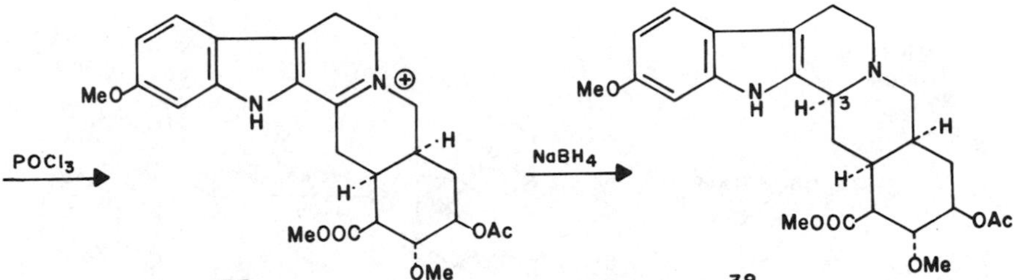

POCl$_3$

NaBH$_4$

37

38

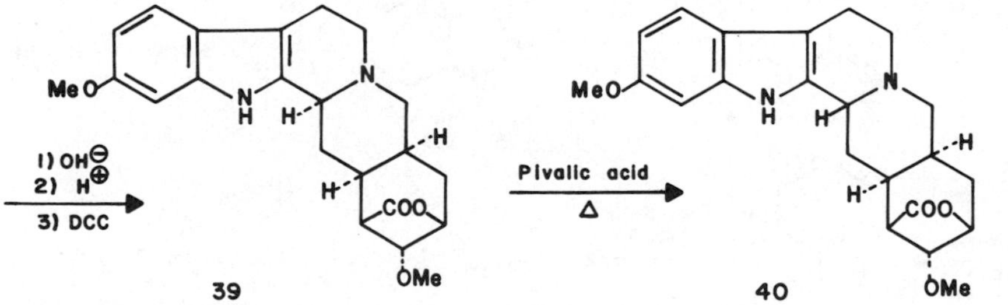

1) OH$^{\ominus}$
2) H$^{\oplus}$
3) DCC

Pivalic acid
Δ

39

40

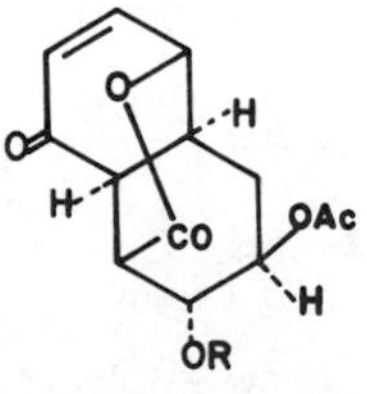

41

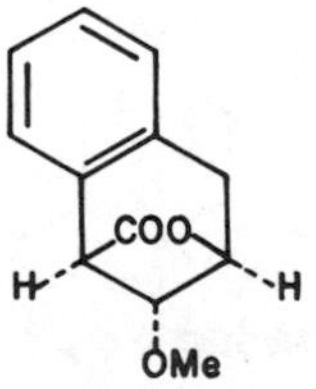

42

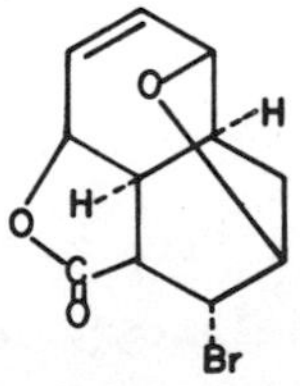

43

sodium methoxide in methanol led to the methoxy ether **27**, stereochemically identical with the product of the earlier synthesis.

Although the double bond of **27** was not attacked by bromine at room temperature, addition of the elements of hypobromous acid was achieved by reaction with *N*-bromsuccinimide at 80°. Attack of the molecule at the convex face gave the di-axial bromohydrin **28**, which was smoothly oxidized to the bromoketone **29**. Where **29** was reduced with zinc in glacial acetic acid, two simultaneous reductive processes occurred (indicated by arrows in **29**), and the unsaturated keto acid **30** was obtained in high yield. The conversion of **30** to the conformationally stable aldehyde ester **34** proceeds in a straightforward manner.

Condensation of **34** with 6-methoxytryptamine provides the expected imine **35**, and the latter upon reduction proceeds to the lactam **36**. Bischler-Napieralski cyclization of **36** completes the formation of the required pentacyclic ring system. Normal reduction of the iminium bond in **37** yields the expected 3-iso-stereochemistry shown in **38**, the more stable axial orientation of the hydrogen atom being formed, as already determined in various structural studies within the yohimbine family. The racemic ester **38** was resolved via its di-*p*-toluyl-ℓ-tartrate salt into ℓ-methyl-*O*-acetylisoreserpate, and the latter was compared with an authentic material prepared from reserpine.

Lactonization (**38** → **39**) of isoreserpic acid hydrochloride, the latter being made available from **38** by saponification, has forced the molecule to adopt the less favored conformation. As inticipated, **39**, is readily, and quantitatively epimerized at C_3 by heating with pivalic acid. The resultant product, reserpic acid lactone **40**, was identical with authentic material. Since the conversion of **40** into reserpine **2** had been achieved previously, the total synthesis of the alkaloid was complete.

In later experiments, d,l-reserpic acid lactone **40** was prepared by omitting the resolution stage. Methanolysis gave d,l-methyl reserpate, which was converted by treatment with 3,4,5-trimethoxybenzoyl chloride, into d,l-reserpine. The latter was then readily resolved via the d-camphor-10-sulfonate salt to yield l-reserpine. A subsequent investigation by Velluz and his group[145] provided some modifications to Woodward's synthesis and also allowed the synthesis of a variety of yohimbine derivatives.

Ernest and Kakac[146] have successfully applied the principles of the Woodward synthesis in preparing appropriate D/E *cis*-fused analogs and then modifying the route for the synthesis of D/E *trans* derivatives in the yohimbane series. Thus the diol **45** available from the known lactone **44** was readily converted to the aldehydo acid **46**, and the latter on equilibration provided a mixture of **46** and **47**. When the latter substance was subjected to the steps outlined below, d,l-apoyohimbane **52** was obtained.

A similar consideration, namely the preponderance of *trans* 1,2-disubstituted cyclohexane derivatives under appropriate equilibrating conditions, led to another synthesis of d,l-yohimbane.[147] Thus the crude acid aldehyde 55, obtained as an equilibrium mixture and assumed to exist mainly in the *trans* stereochemistry, was condensed with tryptamine and the resultant lactam 56 was converted to d,l-yohimbane 57. (+)-Yohimbane was obtained from the racemic mixture by resolution with (−)-dibenzoyltartaric acid.

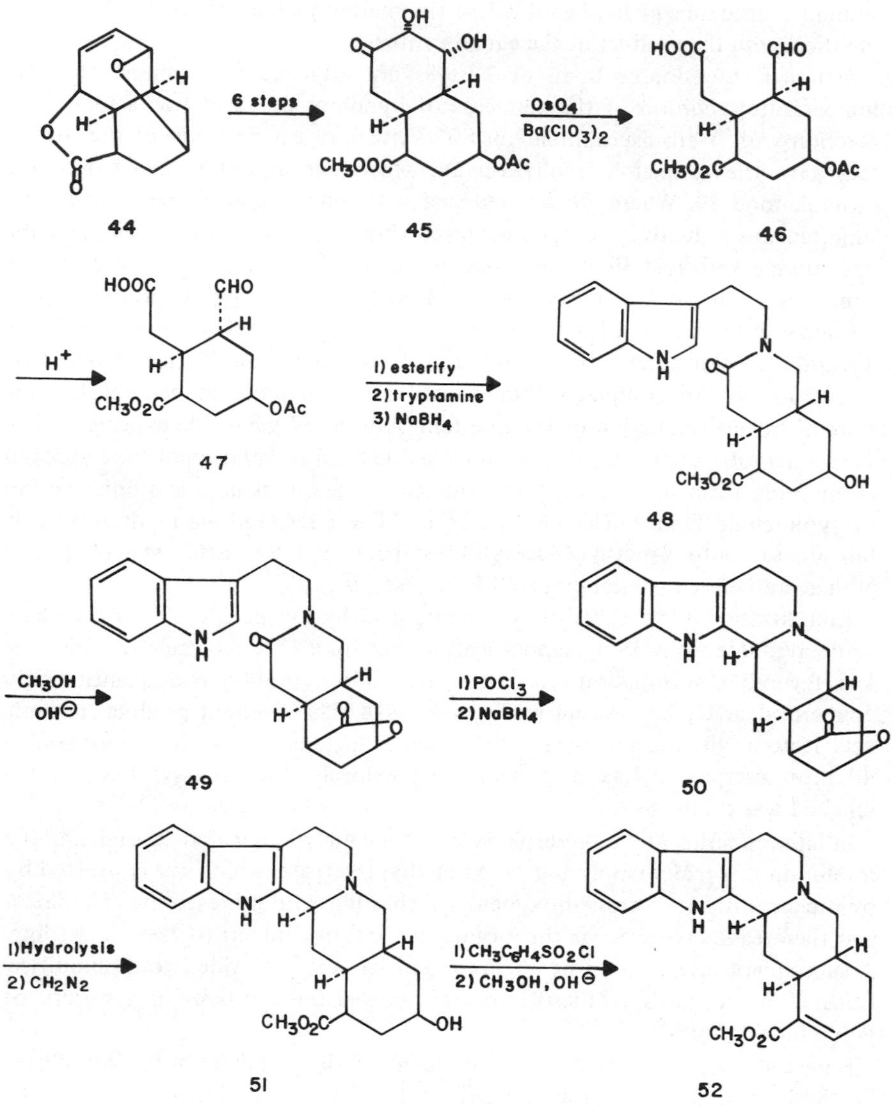

A substantial number of recent syntheses in this family utilize the cyclization of appropriate iminium salts, a method first described by Potts and Robinson[148]

and then subsequently studied extensively by the Wenkert group.[149,150] In this way, d,l-pseudoyohimbane 59 was prepared.[151] Unfortunately, the commonly employed oxidant, mercuric acetate, attacks the decahydroisoquinoline derivative 58 in a rather indiscriminate manner and a mixture of yohimbane isomers is obtained.

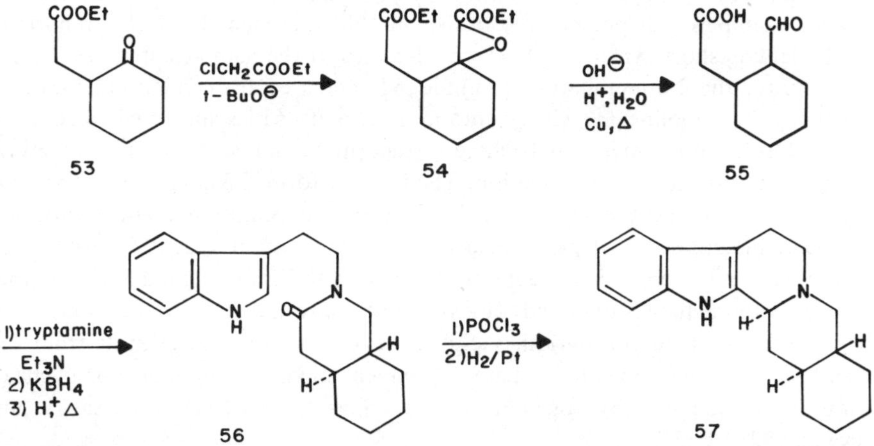

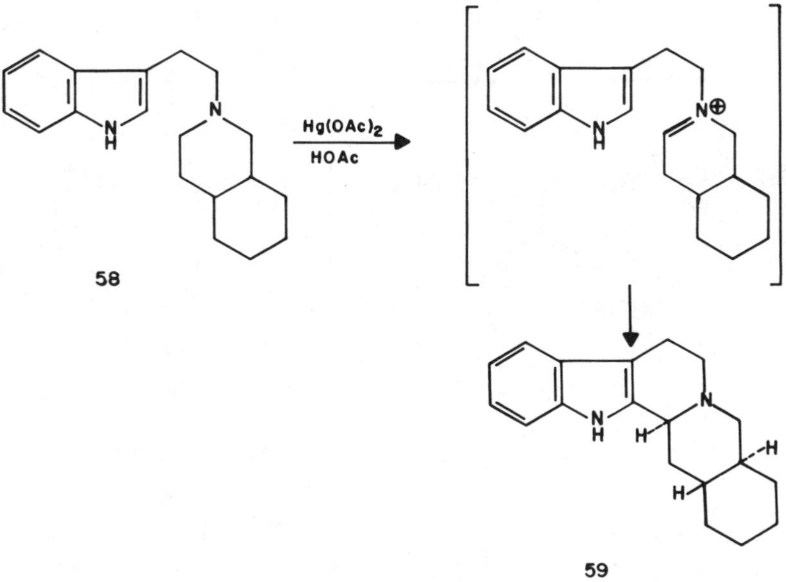

A convenient method for generating the requisite imine systems for subsequent elaboration to the yohimbine and related series is provided via hydride reduction of appropriate pyridinium or isoquinolinium derivatives. Several recent examples will portray the utility of this approach. Beisler[152] reports a high yielding short synthesis of several alkaloids in the gambirtannine series. In this study, the isoquinolinium bromide **62**, prepared by condensing 5-carbomethoxy isoquinoline **60** with tryptophyl bromide **61**, is subjected to sodium borohydride reduction in a multiphase system (methanol-water-ether) containing a high concentration of cyanide ion. The initially formed enamine is trapped by cyanide, and the postulated intermediate **63** receives minimum contact with the reducing medium since it passes into the ether layer as soon as it is formed. In this way, undesirable side reactions are minimized, and a good yield of the desired intermediate is obtained. This procedure was first employed by Fry[153] in the synthesis of dihydropyridines. Subsequent elimination of cyanide from **63** generates the iminium intermediate **64**, which undergoes spontaneous cyclization to provide d,l-dihydrogambirtannine **65** as the hydrochloride salt in an overall yield of 83%.

N H Br + N CH_3O_2C N H N ⊕ Br ⊖ CH_3O_2C

61 60 62

N H NC N CH_3O_2C HCl N H N ⊕ CH_3O_2C

63 64

N H N

65 CH_3O_2C

The synthesis of **65** had been described by Wenkert.[38] This investigation represents an extension of a continuing study on indole alkaloid syntheses in which hydrogenation and cyclization of 1-alkyl-3-acylpyridinium salts occupies a crucial role. In this instance, the reduction of the appropriate isoquinolinium salt was shown to follow a pattern observed previously in the pyridinium series. The successful pathway to **65** follows the sequence **66** → **67** → **68** → **65**.

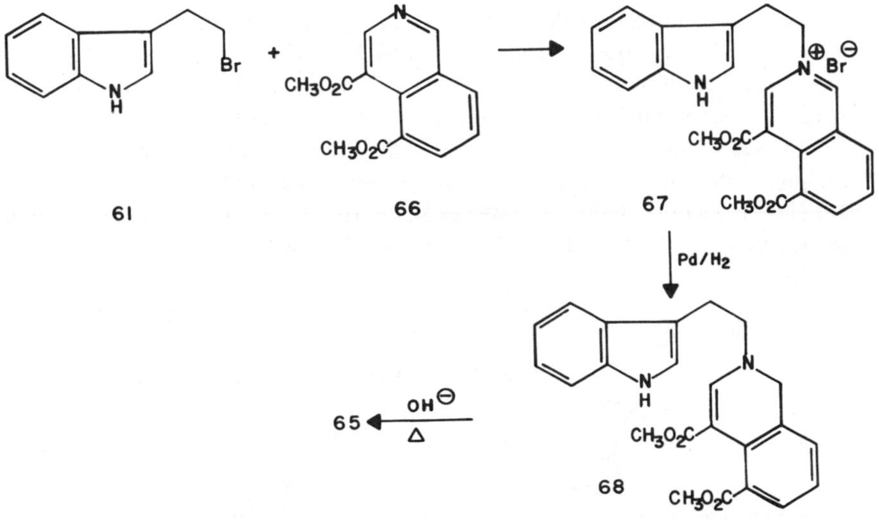

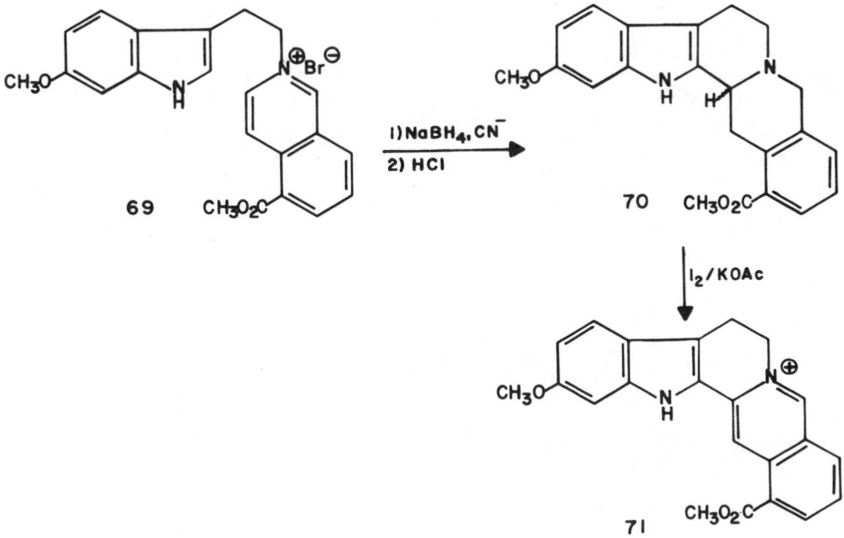

61 66 67 68 65

Pd/H_2

$OH^{\ominus}$ Δ

69 70 71

1) $NaBH_4$, CN^- 2) HCl

I_2/KOAc

A synthesis of the alkaloid alstoniline **71** has been recently described by Beisler[154] in which his approach mentioned previously has been applied.

Another approach to the yohimbine system has been reported by Szantay.[155a] The crucial step in this synthesis involves a Dieckmann cyclization of the appropriate tetracyclic diester **74** or nitrile ester **78**. The tetracyclic keto ester **72** was already available from a previous study. A detailed investigation of the Dieckmann cyclization of **74** was made and a variety of reaction conditions was employed. Unfortunately, the cyclization proceeded predominantly in the undesired direction to provide **75** as the major product. The desired keto ester, yohimbinone **76**, was obtained in only 10 to 15% of the yield. This latter substance on reduction provided a 1:3 mixture of yohimbine **1** and β-yohimbine **77**.

When **72** was reacted with the appropriate phosphonate, the expected nitrile ester **78** was obtained. Saturation of the double bond and ring closure of the resultant product under basic conditions provided the keto nitrile **79** in 52% of the yield. Reduction of the carbonyl group in the latter allowed the isolation of three nitrile alcohols (**80**, **81**, and **82**), two of which were then elaborated to the

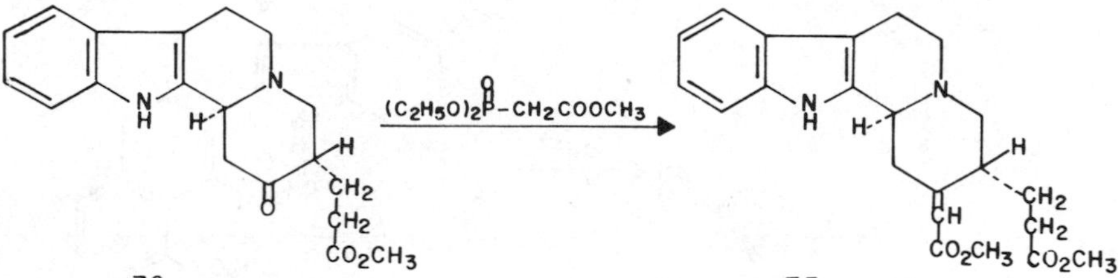

$(C_2H_5O)_2P(O)-CH_2COOCH_3$

72 → 73

Base

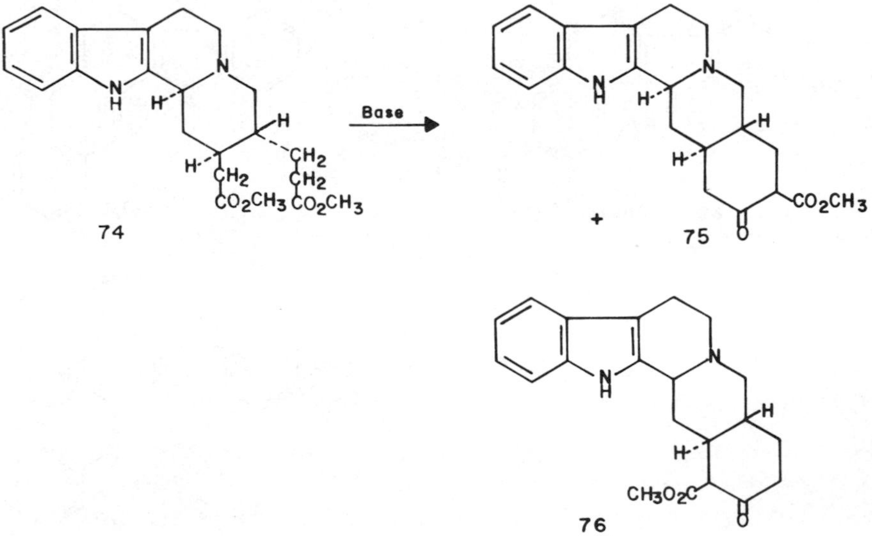

74 → 75 + 76

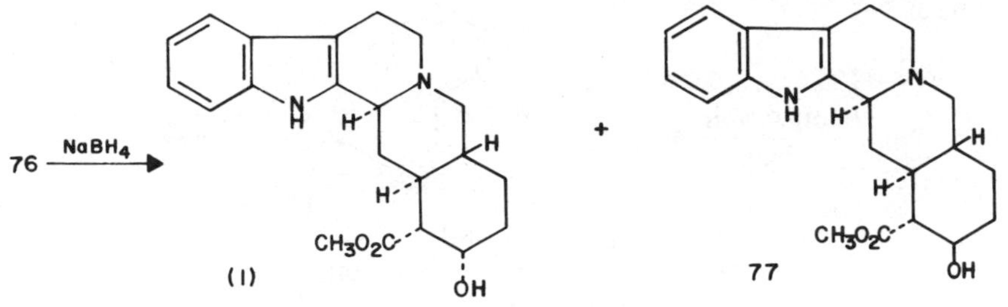

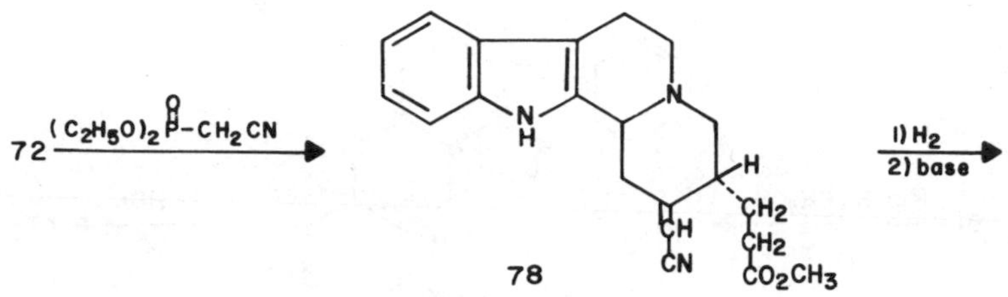

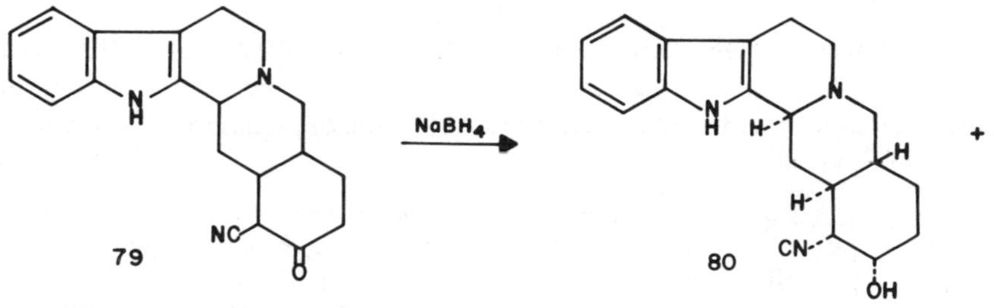

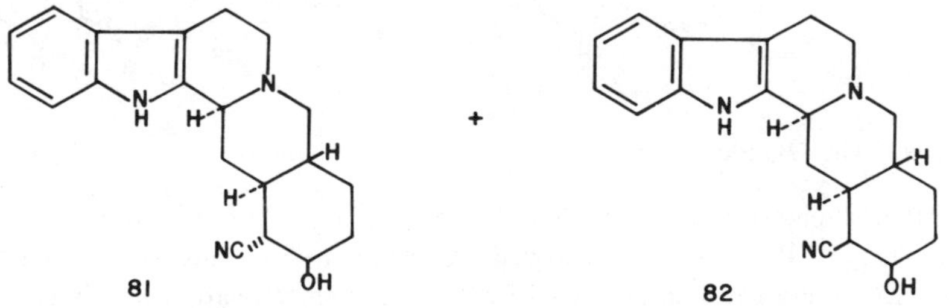

76 $\xrightarrow{NaBH_4}$ (1) + 77

72 $\xrightarrow{(C_2H_5O)_2P(O)-CH_2CN}$ 78 $\xrightarrow{1)\ H_2\ 2)\ base}$

79 $\xrightarrow{NaBH_4}$ 80 +

81 + 82

alkaloid systems (**80** → **1** and **81** → **77**). The synthetic racemic products (**1** and **77**) were resolved by employing *N*-acetyl-L-leucine for d,l-yohimbine and L-camphor sulfonic acid for d,l-β-yohimbine as the resolving agents. In this way, (+)-yohimbine and (−)-β-yohimbine were obtained.

80 $\xrightarrow[H_2O_2]{NaOH, CH_3OH}$ 83 $\xrightarrow[\Delta]{HCl}$ 1

81 $\xrightarrow[H_2O_2]{NaOH, CH_3OH}$ 84 $\xrightarrow[\Delta]{HCl}$ 77

A more recent publication provides an improved synthesis of yohimbine by Dieckmann cyclization of unsaturated diesters.[155b] Other approaches to the yohimbine alkaloids have been reported[156-159] but these have not as yet led to the natural systems.

73 $\xrightarrow{NaH, THF}$ 85 $\xrightarrow{H_2/Pd}$ 86

J. The Oxindole Family

The synthetic achievements in this family have occurred largely within the last decade. The initial approaches involved conversions of tetrahydro-β-carboline alkaloids into oxindoles with the view of revealing the interrelationships between these two families. Such investigations provided structural as well as stereochemical data for the oxindole members and in turn allowed partial syntheses of these compounds.[160,161] The well-known reactivity of the β position of the indole nucleus to electrophilic agents allowed the utilization of this property in the conversion of the appropriate indole alkaloids by means of tert-butyl hypochlorite to the resulting chloroindolenines and subsequent rearrangement of the latter intermediates to the oxindole systems. In this manner, ajmalicine **1** leads to mitraphylline **2**, whereas corynantheine **3** and dihydrocorynantheine **4** give rise to corynoxeine **5** and rhyncophylline **6**. These types of rearrangements have received renewed interest in the last few years, since it has been postulated

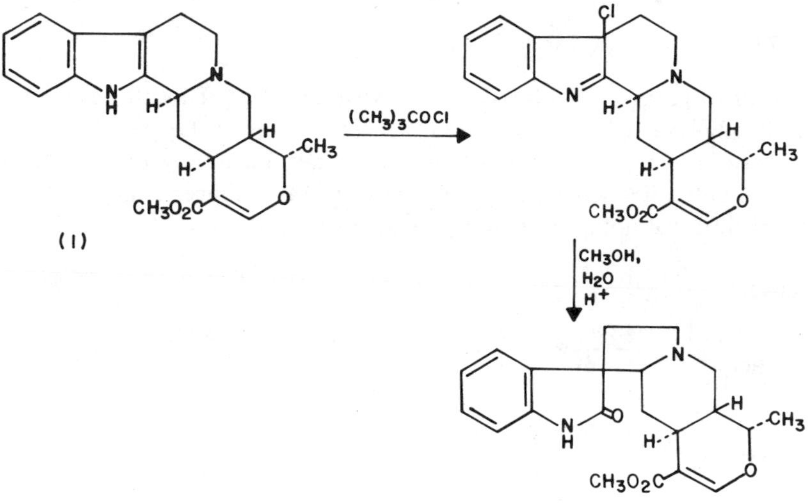

$(CH_3)_3COCl$

(1)

CH_3OH, H_2O, H^+

(2)

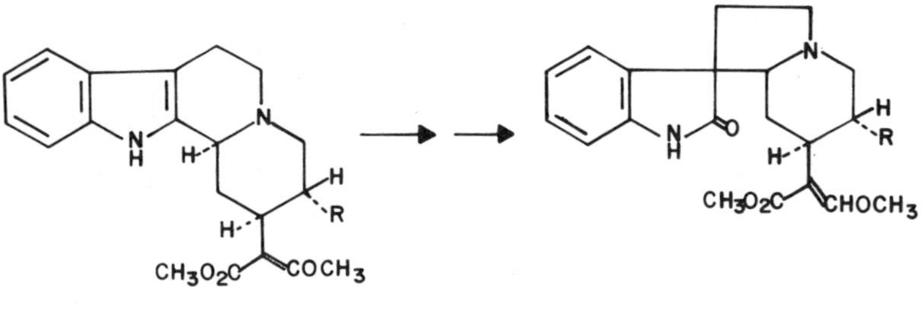

3, R = $CH{=}CH_2$

4, R = Et

5, R = $CH{=}CH_2$

6, R = Et

that the *in vivo* rearrangement of the *Corynanthe* series to the *Strychnos* family in the plants occurs via similar processes.

Several other approaches[162,163] have employed the appropriate hydroxytryptamine derivatives directly to prepare derivatives of the rhyncophylline series. Thus Ban and Oishi[162] were able to condense 1-methyl-2-hydroxytryptamine hydrochloride **7** with the aldehyde **8** to provide the base **9**. A similar sequence

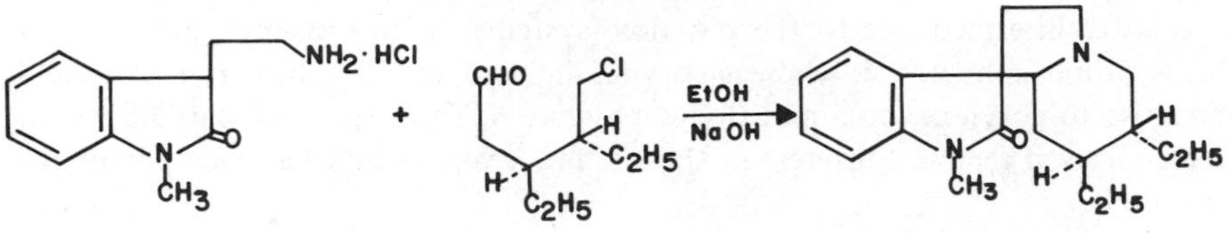

(7) (8) (9)

has served to provide a synthesis of d,l-8-oxovincatine.[164] Thus the reaction of **7** with dimethyl 4-ethyl-4-formylpimelate **10** gave in 70% of the yield the dilactam 8-oxovincatine **11** as a mixture of two racemates apparently differing in configuration at C_5. Reduction of the separated lactams produced a diol **12** that was identical with that obtained from the alkaloid vincatine.

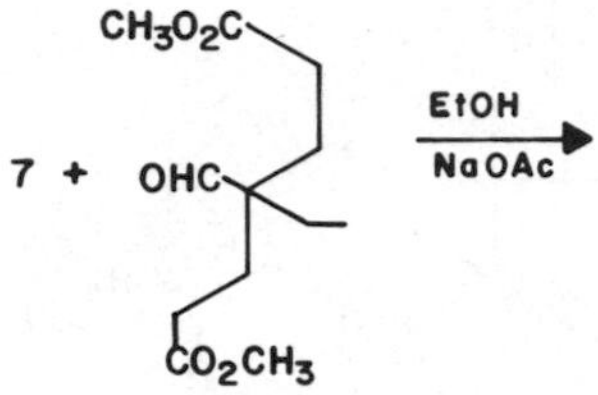

(10) 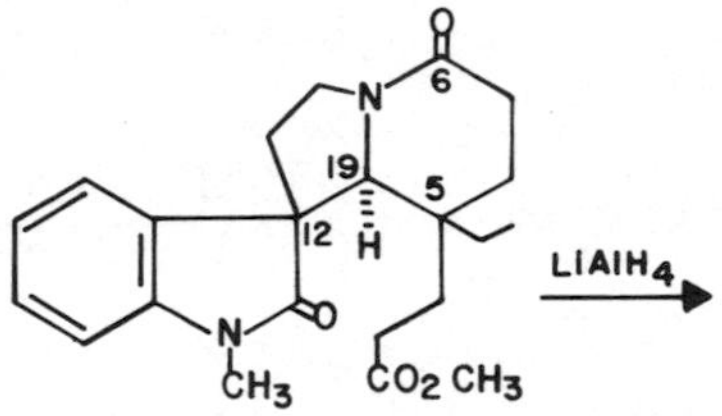 (11) 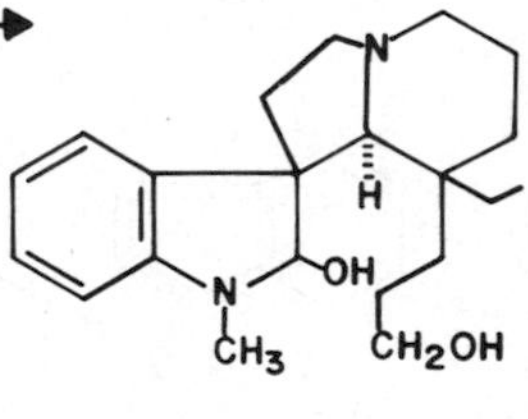 (12)

The extensive investigations of van Tamelen and his group concerning biogenetic-type syntheses have led to a total synthesis of rhyncophyllol **24** and d,l-isorhynocophyllol **29**.[165,166] In the belief that many of the structural types inherent in the numerous indole families are derived by varying modes of cyclization of natural intermediates in which certain reactive sites have been transformed enzymatically to specific oxidation levels, van Tamelen devised a sequence in which the crucial step in the synthesis of **24** involved a cyclization of the appropriate oxindole dialdehyde **22**. A modification of this sequence completed the synthesis of d,l-isorhyncophyllol **29**.

1) B_2H_6
2) H_2O_2, OH^-
3) p-TsCl

OTs

(13) (14)

EtCH CN CO_2Et

CN
C—COOEt
Et

(15)

1) hydrolysis
2) $^-CO_2$

CN
CH_3

(16)

$LiAlH_4$

NH_2
CH_3

(17)

COOH
N H

Δ

O
NH
N H

(18)

1) $LiAlH_4$
2) $C_6H_5CH_2OC(O)-Cl$
K_2CO_3

N-C(O)-$OCH_2C_6H_5$
N H

(19)

OsO_4

N-C(O)-$OCH_2C_6H_5$
N H
HO
HO

(20)

1) NBS
HOAc
2) H_2, Pd

NH
N H O
HO
HO

(21)

$NaIO_4$

NH
N H O
CHO
CHO

(22)

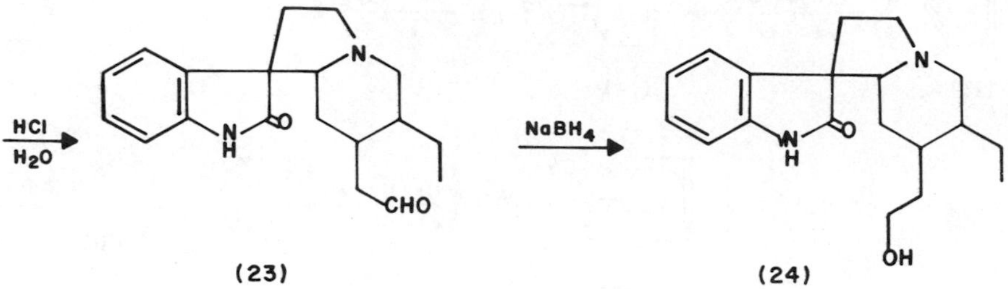
HCl
H2O
CHO
NaBH4
OH

(23) (24)

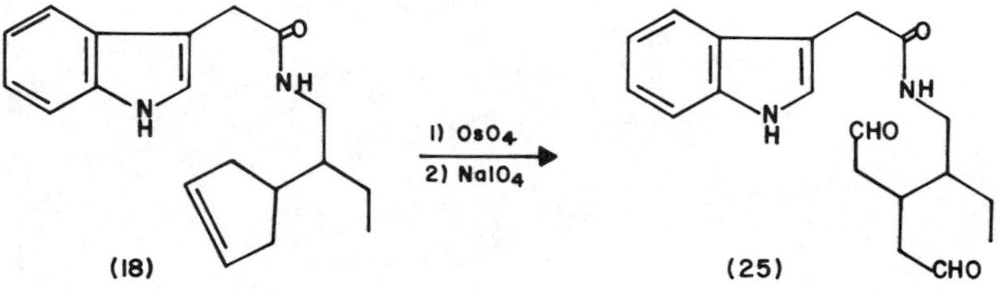
NH
1) OsO4
2) NaIO4
CHO
CHO

(18) (25)

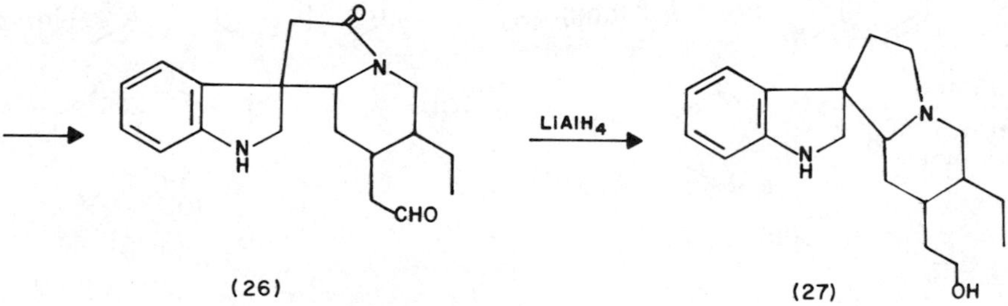
CHO
LiAlH4
OH

(26) (27)

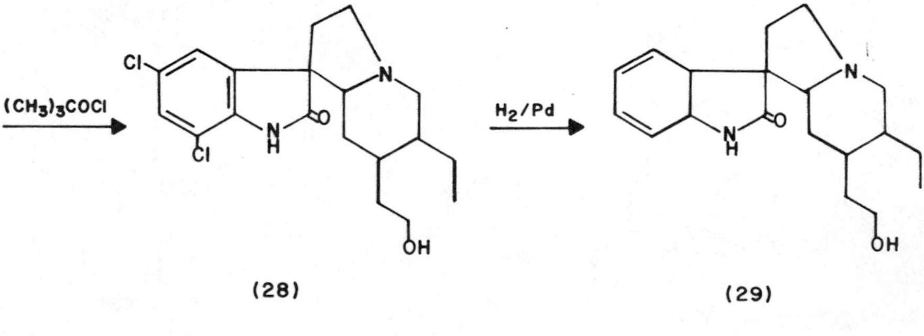
(CH3)3COCl
Cl
Cl
OH
H2/Pd
OH

(28) (29)

K. The *Strychnos* Family

A major synthetic effort by the Woodward group culminated in the first total synthesis of strychnine in 1954.[26] The starting material selected already contains the indole unit; this is then elaborated to the natural system. Thus 2-veratrylindole **1** is prepared via a Fischer indole synthesis, and the required tryptamine side chain is attached by means of a Mannich reaction (**1** → **2**). A frequently employed reaction in indole chemistry, already mentioned earlier in this discussion, is then utilized to elongate the side chain in **2**. Thus the methiodide of **2** provides an excellent leaving group for nucleophilic displacement by cyanide ion and the resulting nitrile, upon reduction, provides **3**. The 3,3-disubstituted indolenine system required for the strychnine skeleton is elaborated via the well-known property of indoles to undergo facile electrophilic substitution at the β-position. Thus the Schiff base **4**, formed initially by reaction of **3** with ethyl glyoxylate, undergoes cyclization in the manner indicated (*p*-toluenesulfonyl chloride and pyridine) to yield **5**. Normal reduction and acetylation leads to **6**, and the latter is then subjected to an exceptional ozonolysis reaction in which the veratrole ring is cleaved to **7**. It is of interest to note that this reaction parallels the postulate known commonly as the "Woodward fission" (dotted line in **6**) to explain the mode of biological cleavage of aromatic rings to the functionality normally found in the corynantheine family of alkaloids. Although subsequent experimentation has excluded such a process in the biosynthesis of these natural products, it is nevertheless a very interesting reaction.

Action of acidic methanol on **7** yields the pyridone **8**. This ring system now provides a stable structure for further construction of the remaining rings necessary and yet plays an important role in the final stages of the synthesis. Thus Dieckmann ring closure of **8** yields the pentacyclic system **9**. Removal of the enolic hydroxyl group in the latter was achieved in an interesting manner. Conversion of **9** to the enol tosylate, and in turn to the thioenolbenzyl ether **9a**, the latter being formed from the former by means of sodium benzylmercaptide, provided the opportunity for reductive removal by means of Raney nickel desulfurization and catalytic hydrogenation of the resulting unsaturated ester. The product (**10**) was resolved and found to be identical with the corresponding product formed by degradation of strychnine. Ring closure of **10** to provide **12** proved unusually difficult, and the only successful method involved conversion of **10** to the methyl ketone **11**, followed by oxidation and cyclization to dehydrostrychninone **12**.

Introduction of the vinyl side chain-involved reaction of **12** with sodium acetylide and reduction of the ethinylcarbinol with Lindlar catalyst to **13**. Hydride reduction of the latter achieves reduction of the pyridone ring as well as the N_b lactam carbonyl group *but* the N_a lactam carbonyl remains intact.

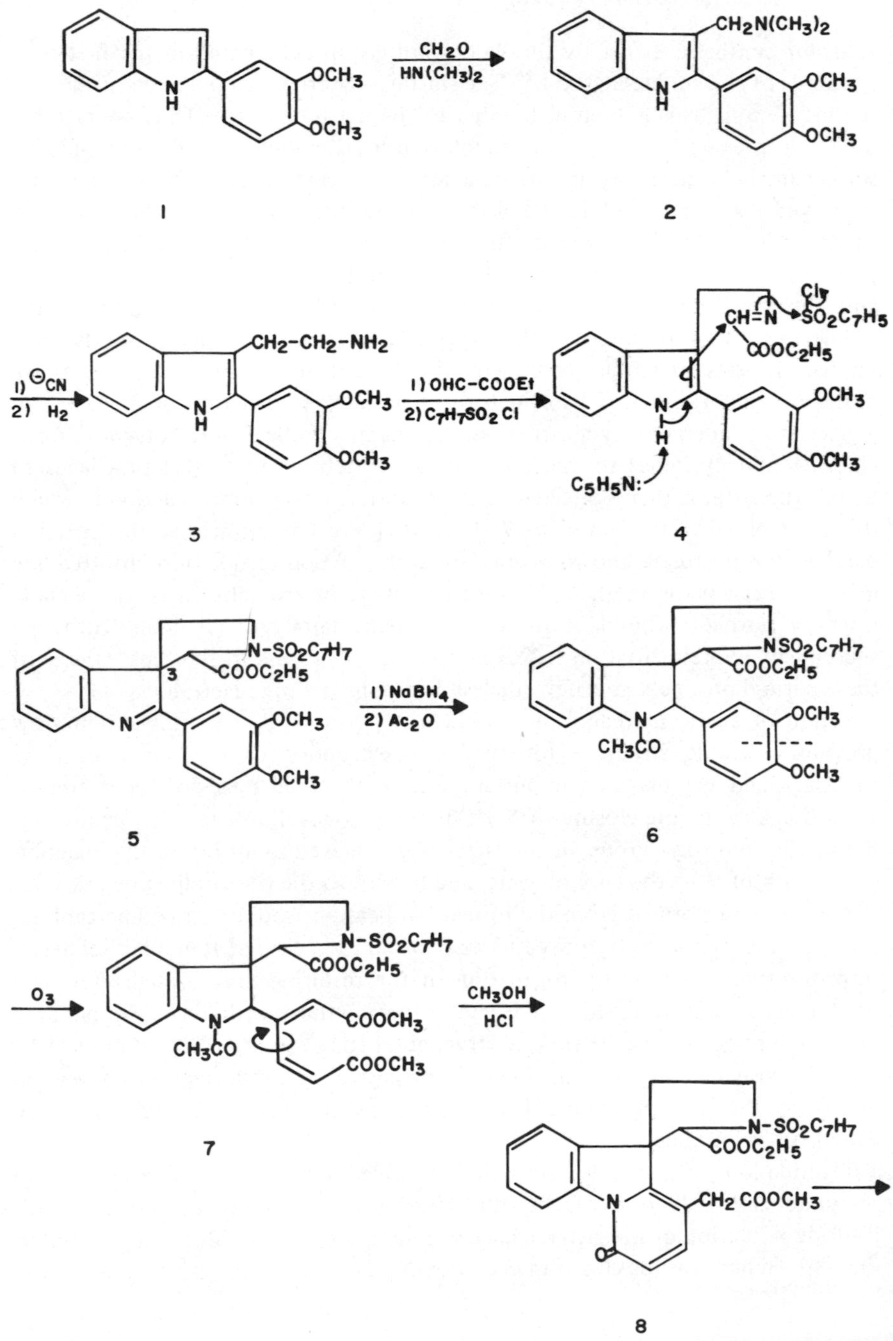

CH_2O, $HN(CH_3)_2$

1 → 2

1) $^{\ominus}CN$, 2) H_2 → 3

1) OHC–COOEt, 2) $C_7H_7SO_2Cl$ → 4

5 → 1) $NaBH_4$, 2) Ac_2O → 6

O_3 → 7

CH_3OH, HCl → 8

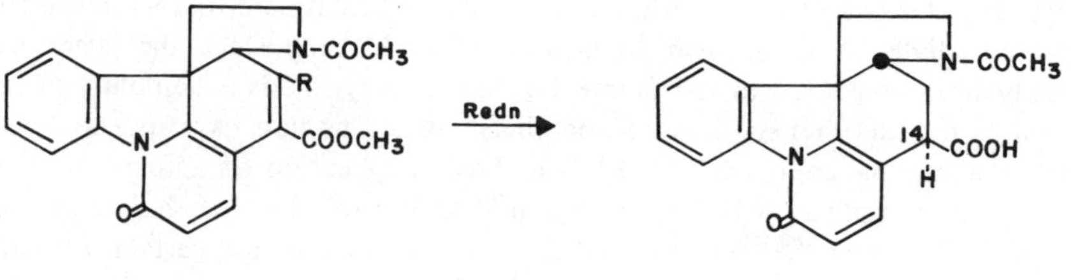

N-COCH3
R
COOCH3
Redn
N-COCH3
14
COOH
H

9, R = OH
9a, R = SC7H7

10

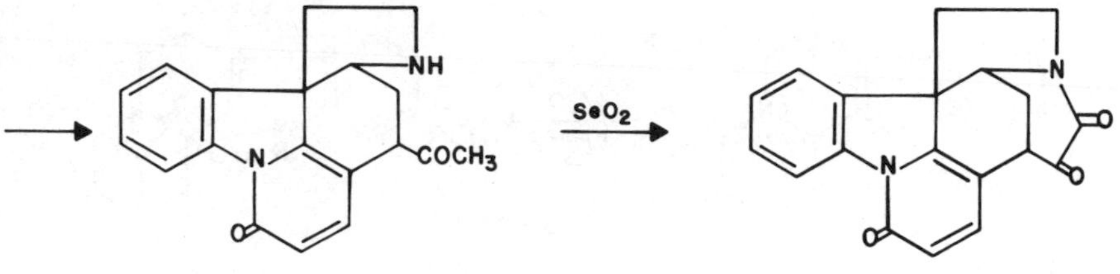

NH
COCH3
SeO2
N

11

12

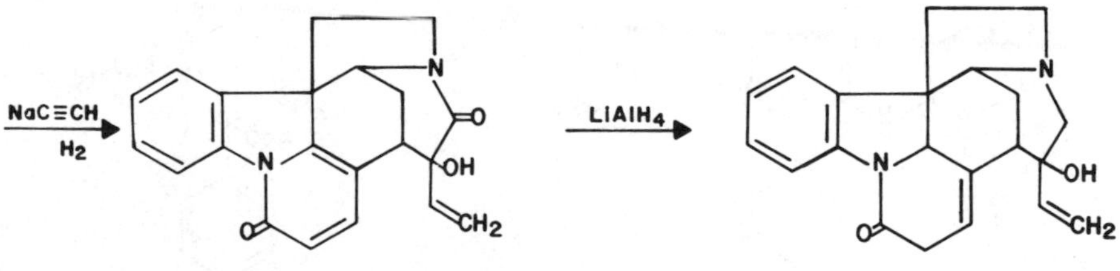

NaC≡CH
H2
N
OH
CH2
LiAlH4
N
OH
CH2

13

14

HBr
N
CH2
OH
N
O

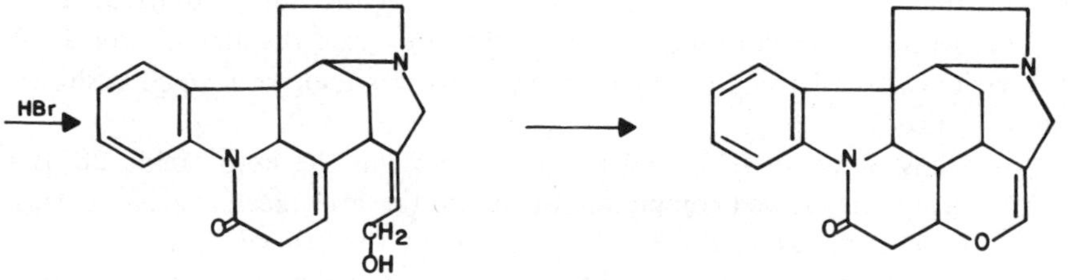

15

16

Obviously this is a rather remarkable reaction. Acid treatment of **14** allowed the required allylic rearrangement to isostrychnine **15**, and since the latter has already been converted to strychnine **16**, the total synthesis is complete. Since that time several other syntheses of members within this class have appeared.

In general, the approach utilized is a direct application of a fragmentation reaction first applied by Dolby and Sakai[167] in the tetrahydrocarboline system and then by Freter.[168,169] Dolby and Sakai demonstrated that certain dihydrocorynantheine derivatives undergo C-D ring cleavage by the action of hot acetic anhydride containing sodium acetate (**17** → **18**). Freter[168] also observed that treatment of 1-phenyl-2-methyltetrahydro-β-carboline **19** with acetic anhydride gave the 1,2-seco compound **20**. Subsequent investigations[169] in his laboratory demonstrated the generality of this approach.

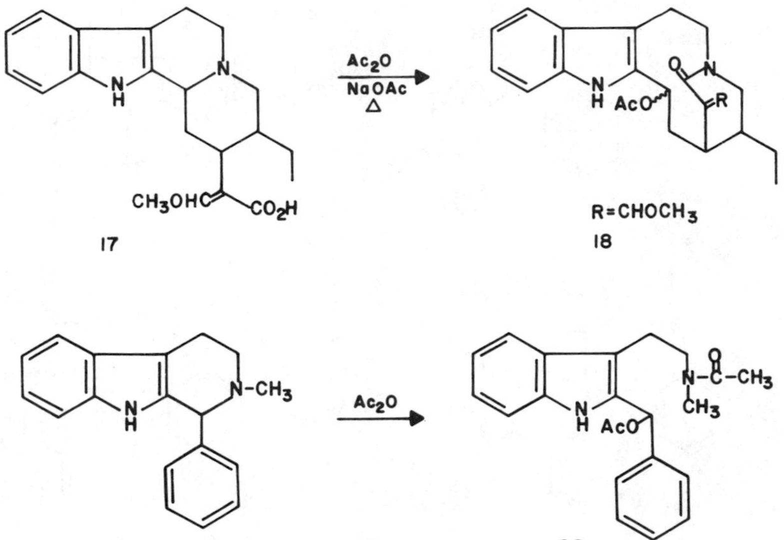

$CH_3OHC{=}C{-}CO_2H$ **17** $\xrightarrow[\Delta]{Ac_2O,\ NaOAc}$ **18** ($R = CHOCH_3$)

19 $\xrightarrow{Ac_2O}$ **20**

Harley-Mason has employed this reaction in providing syntheses of several members of the *Strychnos* family. In the synthesis[170] of racemic tubifoline **25**, tubifolidine **27**, and condyfoline **26**, the hexahydroindoloindolizine **21** was reacted with α,α′-dichlorobutyric anhydride to provide the amide ester **22**. The conversion of the latter to the natural systems was then performed as shown in the sequence.

An extension of these investigations, employing the keto amide **28** as the starting material, allowed completion of the synthesis of racemic geissoschizoline **32**.[171] Here again as in the previous synthesis, the crucial cyclization of the medium-sized ring intermediate (**30**) was accomplished by a procedure previously developed by Schumann and Schmid.[172]

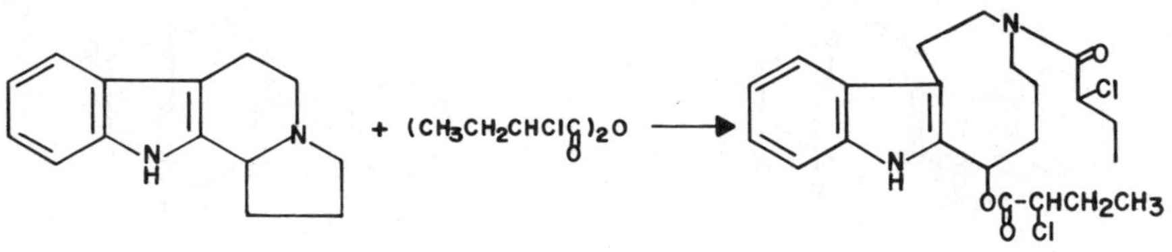

+ $(CH_3CH_2CHClC(O))_2O$ →

21 22

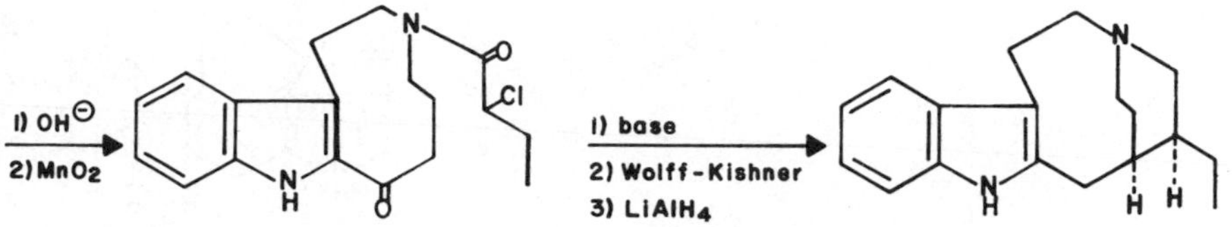

1) $OH^{\ominus}$
2) MnO_2

23

1) base
2) Wolff-Kishner
3) $LiAlH_4$

24

[O]

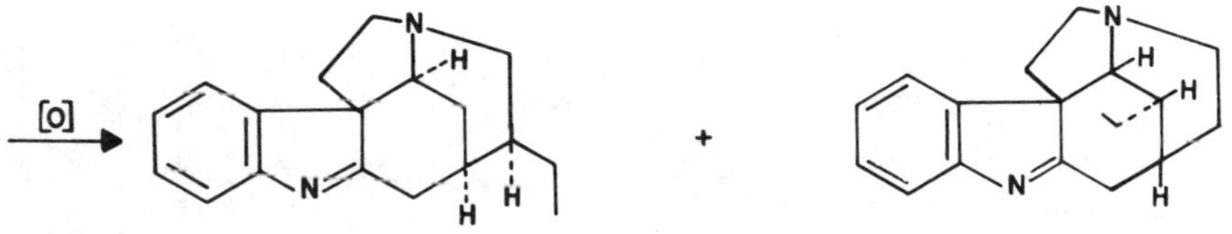

25 + 26

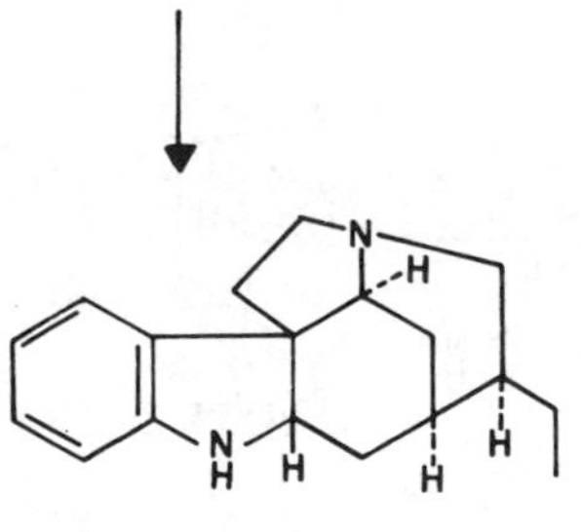

27

Transannular cyclization of the tetracyclic ester amide **30** provided a low yield of racemic tubotaiwine **33**.[173] An improved synthesis of geissoschizoline as well as a synthesis of racemic dihydronorfluorocurarine **37** was subsequently reported by Harley-Mason and Taylor.[174] The approach is identical to those discussed previously. The starting keto amide **28** is converted via routine reactions

28 → 1) $NaBH_4$ 2) Ac_2O → 29

29 → 1) NaCN, DMSO 2) H_2SO_4, CH_3OH → 30

1) $LiAlH_4$ 2) Pt, O_2 → 31 → B_2H_6 → 32

30 → 1) $POCl_3$ 2) base → 33

to the *Strychnos* derivative **37**. Finally, a synthesis of racemic fluorocurarine iodide **41**, a Calabash curare alkaloid, was completed.[175] The authors simply apply the previous reactions with appropriate modifications. The tetracyclic indole derivative **21**, employed earlier, was reacted with the anhydride of 2-bromo-3-methoxybutyric acid, and the resultant product **38** now bears the necessary functionality for subsequent ring closure and elimination to provide the essential ethylidene side chain. The remainder of the sequence is straightforward.

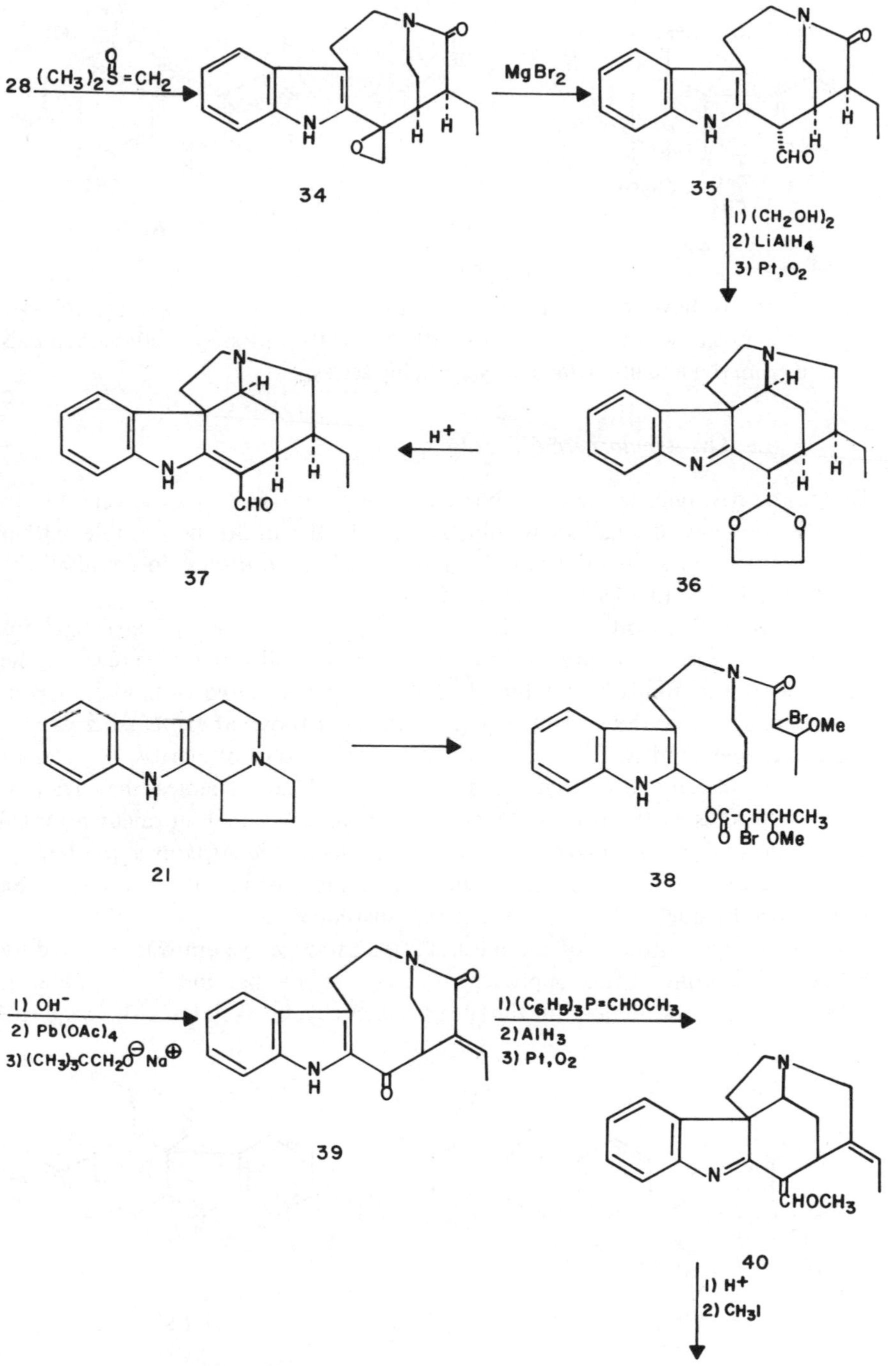

28 $(CH_3)_2\overset{O}{S}=CH_2$ → 34

34 $MgBr_2$ → 35

35 1) $(CH_2OH)_2$ 2) $LiAlH_4$ 3) Pt, O_2 → 36

36 H^+ → 37

CHO

21 → 38

OC-CHCHCH3 / O Br OMe; Br OMe

1) OH^- 2) $Pb(OAc)_4$ 3) $(CH_3)_3CCH_2O^{\ominus}Na^{\oplus}$ → 39

39 1) $(C_6H_5)_3P=CHOCH_3$ 2) AlH_3 3) Pt, O_2 → 40

$CHOCH_3$

40 1) H^+ 2) CH_3I →

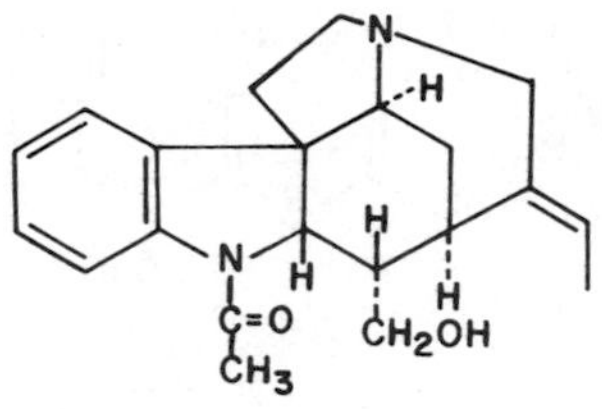

Two groups have reported syntheses of the *Strychnos* alkaloid, retuline 42.[176,177] Both synthetic endeavors started with optically active materials derived from the akuammicine and strychnine series.

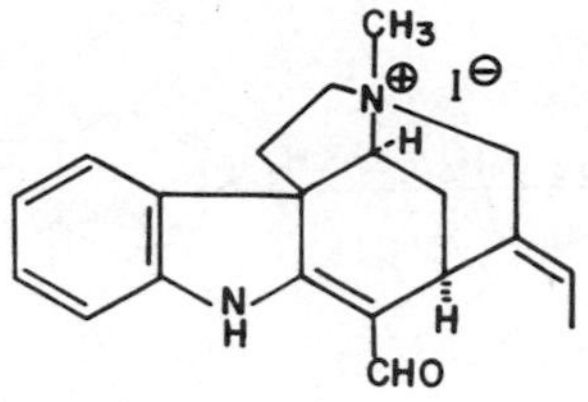

L. The *Aspidosperma* Family

During the past decade there has been an enormous effort by various groups to achieve syntheses of alkaloids within this family. It will not be possible, within the limitations in scope and length of this chapter, to discuss in detail all the synthetic schemes that have been published.

Along with the synthetic endeavors, an extensive series of investigations directed at the isolation and structure elucidation of alkaloids have revealed the presence of the fundamental template normally associated with the *Aspidosperma* family in various plant genera other than those of *Aspidosperma* (for example, *Vinca* and *Kopsia*). Consequently for the sake of clarity, the present discussion concerns itself with the various synthetic schemes that lead to alkaloids possessing the rearranged nontryptophan C_{10} unit **1**, normally regarded as fundamental in the biosynthetic considerations of the *Aspidosperma* family. The two main skeleta contained in the large number of natural systems can be exemplified by quebrachamine 2 and aspidospermidine 3.

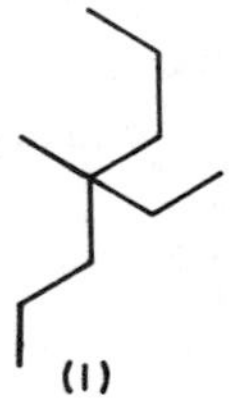

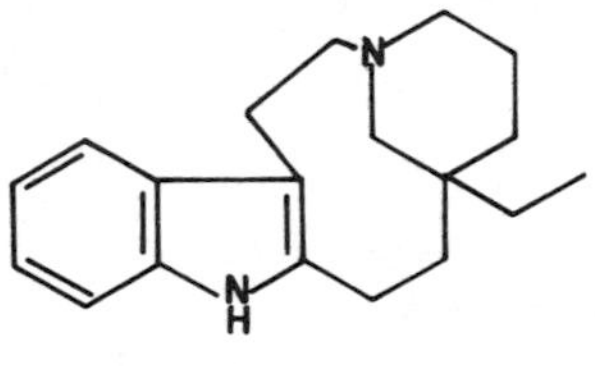

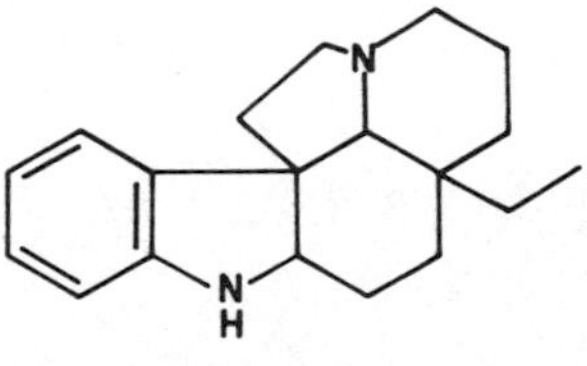

The first total synthesis of the natural *Aspidosperma* system was reported by Stork and Dolfini.[9] This approach utilized the Fischer indole synthesis in achieving the construction of the desired pentacyclic skeleton. The requisite tricyclic intermediate **11** was obtained by effectively employing the pyrrolidine enamine reaction, also developed previously in Stork's laboratory, in the initial stages of the sequence.

The stereochemistry of the various bicyclic and tricyclic intermediates utilized was left undefined since the authors felt that the indolenine **12** being formed under equilibrating conditions would lead to equilibration at the two asterisked

$CH_3CH_2CH_2CHO$ (4) → 1) Pyrrolidine 2) $CH_2{=}CHCO_2CH_3$ 3) HOAc → (5) → 1) Pyrrolidine 2) $CH_3\overset{O}{\overset{\|}{C}}{-}CH{=}CH_2$ 3) HOAc → (6)

1) $(CH_2OH)_2$ 2) NH_4OH → (7) → 1) $LiAlH_4$ 2) H^+, H_2O 3) base → (8) → $ClCH_2\overset{O}{\overset{\|}{C}}{-}Cl$ →

(9) → $\overset{\oplus}{K}\ \overset{\ominus}{O}C(CH_3)_3$ → (10) → 1) $(CH_2OH)_2$ 2) $LiAlH_4$ 3) H^+ → (11)

→ o-methoxyphenylhydrazine, HOAc → (12) → 1) $LiAlH_4$ 2) Ac_2O → (13)

centers via a reverse Mannich reaction. There was precedent in the literature for this situation.[178] Since the conformational expression for aspidospermine **13** denotes the most stable arrangement at the various asymmetric centers, such an equilibration process would be expected to lead eventually to the desired natural stereochemistry. The racemic substance **13** thus obtained was identical with the natural alkaloid.

Fischer indole cyclization of **11** employing phenylhydrazine allowed extension of the synthesis to the quebrachamine system **2**. The reductive cleavage with potassium borohydride had been previously applied by Biemann and Spiteller[179] in a structural elucidation study.

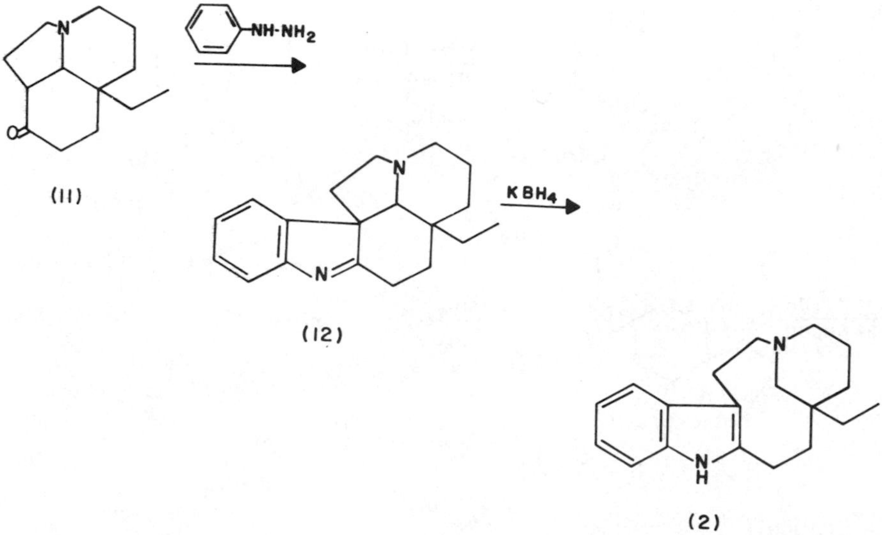

NH-NH$_2$

(11)

KBH$_4$

(12)

(2)

Another total synthesis of aspidospermine was reported by Ban and co-workers.[10] An interesting situation developed during this study, since during their planned objective in similarly utilizing the Fischer indole synthesis in the later steps of the synthesis, these workers developed a different pathway to a bicyclic intermediate **19** identical in gross structure to **8**. However, the physical and chemical data for the compound available from Ban's experiments clearly revealed that this intermediate was different from that obtained by Stork.

Conversion of **19** to **22** and the latter to d,l-aspidospermine **13** in exactly the manner outlined previously in the Stork synthesis (**8** → **13**) provided conclusive evidence that the respective sequences yielded a series of intermediates that were diastereoisomeric. A detailed spectroscopic study allowed the assignment of the conformational structures **10a** and **21a** to the tricyclic keto amides **10** and **21** in the respective studies. The conversion of both **11** and **22** to d,l-aspidospermine supports Stork's contention that equilibration occurs during the Fischer indole

$CH_3CH_2CH_2C(=O)-CH_3$ $\xrightarrow{CH_2=CH-CN}$ (16) $\xrightarrow{H_2SO_4}$ (17)

(15) (16) (17)

17 $\xrightarrow{Pd/H_2, OH^-;\ LiAlH_4}$ (18) $\xrightarrow{K^{\oplus}\ {}^{\ominus}OC(CH_3)_3,\ \text{Cyclohexanone}}$ (19)

17 $\xrightarrow{1)\ H_2,\ Ni;\ 2)\ LiAlH_4}$ (20) $\xrightarrow{K^{\oplus}\ {}^{\ominus}OC(CH_3)_3,\ \text{Cyclohexanone}}$ (19)

(18) (20) (19)

19 ⟶ (20a) ⟶ (21) ⟶ (22) ⟶ 13

(20a) (21) (22)

(10a) (21a)

cyclization; therefore, the stereochemical centers in the various intermediates are of no consequence in determining the end product.

In more recent studies by Ban and Iijima,[180] the tricyclic ketones **21** and **22** have been assigned the conformational structures **21b** and **22a**, whereas Stork's tricyclic ketone **11** has been revised to **11a**. Although the latter apparently bears the necessary stereochemistry for conversion to the natural alkaloid, Ban's ketone (**22** and **22a**) differs from the required stereochemistry. Consequently, a further investigation of the Fischer indole synthesis with **22** was pursued by Ban. These studies provided the synthesis of a racemic stereoisomer of aspidospermine **23**.

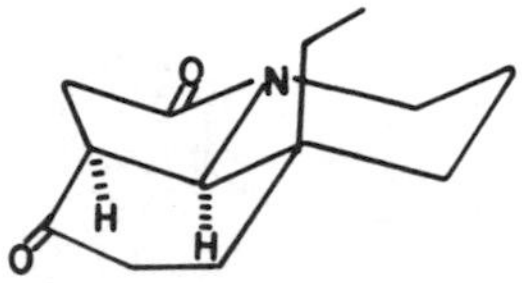

(21b) 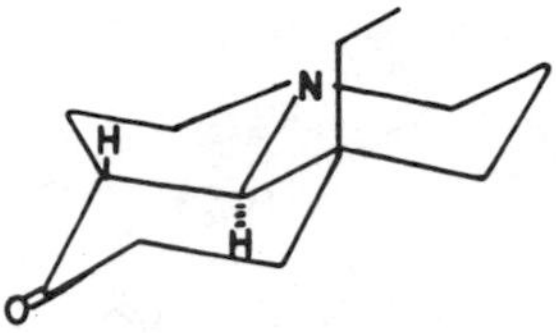 (22a) 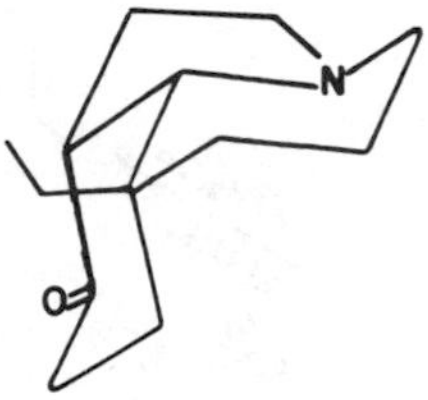 (11a)

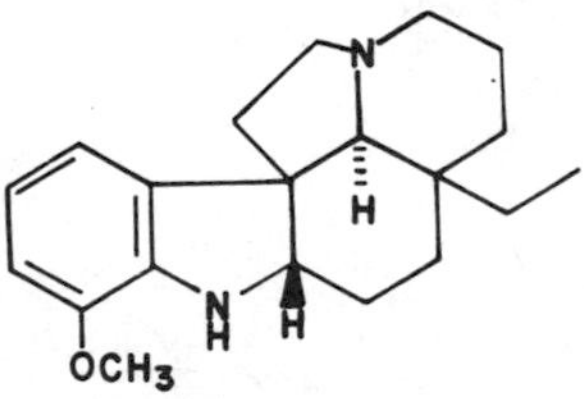

(23)

Another study by Kuehne and Bayha[181] has also provided the tricyclic intermediate of gross structure **11** but possessing the stereochemistry portrayed in **11a** and **30**. The sequence starts with proline ethyl ester **24** and proceeds through the steps shown.

A totally different approach to the synthesis of *Aspidosperma* and related alkaloids was undertaken in our laboratory. Emphasis was placed on generality and versatility whereby it was hoped that subsequent modifications at appropriate stages of the pathway could lead to the synthesis of a large number of natural systems. It was also felt desirable to consider reactions that may have significance in biosynthesis, that is, in the *in vivo* construction of these alkaloids by the living plant systems. For this purpose, we selected the possible utilization of a transannular cyclization reaction of appropriate nine-membered ring inter-

$CH_3CH_2O_2C$ (24) $\xrightarrow{Br(CH_2)_3CO_2CH_2CH_3}$ $CH_3CH_2O_2C$, $(CH_2)_3$, $CO_2CH_2CH_3$ (25) $\xrightarrow{1) NaOH;\ 2) hydrolysis;\ 3) -CO_2}$ (26)

$\xrightarrow{(C_6H_5)_3P{=}CHCH_3}$ (27) $\xrightarrow{1) H_2, Pd;\ 2) Hg(OAc)_2}$ (28) $\xrightarrow{CH_2{=}CH{-}CO_2CH_3}$

(29) $\xrightarrow{LiAlH_4}$ IIa

(29) $\xrightarrow{1) H_2, Pt;\ 2) (i\text{-}PrO)_3Al,\ Acetone}$ (30)

(31) $\xrightarrow{Hg(OAc)_2}$ (32)

(32) $\xrightarrow{HOAc}$ (33)

(33) $\xrightarrow{LiAlH_4}$ (34)

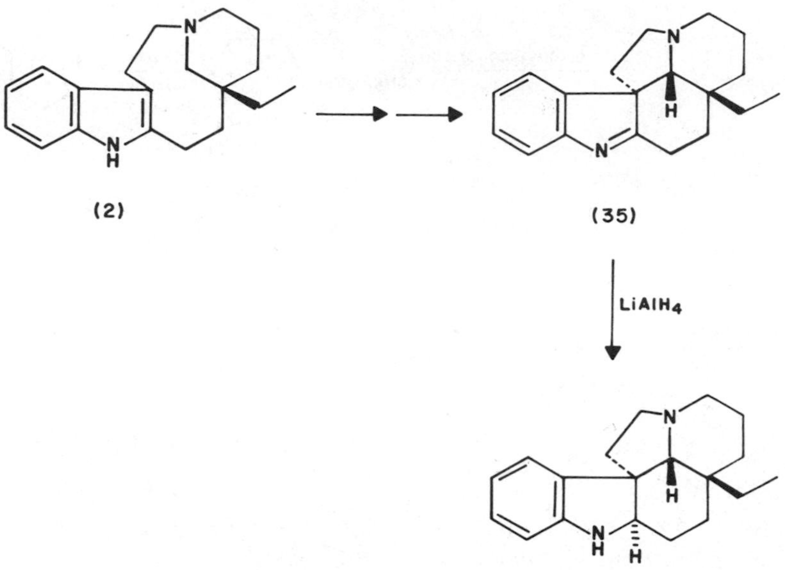

(2)

(35)

LiAlH$_4$

(36)

mediates, since it seemed to satisfy most of our requirements. This study[182-186] revealed the general usefulness of such an approach for the synthesis of a variety of alkaloids. Thus nine-membered ring systems of the cleavamine **31** and quebrachamine **2** series were shown to cyclize in a completely stereospecific manner to generate the necessary asymmetry for the natural systems (**34** and **36**). Utilization of the ester derivatives (**37** and **42**) of the foregoing systems then provided an alteration in the course of the cyclization process and a direct entry into the *Aspidosperma* (**39** and **43**) and *Iboga* (**41**) alkaloids.

The crucial role occupied by the appropriate nine-membered ring intermediates for the total syntheses of a variety of these alkaloids demanded an investigation into their synthesis. The approach selected involved the generation of the medium-sized ring system by means of a reductive cleavage in the last step of the synthesis. A preliminary study in our laboratory had revealed the generality of this reaction (**44** → **45**).

A new total synthesis of d,l-quebrachamine **2** and d,l-aspidospermidine **36** was accomplished by means of this scheme.[187,188] The total synthesis of d,l-quebrachamine **2** also completed the synthesis of d,l-aspidospermidine **36** in view of the previously established transannular cyclization of the former to the latter (see **2** → **35** → **36**). Although the yields in these conversions were generally satisfactory, one of the reactions involving mercuric acetate in the oxidative cycliza-

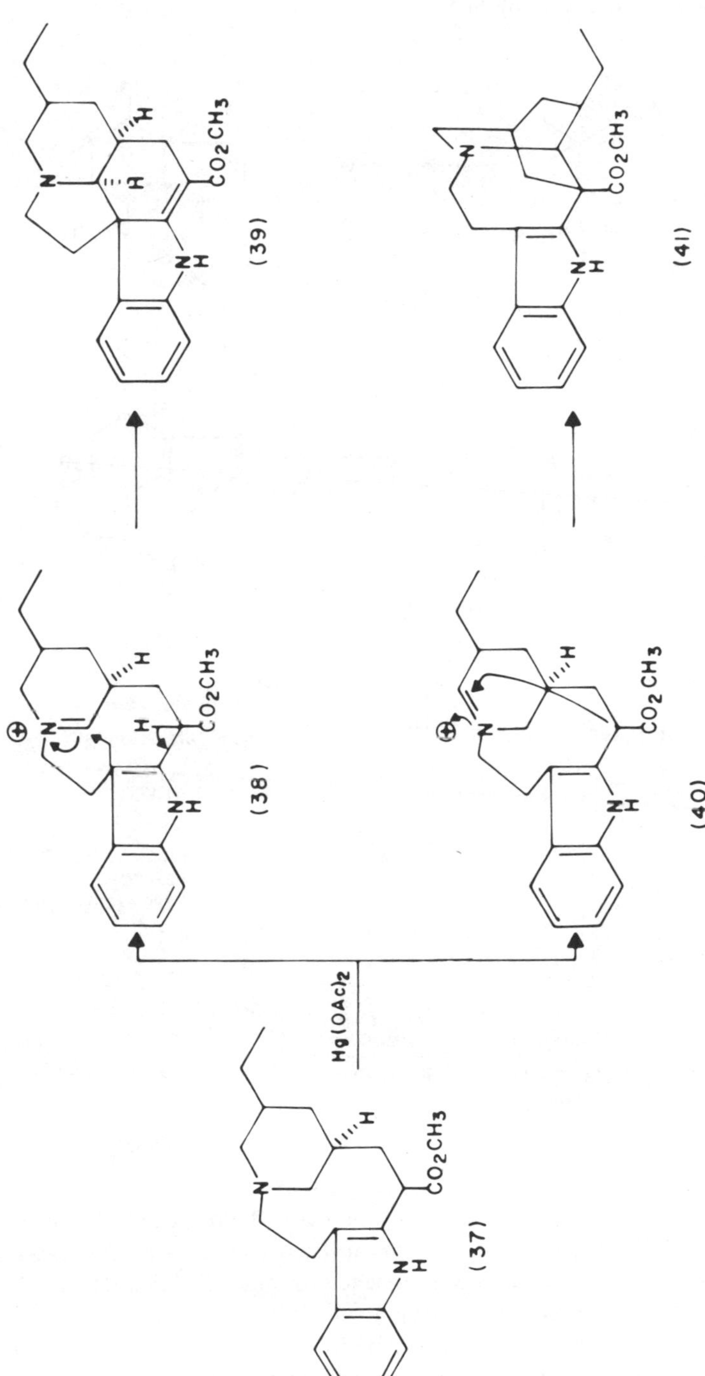

Hg(OAc)2
(37)
(38)
(39)
(40)
(41)
CO2CH3
NH

(42) (43)

(44) (45)

CH_3CH_2I, NaOEt

(46) $C_6H_5CH_2O(CH_2)_3CH(CO_2Et)_2$ → (47) $C_6H_5CH_2O(CH_2)_3C(CH_2CH_3)(CO_2Et)_2$

1) OH^-
2) $-CO_2$
3) CH_3CH_2OH, H^+

(48) $C_6H_5CH_2O(CH_2)_3CH(CH_2CH_3)CO_2Et$

$(C_6H_5)_3C^{\ominus}\ Na^{\oplus}$, $BrCH_2CO_2Et$

(49) $C_6H_5CH_2O-(CH_2)_3C(CH_2CH_3)(CH_2CO_2Et)-CO_2Et$

tion of the tricyclic amine to the tetracyclic base **51** proceeded in only 30 to 40% of the yield. Subsequent modifications in the sequence[12] obviated this difficulty and provided a vastly improved sequence to these alkaloids. The modified scheme involves condensation of tryptamine with the aldehydo ester **55**, whereupon the more reactive aldehyde function directs the cyclization in only one manner. In this way, the desired tetracyclic lactam **56** is formed in 90% of the yield.

Tryptamine →

$CH_2OCH_2C_6H_5$

(50)

(2)

1) $LiAlH_4$
2) $Hg(OAc)_2$
3) $NaBH_4$

Na, NH_3

$CH_2OCH_2C_6H_5$

1) BBr_3
2) CH_3SO_2Cl (pyridine)

$\ominus$ OMes

(51)

(52)

$(C_6H_5)_3C^{\ominus}Na^{\oplus}$

$CH_2{=}CH{-}CH_2Br$

$C_6H_5CH_2O(CH_2)_3{-}CH(CH_2CH_3){-}CO_2Et$

(48)

$C_6H_5CH_2O(CH_2)_3{-}C(CH_2CH_3)(CH_2CH{=}CH_2){-}CO_2Et$

(54)

OsO_4
$NaIO_4$

Tryptamine

$CH_2OCH_2C_6H_5$

(56)

$C_6H_5CH_2O(CH_2)_3{-}C(CH_2CH)(CH_2CHO){-}CO_2Et$

(55)

$\xrightarrow{LiAlH_4}$ 51 → 52 → 2

1) H_2, Pd
2) CH_3SO_2Cl
3) $(CH_3CH_2)_3N$

The versatility of this approach to the synthesis of various alkaloids bearing the *Aspidosperma* template but isolated from *Vinca* plants was now demonstrated. The quaternary mesylate **52** was exposed to nucleophilic displacement by cyanide ion with the resulting generation of the nine-membered ring system possessing the nitrile function at C_3 (**57** and **58**). Subsequent conversion of this function to the ester group, present in many alkaloids, opened a synthetic entry to all of these systems. During reaction at C_3, both possible stereoisomers at this site were obtained. To illustrate the stereochemical consequences in these conversions, the most plausible conformational expressions for these compounds are depicted (**57-60**). The synthetic compounds **59** and **60** represent d,l-vincadine and d,l-epivincadine, respectively. The latter compound was also synthesized from the natural alkaloid **59** by isomerization of the ester function. The synthesis of d,l-vincadine also completes the total synthesis of d,l-vincaminoreine **59** (R = CH_3) and d,l-vincaminorine **60** (R = CH_3) in view of the already known interconversions.[189]

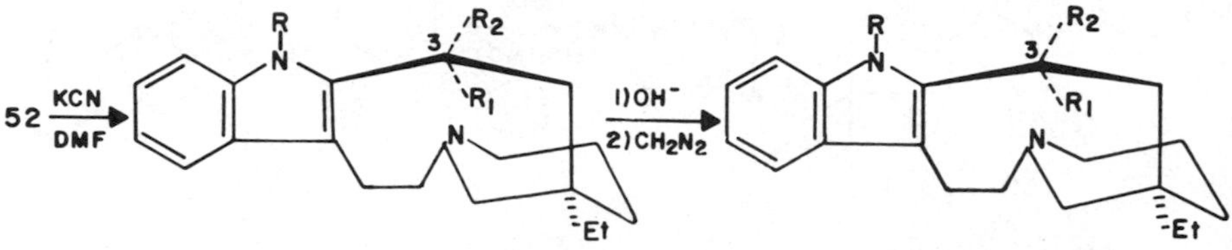

52 —KCN, DMF→ 57/58 —1) OH^-, 2) CH_2N_2→ 59/60

57, R = R_2 = H ; R_1 = CN
58, R = R_1 = H ; R_2 = CN

59, R = R_2 = H ; R_1 = CO_2CH_3
60, R = R_1 = H ; R_2 = CO_2CH_3

The previously established cyclization now made possible the total synthesis of the pentacylcic series exemplified by vincadifformine **61** and minovine **62** from vincadine and vincaminoreine, respectively. There are a number of alkaloids that possess oxygen functions, particularly methoxyl groups, in the aromatic ring, and it was clear that this approach could be expanded to encompass these systems as well. Indeed, reaction of 6-methoxytryptamine **63** with the aldehydo ester **55** provided a 70% yield of the tetracyclic lactam **64**. Further elaboration of this intermediate in the manner already discussed allowed the total synthesis

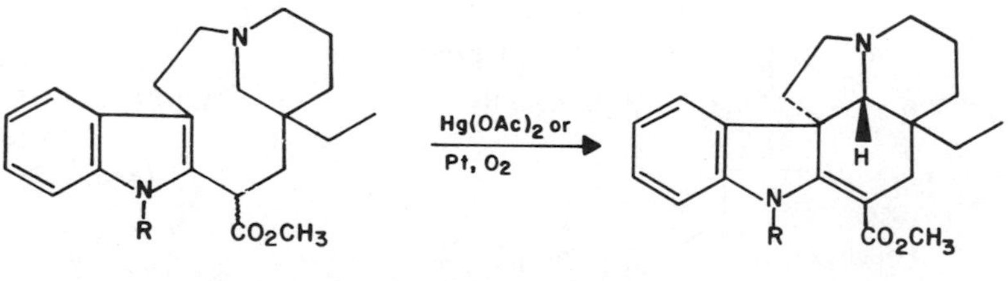

—$Hg(OAc)_2$ or Pt, O_2→

61, R = H
62, R = CH_3

of 16-methoxy vincadine **65** and 16-methoxyepivincadine **66**. These latter substances possess the skeletal features of vincaminoridine, an alkaloid isolated from *Vinca minor* L, but no stereochemical assignment is available from the published data nor could an authentic sample be obtained.

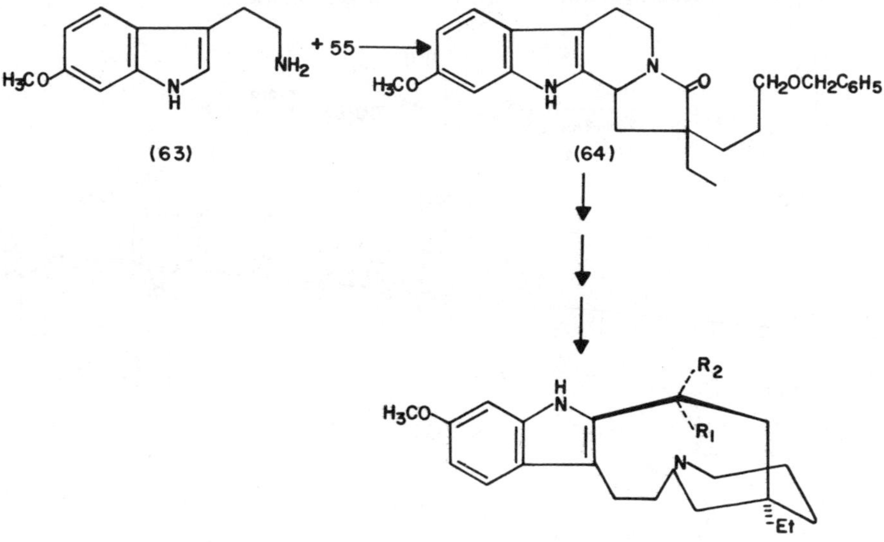

(63) + 55 → (64)

65, $R_1 = COOCH_3$, $R_2 = H$

66, $R_1 = H$, $R_2 = COOCH_3$

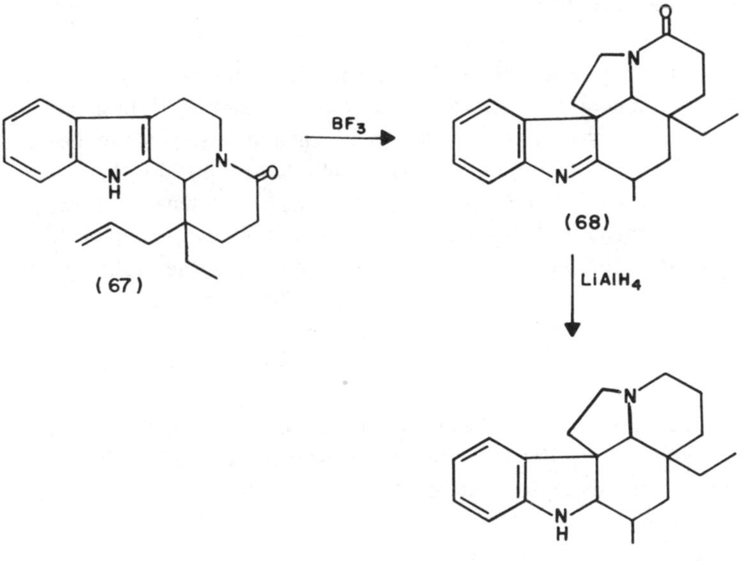

(67) —BF_3→ (68) —$LiAlH_4$→ (69)

An interesting rearrangement reaction has been employed by Harley-Mason to complete a total synthesis of d,l-aspidospermidine **36**. In the initial study,[35] the tetracyclic lactam **67** was shown to rearrange to an *Aspidosperma* skeleton **68**, which upon simple reduction provided 3-methylaspidospermidine **69**. An extension of this work led to synthesis of the natural system **36**.[190]

CH(OCH$_3$)$_2$ CO$_2$CH$_3$ 1) O$_3$ 2) NaBH$_4$ CH(OCH$_3$)$_2$ CO$_2$CH$_3$ Tryptamine HOAc

(70) (71)

(72)

BF$_3$ or H$_2$SO$_4$ LiAlH$_4$

(73) (36)

Still another approach to quebrachamine and the pentacyclic *Aspidosperma* alkaloids was undertaken by Ziegler, Kloek, and Zoretic.[191] In one of the crucial steps of the synthesis, the alkylation of an appropriate enamine, 1-benzyl-3-ethyl-1,4,5,6-tetrahydropyridine **73a**, is invoked. Thus in the quebrachamine synthesis,[191] this enamine, prepared according to the scheme shown, undergoes alkylation with methyl haloacetates to provide **74**, which upon debenzylation and reaction with β-indolylacetyl chloride provided the lactam ester **75**. The cyclization of **75** proceeds in high yield (85%) to **76**, but unfortunately conversion of the latter to quebrachamine **2** was a very low yielding process. The double bond in the accompanying product, 3,4-dehydroquebrachamine **77** was very unreactive to reduction.

The same basic approach has been employed in the synthesis of d,l-minovine **62**.[192] Alkylation of **73a** with the acrylic ester derivative **78** allows the resultant iminium intermediate to undergo facile cyclization at the β-position of the indole nucleus, and the tetracyclic indole derivative **79** is obtained. The two-carbon bridge necessary to complete the synthesis was introduced by means of ethylene dibromide. The synthesis of 6,7-dehydroquebrachamine 83 has also been reported by Ziegler and Bennett.[193]

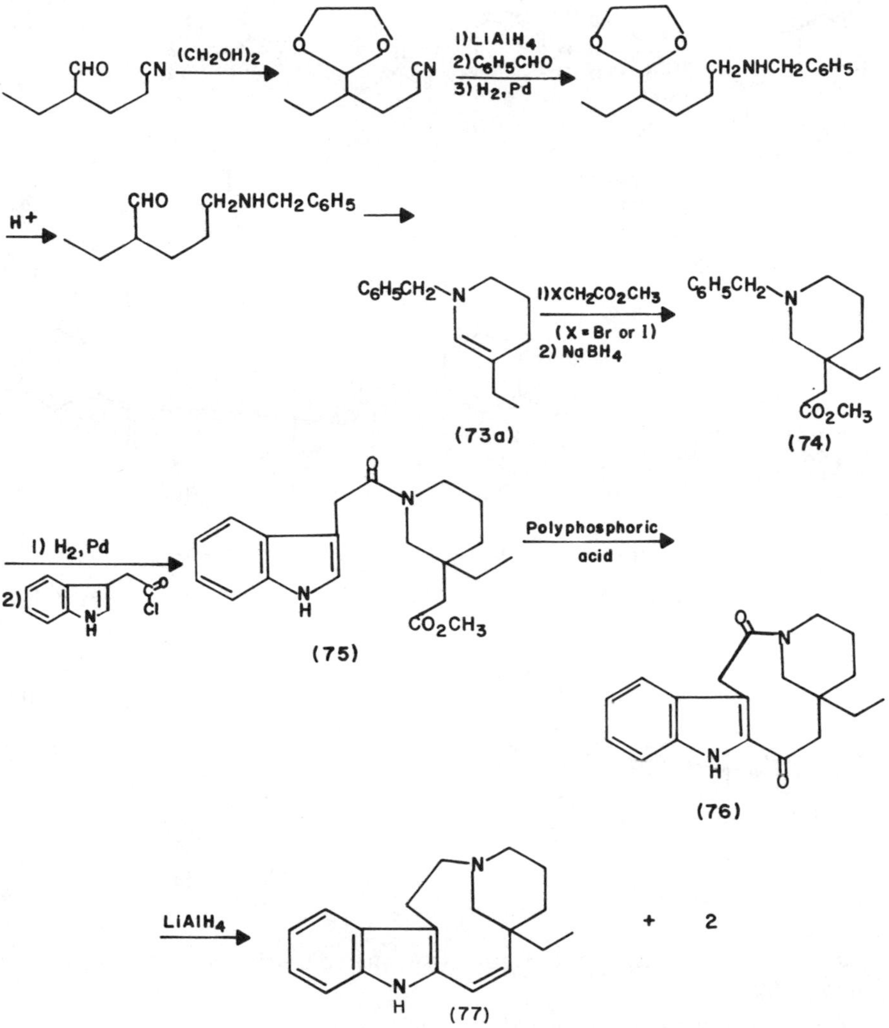

A related approach to the *Aspidosperma* alkaloid nucleus has been investigated in a collaborative study by a French group and Wenkert.[194] Here again, the resulting iminium intermediate generated during the catalytic reduction of the pyridine ring undergoes cyclization at the β-position of the indole nucleus to provide a tetracyclic intermediate 86 similar in structure to that obtained by Ziegler. Introduction of the two-carbon side chain 87 was also performed in a

1) base 2) $(CO_2CH_2CH_3)_2$ 3) OH^-

1) CH_2N_2 2) $(C_6H_5)_3P=CH_2$

(78)

$C_6H_5CH_2N$ (73a) CH_3OH, Δ

1) $LiAlH_4$ 2) H_2, Pd

79, $R_1 = CO_2CH_3$; $R_2 = H$

80, $R_1 = H$, $R = CO_2CH_3$

81, $R_1 = CH_2OH$; $R_2 = H$

82, $R_1 = H$, $R_2 = CH_2OH$

$BrCH_2CH_2Br$ DMF, Na_2CO_3 (on 82)

(62)

(83)

manner similar to Ziegler's. Unfortunately, the conversion **86** → **87** proceeds in only a 10% yield.

A synthesis of d,l-tabersonine **91** by Ziegler and Bennett[195] utilizes, in essence, his previously developed scheme for the construction of the required tetracyclic amino alcohol **88**. The later stages of the synthesis are essentially a repetition of the chemistry previously involved in our synthesis of the monomeric *Vinca* alkaloids, for example, minovine **62** and vincadifformine **61**[188,12,196] (compare the conversion **89** → **90** → **91** with **52** → **57**, **58** → **59**, **60** → **61**,**62**).

Stevens et al.[197] have reported a fundamentally different approach that involves the acid-catalyzed thermal rearrangement of a cyclopropyl imine **92** to an appropriately substituted 2-pyrroline **93** as a crucial stage. This intermediate was then converted according to the scheme shown to the hydrolulolidine **97**,

(84)

1) $NaBH_4$
2) hydrolysis

(85)

H_2, Pt
CH_3OH, HCl

(86)

1) $BrCH_2CH_2Br$
2) DMF, $NaHCO_3$
Δ

(87)

(88)

CH_3SO_2Cl

(89)

KCN

(90)

1) base
2) CH_2N_2
3) Pt, O_2

(91)

which had already been established as a synthetic precursor in the approaches employed by Stork and Ban in their syntheses of d,l-aspidospermine.

Inoue and Ban[198] have made an attempt to extend Ban's previous method to the synthesis of C_{21}-oxygenated *Aspidosperma* alkaloids, as for example, limaspermine **98** and haplocine **99**. These investigations, which employ 5-phenoxypentan-2-one **100** as starting material, proceed via an 11-step sequence to d,l-21-phenoxyisopalosine **101**. Unfortunately, this latter intermediate, when treated with hydrobromic acid with the intention of cleaving the ether linkage, undergoes quaternization to **102** thereby suggesting to the authors that the stereochemistry inherent in their synthetic compounds is not compatible with that of the natural alkaloid.

Buchi, Matsumoto, and Nishinura[199] have recently completed the synthesis of vindorosine **103**, a pentacyclic highly oxygenated *Aspidosperma* alkaloid that bears close resemblance to vindoline **104**, the latter being a major alkaloid in *Vinca rosea* Linn and the monomeric template in the oncolytic alkaloid vincaleukoblastine. The sequence starts with **105**, which undergoes an interesting boron trifluoride-catalyzed cyclization to **106**, the latter process involving a different mechanistic pathway than that envisaged in Harley-Mason's previous rearrangements with this reagent.[35,200] Conversion of **106** to the pentacyclic

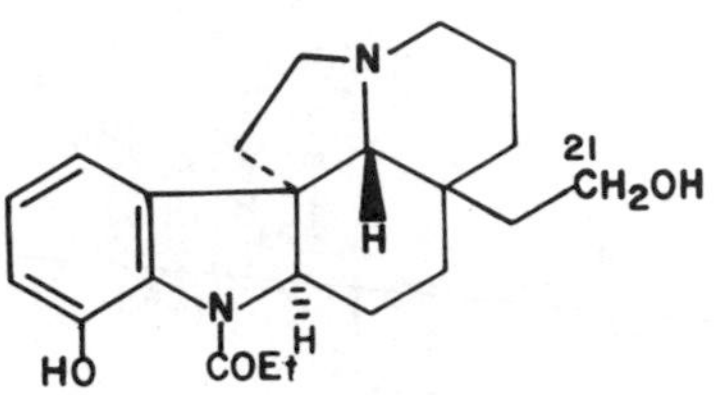

(98)

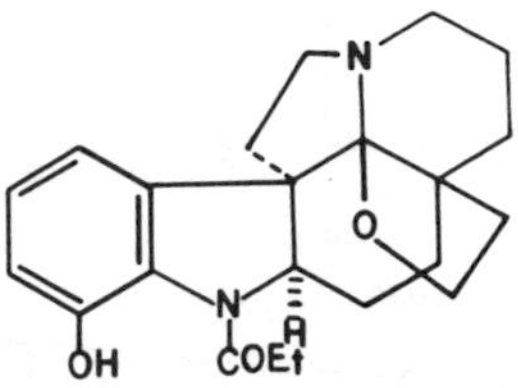

(99)

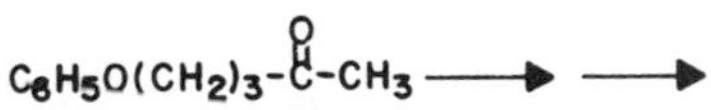

(100)

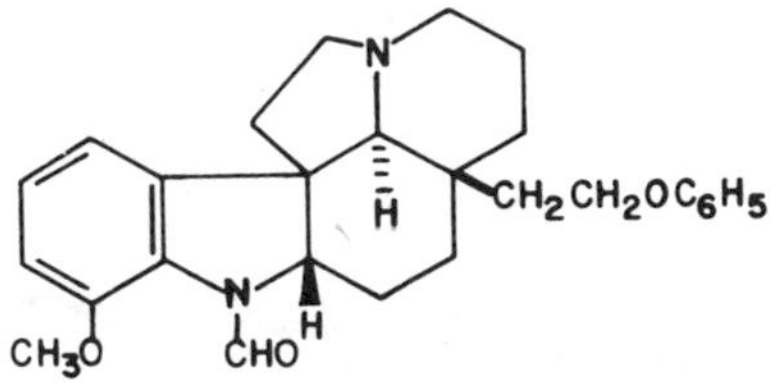

(101)

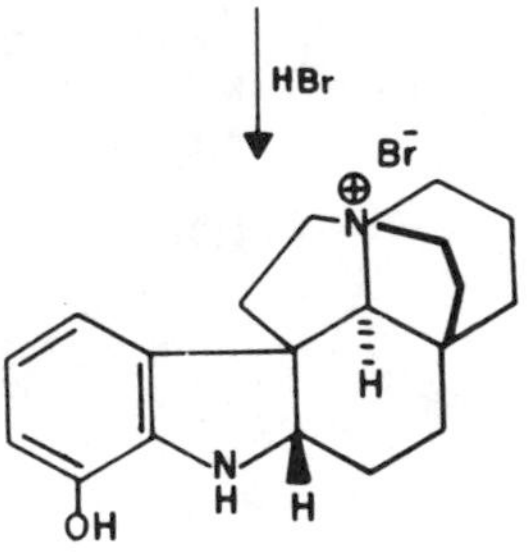

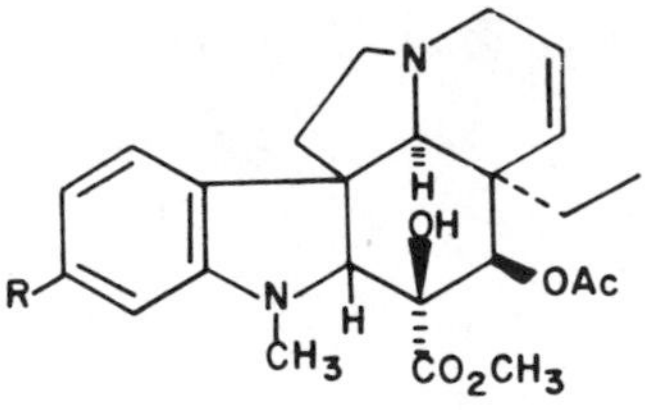

(102)

103, R = H
104, R = OCH_3

systems **107** and **108** follows via appropriate condensations and base-catalyzed alkylation of the resultant conjugated ketone. An interesting hydroxylation of the keto ester **109** is performed by means of an oxygen-hydrogen peroxide treat-

(112) → (113) → (110)

(105) —BF_3→ (106)

(106) —1) HCl; 2) $CH_2=CH-CHO$, $NaOCH_3$; 3) BF_3→ (107)

—CH_3CH_2I, $K^{\oplus}\ {}^{\ominus}OC(CH_3)_3$→ (108) —NaH, $O=C(OCH_3)_2$→ (109)

(109) —$K^{\oplus}\ {}^{\ominus}OC(CH_3)_3$, O_2, H_2O_2→ (110)

(110) —1) $LiAlH_4$, $-70°$→ (111) —Ac_2O→ 103

ment under basic conditions. The authors propose the rationale shown in **112** → **113** → **110**.

There is a series of alkaloids normally grouped under the *Aspidosperma* family, since they occur in *Aspidosperma* species, which do not bear the fundamental structure characterized by the members discussed previously. Examples of these are shown by uleine **114**, dasycarpidone **115**, and apparicine **116**.

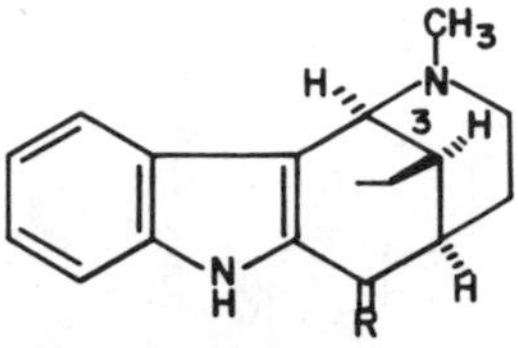

114, R = CH_2
115, R = O

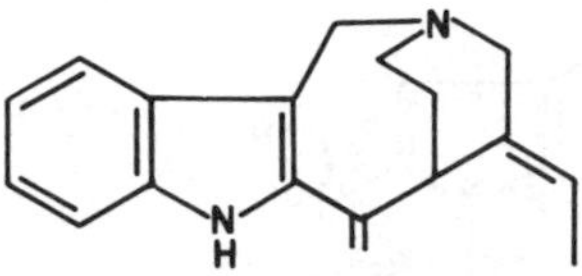

(116)

Several syntheses of these skeletal types are now on hand in the recent literature. Joule and co-workers[201-203] describe syntheses of the racemic forms of dasycarpidone **115**, 3-epidasycarpidone **115** (epimeric ethyl group), uleine **114**, and 3-epiuleine **114** (epimeric ethyl group), according to the indicated scheme.

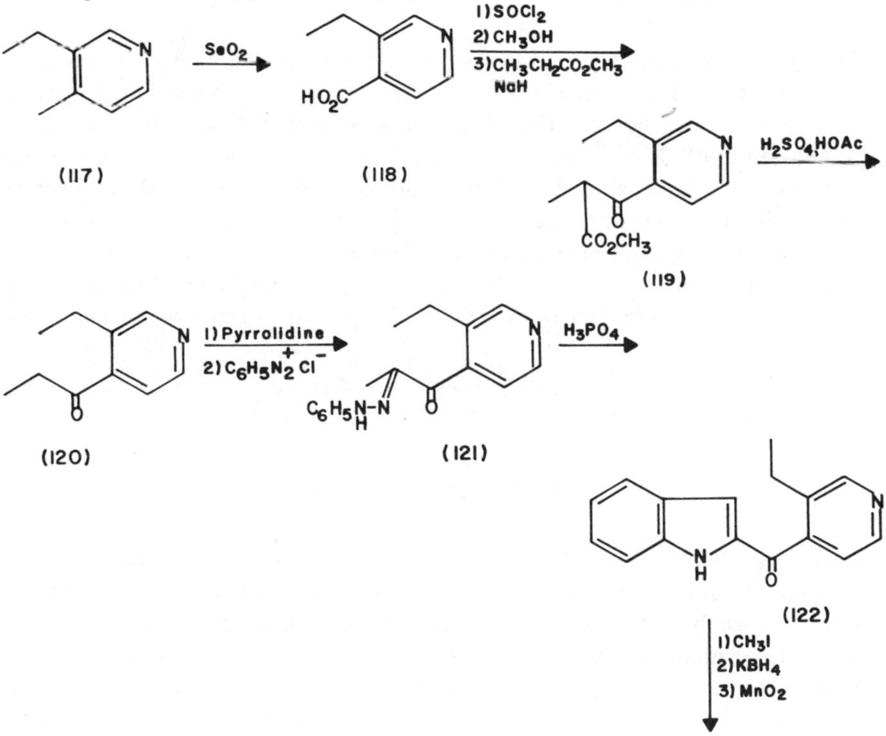

(117) → SeO_2 → (118) → 1) $SOCl_2$; 2) CH_3OH; 3) $CH_3CH_2CO_2CH_3$, NaH → (119) → H_2SO_4, HOAc →

(120) → 1) Pyrrolidine; 2) $C_6H_5\overset{+}{N_2}\ \overset{-}{Cl}$ → (121) → H_3PO_4 → (122)

(122) → 1) CH_3I; 2) KBH_4; 3) MnO_2 →

(124) ← DMSO, Na — (123)

HOAC → 126, R_1=Et; R_2=H

125, R =Et; R =H

epimerizes at C-3

115

3-epi-143

The authors have employed a Fischer indole synthesis to construct the indole system (**121** → **122**) and the latter, upon further elaboration within the pyridine unit, allows conversion to **123**. The crucial step in the scheme involved the isomerization of **123** to the intermediate enamine **124**, which immediately undergoes ring closure to **125**. The latter is shown to proceed to the final systems by reaction with acetic acid. Ring closure at the β-position of the indole nucleus is visualized as proceeding via intermediates such as **126**.

Extension of the foregoing to the uleine series **114** simply involved converting the two ketones (**115** and its 3-epiisomer) into the corresponding methylene

$$115 \xrightarrow{(C_6H_5)_3P = CH_2} 114$$

$$3\text{-epi-}115 \xrightarrow[CH_2I_2]{Mg(Hg)} 3\text{-epi-}114$$

derivatives. Dolby and Biere[204,205] also succeeded in synthesizing these alkaloids at about the same time and have employed an entirely different approach. Dolby and Biere's synthetic scheme takes advantage of the high reactivity of indole systems to electrophilic attack at both the α and β positions. Thus an appropriately substituted piperidine derivative **127** is attached at these positions to the indole itself. Vilsmeier condensation followed by reduction provides **128** and the latter after saponification and ring closure affords **115** and 3-epi-**115**. The conversion to the uleine system **114** also followed a different pathway from that established by Joule.

(127) (128) 1) $POCl_3$ 2) $NaBH_4$; 1) $Ba(OH)_2$ 2) polyphosphoric acid; 115 + 3-epi-115; CH_3Li; (129); H_3PO_4 or Al_2O_3; 3-epi-114

Stereospecific syntheses of uleine **114** and epiuleine (3-epi-**114**) have also been reported by Buchi, Gould, and Naf.[206] The scheme for epiuleine starts with a Mannich condensation of 3-formylindole **130** with 1-aminohexan-3-one **131**. The resulting *trans*-disubstituted piperidone **132** is then elaborated via the acetylenic alcohol **133** to the final product, the crucial cyclization to the required exocyclic olefin **136** being performed by means of boron trifluoride.

(130) + (131) NaOH (132) 1) H–C(=O)–OAc 2) $K^+ \bar{C}{\equiv}CH$

(133) $Hg(OAc)_2$ (134) Li/NH_3

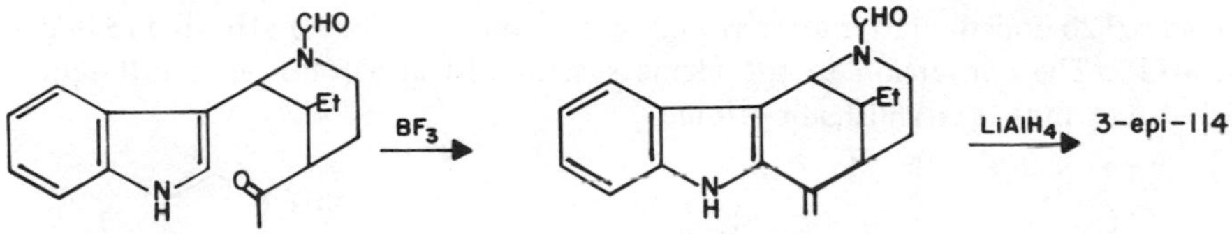

(135) —BF_3→ (136) —$LiAlH_4$→ 3-epi-114

In the uleine sequence, the acetate **134** was chosen as the starting material.

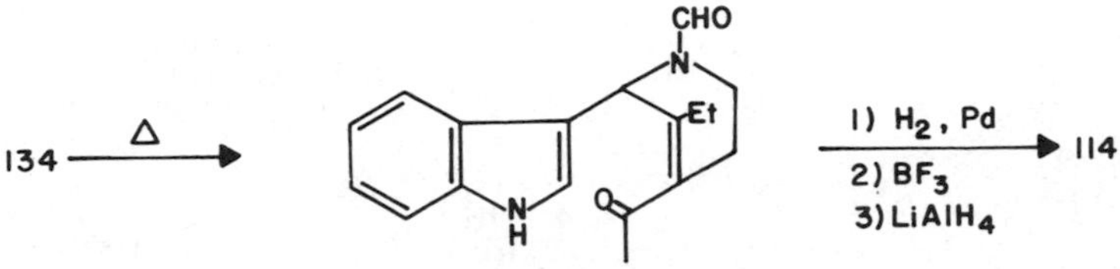

134 —Δ→ (137) —1) H_2, Pd; 2) BF_3; 3) $LiAlH_4$→ 114

Kametani and his group have also succeeded in synthesizing several members of the dasycarpidone-uleine series. The approach employed[207-209] involves, in the key step, condensation of indolylmagnesium bromide **138** with the appropriate

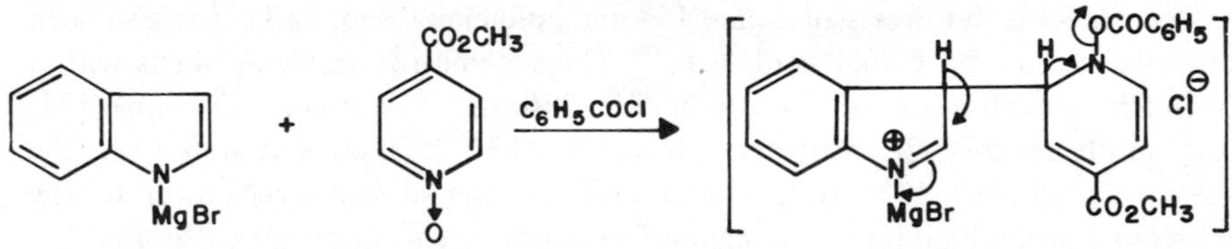

(138) + (139) —C_6H_5COCl→ [intermediate]

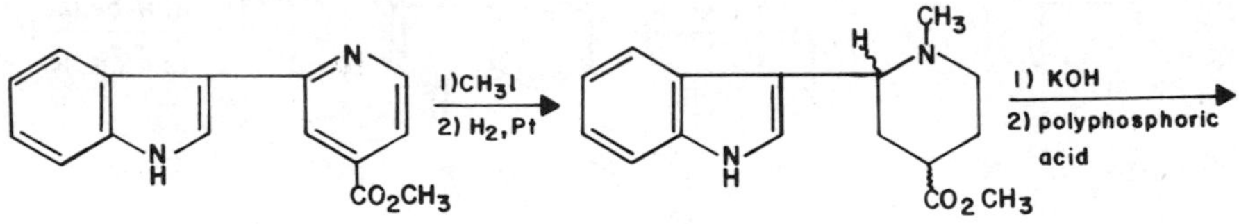

(140) —1) CH_3I; 2) H_2, Pt→ (141) —1) KOH; 2) polyphosphoric acid→

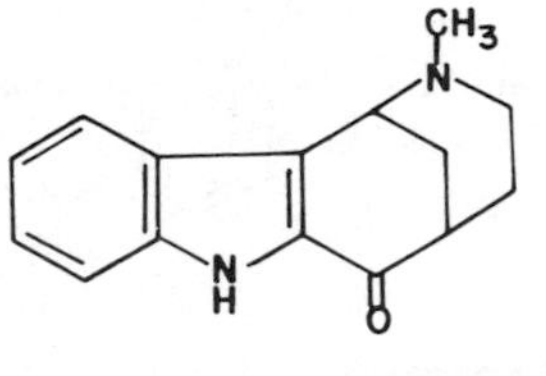

(142)

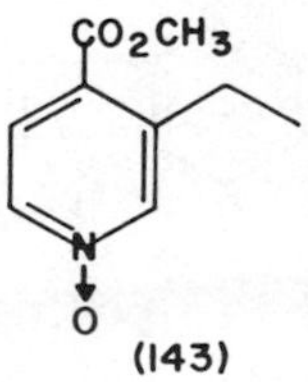

(143)

nicotinate *N*-oxide (for example, **139**) and subsequent conversion of the pyridine unit in the latter (for example, **140**) to the reduced piperidine system. Final closure of this intermediate with polyphosphoric acid yields the desired system. The synthesis of d,l-de-ethyldasycarpidone **142** is typical to the approach.

Extension of this scheme, employing methyl 3-ethylisonicotinate 1-oxide **143**, completed the syntheses of dasycarpidone **115**, 3-epidasycarpidone, uleine **114**, 3-epiuleine, as well as the various other isomers in this series.

Finally, there is yet another family of alkaloids that in their fundamental structure lacks the normal tryptamine unit but that is best considered under the *Aspidosperma* series, since it occurs in *Aspidosperma,* as well as other species. This family is best exemplified by the alkaloids olivacine **144** and ellipticine **145**.

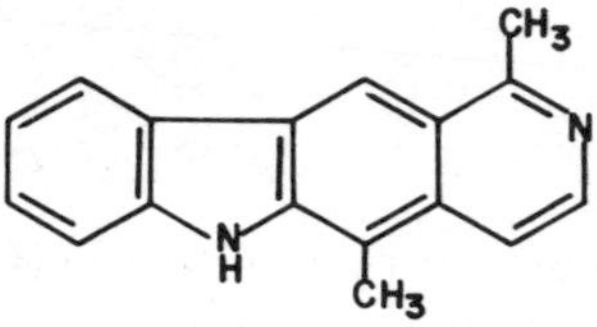

(144) 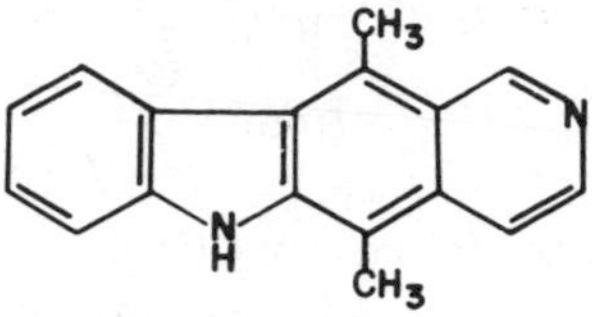 (145)

Both of these alkaloids and their respective derivatives attained interest beyond their chemistry, since in appropriate biological evaluation in various tumor systems they exhibited some antitumor activity.

The first synthesis of ellipticine by Woodward, Iacobucci, and Hochstein[210] was followed by several other successful syntheses.[211-214] The five-step scheme employed by Cranwell and Saxton[212] has been improved in the more recent investigations of the Australian group,[214] and this approach has been adapted for large scale preparation of a wide variety of ellipticine analogs for purposes of antitumor evaluation. It is, therefore, appropriate to discuss this pathway. In essence, versatility is provided in the initial phases since condensation of indole, or any of its substituted derivatives **146** ($R_1 \neq R_2 \neq H$), with hexane-2,5-dione **147** yields a variety of 1,4-dimethylcarbazole analogs **148**, which are then elaborated to the final product.

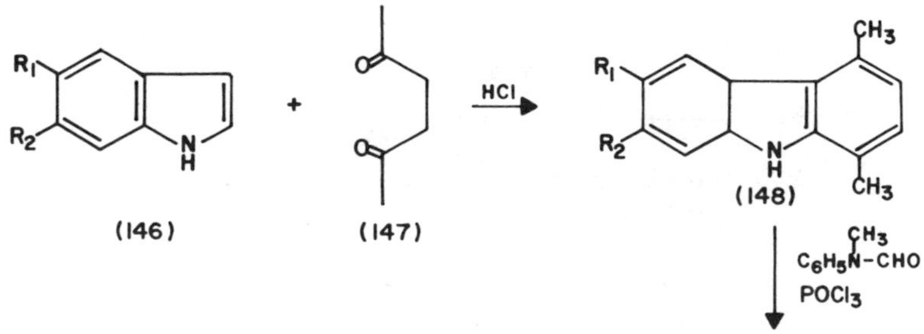

(146) (147) HCl → (148)

$C_6H_5N(CH_3)-CHO$, $POCl_3$

(150) $\xleftarrow{(EtO)_2CHCH_2NH_2}$ (149)

1) H_2, Ni
2) HCl
3) Pd on C

(145)

In the initial work Cranwell and Saxton attempted to cyclize the azomethine **150** under conditions normally employed in the Pomeranz-Fritsch isoquinoline synthesis (sulfuric acid, arsenic pentoxide, polyphosphoric acid) but were unable to succeed. It was, therefore, necessary to reduce **150**, then cyclize and finally dehydrogenate the latter to the final product.

By utilizing orthophosphoric acid, the Australian workers[214] were able to effect the direct cyclization of **150**, thus eliminating two steps in the synthesis and improving the yield.

Olivacine **144** has also been synthesized by several groups.[215-217] Again the latest investigations by the group at the Stanford Research Institute[217]

(151) $\xrightarrow{1) CH_3NO_2, NH_4OAc;\ 2) LiAlH_4}$ (152) $\xrightarrow{1) Ac_2O;\ 2) POCl_3}$

(153) $\xrightarrow{Pd/C}$ 144

constitute an improvement in the Swiss synthesis[215] for purposes of large-scale preparation for antitumor testing. A brief outline of the synthesis follows.

M. The Eburnamine-Vincamine Family

This family of alkaloids possesses the same fundamental C_{10} nontryptophan unit 1 already noted in the *Aspidosperma* series. It is, however, incorporated into the alkaloid system in a rather different fashion, and it is best to discuss it separately. The best known alkaloids in this family are probably eburnamine 2 and vincamine 3, and they are cited as examples.

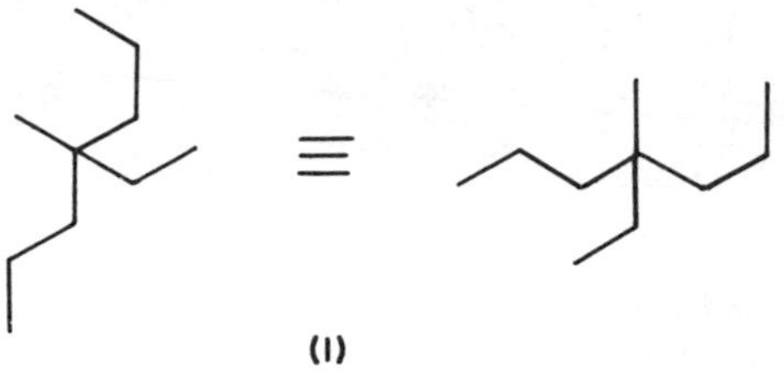

(1)

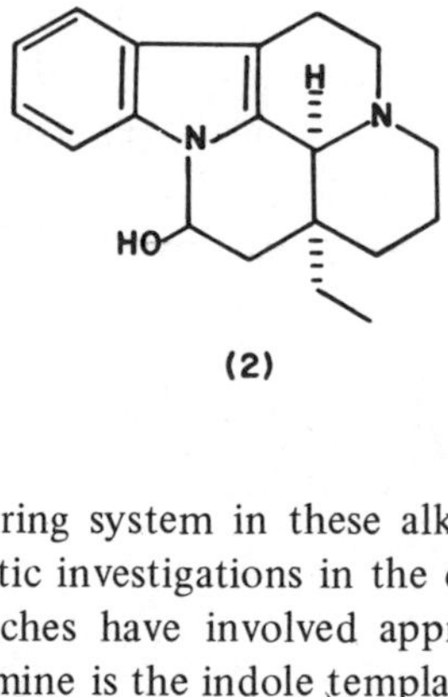

HO N N H

(2)

CH$_3$OOC HO N N H

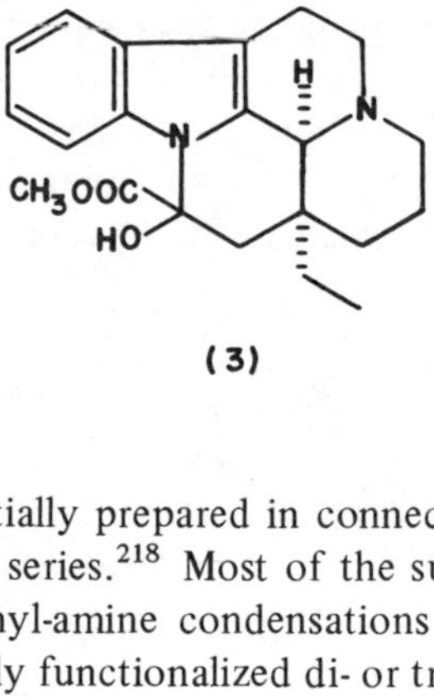

(3)

The ring system in these alkaloids was initially prepared in connection with synthetic investigations in the curare alkaloid series.[218] Most of the subsequent approaches have involved appropriate carbonyl-amine condensations in which tryptamine is the indole template and a suitably functionalized di- or tricarbonyl system is employed for the nontryptophan unit.

Bartlett and Taylor[219] in their structural studies in this series provided a simple seven-step synthesis of d,l-eburnamonine 6. The key step involves condensation of β-ethyl-β-formyladipic acid 4 with tryptamine with the resultant formation of d,l-eburnamonine lactam 5. The latter substance is then converted to 6 in a straightforward manner.

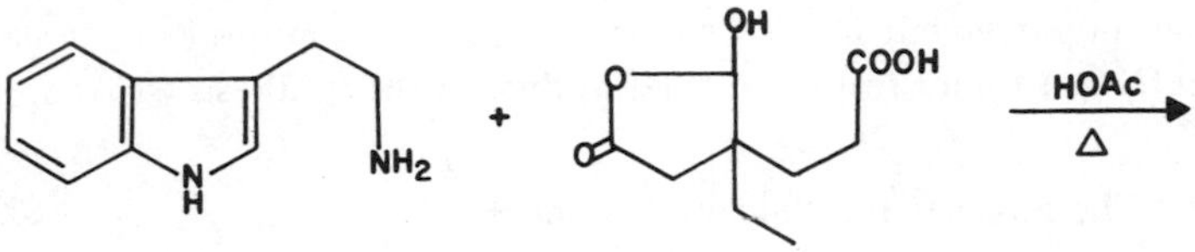

(4)

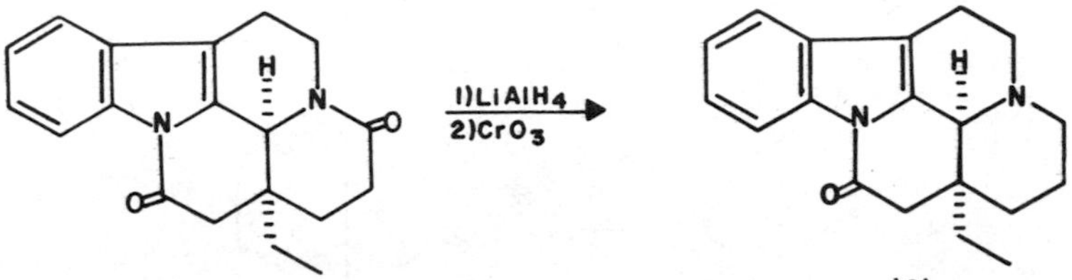

(5) (6)

Kuehne[36] has provided a synthesis of d,l-vincamine **3** by condensing tryptamine with 4-ethyl-4-formylpimelate **7** in the initial phases of the scheme. The resultant product **9** was then converted to the thiolactam **10** with the objective of removing the amide carbonyl but retaining the ester function **11**. This latter product, a mixture of isomers, was equilibrated through the immonium intermediate **12** to the two amino esters to which the relative configuration shown in **11a** and **11b** could be assigned. One of these (**11a**) is then converted in low yield (3%) to d,l-vincamine by the sequence shown.

Harley-Mason and co-workers[35,220] have employed a similar approach in the condensation of **13** with tryptamine and subsequent elaboration of the lactam **14** to d,l-eburnamine **2**. Subsequently,[221] it was found that the lactam **14** was

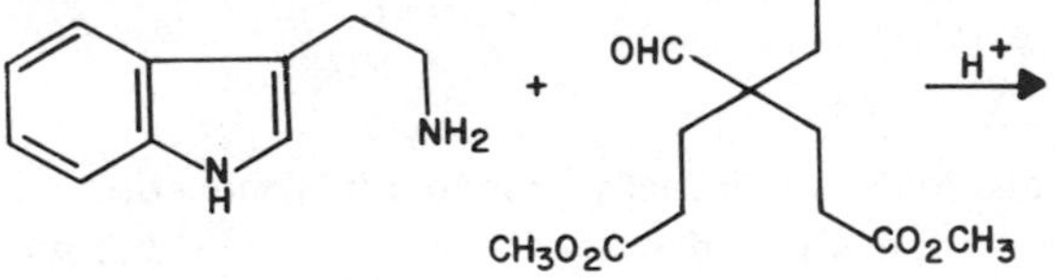

(7)

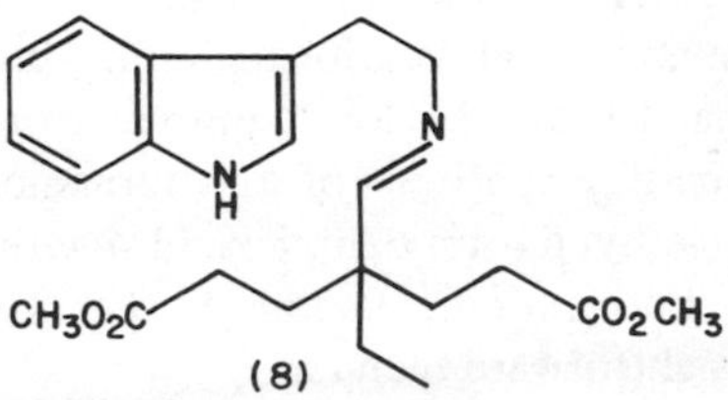

(8)

$\xrightarrow{\Delta}$ (9) $\xrightarrow{P_2S_5}$ (10) $\xrightarrow[\text{Ni}]{\text{Raney}}$

(11) $\xrightarrow{Hg(OAc)_2}$ (12) $\xrightarrow{NaBH_4}$

(11a) + (11b) $\xrightarrow[(C_6H_5)_3C^{\ominus}\ Na^{\oplus}]{ON-C_6H_4-N(CH_3)_2}$ 3

(13) + $\longrightarrow$

(14) $\xrightarrow[\text{2) } NaIO_4]{\text{1) } OsO_4}$ (15) $\xrightarrow{LiAlH_4}$ 2

not homogeneous and in the reaction with osmium tetroxide yielded a mixture of diols that could be separated into the four possible racemates. Either of the *cis* diols **16** on oxidation, cyclization, and reduction provides d,l-vincaminol **17**.

Another synthesis of d,l-eburnamine has been reported more recently by Gibson and Saxton.[222] In this scheme, the ester diol **18** is the starting material. An extension of this scheme provided a new synthesis of d,l-vincamine.[223] d,l-Homoeburnamenine **25** available from this sequence is converted in the manner shown to the alkaloid in racemic form.

Wenkert and Wickberg in their general study on the utilization of piperideine intermediates in indole alkaloid syntheses have published a completely different scheme to d,l-eburnamonine **6**.[48] The crucial conversion involves the hydrogenation of a pyridinium salt 27 to a Δ^2-piperideine 28, and the latter is then

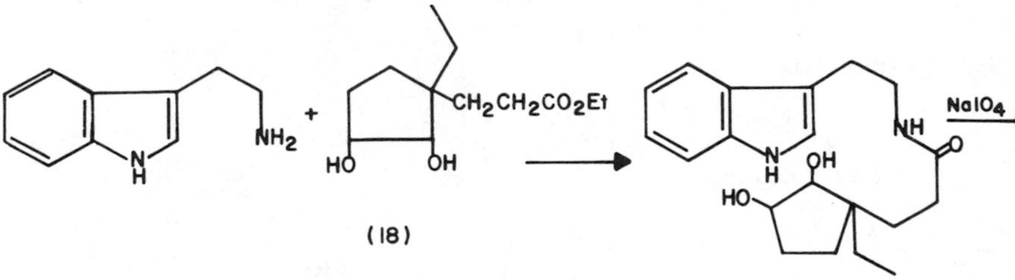

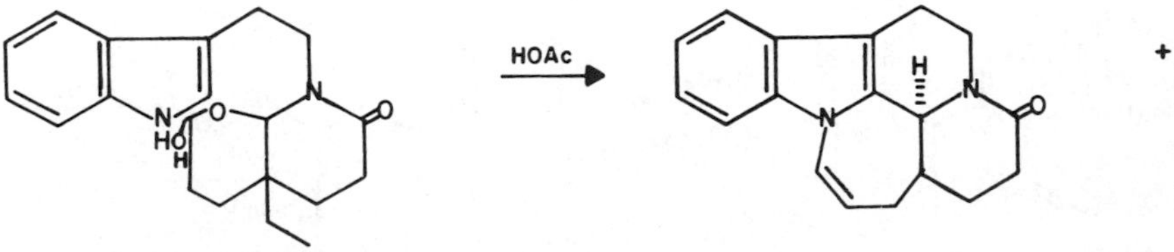

OsO_4 (on 21)

(22) (23)

1) $Pb(OAc)_4$
2) K_2CO_3

$LiAlH_4$

2

(24)

1) OsO_4
2) DMSO:NEt_3: pyridine:SO_3: H_2O

1) $OH^{\ominus}$
2) CH_2N_2
3) DMSO:NEt_3: pyridine.SO_3: H_2O

3

(25) (26)

+

Δ

$Br^{\ominus}$

H_2, Pd
Et_3N

(27)

HCl

H_2N-NH_2
KOH

(28) (29)

1) $Hg(OAc)_2$
2) $NaClO_4$

(30) 31 $ClO_4^{\ominus}$ NaOH

$ICH_2COOCH_2CH_3$

(32) (33) CH_3H_2OOC

1) H_2 or $NaBH_4$
2) $\bar{O}Et$

(6) + (34)

exposed to a Pictet-Spengler cyclization. The resulting product **29** is reduced and converted to an immonium salt **31**, which then undergoes the expected base-catalyzed isomerization to the enamine **32**. The latter is then alkylated to **33**, which after reduction and treatment with base affords d,l-eburnamonine **6** and d,l-epieburnamonine **34**, a compound isomeric with the natural alkaloid.

N. The *Iboga* Family

The *Iboga* family has been extensively pursued from a structural as well as a synthetic standpoint. Its members also reveal within the structural framework a different variant of the C_{10} nontryptophan unit **1** from that portrayed in the two previous classes. Since a number of these alkaloids occur along with the *Aspidosperma* members in the same plant species (for example, *Vinca rosea* Linn), they have also received a great deal of attention in biosynthetic considerations and in medicinal applications. The dimeric *Vinca* alkaloids, of which vin-

(1)

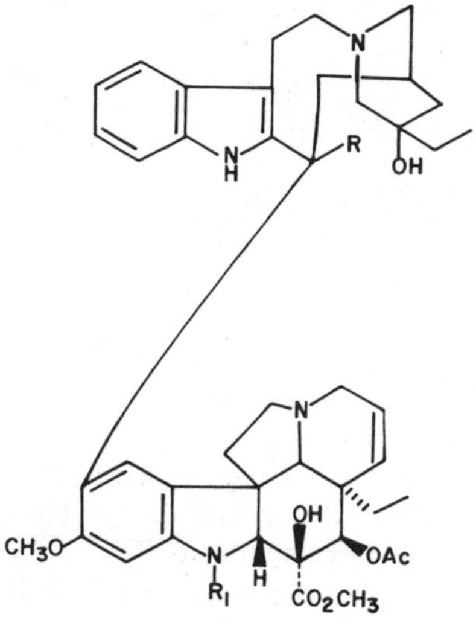

2, R = $COOCH_3$; R_1 = CH_3

3, R = $COOCH_3$; R_1 = CHO

blastine 2 and vincristine 3 may be cited as examples, are important in the clinical treatment of cancer in humans.

These dimers portray an interesting combination of the *Iboga* and *Aspidosperma* systems. Thus the dihydroindole unit represents the highly oxygenated *Aspidosperma* alkaloid, vindoline, already mentioned in Section L, whereas the indole unit reveals one possible natural variant of the *Iboga* family. This unit, with its nine-membered ring system, has been assigned the name velbanamine 4, and it can be obtained during acidic cleavage of the dimers, whereas the parent system (5) also available from 4 or the dimers has been given the name cleavamine. The cleavamine-velbanamine series has not been isolated from a natural source other than in the dimeric family exemplified previously. The more common *Iboga* members bear the more rigid cyclic system exemplified by ibogamine 6 and catharanthine 7.

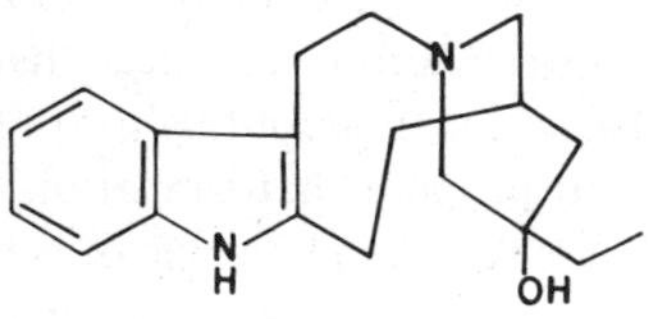

(4)

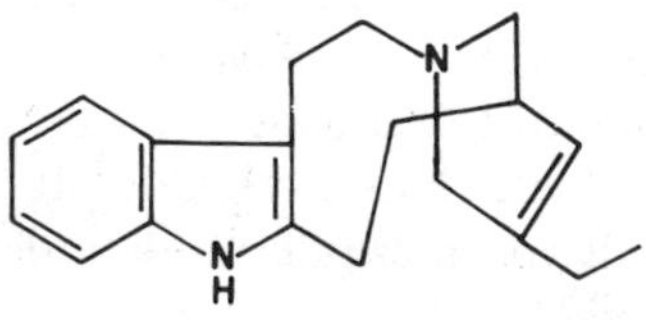

(5)

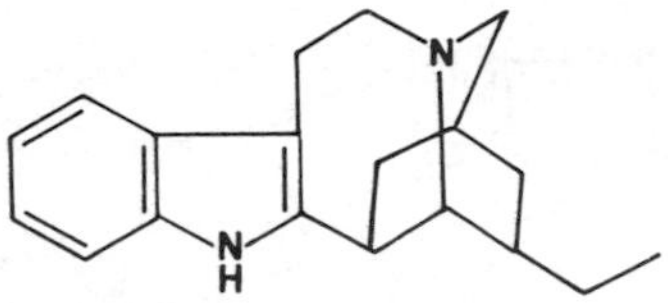

(6)

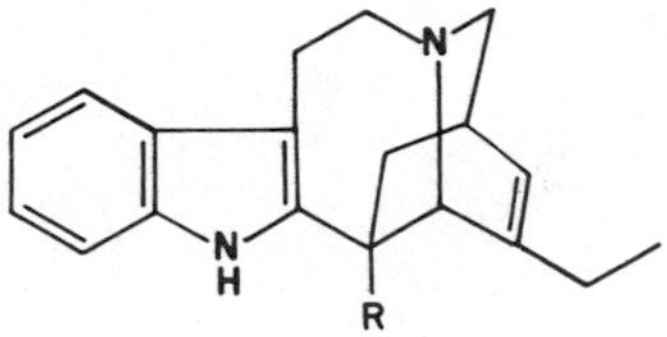

7, R = $COOCH_3$

The first total synthesis of ibogamine 6 and epi-ibogamine 19 was reported by Buchi et al.[27,28] The approach involved first the synthesis of the isoquinuclidine ring characteristic of the *Iboga* alkaloids and then subsequent attachment of the indole portion to this ring system. Preparation of the functionalized isoquinuclidine intermediate 9 required for the synthesis proceeded via a Diels-Alder condensation of the dihydropyridine 8 with methyl vinyl ketone. Subsequent elaboration of 9 to the key intermediate 12 followed the pathway indicated. The most interesting conversion of note in proceeding from 9 to the isoquinuclidone 12 is the Hofmann reaction, which allows isolation of the intermediate urethan **10**. The authors feel that formation of the latter proceeds through an initially formed vinyl isocyanate, which then rearranges to the enamine **20** or the conjugated imine **21**. Reaction of **12** with β-indolylacetyl chloride proceeded in the expected fashion to provide **13**. However, the subsequent reactions in the scheme involve a series of interesting rearrangements and some comments are, therefore, appropriate.

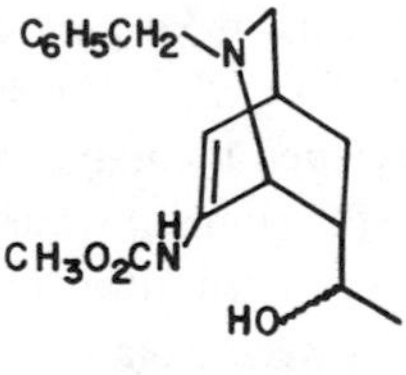

(20)

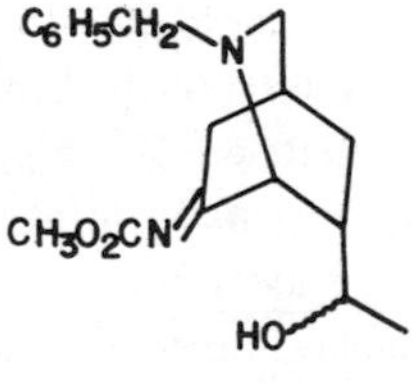

(21)

In the initial publication of 1965, the authors postulated that in the acid-catalyzed cyclization of **13**, the cyclic diol monoacetate thus obtained possesses the structure **22**. This would be the expected structure if a normal cyclization process were to prevail. However, further investigations revealed that the structure of this product must be altered to **14**, and it is interesting to speculate on the formation of the latter. It is tempting to propose that formation of **14** proceeds via an aziridine intermediate **23** and as is seen later in some of Nagata's approaches to the *Iboga* series, such intermediates can be effectively employed for this purpose. The aziridine system does rationalize the isolation of **14**, but as

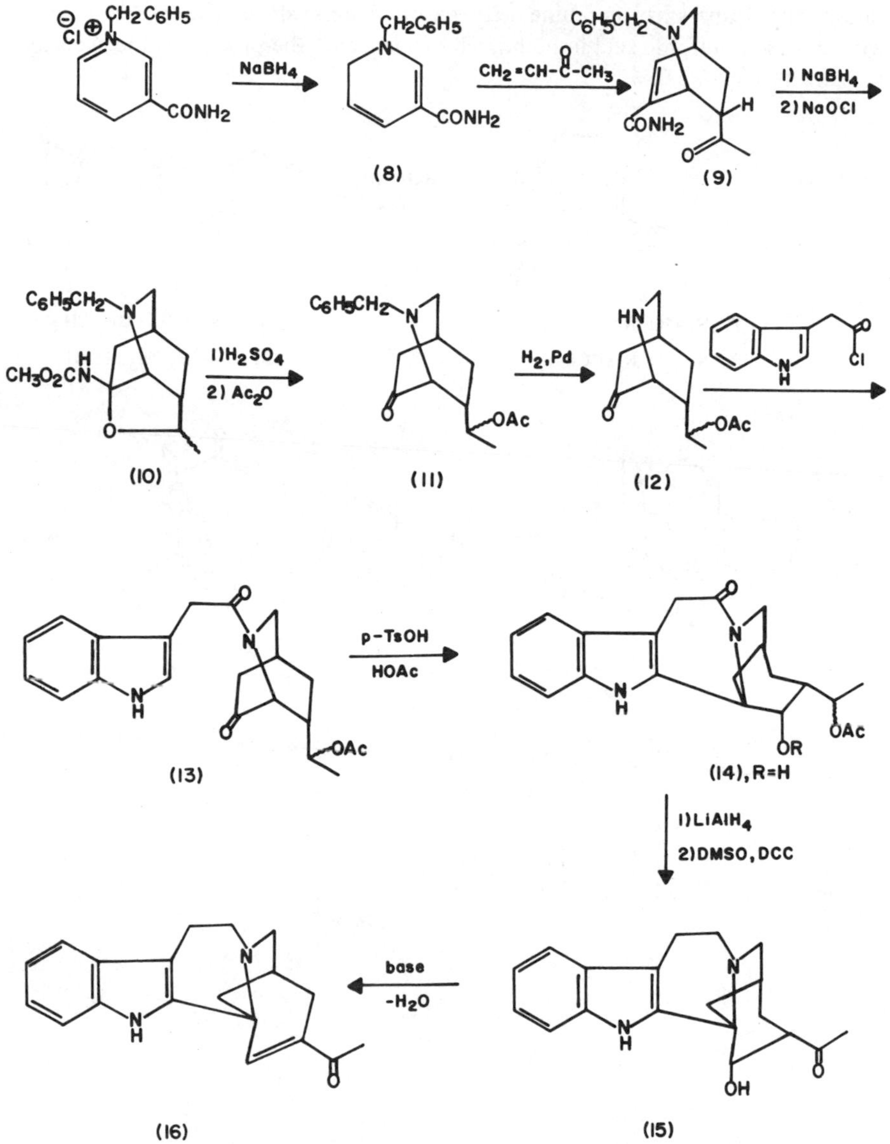

(8)

(9)

(10)

(11)

(12)

(13)

(14), R=H

(16)

(15)

the authors point out, it should be noted that cyclization of **13** gives *mainly* the monoacetate **14** (R = H) rather than the diacetate **14** (R = Ac), although the reaction is performed in acetic acid. This product, therefore, seems to result from internal return within ion pairs and demands that leaving and entering hydroxyl groups be located on the same face of the molecule. The authors propose a cation **24** for this purpose. Under more strenuous conditions, the cation does apparently combine with external ions, since when **13** is reacted with *p*-toluenesulfonic acid in boiling chlorobenzene, the product is the acetoxy-tosylate **14** (R = Tos).

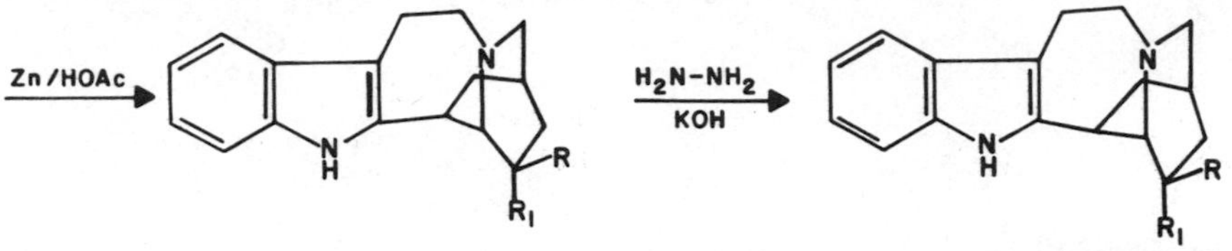

Zn/HOAc

H_2N-NH_2 KOH

17, R_1 = H; R = $COCH_3$
18, R = H; R_1 = $COCH_3$

6, R_1 = H; R = CH_2CH_3
19, R_1 = CH_2CH_3; R = H

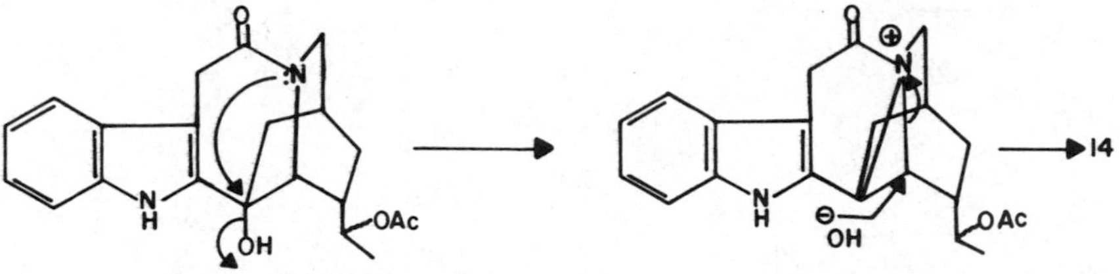

(22)

(23)

14

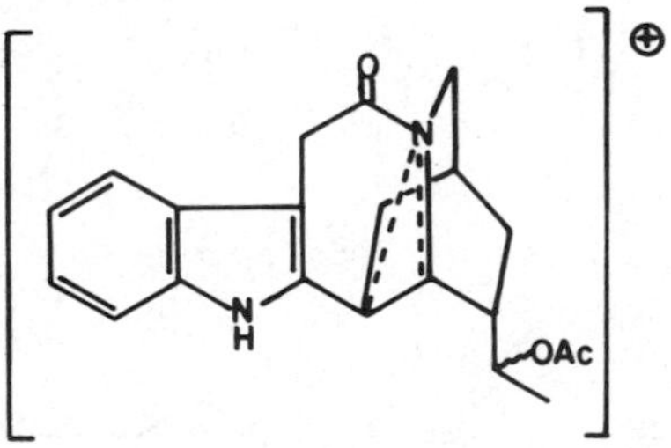

(24)

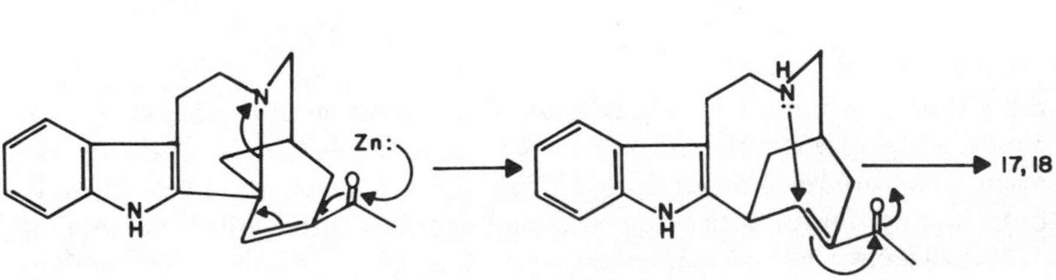

(16)

(25)

17, 18

Another reaction of interest in the later stages of the pathway concerns the rearrangement of the azabicyclo[1,2,3]octane system portrayed in **16** to the isoquinuclidine skeleton shown in **17** and **18**. This rearrangement prevailed during zinc-acetic acid reduction of the α,β-unsaturated ketone **16** and the mechanistic rationale given by the authors is portrayed in the scheme (**16** → **25** → **17**, **18**). Extensions of this scheme allowed the synthesis of ibogaine **27** and epi-ibogaine **28**. Thus condensation of 3-(5-methoxyindolyl)acetyl chloride with the amine **12** provided **26**, which was then elaborated to the alkaloid in the manner already described.

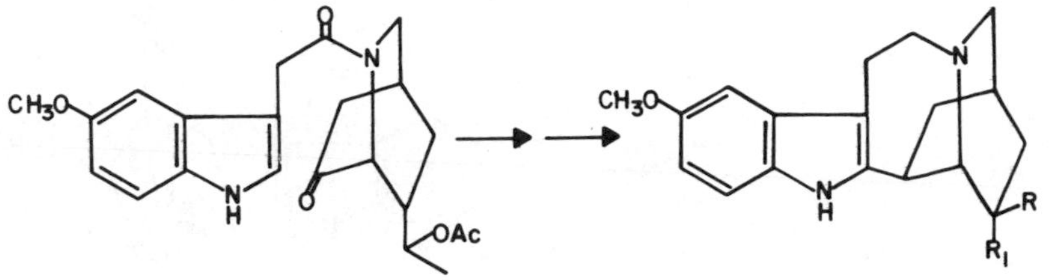

(26)

27, R_1 = H; R = CH_2CH_3
28, R_1 = CH_2CH_3; R = H

Investigations in our laboratory,[183,196,224] provided a general entry into the *Iboga* family. The approach employed is totally different from that already described and is based on the transannular cyclization of appropriate nine-membered ring intermediates in the cleavamine series for the final elaboration of the rigid cyclic systems revealed in ibogamine **6**, catharanthine **7**, and so on.

The initial investigations published in 1964[183] provided the approach involving the cleavamine derivatives. Thus carbomethoxydihydrocleavamine **29**, upon reaction with mercuric acetate, generates as one of the intermediates an iminium system, which cyclizes spontaneously to the *Iboga* structures, coronaridine **34** and dihydrocatharanthine **33**. The isolation of both of the latter compounds demands that the iminium species initially formed (**30**) undergo equilibration via the enamine **31** prior to transannular cyclization.

On this basis, it was now clear that the synthesis of the requisite cleavamine derivatives would complete the total synthesis of these alkaloids. The pathway selected[224,196] paralleled one already discussed in the *Aspidosperma* alkaloid section. The key step involved the reductive cleavage of the appropriate quaternary system to the nine-membered ring in the final stage (for example, **35** → **36**). The successful sequence that allows the synthesis of **35** and, in turn, the 4α- and 4β-dihydrocleavamines (**42** and **43**, respectively) now follows.

It should be noted that the stereochemical problems associated with this synthesis are simplified markedly by the fact that the transannular cyclization

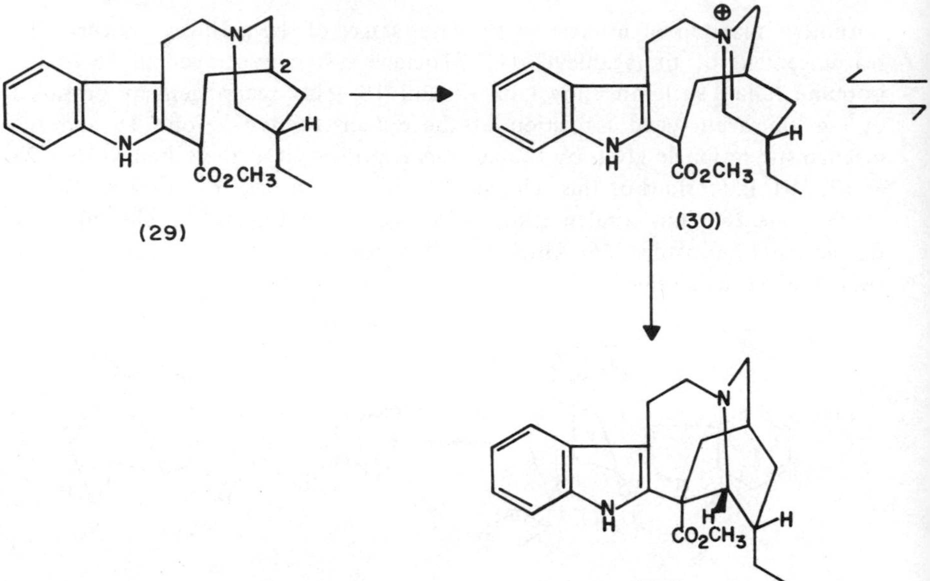

(29)

(30)

(33)

process is completely stereospecific. The previous results had already indicated that the asymmetric center at C_2 (see 29) in the cleavamine series completely controls the steric course of this cyclization. Thus it was not necessary to give serious consideration to the stereochemistry at the various stages of the pathway. Since the absolute configuration of the 4α- and 4β-dihydrocleavamines (42 and 43) is known with certainty from our previous X-ray work, it becomes clear from this sequence that C_2 and C_4 (asterisked centers in 42 and 43), the two

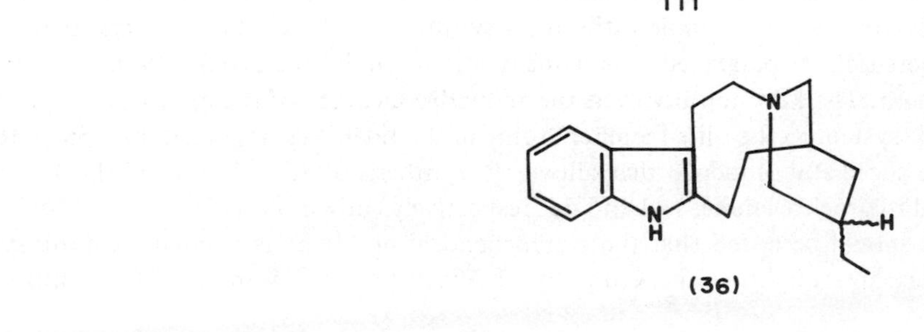

(35)

(36)

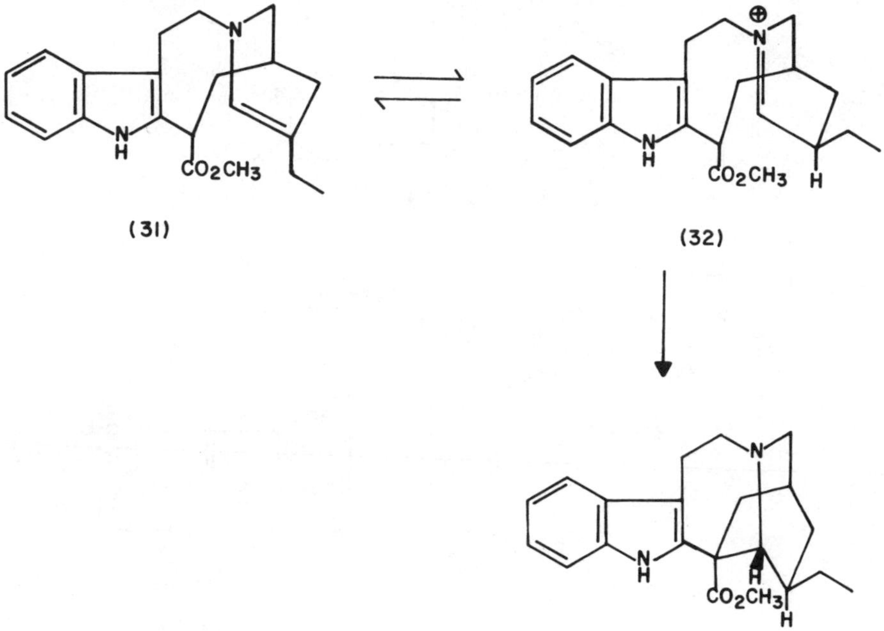

(31) ⇌ (32) → (34)

asymmetric centers are derived directly from the two relevant centers in the succinate ester (see asterisked centers in 37). Therefore, the appropriate stereo-

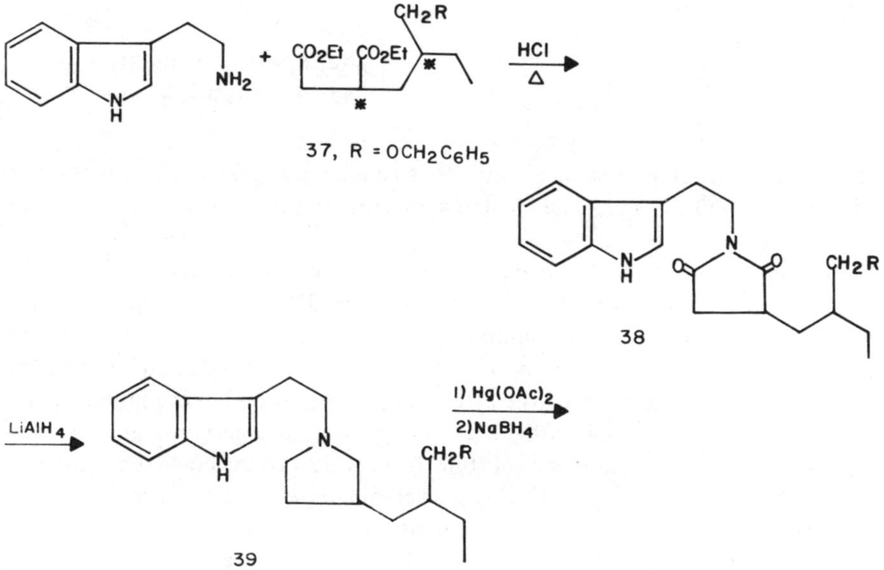

37, R = $OCH_2C_6H_5$

HCl, Δ → 38

$LiAlH_4$ → 39

1) $Hg(OAc)_2$
2) $NaBH_4$

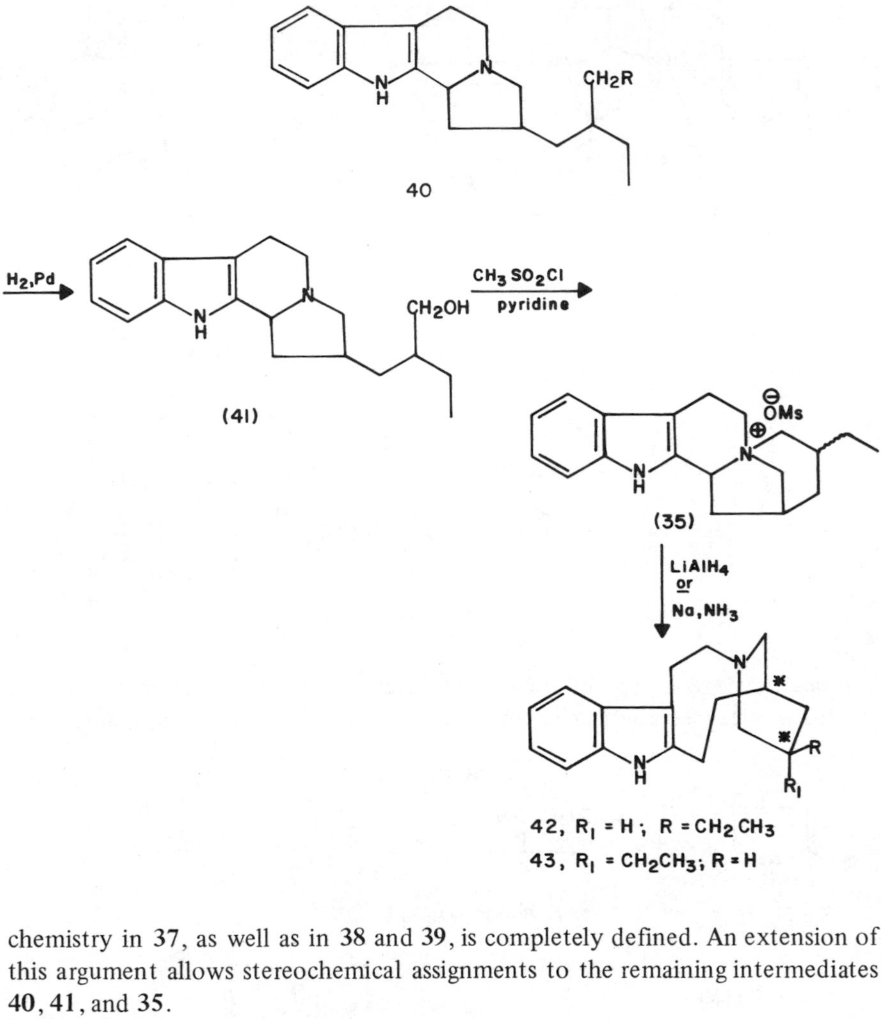

42, R_1 = H; R = CH_2CH_3
43, R_1 = CH_2CH_3; R = H

chemistry in **37**, as well as in **38** and **39**, is completely defined. An extension of this argument allows stereochemical assignments to the remaining intermediates **40**, **41**, and **35**.

The completion of the *Iboga* alkaloid synthesis made it necessary to introduce a carbomethoxy group into the C_{18} position of the dihydrocleavamine molecule. For this purpose, the chloroindolenine of 4β-dihydrocleavamine **44** was prepared and an extensive series of investigations were conducted to optimize conditions for introduction of the ester group. It was found in the initial experiments that reaction of **44** with potassium cyanide in a mixture of methanol-water-ether provided a complex mixture from which 18-cyano-4β-dihydrocleavamine **45** and 18α-methoxy-4β-dihydrocleavamine were isolated in low yield. However, when **44** was reacted with anhydrous sodium acetate in glacial acetic acid, an unstable 18-acetoxy-4β-dihydrocleavamine **46** was formed, which in

(43) $\xrightarrow{(CH_3)_3COCl}$ (44)

KCN or NaOAc

45, R = CN
46, R = OAc

on 46 only → (47)

KCN → (48)

1) KOH
2) CH_2N_2

(49)

33 + 34

time converted to a quaternary species **47** and the latter, without isolation, provided 18β-cyano-4β-dihydrocleavamine **48** upon reaction with potassium cyanide. Hydrolysis of the latter affords 18β-carbomethoxy-4β-dihydrocleavamine **49**. The conversion (**44** → **46** → **47** → **48**) could be performed without isolation of intermediates and in optimum yields.

Conversion of **49** to dihydrocatharanthine **33** and coronaridine **34** has already been discussed. Since the conversion of coronaridine to ibogamine has also been reported,[225] this sequence also completes a total synthesis of ibogamine **6**.

An approach similar to the one already described for the cleavamine series has been investigated by Harley-Mason and co-workers.[226-229] The triester acetal **50**, upon reaction with tryptamine, afforded the amide ester **51**. Conversion of the latter to the quaternary salt **53** and reductive cleavage to **42** and **43** followed the

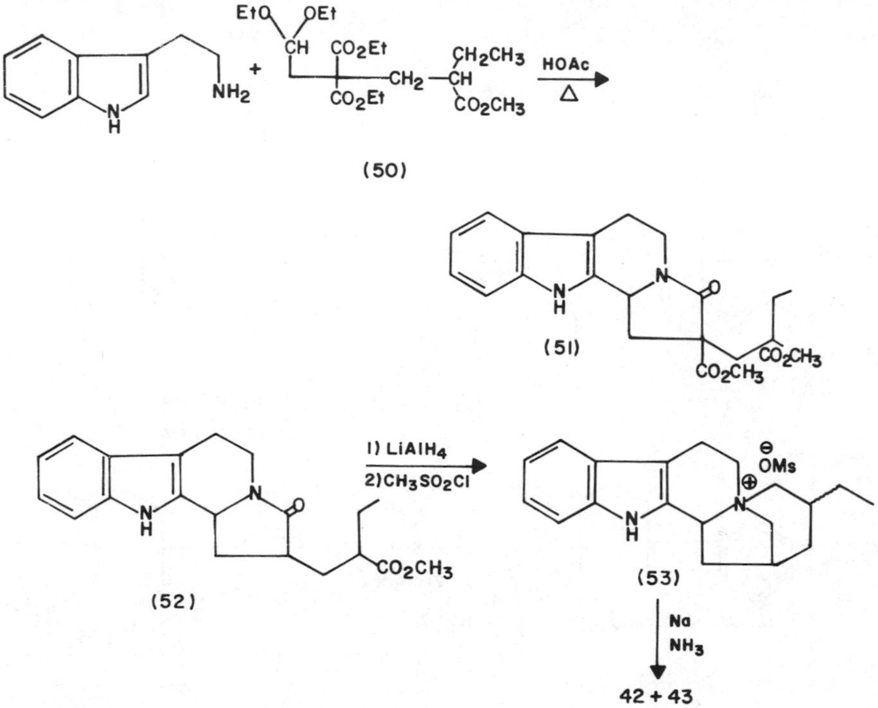

HOAC, Δ

(50)

(51)

1) $LiAlH_4$
2) CH_3SO_2Cl

(52)

(53)

Na, NH_3

42 + 43

procedures already mentioned in our sequence. Similarly, the conversion of **53** to the 18-cyanocleavamine series (see **45**) and subsequently to the 18-carbomethoxy compounds (see **49**) followed a route described previously. An extension of this investigation provided a preparation of the 18-hydroxycleavamine derivatives **56** (two isomers at ethyl group).

The most recent publication[229] describes an improvement in the synthesis of **52**. In this modification, a procedure developed elsewhere[230] and employing sodium cyanide in dimethylsulfoxide to remove an ester group was utilized (**50** → **57**). The latter, upon reaction with tryptamine, affords **52** in 85% of the yield.

A fundamentally different approach to the *Iboga* alkaloid skeleton was developed by Nagata and his group[231-232] The crucial reaction involves a new method for isoquinuclidine synthesis in which aziridine intermediates are employed. Thus when the bridged aziridines **58a** through **c** are treated with an acylating agent (acyl halide or acid anhydride) in an appropriate solvent, the

52 → (1) P_2S_5 2) Raney Ni 3) $OH^{\ominus}$) → (54) → (Ac_2O, NaOAc) → (55) → (1) $OH^{\ominus}$ 2) $LiAlH_4$) → (56)

50 → (NaCN, DMSO) → $(CH_3CH_2O)_2CH_2C(H)(CO_2CH_2CH_3)-CH_2CH(CH_2CH_3)CO_2CH_3$ (57) → (tryptamine) → 52

isoquinuclidine system **59** is generated in excellent yield. This reaction initially allowed a synthesis of desethylibogamine[231] and subsequently[232] provided a total synthesis of ibogamine **6** and epiibogamine **19**. The sequence leading to

58 → (RX) → (59) + (60)

58a, $R_1 = R_2 = H$
b, $R_1 = CH_3$; $R_2 = H$
c, $R_1R_2 = -CH_2(CH_2)_2CH_2-$

(61) $\xrightarrow{Pb(OAc)_4}$ (62) $\xrightarrow[\text{acetic anhyhydride}]{\beta\text{-indolyl}}$ 63, R = β-indolyl- acetyl

$\xrightarrow[\text{2) } Ac_2O \text{ DMSO}]{\text{1) } OH^{\ominus}}$ (64) $\xrightarrow[\Delta]{p\text{-TsOH}}$ (65) → (66) $\xrightarrow[\text{2) } Al_2O_3]{\text{1) } LiAlH_4}$ (67) $\xrightarrow{(H)}$ 6

these latter systems is now described. It is of interest to note that in the cyclization (**64** → **65**) no rearrangement of the type observed by Buchi (compare **13** → **14**) was noted.

A parallel series of experiments employing the isomer **68** led to epi-ibogamine **19**. Sallay[233] has also reported a synthesis of d,l-ibogamine. In this study, the requisite stereochemistry in the isoquinuclidine ring system is developed by

(68) → → 19

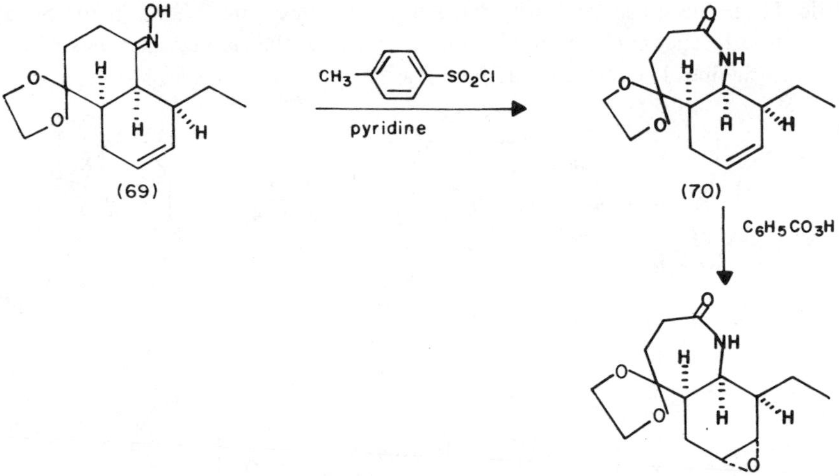

CH3—C6H4—SO2Cl, pyridine

(69) (70)

$C_6H_5CO_3H$

(71)

employing stereochemically controlled reactions in the initial phases. The final stage of the synthesis involves a Fischer indolization of the tricyclic ketone **75** to provide the alkaloid system.

Ikezaki, Wakamatsu, and Ban[234] have succeeded in the synthesis of d,l-ibogamine via a seqeunce that involves the Ziegler cyclization of an appropriate tri-

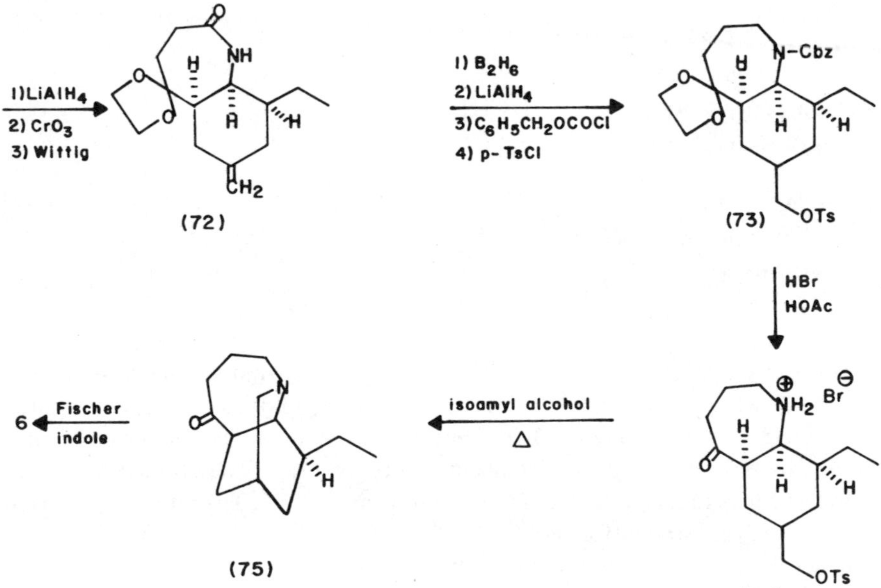

1) $LiAlH_4$ 2) CrO_3 3) Wittig

(72)

1) B_2H_6 2) $LiAlH_4$ 3) $C_6H_5CH_2OCOCl$ 4) p-TsCl

(73)

HBr HOAc

(74)

isoamyl alcohol Δ

(75)

Fischer indole

6

nitrile 76 in the construction of the isoquinuclidine ring 77. As in the Sallay synthesis, a Fischer cyclization of the hydrazones of the isomeric ketones 81 and 82 forms the indole nucleus in the final stages.

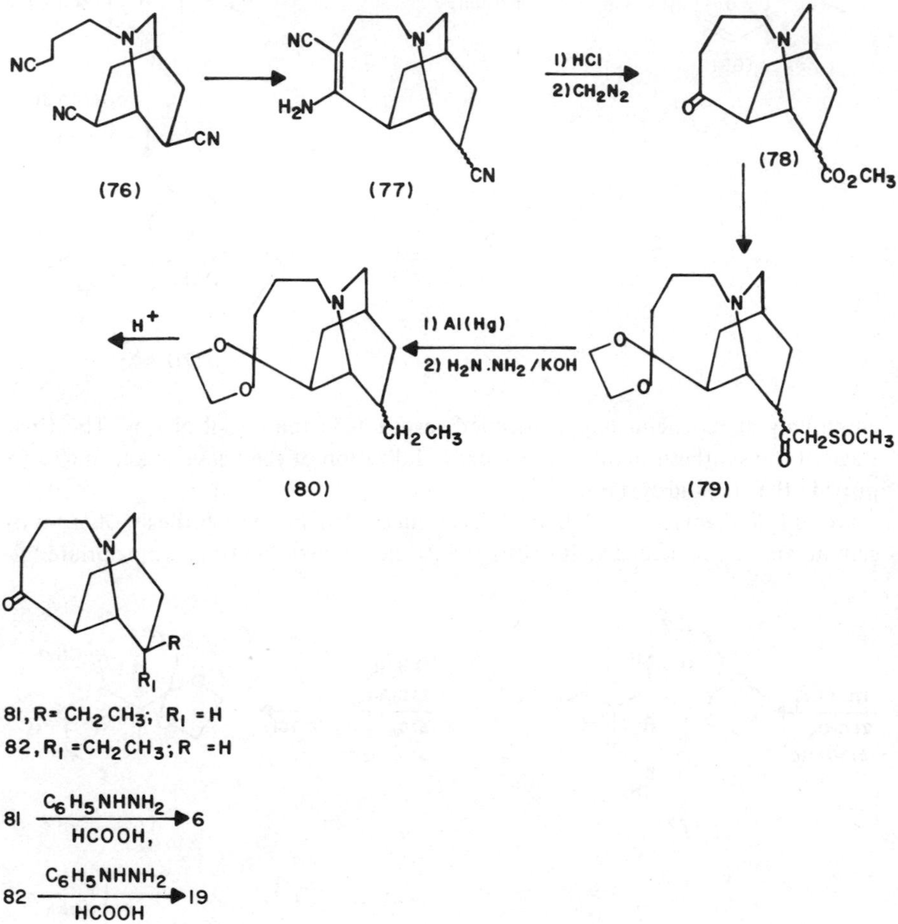

(76) → (77) —1) HCl, 2) CH_2N_2→ (78) → (79) —1) Al(Hg), 2) $H_2N.NH_2$/KOH→ (80) —H^+→

81, R = CH_2CH_3; R_1 = H
82, R_1 = CH_2CH_3; R = H

81 —$C_6H_5NHNH_2$, HCOOH,→ 6

82 —$C_6H_5NHNH_2$, HCOOH→ 19

Another group of investigators[235] have described a total synthesis of epi-ibogamine in which a new indole synthesis has been employed. In this approach, a good yield of the required 1,3-diketone 84 is obtained when an enamine of a ketone 83 is reacted with *O*-nitro phenyl acetyl chloride. The product 84 is then converted in a reducing medium to the indole system 85. The remaining steps in the sequence are straightforward.

A synthesis of desethylibogamine 91 in several steps from the epoxide of methyl 3-cyclohexene-1-carboxylate 88 has been described by Huffman, Rao,

(83)

(84)

1) H_2, Pd
2) $H_2N{\cdot}OH$

(85)

1) H_2, Raney Ni
2) Δ

(86)

1) $LiAlH_4$
2) $BrCH_2CO_2R$
3) polyphosphoric acid

(87)

$LiAlH_4$ → 19

(88)

tryptamine

(89)

Δ

(90)

1) p-TsCl
2) $AlCl_3$
3) $LiAlH_4$

(91)

and Kamiya.[236] This isoquinuclidine nucleus **90**, as in the previous synthesis, is formed by pyrolysis of an appropriate amino ester **89**.

Augustine and Pierson[237] found that catalytic reduction of **92** proceeded directly to the tricyclic lactam **94**. This facile ring closure may be facilitated by the favorable stereochemistry of the carboxy and amino groups in the *cis*-fused intermediate **93**, which would be expected from the reduction process. Now that

H_2,Ru

(92) (93) (94)

the facile formation of the isoquinuclidone **94** was demonstrated, it only remained to prepare the appropriate compound possessing an oxygen function for attachment of the indole ring. The authors selected **96** for this purpose, and the successful sequence is outlined.

$KMnO_4$; 1) $OH^{\ominus}$ 2) H_2, Ru ; CrO_3

(59) (96) (97)

1) $C_6H_5NHNH_2$ 2) polyphosphoric acid ; $LiAlH_4$ → 91

(98) (99)

In some of the most recent studies three groups have reported successful syntheses of velbanamine **4**, the monomeric indole unit of the oncolytic *Vinca* alkaloids; vinblastine **2**, and vincristine **3**.

Buchi and co-workers[238,239] reported the first synthesis of velbanamine employing as starting material the isoquinuclidine **9**, previously utilized in the *Iboga* alkaloid synthesis. Conversion of **9** to the appropriately functionalized isoquinuclidone **103** proceeded according to the sequence of reactions shown. This intermediate was then condensed with sodium indole acetate to provide the

O_2, $(CH_3)_3CO^{\ominus}K^{\oplus}$, −20°

(9) (100)

1) $NaBH_4$
2) $NaIO_4$

(101)

1) $HC(OCH_3)_3$, H^+
2) NaOCl
3) Na_2CO_3

(102)

H_2

(103)

sodium indole acetate

(104)

p-TsOH

(105)

$HClO_4$
HOAc

(106)

$(CH_3)_3CO^{\ominus}K^{\oplus}$
$(CH_3)_3COH$

(107)

1) $NaBH_4$
2) $SnCl_2$, Sn
p-TsOH

(108)

1) DMSO, DCC
2) CH_3CH_2MgBr
3) $LiAlH_4$

(4)

amide **104**, which is finally elaborated to velbanamine. One of the interesting reactions of note in the final stages is the retroaldolization process revealed in the conversion (**106** → **107**).

An extension of this work led to the total synthesis of catharanthine 7. The starting material for this synthesis was the cyclic ketone **105**, which on exposure to a Grignard reaction with vinylmagnesium bromide afforded the vinylcarbinol **109**. This latter substance could be readily converted to **110**, which now bears the ethyl side chain necessary for the catharanthine system. The chloroindolenine method for introducing the ester function (**111** → 7) had already been known from previous studies.

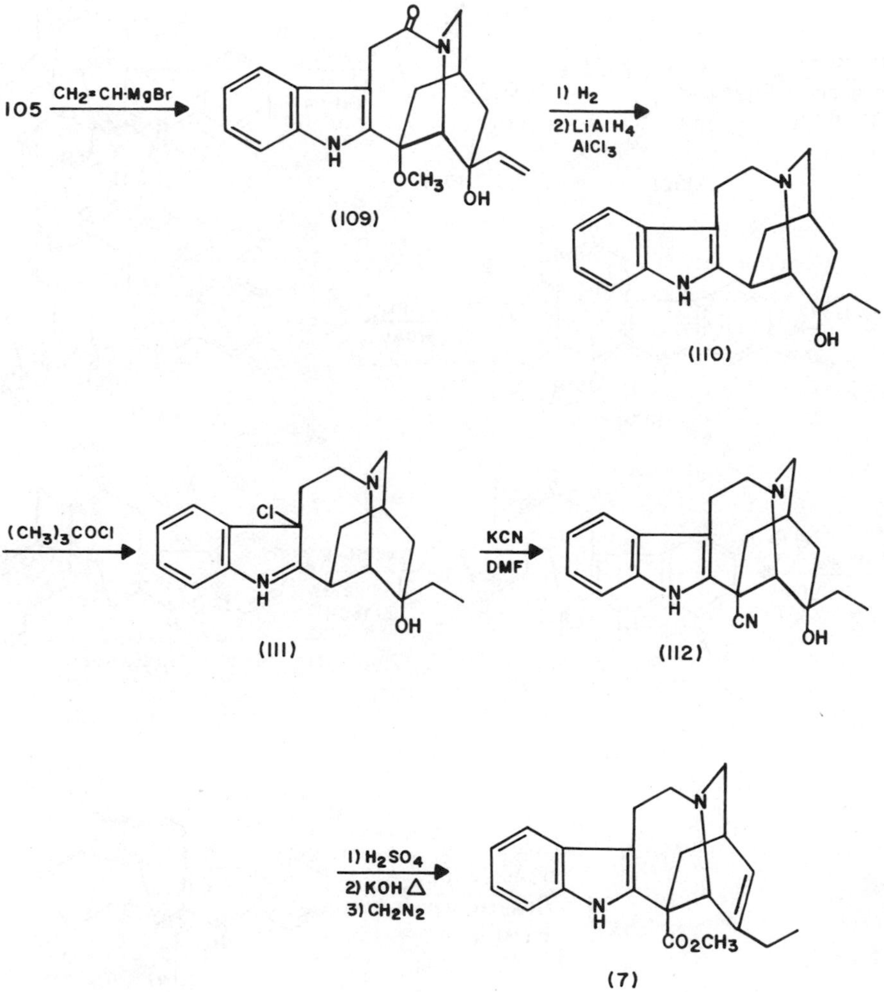

105 $\xrightarrow{CH_2{=}CH\text{·}MgBr}$ (109) $\xrightarrow[2)\ LiAlH_4,\ AlCl_3]{1)\ H_2}$ (110)

$\xrightarrow{(CH_3)_3COCl}$ (111) $\xrightarrow[DMF]{KCN}$ (112)

$\xrightarrow[2)\ KOH\ \Delta;\ 3)\ CH_2N_2]{1)\ H_2SO_4}$ (7)

A completely different approach to the cleavamine-velbanamine series was developed in our laboratory.[240] With the view of trying to obtain generality and versatility in a pathway that would lead to a variety of C_3-C_4-functionalized derivatives in this series, we turned our attention to the fragmentation reaction that we hoped would provide these compounds from the readily available catharanthine system. It was proposed that the catharanthine molecule possesses the necessary stereochemistry for the fragmentation process shown in **113** → **114** → **115**. If such a conversion were to succeed, the tetracyclic system generated would possess the necessary activation for functionalization at the 3,4 positions. The sequence shown here reveals the successful achievement of this goal.

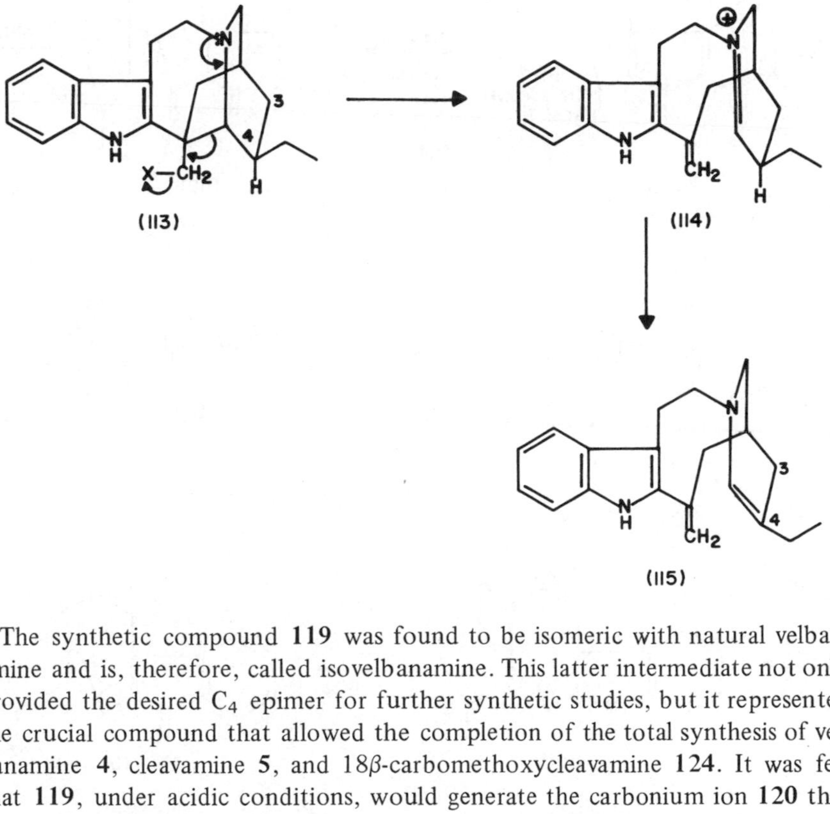

(113)

(114)

(115)

The synthetic compound **119** was found to be isomeric with natural velbanamine and is, therefore, called isovelbanamine. This latter intermediate not only provided the desired C_4 epimer for further synthetic studies, but it represented the crucial compound that allowed the completion of the total synthesis of velbanamine **4**, cleavamine **5**, and 18β-carbomethoxycleavamine **124**. It was felt that **119**, under acidic conditions, would generate the carbonium ion **120** that would be expected to eliminate a proton to provide cleavamine or hydrate to generate the thermodynamically more favorable velbanamine system. Indeed, when isovelbanamine was treated with concentrated sulfuric acid, the product was cleavamine **5**, but when aqueous sulfuric acid was utilized, velbanamine **4** was isolated. The conversion of cleavamine to the 18β-carbomethoxy derivative **124** was accomplished by the chloroindolenine approach already described. The

(33) $\xrightarrow{\text{1) LiAlH}_4\text{; 2) p-TsCl}}$ (116) $\xrightarrow{(CH_3CH_2)_3N}$ (114) $\xrightarrow{\text{base}}$ 115

$\xrightarrow{OsO_4}$ (117) $\xrightarrow{\text{1) NaBH}_4\text{; 2) NaIO}_4}$ (118) $\xrightarrow{LiAlH_4}$

(119) $\xrightarrow{H_2SO_4,\ H_2O}$ (4)

119 $\xrightarrow{H^+,\ -H_2O}$ (120) $\xrightarrow{\text{conc. } H_2SO_4}$ (5)

$\xrightarrow{(CH_3)_3COCl}$ (121) $\xrightarrow{NaOAc,\ HOAc}$ (122)

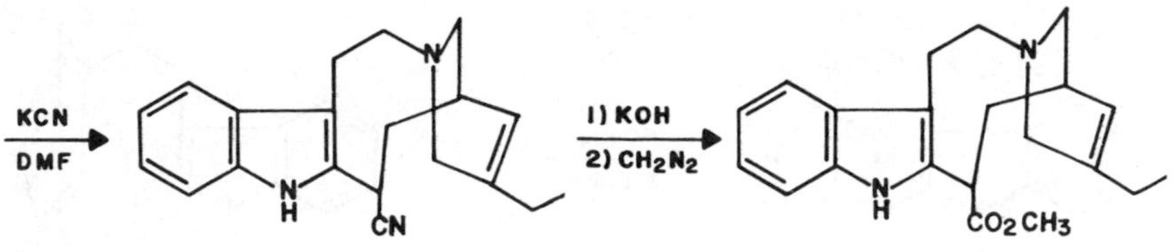

KCN DMF (123) 1) KOH 2) CH_2N_2 (124)

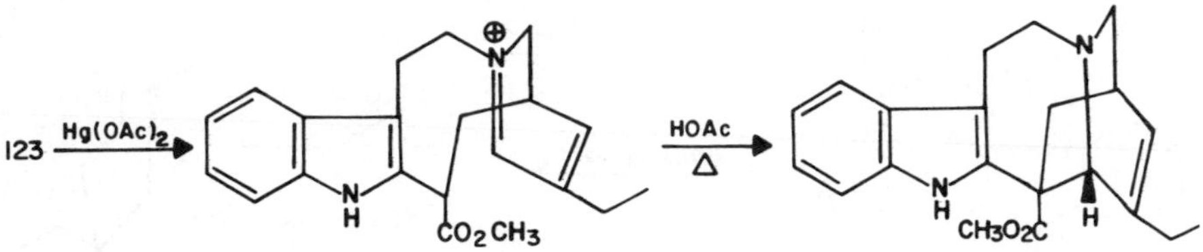

123 $Hg(OAc)_2$ (125) HOAc Δ (7)

transannular cyclization of **124** to catharanthine **7** was also performed according to the methods previously developed in our laboratory and discussed earlier in this section. This sequence, therefore, completed the total synthesis of this *Vinca* alkaloid.

Narisada, Watanabe, and Nagata[241] have achieved a synthesis of velbanamine **4** and isovelbanamine **119** utilizing an oxidative fragmentation in the key step of the synthetic pathway. The hydroxylactam **126**, available from previous studies in the *Iboga* series on reaction with lead tetraacetate, undergoes a fragmentation to the tetracyclic skeleton shown in **127**, and this compound is then converted to the final systems by the sequence of reactions shown.

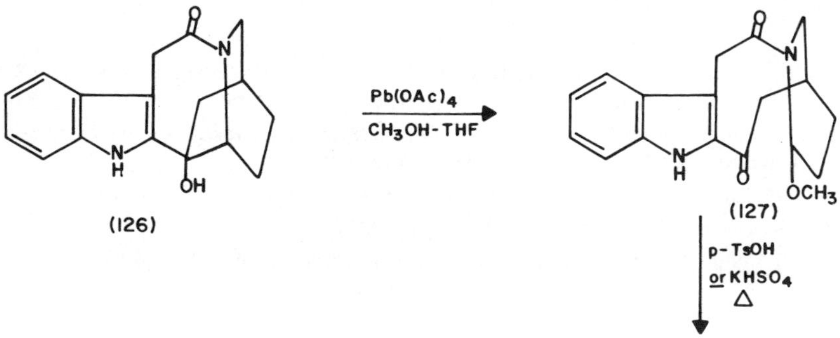

(126) $Pb(OAc)_4$ CH_3OH-THF (127) p-TsOH or $KHSO_4$ Δ

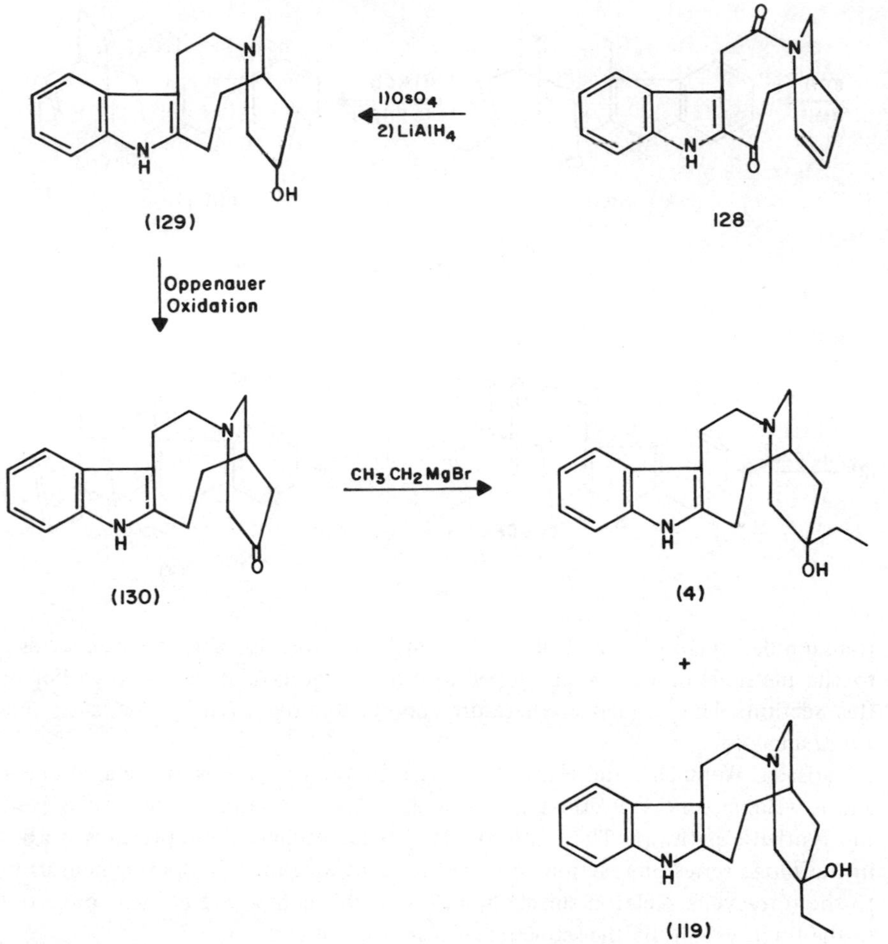

(129) ← 1) OsO_4 2) $LiAlH_4$ — 128

Oppenauer Oxidation

(130) → CH_3CH_2MgBr → (4) + (119)

REFERENCES

1. P. L. Julian, E. W. Meyer, and H. C. Printy, *Heterocyclic Compounds,* Vol. 3, (R. C. Elderfield, Ed.), Wiley, New York, 1952, p. 1.
2. W. C. Sumpter and F. M. Miller, *The Chemistry of Heterocyclic Compounds,* Vol. 8, (A. Weissberger, Ed.), Interscience, New York, 1954.
3. R. J. Sundberg, *The Chemistry of Indoles,* Academic Press, New York, 1970.
4. B. Robinson, *Chem. Rev.,* **63**, 373 (1963); **69**, 227 (1969).
5. H. J. Shine, *Aromatic Rearrangements,* Elsevier, Amsterdam, 1967, p. 190.
6. R. A. Abramovitch, *J. Chem. Soc.,* 4593 (1956).

7. R. A. Abramovitch and D. Shapiro, *J. Chem. Soc.*, 4589 (1956).
8. G. R. Clemo and G. A. Swan, *J. Chem. Soc.*, 617 (1946).
9. G. Stork and J. E. Dolfini, *J. Amer. Chem. Soc.*, **85**, 2872 (1963).
10. Y. Ban, Y. Sato, I. Inoue, M. Nagai (nee Seo), T. Oishi, M. Terashima, O. Yonemitsu, and Y. Kanaoka, *Tetrahedron Lett.*, 2261 (1965).
11. R. B. Woodward, F. E. Bader, H. Bickel, A. J. Frey, and R. W. Kierstead, *J. Amer. Chem. Soc.*, 78, 2023, 2657 (1956); *Tetrahedron*, **2**, 1 (1958).
12. J. P. Kutney, K. K. Chan, A. Failli, J. H. Fromson, C. Gletsos, and V. R. Nelson, *J. Amer. Chem. Soc.*, **90**, 3891 (1968).
13. J. V. Braun, K. Heider, and E. Muller, *Berichte*, **50**, 1637 (1917).
14. P. L. Julian and J. Pikl, *J. Amer. Chem. Soc.*, **57**, 539 (1935).
15. P. L. Julian and J. Pikl, *J. Amer. Chem. Soc.*, **57**, 563 (1935).
16. P. L. Julian and J. Pikl, *J. Amer. Chem. Soc.*, **57**, 755 (1935).
17. R. J. Sundberg, *The Chemistry of Indoles*, Academic Press, New York, 1970, Chapter III.
18. M. E. Speeter and W. C. Anthony, *J. Amer. Chem. Soc.*, **76**, 6208 (1954).
19. H. Kuhn and O. Stein, *Berichte*, **70**, 567 (1937).
20. T. A. Geissman and A. Armen, *J. Amer. Chem. Soc.*, **74**, 3916 (1952).
21. H. R. Snyder, C. W. Smith, and J. M. Stewart, *J. Amer. Chem. Soc.*, **66**, 200 (1944).
22. C. Hausch and J. C. Godfrey, *J. Amer. Chem. Soc.*, **73**, 3518 (1951).
23. E. L. Eliel and N. J. Murphy, *J. Amer. Chem. Soc.*, **75**, 3589 (1953).
24. M. E. Speeter, R. V. Heinzelmann, and D. I. Weisblat, *J. Amer. Chem. Soc.*, **73**, 5515 (1951).
25. K. E. Hamlin and F. E. Fisher, *J. Amer. Chem. Soc.*, **73**, 5007 (1951).
26. R. B. Woodward, M. P. Cava, W. D. Ollis, A. Hunger, H. U. Daeniker, and K. Schenker, *J. Amer. Chem. Soc.*, **76**, 4749 (1954); *Tetrahedron*, **19**, 247 (1963).
27. G. Buchi, D. L. Coffen, K. Kocsis, P. E. Sonnet, and F. E. Ziegler, *J. Amer. Chem. Soc.*, 87, 2073 (1965).
28. G. Buchi, D. L. Coffen, K. Kocsis, P. E. Sonnet, and F. E. Ziegler, *J. Amer. Chem. Soc.*, 88, 3099 (1966).
29. A. H. Jackson and A. E. Smith, *Tetrahedron*, **24**, 403 (1968).
30. A. H. Jackson and P. Smith, *Tetrahedron*, **24**, 2227 (1968).
31. A. H. Jackson, B. Naidoo, and P. Smith, *Tetrahedron*, **24**, 6119 (1968).
32. K. M. Biswas and A. H. Jackson, *Tetrahedron*, **25**, 227 (1969).
33. A. Pictet and T. Spengler, *Berichte*, **44**, 2030 (1911).
34. W. M. Whaley and T. R. Govindachari, *Organic Reactions*, Vol. 6, Wiley, New York, 1951, p. 151.
35. J. E. D. Barton and J. Harley-Mason, *Chem. Comm.*, **298** (1965).
36. M. E. Kuehne, *J. Amer. Chem. Soc.*, **86**, 2946 (1964); *Lloydia*, **27**, 435 (1964).
37. E. Wenkert, *Acct. Chem. Res.*, **1**, 78 (1968).
38. E. Wenkert, K. G. Dave, C. T. Gnewuch, and P. W. Sprague, *J. Amer. Chem. Soc.*, **90**, 5251 (1968).
39. E. Wenkert, K. G. Dave, and F. Haglin, *J. Amer. Chem. Soc.*, **87**, 5461 (1965).

40. E. E. van Tamelen, M. Shamma, A. W. Burgstahler, J. Wolinsky, R. Tamm, and P. E. Aldrich, *J. Amer. Chem. Soc.,* **80**, 5006 (1958).
41. E. E. van Tamelen, M. Shamma, A. W. Burgstahler, J. Wolinsky, R. Tamm, and P. E. Aldrich, *J. Amer. Chem. Soc.,* **91**, 7315 (1969).
42. A. Bischler and B. Napieralski, *Berichte,* **26**, 1903 (1893).
43. W. M. Whaley and T. R. Govindachari, *Org. Reactions,* **6**, 74 (1951).
44. R. A. Abramovitch and I. D. Spenser, *Adv. Heterocyclic Chem.,* **3**, 79 (1964).
45. E. E. van Tamelen and I. G. Wright, *J. Amer. Chem. Soc.,* **91**, 7349 (1969).
46. E. E. van Tamelen and J. B. Hester, *J. Amer. Chem. Soc.,* **91**, 7342 (1969).
47. E. E. van Tamelen, C. Placeway, G. P. Schiemenz, and I. G. Wright, *J. Amer. Chem. Soc.,* **91**, 7359 (1969).
48. E. Wenkert and B. Wickberg, *J. Amer. Chem. Soc.,* 87, 1580 (1965).
49. L. Marion, in *The Alkaloids,* Vol. I (R. H. F. Mausk, Ed.), Academic, New York, 1952, p. 369.
50. J. E. Saxton, in *The Alkaloids,* Vol. VII (R. H. F. Mausk, Ed.), Academic, New York, 1960, p. 4.
51. J. E. Saxton, in *The Alkaloids,* Vol. VIII (R. H. F. Mausk, Ed.), Academic, New York, 1965, p. 4.
52. J. E. Saxton, in *The Alkaloids,* Vol. X (R. H. F. Mausk, Ed.), Academic New York, 1968, p. 491.
53. H. R. Snyder and V. W. Smith, *J. Amer. Chem. Soc.,* **66**, 350 (1944).
54. E. E. Howe, A. J. Zambito, H. R. Snyder, and M. Tishler, *J. Amer. Chem. Soc.,* **67**, 38 (1945).
55. N. F. Albertson, S. Archer, and C. M. Suter, *J. Amer. Chem. Soc.,* **66**, 500 (1944).
56. N. F. Albertson, S. Archer and C. M. Suter, *J. Amer. Chem. Soc.,* **67**, 36 (1945).
57. D. A. Lyttle and D. I. Weisblat, *J. Amer. Chem. Soc.,* **69**. 2118 (1947).
58. D. I. Weisblat and D. A. Lyttle, *J. Amer. Chem. Soc.,* **71**, 3079 (1949).
59. V. L. Stromberg, *J. Amer. Chem. Soc.,* **76**, 1707 (1954).
60. M. S. Fish, N. M. Johnson, and E. F. Horning, *J. Amer. Chem. Soc.,* **77**, 5892 (1955).
61. J. Harley-Mason and A. H. Jackson, *J. Chem. Soc.,* 1165 (1954).
62. A. Stoll, F. Troxler, J. Peyer, and A. Hofmann, *Helv. Chim. Acta,* **38**, 1452 (1955).
63. A. Hofmann, R. Heim, A. Brack, and H. Kobel, *Experientia,* **14**, 107 (1958).
64. A. Hofmann, R. Heim, A. Brack, H. Kobel, A. Frey, H. Ott, T. Petrzilka, and F. Troxler, *Helv. Chim. Acta,* **42**, 1557 (1959).
65. A. Hofmann, A. Frey, H. Ott, T. Petrzilka, and F. Troxler, *Experientia,* **14**, 397 (1958).
66. A. Hofmann and F. Troxler, *Experientia,* **15**, 101 (1959).
67. G. P. Lewis, ed., *5-Hydroxytryptamine.* Permagon Press, New York, 1958; Symposium on 5-Hydroxytryptamine, *Ann. N.Y. Acad. Sci.,* **66**, 592 (1957).
68. S. Udenfriend, P. A. Shore, D. F. Bogdanski, H. Weisback, and B. B. Brodie, *Recent Progr. Hormone Res.,* **13**, 1 (1957).
69. E. H. P. Young, *J. Chem. Soc.,* 3493 (1958).
70. A. S. F. Ash and W. R. Wragg, *J. Chem. Soc.,* 3887 (1958).
71. W. E. Noland and R. A. Hovden, *J. Org. Chem.,* **24**, 894 (1959).

72. T. Kametani and K. Fulumoto, *Japan J. Pharm. Chem.*, **33**, 83 (1961); *Chem. Abstr.*, **55**, 19897 (1961).
73. R. A. Abramovitch and D. Shapiro, *Chem. Ind. (London)*, 1255 (1955).
74. R. Justoni and R. Pessina, *Farmaco (Pavia) Ed. Sci.*, **10**, 356 (1955); *Chem. Abstr.*, **49**, 13986 (1955).
75. J. D. Crum and P. W. Sprague, *Chem. Comm.*, 417 (1966).
76. D. P. Chakraborty, K. C. Das, and B. K. Chowdhury, *Chem. Ind. (London)*, 1684 (1966).
77. W. F. Gannon, J. D. Benigni, J. Suzuki, and J. W. Daly, *Tetrahedron Lett.*, 1531 (1967).
78. L. Marion, in *The Alkaloids, II*, 393 (1952); R. H. F. Manske, in *The Alkaloids, VIII*, 47 (1965).
79. R. H. F. Manske, W. H. Perkin, Jr., and R. Robinson, *J. Chem. Soc.*, 1 (1927).
80. E. Spath and E. Lederer, *Berichte*, **63**, 120 (1930).
81. I. D. Spencer, *Can. J. Chem.*, **37**, 1851 (1959).
82. A. Stoll and A. Hofmann, in *The Alkaloids*, Vol. VIII. 1965, p. 725.
83. F. C. Uhle and W. A. Jacobs, *J. Org. Chem.*, **10**, 76 (1945).
84. A. Stoll, J. Rutschmann, and W. Schlientz, *Helv. Chim. Acta*, **33**, 375 (1950).
85. A. Stoll and J. Rutschmann, *Helv. Chim. Acta*, **33**, 67 (1950).
86. A. Stoll and J. Rutschmann, *Helv. Chim. Acta*, **35**, 1512 (1953).
87. A. Stoll and J. Rutschmann, *Helv. Chim. Acta*, **37**, 814 (1954).
88. A. Stoll, Th. Petrzilka, and J. Rutschmann, *Helv. Chim. Acta*, **35**, 1249 (1952).
89. A. Stoll and Th. Petrzilka, *Helv. Chim. Acta*, **36**, 1125 (1953).
90. A. Stoll and Th. Petrzilka, *Helv. Chim. Acta*, **36**, 1137 (1953).
91. F. R. Atherton, F. Bergel, A. Cohen, B. Heath-Brown, and A. H. Rees, *Chem. Ind. (London)*, 1151 (1953).
92. E. C. Kornfeld, E. J. Fornefeld, G. B. Kline, M. J. Mann, R. G. Jones, and R. B. Woodward, *J. Amer. Chem. Soc.*, **76**, 5256 (1954).
93. K. C. Kornfeld, E. J. Fornefeld, G. B. Kline, M. N. Mann, D. E. Morrison, R. G. Jones, and R. B. Woodward, *J. Amer. Chem. Soc.*, **78**, 3987 (1956).
94. M. M. Kolosov, L. I. Metreveli, and N. A. Preabrazhensky, *J. Gen. Chem. USSR*, **23**, 2143 (1953).
95. S. Sugasawa and M. Murayama, *Chem. Pharm. Bull. (Tokyo)*, **6**, 194 (1958).
96. S. Sugasawa and M. Murayama, *Chem. Pharm. Bull. (Tokyo)*, **6**, 200 (1958).
97. J. Harley-Mason and A. H. Jackson, *J. Chem. Soc.*, 3651 (1954).
98. B. Witkop and R. K. Hill, *J. Amer. Chem. Soc.*, **77**, 6592 (1955).
99. P. Roesnmund and A. Sotiriou, *Angew. Chem.*, **76**, 787 (1964).
100. R. B. Longmore and B. Robinson, *Chem. Ind. (London)*, 1297 (1965).
101. R. Robinson and H. J. Teuber, *Chem. Ind. (London)*, 783 (1954).
102. R. B. Woodward, W. C. Yang, T. J. Katz, V. M. Clark, J. Harley-Mason, R. J. F. Ingleby, and N. Sheppard, *Proc. Chem. Soc.*, 76 (1960).
103. J. B. Hendrickson, R. Rees, and R. Goschke, *Proc. Chem. Soc.*, 393 (1962); *Tetrahedron*, **20**, 565 (1964).

104. A. I. Scott, F. McCapra, and E. S. Hall, *J. Amer. Chem. Soc.,* **86**, 302 (1964); E. S. Hall, F. McCapra, and A. I. Scott, *Tetrahedron,* **23**, 4131 (1967).
105. M. G. Reinecke, H. W. Johnson, and J. F. Sebastian, *Tetrahedron Lett.,* 1183 (1963).
106. G. Champetier, *Bull. Chim. Soc.,* **47**, 1131 (1930).
107. T. Hino and S. Yamada, *Tetrahedron Lett.,* 1757 (1963).
108. T. Hino, *Chem. Pharm. Bull.,* **9**, 979, 988 (1961).
109. R. H. F. Manske and H. L. Holmes, *The Alkaloids,* Vol. II (R. H. F. Mausk, Ed.), Academic, New York, 1952, p. 420.
110. R. H. F. Manske, *The Alkaloids,* Vol. VII (R. H. F. Mausk, Ed.), Academic, New York, 1960, p. 37.
111. A. R. Battersby, *Pure Appl. Chem.,* **14**, 117 (1967).
112. E. Leete, *Acct. Chem. Res.,* **2**, 59 (1969).
113. I. D. Spenser, in *Comprehensive Biochemistry,* Vol. 20, (M. Florkin and E. H. Stotz, Eds., Elsevier, New York, Chapter 6 (1968).
114. A. I. Scott, *Acct. Chem. Res.,* **3**, 151 (1970).
115. J. P. Kutney, J. F. Beck, C. Ehret, G. Poulton, R. S. Sood, and N. D. Westcott, *Bioorg. Chem.,* **1**, 194 (1971).
116. K. T. D. De Silva, G. N. Smith, and K. E. H. Warren, *Chem. Comm.,* 905 (1971).
117. W. P. Blackstock, R. T. Brown, and G. K. Lee, *Chem. Comm.,* 910 (1971).
118. N. A. Preobrazhenskii, *J. Gen. Chem. USSR,* **22**, 1467 (1952).
119. E. E. van Tamelen, P. E. Aldrich, and J. B. Hester, *J. Amer. Chem. Soc.,* **79**, 4817 (1957).
120. E. E. van Tamelen and I. G. Wright, *Tetrahedron Lett.,* 295 (1964).
121. J. W. Powell and M. C. Whiting, *Tetrahedron,* **12**, 168 (1961).
122. R. L. Autrey and P. W. Scullard, *J. Amer. Chem. Soc.,* **90**, 4917 (1968).
123. G. A. Swan, *J. Chem. Soc.,* 1534 (1950).
124. J. A. Weisbach, J. L. Kirkpatrick, K. R. Williams, E. L. Anderson, N. C. Yim, and B. Douglas, *Tetrahedron Lett.,* 3457 (1965).
125. Cs. Szantay and M. Barczai-Beke, *Tetrahedron Lett.,* 1405 (1968).
126. Cs. Szantay and M. Barczai-Beke, *Chem. Ber.,* **102**, 3963 (1969).
127. E. Wenkert, K. G. Dave, R. G. Lewsi, and P. W. Sprague, *J. Amer. Chem. Soc.,* **89**, 6741 (1967).
128. F. E. Ziegler and J. G. Sweeney, *Tetrahedron Lett.,* 1097 (1969).
129. A. E. Wick, D. Felix, K. Steen, and A. Eschenmoser, *Helv. Chim. Acta,* **47**, 2425 (1964).
130. Y. K. Sawa and H. Matsumura, *Tetrahedron,* **25**, 5319, 5329 (1969); **26**, 2931 (1970).
131. E. E. van Tamelen and C. Placeway, *J. Amer. Chem. Soc.,* **83**, 2594 (1961).
132. F. Korte and H. Machleidt, *Chem. Ber.,* **88**, 136 (1955).
133. W. I. Taylor in *The Alkaloids,* Vol. 8 (R. H. F. Manske, Ed.), Academic, New York, 1965, Chapter 22.
134. W. I. Taylor in *The Alkaloids,* Vol. II (R. H. F. Manske, Ed.), Academic, New York, (1968), Chapter 2.
135. S. Masamune, Sining K. Ang, C. Egli, N. Nakatsuka, S. K. Sarkar, and Y. Yasunari, *J. Amer. Chem. Soc.,* **89**, 2506 (1967).

136. F. A. L. Anet, D. Chakravarti, R. Robinson, and E. Schlittler, *J. Chem. Soc.*, 1242 (1954).

137. E. E. van Tamelen and L. K. Oliver, *J. Amer. Chem. Soc.*, **92**, 2136 (1970).

138. M. F. Bartlett, B. F. Lambert, H. H. Werblood, and W. I. Taylor, *J. Amer. Chem. Soc.*, **85**, 475 (1963).

139. J. D. Hobson and J. G. McCluskey, *J. Chem. Soc.*, 2015 (1967).

140. K. Mashimo and Y. Sato, *Tetrahedron Lett.*, 901 (1969); *Tetrahedron,* **26**, 803 (1970).

141. N. Yoneda, *Chem. Pharm. Bull. Japan,* **13**, 1231 (1965).

142. K. Mashimo and Y. Sato, *Tetrahedron Lett.*, 905 (1969).

143. J. A. Weisbach and B. Douglas, *Chem. Ind. (London),* 623 (1965); 233 (1966).

144. T. Shiori and S. Yamada, *Tetrahedron,* **24**, 4159 (1968).

145. L. Velluz, G. Muller, R. Joly, G. Nomine, J. Mathieu, A. Allais, J. Warnant, J. Valls, R. Bucoust, and J. Joly, *Bull. Soc. Chim. France,* 673 (1958).

146. I. Ernest and B. Kakac, *Chem. Ind. (London),* 513 (1965).

147. G. C. Morrinson, W. A. Cetenko, and J. Shavel, Jr., *J. Org. Chem.*, **31**, 2695 (1966).

148. K. T. Potts and R. Robinson, *J. Chem. Soc.*, 2675 (1955). See also P. L. Julian and A. Magnani, *J. Amer. Chem. Soc.*, **71**, 3207 (1949); B. Belleau, *Chem. Ind. (London),* 229 (1955).

149. E. Wenkert, R. A. Massy-Westropp, and R. G. Lewis, *J. Amer. Chem. Soc.*, **84**, 3732 (1962).

150. E. Wenkert and B. Wickberg, *J. Amer. Chem. Soc.*, **84**, 4914 (1962).

151. G. C. Morrinson, W. Cetenko, and J. Shavel, Jr., *J. Org. Chem.*, **32**, 4089 (1967).

152. J. A. Beisler, *Tetrahedron,* **26**, 1961 (1970).

153. E. M. Fry, *J. Org. Chem.*, **28**, 1869 (1963); **29**, 1647 (1964).

154. J. A. Beisler, *Chem. Ber.*, **103**, 3360 (1970).

155. (a) L. Toke, K. Honty, and Cs. Szantay, *Chem. Ber.*, **102**, 3248 (1969). (b) Cs. Szantay, K. Honty, L. Toke, A. Buzas, and J. P. Jacquet, *Tetrahedron Lett.*, 4871 (1971).

156. W. Meise and F. Zymalkowski, *Angew. Chem. Int. Ed.*, **8**, 445 (1969).

157. F. E. Ziegler and J. G. Sweeny, *J. Org. Chem.*, **34**, 3545 (1969).

158. E. Winterfeldt and H. Riesner, *Synthesis,* 261 (1970).

159. F. V. Brutcher, Jr., W. D. Vanderwerff, and B. Dreikorn, *J. Org. Chem.*, **37**, 297 (1972).

160. N. Finch and W. I. Taylor, *J. Amer. Chem. Soc.*, **84**, 1318 (1962).

161. J. Shavel and H. Zinnes, *J. Amer. Chem. Soc.*, **84**, 1320 (1962).

162. Y. Ban and T. Oishi, *Chem. Pharm. Bull. Japan,* **11**, 451 (1963).

163. J. B. Hendrickson and R. A. Silva, *J. Amer. Chem. Soc.*, **84**, 643 (1962).

164. L. Castedo, J. Harley-Mason, and M. Kaplan, *Chem. Comm.*, 1444 (1969).

165. E. E. van Tamelen, J. P. Yardley, and M. Miyano, *Tetrahedron Lett.*, 1011, (1963).

166. E. E. van Tamelen, J. P. Yardley, M. Miyano, and W. B. Hinshaw, *J. Amer. Chem. Soc.*, **91**, 7333 (1969).

167. L. J. Dolby and S. Sakai, *J. Amer. Chem. Soc.*, **86**, 1890 (1964).

168. K. Freter, H. H. Hubner, H. Merz, H. D. Schroeder, and K. Zeile, *Annalen,* **684**, 159

(1965).

169. K. Freter and K. Zeile, *Chem. Comm.*, 416 (1967).
170. B. A. Dadson, J. Harley-Mason, and G. H. Foster, *Chem. Comm.*, 1233 (1968).
171. B. A. Dadson, J. Harley-Mason, *Chem. Comm.*, 665 (1969).
172. D. Schumann and H. Schmid, *Helv. Chim. Acta*, **46**, 1966 (1963).
173. B. A. Dadson and J. Harley-Mason, *Chem. Comm.*, 665 (1969).
174. J. Harley-Mason and C. G. Taylor, *Chem. Comm.*, 812 (1970).
175. G. C. Crawley and J. Harley-Mason, *Chem. Comm.*, 685 (1971).
176. J. R. Hymon and H. Schmid, *Helv. Chim. Acta*, **49**, 2067 (1966).
177. E. Wenkert and R. Sklar, *J. Org. Chem.*, 2689 (1966).
178. G. F. Smith and J. T. Wrobel, *J. Chem. Soc.*, 792 (1960).
179. K. Bieman and G. Spiteller, *Tetrahedron Lett.*, 299 (1961).
180. Y. Ban and I. Iijima, *Tetrahedron Lett.*, 2523 (1969).
181. M. E. Kuehne and C. Bayha, *Tetrahedron Lett.*, 1311 (1966).
182. J. P. Kutney and E. Piers, *J. Amer. Chem. Soc.*, **86**, 953 (1964).
183. J. P. Kutney, R. T. Brown, and E. Piers, *J. Amer. Chem. Soc.*, **86**, 2286, 2287 (1964).
164. A. Camerman, N. Camerman, J. P. Kutney, E. Piers, and J. Trotter, *Tetrahedron Lett.*, 637 (1965).
185. J. P. Kutney, R. T. Brown, and E. Piers, *Can. J. Chem.*, **44**, 637 (1966).
186. J. P. Kutney, E. Piers, and R. T. Brown, *J. Amer. Chem. Soc.*, **92**, 1700 (1970).
187. J. P. Kutney, N. Abdurahman, P. LeQuesne, E. Piers, and I. Vlattas, *J. Amer. Chem. Soc.*, 88, 3656 (1966).
188. J. P. Kutney, N. Abdurahman, C. Gletsos, P. LeQuesne, E. Piers, and I. Vlattas, *J. Amer. Chem. Soc.*, **92**, 1927 (1970).
189. J. Mokry and I. Kompis, *Lloydia*, **27**, 428 (1964).
190. J. Harley-Mason and M. Kaplan, *Chem. Comm.*, 915 (1967).
191. F. E. Ziegler, J. A. Kloek, and P. A. Zoretic, *J. Amer. Chem. Soc.*, **91**, 2343 (1969).
192. F. E. Ziegler, and E. B. Spitzner, *J. Amer. Chem. Soc.*, **92**, 3492 (1970).
193. F. E. Ziegler and G. B. Bennett, *Tetrahedron Lett.*, 2545 (1970).
194. H. P. Husson, C. Thal, P. Potier, and E. Wenkert, *Chem. Comm.*, 480 (1970).
195. F. E. Ziegler and G. B. Bennett, *J. Amer. Chem. Soc.*, **93**, 5930 (1971).
196. J. P. Kutney, W. J. Cretney, P. LeQuesne, B. McKauge, and E. Piers, *J. Amer. Chem. Soc.*, **92**, 1712 (1970).
197. R. V. Stevens, J. M. Fitzpatrick, M. Kaplan, and R. L. Zimmerman, *Chem. Comm.*, 857 (1971).
198. I. Inoue and Y. Ban, *J. Chem. Soc. (C)*, 602 (1970).
199. G. Buchi, K. E. Matsumoto, and H. Nishimura, *J. Amer. Chem. Soc.*, 93, 3299 (1971).
200. J. Harley-Mason and W. R. Waterfield, *Tetrahedron*, **19**, 65 (1963).
201. A. Jackson, A. J. Gaskell, N. D. V. Wilson, and J. A. Joule, *Chem. Comm.*, 364 (1968).
202. N. D. V. Wilson, A. Jackson, A. J. Gaskell, and J. A. Joule, *Chem. Comm.*, 584 (1968).
203. A. Jackson, N. D. V. Wilson, A. J. Gaskell, and J. A. Joule, *J. Chem. Soc. (C)*, 2738 (1969).

204. L. J. Dolby and H. Biere, *J. Amer. Chem. Soc.*, **90**, 2699 (1968).
205. L. J. Dolby and H. Biere, *J. Org. Chem.*, **35**, 3843 (1970).
206. G. Buchi, S. J. Gould, and F. Naf, *J. Amer. Chem. Soc.*, **93**, 2492 (1971).
207. T. Kametani and T. Suzuki, *J. Chem. Soc. (C)*, 1053 (1971).
208. T. Kametani and T. Suzki, *J. Org. Chem.*, **36** 1291 (1971).
209. T. Kametani and T. Suzuki, *Chem. Pharm. Bull. Japan*, **19**, 1424 (1971).
210. R. B. Woodward, G. A. Iacubucci, and F. A. Hochstein, *J. Amer. Chem. Soc.*, **81**, 4434 (1959).
211. G. Buchi, D. W. Mayo, and F. A. Hochstein, *Tetrahedron*, **15**, 167 (1961).
212. P. A. Cranwell and J. E. Saxton, *J. Chem. Soc.*, 3482 (1962).
213. T. R. Govindachari, S. Rajappa, and V. Sudarsan, *Indian J. Chem.*, **1**, 247 (1963).
214. L. K. Dalton, S. Demarac, B. C. Elmes, J. W. Loder, J. M. Swan, and T. Teitei, *Austral J. Chem.*, **20**, 2715 (1967).
215. J. Schmutz and H. Wittwer, *Helv. Chim. Acta*, **43**, 793 (1960).
216. E. Wenkert and K. G. Dave, *J. Amer. Chem. Soc.*, **84**, 94 (1962).
217. C. W. Mosher, O. P. Crews, E. M. Acton, and L. Goodman, *J. Org. Chem.*, **9**, 237 (1966).
218. T. Wieland and E. Neeb, *Annalen*, **600**, 161 (1956).
219. M. F. Bartlett and W. I. Taylor, *J. Amer. Chem. Soc.*, **82**, 5491 (1960).
220. J. E. D. Barton, J. Harley-Mason, and K. C. Yates, *Tetrahedron Lett.*, 3669, (1965).
221. L. Castedo, J. Harley-Mason, and T. J. Leeny, *Chem. Comm.*, 1186 (1968).
222. K. H. Gibson and J. E. Saxton, *Chem. Comm.*, 799 (1969).
223. K. H. Gibson and J. E. Saxton, *Chem. Comm.*, 1490 (1969).
224. J. P. Kutney, R. T. Brown, E. Piers, and J. R. Hadfield, *J. Amer. Chem. Soc.*, **92**, 1708 (1970).
225. M. Gorman, N. Neuss, N. J. Cone, and J. A. Deyrup, *J. Amer. Chem. Soc.*, **82**, 1142 (1960).
226. J. Harley-Mason, A. Rahman, and J. A. Beisler, *Chem. Comm.*, 743 (1966).
227. J. Harley-Mason and A. Rahman, *Chem. Comm.*, 208 (1967).
228. J. Harley-Mason and A. Rahman, *Chem. Comm.*, 1048 (1967).
229. J. Harley-Mason and A. Rahman, *Chem. Ind. (London)*, 1845 (1968).
230. A. P. Krapcho, G. A. Glynn, and B. J. Grenon, *Tetrahedron Lett.*, 215 (1967).
231. W. Nagata, S. Hirai, K. Kawata, and T. Aoki, *J. Amer. Chem. Soc.*, **89**, 5045 (1967).
232. W. Nagata, S. Hirai, T. Okumura, and K. Kawata, *J. Amer. Chem. Soc.*, **90**, 1650 (1968).
233. S. I. Sallay, *J. Amer. Chem. Soc.*, **89**, 6762 (1967).
234. M. Ikezaki, T. Wakamatsu, and Y. Ban, *Chem. Comm.*, 88 (1969).
235. P. Rosenmund, W. H. Haase, and J. Bauer, *Tetrahedron Lett.*, 4121 (1969).
236. J. W. Huffman, C. B. S. Rao, and T. Kamiya, *J. Org. Chem.*, **32**, 697 (1967).
237. R. L. Augustine and W. G. Pierson, *J. Org. Chem.*, **34**, 1070 (1969).
238. G. Buchi, P. Kulsa, and R. L. Rosati, *J. Amer. Chem. Soc.*, **90**, 2448 (1968).
239. G. Buchi, P. Kulsa, K. Ogasawara, and R. L. Rosati, *J. Amer. Chem. Soc.*, **92**, 999 (1970).

240. J. P. Kutney and F. Bylsma, *J. Amer. Chem. Soc.*, **92**, 6090 (1970).

241. M. Narisada, F. Watanabe, and W. Nagata, *Tetrahedron Lett.*, 3681 (1971).

APPENDIX

In order to update this chapter from the time of its writing, the following list of references is provided. This list is not intended to be exhaustive but rather to provide more recent references in the areas that have been either mentioned or that are rather closely related to those discussed.

1. C. Thal, T. Sevenet, H. P. Husson, and P. Potier, Synthesis of (±)-deethylvincamine and (±)-vincamine. *C. R. Acad. Sci. Ser. (C)*, **275**, 1295 (1972).
2. K. Mori, I. Takemoto, and M. Matsui, Synthesis of (±)-yohimbone. *Arg. Biol. Chem.*, **36**, 2605 (1972).
3. Y. Ban, T. Ohnuma, N. Nagai, Y. Sendo, and T. Oishi, Conversion of oxidoles to the aspidosperma skeleton. Synthesis of dl-*N*(a)-acetyldeethylaspidospermidine. *Tetrahedron Lett.*, 5023 (1972).
4. Y. Ban, Y. Sendo, M. Nagai, and T. Oishi, Total synthesis of dl-*N*(a)-acetyl-7β-ethyl-5-deethylaspidospermidine. *Tetrahedron Lett.*, 5027 (1972).
5. T. Kametani, T. Yamanaka, and K. Nyu, Synthesis of heterocyclic compounds. CDXCIX. Synthesis of pseudorutecarpine with triethyl phosphite. *J. Heterocyclic Chem.*, 9, 1281 (1972).
6. Cs Szantay, L. Szabo, and Gy. Kalaus, Stereoselective total synthesis of (+)-vancamine. *Tetrahedron Lett.*, 191 (1973).
7. S. G. Agbalyan and V. V. Darbinyan, Syntheses based on harmine and tetrahydroharmine. IV. Esters of tetrahydroharmylethanol. *Arm. Khim. Zh.*, **25**, 693 (1972).
8. S. G. Agbalyan and V. V. Darbinyan, Syntheses based on harmine and tetrahydroharmine. II. Methoxy derivatives of 1,2,3,4-tetrahydro-1-(2′-phenylethyl)norharmine. *Arm. Khim. Zh.*, **25**, 689 (1972).
9. P. Rosenmund, W. H. Haase, J. Bauer, and R. Frische, Chemistry in indole. IV. Syntheses in the iboga series. I. (d,l)-Deethylibogamine. *Chem. Ber.*, **106**, 1459 (1973).
10. P. Rosenmund, W. H. Haase, J. Bauer, and R. Frische, Chemistry of indole. V. Synthesis in the iboga series. II. 8-Oxodeethylibogaine and deethylisoibogaine. *Chem. Ber.*, **106**, 1474 (1973).
11. L. Toke, Z. Gombos, G. Blasko, K. Honty, L. Szabo, J. Tamas, and C. Szantay, Synthesis of yohimbines. II. Alternative route to alloyohimbine alkaloids. *J. Org. Chem.*, **38**, 2501 (1973).
12. L. Toke, K. Honty, L. Szabo, G. Blasko, and C. Szantay, Synthesis of yohimbines. I. Total synthesis of alloyohimbine and α-yohimbine and their epimers. Revised structure of natural alloyohimbine. *J. Org. Chem.*, 38, 2496 (1973).
13. L. J. Dolby and S. J. Nelson, Model studies of the synthesis of echitamine and related indole alkaloids. II. *J. Org. Chem.*, **38**, 2882 (1973).
14. E. Wenkert and G. D. Reynolds, Synthesis of indole alkaloids. XI. Vallesiachotamine models. *Syn. Comm.*, **3**, 241 (1973).

15. N. Aimi, E. Yamanaka, J. Endo, S. Sakai, and J. Haginiwa, Transformation of indole alkaloids. I. Conversion of oxindole alkaloids into indole alkaloids. *Tetrahedron,* **29**, 2015 (1973).

16. S. Sakai, A. Kubo, K. Katano, N. Shinma, and K. Sasago, Transformation of indole alkaloids. II. C/D ring opening and closing reactions of indole alkaloids and the syntheses of vobasine type alkaloids. *Yakugaku Zasshi,* **93**, 1165 (1973).

17. H. P. Husson, T. Imbert, C. Thal, and P. Potier, Indole series. IV. Stereospecific synthesis of (±)-deethyleburnamonine and of (±)-21-epideethyleburnamonine from a common intermediate. *Bull. Soc. Chim. France,* 2013 (1973).

18. C. Thal, T. Imbert, H. P. Husson, and P. Potier, Indole series. III. Synthesis of indoloquinolizidines by hydrogenation-cyclization of indolethylpyridinium salts in acidic medium. 21-Epideethyleburnamonine. *Bull. Soc. Chim. France.,* 2010 (1973).

19. E. Wenkert, P. W. Sprague, and R. L. Webb, General methods of synthesis of indole alkaloids. XII. Syntheses of di-18,19-dihydroantirhine and methyl demethylilludininate. *J. Org. Chem.,* **38**, 4305 (1973).

20. F. E. Ziegler and E. B. Spitzner, Biogenetically modeled synthesis via an indole acrylic ester Total synthesis of (±)-minovine. *J. Amer. Chem. Soc.,* **95**, 7146 (1973).

21. F. E. Aiegler and G. B. Bennet, Claisen rearrangement in indole alkaloid synthesis. Total synthesis of (±)-tabersonine. *J. Amer. Chem. Soc.,* **95**, 7458 (1973).

22. H. Rischke, J. D. Wilcock, and E. Winterfeldt, Reactions with indole derivatives. XIX. Stereoselective cyclizations in the indole series. *Chem. Ber.,* **106**, 3106 (1973).

23. A. Husson, Y. Langlois, C. Riche, H. P. Husson, and P. Potier, Indoles. VI. Transformation of vobasine alkaloids to dehydroervatamine alkaloids. X-ray analysis of ervatamine. *Tetrahedron,* **29**, 3095 (1973).

24. I. Ninomiya, H. Takasugi, and T. Naito, Total synthesis of yohimbine-type alkaloids. The yohimbine skeleton and angustidine. *J. Chem. Soc. Chem. Comm.,* 732 (1973).

25. T. Imbert, C. Thal, H. P. Husson, and P. Potier Indole series. V. Synthesis of eburnane and homoeburnane derivatives oxygenated at C-15. *Bull. Soc. Chim. France.,* 2705 (1973).

26. M. Koch, M. Plat and N. Preaux, Partial synthesis and stereochemistry at ochrolifuanines A and B, alkaloids from Ochrosia lifuana (Apocynaceae). *Bull. Soc. Chim. France,* 2868 (1973).

27. Y. Ban, N. Taga, and T. Oishi, Synthesis of 3-spirooxindole derivatives. Total syntheses of dl-formosanine, dl-isoformosanine, dl-mitraphylline and dl-isomitraphylline. *Tetrahedron Lett.,* 187 (1974).

28. E. Wenkert, J. S. Bindra, C. J. Chang, D. W. Cochran, and D. E. Rearick, General methods of synthesis of indole alkaloids. XIII. Oxindole alkaloid models. *J. Org. Chem.,* **39**, 1662 (1974).

29. C. Kan-Fan, G. Massiot, A. Ahond, B. C. Das, H. P. Husson, P. Potier, A. I. Scott, and C. C. Wei, Structure and biogenetictype synthesis of andranginine. Indole alkaloid of a new type. *J. Chem. Soc. Chem. Comm.,* 164 (1974).

30. J. L. Herrmann, R. J. Cregge, J. E. Richman, C. L. Semmelhack, and R. H. Schlessinger, High yield stereospecific total synthesis of vincamine. *J. Amer. Chem. Soc.,* **96**, 3702 (1974).

31. D. L. Coffen, D. A. Katonak, and F. Wong, Rearrangement of benzylidenequinuclidinones to tetrahydropyridoindoles. Novel synthesis of indole alkaloids of the eubrnamine type. *J. Amer. Chem. Soc.,* **96**, 3966 (1974).

32. S. Nakatsuka, T. Fukuyama, and Y. Kishi, Total synthesis of d,l-sporidesmin B. *Tetrahedron Lett.*, 1549 (1974).

33. T. Ohnuma, K. Seki, T. Oishi, and Y. Ban, Total Synthesis of the alkaloid (±)-deoxy-aspidodispermine. *J. Chem. Soc. Chem. Comm.*, 296 (1974).

34. T. Kametani, M. Kajiwara, and K. Fukumoto, Syntheses of heterocyclic compounds. DXLVII. Synthesis of a yohimbane derivative by thermolysis. *Tetrahedron,* **30**, 1053 (1974).

35. K. Dora-Horvath and O. Clauder, Alkaloids derived from indolo[2,3-c]quinazolino[3,2-a]pyridine. II. Synthesis and examination of rutecarpine-carboxylic acid. *Acta Pharm Hung. Suppl,* **44**, 80-2 (1974).

36. D. Cohylakis, G. J. Hignett, K. V. Lichman, and J. A. Joule, Synthesis of 5-ethyl-11H-pyrido[3,4-a]carbazole by two routes and conversion of uleine into 5-ethyl-1,2,3,4-tetrahydro-2-methyl-11H-pyrido[3,4-a]carbazole. *J. Chem. Soc. Perkin Trans. I,* 1518 (1974).

37. M. Sainsbury and B. Webb, Synthesis of 9-aminoellipticine (9-amino-5,11-dimethyl-6H-pyrido[4,3-b]carbazole) and related compounds. *J. Chem. Soc. Perkin Trans. I,* 1580 (1974).

38. K. B. Prasad and S. C. Shaw, Synthesis of indole alkaloids and related compounds. II. Synthesis of some 12H-benz[f]indolo[2,3-a]-pyridocolinium salts. *Indian J. Chem.*, **12**, 344 (1974).

39. E. Winterfeldt, Stereoselective total synthesis of indole alkaloids. *Fortschr. Chem. Org. Naturst.*, **31**, 469 (1974).

40. T. Kametani, T. Susuki, K. Takahashi and K. Fukumoto, Synthesis of heterocyclic compounds. DLIX. Synthesis of benzocarbazole derivatives by thermolysis. *Tetrahedron,* **30**, 2207 (1974).

41. I. Ninomiya and T. Naito, Total synthesis of (±)-angustoline. *Heterocycles,* **2**, 607 (1974).

42. R. T. Brown, C. L. Chapple, and A. A. Charalambides, Biomimetic synthesis of indole alkaloids. Dihydromancunine. *J. Chem. Soc. Chem. Commun.*, 756 (1974).

43. R. T. Brown and C. L. Chapple, Biomimetic conversion of vincoside into heteroyohimbine alkaloids. *J. Chem. Soc. Chem. Comm.*, 740 (1974).

44. T. Kametani, H. Takeda, Y. Hirai, F. Satoh, and K. Fukumoto, Simple route to spiro-[indene-2,1′-isoquinoline]s, a spiro[indene-2,1′-β-carboline]. and hexadehydroyohimbane. *J. Chem. Soc. Perkin Trans. I,* 2141 (1974).

45. H. Akimoto, K. Okamura, M. Yui, T. Shioiri, M. Kuramoto, Y. Kikugawa, and S. Yamada, Amino acid and peptides. XIII. New approach to the biogenetic-type, asymmetric synthesis of indole and isoquinoline alkaloids by 1,3-transfer of asymmetry. *Chem. Pharm. Bull.*, **22**, 2614 (1974).

46. V. Zikan, K. Rezabek, and M. Semonsky, Synthesis of 1,1-diethyl-4-(D-6-methyl-8-ergolin-I-yl)-semicarbazide and its isoergolin-I-yl and isoergolin-II-yl isomers. *Collection Czech. Chem. Comm.*, **39**, 3144 (1974).

47. A. I. Scott and A. A. Qureshi, Regio- and stereospecific models for the biosynthesis of the indole alkaloids. I. Development of strategic approaches and preliminary experiments. *Tetrahedron,* **30**, 2993 (1974).

48. A. I. Scott, P. C. Cherry, and C. C. Wei, Regio- and stereospecific models for the biosynthesis of the indole alkaloids. III. Aspidosperma-Iboga-secodine relation. *Tetrahedron,* **30**, 3013 (1974).

49. N. J. Bach and E. C. Kornfeld, Conversion of lysergic acid to the clavine alkaloids, penniclavine and elymoclavine, and to decarboxylysergic acid. *Tetrahedron Lett.*, 3225 (1974).

50. S. Sakai and N. Shinma, Transformation of indole alkaloids. Chemical transformation of corynantheine type alkaloids to C-mavacurine type alkaloids. *Chem. Pharm. Bull.*, **22**, 3013 (1974).

51. K. Yamada, K. Aoki, T. Kato, D. Uemura, and E. E. Van Tamelen, Total synthesis of the alkaloid (±)-geissoschizine. *J. Chem. Soc. Chem. Comm.*, 908 (1974).

52. R. T. Brown, C. L. Chapple, R. Platt, and H. Spencer, Biomimetic inversion of C-3 in monoterpenoid indole alkaloids. *J. Chem. Soc. Chem. Comm.*, 929 (1974).

53. G. Benz, H. Riesner, and E. Winterfeldt, Reactions with indole derivatives. XXV. The stereoselective total synthesis of roxburghin D. *Chem. Ber.*, **108**, 248 (1975).

54. T. Kametani and K. Fukumoto, Total synthesis of certain isoquinoline and indole alkaloids by thermolysis. *Heterocycles*, **3**, 29 (1975).

55. J. P. Kutney and D. S. Grierson, Improved synthesis of the pyridocarbazole indole alkaloid olivacine. *Heterocycles*, **3**, 171 (1975).

56. T. Kametani, M. Kajiwara, T. Takahashi, and K. Fukumoto, Total synthesis of (±)-yohinbine. *Heterocycles*, **3**, 179 (1975).

57. F. R. Shiroyan, V. T. Avertyan, and G. T. Tatevosyan, Indole derivatives. 3-methyl-15,16,17,18,19,20-hexadehydroyohimban. *Arm. Khim. Zh.*, **27**, 978 (1974).

58. H. Riesner and E. Winterfeldt, Reactions with indole derivatives. XXIV. Yohimbone derivatives by enamine cyclization. *Chem. Ber.*, **108**, 243 (1975).

59. L. A. Mitscher, M. Shipchandler, H. D. Showalter, and M. S. Bathala, Antimicrobial agents from higher plants. Synthesis in the canthin-6-one (6H-indolo[3,2,1-de] [1,5] naphthyridine-6-one) series. *Heterocycles*, **3**, 7 (1975).

60. Y. Morita, S. Savaskan, K. A. Jaeggi, M. Hesse, U. Renner, and H. Schmid, Alkaloids. 154. Tranformations of the Iboga alkaloids, voacangine and conopharyngine. *Helv. Chim. Acta*, **58**, 211 (1975).

61. K. Horvath-Dora and O. Clauder, Alkaloids containing the indolo[2,3-c] quinazolino [3,2-a] pyridine skeleton. III. 3,14-dihydrorutecarpine. *Acta Chim Acad. Sci. Hung.*, **84**, 93 (1975).

62. H. Irie, J. Fukudome, T. Ohmori, and J. Tanaka, Synthesis of the alkaloid, oxogambirtannine. *J. Chem. Soc. Chem. Comm.*, 63 (1975).

63. R. Besselievre, C. Thal, H. P. Husson, and P. Potier, Novel synthesis of the indole alkaloid ellipticine. *J. Chem. Soc. Chem. Comm.*, 90 (1975).

64. K. Seki, T. Ohnuma, T. Oishi, and Y. Ban, Total synthesis of the alkaloid (±)-1-acetylaspidospermidine. *Tetrahedron Lett.*, 723, (1975).

65. Y. Ban, T. Ohnuma, K. Seki, and T. Oishi, Total synthesis of the alkaloid (±)-acetylaspidoalbidine. *Tetrahedron Lett.*, 727 (1975).

66. J. P. Kutney and G. B. Fuller, Total synthesis of akuammicine and 16-epistemmadenine. Absolute configuration of stemmadenine. *Heterocycles*, **3**, 197 (1975).

67. Y. Langlois, N. Langlois, and P. Potier, Application of a biogenetic scheme to total synthesis. Ellipticine. *Tetrahedron Lett.*, 955 (1975).

68. Y. Langlois and P. Potier, Indoles. VIII. Total synthesis of (±)-16-de(methoxycarbonyl)vobasine. *Tetrahedron*, **31**, 419 (1975).

69. Y. Langlois and P. Potier, Indoles IX. Total synthesis of (±)-deethylervatamine and

(±)-15,20-dehydroervatamine. *Tetrahedron,* **31**, 423 (1975).

70. P. Mangeney, Y. Langlois, and P. Potier, Indoles. X. Neighboring group reactions and demethylation in the dregamine series. *Tetrahedron,* **31**, 429 (1975).

71. P. Pfaeffli, W. Oppolzer, R. Wenger, and H. Hauth, Stereoselective synthesis of optically active vincamine. *Helv. Chim. Acta,* **58**, 1131 (1975).

72. P. Rosenmund, W. H. Haase, J. Bauer, and R. Frische, Chemistry of indole. VII. Synthesis in the iboga series. III. Ibogamine, ibogaine and epiibogamine. *Chem. Ber.,* **108**, 1871 (1975).

73. H. Plieninger, W. Lehnert, D. Mangold, D. Schmalz, A. Voelkl, and J. Westphal, Total synthesis of (±)-chanoclavine I. *Tetrahedron Lett.,* 1827 (1975).

74. L. Merlini, G. Nasini, and M. Palamareva, Models for the synthesis of indole alkaloids. *Gazz. Chim. Ital.,* **105**, 339 (1975).

75. A. Wu and V. Snieckus, New synthesis of a stemmadenine model. *Tetrahedron Lett.,* 2057 (1975).

76. W. Mueller, R. Preuss, and E. Winterfeldt, Simple access to 1,4-disubstituted β-carboline derivatives. Total synthesis of N^a-methylbrevicolline. *Angew. Chem.,* **87**, 385 (1975).

77. J. P. Kutney, J. Cook, K. Fuji, A. M. Treasurywala, J. Clardy, J. Fayos, and H. Wright, Synthesis of bisindole alkaloids. Synthesis, structure, and absolute configuration of 18′-epi-4′-deoxo-4′-epivinblastine, 18′-decarbomethoxy-18′-epi-4-epivinblastine, and 18′-epi-3′-4′-dehydrovinblastine. *Heterocycles,* **3**, 205 (1975).

78. T. Kametani, M. Kajiwara, T. Takahashi, and K. Fukumoto, Tetra- and hexadehydroyohimban synthesis by an intermolecular cycloaddition of o-quinodimethan. *J. Chem. Soc. Perkin Trans. I,* 737 (1975).

79. T. Kametani, Y. Ichikawa, T. Suzuki, and K. Fukumoto, Fascinating synthesis of olivacine. *Heterocycles,* **3**, 401 (1975).

80. H. Ishii and Y. Murakami, Fischer indolization and its related compounds. X. Application of the advanced Fischer indolization of a 2-methoxyphenylhydrazone derivative to syntheses of naturally occurring 6-substituted indoles. *Tetrahedron,* **31**, 933 (1975).

81. G. H. Buechi, Synthesis of naturally occurring indole derivative. *Chimia,* **29**, 172 (1975).

82. M. Ono, M. Shimamine, K. Takahasi, and T. Inoue, Hallucinogens. V. Synthesis of psilocybin. *Eisei Shikenjo Hokoku,* **92**, 41 (1974).

83. J. R. Knox and J. Slobbe, Indole alkaloids from Ervatamia orientalis. III. Configurations of the ethyl side chains of dregamine and tabernaemontanine and some further chemistry of the vobasine group. *Austral. J. Chem.,* **28**, 1843 (1975).

84. K. Yamada, K. Aoki, and D. Uemura, Synthesis and stereochemistry of (±)-3′,4′-dihydrousambarensine. *J. Org. Chem.,* **40**, 2572 (1975).

85. C. Dieng, C. Thal, H. P. Husson, and P. Potier, Indole series. VII. Synthesis of new pyridocarbazoles obtained by photocyclization of 1-(β-indolyl)-2-(pyridyl)acrylonitriles. *J. Heterocyclic Chem.,* **12**, 455 (1975).

86. T. Kametani, M. Takeshita, M. Ihara, and K. Fukumoto, Total synthesis of angustine. *Heterocycles,* **3**, 627 (1975).

87. P. Potier, N. Langlois, Y. Langlois and F. Gueritte, Partial synthesis of vinblastine-type alkaloids. *J. Chem. Soc. Chem. Commun.,* 670 (1975).

88. J. P. Kutney, A. H. Ratcliffe, A. M. Treasurywala, and S. Wunderly, Synthesis of bis-

indole alkaloids. II. Synthesis of 3′-4′-dehydrovinblastine, 4′-deoxovinblastine, and related analogs. Biogenetic approach. *Heterocycles,* **3**, 639 (1975).

89. J. Harley-Mason, Synthetic studies in the strychnos-type alkaloid field. *Pure Appl. Chem.,* **41**, 167 (1975).

90. S. S. Klioze and F. P. Darmory, Total synthesis of (±)-6,7-didehydroaspidospermine. *J. Org. Chem.,* **40**, 1588 (1975).

91. T. Kametani, T. Suzuki, C. Otsuka, K. Takahashi, and K. Fukumoto, Syntheses of heterocyclic compounds. DCVI. Decarboxylation of tryptophan and a synthesis of harman. *Yakugaku Zasshi,* **95**, 363 (1975).

92. K. Krohn and E. Winterfeldt, Reactions with indole derivatives. XXVIII. Alkylation of camptothecin intermediates. New appraoch to camptothecin. *Chem. Ber.,* **108**, 3030 (1975).

93. T. Kametani, T. Takahashi, M. Kajiwara, and K. Fukumoto, Syntheses of heterocyclic compounds. DLXXVIII. Synthesis of the Yohimbane ring system by an electrocyclic reaction of o-quinodimethane. *An. Quim.,* **70**, 1000 (1974).

94. R. Z. Andriamalisoa, L. Diatta, P. Rasoanaivo, N. Langlois, and P. Potier, Vindoline. II. Partial synthesis of andranginine. *Tetrahedron,* **31**, 2347 (1975).

95. M. Ando, G. Buechi, and T. Ohnuma, Total synthesis of (±)-vindoline. *J. Amer. Chem. Soc.,* **97**, 6880 (1975).

96. S. Sakai, N. Aimi, K. Kato, H. Ido, K. Masuda, Y. Watanabe, and J. Haginiwa, Plants containing indole alkaloids. 5. Identification of tetrahydrosecamine in Amsonia elliptica and the synthesis of secamine and presecamine skeletons. *Yakugaku Zasshi,* **95**, 1152 (1975).

97. L. Chevolot, H. P. Husson, and P. Potier, Indoles. XI. Synthesis of dl-18,19-dihydroantirhine. *Tetrahedron,* **31**, 2491 (1975).

98. L. A. Djakoure, F. X. Jarreau, and R. Goutarel, Indole alkaloids. CI. Preparation of heteroyohimban derivatives from Corynantheine. Activiation of the vinylic double bond by mercuration. Synthesis of ajmalicine, epi-19 ajmalicine and an abeo-18(17 → 16)yohimban. *Tetrahedron,* **31**, 2695 (1975).

99. M. Lounasmaa and C. J. Johansson, Synthetic studies in the alkaloid field. II. Preparation of N-(β-indolylethyl)dihydro- and N-(β-indolylethyl)tetrahydropyridines by catalytic hydrogenation, and their acid-induced cyclization. *Acta Chem. Scand. Ser. B,* **B29**, 655 (1975).

100. M. Nakagawa, Y. Okajima, K. Kobayashi, T. Asaka, and T. Hino, Biomimetic transformation of indoloquinolizidine derivative to 2-acylindole. *Heterocycles,* **3**, 799 (1975).

101. T. Kametani, T. Suzuki, Y. Ichiikawa, and K. Fukumoto, Synthesis of heterocyclic compounds. CDXXXIV. Novel total synthesis of olivacine (1,5-dimethyl-6H-pyrido [4,3-b] carbazole). *J. Chem. Soc. Perkin Trans. I,* 2120 (1975).

102. J. P. Kutney, K. K. Chan, A. Failli, J. M. Fromson, C. Gletsos, A. Leutwiler, V. R. Nelson, and J. P. de Souza, Total synthesis of indole and dihydroindole alkaloids. VI. Total synthesis of some monomeric vinca alkaloids, dl-vincadine, dl-vincaminoreine, dl-vincaminorine, dl-vincadifformine, dl-minovine and dl-vincaminoridine. *Helv. Chim. Acta,* **58**, 1648 (1975).

103. J. P. Kutney and F. Bylsma, Total synthesis of indole and dihydroindole alkaloids. VII. Total synthesis of isovelbanamine, velbanamine, cleavamine, 18β-carbomethoxycleavamine and catharanthine. *Helv. Chim. Acta,* **58**, 1672 (1975).

104. J. P. Kutney, J. Beck, F. Bylsma, J. Cook, W. J. Cretney, K. Fuji, R. Imhof, and A. M.

Treasurywala, Total synthesis of indole and dehydroindole alkaloids. VIII. Synthesis of bisindole alkaloids in the vinblastine-vincristine series. Chloroindolenine approach. *Helv. Chim. Acta,* 58, 1690 (1975).

105. J. E. Saxton, A. J. Smith, and G. Lawton, Aspidosperma akaloids. Total synthesis of (±)-N,O-diacetylindocarpinol, (±)-cylindrocarine, (±)-cylindrocarpine, (±)-cylindrocarpidine, *Tetrahedron Lett.*, 4161 (1975).

106. Y. Ban, M. Seto, and T. Oishi, Synthesis of 3-spirooxindole derivatives. VII. Total synthesis of alkaloids (±)-rhynchophylline and (±)-isorhynchophylline. *Chem. Pharm. Bull.*, **23**, 2605 (1975).

107. T. Kametani, Y. Hirai, M. Kajiwara, T. Takahashi, and K. Fukumoto, Syntheses of heterocyclic compounds. DCXLI. Convenient synthesis of hexadehydroyohimbine and a total synthesis of yohimbine. *Chem. Pharm. Bull.*, **23**, 2634 (1975).

108. T. Fehr and P. A. Stadler, Ergot alkaloids. 80. Stereospecific, light- and acid-catalyzed allylic rearrangement. Synthesis of paliclavine. *Helv. Chim. Acta,* 58, 2484 (1975).

109. R. T. Brown, C. L. Chapple, D. M. Duckworth, and R. Platt, Conversion of secologanin into elenolic acid and 18-oxayohimban alkaloids. *J. Chem. Soc. Perkin Trans. I,* 160 (1976).

110. T. Kametani, Y. Hirai, and K. Fukumoto. Alternative synthesis of (±)-yohimbine. *Heterocycles,* 4, 29 (1976).

111. T. Kametani, T. Higa, K. Fukumoto, and M. Koizumi, One-step synthesis of evodiamine and rutecarpine. *Heterocycles,* 4, 23 (1976).

112. T. Kametani, M. Takeshita, and M. Ihara, A total synthesis of nauclefine. *Heterocycles,* 4, 247 (1976).

113. A. Rahman, N. Waheed, and M. Ghazala, Synthetic studies towards vinblastine and its analogs. VI. The Markownikoff addition of acetic acid to catharanthine by a modification of the Prevost reaction. *Z. Naturforsch. B. Anorg. Chem. Org. Chem.*, **31b**, 264 (1976).

114. D. G. I. Kingston, B. B. Gerhart, and F. Ionescu, Isolation, structural elucidation, and synthesis of tabernamine, a new cytotoxic bis-indole alkaloid from Tabernamontana Johnstonii. *Tetrahedron Lett.*, 649 (1976).

115. S. Yamada, K. Murato, and T. Shioiri, A new entry into cinchona alkaloids via biomimetic pathway. *Tetrahedron Lett.*, 1605 (1976).

116. Y. Morita, M. Hesse, U. Renner, and H. Schmid, Transformation of the Iboga alkaloid voacangine into voaketone, a derivative of β-carboline, *Helv. Chim. Acta,* **59**, 532 (1976).

117. S. Takano, S. Hatakeyama, and K. Ogasawara, Synthesis of the non-tryptamine moiety of the Aspidosperma-type indole alkaloids via cleavage of cyclic α-diketone monothioketal. An efficient synthesis of (dl)-quebrachamine and a formal synthesis of (dl)-tabersonine. *J. Amer. Chem. Soc.*, 98, 3022 (1976).

118. J. P. Kutney, J. Balsevich, G. H. Bokelman, T. Hibino, I. Itoh, and A. H. Ratcliffe, Studies on the synthesis of bisindole alkaloids. III. The synthesis of leurosine and 3′-hydroxyvinblastine. *Heterocycles,* 4, 997 (1976).

Alkaloid Synthesis

R. V. STEVENS
Department of Chemistry
University of California
Los Angeles, California

1. INTRODUCTION AND GENERAL PRINCIPLES

This chapter is primarily concerned with the theory and practice of organochemical synthesis as applied to a variety of natural (and some unnatural) products. Therefore, perhaps it is appropriate to begin with a brief outline of what some of the broad objectives of organic synthesis are. In certain respects, the answer to this inquiry is that they defy precise definition, for as many of the Grand Masters have made abundantly clear in both word and deed, there is an incalculable element of creative art and imagination in organic synthesis. Indeed, to a large extent, one might say that it is only the imaginations of those engaged in the profession that place any limits on its potential accomplishments.

As a major partner of the science as a whole, organic synthesis has contributed impressively to at least two of the most essential elements of any chemical activity: mainly, in revealing new chemistry and in understanding, Thus one certainly quite defensible view of the role of organic synthesis is that it is an effective means for the discovery of new chemical transformations and that it fosters and improves upon old ones. Insofar as the major emphasis of chemistry continues to be on chemical reactions, the necessity of discovering new or uncovering unknown aspects of previously defined reactions seems apparent. It is this role of being able to provide the means within which discovery can be made that is certainly one of the most important functions of organic synthesis. And, of course, ultimately, discovery leads to a better understanding, As such, we find in synthesis an inviting and perhaps unparalleled opportunity for expanding our knowledge of the environment in which we toil.

Although there have been impressive and vast advances in the precision of our knowledge and understanding or organic chemistry, it is still largely an experimental science and is likely to remain so for some time. Indeed, one discovers this fact rapidly and not infrequently brutally when attempting to reduce to practice even a well-documented series of transformations aimed at a relatively modest synthetic target. But all synthetic objectives require a plan, and the greatest limitation on the art and science of organic synthesis presently is the degree to which sound planning is possible. Now, this is not to malign the importance of purely speculative planning, on the contrary. Nevertheless, the monumental edifice of chemical theory that we now bear witness to enables us to predict with some confidence the most likely outcome of a vast number of transformations, often without prior art. Furthermore, any deviations from such predictions can always be explained provided the chemist is willing to pursue such problems. Thus modern theory allows synthetic planning and the subsequent reduction of such plans to practice on an ever more impressive scale. Such is the nature of the science.

A. General Principles That Guide Synthetic Planning

Before commencing actual discussion of alkaloid synthesis, an outline of the general principles that guide synthetic planning and the execution thereof is in order. The considerable task of analyzing and understanding organic synthesis as a scholarly endeavor may appear rather formidable at first. The impressive and ever growing number of reactions that are at our disposal and the often justified uncertainty as to their generality may create the impression that the decisive formulation of a synthetic plan is frequently problematical. This uncertainty is largely a consequence of the fact that although a number of definite operations may be identified in the synthesis of intricate organic molecules, they are not strictly independent of one another. Nevertheless, the impressive quantity and quality of successful syntheses that have succumbed to the chemist's imagination and skill serve as a monument to the fact that worthy and reliable synthetic plans can indeed be devised. There can be no doubt that such efforts have already led to a more meaningful comprehension of synthesis and provided the means for its continued growth and development, which leads us to the main topic of this discussion: How does a chemist arrive at a particular outline for the synthesis of complex organic molecules such as those considered herein? Of course, this is in part a function of the particular chemist involved, but rather surprisingly this important question had not been extensively or systematically dealt with until quite recently when an important and most illuminating contribution by Corey[1] appeared (the interested reader is referred to this source). Some of the broad general principles that served to guide the synthetic planning outlined in this chapter are dealt with explicitly and in the sequel implicitly.

1. The synthesis of complex molecules (and even some that appear deceptively simple) most often require the employment of carefully selected sequences of chemical transformations defined in a rather specific order. A relatively large number of possible approaches to the synthesis of such substances can usually be derived, and the number of possible approaches can be anticipated to increase with increasing molecular size and complexity. Obviously, failure of even one of these synthetic steps may doom an entire synthetic plan.
2. An integral part of the planning stage is a thorough evaluation of the molecular history of the target molecule, since the results of earlier investigations may play an important role in the ultimate selection of a particular scheme. For example, during the elucidation of structure of many natural products or as a result of previous synthetic efforts, certain decisive molecular characteristics may reveal themselves. Such history often will be useful in planning the attack, since prospective problems may be anticipated before they arise in the actual execution of the synthesis. Also, occasionally as part of degradative structural studies, partial syntheses will have been achieved, thus providing an attractive alternative objective.
3. Synthetic organic chemistry, perhaps more than any other subdiscipline, demands the application of the knowledge and technical developments of the science as a whole. There can be no doubt that an intimate and broad

knowledge of organic reactions and their mechanisms is a primary importance in conceiving of and executing organic syntheses. The application and continued acquisition of powerful physical tools and refined methods of analysis and purification have armed us for the attack as never before. In fact, these exciting developments are modifying the very nature of organic synthesis. Truly signficant synthetic programs must now either cope with problems not amenable to trivial solution or accept the responsibility and challenge of developing and/or employing new methods of approach that test and expand the very principles upon which the science is founded. Modern physical methods permit the examination of extremely minute quantities of material and allow us to penetrate further and further into the environment. In an ever increasing number of examples, they have partially or completely displaced degradative methods of analysis. For example, single-crystal, X-ray structural analysis may be regarded presently as the ultimate weapon, since it requires relatively small amounts of material (provided that a suitable crystal or crystalline derivative can be secured) and imparts little or no change in the substance. In the past, the discovery of the specific chemistry of new substances was most often the result of degradative studies; the present situation affords the synthetic chemist an almost unique challenge to secure knowledge on a strictly *de novo* basis. One can anticipate that with the continued use and refinement of such powerful physical tools, the synthetic chemist will play a dominant role in the development of the chemistry of new structural types. Although it is beyond the scope of this inquiry to consider all of the factors that play a role in organic synthesis, an even cursory inspection of these pages reveals the broad applications and implications of related disciplines.

4. There are distinct but not unequivocal criteria that may be employed to evaluate the utility or virtues of alternative contemplated syntheses. Quite often the selection of a preferred synthetic approach is forced to be arbitrary if there are a number of unknown factors, However, even in such cases these criteria allow the dismissal of inferior possibilities. The diagnostic criteria by which potentially competitive approaches may be evaluated are of a routine nature to the practicing organic chemist; nevertheless, the following comments are included for thoughtful perusal.

   a. The most satisfactory solutions to even complex synthetic problems are usually, at least in retrospect, relatively simple. Excessively lengthy possible routes should be suspiciously and critically evaluated before their execution and shorter possibilities sought.
   b. It is often desirable to employ and/or develop new reactions or applications in chemical synthesis, especially if it is possible that this will

enhance the efficiency of the synthesis or provide new chemical principles. However, the number of such speculative transformations and their position in the overall plan must obviously be carefully considered. The execution of doubtful steps should, in general, be reserved for the first stages of the synthesis and conversely abhorred in the latter stages. Of course, the proper selection of known chemical transformations of proved dependability and mechanistically sound interpretation occupies a central role in sound synthetic planning and increases the probability of affecting each stage of the synthesis.

c. Alternative means of achieving required transformations in the synthetic plan are desirable because they enhance the chances of success, especially if such steps are the subject of some doubt.

2. THE MESEMBRINE ALKALOIDS*,†

At this time, the mesembrine alkaloids constitute a relatively homogeneous group of substances that have only recently been the subject of serious investigation. Interest in these bases has been catalyzed by their isolation from a drug preparation known as "Channa" or Koegoed," which is used by the natives of Southwest Africa as a narcotic and which is claimed to exert an effect similar to that of cocaine. The structures and names‡ of those alkaloids that are currently known are portrayed together with key literature references.

Synthesis of the mesembrine alkaloids has witnessed considerable activity in recent years. This may be ascribed, in part, to their relative simplicity, which

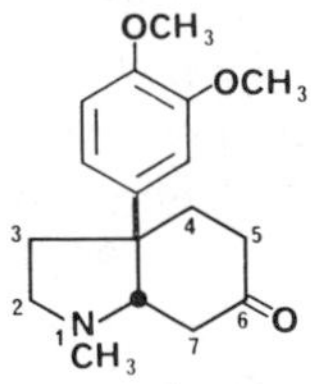

1: mesembrine[2,4]

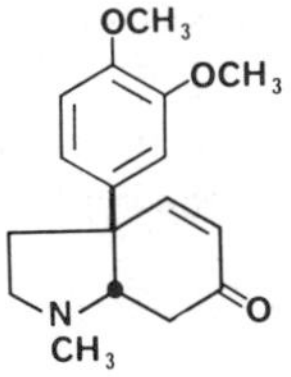

2: mesembrinine[2]

*In the older literature, reference is made to the isolation of many of these alkaloids from the genus *Mesembryanthemum* Dill from which the names of several of these bases were derived. More recently, however, this classification has been revised to the genus *Sceletium* N. E. Brown (*Ficoidaceae* or *Aizoaceae*). Therefore, reference to these bases as mesembrine alkaloids is technically a misnomer. Nevertheless, the term appears to have sufficient currency to warrant its retention.

†For a review of these alkaloids see Popelak and Lettenbauer.[2]

‡The reader should be alerted to certain ambiguities found in the naming of these alkaloids and is referred to footnote 11 of Jeffs, Hawks, and Farrier[3] for a discussion of this matter.

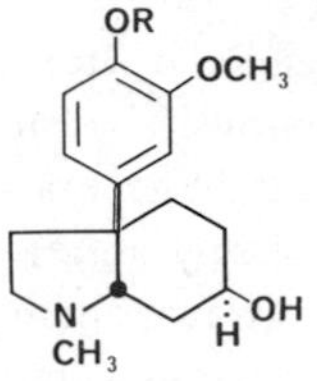

3: mesembranol[4]
R = CH_3

4: 4'-O-demethylmesembranol[5]
R = H

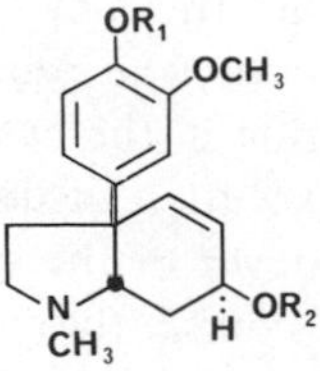

5: mesembrenol[5]
R_1 = CH_3, R_2 = H

6: O-acetylmesembrenol[5]
R_1 = CH_3, R_2 = Ac

7: 4'-O-demethylmesembrenol[5]
R_1 = R_2 = H

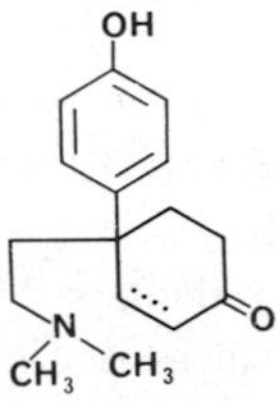

8: joubertiamine (····)[6]

9: dihydrojoubertiamine[6]

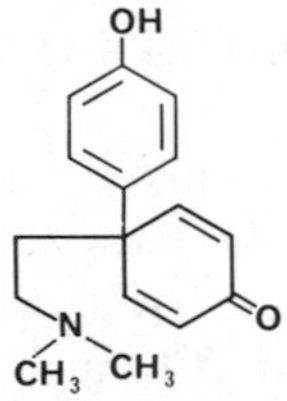

10: dehydrojoubertiamine[6]

provides the synthetic chemist the relative luxury of being able to probe the frontiers of the science with the hope that new synthetic methods may be developed. Furthermore, the basic mesembrane skeleton **11** bears a rather close structural and stereochemical relationship with the somewhat more complex 5,10b-ethanophenanthridine (crinane) group **12** of *Amaryllidaceae* alkaloids and thus provides an additional outlet upon which to apply any new methods of approach revealed in the synthesis of the mesembrine bases (and vice versa).

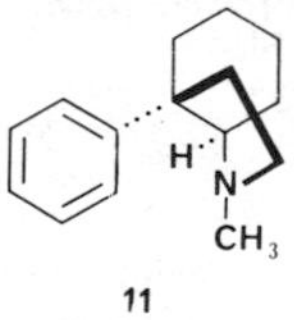

11

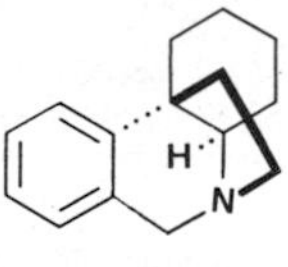

12

The first synthetic study in this family was initiated to provide final and definitive proof concerning the skeleton of mesembrine. This was accomplished by construction of the degradation product (±)-mesembrane. The basic method of approach was the same as that employed previously by Wildman in the synthesis of (±)-crinane (see Chart 2-1).

Chart 2-1.[6]

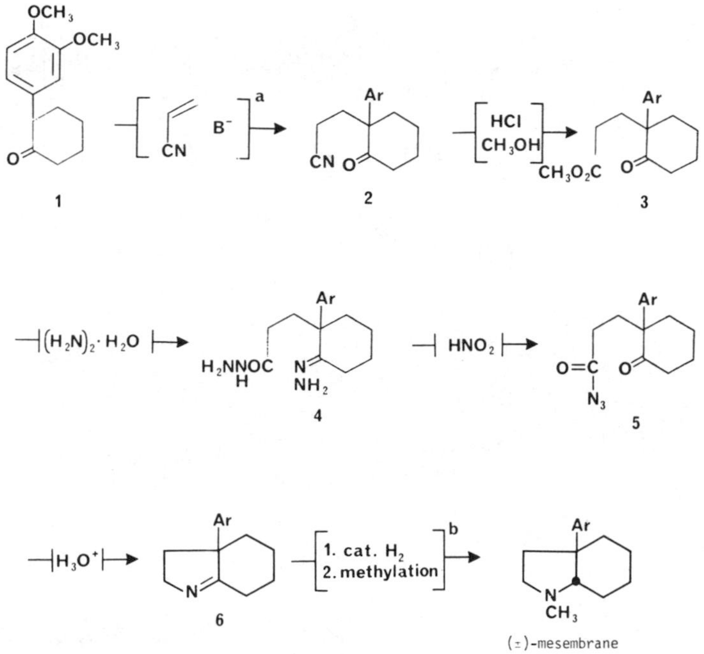

[a]The employment of the venerable cyanoethylation reaction as a reliable method for introducing the crucial quaternary center is perhaps worthy of comment. Although a subsequent three-stage modification of this chain is required to provide a suitable precursor (5) to the hexahydroindole, reliability is a desirable virtue in any synthetic plan and should never be regarded lightly. Since the objective in the present case was proof of structure via synthesis, this consideration was justifiably raised to a premium. In view of the subsequent difficulties encountered by Shamma and Rodriguez[7] in attempting a very similar alkylation with somewhat more novel reagents, this decision is amply justified.

[b]In this paper the conversion of indolenine 6 to mesembrane was accomplished by methylation and subsequent reduction (conditions unspecified). By this method, a mixture of the *cis* and *trans* isomers was produced. However, in a patent (cited in Ref. 2, p. 473), these same authors observed that reversing the order in which these steps are carried out leads almost exclusively to the desired *cis* isomer. The stereochemical course of the catalytic hydrogenation of 6 could be predicted with some confidence from the analogous step in Wildman's crinane synthesis[8] and has subsequently been discussed and exploited by Whitlock and Smith[9] as a key step in the synthesis of crinine.

The first total synthesis of mesembrine was achieved by the following rather lengthy route. However, certain of the individual steps are instructive. This initial effort was followed by a number of quite different and certainly more expeditious syntheses. One such approach was developed almost simultaneously

Chart 2-2.[7]

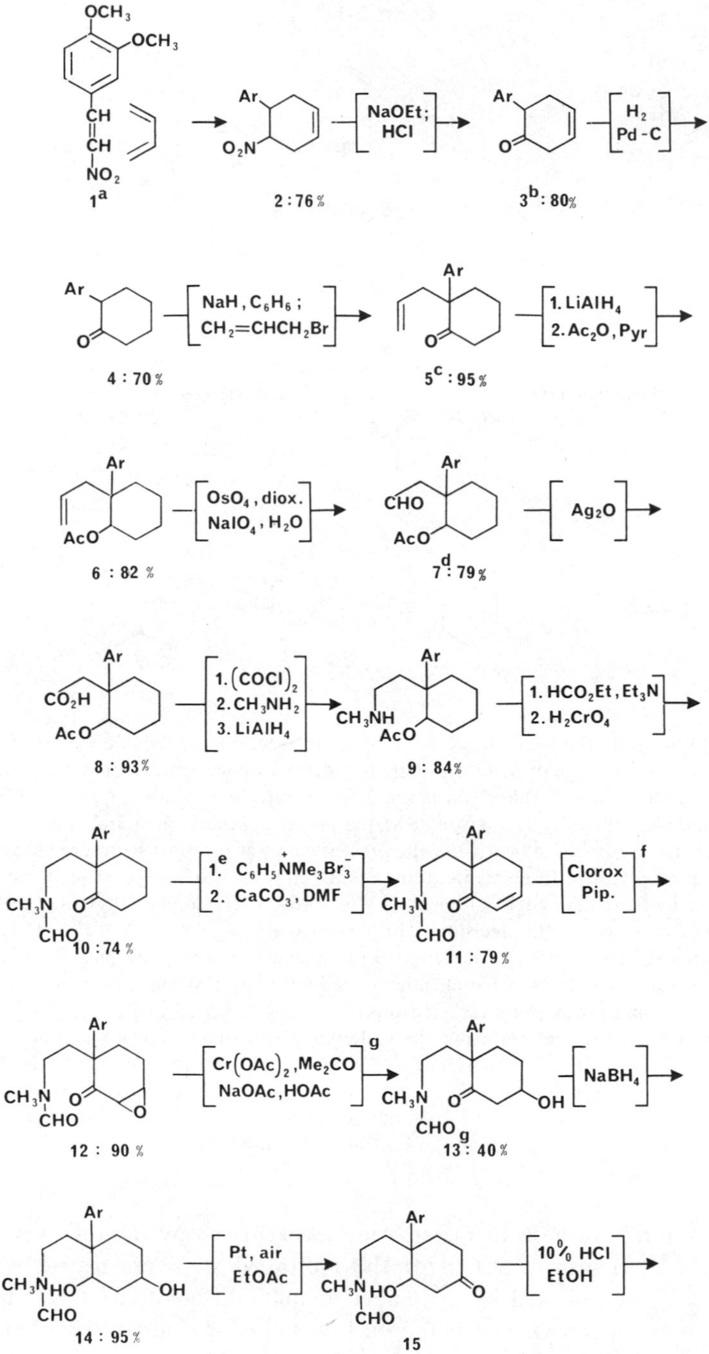

OCH_3, OCH_3, CH=CH, NO_2

1^a

→ Ar, O_2N

$2 : 76\%$

[NaOEt; HCl] → Ar, O

$3^b : 80\%$

[H_2, Pd-C] →

Ar, O

$4 : 70\%$

[NaH, C_6H_6; $CH_2{=}CHCH_2Br$] → Ar, O

$5^c : 95\%$

[1. $LiAlH_4$ 2. Ac_2O, Pyr] →

Ar, AcO

$6 : 82\%$

[OsO_4, diox. $NaIO_4$, H_2O] → Ar, CHO, AcO

$7^d : 79\%$

[Ag_2O] →

Ar, CO_2H, AcO

$8 : 93\%$

[1. $(COCl)_2$ 2. CH_3NH_2 3. $LiAlH_4$] → Ar, CH_3NH, AcO

$9 : 84\%$

[1. HCO_2Et, Et_3N 2. H_2CrO_4] →

Ar, CH_3N, CHO, O

$10 : 74\%$

[1.[e] $C_6H_5\overset{+}{N}Me_3Br_3^-$ 2. $CaCO_3$, DMF] → Ar, CH_3N, CHO, O

$11 : 79\%$

[Clorox Pip.][f] →

Ar, CH_3N, CHO, O, O

$12 : 90\%$

[$Cr(OAc)_2$, Me_2CO NaOAc, HOAc][g] → Ar, CH_3N, CHO, O, OH

$13^g : 40\%$

[$NaBH_4$] →

Ar, CH_3N, CHO, HO, OH

$14 : 95\%$

[Pt, air EtOAc] → Ar, CH_3N, CHO, HO, O

15

[10% HCl EtOH] →

Chart 2-2 (Continued).

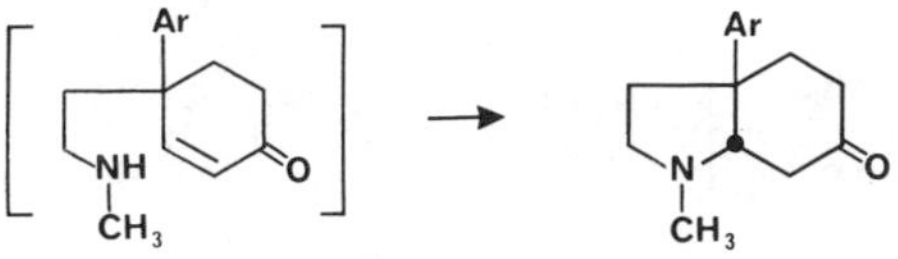

Ar NH CH_3 O → Ar N O CH_3

16[h] (±)-mesembrine

~50% from 14

[a]Prepared from veratraldehyde and nitromethane according to the procedure of Raiford and Fox.[10]
[b]This transformation is based on the original work of Wildman and Wildman[11] and has been employed frequently. In the present study, "considerable difficulty was originally encountered with the Nef reaction ···. However, after it was realized that solutions of the sodium salt of **2** were sensitive to both oxygen and heat considerable improvement of the yield was achieved."
[c]The employment of allyl bromide and the subsequent degradation to a two-carbon side chain was necessitated by the fact that more desirable candidates such as chloroacetonitrile, bromoethyl acetate, bromoacetic acid, *N*-methyleneimine, and bromoethylamine failed to yield useful products.
[d]The procedure involved a modification of the method of Pappo et al.[12]
[e]This selective bromination in the presence of a highly reactive aromatic nucleus is worthy of note, see Marwuet and Jacques.[13]
[f]According to the procedure of Marmor.[14]
[g]cf. Julian et al.[15]
[h]Of intrinsic interest in this synthesis is the transposition of enones **11** and **16**. This task was accomplished in five stages and employed chromous acetate reduction[g] of epoxyketone **12** and selective oxidation of the least-hindered alcohol of diol **14** as key steps.

in three different laboratories and featured the acid-catalyzed, thermally induced rearrangement of cyclopropyl imines and methyl vinyl ketone annelation of the resultant 2-pyrrolines as key steps. Employed individually or in combination, these two synthetic methods have evolved into powerful general methods, and the reader is referred to Section 6 for a further discussion of how these two synthetic methods have been developed and applied to the synthesis of mesembrine and other alkaloid families.

Chart 2-3.[16]

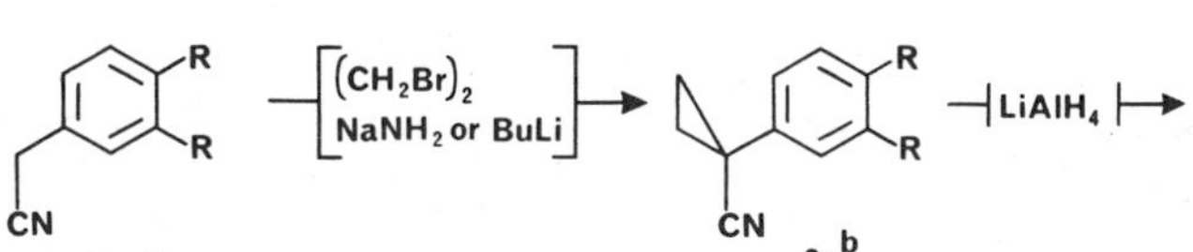

R R CN → $(CH_2Br)_2$ $NaNH_2$ or BuLi → R R CN → $LiAlH_4$ →

1a R = H
b R = OCH_3

2a[b]

Chart 2-3 (Continued).

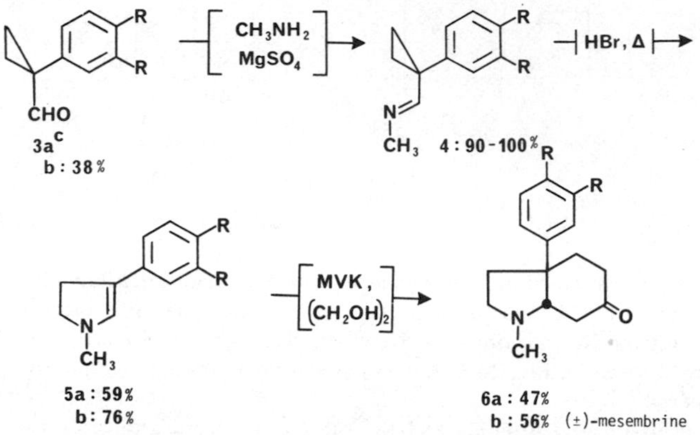

3a[c] b : 38% → [CH_3NH_2, $MgSO_4$] → 4 : 90–100% → [HBr, Δ] →

5a : 59%, b : 76% → [MVK, $(CH_2OH)_2$] → 6a : 47%, b : 56% (±)-mesembrine

[a]Many of the yields reported in this paper have subsequently been substantially improved in closely related systems. An extensive discussion of each of these transformations is included in Section 6.

[b] Nitrile **2a** was prepared according to the procedure of Dupin and Fraisse-Julien.[17]

[c]As described by Schuster and Roberts.[18]

Chart 2-4.[a,19]

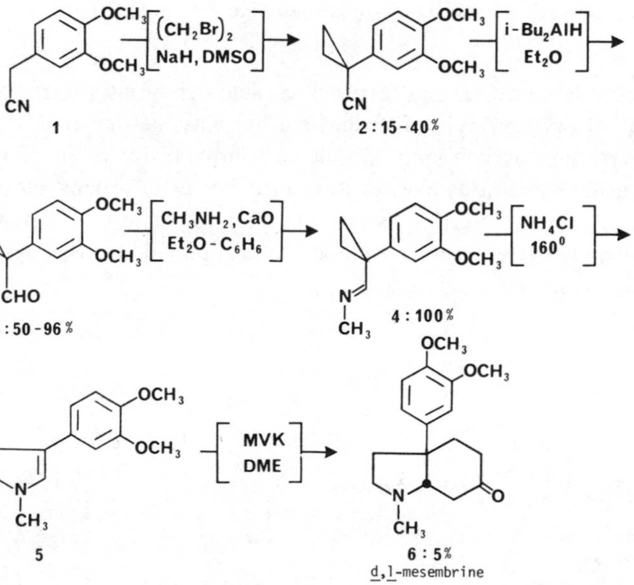

1 → [$(CH_2Br)_2$, NaH, DMSO] → 2 : 15–40% → [i-Bu_2AlH, Et_2O] →

3 : 50–96% → [CH_3NH_2, CaO, Et_2O-C_6H_6] → 4 : 100% → [NH_4Cl, 160°] →

5 → [MVK, DME] → 6 : 5% d,l-mesembrine

[a]A detailed discussion of these transformations is included in Section 6.

Chart 2-5.[a,20]

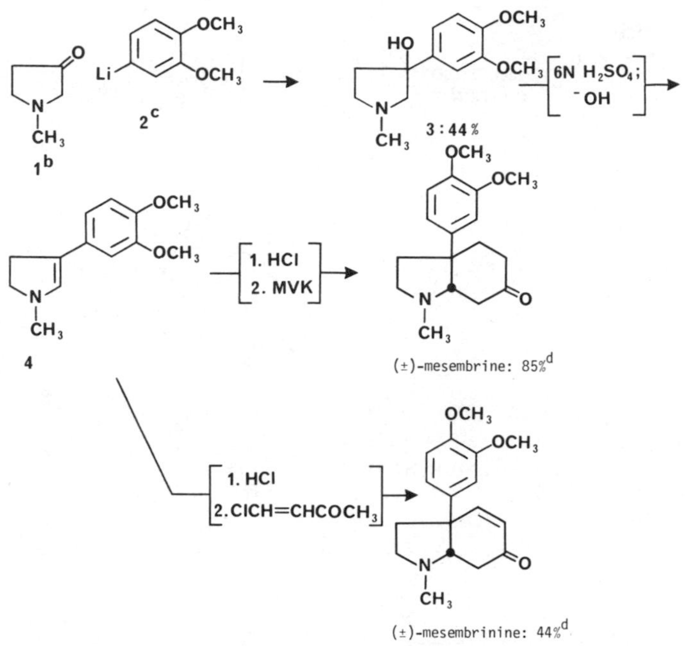

1[b] + 2[c] (Li) → 3: 44% → [6N H_2SO_4; ^-OH] → 4

4 → [1. HCl; 2. MVK] → (±)-mesembrine: 85%[d]

4 → [1. HCl; 2. $ClCH{=}CHCOCH_3$] → (±)-mesembrinine: 44%[d]

[a]A more detailed discussion of this work is included in Section 6.
[b]This ketone was prepared by Dieckmann condensation of **5** and subsequent hydrolysis and decarboxylation as described by Prill and McElvain.[21]

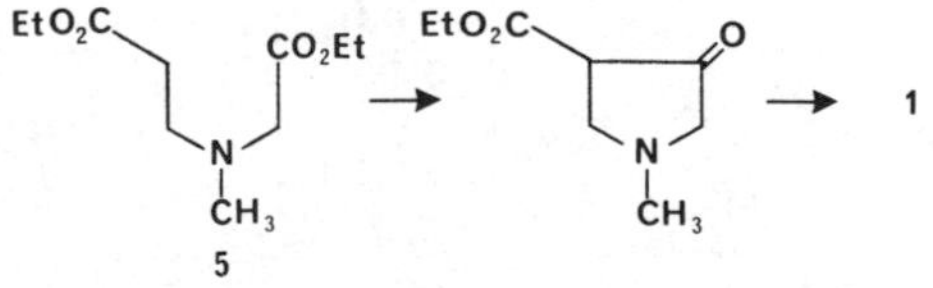

5 → → **1**

[c]cf. Bruce[22] and Howell and Taylor.[23]
[d]"When **4** was condensed with methyl vinyl ketone in refluxing water a 14% yield of d,l-mesembrine was obtained. On the other hand, reaction of **4** with methyl ethynyl ketone, methyl β-chlorovinyl ketone, or β-acetyl trimethylammonium chloride under the same conditions failed to yield any detectable (by TLC) amount of mesembrine. Reactions attempted in solvents other than water led to no fruitful results. Closer scrutiny of the reaction between **4** and methyl vinyl ketone, however, showed that it was strongly subject to acid catalysis."

Yet another fundamentally different approach to the mesembrine alkaloids has been provided by Kugita and his collaborators and is recorded as follows.

Chart 2-6.[24]

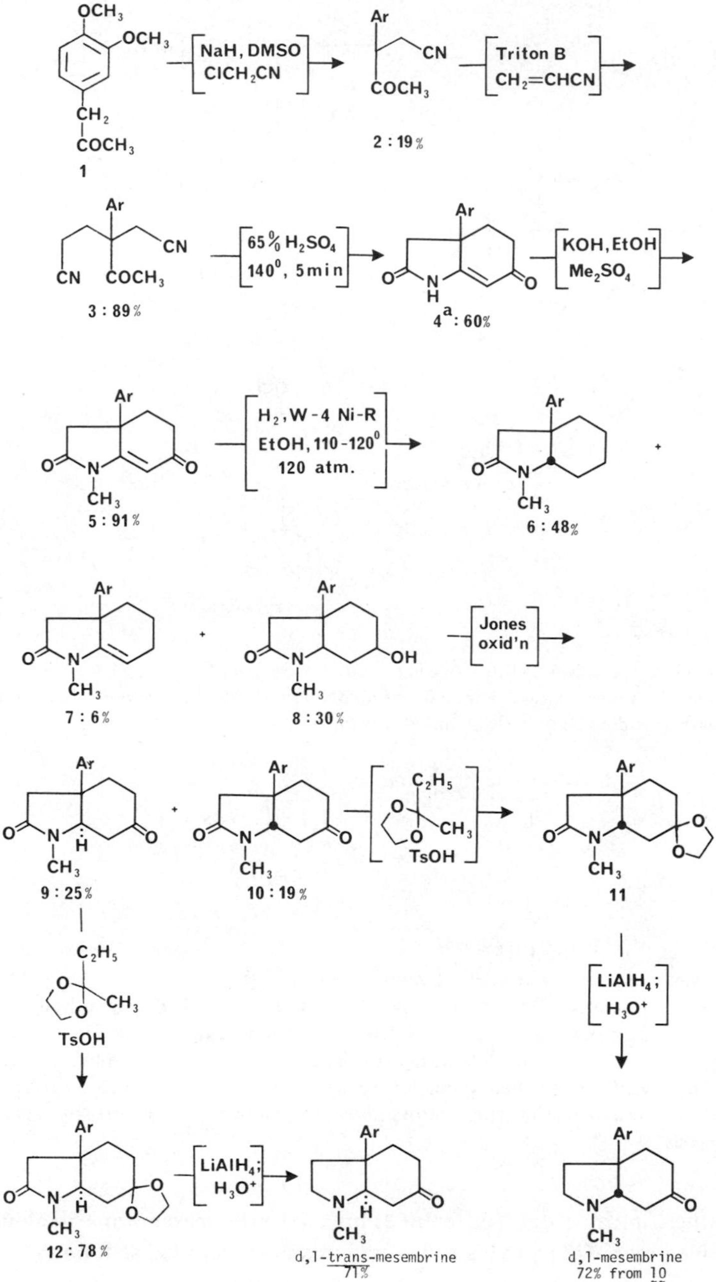

OCH_3
OCH_3
CH_2
$COCH_3$
1

[NaH, DMSO; $ClCH_2CN$]

Ar
CN
$COCH_3$
2 : 19%

[Triton B; $CH_2{=}CHCN$]

Ar
CN
CN $COCH_3$
3 : 89%

[65% H_2SO_4; 140°, 5 min]

Ar
O N H O
4[a] : 60%

[KOH, EtOH; Me_2SO_4]

Ar
O N O
CH_3
5 : 91%

[H_2, W-4 Ni-R; EtOH, 110-120°; 120 atm.]

Ar
O N
CH_3
6 : 48%

+

Ar
O N
CH_3
7 : 6%

+

Ar
O N OH
CH_3
8 : 30%

[Jones oxid'n]

Ar
O N H O
CH_3
9 : 25%

+

Ar
O N O
CH_3
10 : 19%

[C_2H_5, O, O, CH_3; TsOH]

Ar
O N O O
CH_3
11

9 → [C_2H_5, O, O, CH_3; TsOH]

Ar
O N H O O
CH_3
12 : 78%

[$LiAlH_4$; H_3O^+]

Ar
N H O
CH_3
d,l-trans-mesembrine
71%

11 → [$LiAlH_4$; H_3O^+]

Ar
N O
CH_3
d,l-mesembrine
72% from 10

[a]This interesting transformation was first established by Koelsch et al.[25,26] who studied the hydrolytic cyclization of γ-acetylpimelonitriles using 65 to 80% H_2SO_4 and was later emloyed by Ban et al.[27] as a crucial step in the total synthesis of aspidospermine:

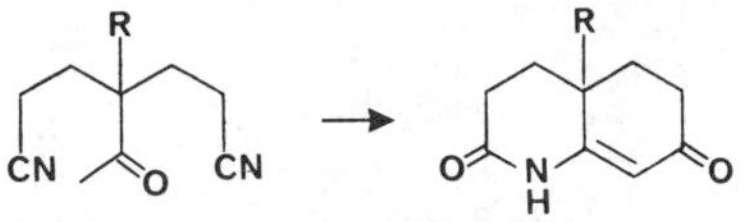

The important and somewhat surprising additional observation, established in the present study, that intermediates **3**, **14**, and **15** are all suitable precursors of **4** adds considerably to the utility of this process. A consideration of the mechanism of these reactions is provided in the first chapter of this book.

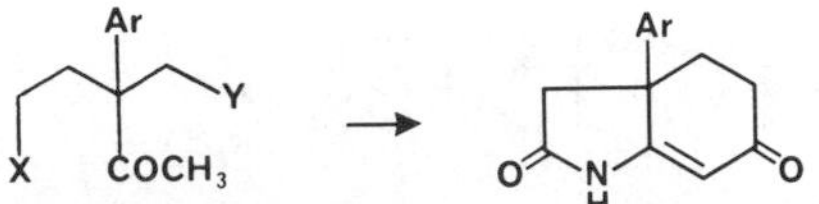

3 : X = Y = CN
14 : X = CN, Y = CO_2CH_3
15 : X = CO_2CH_3, Y = CN

4

Chart 2-7 outlines an approach to the hydroindolone nucleus and its incorporation into the crinine skeleton. The synthesis of joubertiamine and dihydrojoubertiamine are illustrated in Chart 2-8.

Chart 2-7.[28]

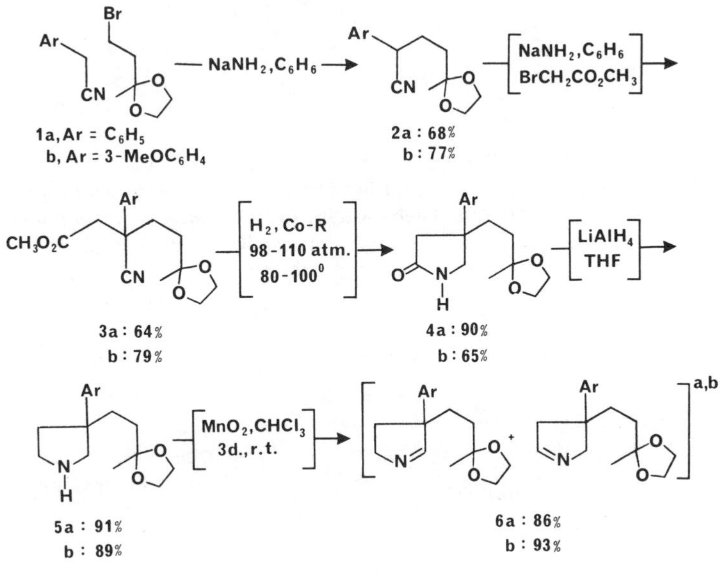

1a, Ar = C_6H_5
b, Ar = 3-$MeOC_6H_4$

$NaNH_2, C_6H_6$

2a : 68%
b : 77%

$NaNH_2, C_6H_6$
$BrCH_2CO_2CH_3$

3a : 64%
b : 79%

H_2, Co-R
98-110 atm.
80-100°

4a : 90%
b : 65%

$LiAlH_4$
THF

5a : 91%
b : 89%

$MnO_2, CHCl_3$
3d., r.t.

6a : 86%
b : 93%

a,b

Chart 2-7 (Continued).

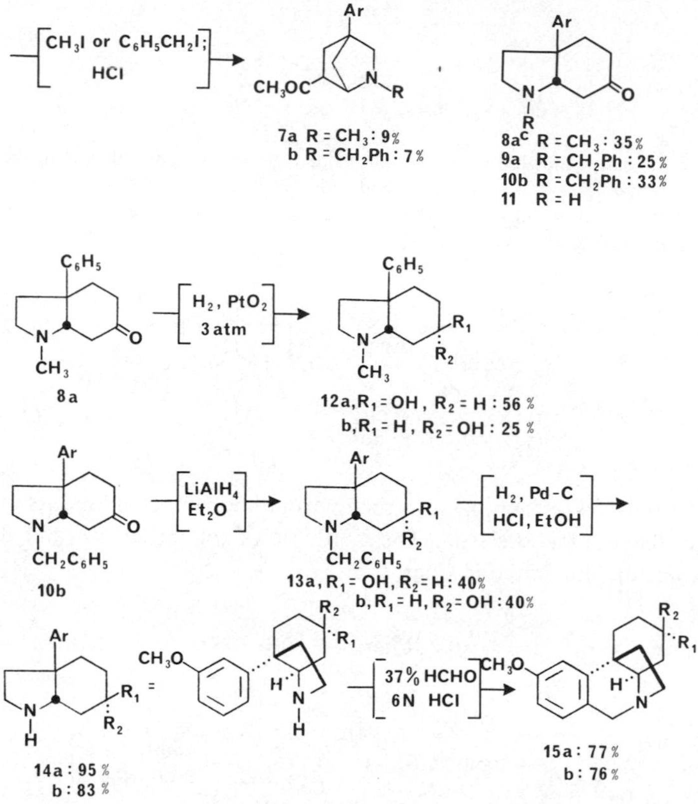

[CH_3I or $C_6H_5CH_2I$; HCl]

7a R = CH_3 : 9%
b R = CH_2Ph : 7%

8a[c] R = CH_3 : 35%
9a R = CH_2Ph : 25%
10b R = CH_2Ph : 33%
11 R = H

8a [H_2, PtO_2 3 atm]

12a, R_1 = OH, R_2 = H : 56%
b, R_1 = H, R_2 = OH : 25%

10b [$LiAlH_4$ Et_2O] 13a, R_1 = OH, R_2 = H : 40%
b, R_1 = H, R_2 = OH : 40%

[H_2, Pd-C HCl, EtOH]

14a : 95%
b : 83%

[37% HCHO 6N HCl]

15a : 77%
b : 76%

[a]The isomeric mixture of pyrrolines **6a** and **6b** was used directly in the cyclization step.
[b]Inexplicably, direct acid-catalyzed cyclization of **6a** to the desired octahydroindole system **11** proceeded in only 3% of the yield necessitating the alternative procedure of preparation and catalytic debenzylation of **9a**. The structural and stereochemical features of **11** were verified by methylation of the known[c] *N*-methyl series **8a**.
[c] Ref. 16.

Chart 2-8.[a,29]

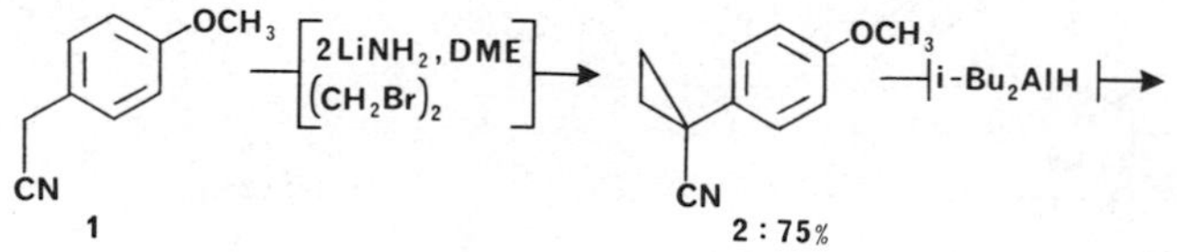

[2 $LiNH_2$, DME $(CH_2Br)_2$]

1

2 : 75%

[i-Bu_2AlH]

CHO

3 : 86%

$-|CH_3NH_2|\rightarrow$

4 : 90%

$-|NH_4Cl, \Delta|\rightarrow$

5 : 100%

$-[HCl; MVK, CH_3CN]\rightarrow$

6 : 83%

$-[CH_3I, \Delta; {}^-OH]\rightarrow$

7 : 62%

$-|HBr, \Delta|\rightarrow$

Joubertiamine: 80%

$-[H_2, Pd-C]\rightarrow$

Dihydrojoubertiamine

[a]cf. Section 6.

3. THE *AMARYLLIDACEAE* ALKALOIDS*

The *Amaryllidaceae* family has been known for some time now to be a rich source of a number of complex and intriguing alkaloids. More than 100 bases have been isolated to date, and the structural diversity that characterizes these interesting substances necessitates their classification into several skeletally homogeneous subgroups. The structures of the parent alkaloid of each of the major subgroups is provided here.

Lycorine

Lycorenine

*Excellent comprehensive reviews of these alkaloids have been provided by Wildman.[30]

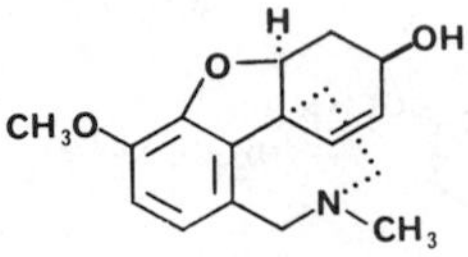

Galanthamine

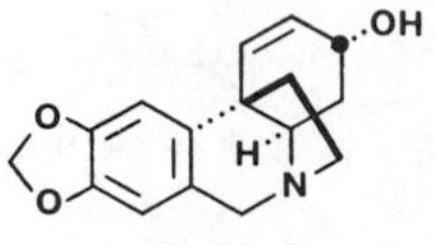

Crinine

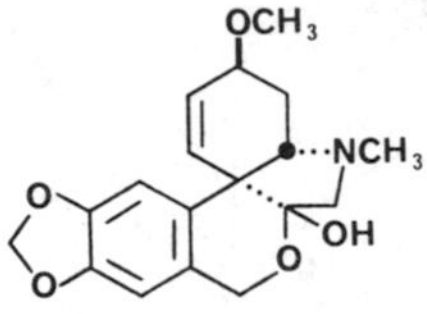

Tazettine

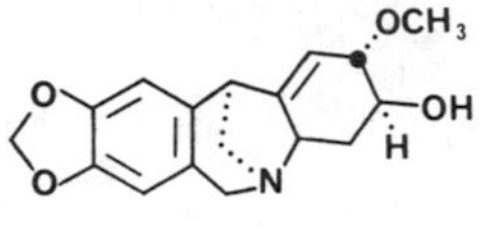

Montanine

Synthesis of the *Amaryllidaceae* alkaloids has been inexplicably rather slow in developing. However, recent developments suggest that this situation is likely to change dramatically. Thus far, attention has been focused mainly on the crinine-, galanthamine-, and lycorine-types of alkaloids, and these will be treated separately in the sequel. Several biogenetic-type syntheses have been reported and for convenience, and comparative purposes are also dealt with in a separate section.

Chart 3-1.[a,8]

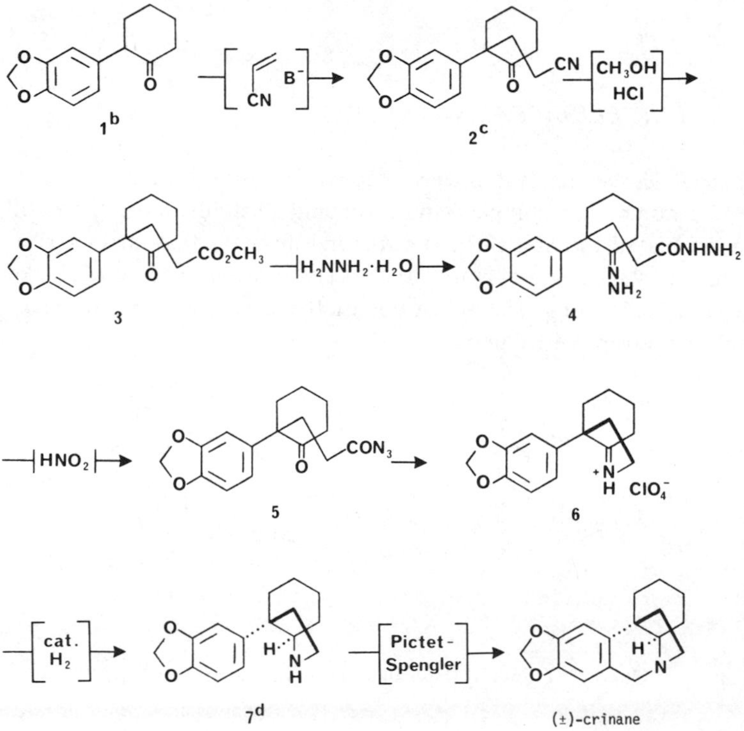

1[b] —[CN, B⁻]→ 2[c] —[CH_3OH, HCl]→ 3 —[$H_2NNH_2 \cdot H_2O$]→ 4 —[HNO_2]→ 5 → 6 (ClO_4^-) —[cat. H_2]→ 7[d] —[Pictet-Spengler]→ (±)-crinane

[a]The precise reagents and conditions employed and the yields obtained are not given in this brief communication.

[b]This starting material had been synthesized previously by Wildman and Mason[31] and Wildman and Wildman.[11] The basic method, which is outlined here, has subsequently been extensively employed as a reliable method for the synthesis of 2-aryl cyclohexanones (cf. Charts 2-1, 2-2, and 3-11):

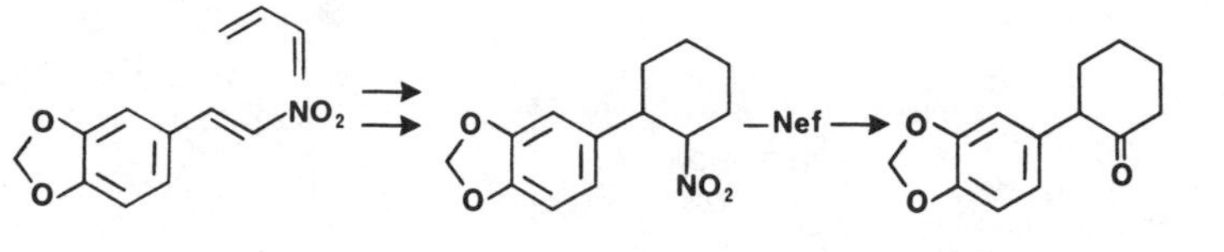

NO_2 — NO_2 — Nef — O

[c]cf. also Charts 2-1, 2-2, and 3-11.

[d]The stereochemical course of this reduction was uncertain until it was established later that crinine possesses a *cis* ring fusion. This stereoselective reduction has subsequently been employed in other syntheses (cf. Charts 2-1 and 3-3).

Chart 3-2.[32]

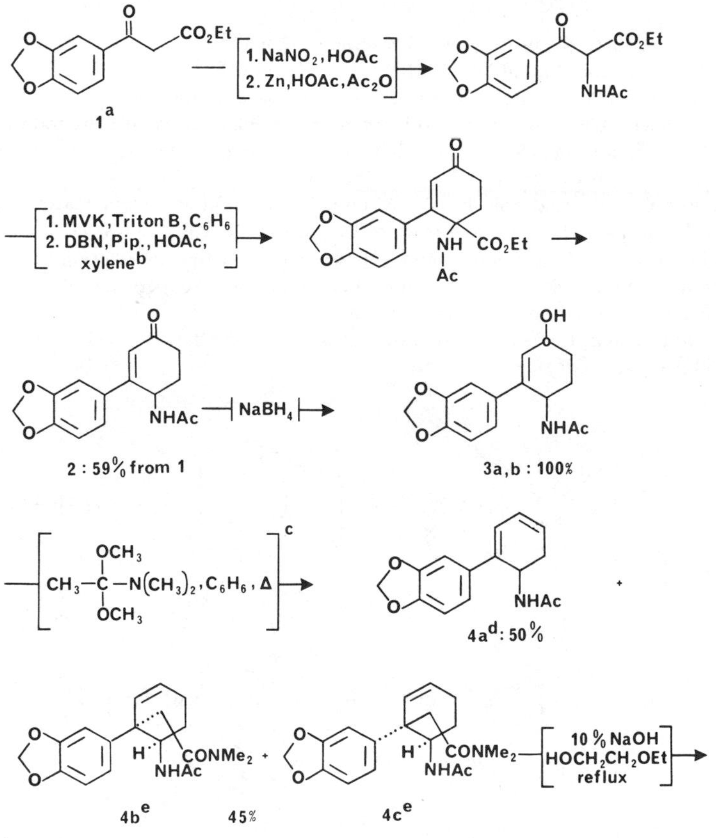

1[a] (CO_2Et) — [1. $NaNO_2$, HOAc; 2. Zn, HOAc, Ac_2O] → (CO_2Et, NHAc)

— [1. MVK, Triton B, C_6H_6; 2. DBN, Pip., HOAc, xylene[b]] → (NH Ac, CO_2Et) →

2 : 59% from 1 (NHAc) — [$NaBH_4$] → 3a,b : 100% (OH, NHAc)

— [$CH_3-C(OCH_3)_2-N(CH_3)_2$, C_6H_6, Δ][c] → 4a[d]: 50% (NHAc) +

4b[e] (H, $CONMe_2$, NHAc) + 4c[e] (H, $CONMe_2$, NHAc) 45% — [10% NaOH, $HOCH_2CH_2OEt$, reflux] →

Chart 3-2 (Continued).

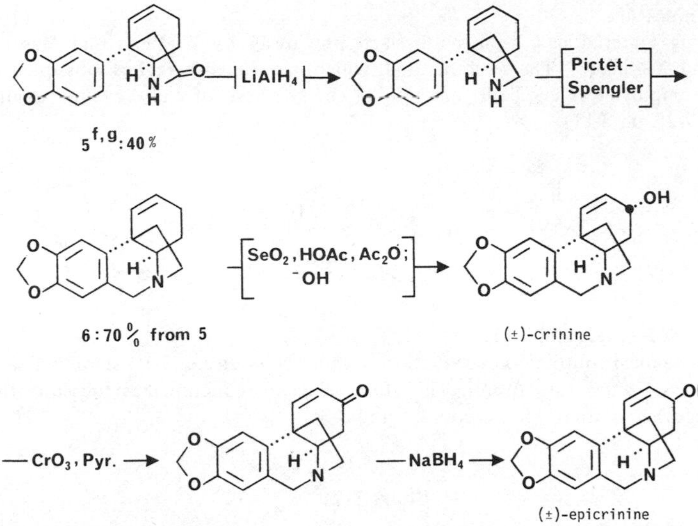

$5^{f,g}$: 40%

LiAlH$_4$ → Pictet-Spengler →

6 : 70% from 5

SeO$_2$, HOAc, Ac$_2$O; $^-$OH → (±)-crinine

CrO$_3$, Pyr. → NaBH$_4$ → (±)-epicrinine

[a]Prepared according to the procedure of Kametani and Ida.[33]
[b]According to the procedure of Muxfeldt.[34]
[c]The mixture of epimeric alcohols **3a** and **3b** was employed in this step.
[d]Although undesirable in this series, this extremely mild method of dehydration is truly noteworthy. Application to systems that cannot undergo the Meerwein-Eschenmoser reaction[e] are under investigation.
[e]The authors propose that this ingenious modification of the Claisen rearrangement be known as the Meerwein-Eschenmoser reaction (cf. Meerwein et al.[35] and Wick et al.[36]
[f]In addition to lactam **5**, unreacted amide **4b** was secured from the reaction mixture when the isomeric mixture of **4b** and **4c** was employed. In a separate experiment, **4b** could not be transformed into a lactam under these conditions.
[g]Unambiguous demonstration of the configuration of this crucial intermediate was obtained by the following series of transformations:

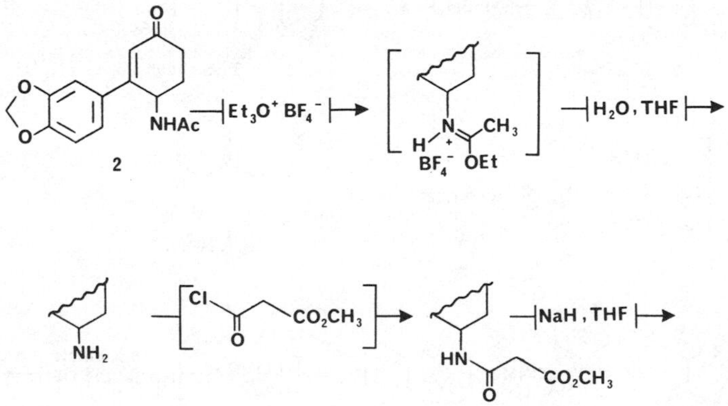

2 (NHAc) → $Et_3O^+ BF_4^-$ → [BF_4^-, OEt, CH_3] → H_2O, THF →

NH_2 → [Cl–CO–CH$_2$–CO_2CH_3] → HN–CO–CH$_2$–CO_2CH_3 → NaH, THF →

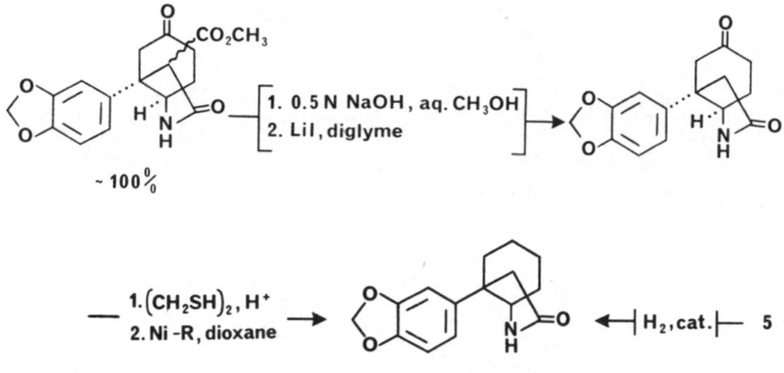

CO_2CH_3

[1. 0.5 N NaOH, aq. CH_3OH / 2. LiI, diglyme]

~100%

[1. $(CH_2SH)_2$, H^+ / 2. Ni-R, dioxane] → 7 ← [H_2, cat.] — 5

Further confirmation of the structural and stereochemical features of 7 (and hence 5) was obtained by the following conversion to the known (±)-crinane:[37]

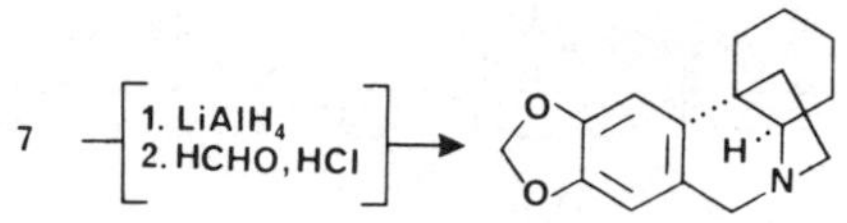

7 — [1. $LiAlH_4$ / 2. HCHO, HCl] →

Chart 3-3.[9]

EtO_2C, CN, CO_2Et — [NaOEt / EtOH][a] → Ar, CN, CO_2Et (1) — [CH_3OH / HCl][b] →

Ar, CO_2CH_3, CO_2CH_3 (2) — [HCl / HOAc][b] → Ar, CO_2H (3) — [CH_3OH / HCl][c,d] → Ar, CO_2CH_3 (4 : 35% overall)

— [$NaOCH_3$ / C_6H_6][e] → Ar, OH (5 : 80%) — [PCl_3 / $CHCl_3$][f] → Ar, Cl (6 : 63%) — [aziridine / TEA, THF] →

7 : 80% — [NaI / diglyme / 145°][g] → 8 : 55% — [H_2-Pt / EtOH, H^+][h] →

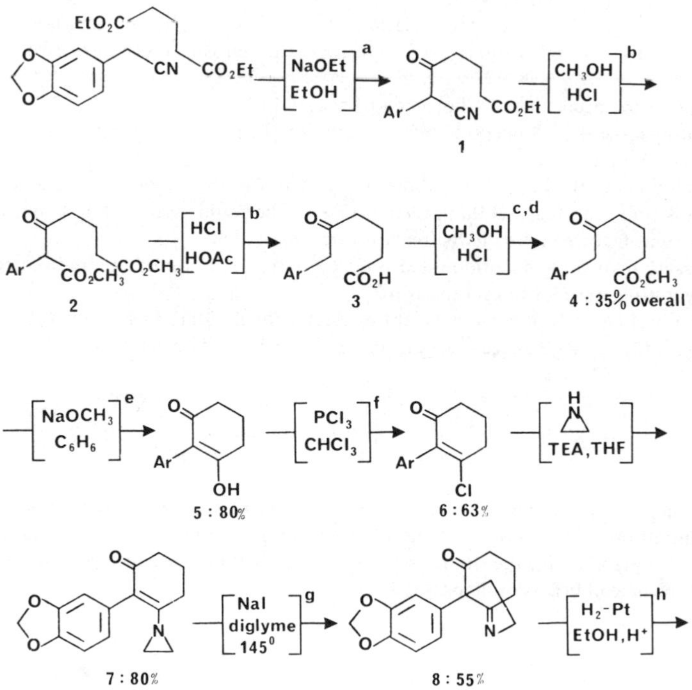

Chart 3-3 (Continued).

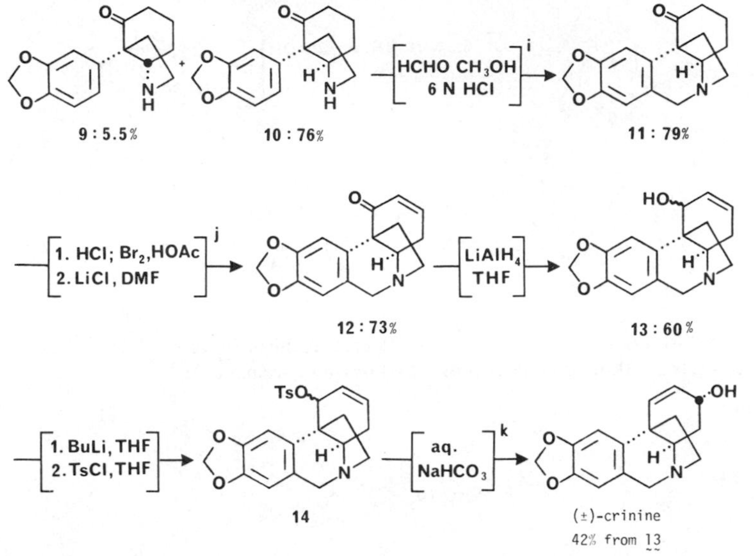

9 : 5.5%   10 : 76%   i: HCHO CH_3OH, 6 N HCl   11 : 79%

j: 1. HCl; Br_2,HOAc 2. LiCl, DMF   12 : 73% $LiAlH_4$, THF   13 : 60%

1. BuLi, THF 2. TsCl, THF   14   k: aq. $NaHCO_3$ (±)-crinine 42% from 13

[a]The synthesis of **5** was a modification of the procedure for 2-phenyl-1,3-cyclohexanedione by Born, Pappo, and Szmuszkovicz.[38]

[b]The direct hydrolysis and decarboxylation of **1** to **3** with refluxing HCl led to extensive hydrolysis of the methylenedioxy group and necessitated this alternative, two stage modification.

[c]According to the procedure of Clinton and Laskowski.[39]

[d]The hydrolysis of **4** to **8** occurred very readily suggesting neighboring-group participation of the ketone hydrate.

[e]The employment of CH_3OH or ethers as solvent for this reaction provided very poor yields of diketone and suggests that precipitation of the sodium-salt of **5** in benzene solution is an important feature in driving the reaction to completion.

[f]Direct reaction of **5** with aziridine led only to polymerization of the amine necessitating this indirect procedure for the synthesis of **7**.

[g]This key novel heterolytic rearrangement of an *N*-vinylaziridine to a Δ^1-pyrroline had some analogy with the iodide induced rearrangement of *N*-acylaziridines (cf. Heine[40]).

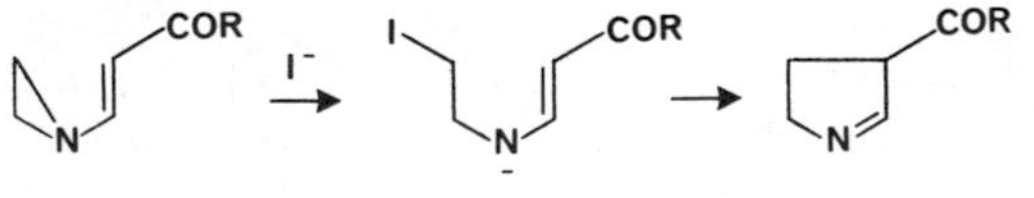

Several models of the desired rearrangement were investigated to determine the utility and/or limitations of this process. Thus whereas ethyl β-aziridinocrotonate **i** failed to yield the desired pyrroline **ii**, the more relevant model (**iii**) provided hydroindole **iv** when subjected to the conditions outlined:

The isomerization of 7 under these conditions provided only poor yields of 8, presumably due to its decreased volatility (**iv** was obtained by bubbling N_2 through the NaI melt subliming it into an attached condenser). However, by executing the reaction in diglyme at 145°, the 55% yield indicated was obtained.

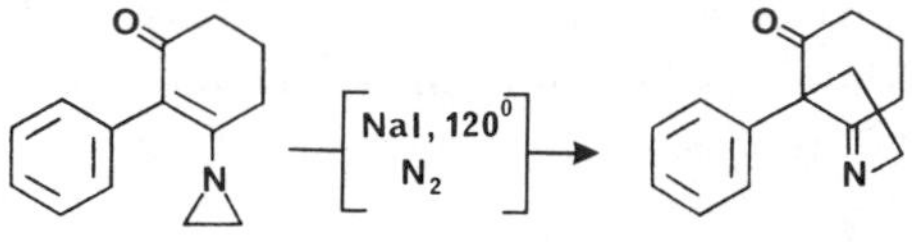

[h]Assignment of a *cis* ring fusion to the major product (**10**) could be made with some confidence based on Wildman's earlier observation on a closely related system (cf. Chart 3-1).

[i]The conditions of this Pictet-Spengler cyclization have been employed extensively in other syntheses of the crinine-type bases discussed in this section. In the present case, the product **11** was identified as (±)-dehydrodihydrodesmethoxybuphanamine by direct comparison with a sample of this optically active degradation product of buphanamine.[41]

[j]Direct introduction of the conjugated double bond with DDQ yielded either starting material or decomposition products. Reaction of **11** with Ph_3CK produced the enolate ion as demonstrated by its reaction with Ac_2O to yield the enol acetate, but reaction of either the enolate or the enol acetate with DDQ afforded only recovered starting materials or decomposition products. Employment of the morpholine enamine to achieve this result was precluded by the inability to generate the enamine.

[k]The conversion of **14** to crinine is an interesting application of the observations made by Goering[42] that cyclohexenyl cations prefer to add nucleophiles in a quasiaxial manner:

In the present case there are two possible modes of quasiaxial attack. However, formation of the isomeric allylic alcohol **vii** should be suppressed for steric reasons.

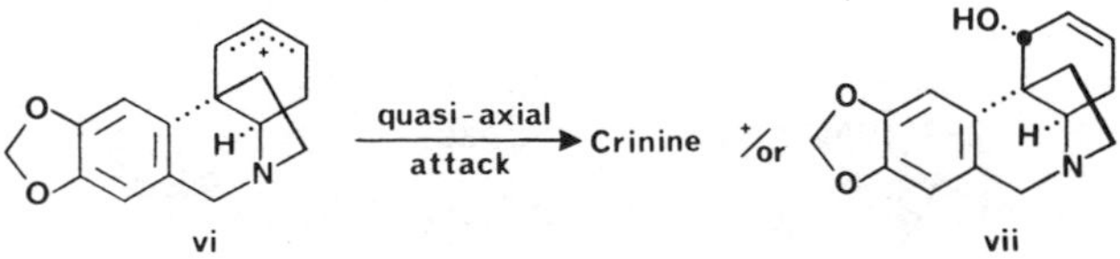

A. Crinine-type Alkaloids

The first synthetic effort in the *Amaryllidaceae* family was initiated by Wildman to provide definitive proof concerning the basic skeleton of crinine, and this was achieved by an unambiguous synthesis and comparison with one of its degradation products–crinane. It is interesting to note that several of the stages employed in this initial effort have been adopted by others in a number of subsequent synthetic investigations in both the *Amaryllidaceae* and mesembrine families as inspection of the footnotes adequately reveals.

Chart 3-4.[43]

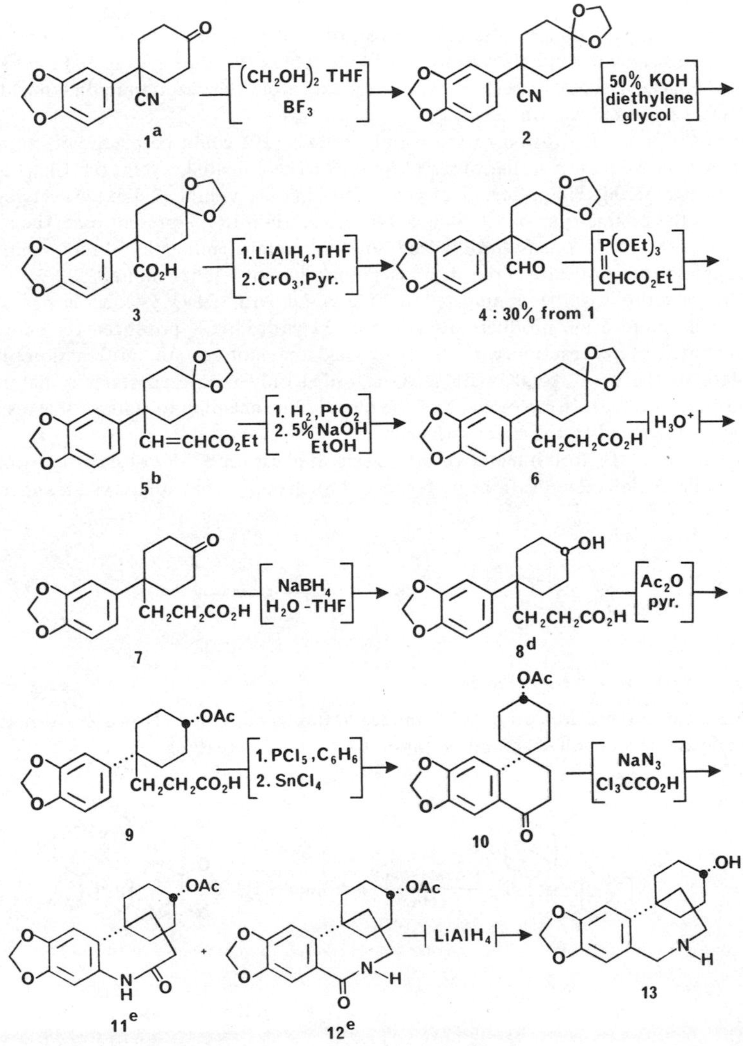

Chart 3-4 (Continued).

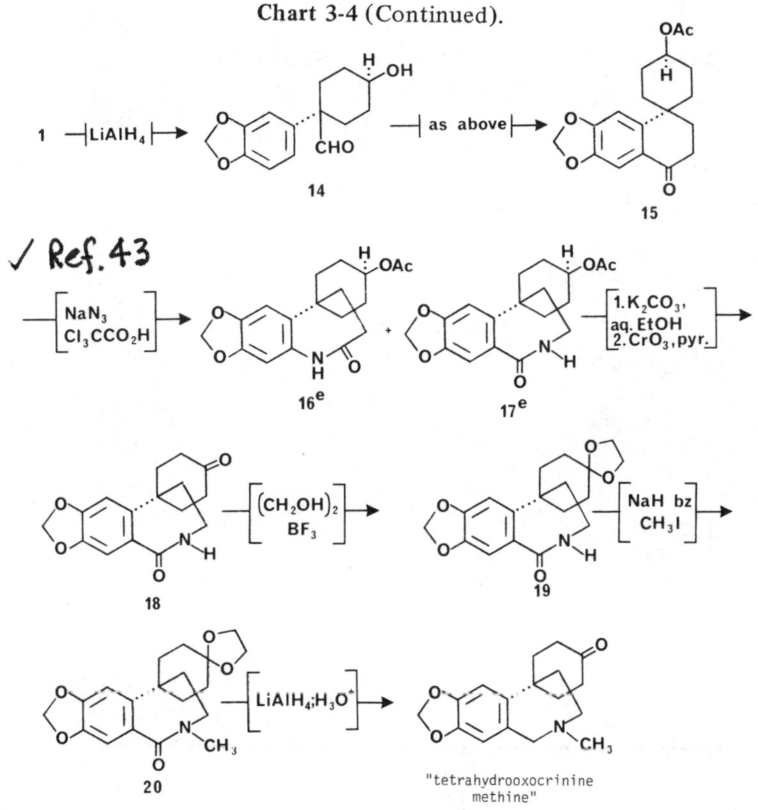

1 —LiAlH$_4$→ **14** —as above→ **15**

✓ Ref. 43

—[NaN_3, Cl_3CCO_2H]→ **16**[e] + **17**[e] —[1. K_2CO_3, aq. EtOH 2. CrO_3, pyr.]→

18 —[$(CH_2OH)_2$, BF_3]→ **19** —[NaH bz, CH_3I]→

20 —[$LiAlH_4$;H_3O^+]→ "tetrahydrooxocrinine methine"

[a]The synthesis of cyano ketone **1** was described previously by these authors: Irie, Tsuda, and Uyeo.[44]

[b]Only the Wadsworth-Emmons modification of the Wittig reaction provided the acrylate **5** in high yield. "The Cope, Knoevenagel, and Reformatsky reactions did not give satisfactory results."

[c]Wadsworth and Emmons.[45]

[d]A mixture of epimeric alcohols in which the *trans* isomer predominated was obtained from this reduction. These were separated and the *trans* alcohol acetylated to provide **9**.

[e]The Schmidt reaction provided lactams **11** and **12** in about equal quantities. This result should be contrasted with the behavior of tetralone itself, which yields only the benzazepin system:

However, prior work by this group had established that placement of the methoxy group *para* to the carbonyl function can substantially alter the course of this reaction:

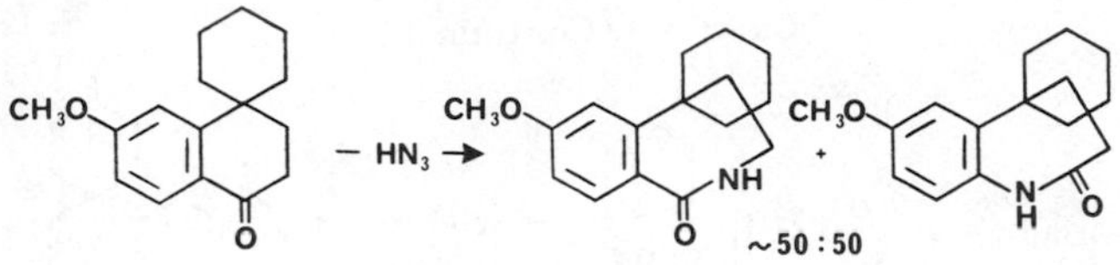

CH_3O — HN_3 → CH_3O ... NH + CH_3O ... N H O ~50 : 50

Chart 3-5.[46]

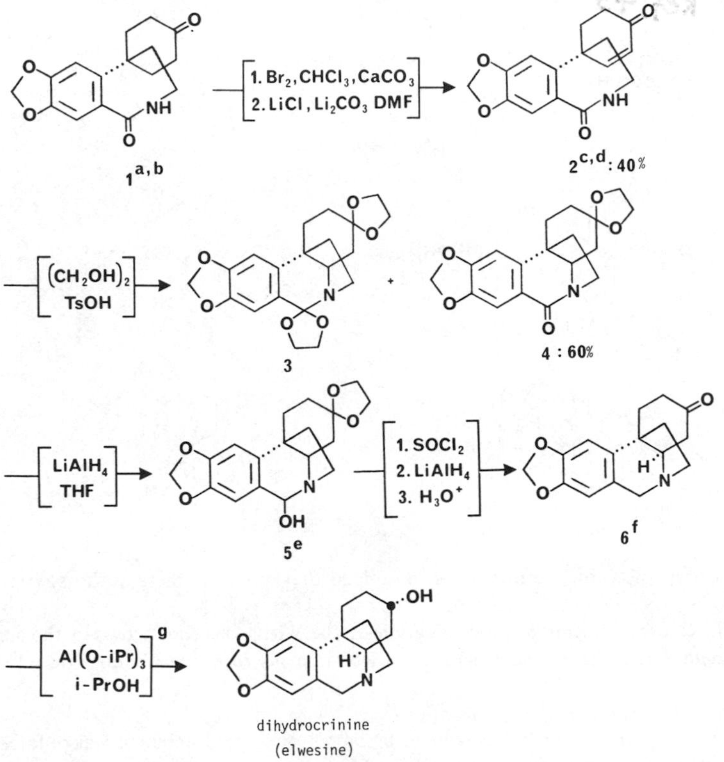

$1^{a,b}$ — [1. Br_2, $CHCl_3$, $CaCO_3$; 2. LiCl, Li_2CO_3 DMF] → $2^{c,d}$: 40%

— [$(CH_2OH)_2$, TsOH] → 3 + 4 : 60%

— [$LiAlH_4$, THF] → 5^e — [1. $SOCl_2$; 2. $LiAlH_4$; 3. H_3O^+] → 6^f

— [$Al(O\text{-}iPr)_3$, i-PrOH]^g →

dihydrocrinine
(elwesine)

[a]The synthesis of this key intermediate was described in a previous communication (cf. Chart 3-4).
[b]Attempts to reduce the amide function at this stage to the corresponding 2°-amine with $LiAlH_4$ proceeded only in poor yield despite the fact that the corresponding *N*-methyl lactam was easily transformed to the 3°-amine.
[c]Attempts to introduce this double bond by employing DDQ gave only starting material even after prolonged heating in various solvents.
[d]The intramolecular, base-induced conjugate addition of the amide nitrogen to the enone proceeded in less than 5% of the yield when NaH was employed as the base. Mainly starting material was recovered.
[e]Incomplete reduction of the "amide" function may be accounted for by observing that the required intermediate for complete reduction would violate Bredt's rule.

[f]The racemic dihydro-oxocrinine **6** was resolved by means of di-(*p*-toluoyl)-D- and -L-tartaric acids to give optically active dihydro-oxocrine and its enantiomer dihydro-oxovittatine. Each of these enantiomers, in turn, was subjected to Meerwein-Ponndorf reduction to afford dihydrocrinine (elwesine) and dihydrovittatine, respectively.
[g]Employment of $LiAlH_4$ in this reduction afforded only the epimeric alcohol.

Chart 3-6.[47]

haemanthidine

[a]The reagents, conditions, and yields for many of the steps outlined in this synthesis were omitted from this brief communication (see, however, Chart 3-7, footnote a).

[b]This transformation assures ring C of correct stereochemistry as the subsequent Curtius reaction is known to proceed with retention of configuration.

[c]The selectivity observed in this opening may possibly be attributed to the powerful electron withdrawing effect of the adjacent 3,4-methylene-dioxyphenyl group.

[d]Two alcohols epimeric at C_{11} were produced in approximately equivalent amounts. Formation of the carbinolamine moiety is also of interest in this reduction and occurs as a consequence of the fact that normal amide resonance would constitute a violation of Bredt's rule.

Chart 3-7.[48]

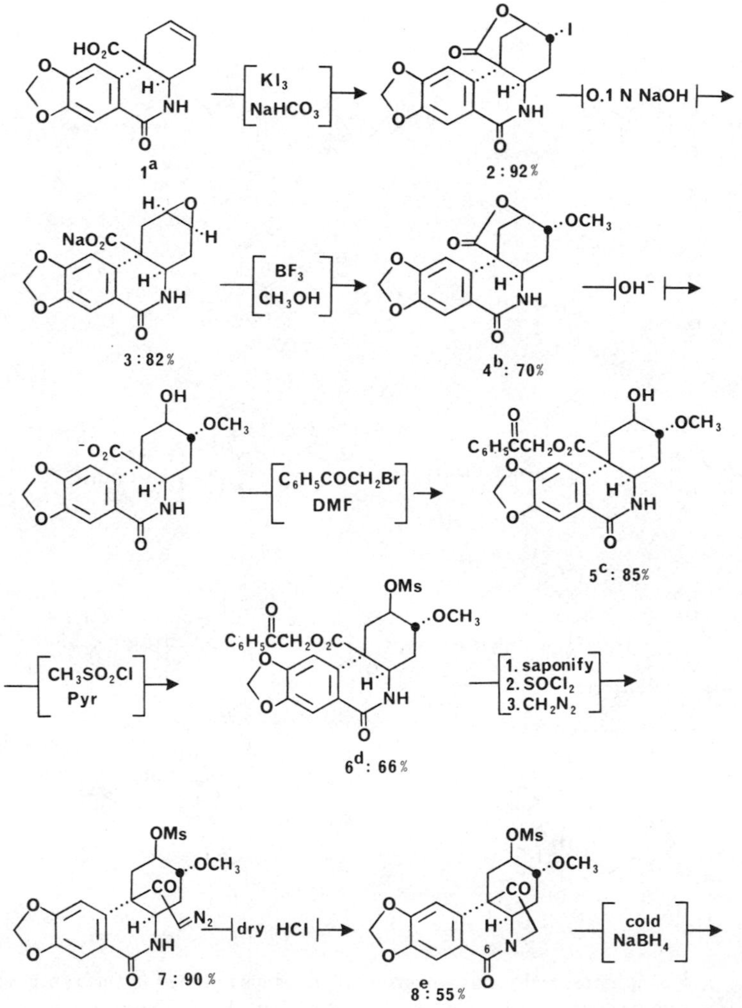

1^a —[KI_3, $NaHCO_3$]→ 2 : 92% —[0.1 N NaOH]→

3 : 82% —[BF_3, CH_3OH]→ 4^b: 70% —[OH^-]→

—[$C_6H_5COCH_2Br$, DMF]→ 5^c: 85%

—[CH_3SO_2Cl, Pyr]→ 6^d: 66% —[1. saponify 2. $SOCl_2$ 3. CH_2N_2]→

7 : 90% —[dry HCl]→ 8^e: 55% —[cold $NaBH_4$]→

Chart 3-7 (Continued).

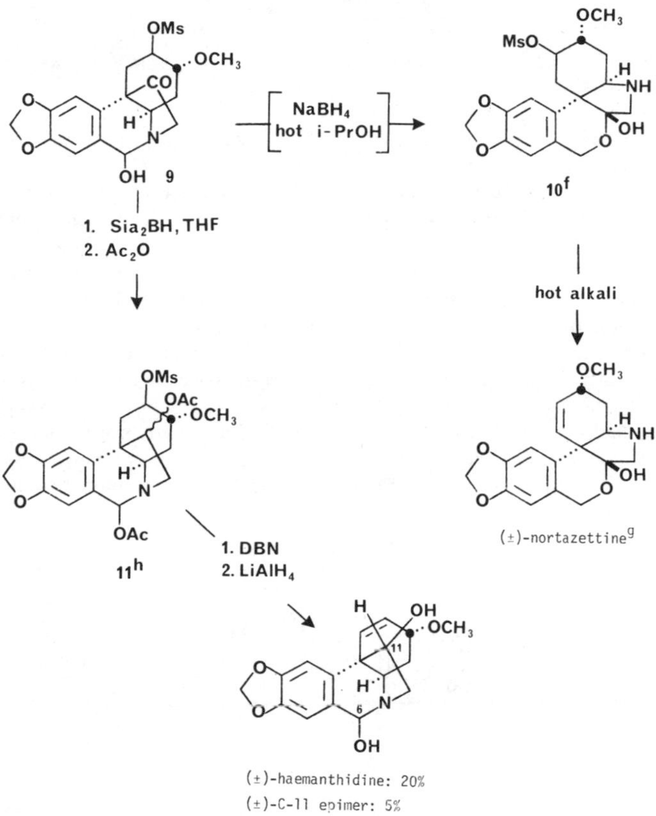

OMs
..OCH$_3$
CO
H
N
OH 9

NaBH$_4$
hot i-PrOH

OCH$_3$
MsO
H
NH
OH
O
10[f]

1. Sia_2BH, THF
2. Ac_2O

hot alkali

OMs
OAc
..OCH$_3$
H
N
OAc
11[h]

OCH$_3$
H
NH
OH
O
(±)-nortazettine[g]

1. DBN
2. $LiAlH_4$

H
OH
..OCH$_3$
11
H
6
N
OH

(±)-haemanthidine: 20%
(±)-C-11 epimer: 5%

[a]A system of synthetic design (to be published by Professor Hendrickson) identifies the creation of the quaternary carbons found in these alkaloids (and many other substances for that matter) as the operation most limited by known synthetic methods. The venerable Diels-Alder reaction has frequently been called upon in such cases, and, indeed, constituted a key step in the synthesis of intermediate **1**. It is reported in this communication that this substance can be prepared in eight-stages and 14% overall yield by improvements in the original method outlined in Chart 3-6. The rigid *trans*-decalin system and dominant axial carboxyl function found in **1** provides a powerful method for control of the remaining asymmetric centers.

[b]Nucleophilic (S_N2) opening of three-membered rings is known to prefer back-side attack. In the present case, there are two possible sites for such attack. However in a conformationally rigid cyclohexanoid system, antiparallel attack, which leads through a chair-like transition state, is usually favored over parallel attack, which must proceed through a boat-like transition state, unless additional steric or electronic considerations are involved in the system in question.

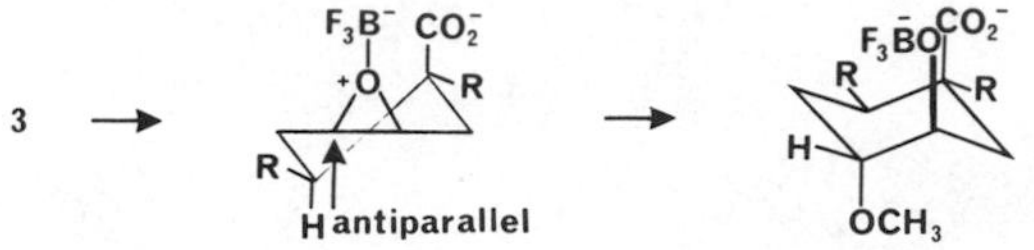

3 → [scheme] → [scheme]

H antiparallel

It is of additional interest to note that these considerations require the methoxyl and hydroxyl groups to assume the less stable axial orientation.

[c]Normal methods of esterification resulted in preferential relactonization thus requiring this ingenious and, no doubt, general alternative method or esterification in order to free the hydroxyl group for further reaction.

[d]The antiparallel opeining of epoxide **3** also insures an axial orientation of the mesylate function, which should permit subsequent facile *trans*-diaxial β-elimination. Attempts to induce elimination at this stage were frustrated by concomitant decarboxylation.

[e]Normal amide resonance between the nitrogen and the carbonyl group at C_6 constitutes a violation of Bredt's rule in this intermediate and permits selective reduction to the carbinolamine with cold $NaBH_4$.

[f]It had been established previously that haemanthidine is readily converted into nortazettine upon treatment with base. The formation of **10** from the borohydride reduction of **9** is therefore not as surprising as it might at first appear and can be accounted for in the following manner:

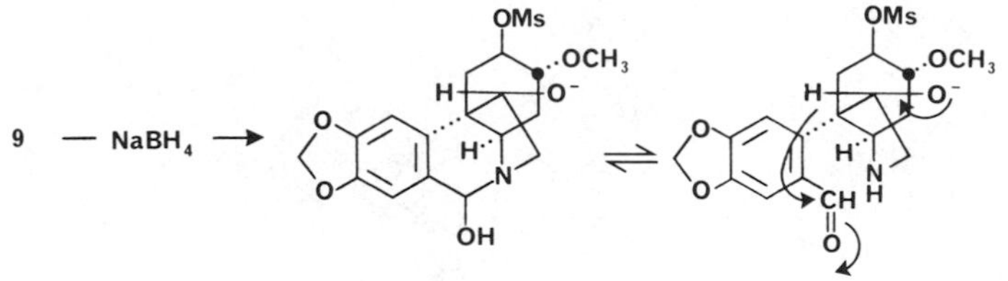

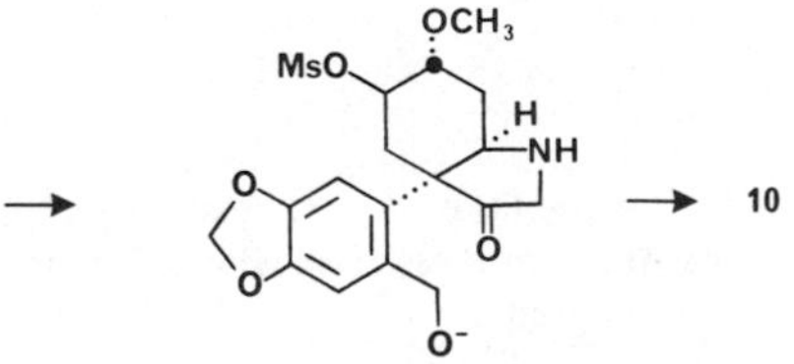

9 — $NaBH_4$ → [scheme] ⇌ [scheme]

→ [scheme] → 10

[g]*N*-Methylation of nortazettine yields tazettine.

[h]The conversion of **9** into haemanthidine obviously requires some means of the preventing the intramolecular Cannizzaro hydride transfer discussed previously. This was achieved by employing disiamylborane as the reducing agent and trapping the intermediate diol by acylation. Reduction of the C_{11} carbonyl group in natural 11-oxo derivatives and previously been shown to yield predominantly the wrong 11-epimer. In this study, it was hoped that this result would be reversed by the presence of the axial mexylate function at C_2. The prominence of haemanthidine over its C_{11} epimer (4:1) demonstrates the wisdom of postponing elimination of this group until after reduction with the hindered boron hydride.

Chart 3-8.[49]

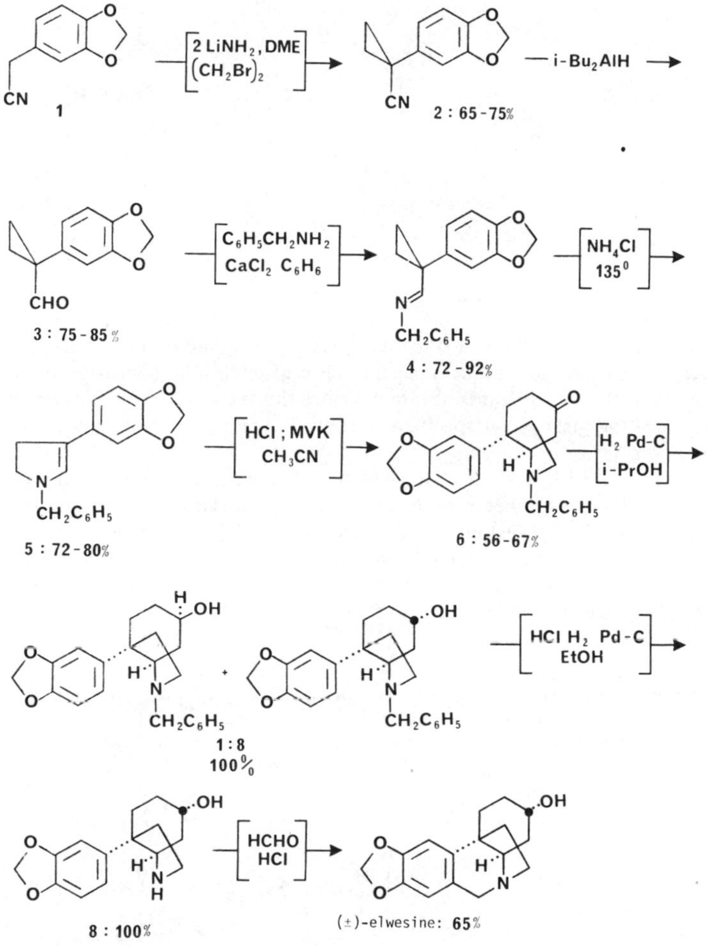

$\xrightarrow{2\ LiNH_2,\ DME;\ (CH_2Br)_2}$ 2 : 65-75% — i-Bu_2AlH →

CHO 3 : 75-85% $\xrightarrow{C_6H_5CH_2NH_2;\ CaCl_2\ C_6H_6}$ N $CH_2C_6H_5$ 4 : 72-92% $\xrightarrow{NH_4Cl;\ 135^0}$

N $CH_2C_6H_5$ 5 : 72-80% $\xrightarrow{HCl;\ MVK;\ CH_3CN}$ H N $CH_2C_6H_5$ 6 : 56-67% $\xrightarrow{H_2\ Pd\text{-}C;\ i\text{-}PrOH}$

H OH H N $CH_2C_6H_5$ + OH H N $CH_2C_6H_5$ 1:8 100% $\xrightarrow{HCl\ H_2\ Pd\text{-}C;\ EtOH}$

OH H N H 8 : 100% $\xrightarrow{HCHO;\ HCl}$ OH H N (±)-elwesine: 65%

[a]A detailed discussion of several of these transformations is provided in Section 6.

Chart 3-9.[50]

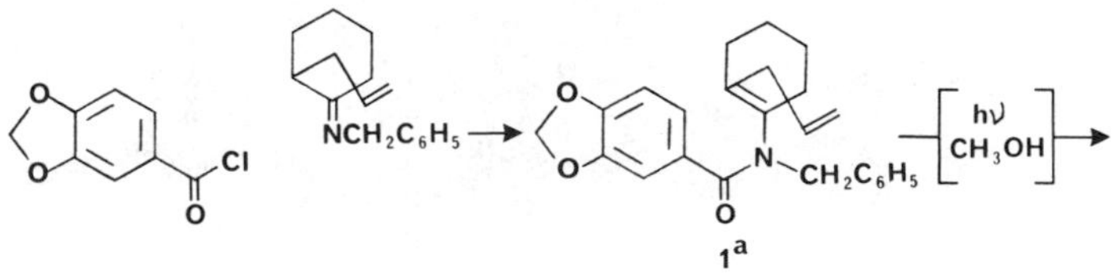

Cl O + $NCH_2C_6H_5$ → N $CH_2C_6H_5$ O 1[a] $\xrightarrow{h\nu;\ CH_3OH}$

Chart 3-9 (Continued.)

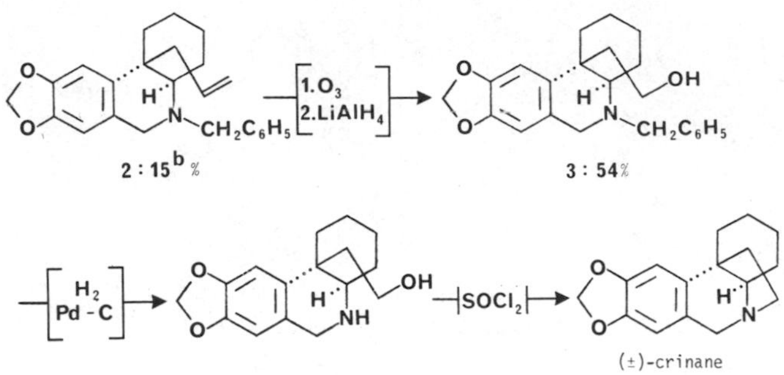

2 : 15[b] %

3 : 54%

(±)-crinane

[a]Unfortunately, many of the yields in this brief communication are not recorded so that it is difficult to draw any definite conclusions. For example, the formation of enamide **1** in the direction of the substituent is of interest since this is contrary to the usual course of this reaction. Since the yield is not specified, however, it may be only a minor product in this transformation.

[b]Even though formed in very modest yield the stereochemical course of this photocyclization is of considerable synthetic and mechanistic interest. This observation is in accord with those recorded earlier by Chapman and Cleveland[51] on the nonoxidative photocyclization of alkyl-substituted acrylic acid anilides to dihydrocarbostyrils in which the initial electrocyclic ring closure was shown by labeling experiments to be followed by symmetry-allowed [1,3] and [1,5] sigmatropic hydrogen shifts:

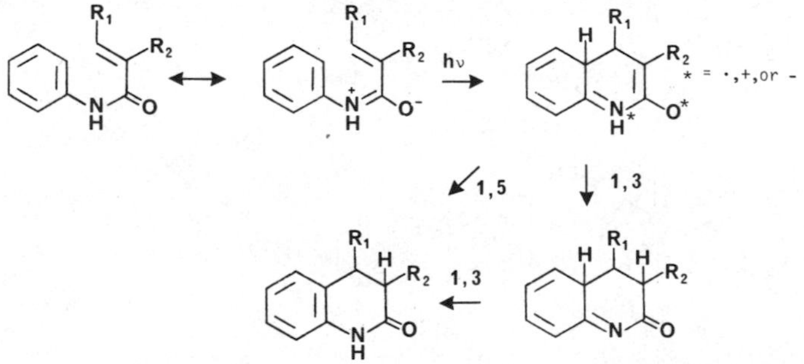

* = ·,+,or -

By analogy, in this series, if the initial electrocyilc reaction follows the usual conrotatory course then a subsequent symmetry-allowed suprafacial [1,5] hydrogen shift would explain the observed stereochemical course:

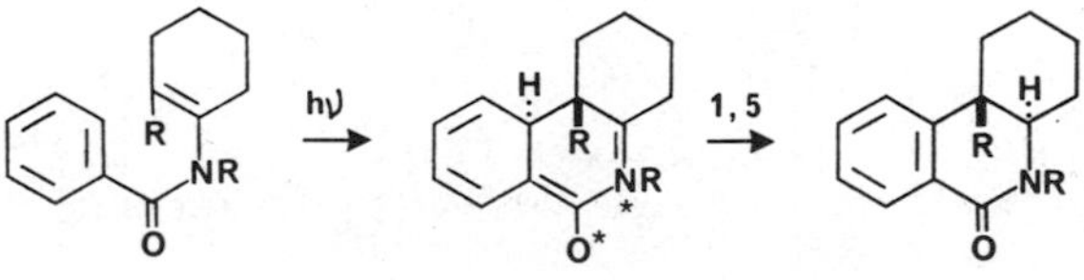

[c]Chapman and Cleveland[51] and Chapman and Eian.[52]

Chart 3-10.[53]

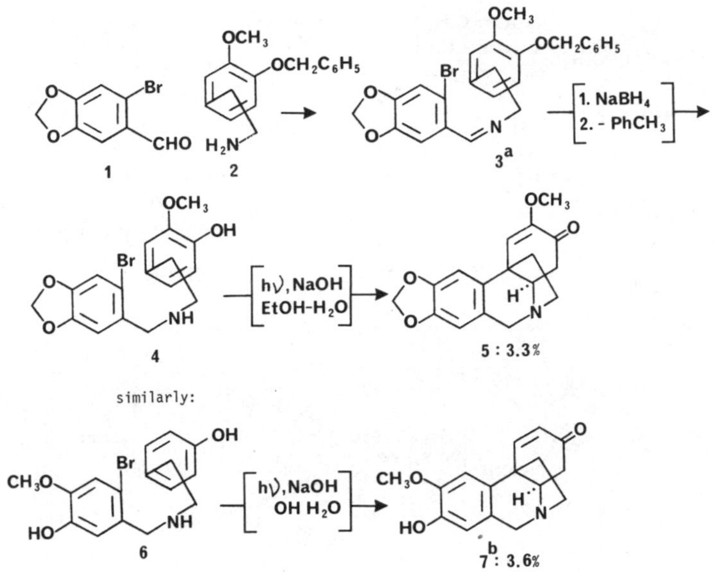

similarly:

[a]Except for the photocyclization products, yields, reagents, and conditions were omitted in this brief communication.
[b]Enone 7 was compared spectroscopically with Schwartz's intermediate in the synthesis of maritidine (cf. Chart 3-23).

B. Galanthamine-type Alkaloids

The galanthamine-type of *Amaryllidaceae* alkaloids have also been the subject of several investigations. Many of these have been executed as biosynthetic models, and the reader is referred to Section 3 for additional synthetic approaches.

Chart 3-3.11.[54]

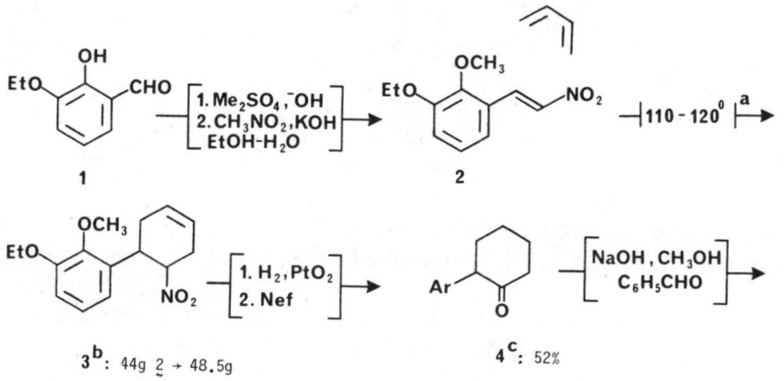

Chart 3-11 (Continued).

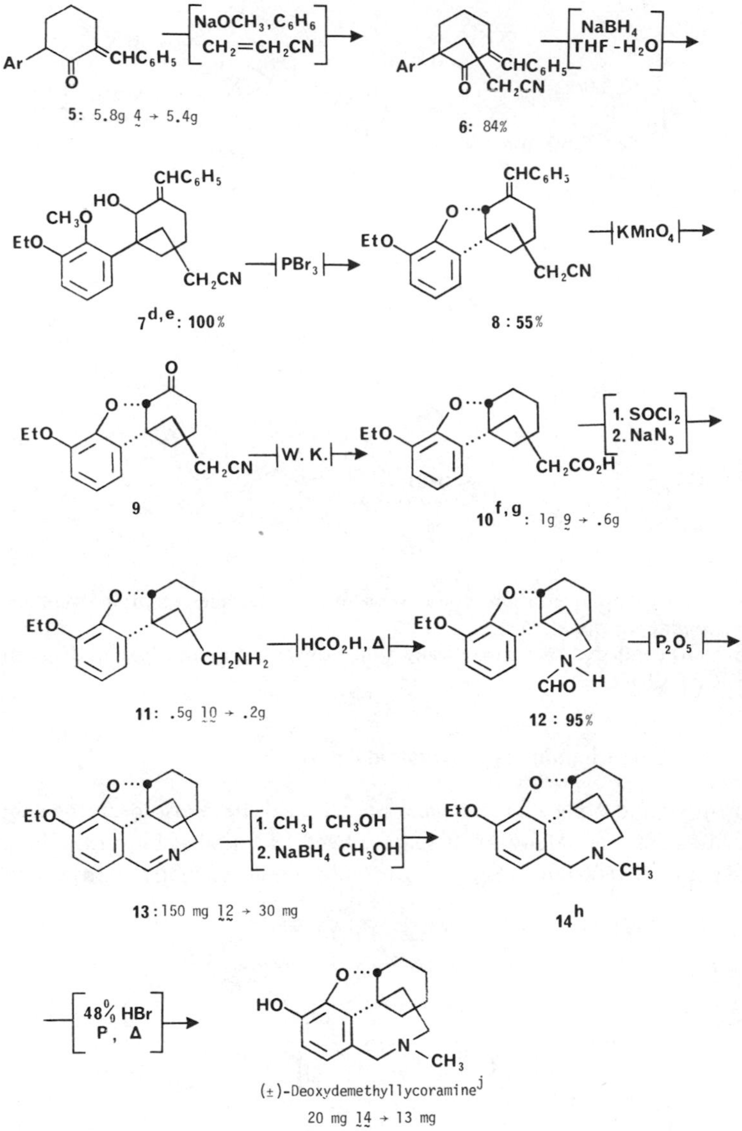

5: 5.8g 4 → 5.4g

[$NaOCH_3$, C_6H_6; $CH_2{=}CH_2CN$]

6: 84%

[$NaBH_4$; THF - H_2O]

$7^{d,e}$: 100%

[PBr_3]

8 : 55%

[$KMnO_4$]

9

[W. K.]

$10^{f,g}$: 1g 9 → .6g

[1. $SOCl_2$ 2. NaN_3]

11: .5g 10 → .2g

[HCO_2H, Δ]

12 : 95%

[P_2O_5]

13: 150 mg 12 → 30 mg

[1. CH_3I CH_3OH 2. $NaBH_4$ CH_3OH]

14^h

[48% HBr P, Δ]

(±)-Deoxydemethyllycoramine[j]

20 mg 14 → 13 mg

[a]According to the procedure of Wildman and Wildman[11]

[b]A Nef reaction of this compound did not proceed smoothly and gave only a 20% yield (cf. Charts 2-2 and 3-1).

[c]Preliminary experiments indicated that direct cyanoethylation of **5** yielded a mixture of mono- and disubstituted products, from which mono-alkylated material was difficult to

separate necessitating the employment of the benzylidine blocking group. Compare with Charts 2-2 and 3-1.
[d]Some of the saturated alcohol was also produced in this reduction.
[e]Dealkylation of 7 by heating with pyridine hydrochloride yielded the desired hydroxyhexahydrodibenzofuran **15**. However, the low yield (12%) encountered and a tedious isolation procedure prevented further deployment.

7 —[Pyr·HCl, 200°]→ **15**: 12%

[f]A by-product of this Wolff-Kishner reduction was the corresponding phenol. Precedence for this dealkylation during Wolff-Kishner reduction may be found in the work of Gates and Tschudi.[55]
[g]The original plan for deployment of **10** involved its conversion to **16** and subsequent transformation to the α-oxime **1**. However, attempted Beckmann rearrangement of the latter substance resulted in a cleavage to **18** rather than rearrangement.

10 —[H_3PO_4, P_2O_5]→ **16** —[KO-t-Bu, i-AmONO]→ **17** —[$SOCl_2$, CH_3OH]→ **18**

[h]This alternative route to the benzazepine system was based on the model transformations outlined below. An unsuccessful attempt at Pictet-Spengler cyclization of amine **19** had been reported earlier by Tomita and Minami.[56] However, "treatment of the formate of this amine with P_2O_5 in boiling toluene gave the expected benzazepine **20** in 12.3% yield···. In view of this successful model experiment, we carried out the Bischler-Napieralski reaction on the amine **11**···."!!!

19[i] → **20**: 12.3% →

[i]The preparation of this intermediate was described by Takahashi, Yori, and Kanbara.[57]
[j]Confirmation of the structural and stereochemical features of this product was achieved by the following degradation of natural (−)-galanthamine:

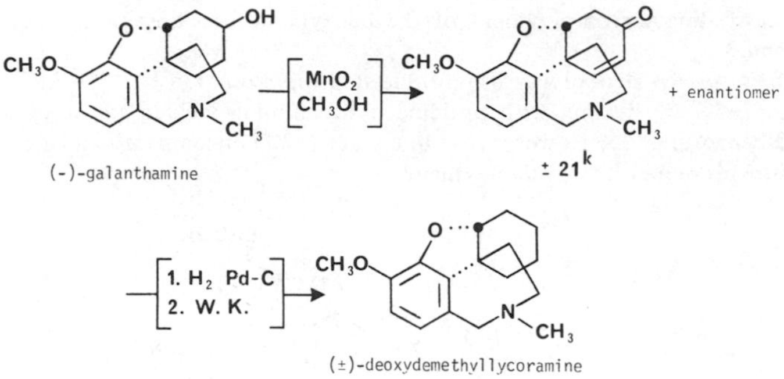

CH_3O OH N CH_3 — [MnO_2 / CH_3OH] → CH_3O O N CH_3 + enantiomer

(-)-galanthamine ± 21[k]

— [1. H_2 Pd-C / 2. W. K.] → CH_3O N CH_3

(±)-deoxydemethyllycoramine

[k]The production of the racemic enone **21** during the oxidation of (−)-galanthamine had been established previously by Koizumi, Kobayashi, and Uyeo.[58] This unusual racemization undoubtedly results from a facile β-elimination of the initially formed enone **21a** to the symmetrical dienone **22**. Subsequent reclosure would then provide **21a** or its enantiomer **21b**.

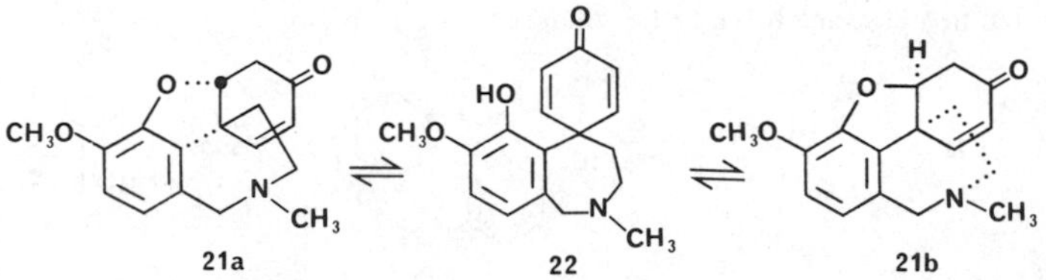

CH_3O O N CH_3 **21a** ⇌ HO CH_3O O N CH_3 **22** ⇌ H CH_3O O N CH_3 **21b**

Chart 3-12.[59]

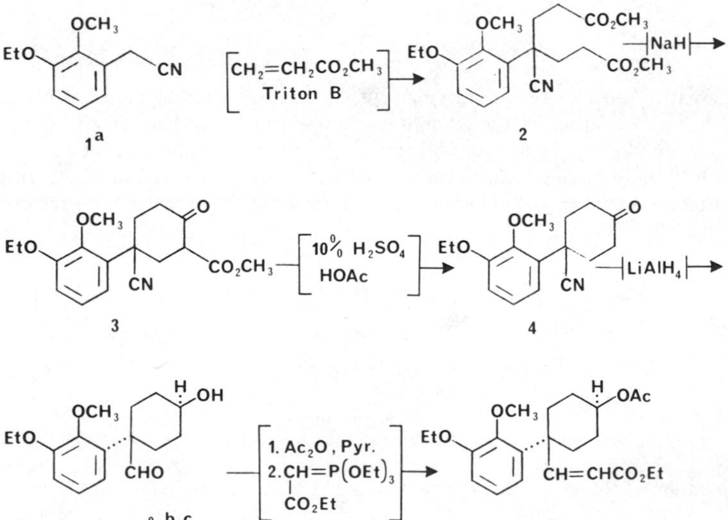

EtO OCH_3 CN **1**[a] [CH_2=$CH_2CO_2CH_3$ / Triton B] → EtO OCH_3 CO_2CH_3 CO_2CH_3 CN **2** —[NaH]→

EtO OCH_3 O CN CO_2CH_3 **3** —[10% H_2SO_4 / HOAc] → EtO OCH_3 O CN **4** —[$LiAlH_4$]→

EtO OCH_3 H OH CHO **5** : 30%[b,c] —[1. Ac_2O, Pyr. / 2. CH=P(OEt)$_3$ / CO_2Et] → EtO OCH_3 H OAc CH=CHCO$_2$Et **6**

Chart 3-12 (Continued).

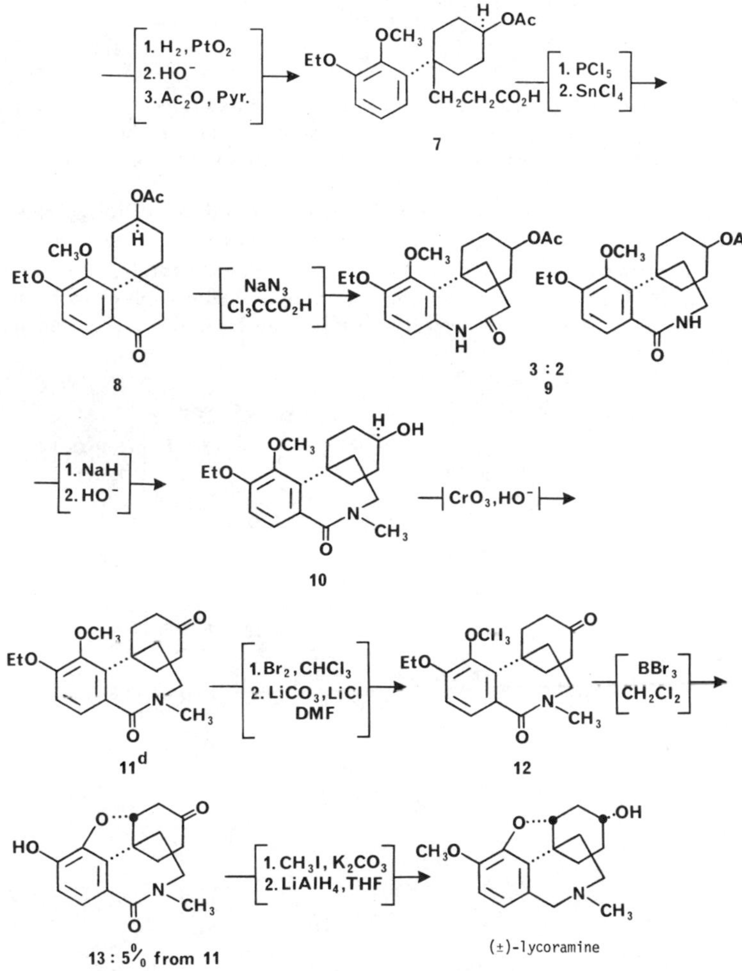

[a]This substance was prepared from 3-ethoxy-2-hydroxybenzaldehyde by (a) methylation of the phenol, (b) crossed Cannizzaro reduction of the aldehyde with formaldehyde, (c) conversion of the resultant benzylic alcohol to the corresponding chloride, and (d) displacement with cyanide.

[b]In addition to the desired aldehyde **5**, a mixture consisting of the corresponding hydroxy-nitrile and hydroxy-amine was also produced in this reduction. This result is in contrast to the behavior of keto-nitrile **14**, which provided good yields of the corresponding alkdehyde upon reduction with $LiAlH_4$ and reflects perhaps the high degree of congestion about the nitrile function in intermediate **4**. Other reagents were no more successful.

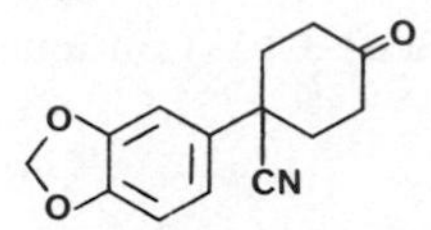

14

[c] The authors candidly point out that the low yield of hydroxy-aldehyde **5** is the chief disadvantage of this synthesis. One might add that the transformation of **8** into the unfavorable product ratio **9** and the extremely low overall yield ovserved in the conversion of **11** to **13** fully corroborate this appraisal.

[d] "Although this compound possesses the ABC ring system comparable to lycoramine, the amount of the material obtained after the many steps required was so small that it would not have been possible to proceed if we had not found that the dihydrofuran ring of natural oxolycoraminone (**15**)···is cleaved smoothly by catalytic hydrogenolysis in alkali." The conversion of oxolycoraminone obtained from natural sources into the relay compound **11** is outlined here.

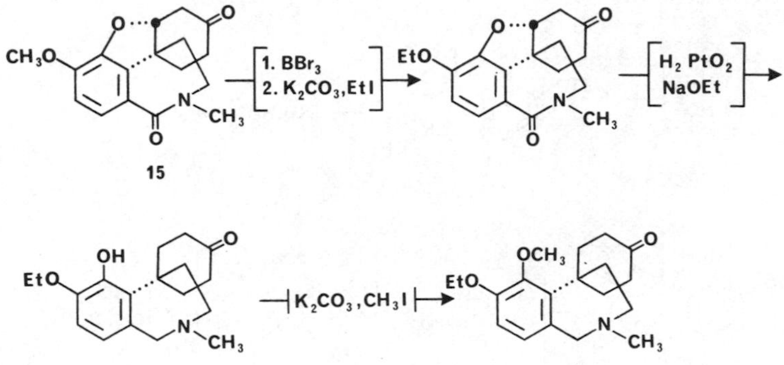

15 —[1. BBr_3; 2. K_2CO_3, EtI]→ —[H_2 PtO_2, NaOEt]→ —[K_2CO_3, CH_3I]→ 11

Chart 3-13.[60]

1 —[$CH_3CH_2NO_2$, n-$BuNH_2$]→ 2 : 72% —[Fe HOAc]→ 3 : 62% —[CH_2=CHCN, t-BuOK]→ 4 : 65% —[1. $^-$OH; 2. EtOH, TsOH]→

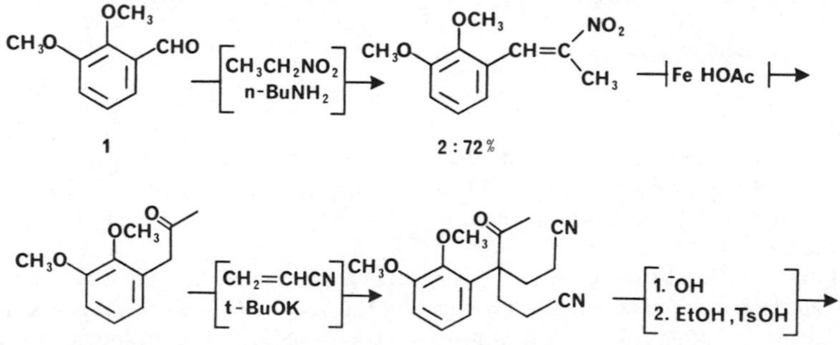

Chart 3-13 (Continued).

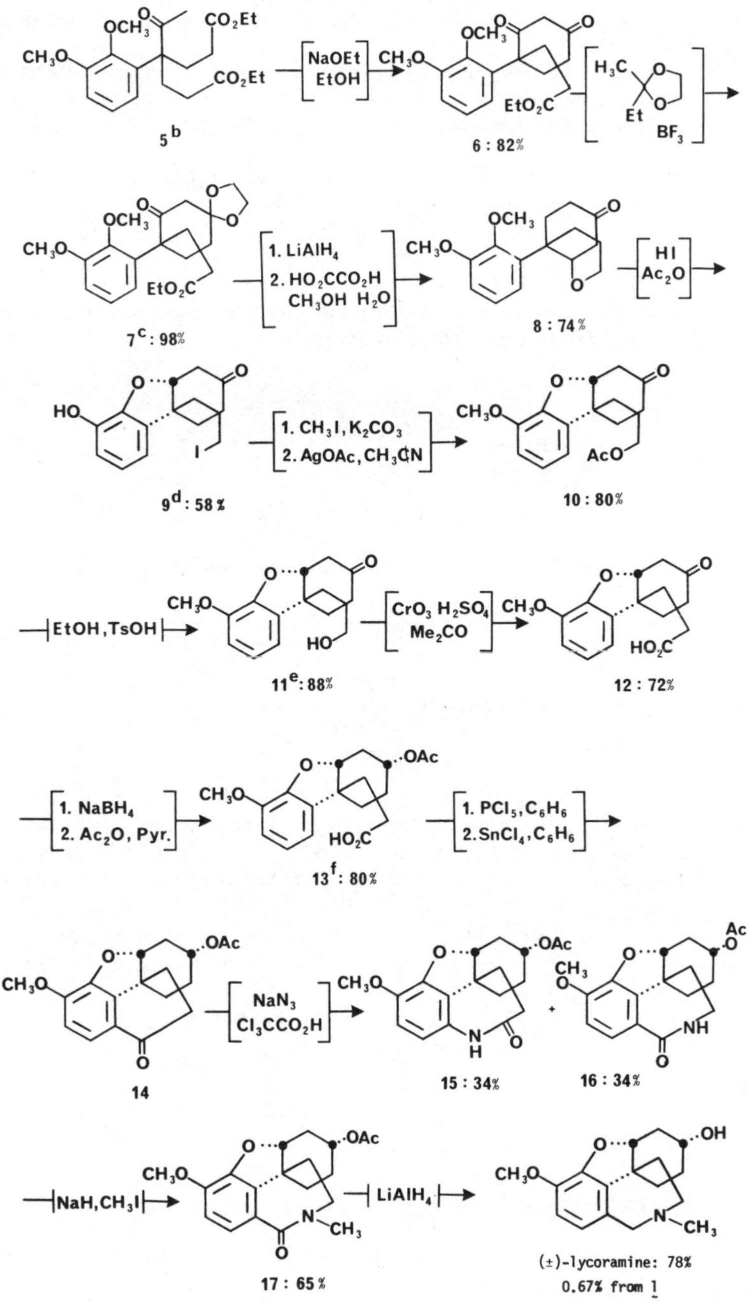

5[b] → [NaOEt, EtOH] → 6 : 82% → [H_3C, Et (2-ethyl-2-methyl-1,3-dioxolane), BF_3] →

7[c] : 98% → [1. $LiAlH_4$; 2. HO_2CCO_2H, CH_3OH, H_2O] → 8 : 74% → [HI, Ac_2O] →

9[d] : 58% → [1. CH_3I, K_2CO_3; 2. AgOAc, CH_3CN] → 10 : 80%

→ [EtOH, TsOH] → 11[e] : 88% → [CrO_3, H_2SO_4, Me_2CO] → 12 : 72%

→ [1. $NaBH_4$; 2. Ac_2O, Pyr.] → 13[f] : 80% → [1. PCl_5, C_6H_6; 2. $SnCl_4, C_6H_6$] →

14 → [NaN_3, Cl_3CCO_2H] → 15 : 34% + 16 : 34%

→ [NaH, CH_3I] → 17 : 65% → [$LiAlH_4$] → (±)-lycoramine: 78%
0.67% from 1

[a]The overall yield (0.67%) of this 19-stage synthesis represents a substantial improvement over the alternative approach provided by these same authors (cf. Chart 3-12).
[b]Attempts to directly convert **3** to **5** by alkylation with methyl acrylate provided only a monoalkylated product.
[c]The selectivity observed in this transdioxonalation procedure is undoubtedly steric in origin. It is of interest to note that direct ketalization with ethylene glycol and BF_3 catalyst gave a mixture of **7** and its enol ether **18**:

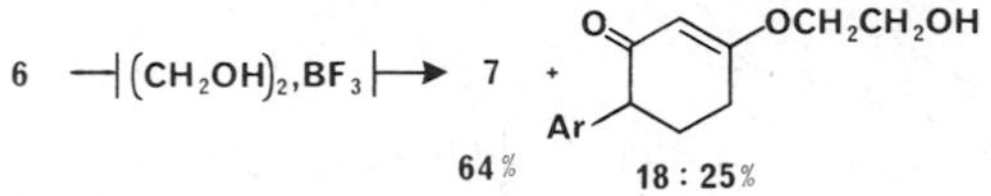

[d]Although the exact timing of the various transformations involved in this process remains obscure, it is clear that **19** or its equivalent is involved:

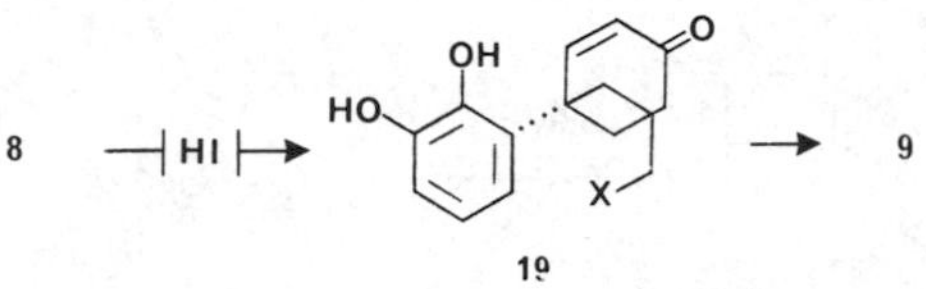

[e]An attempt to convert **9** directly into **11** by diplacement with ¯OH provided **20** instead necessitating these additional steps.

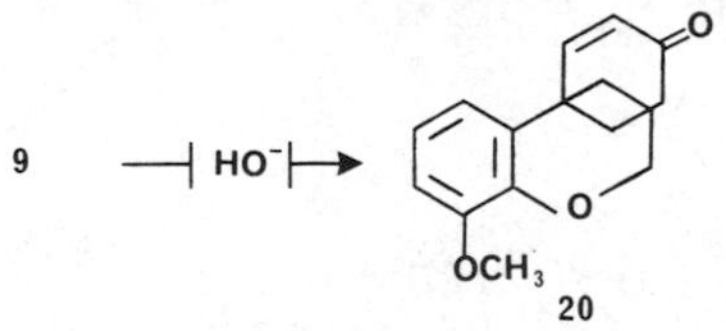

[f] The selection of $NaBH_4$ as the reducing agent was dictated by the fact that lycoraminone **21** had previously been demonstrated by these workers (cf. Hakama et al.[59]) to provide lycoramine. The stereochemical features of **12** and **21** are sufficiently similar to warrant this decision.

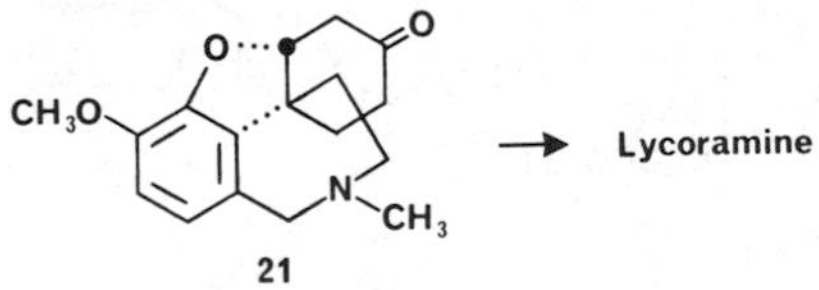

C. Lycorine-type Alkaloids

Lycorine itself is the most abundant and widespread of all of the *Amaryllidaceae* alkaloids. It is, therefore, surprising to discover that synthetic activity in this area has been rather limited, and that, to date, no total syntheses have been recorded.

Chart 3-14.[a,61]

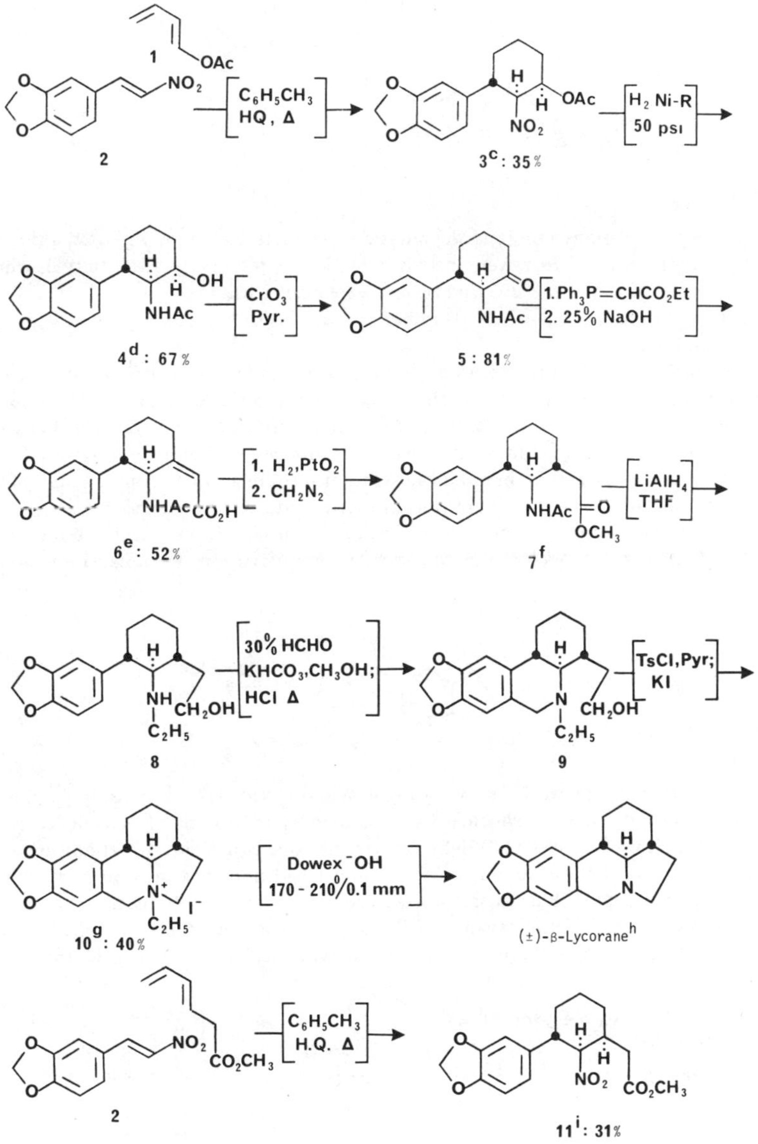

1
OAc
NO_2
$C_6H_5CH_3$
HQ, Δ
2
H
H
OAc
NO_2
3^c: 35%
H_2 Ni-R
50 psi

H
H
OH
NHAc
4^d: 67%
CrO_3
Pyr.
H
O
NHAc
5 : 81%
1. $Ph_3P{=}CHCO_2Et$
2. 25% NaOH

H
$NHAc\ CO_2H$
6^e: 52%
1. H_2,PtO_2
2. CH_2N_2
H
NHAc
O
OCH_3
7^f
$LiAlH_4$
THF

H
NH
CH_2OH
C_2H_5
8
30% HCHO
$KHCO_3$,CH_3OH;
HCl Δ
H
N
CH_2OH
C_2H_5
9
TsCl,Pyr;
KI

H
N^+
I^-
C_2H_5
10^g: 40%
Dowex $^-$OH
170 - 210°/0.1 mm
H
N
(±)-β-Lycorane[h]

NO_2
CO_2CH_3
2
$C_6H_5CH_3$
H.Q. Δ
H
H
NO_2
CO_2CH_3
11^i: 31%

Chart 3-14 (Continued).

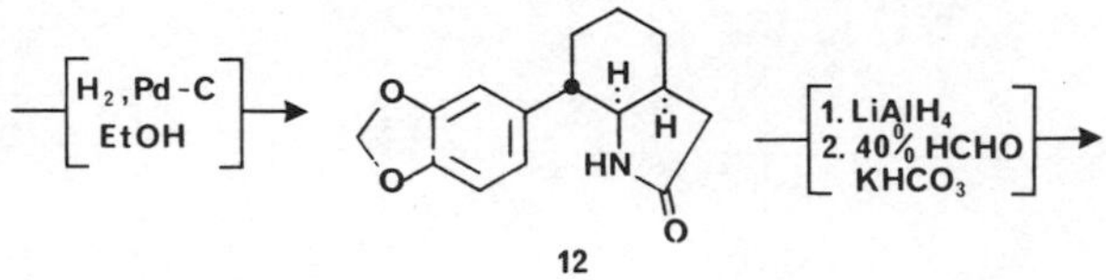

$\left[\begin{array}{c}H_2, Pd\text{-}C \\ EtOH\end{array}\right] \rightarrow$ 12 $\left[\begin{array}{l}1.\ LiAlH_4 \\ 2.\ 40\%\ HCHO \\ \quad KHCO_3\end{array}\right] \rightarrow$

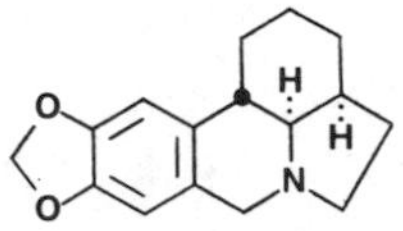

(±)-α-Lycorane

[a]The synthesis of various aromatic degradation products had been reported prior to this study by several groups.[b] However, this investigation represents the first study in which the stereochemical features of the B/C ring fusion were dealt with.

[b]Kelly, Taylor, and K. Wiesner;[62] Humber et al.,[63] Kondo, Takagi and Uyeo;[64] Fales Warnhoff, and Wildman.[65]

[c]The structural and especially the stereochemical features of **3** were assigned on the basis of several arguments. These include: a the orientation is precisely that which one would predict on polar or radical grounds; b the *trans* geometry of β-nitrostyrenes is known to be preserved during the cycloaddition process; and c the stereochemisty corresponds to that predicted by the Alder and Stein rule of maximum accumulation of double bonds. The subsequent reactions of this adduct are in full agreement with these conclusions.

[d]This amide results from O–N acyl migration of the initially formed amino-ester **13**, which could be secured in low yield from the mother liquors. Upon warming, **13** is converted smoothly into **4**.

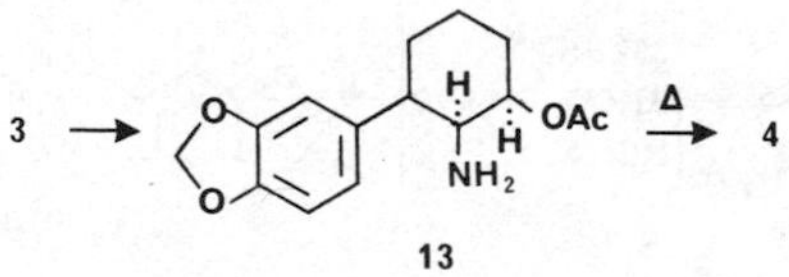

3 → 13 (OAc, NH_2) $\xrightarrow{\Delta}$ 4

The facile rearrangement of **13** to **4** strongly supports the 1,2-relationship of these two functional groups. Conclusive evidence was obtained by hydrolysis of **4** to the corresponding aminoalcohol, which consumed slightly more than one equivalent of periodic acid.

[e]Application of the Wittig reaction to α-ketoamides had not been previously examined. 2-Acetamidocyclohexanone was, therefore, selected as a model. Reduction of the Wittig product (**14**) with either Pd in ethanol or Pt in acetic acid gave the same dihydroester **15**, which was hydrolyzed to the known *trans*-2-acetamidocyclohexaneacetic acid **16**:

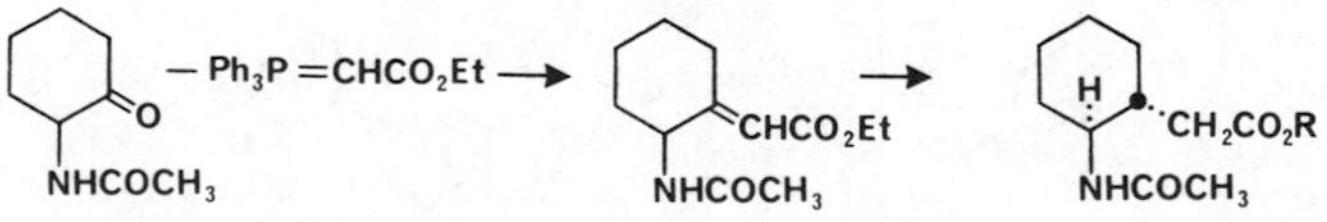

$NHCOCH_3$ — $Ph_3P{=}CHCO_2Et$ → 14 ($CHCO_2Et$, $NHCOCH_3$) → (CH_2CO_2R, $NHCOCH_3$)

15 : R = Et
16 : R = H

[f]At this stage the reasonable assumption was made that the stereochemical course of this hydrogenation should parallel that observed in the closely related model **14**, thus generating the all-equatorial isomer depicted. The somewhat more circuitous route followed in converting 7 into the tetracyclic system was necessitated by the fact that attempts to deacylate 7 were uniformly unsuccessful.

[g]Hofmann elimination is particularly suitable here since, except for ring D, only the ethyl group contains β-hydrogens, which are able to satisfy the usual *anti*-coplanar stereoelectronic requirements of this reaction. The reluctance of the β-hydrogens of ring D to participate in a Hofmann elimination had been established previously by the failure of dihydrolycorine methiodide to undergo this reaction.

[h]This synthesis confirms the gross structural features of β-lycorane and provides unambiguous proof of the *trans* B/C ring fusion found in lycorine and its relatives.

[i]The orientation and stereochemistry assigned this intermediate follows from the knowledge gained in the previous series.

Chart 3-15.[66]

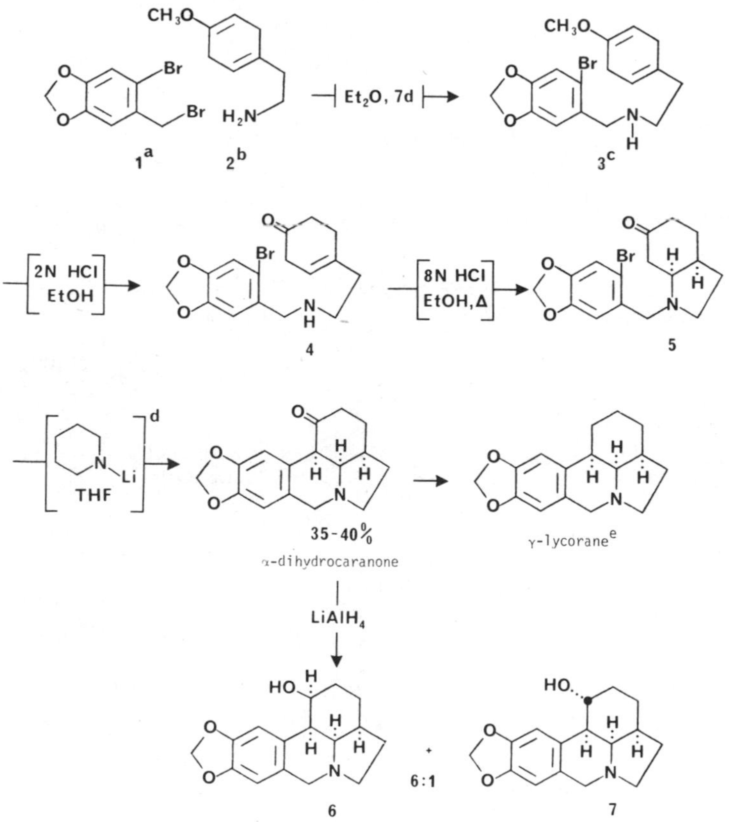

CH_3O Br Br H_2N

1[a] 2[b]

Et_2O, 7d

CH_3O Br N H

3[c]

2N HCl EtOH

Br N H

4

8N HCl EtOH,Δ

Br H H N

5

N Li THF [d]

H H H N

35-40%

α-dihydrocaranone

H H H N

γ-lycorane[e]

$LiAlH_4$

HO H H H H N

6

6:1

HO H H H N

7

[a] Prepared from piperonal according to the procedure of Naik and Wheeler.[61]

[b] This compound was prepared by a method similar to that described by Clark and Pinder.[68]

[c] The corresponding dialkylated amine was also obtained in this alkylation in a ratio of 1:3 favoring **3**.

[d] For other ring-closures involving benzyne intermediates see Bunnett and Hautfiord,[69] Bunnett and Skorcz,[70] Bunnett et al.[71] The reaction of **5** with $NaNH_2$ or KNH_2 in liquid NH_3 did not yield **6**, rather gave amine **8**.

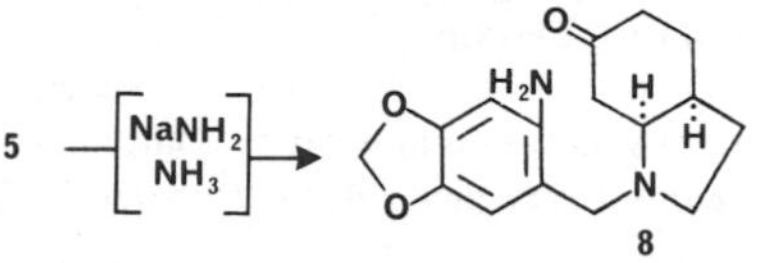

5 —[$NaNH_2$, NH_3]→ 8

[e] The conversion of **6** to γ-lycorane was achieved by formation of the thioketal and desulfurization. The stereochemical features of this substance were confirmed by comparison with an authentic sample obtained from lycorine itself.[74]

[f] All four of the possible stereoisomers of lycorane had been established previously by degradation of lycorine: Takeda et al.,[72] Kotera,[73] and Koreta.[74]

Chart 3-16.[75]

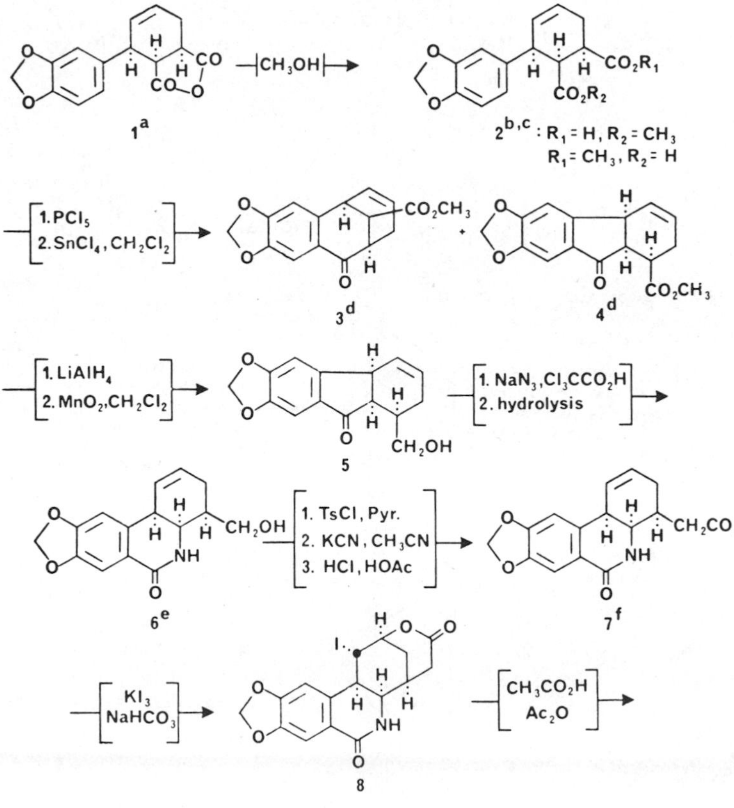

1^a —[CH_3OH]→ $2^{b,c}$: $R_1 = H, R_2 = CH_3$; $R_1 = CH_3, R_2 = H$

—[1. PCl_5; 2. $SnCl_4, CH_2Cl_2$]→ 3^d (CO_2CH_3) + 4^d (CO_2CH_3)

—[1. $LiAlH_4$; 2. MnO_2, CH_2Cl_2]→ 5 (CH_2OH) —[1. NaN_3, Cl_3CCO_2H; 2. hydrolysis]→

6^e (CH_2OH, NH) —[1. TsCl, Pyr.; 2. KCN, CH_3CN; 3. HCl, HOAc]→ 7^f (CH_2CO_2H, NH)

—[KI_3, $NaHCO_3$]→ 8 —[CH_3CO_2H, Ac_2O]→

Chart 3-16 (Continued).

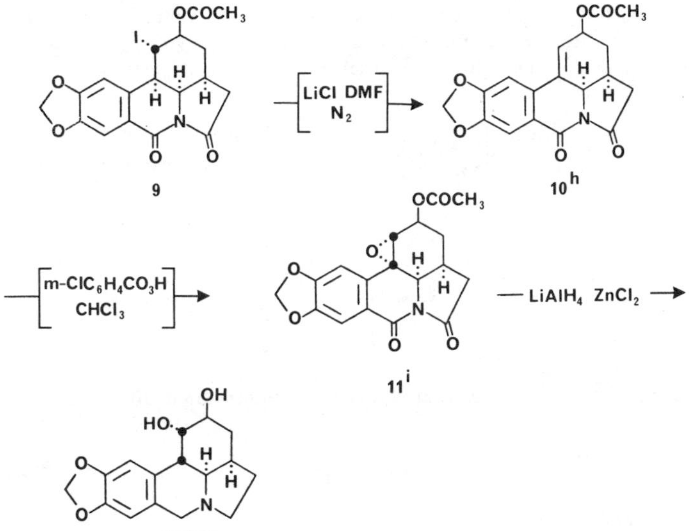

OCOCH$_3$ ... 9 —[LiCl DMF, N$_2$]→ 10[h] —[m-ClC$_6$H$_4$CO$_3$H, CHCl$_3$]→ 11[i] — LiAlH$_4$ ZnCl$_2$ → 12 : 5%

(±)-dihydrolycorine

[a]cf. Arnold and Coyner,[76] Quelet, Dran, and Lukacs,[77] and Dran and Prang.[78]
[b]The isomeric mixture was used directly in the next step without separation.
[c]The synthesis, as outlined in this brief communication, suffers from the fact that isomeric mixtures and/or low yields are encountered in many of the stages. Also, nearly all of the yields are unspecified; therefore, it is difficult to arrive at any definite conclusions with regard to the reliability of the steps as they are outlined.
[d]The isomeric mixture of **3** and **4** was separated. "Cyclization of **1** or the corresponding diacid under Friedel-Crafts conditions afforded exclusively the tetralone-carboxylic acid (of **3**) and not the indanone-carboxylic acid **4** as claimed by previous authors." "The indanone **4** was submitted to the conditions of the Schmidt reaction (with NaN$_3$ in trichloroacetic acid) without success."
[e]The product of this reaction (yield unspecified) was a mixture of two isomeric lactams (ratio unspecified).
[f]Confirmation of the structural and stereochemical features of these intermediates was established by the conversion of **7** into (±)-γ-lycorane whose stereochemistry had been previously defined[g]:

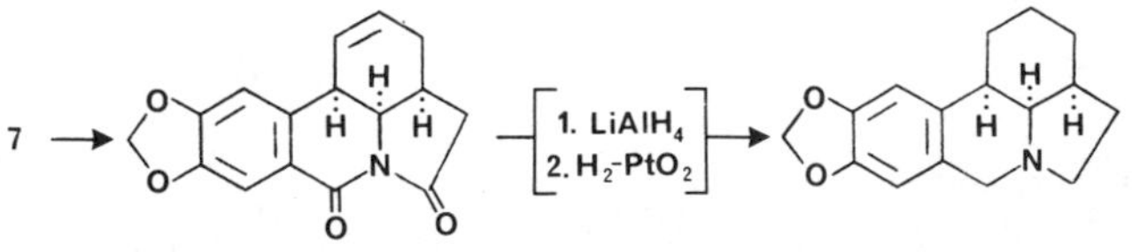

7 → —[1. LiAlH$_4$, 2. H$_2$-PtO$_2$]→

[g]Kotera[79] and Ueda, Tokuyama, and Sakan.[80]
[h]Although formation of the conjugated olefin undoubtedly provides considerable driving force, it is also of interest to note the overall transformation corresponds to a *cis* dehydrohalogenation.

[i]Peracid epoxidations normally proceed by attack of the relatively bulky peracid on the least hindered (in this case convex) side. However, the direction of attack and even its rate way be influenced by neighboring polar substituents. By virtue of the fact that it can hydrogen bonded with the attacking peracid, a neighboring hydroxyl group is particularly effective in this regard as the following example amply demonstrates:

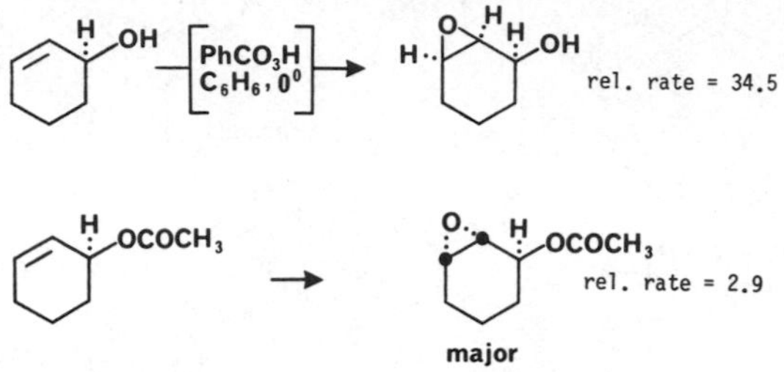

H OH — [$PhCO_3H$, C_6H_6, 0^0] → H.. O H H OH rel. rate = 34.5

H $OCOCH_3$ → O H $OCOCH_3$ rel. rate = 2.9

major

The merit of protecting the allylic hydroxyl group in intermediate **10** is, therefore, obvious.

Chart 3-17.[81]

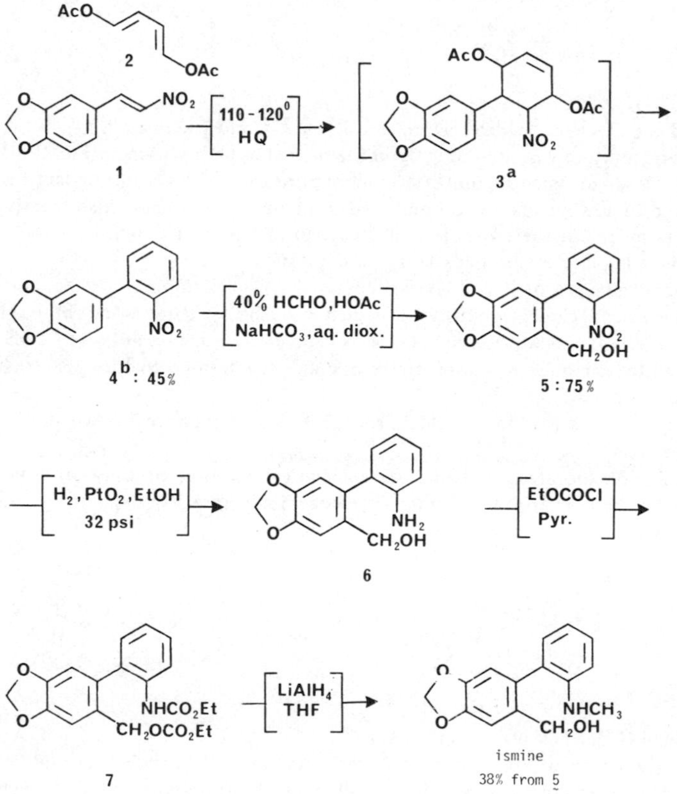

AcO 2 OAc

O O NO_2 1 — [110 - 120^0, HQ] → [AcO O O OAc NO_2] 3^a →

O O NO_2 4^b: 45% — [40% HCHO,HOAc, $NaHCO_3$,aq. diox.] → O O NO_2 CH_2OH 5 : 75%

— [H_2, PtO_2, EtOH, 32 psi] → O O NH_2 CH_2OH 6 — [EtOCOCl, Pyr.] →

O O $NHCO_2Et$ CH_2OCO_2Et 7 — [$LiAlH_4$, THF] → O O $NHCH_3$ CH_2OH ismine 38% from 5

[a]The diacetoxyolefin is not stable under the conditions of the reaction and eliminates spontaneously the elements of two equivalents of acetic acid. The reaction appears to be a general one and provides a useful alternative to more conventional methods for the preparation of unsymmetrical biphenyls such as the Bachman-Gomberg arylation or the mixed Ullmann coupling, which suffers from orientation problems and not infrequently unsatisfactory yields and/or tedious separations.

[b]Alternatively, in principle, **4** could be converted into **6** by (a) bromination of **4** with pyridinium bromide perbromide (b) replacement of the bromo substituent with a cyano function by means of CuCN, (c) complete hydrolysis to the nitro acid, and (d) complete reduction. In practice, the yields encountered in the hydrolysis step (c) even after several days reflux with concentrated HCl were impractically low, and this approach was abandoned.

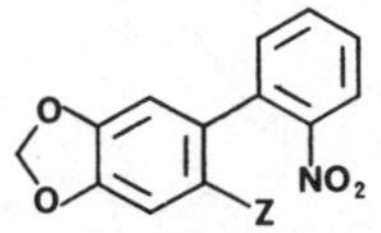

NO_2 Z

a. Z = Br 98%
b. Z = CN 33%
c. Z = COOH low

D. Biogenetic-type Syntheses

Chart 3-18.[82]

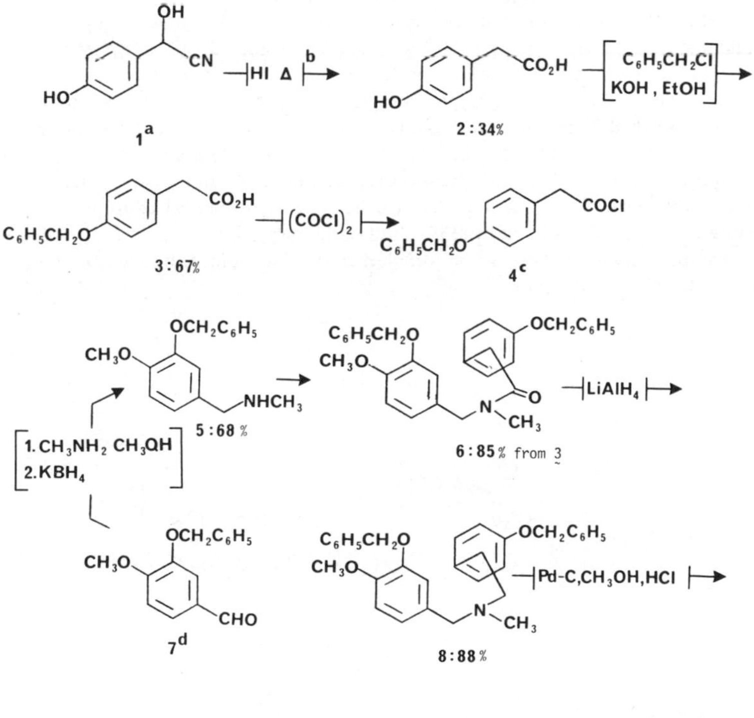

OH CN HO $\dashv$ HI Δ [b] $\mapsto$ HO CO_2H $\dashv$ [$C_6H_5CH_2Cl$ KOH, EtOH] $\to$

1[a] 2 : 34%

$C_6H_5CH_2O$ CO_2H $\dashv (COCl)_2 \mapsto$ $C_6H_5CH_2O$ COCl

3 : 67% 4[c]

CH_3O $OCH_2C_6H_5$ $NHCH_3$ $\to$ $C_6H_5CH_2O$ CH_3O $OCH_2C_6H_5$ N O CH_3 $\dashv LiAlH_4 \mapsto$

5 : 68 % 6 : 85% from 3

[1. CH_3NH_2 CH_3OH 2. KBH_4]

CH_3O $OCH_2C_6H_5$ CHO $C_6H_5CH_2O$ CH_3O $OCH_2C_6H_5$ N CH_3 $\dashv$ Pd-C, CH_3OH, HCl $\mapsto$

7[d] 8 : 88 %

Chart 3-18 (Continued).

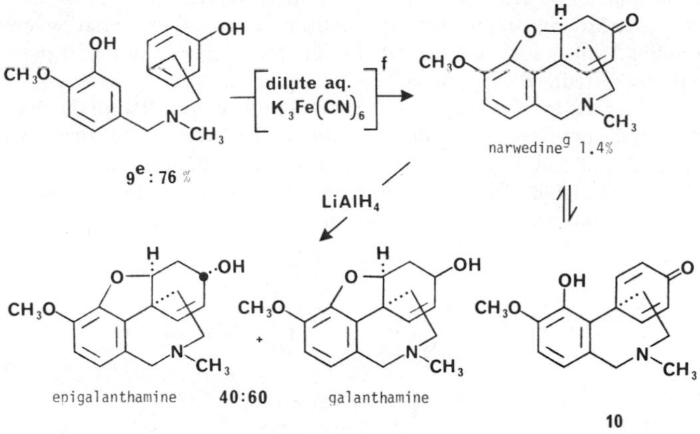

9^e: 76 %

dilute aq. $K_3Fe(CN)_6$ [f]

narwedine[g] 1.4%

$LiAlH_4$

epigalanthamine 40:60 galanthamine

10

[a]This cyanohydrin was prepared according to the procedure of Ladenburg, Folkers, and Major[83]

[b]As recorded by Czapliski, Kostanecki, and Lampe.[84]

[c]This acid chloride was not isolated but treated directly with two equivalents of amine 5.

[d]*O*-Benzylisovanillin had been previously prepared by Späth, Orechoff, and Kuffper.[85]

[e]Confirmation of this structure was achieved by treatment with diazomethane. The resultant trimethyl ether (belladine) had been synthesized previously by Surrey et al.[86]

[f]The alkaline ferricyanide oxidation of 9 was studied in some detail. In every case, the major product consisted of a $CHCl_3$-insoluble polymer whose formation could not be suppressed even in very dilute solutions. Attempts were also made to oxidize the diphenol on solid surfaces since absorbed phenoxide radicals might have a better chance of coupling intramolecularly than those freely diffusing in solvent. Active manganese and lead dioxide were tested but once again polymerization predominated. Anodic oxidation was equally discouraging. Since this pioneering work, numerous other phenolate radical coupling reactions have found use in the synthesis of natural products and many new reagents and procedures have been developed. Some of these are chronicled here. For an informative and comprehensive review see Battersby and Taylor.[87]

Chart 3-19.[88]

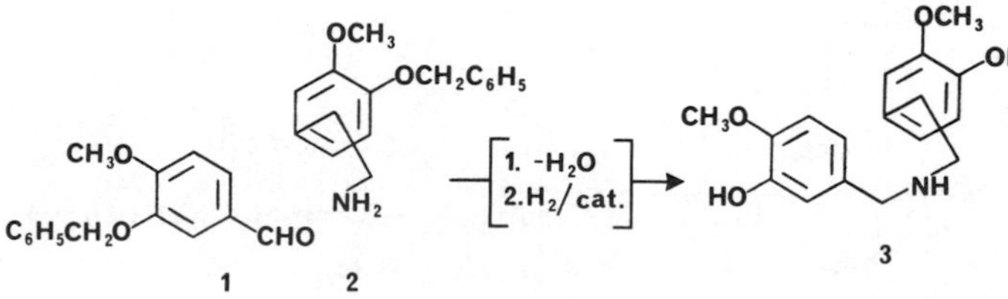

1. $-H_2O$
2. H_2/cat.

1 2 3

Chart 3-19 (Continued).

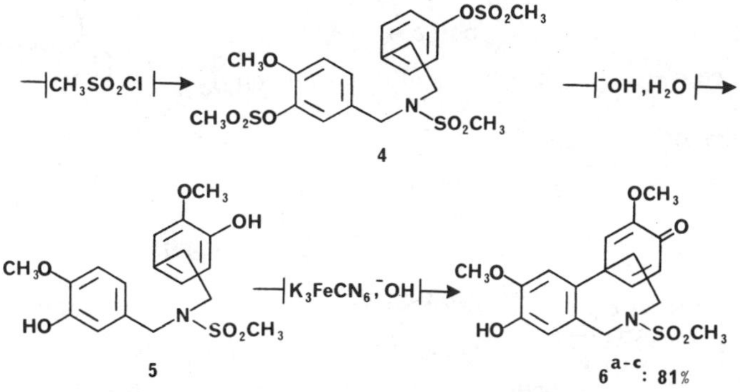

$—|CH_3SO_2Cl|→$ CH_3O CH_3O_2SO OSO_2CH_3 N SO_2CH_3 4 $—|^-OH, H_2O|→$

OCH_3 OH CH_3O HO N SO_2CH_3 5 $—|K_3FeCN_6, ^-OH|→$ OCH_3 O CH_3O HO N SO_2CH_3 6^{a-c}: 81%

[a]The extremely high yield of **6** should be compared with other similar oxidations (see especially Chart 3-20, which reports a substantially lower yield for the same reaction reported here). It is clear, however, that tying up the nitrogen lone electron pair has a dramatic effect upon improving the intramolecular coupling process by preventing indole formation, and this has subsequently been taken great advantage of.

[b]The alternative *ortho-para* coupled products (7 and 8) were excluded on the basis of spectral properties and the fact that 8 would be expected to undergo internal ether formation by analogy with Barton's study (Chart 3-18).

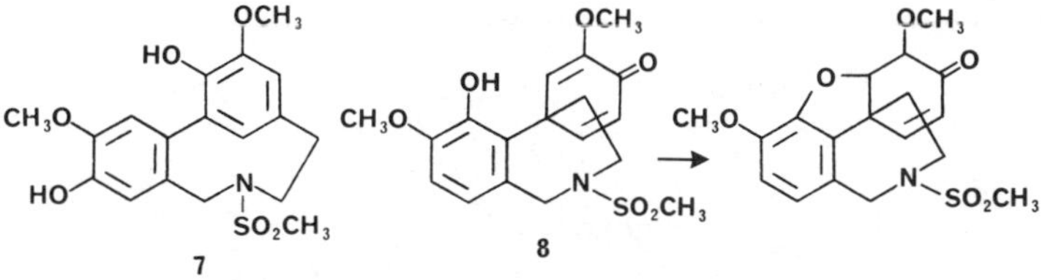

OCH_3 HO CH_3O HO N SO_2CH_3 7 OCH_3 OH O CH_3O N SO_2CH_3 8 → OCH_3 O O CH_3O N SO_2CH_3

[c]Various attempts to effect *N*-demesylation of **6** under mild conditions failed, as did attempts to isolate any pure products from the oxidation of 9 and **10**.

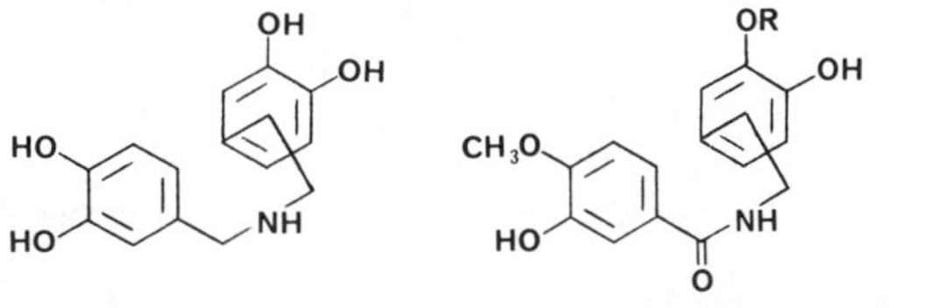

OH OH HO HO NH 9 OR OH CH_3O HO NH O 10 : R = H or SO_2CH_3

Chart 3-20.[89]

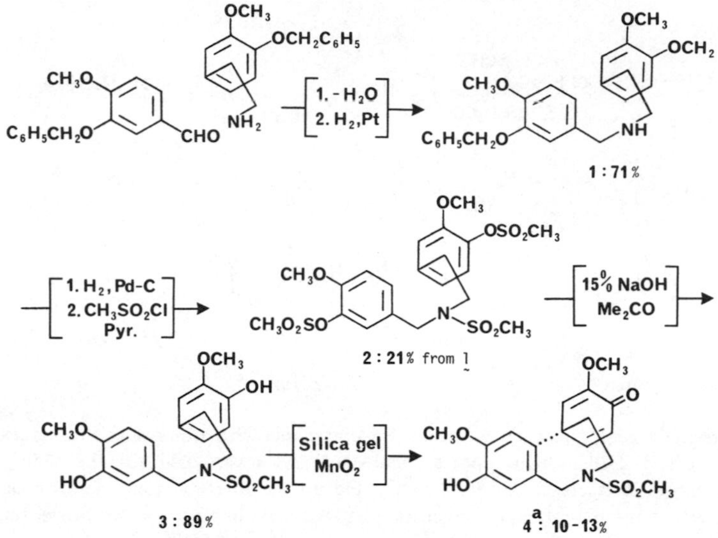

CH₃O C₆H₅CH₂O CHO NH₂ OCH₃ OCH₂C₆H₅ [1. −H₂O 2. H₂,Pt] → 1 : 71%

[1. H₂, Pd-C 2. CH₃SO₂Cl Pyr.] → 2 : 21% from 1 → [15% NaOH Me₂CO] →

3 : 89% → [Silica gel MnO₂] → 4[a] : 10–13%

[a]Oxidation of **3** with alkaline $K_3Fe(CN)_6$ gave only a complex mixture (compare with Chart 3-19).

Chart 3-21.[90]

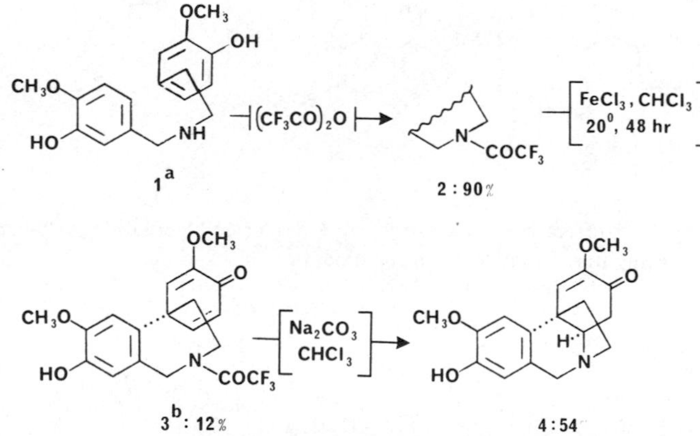

1[a] → [$(CF_3CO)_2O$] → 2 : 90% → [$FeCl_3$, $CHCl_3$ 20°, 48 hr] →

3[b] : 12% → [Na_2CO_3 $CHCl_3$] → 4 : 54

[a]Prepared according to the procedure of Abramovitch and Takahaski[88] (Chart 3-19) and Schiebel.[91]

[b]The advantage of employing the more labile trifluoroacetamide protecting group rather than the mesylate employed previously by Abramovitch and Takahashi[88] (Chart 3-19) becomes apparent in the **3** to **4** transformation.

Chart 3-22.[92]

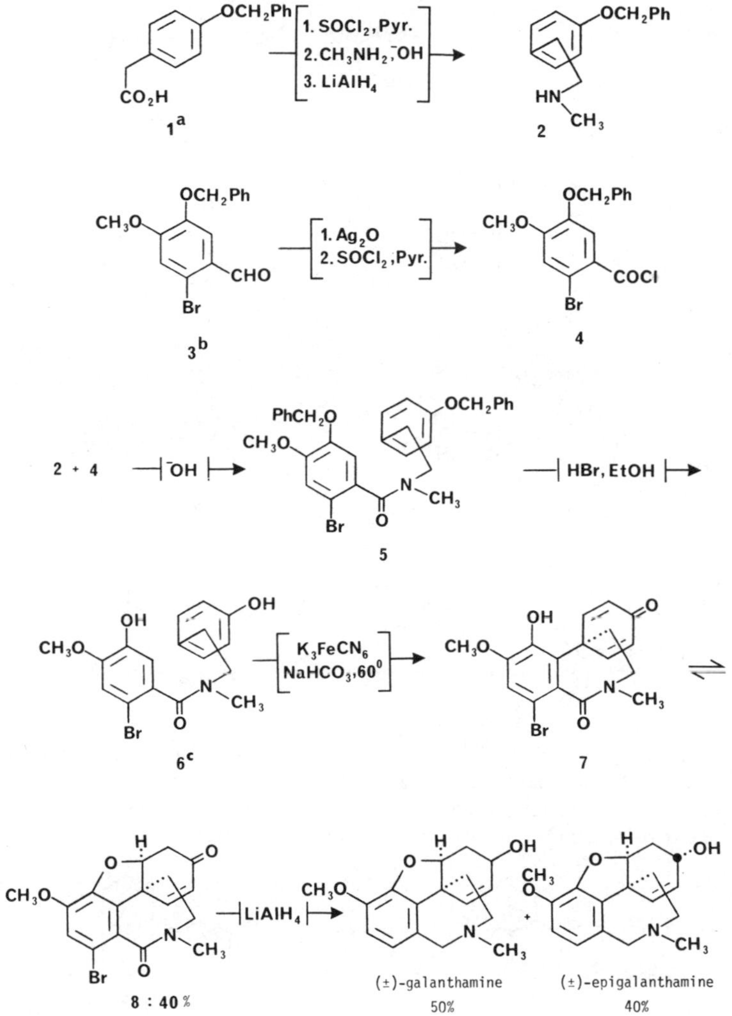

[a]Prepared according to the procedure of Vaghani and Merchant.[93]

[b]2-Bromo-*O*-benzylisovanillin **3** was obtained by the method of Jackson and Martin.[94]

[c]The selection of diphenol **6** as a suitable candidate for oxidative cyclization follows from the previous work cited in this section. Thus an amide was selected to prevent conjugate addition of the nitrogen atom to the α,β-unsaturated ketone formed in the first step of the oxidation to give an indole system, which could not undergo coupling of the aromatic nuclei. The bromine substituent was included to inhibit undesirable coupling *para* to the hydroxyl group and to favor the desired *ortho* coupling process.

Chart 3-23.[95]

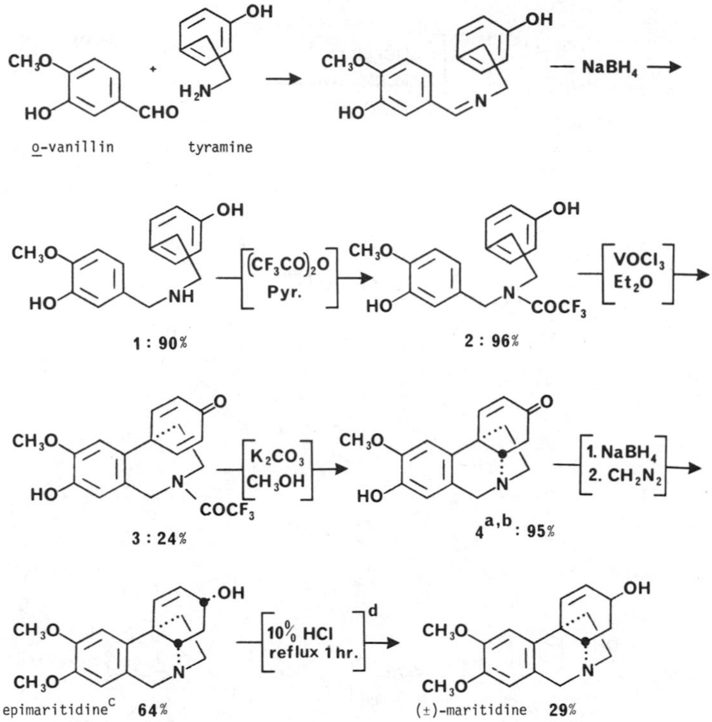

CH$_3$O, HO, CHO, o-vanillin + H$_2$N, OH, tyramine → CH$_3$O, HO, N, OH — NaBH$_4$ →

CH$_3$O, HO, NH, OH, **1 : 90%** —[$(CF_3CO)_2O$, Pyr.]→ CH$_3$O, HO, N, COCF$_3$, OH, **2 : 96%** —[VOCl$_3$, Et$_2$O]→

CH$_3$O, HO, N, COCF$_3$, O, **3 : 24%** —[K$_2$CO$_3$, CH$_3$OH]→ CH$_3$O, HO, N, O, **4**[a,b] **: 95%** —[1. NaBH$_4$, 2. CH$_2$N$_2$]→

CH$_3$O, CH$_3$O, N, OH, epimaritidine[c] **64%** —[10% HCl, reflux 1 hr.][d]→ CH$_3$O, CH$_3$O, N, OH, (±)-maritidine **29%**

[a]The spontaneous cyclization to **4** upon removal of the *N*-trifluoroacetyl group from **3** has ample precedence: Franck and Lubs[96] (Charts 3-21), and Goosen et al.[97]
[b]Confirmation of the structural and stereochemical features of **4** was established by conversion into oxomaritidine and comparison with an authentic sample:

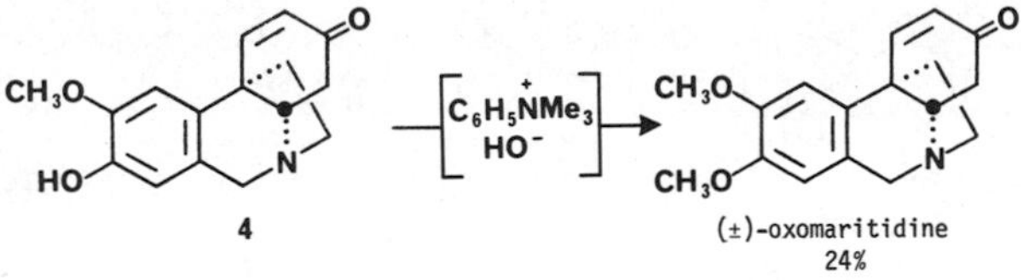

CH$_3$O, HO, N, O, **4** —[$C_6H_5\overset{+}{N}Me_3$, HO$^-$]→ CH$_3$O, CH$_3$O, N, O, (±)-oxomaritidine 24%

The yield of this methylation reaction[98] was low. However, a variety of other methods on either enone **4** or its precursor (**3**) were even less satisfactory.
[c]Assignment of the relative stereochemistry of the allylic alcohol was made by comparison of its pmr spectrum with that of *epi*-crinine whose stereochemical features were well established. Similarly, the relative stereochemistry of maritidine was defined by comparison with crinine itself.
[d]This equilibration had been employed previously by Whitlock and Smith[99] in the synthesis of crinine (Chart 3-3).

4. THE *LYCOPODIUM* ALKALOIDS[100]

In spite of the fact that investigation of the *Lycopodium* alkaloids has a very long history,[100] progress in isolation and structure illucidation was slow to develop until 1957 when Wiesner and his collaborators[101] reported the structure of annotinine. Rapid progress followed this initial breakthrough, and activity in this area continues to be impressive. Indeed, the number of fully characterized *Lycopodium* alkaloids presently known practically dwarfs those described only a few years ago, and one can only anticipate a further blossoming of new and novel structures of which the list portrayed below may be regarded as typical. The synthetic chemist has found this fertile area impossible to resist, and some of the sciences most notable achievements have been recorded therein.

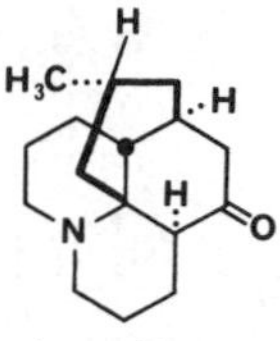

Lycopodine 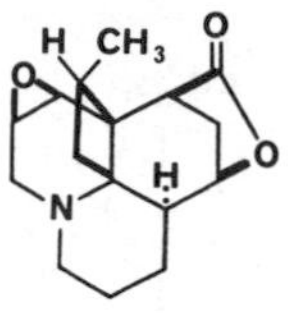 Annotinine 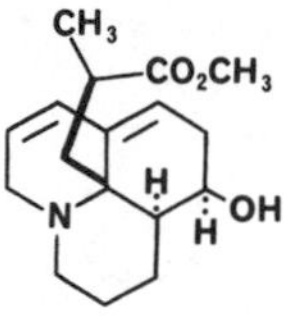 Lyconnotine

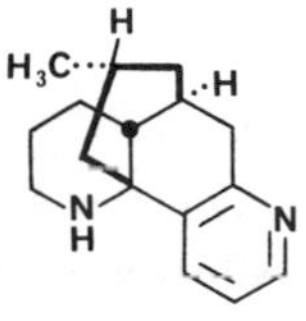

Lycodine 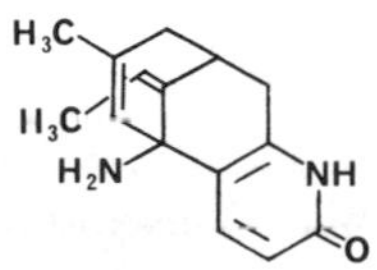 Selagine 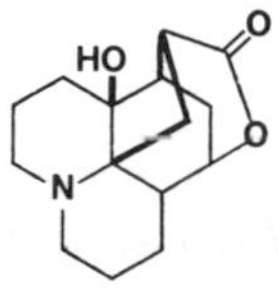 Annotine

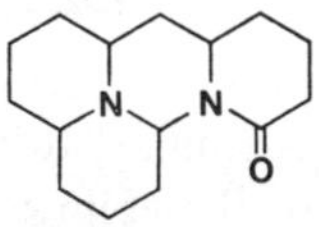

Cernuine 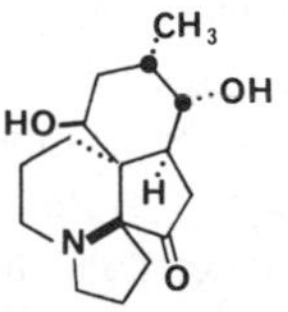 Serratinine

We initiate the present discussion by considering certain structural features, stereochemical traits, and synthetic interconversion of potential or established utility to synthetic investigations. By far the most common structural pattern that occurs throughout the *Lycopodium* family is that found in lycopodine. As might be expected from an examination of their structures, several members of this group have been chemically interrelated. An exceedingly interesting example is provided by the transformation of lycopodine into annofoline.[102] Permanganate oxidation of lycopodine provided a keto-lactam in 45% of the yield, which was reduced with sodium borohydride. Exposure of the resultant alcohol to a refluxing benzene solution of lead tetraacetate gave ether **1** in a 90% yield.

Boron trifluoride-catalyzed solvolysis of this ether in the presence of acetic anhydride led to olefinic acetate 2. Antimarkovnikov hydration of the double bond was accomplished by means of diborane followed by alkaline hydrogen peroxide. Of course, the lactam function was simultaneously reduced also. The final objective, annofoline, was reached by oxidation of the alcohol function and saponification of the acetate.

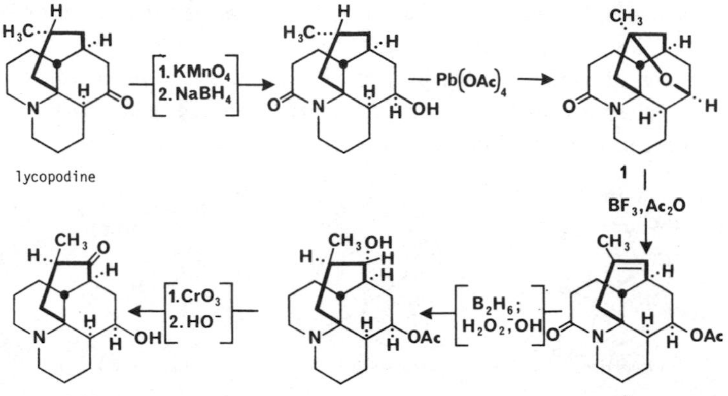

lycopodine

1

2

annofoline

The orientation of the methyl function in annofoline is opposite that found in most other *Lycopodium* alkaloids. This distinctive feature was uncovered by Anet[103] in an elegant conformational study wherein the thermodynamic preference of ring D to assume a boat-like conformation was also demonstrated. Examination of Dreiding models of the ring D chair and boat conformations of each C_{15} epimer reveals that a severe nonbonded interaction between the C_{15} methyl, and C_5 hydroxyl is present in the chair conformation **3b** and is, therefore, excluded from further consideration. A similar, although less severe, interaction is found in the chair conformation of the epimer **(4b)**, and inversion to the corresponding boat conformation **(4a)** is plagued by a bowsprit-flagpole interaction involving the C_{15} methyl and C_{12} hydrogen. The remaining isomer **(3a)** contains a similar type of interaction. However, only two hydrogens are involved in this case, and it, therefore, becomes less objectionable by contrast.

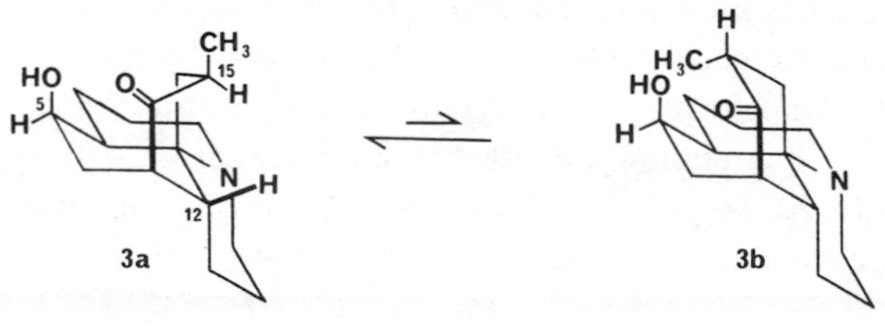

3a

3b

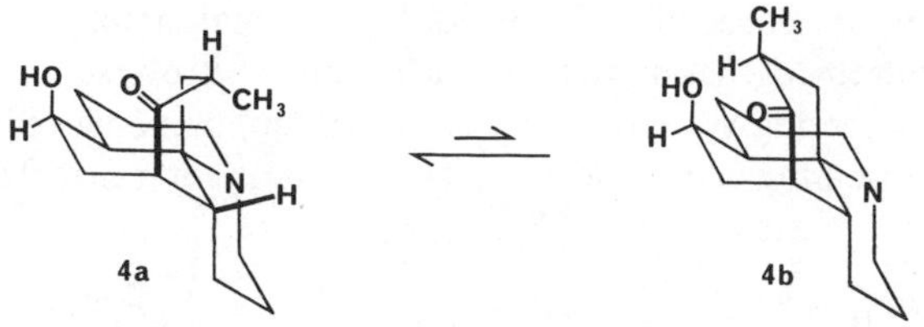

Yet another interesting interconversion is illustrated by the transformation of lycopodine into its C_6 hydroxyl derivative-alkaloid L-20.[104] Bromination of lycopodine gave the C_6 axial bromide, epimerization of which occurred readily by warming in acetic acid. Surprisingly, when either epimer was exposed to a mild base treatment (aqueous sodium bicarbonate), the same α-hydroxyketone ≡ alkaloid L-20 was obtained. The apparent contradiction of this result with the well-established stereoelectronic requirements for backside attack in such displacements is rendered more palatable by assuming epimerization at the reaction site either prior to or after displacement. Somewhat more drastic conditions (aqueous sodium hydroxide) altered the reaction course entirely, yielding an enone **5** and an enolic α-diketone **6**, which was identical to the product of selenium oxide oxidation of lycopodine itself. It is possible to rationalize the formation of each of these products by postulating a dipolar (cyclopropanone) intermediate as illustrated here.

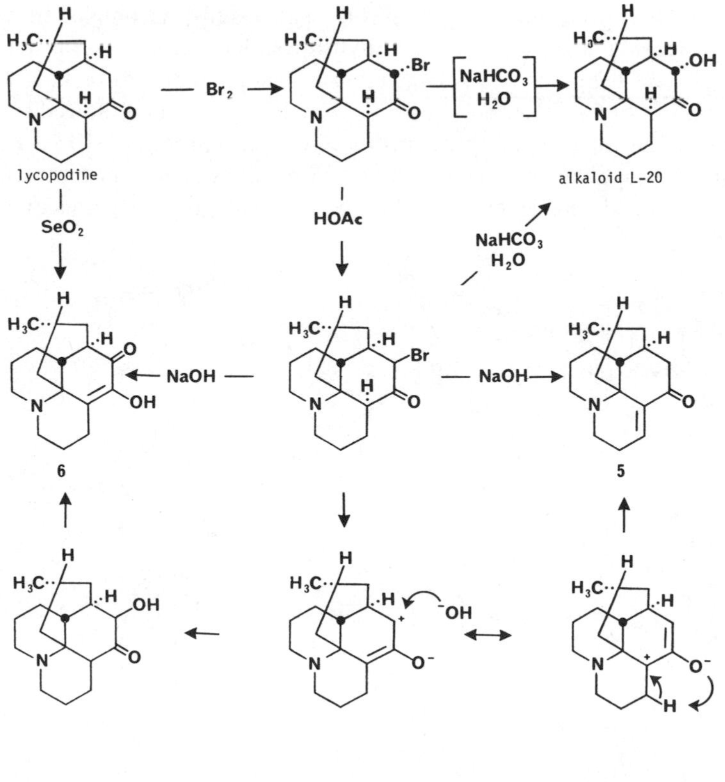

Bromination of α-obscurine followed by chromatography on basic alumina induced dehydrohalogenation to β-obscurine, and the reverse transformation was realized by subjecting the latter base to a lithium/ammonia reduction.[105] Conversion of β-obscurine and lycodine to the common intermediate **7** brought into sharp focus their great similarity.[106]

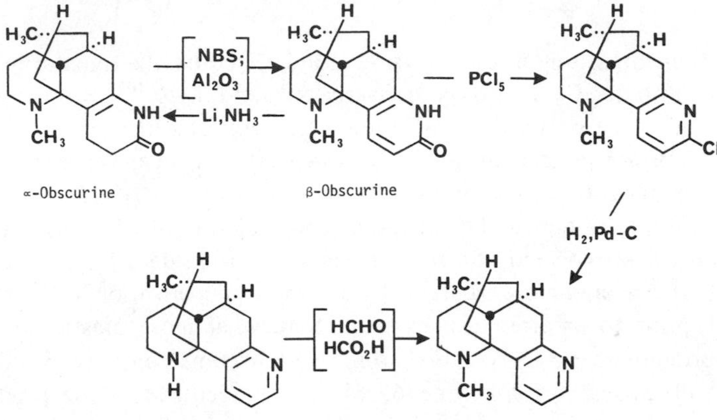

α-Obscurine β-Obscurine

Lycodine **7**

α-Obscurine has also served as the point of departure for the dramatic conversion of this structural type of *Lycopodium* alkaloids into those of lycopodine class. Thus demethylation of α-obscurine was readily accomplished with a nitrous-acetic acid mixture, and the dihydropyridine nucleus of the resultant *des*-methyl base **8** hydrolyzed under vigorous acidic conditions to keto acid **9**. Transformation of the latter substance to dihydrolycopodine followed standard procedures. The identity of this substance with the product **(10)** of hydride reduction of lycopodine revealed for the first time the stereochemical features of the obscurines and related bases, as lycopodine had previously succumbed to such an analysis.[107]

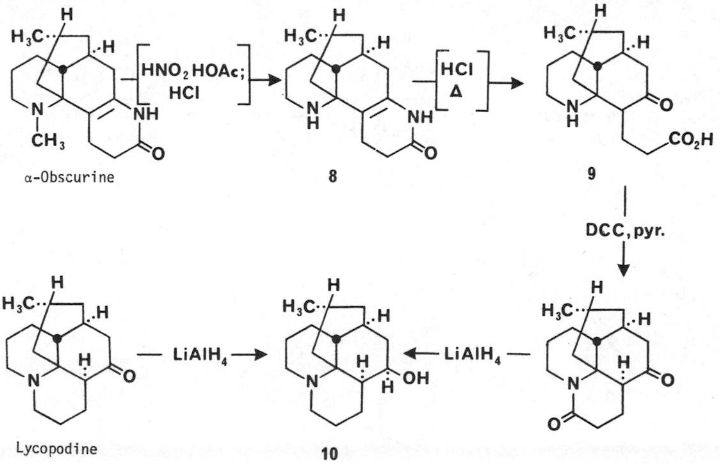

α-Obscurine **8** **9**

Lycopodine **10**

In an equally interesting study, the order of the previous steps was, in effect, reversed.[108] Thus the minor product (**11**) of von Braun degradation of lycopodine was treated with sodium azide and then reduced in acidic ethanol. This yielded tetrahydropyridine **12**, which underwent smooth dehydrogenation with Pd—C in refluxing *p*-cymene. Hydrolysis of the cyanamide provided lycodine.

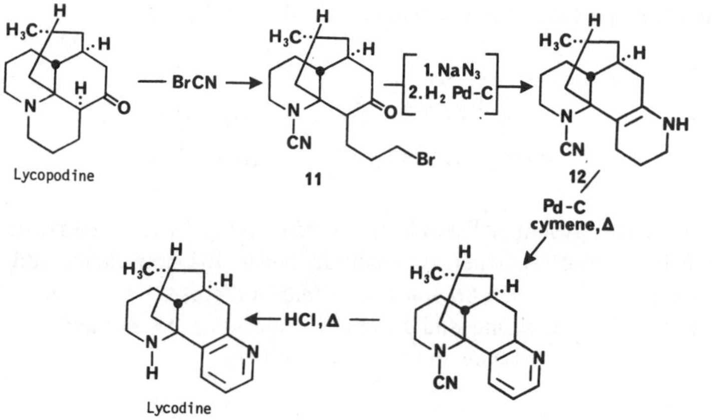

Lycopodine — BrCN → 11 — 1. NaN_3 2. H_2, Pd-C → 12 — Pd-C, cymene, Δ → — HCl, Δ → Lycodine

Other, less spectacular, interconversions include the correlation of fawcettiine, clavolonine, and annofoline;[109] that of lycofoline, base M, and acrifoline;[110] and, finally, that of flabelline and lycopodine.[111] Prior to turning our attention to the total synthesis of some of these alkaloids, we find it expeditious to consider a few fundamental features of their basic ring nucleus—the hexahydrojulolidine system.

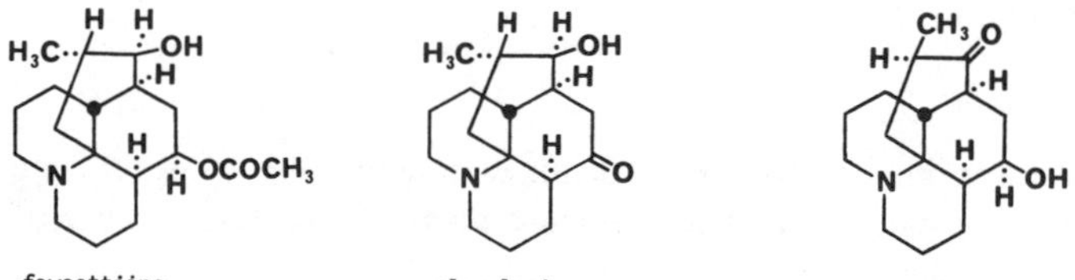

fawcettiine

clavolonine

annofoline

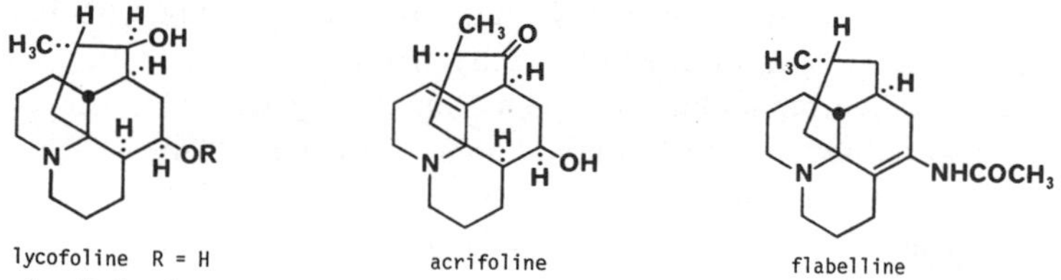

lycofoline R = H
Base M R = Ac

acrifoline

flabelline

A. Selected Aspects of Julolidine and Hexahydrojulolidine Chemistry

The so-called julolidine ring system has been known since 1891 when Reissert[112] reported that heating various acetoacetic esters with tetrahydroquinoline gave a low yield of a julolidine. Since that time, several synthetic and chemical studies have appeared in the literature dealing with these substances. Some of these accomplishments have been reviewed by Mosby.[113]

This basic ring system is listed in the *Revised Ring Index* as benzo-[ij]-quinolizine and its perhydro-partner as hexahydrobenzo-[ij]-quinolizine and numbered as illustrated. Even though *Chemical Abstracts* adheres to this usage, retention of the trivial names julolidine and hexahydrojulolidine by numerous authors has gained considerable currency and is employed here.

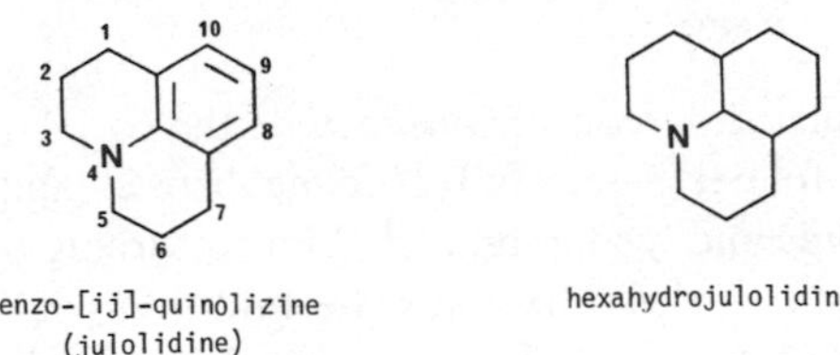

benzo-[ij]-quinolizine (julolidine) hexahydrojulolidine

B. Methods of Synthesis

Julolidines

It is not our intention to provide here an exhaustive list of general methods. However, careful inspection of the literature reveals a modest number of highly useful basic approaches.

The employment of tetrahydroquinoline(s) and various electrophilic three-carbon units based on the historical Reissert synthesis has been developed into generally useful procedures. For example, condensation of tetrahydroquinoline with acrylonitrile or ethyl acrylate and subsequent intramolecular Friedel-Crafts acylation provided a highly efficient synthesis of 1-keto-julolidine.[114] Julolidine itself can be obtained from this substance by Wolff-Kishner reduction[115] or by acylation of tetrahydroquinoline with β-chloropropionyl chloride, Friedel-Crafts alkylation, and reduction with lithium aluminum hydride.[116] Perhaps the most convenient synthesis of julolidine consists of merely boiling a solution of trimethylene chlorobromide and tetrahydroquinoline.[117]

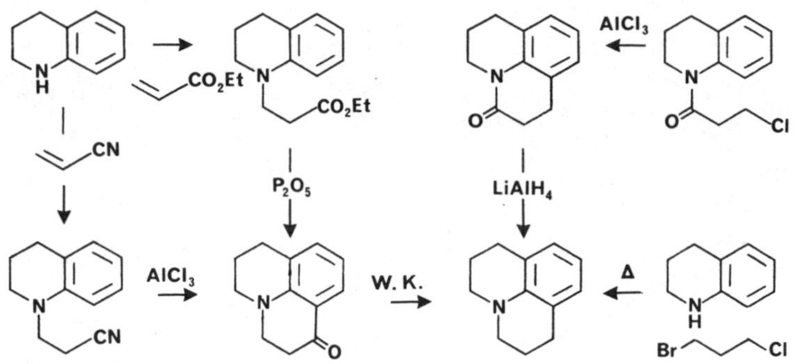

Other useful syntheses of julolidine(s) are illustrated by the *bis*-C_3-annelation of various anilines such as aniline itself[118] or *m*-methoxyaniline.[119]

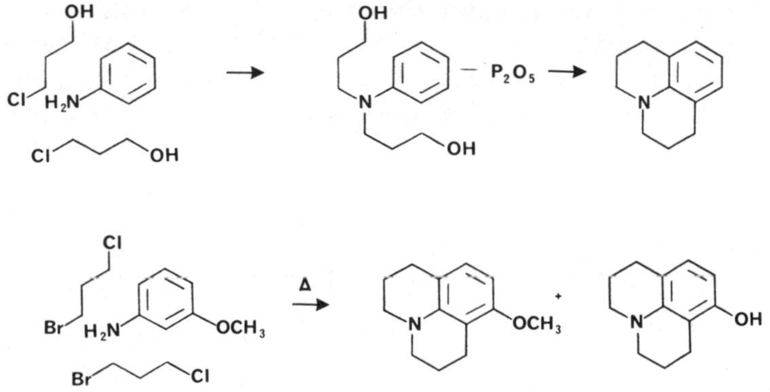

Hexahydrojulolidines

The hexahydrojulolidine system is endowed with three asymmetric centers for which there are, in principle, eight possible stereoisomers. Of the four predicted d,l-pairs two are characterized by a plane of symmetry and one by a C_2-axis, thus reducing the original estimate to two *meso* and one d,l-form. These are portrayed here.

The first report of the hexahydrojulolidine system was made by Boekelheide and Quinn,[120] who obtained an oily product from the Raney nickel-catalyzed hydrogenation of julolidine. This material analyzed correctly as an amorphous picrate. Protiva and Prelog[121] repeated this work but succeeded in separating the amorphous picrate into two stereoisomers by fractional crystallization. Regardless of the precise conditions of the reduction the same two isomers were obtain-

meso meso d,l-

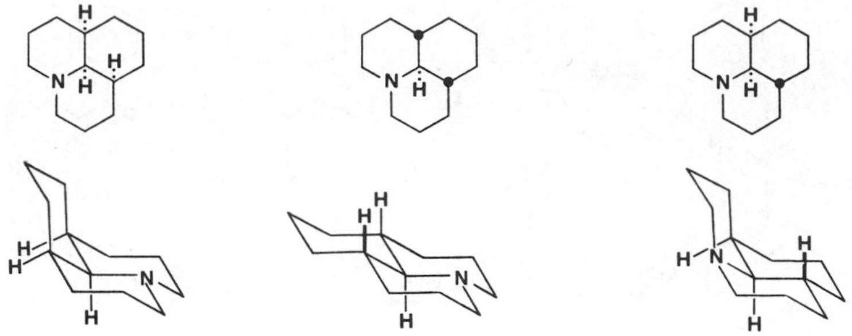

ed, albeit in varying ratios. Protiva and Prelog assigned the labels A and B to these two substances, but no attempt was made to define their stereochemistry.

Later Leonard and Middleton,[122] disclosed a new synthesis of isomer A. Ketodiester **13** was prepared by a standard series of reactions. The corresponding oxime was reduced over a copper chromite catalyst. A single basic product resulted that formed a crystalline picrate identical with that from isomer A. On the basis of its method of synthesis, ketodiester **13** was assigned *trans* stereochemistry; consequently, the hexahydrojulolidine derived therefrom could only be the racemic *cis-trans* isomer.

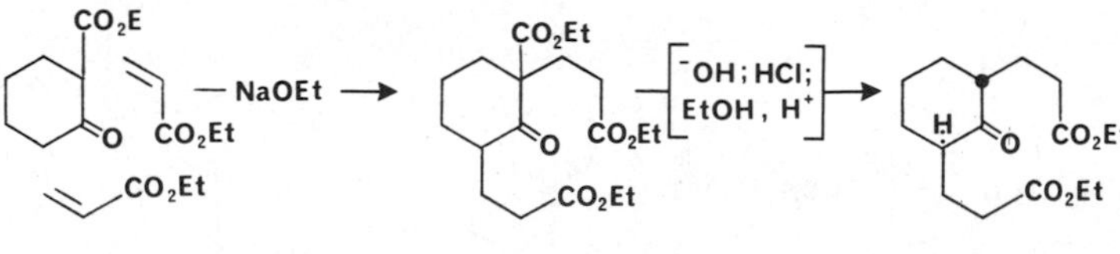

CO_2E — NaOEt → CO_2Et CO_2Et CO_2Et [^-OH; HCl; EtOH, H^+] → CO_2Et CO_2Et

13

→ NOH CO_2Et CO_2Et — H_2, $CuCr_2O_4$ →

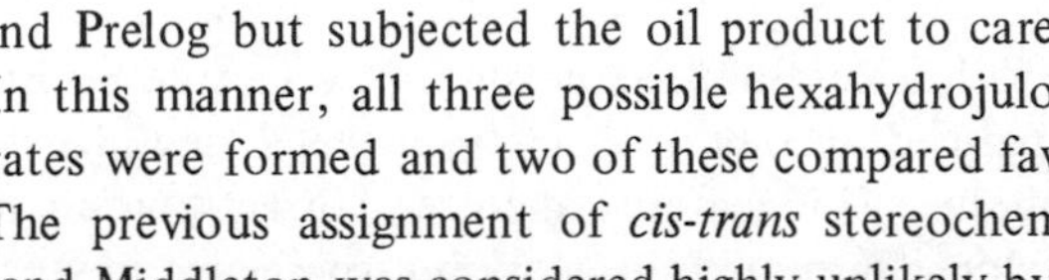

In 1958 Bohlmann and Arndt[123] repeated the hydrogenation procedure of Protiva and Prelog but subjected the oil product to careful column chromatography. In this manner, all three possible hexahydrojulolidines were obtained, their picrates were formed and two of these compared favorably with isomers A and B. The previous assignment of *cis-trans* stereochemistry to isomer A by Leonard and Middleton was considered highly unlikely by the German chemists. They reasoned that the method of synthesis of ketodiester **13** had provided ample opportunity for equilibration of the asymmetric centers and that the *cis*

isomer, with both bulky propionic ester side-chains disposed equatorially, was the most likely fomulation of this substance. Modification of the stereochemistry of this crucial intermediate also required revision of the stereochemistry of isomer A. The all-*trans* configuration was selected and set the stage for an unambiguous structure proof.

Examination of the infrared spectra of the three isomeric hexahydrojulolidines provided the first clue, which ultimately led to their demise. Bohlmann and Arndt noticed that isomers A and B exhibited a characteristic absorption in the 2700 to 2800-cm^{-1} region. A previous lengthy study in Bohlman's laboratory[124] had defined certain empirical stereoelectronic requirements for this unusual absorption in various quinolizidine systems. Wenkert[125] had simultaneously arrived at the same conclusion by examining the infrared spectra of various indole alkaloids containing the quinolizidine skeleton. These studies revealed that the presence or absence of absorption in this region was dictated by a rather precise stereoelectronic relationship between the nonbonding nitrogen electron pair and the adjacent protons. Thus the requirements for absorption are fulfilled only if two (or more) of the adjacent protons are disposed in an anti- and coplanar fashion with respect to the lone electron pair on nitrogen. For example, inspection of the preferred conformations of the hypothetically distinct *cis* and *trans* forms of quinolizidine itself reveals the presence of three protons that can satisfy the stereoelectronic requirement in the *trans* isomer but only one in its *cis* counterpart. One would, therefore, anticipate the unique 2700 to 2800-cm^{-1} absorption in the former but not in the latter.

A similar analysis of the preferred conformations of the three isomeric hexahydrojulolidines (cf. following diagram) revealed that only the all-*cis* and all-*trans* isomers were capable of fulfilling this requirement. The remaining *cis-trans* isomer cannot and, therefore, was assigned by Bohlmann and Arndt to isomer C, which exhibited no such absorption. Assignment of the all-*trans*

arrangement to isomer A followed from the previously described reinterpretation of Leonard and Middleton's data leaving the all-*cis* stereochemistry to isomer B.

Further corroboration of these assignments was provided by a determination of the relative rates of mercuric acetate oxidation of each isomer. Leonard and co-workers[126] had established previously that facile elimination can be achieved only when the participating centers can achieve an *anti* and coplanar arrangement in the transition state. Furthermore, tertiary hydrogens were shown to be eliminated several times faster than secondary ones, which in turn reacted much more readily than their primary counterparts.

Oxidation of each of the isomeric hexahydrojulolidines with mercuric acetate proceeded at three distinctly different rates.[123] Thus isomer C reacted very slowly. This is in agreement with its formulation as the *cis-trans* isomer, which possesses but a single (secondary) hydrogen capable of satisfying the stereoelectronic requirements of the reaction. A distinction between the remaining two isomers was made on the basis of a fivefold difference in the rate of oxidation of isomer B over A. Bohlmann and Arndt argue that this difference supports the all-*cis* configuration of isomer B, since elimination of the (tertiary) hydrogen would result in relief of steric strain in the transition state, thus accelerating the rate of reaction relative to isomer A in which no such factor can intervene. The rates of oxidation are, therefore, in consonance with the infrared assignments.

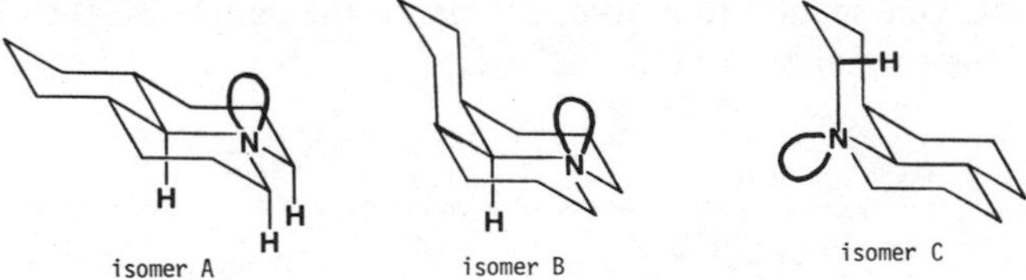

Further support for these assignments was provided by Tsuda and Saeki[127] who measured the relative rates of salt formation of isomers A and B with methyl iodide in refluxing ether. Under these conditions, isomer A was readily methylated in contrast to isomer B, which showed no evidence of reaction even after 10 hr. A final demonstration of the unfavorable nonbonded interactions that plague isomer B was made by the Japanese chemists who subjected this substance to an aluminum chloride-catalyzed isomerization. Only the thermodynamically most favorable all-*trans* isomer A resulted from this isomerization.

In a more recent elegant exploitation of the Leonard reductive cyclization already cited, Mandell and his associates[128] exploited the ready availability of ketodinitrile **14** and ketoester nitrile **15**. Each of these substances was prepared by α,α'-alkylation of the pyrrolidine enamine of cyclohexanone with acrylonitrile and/or ethyl acrylate. Upon reduction, **14** afforded a mixture of isomers A and B in a ratio of 2:7. The latter substance was the sole product when the hydrogenation was performed in acetic acid as solvent. Reductive cyclization of **15** yielded a mixture of 3-ketohexahydrojulolidines whose further reduction with lithium aluminum hydride provided isomers A and B in a ratio of 2:3.

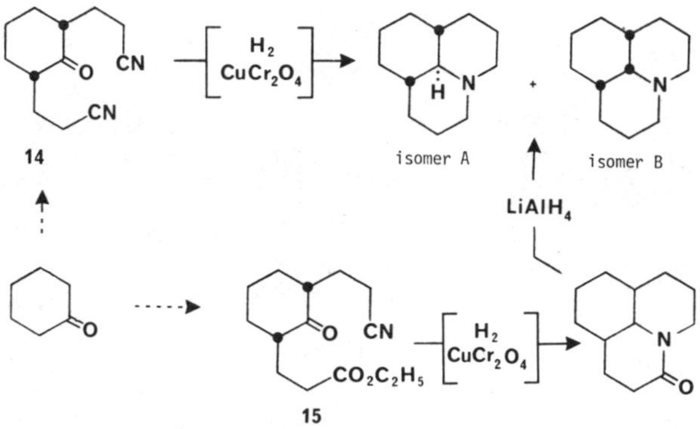

A basically different approach to the hydrojulolidine nucleus, and one of obvious interest to the synthesis of *Lycopodium* alkaloids, was recently reported by Wenkert an his collaborators. The method of approach adopted in this study was a logical extension of a fundamental and increasingly important general method of alkaloid synthesis that features the selective reduction of certain types of β-substituted pyridines and pyridinium salts to tetrahydropyridines[129] and was first employed as a key step in the synthesis of the indoloquinolizidine skeleton of eburnamonine.[130]

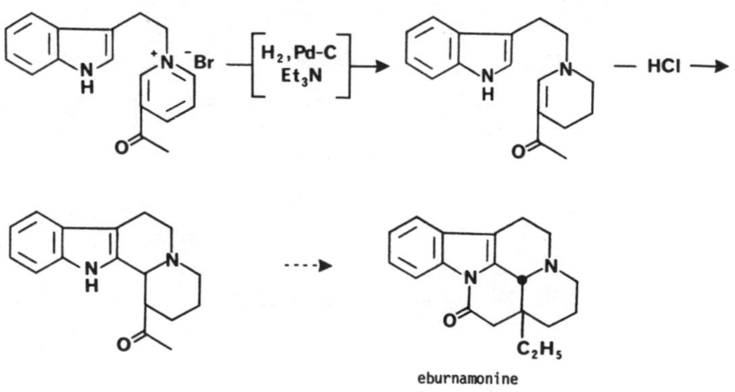

The novelty of this reduction process and the potentially profound utility of the resulting tetrahydropyridine intermediates prompted a broad study of its general utility. The reader is referred to Wenkert's stimulating account[131] of how these intermediates have been exploited as a general devise in alkaloid synthesis. Of special interest to this discussion was the additional observation[132] that upon treatment with anhydrous HCl β-methoxycarbonyl enamine **16** (prepared by selective reduction of the corresponding pyridinium salt) underwent a novel cyclization undoubtedly involving the enol form of the ketal **17**. The ultimate product of this transformation (**18**) was admirably suitable for subsequent conversion into lupinine.

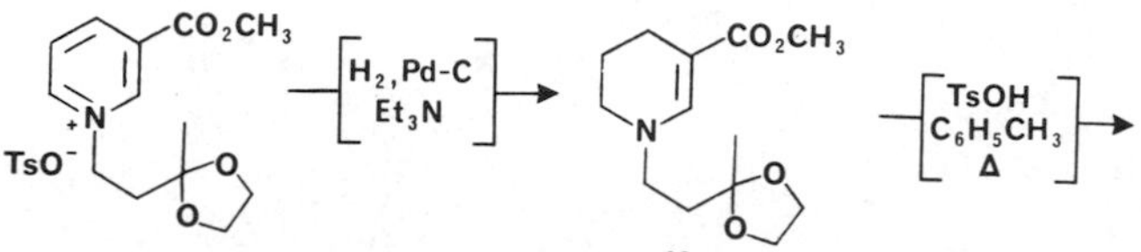

CO_2CH_3 TsO⁻ H_2, Pd-C Et_3N CO_2CH_3 TsOH $C_6H_5CH_3$ Δ

16

CO_2CH_3 CO_2CH_3 OH

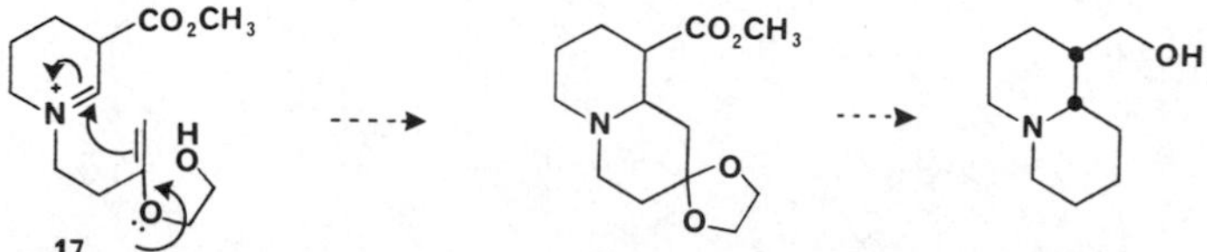

17 18 lupinine

In light of these results, examination of the structures of various *Lycopodium* alkaloids virtually dictated the following course of action. Reasoning that hydrogenation of pyridinium salt **19** should provide tetrahydropyridine **20**, subsequent acid-catalyzed cyclization would then yield quinolizidine **21**. Transformation of this interemediate to hydrojulolidine **22** could be anticipated with considerable confidence. By judicious selection of the R function in these substances, certain of the *Lycopodium* alkaloids could be anticipated. These transformations have now been realized in a model study (R = H).

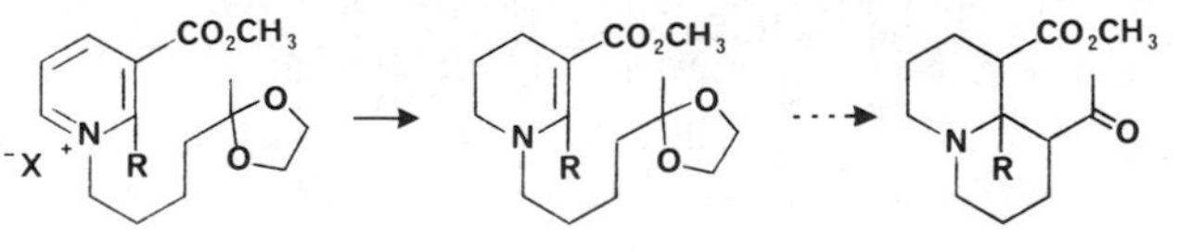

CO_2CH_3 CO_2CH_3 CO_2CH_3

19 20 21

Lycopodium alkaloids

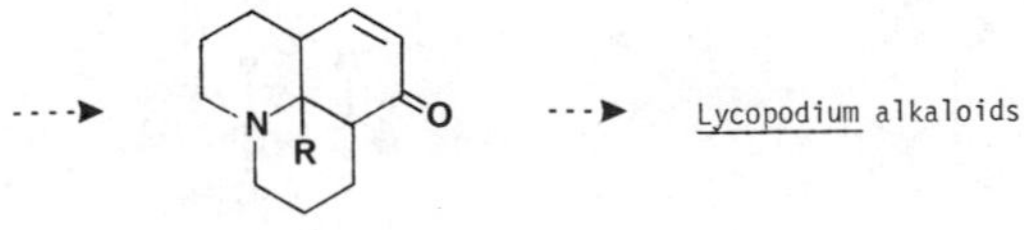

22

Chart 4-1.[133]

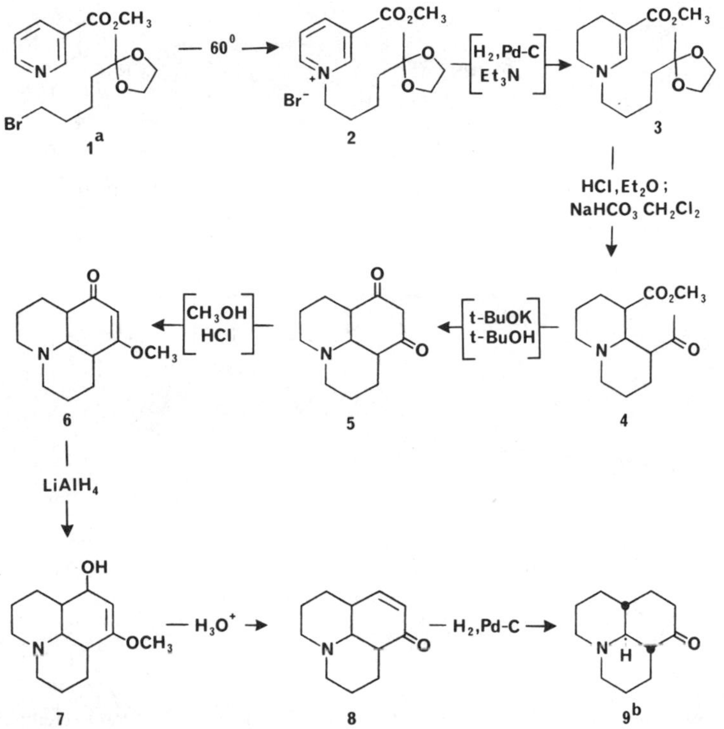

— 60° →

H$_2$,Pd-C
Et$_3$N

HCl,Et$_2$O;
NaHCO$_3$ CH$_2$Cl$_2$

t-BuOK
t-BuOH

CH$_3$OH
HCl

LiAlH$_4$

— H$_3$O$^+$ →

— H$_2$,Pd-C →

^{a}Ketal bromide **1** was prepared from acetylbutyrolactone according to the procedure of Anderson, Crawford, and Sherrill.[134]

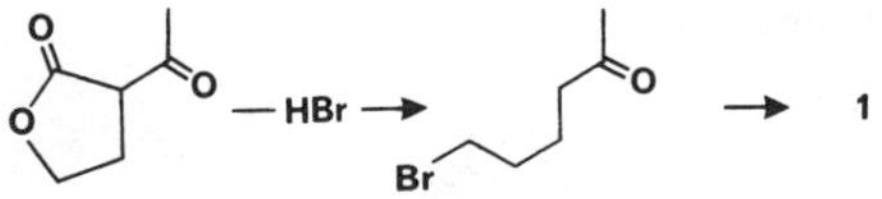

— HBr → → 1

^{b}Assignment of the all-*trans* stereochemistry to these intermediates was based on the method of synthesis and the appearance of *trans*-quinolizidine absorption in their infrared spectra. Confirmation was provided by deoxygenation of **9** to the known all-*trans* hexahydrojulolidine. This was accomplished by Wolff-Kishner reduction of the semicarbazone or by sodium borohydride reduction, tosylation, and lithium aluminum hydride reduction.

Chart 4-2.[a,135]

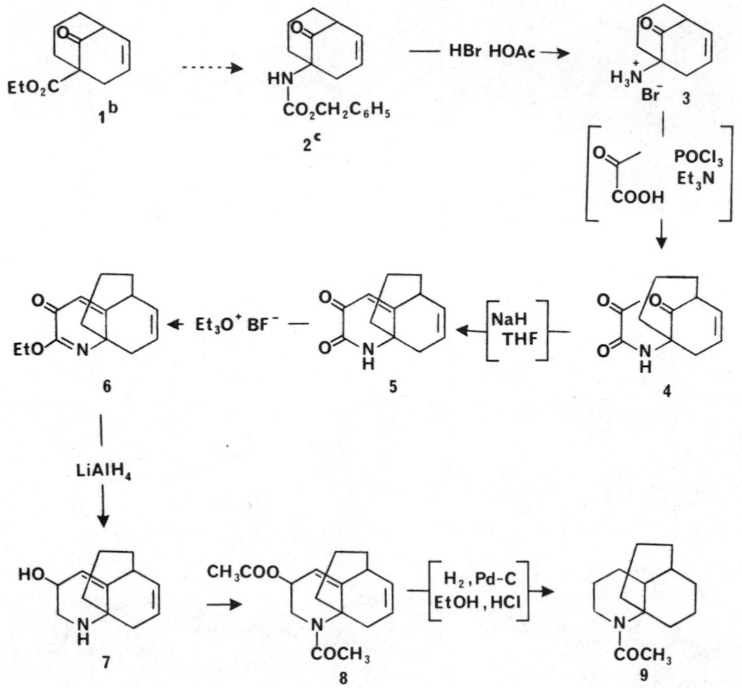

[a]For a closely related but more extensive study, see Horrii et al.[136] and references cited therein.
[b]As described by Cope and Synerholm[137] and Colvin and Parker.[138]
[c]The transformation of **1** to **2** was achieved by conversion of **1** to the corresponding acid, acid azide, isocyanate, and finally the benzylcarbamate **2**. Yields and procedures are not given in this brief communication.

Chart 4-3.[139]

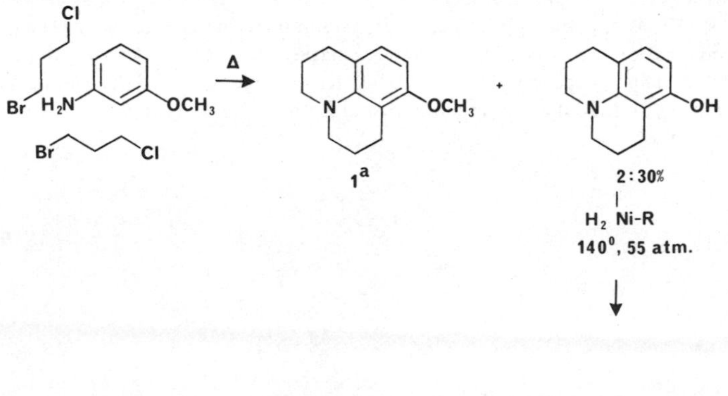

Chart 4-3 (Continued).

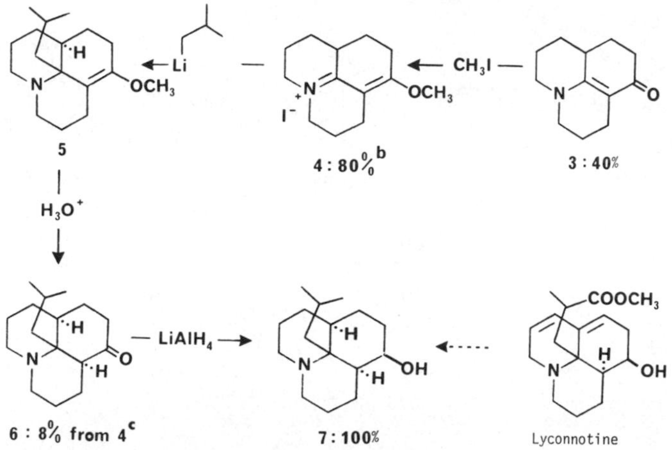

5

H_3O^+

4 : 80%[b]

3 : 40%

— $LiAlH_4$ →

6 : 8% from 4[c]

7 : 100%

Lyconnotine

[a]Based on the synthesis of julolidine by Pinkus.[140]
[b]For a comparable *O*-alkylation of an enamide see Leonard and Adamcik.[141]
[c]Higher yields (up to 35%) could be obtained with less bulky alkyl lithiums and Grignard reagents (see Chart 4-4). None of the isomeric isobutyl ketone could be detected in the reaction mixture. Therefore, both the alkylation and enol ether hydrolysis steps are stereospecific.

Chart 4-4.[142]

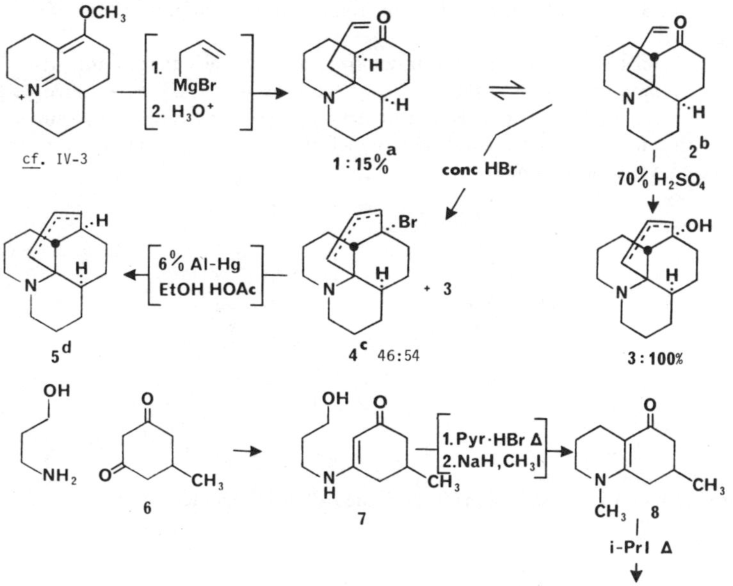

1. MgBr
2. H_3O^+

cf. IV-3

1 : 15%[a]

conc HBr

2^b

70% H_2SO_4

6% Al-Hg
EtOH HOAc

5^d

4^c 46:54

+ 3

3 : 100%

6

7

1. Pyr·HBr Δ
2. NaH, CH_3I

8

i-PrI Δ

Chart 4-4 (Continued).

11: 70%[f] ← 75% H_2SO_4 — 10: 50% ← [allyl MgBr] — 9

— $SOCl_2$ → 12 — Na NH_3 → 13

[a]The stereochemistry of **1** follows from the presence of *trans*-quinolizidine bands in its infrared spectrum, and the fact that the axial alcohol formed from **1** by reduction with $LiAH_4$ forms a cyclic bromoether upon exposure to NBS and from its subsequent transformations.

[b]Extended absorption of **1** on basic alumina or treatment with KOH in methanol provides an equilibrium mixture (1:1) of **1** and **2**. An almost complete conversion of **1** to **2** can be affected by repeated equilibration. The *cis-trans* stereochemistry of **2** follows from the absence of pronounced *trans*-quinolizidine bands in its infrared spectrum.

[c]The position of the double bond in each of the intermediates cited in this paper remains undetermined.

[d]Alternatively, **5** can be obtained from **3** by treatment with thionyl chloride and reduction of the resultant tertiary chloride.

[e]The stereochemistry of **3**, **4**, and **5** follows from the discovery that intermediates **1** and **2** cannot be equilibrated under strongly acidic conditions. This may be ascribed to the fact that such an epimerization would require inversion at *both* the carbon adjacent to the carbonyl and the nitrogen. Under conditions which insure virtually complete protonation of the nitrogen atom is clearly impossible:

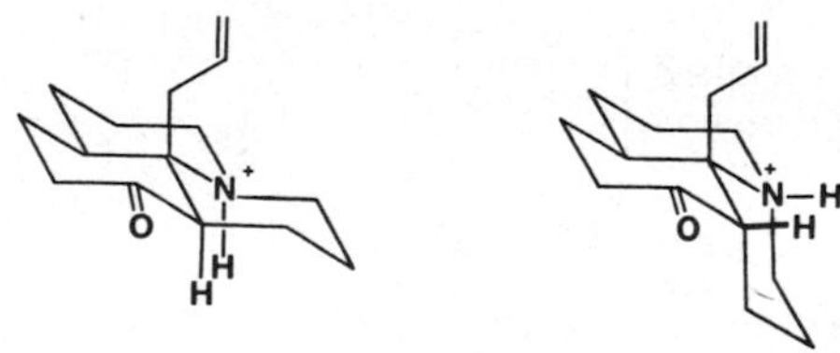

[f]The formation of **11** may be rationalized in the following manner:

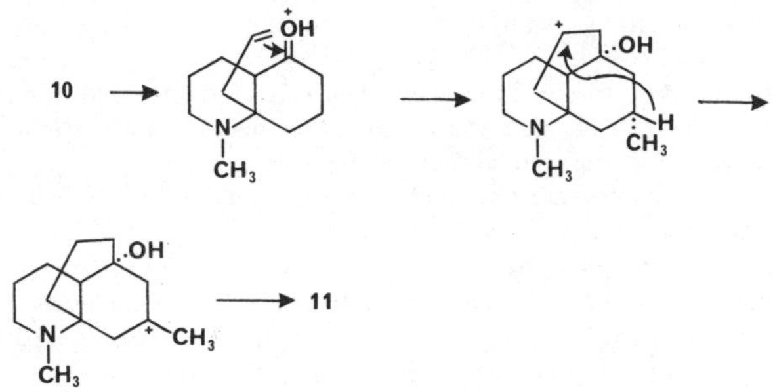

10 → [structure] → [structure] → [structure] → 11

The occurrence of the hydride transfer during this cyclization reaction established conclusively the stereochemistry of the Grignard addition.

Chart 4-5.[143]

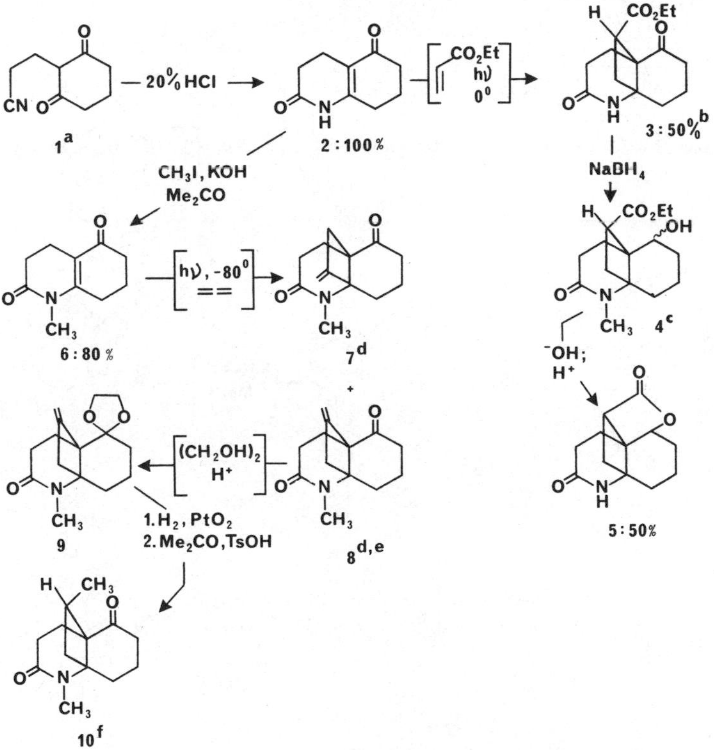

1[a] — 20% HCl → 2 : 100% — [CO_2Et, hν, 0°] → 3 : 50%[b]

3 — $NaBH_4$ → 4[c] — ^-OH; H^+ → 5 : 50%

2 — CH_3I, KOH, Me_2CO → 6 : 80% — [hν, −80°, ==] → 7[d] + 8[d,e]

8 — [$(CH_2OH)_2$, H^+] → 9 — 1. H_2, PtO_2; 2. Me_2CO, TsOH → 10[f]

[a]Prepared by cyanoethylation of dihydroresorcinol according to the procedure of Nazarov and Zar'valov.[144]
[b]The photoaddition was performed in neat ethyl acrylate and appeared to be stereospecific.
[c]A mixture of two epimeric alcohols was produced in this reduction. The mixture was used in the subsequent saponification and lactonization sequence.
[d]The photoadducts 7 and 8 were obtained in about equal amounts in 50% of the yield. They were separated by preparative vpc.
[e]Direct reduction of 8 with $Pd/BaSO_4$ in ethanol produced a mixture of **10** and its methyl-epimer. Prior conversion of 8 to the ketal 9 effectively blocks this undesired mode of attack.
[f]An alternative approach to this system was aborted when it was discovered that the crucial Beckmann rearrangement step proceeded in poor yield.

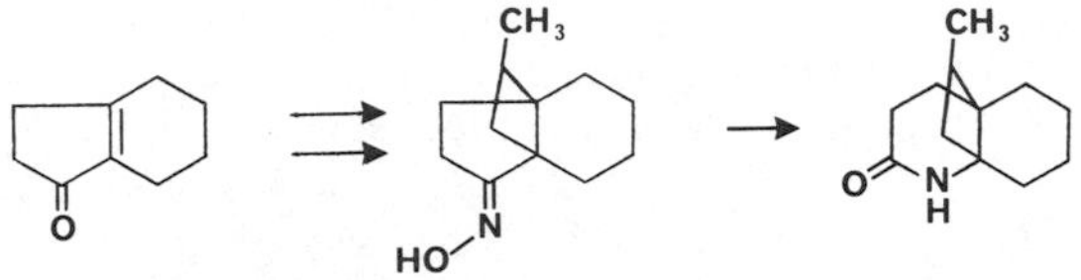

Chart 4-6.[145]

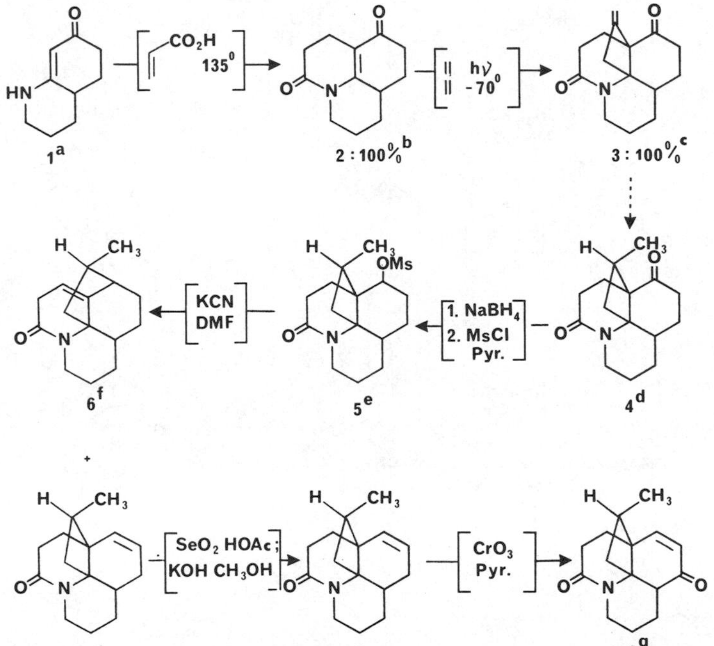

[a]Prepared according to the procedure of Grob and Wilkens.[146]
[b]Attempts to alkylate **1** in basic solution with ethyl acrylate or acrylonitrile gave only very low yields of **2**. This interesting alternative procedure probably proceeds in the following manner:

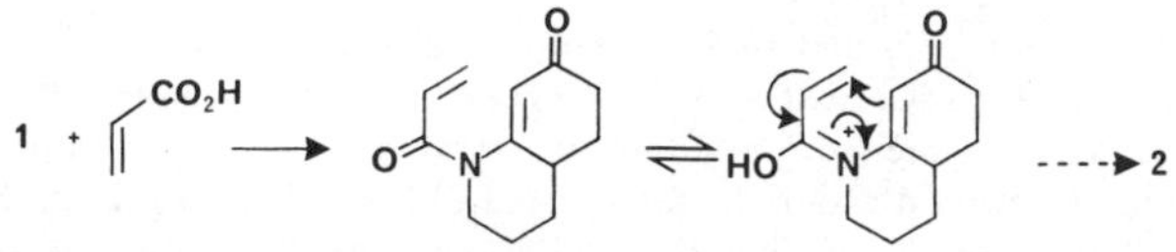

1 + CO_2H ⟶ ⇌ HO ---→ 2

[c]See Chart 4-5 for a discussion of this important transformation.
[d]The stereospecific conversion of **3** to **4** was achieved by ketalization, hydrogenation, and transdioxonolation as described in Chart 4-5.
[e]Mesylate **5** was eliminated only with great difficulty and under a variety of conditions always produced a mixture of **7** and the rearranged material **6**.
[f]The isomeric mixture of **6** and **7** was employed in the allylic oxidation with SeO_2, subsequent saponification, and CrO_3 oxidation. At this stage the desired enone **9** was separated from ketolactam **10** by chromatography.

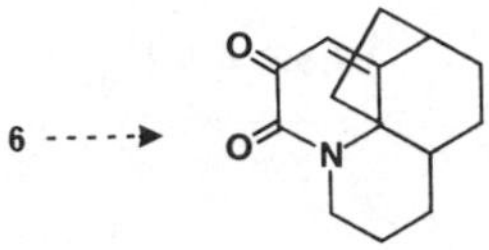

6 ----→ **10**

[g]This substance was identical with an optically active radation product of annotinine and constituted the first chemical proof of the orientation of the methyl function.

Chart 4-7.[147]

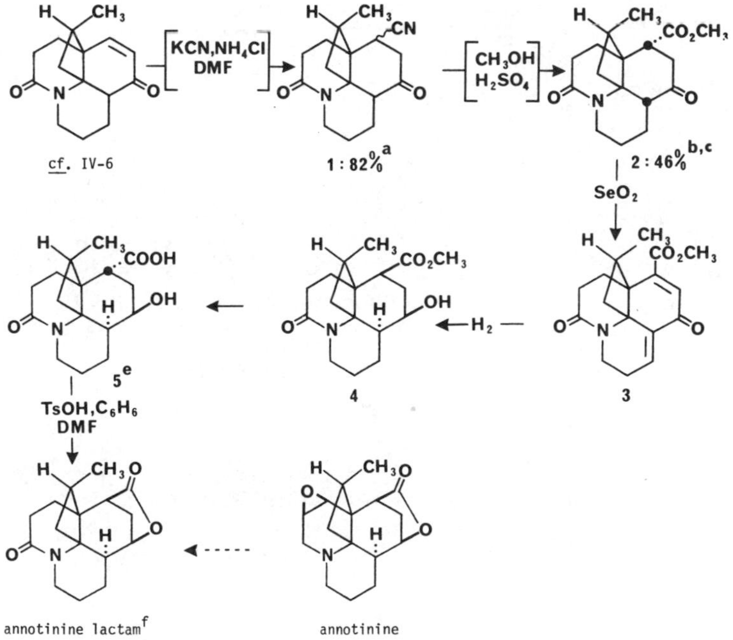

cf. IV-6 —[KCN, NH_4Cl, DMF]→ 1 : 82%[a] —[CH_3OH, H_2SO_4]→ 2 : 46%[b,c]

SeO_2 → 3 —H_2→ 4 → 5[e]

$TsOH, C_6H_6$, DMF → annotinine lactam[f] ←---- annotinine

[a]The conjugate addition afforded two epimeric cyanoketones, which were not separated but, rather used directly in the next step. The general procedure was that of Nigata et al.[148]
[b]Considerable amounts of the corresponding acid were secured from the aqueous layer upon workup. Esterification with diazomethane provided additional quantities of **2**.
[c]The synthetic racemate **2** was identical with the optically active substance obtained previously[d] from annotinine. The selenium dioxide oxidation and subsequent reduction to **4** had also been executed previously on this degradation product.
[d]Wiesner et al.[144]
[e]Saponification of the ester function and epimerization had also been described previously by Valenta et al.[150]
[f]An alternative more efficient synthesis of this lactam was subsequently reported (see Chart 4-9). The conversion of annotinine to annotinine lactam involves three stages and was first described by MacLean and Prime.[151]

Chart 4-8.[152]

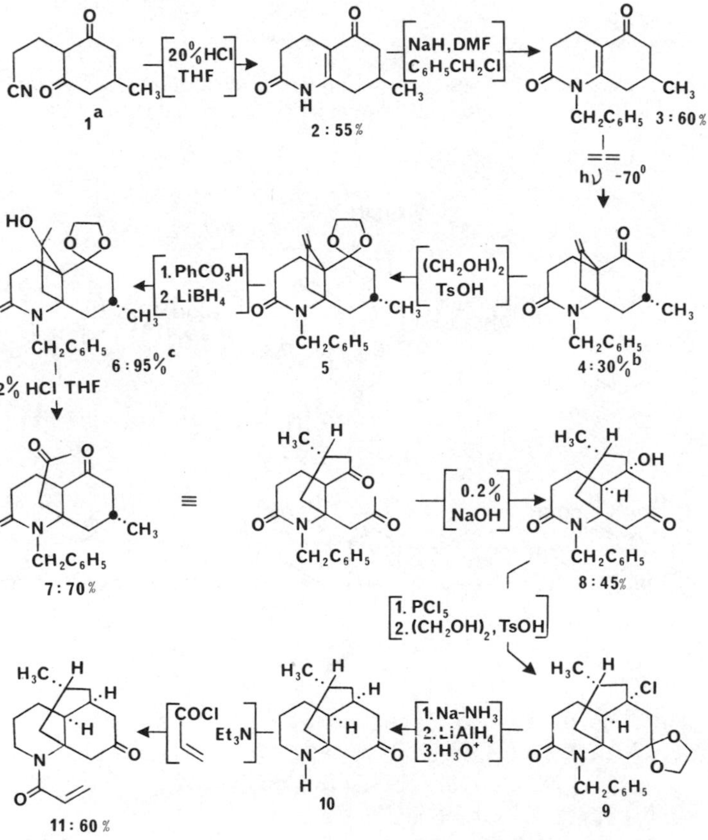

Chart 4-8 (Continued).

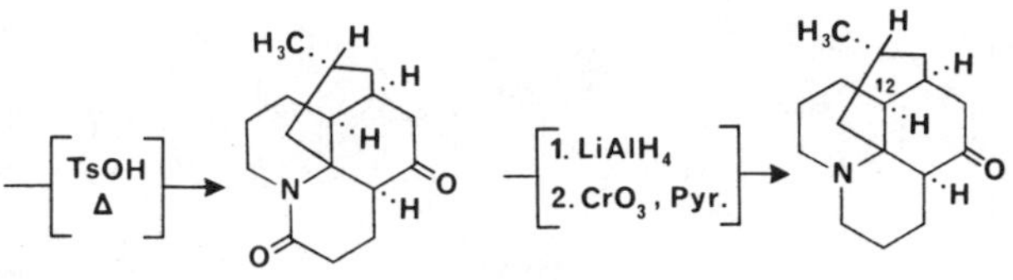

TsOH
Δ

12

1. $LiAlH_4$
2. CrO_3, Pyr.

12-*epi*-lycopodine

[a]The starting material is the cyanoethylation product of dihydroorcinol.[153]
[b]The low yield is due to the formation of the reverse addition product. The orientation observed in the cycloaddition of allene to these vinylogous amides is strongly influenced by the nature of the group on the nitrogen atom (compare with Charts 4-5 and 4-6).
[c]Epoxidation of **5** clearly proceeded by attack on the least hindered face of the exocyclic double bond.

Chart 4-9 [154]

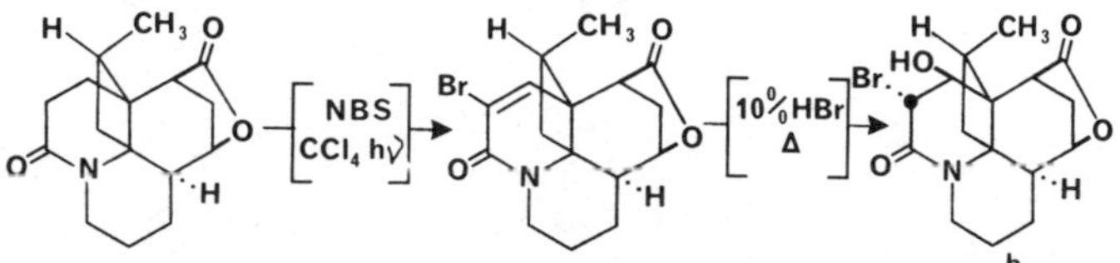

cf. IV-7

NBS
CCl_4, hν

1[a]

10% HBr
Δ

2[b]

$NaHCO_3$
aq. Me_2CO

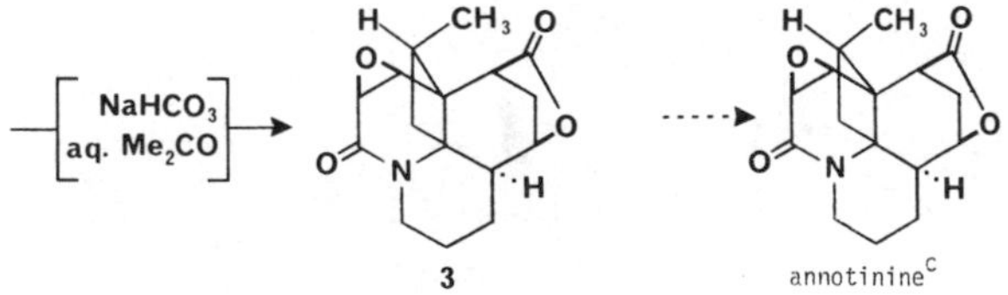

3

annotinine[c]

[a]All of the intermediates described in this synthesis were known transformation products of annotinine.
[b]Other products were formed during this hydration. A potentially tedious purification procedure was avoided by virtue of the fact that the desired bromohydrin **2** crystallized directly from the reaction mixture.
[c]The conversion of oxoannotinine **3** to annotinine had already been described by Betts and MacLean.[155]

Chart 4-10.[156]

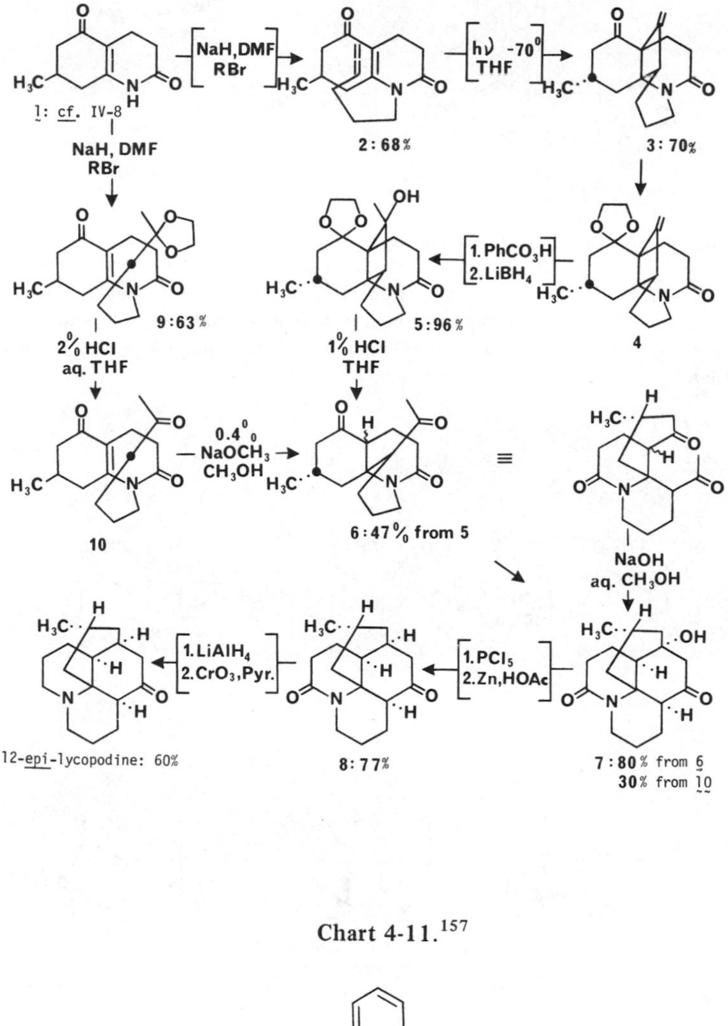

1: cf. IV-8

NaH, DMF / RBr

2: 68%

hν −70° / THF

3: 70%

4

1. $PhCO_3H$ / 2. $LiBH_4$

5: 96%

9: 63%

2% HCl / aq. THF

1% HCl / THF

10

0.4% $NaOCH_3$ / CH_3OH

6: 47% from 5

NaOH / aq. CH_3OH

7: 80% from 6 / 30% from 10

1. PCl_5 / 2. Zn, HOAc

8: 77%

1. $LiAlH_4$ / 2. CrO_3, Pyr.

12-epi-lycopodine: 60%

Chart 4-11.[157]

1. NaOEt / 2. 15% aq. K_2CO_3

1: 36%

EtOH H^+

2

$LiAlH_4$; H_3O^+

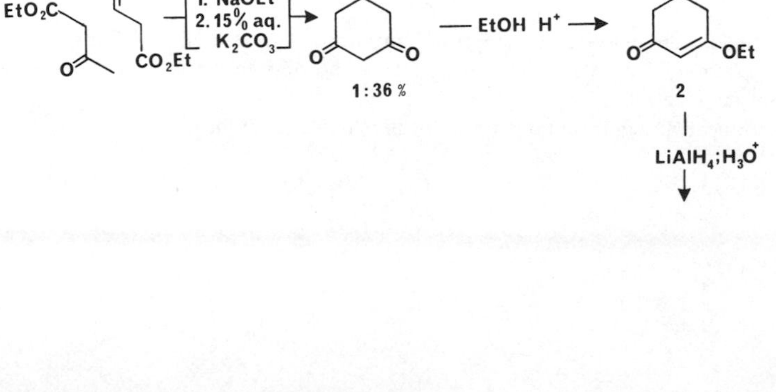

Chart 4-11 (Continued).

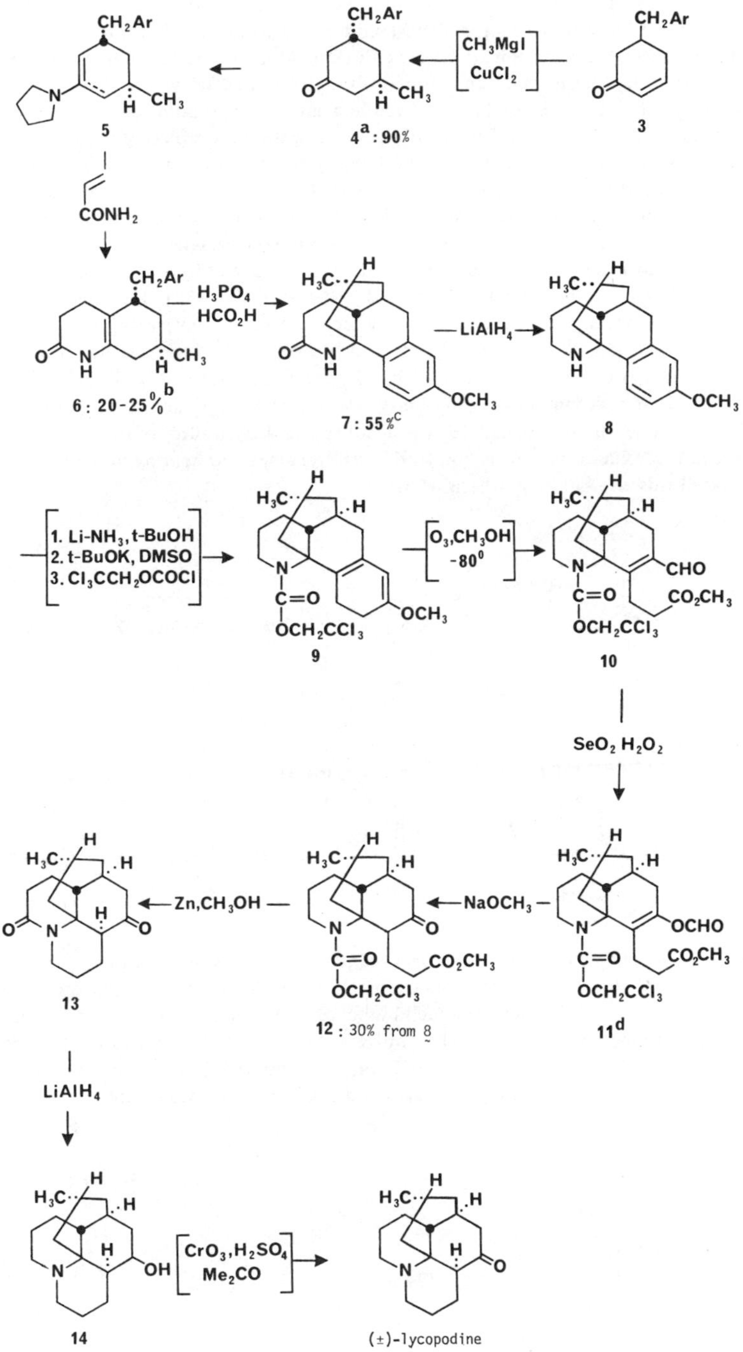

CH_2Ar / 3 → [CH_3MgI, $CuCl_2$] → 4^a : 90% → 5 → $CONH_2$ → 6 : 20-25%[b] → H_3PO_4, HCO_2H → 7 : 55%[c] → $LiAlH_4$ → 8

→ [1. Li-NH_3, t-BuOH; 2. t-BuOK, DMSO; 3. Cl_3CCH_2OCOCl] → 9 → [O_3, CH_3OH, -80^0] → 10 (CHO, CO_2CH_3, OCH_2CCl_3) → SeO_2 H_2O_2 → 11[d] (OCHO) → $NaOCH_3$ → 12 : 30% from 8 → Zn, CH_3OH → 13 → $LiAlH_4$ → 14 → [CrO_3, H_2SO_4, Me_2CO] → (±)-lycopodine

[a]The conjugate addition proceeded best under relatively high dilution (1% in ether at 0°) employing a catalytic amount of cupric chloride. Axial introduction of the methyl group in such reactions had been established previously Allinger and Riew.[158]

[b]This simple annelation procedure involved heating the pyrrolidine enamine with acrylamide in dioxane solution followed by the addition of water and refluxing for 1 hr. In the case of cyclohexanone itself, a 70% yield of the hexahydroquinolone was obtained. Application of this "aza-annelation" to the conversion of **4** to **6** is plagued by the fact that both possible isomeric enamines **5** are produced. Although the desired hexahydroquinolone (**6**) could be separated from its isomer, an alternative method to overcome this lack of selectivity was developed. Stork reasoned that conjugate addition of *m*-methoxybenzylmagnesium bromide (catalyzed cuprous chloride) would provide the proper enolate **15**. This enolate could be trapped as a trimethylsilyl ether **16** from which the more relative lithium enolate **17** could be regenerated and alkylated (cf. Stork and Hudrlik[159]). Alternatively, it was discovered that the magnesium enolate **15** could be alkylated directly and without loss of structural specificity by employing hexamethylphosphoramide (HMPA) as solvent. Conversion of the allylated ketone **18** to ketoacid **19** was accomplished by hydroboration of the corresponding ketal and oxidation. Conversion to the methyl ester and heating in methanolic ammonia completed this alternative synthesis of **6**.

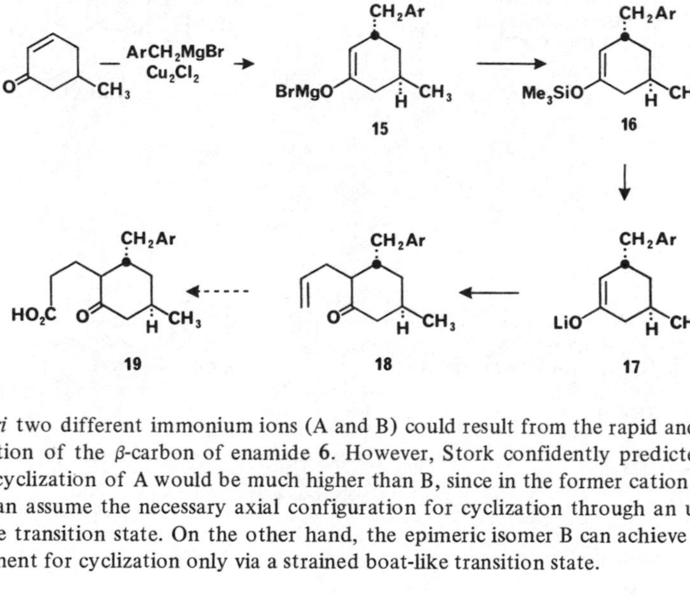

$ArCH_2MgBr$, Cu_2Cl_2; CH_3; CH_2Ar; $BrMgO$; H; **15**; Me_3SiO; **16**; LiO; **17**; **18**; HO_2C; **19**

[c]*A priori* two different immonium ions (A and B) could result from the rapid and reversible protonation of the β-carbon of enamide **6**. However, Stork confidently predicted that the rate of cyclization of A would be much higher than B, since in the former cation the benzyl group can assume the necessary axial configuration for cyclization through an undistorted chair-like transition state. On the other hand, the epimeric isomer B can achieve the proper arrangement for cyclization only via a strained boat-like transition state.

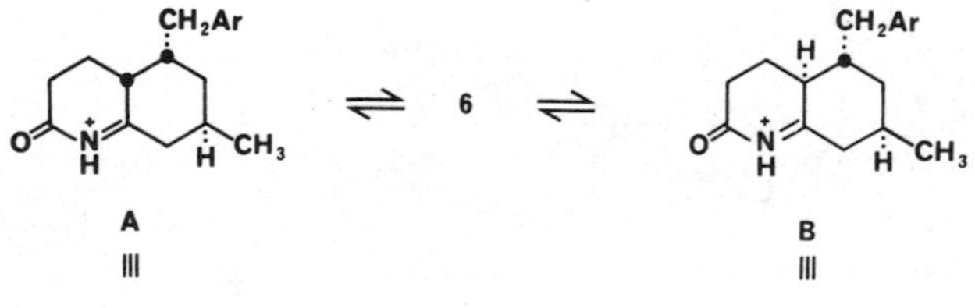

CH_2Ar; $\overset{+}{N}$; H; CH_3; **6**; A; B; III

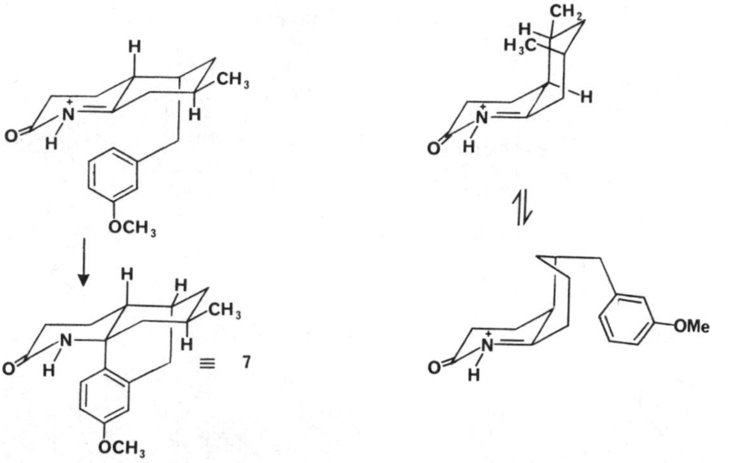

[d]It was anticipated that Birch reduction of 8 followed by hydrolysis of the dihydroanisole would provide the α,β-unsaturated ketone **20**. However, upon execution of this experiment only the β,γ-isomer **21** was isolated, and this substance resisted all attempts at isomerization. This result necessitated this ingenious modification of the original plan.

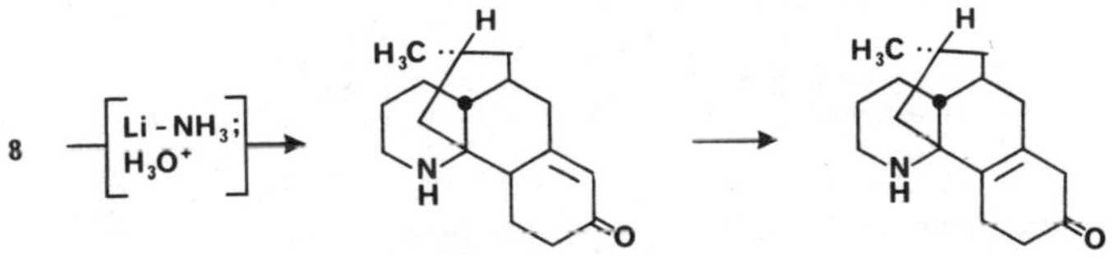

Chart 4-12.[160]

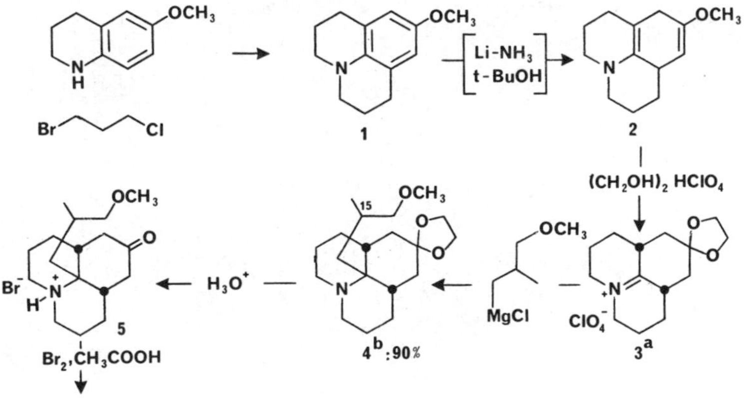

Chart 4-12 (Continued).

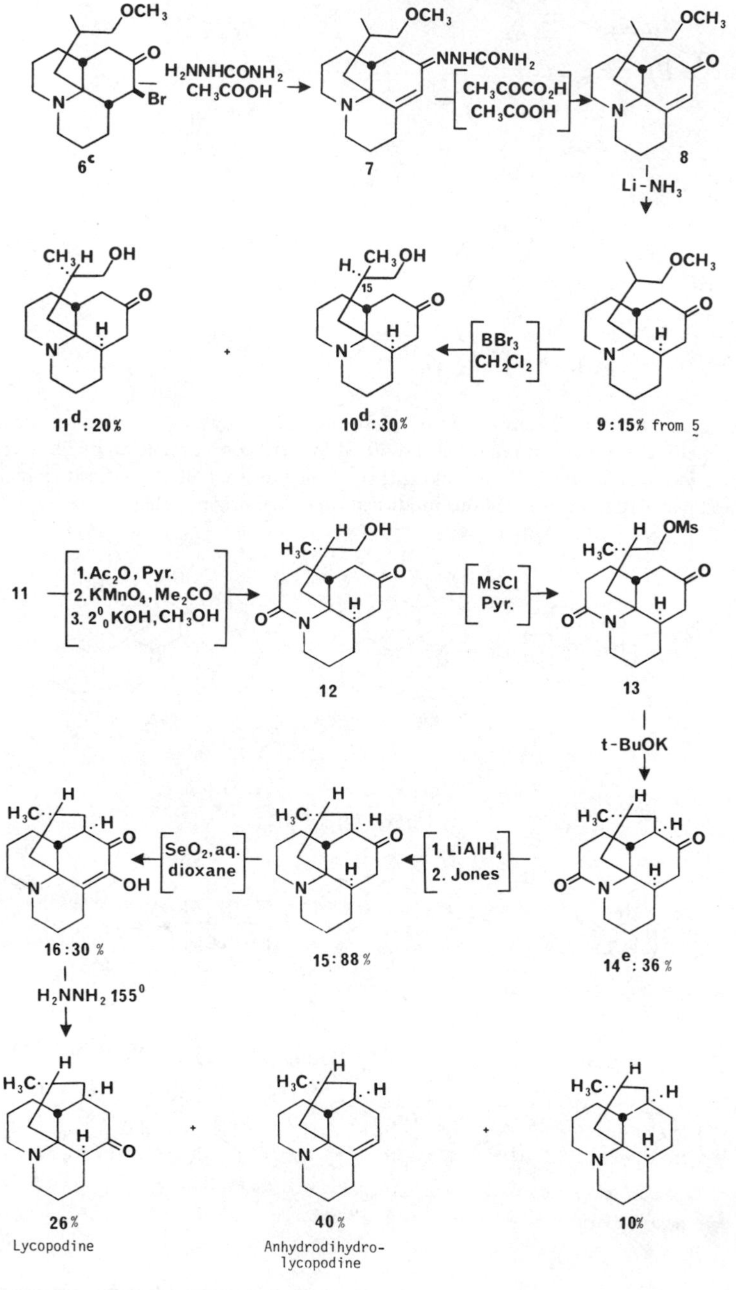

$H_2NNHCONH_2$
CH_3COOH

OCH_3 O Br N
6^c

NNHCONH$_2$
7

[CH_3COCO_2H CH_3COOH]

8

Li-NH_3

9 : 15% from 5

[BBr_3 CH_2Cl_2]

10^d : 30%

11^d : 20%

11 [1. Ac_2O, Pyr. 2. $KMnO_4$, Me_2CO 3. 2^0_0 KOH, CH_3OH]

12

[MsCl Pyr.]

13

t-BuOK

14^e : 36%

[1. $LiAlH_4$ 2. Jones]

15 : 88%

[SeO_2, aq. dioxane]

16 : 30%

H_2NNH_2 155^0

26%
Lycopodine

40%
Anhydrodihydro-lycopodine

10%

[a]The synthesis of this intermediate and a model study of its reactions with certain Grignard reagents was reported earlier by this group (cf. Ayer et al.[161]).
[b]The *cis-cis* stereochemistry of this substance was assigned by analogy with the model studies cited in footnote a; by the fact that it exhibited Bohlmann bands in the infrared spectrum; and on theoretical grounds since this corresponds to addition of the Grignard reagent to the less hindered face of immonium salt **3**.
[c]Dehydrobromination of this bromoketone proved to be difficult by conventional methods. This fact may be ascribed to the equatorial nature of the bromine substituent and necessitated this slightly more circuitous method developed previously by Green and Long.[162]
[d]The mixture of C_{15} epimers was separated by chromatography in alumina.
[e]Optically active ketolactam **14** was secured by an extensive degradation of lycopodine itself, and this material was employed as a relay in all subsequent steps.

5. THE PYRROLIZIDINE ALKALOIDS

The widespread occurrence of the basic pyrrolizidine system (**1**) in several botanical families (such as *Compositae, Santalaceae, Gramineae, Boraginaceae,* and *Leguminosae*) has attracted considerable attention over the past several decades. Although the title pyrrolizidine alkaloids is most descriptive, frequent-

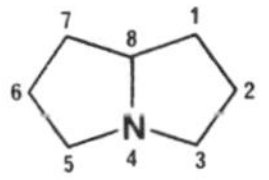

7 8 1 6 2 N 5 4 3

1

ly, members of this group have been designated as *Senecio* alkaloids derived from the fact that various alkaloids of this group were first discovered in the generea *Senecio* of the family *Compositae*). This practice appears to be losing support, however. Comprehensive reviews of the chemistry of these alkaloids have appeared periodically in the recent literature.[163-165]

Representative pyrrolizidine alkaloids are portrayed in the accompanying table (absolute stereochemistry as indicated). In the case of the 1-hydroxymethyl alkaloids (e.g., lindelofidine **2**), these bases occur in nature as esters of various acids, thus giving rise to a host of additional alkaloids. However, in some cases (e.g., lindelofidine), the unesterified base has been detected in the plant. We are concerned here only with the synthesis of the basic portion. Trachelanthamidine **4**; supinidine **6, 7**; platynecine **9**; heliotridine **12**; retonecine **11**; and rosmarinecine **14** are also found in the form of their *N*-oxides, and retronecine and supinidine methyl ethers also occur as the free bases.

The 1-hydroxymethyl pyrrolizidine alkaloids (lindelofidine, (–)-isoretronecanol, trachelanthamidine, and laburnine) have been the subject of several investigations. Lindelofidine 2 was first isolated by Labenskii and Men'shikov,[166]

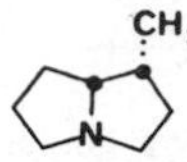

Lindelofidine (+)-Isoretronecanol [2]

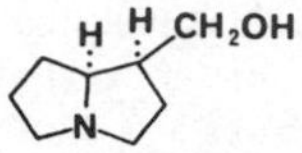

(-)-Isoretronecanol [3]

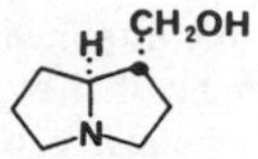

(-)-Trachelanthamidine [4]

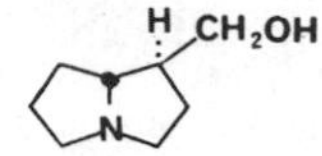

Laburnine (+)-Trachelanthamidine [5]

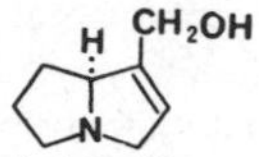

(-)-Supinidine [6]

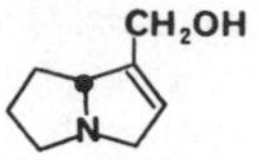

(+)-Supinidine [7]

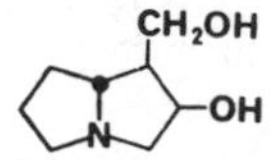

(+)-Macronecine [8]

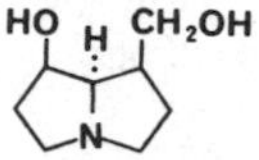

Platynecine [9]

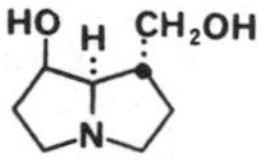

Hastanecine [10]

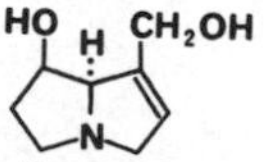

Retronecine [11]

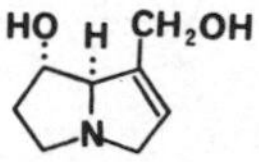

Heliotridine [12]

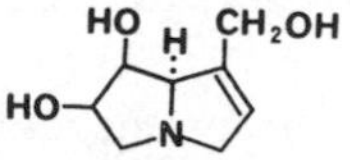

Crotanecine [13]

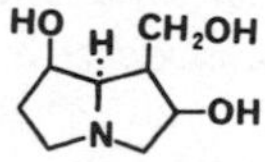

Rosmarinecine [14]

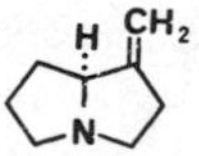

[15]

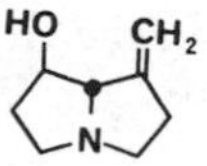

[16]

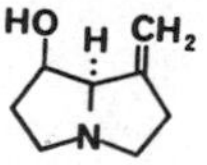

[17]

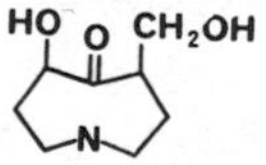

[18]

whereas isoretronecanol 3 was first encountered by Adams and Hamlin[167] during the course of a structural investigation on retronecine **11** and has only recently been identified from natural sources by Hart, Johns, and Lamberton.[168] Trachelanthamidine **4** was isolated and identified by Men'shikov and Borodina,[169] and laburnine **5**, by Galinovsky, Goldberger, and Pöhm.[170] The relative stereochemistry at C_1 and C_8 of these four stereoisomers was deduced by Labenskii, Serova, and Men'shikov[171] and Likhosherstor, Kulakov, and Kochetkov.[172] The assignment of absolute stereochemistry to these four alkaloids was made by Warren and von Klemperer[173] who correlated (−)-heliotridane **19** with (+)-3-methylheptane **20** of known absolute stereochemistry. Since the former substance had previously been related to platynecine **9** by Konovalova and Orekhov[174] and since Adams and Hamlin[167] had demonstrated the relationship between platynecine **9** and (−)-isoretronecanol **3**, the absolute stereochemistry of each of these bases was established.

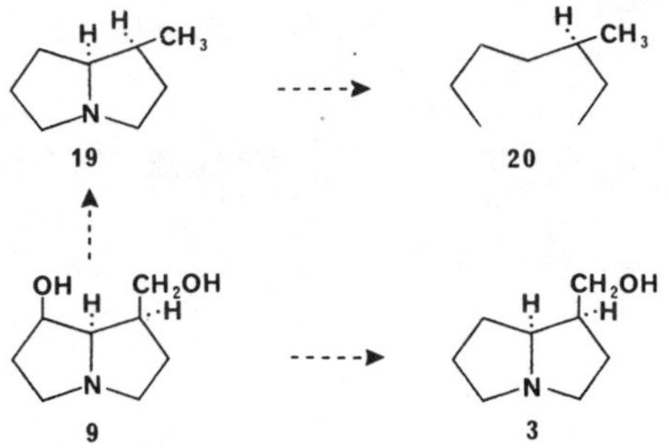
H
H
CH3
N
19
H
CH3
20
OH
H
CH2OH
H
N
9
CH2OH
H
H
N
3

Each of the 1-hydroxymethylpyrrolizidine alkaloids has been synthesized.

Chart 5-1.[175]

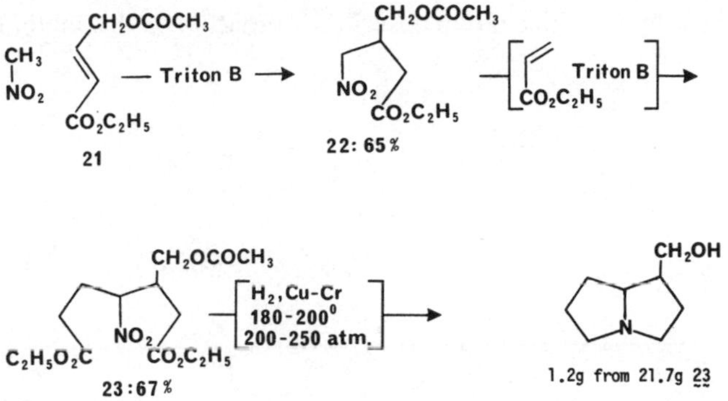
CH2OCOCH3
CH3
NO2
CO2C2H5
21
Triton B
CH2OCOCH3
NO2
CO2C2H5
22: 65%
Triton B
CO2C2H5
CH2OCOCH3
NO2
C2H5O2C
CO2C2H5
23:67%
H2, Cu-Cr
180-200°
200-250 atm.
CH2OH
N
1.2g from 21.7g 23

This paper reports the first total synthesis of a 1-hydroxymethylpyrrolizidine but proceeded in low yield. Thus conjugate addition of nitromethane to ethyl γ-acetoxycrotonate gave nitrodiester **22**. Michael addition of this intermediate to ethyl acrylate provided the triester **23**, which was reduced over a copper chromite catalyst. The stereochemical features of the resultant amine remain uncertain but is probably (±)-trachelanthamidine.

A much more convenient synthesis is described in this paper.

Chart 5-2.[176]

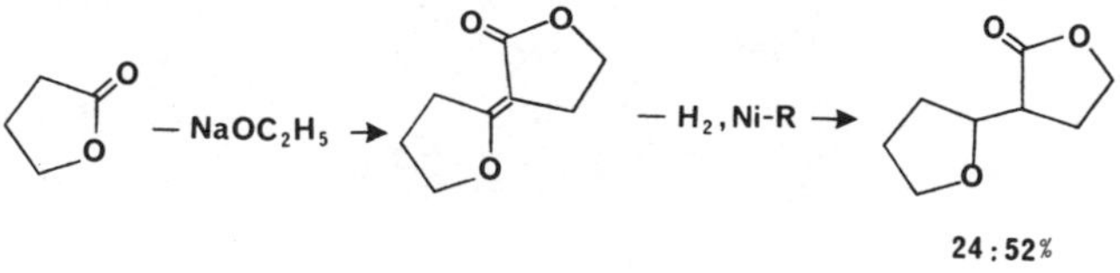
O
O
NaOC2H5
O
O
O
H2, Ni-R
O
O
O
24:52%

Chart 5-2 (Continued).

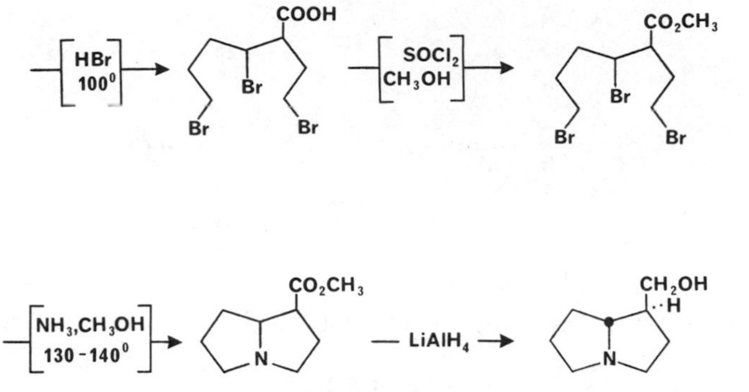

HBr, 100° → (COOH, Br, Br, Br) → $SOCl_2$, CH_3OH → (CO_2CH_3, Br, Br, Br)

NH_3, CH_3OH, 130-140° → (CO_2CH_3, N) — $LiAlH_4$ → (CH_2OH, H, N)

25: 32% from 24 (±)-Trachelanthamidine

A later paper by this same group describes a variation of this method.

Chart 5-3.[177]

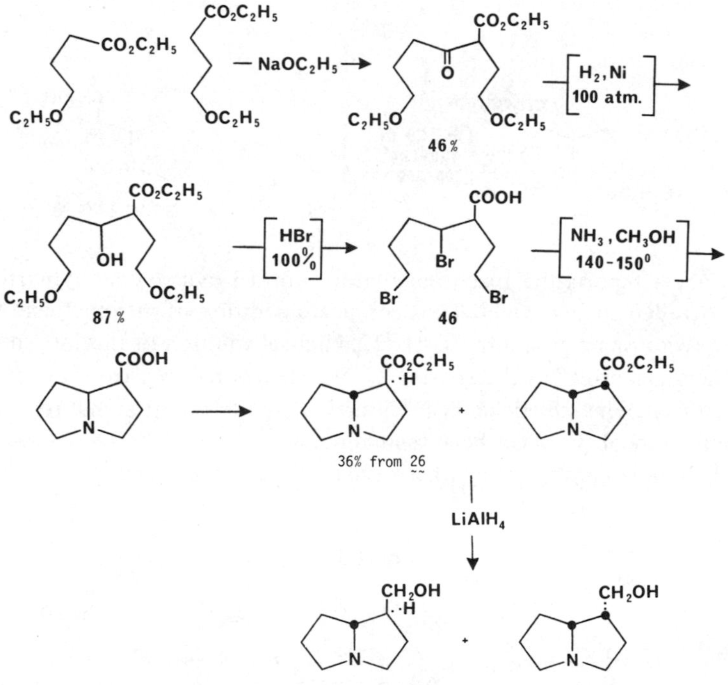

($CO_2C_2H_5$, C_2H_5O) ($CO_2C_2H_5$, OC_2H_5) — $NaOC_2H_5$ → ($CO_2C_2H_5$, O, C_2H_5O, OC_2H_5) 46% → H_2, Ni, 100 atm. →

($CO_2C_2H_5$, OH, C_2H_5O, OC_2H_5) 87% → HBr 100% → (COOH, Br, Br, Br) 46 → NH_3, CH_3OH, 140-150° →

(COOH, N) → ($CO_2C_2H_5$, H, N) + ($CO_2C_2H_5$, N)

36% from 26

$LiAlH_4$

(CH_2OH, H, N) + (CH_2OH, N)

(±)-Trachelanthamidine (±)-Isoretronecanol

Chart 5-4.[179]

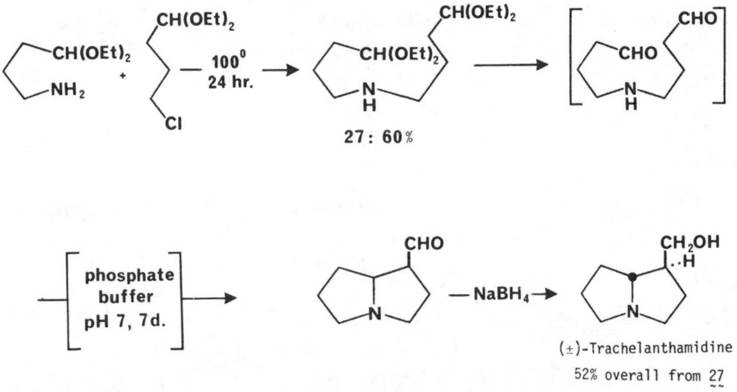

The cyclization of diacetal at pH 4.2 has been reported to yield (±)-isoretronecanol.[179]

Based upon the original observation of Mandell and Blanchard,[181] a mixture of *N*-benzylpyrrole and acetylene dicarboxylic acid was refluxed in ether for 24 hr from which three products were isolated. Direct reduction of fumarate **28** over Raney nickel catalyst was accompanied by decarboxylation and cyclization to 3-oxopyrrolizidine, which could be further reduced to pyrrolizidine itself. On the other hand, controlled reduction of **28** over a carefully prepared palladium catalyst yielded the succinate **30**, which was esterified and subjected to a futher reduction. When exposed to lithium, aluminum hydride oxopyrrolizidine **31** provided a diastereoisomeric mixture of 1-hydroxymethylpyrrolizidines **32**, which was shown by gas chromatography to consist of 90% (±)-trachelanthamidine and 10% (±)-isoretronecanol. The mixture could be separated by means of the corresponding picrates and laburnine **5** obtained from the racemate.

Chart 5-5.[180]

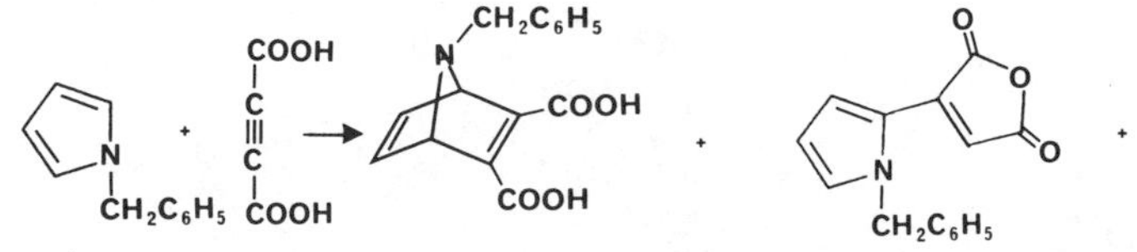

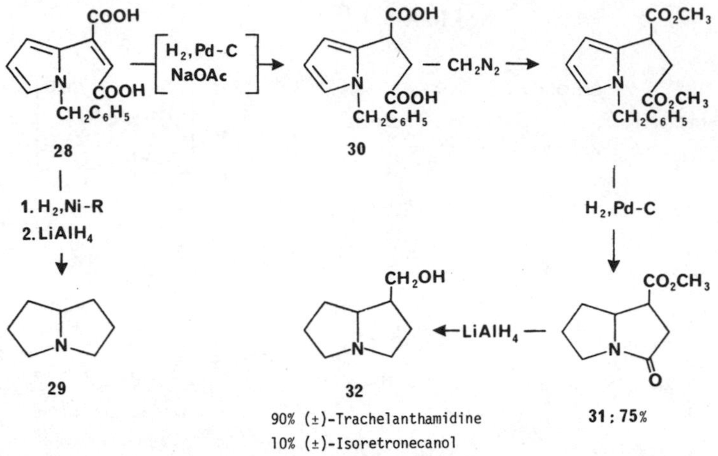

COOH
H_2,Pd-C
NaOAc
COOH
$CH_2C_6H_5$
28
COOH
COOH
$CH_2C_6H_5$
30
— CH_2N_2 →
CO_2CH_3
CO_2CH_3
$CH_2C_6H_5$
1. H_2,Ni-R
2. $LiAlH_4$
29
H_2,Pd-C
CO_2CH_3
O
31 ; 75%
CH_2OH
← $LiAlH_4$ —
32
90% (±)-Trachelanthamidine
10% (±)-Isoretronecanol

Chart 5-6.[182]

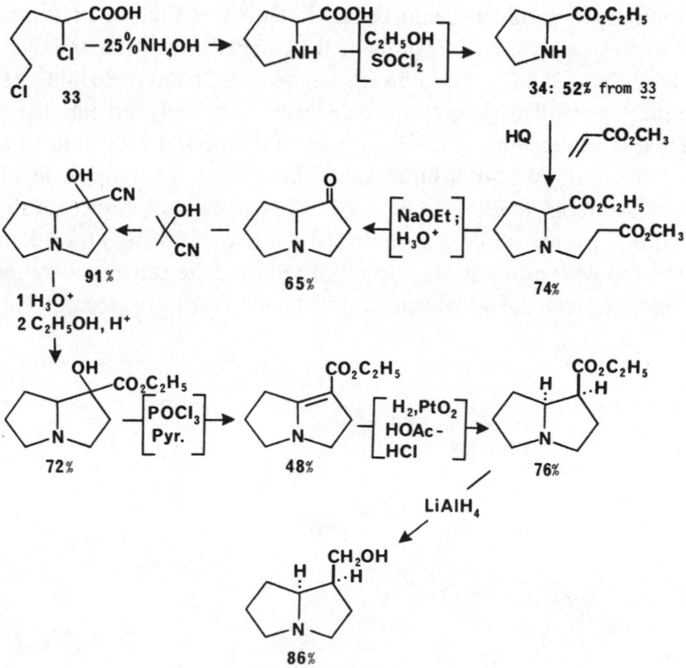

COOH
Cl
Cl
33
— 25% NH_4OH →
COOH
NH
C_2H_5OH
$SOCl_2$
$CO_2C_2H_5$
NH
34: 52% from 33
HQ
CO_2CH_3
$CO_2C_2H_5$
CO_2CH_3
74%
NaOEt;
H_3O^+
O
65%
OH
CN
OH
CN
91%
1 H_3O^+
2 C_2H_5OH, H^+
OH
$CO_2C_2H_5$
72%
$POCl_3$
Pyr.
$CO_2C_2H_5$
48%
H_2,PtO_2
HOAc-
HCl
H
$CO_2C_2H_5$
..H
76%
$LiAlH_4$
H
CH_2OH
..H
86%
(±)-Isoretronecanol

An additional stereospecific synthesis of racemic isoretronecanol is reported. Condensation of diethyl oxylate and ethyl 2-pyrrolidylacetate provided pyrrolizidine **35**, which was reduced catalytically over rhodium on alumina to the alcohol **36** and then dehydrated to **37**. The catalytic hydrogenation of **37** predictably proceeded by *cis* addition to the less hindered convex face of the pyrrolizidine nucleus. Concommitant hydride reduction of the lactam and ester functions completed the synthesis of (±)-isoretronecanol.

Chart 5-7.[183]

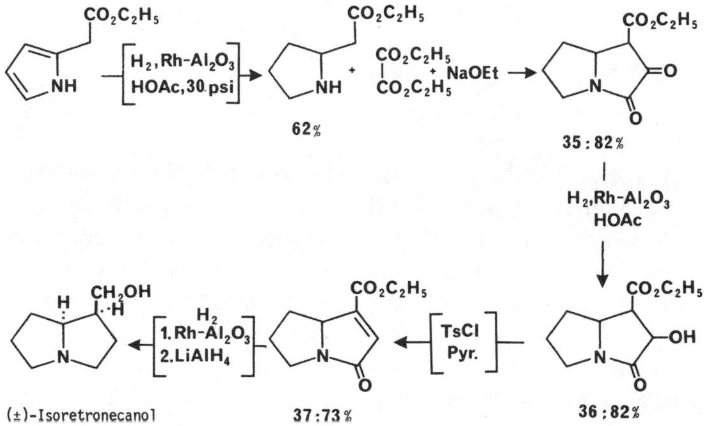

$CO_2C_2H_5$; NH ; H_2,Rh-Al_2O_3 HOAc,30 psi ; 62% ; $CO_2C_2H_5$; $CO_2C_2H_5$; NaOEt ; 35 : 82% ; H_2,Rh-Al_2O_3 HOAc ; 36 ; 82% ; OH ; TsCl Pyr. ; 37 : 73% ; H_2 1. Rh-Al_2O_3 2. $LiAlH_4$; CH_2OH ; (±)-Isoretronecanol

The discovery that a retusamine salt exists in the *O*-protonated transannular form and that otonecine-derived alkaloids show a strong N—C_{CO} interaction prompted application of the knowedge to the synthesis of pyrrolizidine bases.

Chart 5-8.[184]

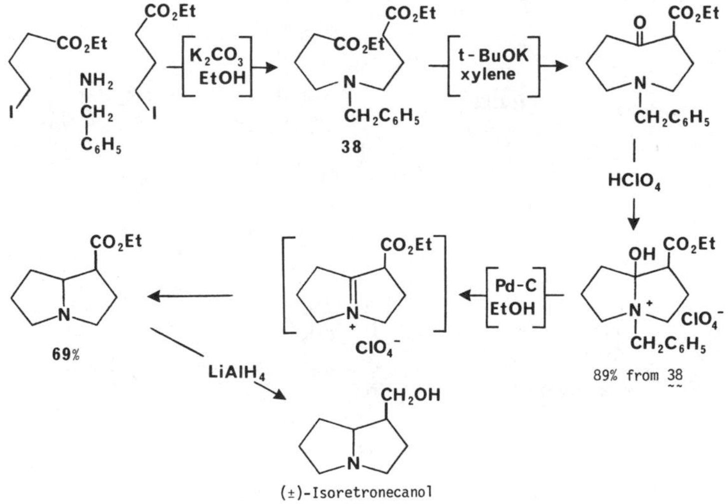

CO_2Et ; NH_2 ; CH_2 ; C_6H_5 ; I ; K_2CO_3 EtOH ; 38 ; $CH_2C_6H_5$; t-BuOK xylene ; $HClO_4$; OH ; ClO_4^- ; 89% from 38 ; Pd-C EtOH ; $LiAlH_4$; 69% ; CH_2OH ; (±)-Isoretronecanol

In addition to this 1-hydroxymethylpyrrolizidine synthesis, Jezŏ and Kaláč[185] reported the synthesis of (apparently) (±)-isoretrone. Total synthesis of pyrroliziidine alkaloids containing more than one hydroxyl function are not nearly as plentiful. In fact, only a few have been disclosed.

Chart 5-9.[186]

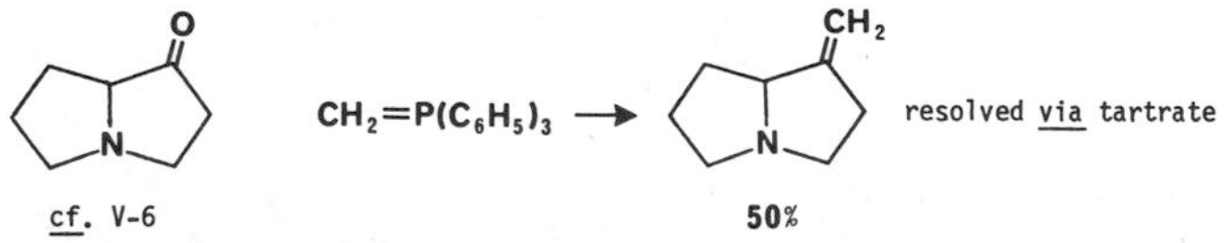

cf. V-6 $CH_2{=}P(C_6H_5)_3$ → CH_2 50% resolved via tartrate

In this synthesis 3-hydroxy-4-methoxybutyronitrile **39** (X = OH) was converted to the chloride **39** (X = Cl) with $SOCl_2$. Alkylation with diethylmalonate provided **40**, which could be reduced catalytically to **41** (under pressure with PtO_2 at room temperature) or **42** (with Raney nickel at 120-130°/80 atm). Upon hydrolysis with aqueous base and acylation with 3,5-dinitrobenzoyl chloride, both **41** and **42** gave the same diacid **43**. The latter substance could be converted to the pyrrolidine **46** by two routes. In one, diacid **43** was brominated in glacial acetic acid, and the resultant bromide **44**, cyclized in alkali to pyrrolidine **45** (not isolated). Subsequent decarboxylation and esterification then gave amino ester **46**. Conjugate addition of this substance to ethyl acrylate followed by Dieckmann condensation, hydrolysis, and decarboxylation yielded keto-alcohol **48**. Catalytic hydrogenation of this intermediate completed the synthesis of (±)-platynecine.

Chart 5-10.[187]

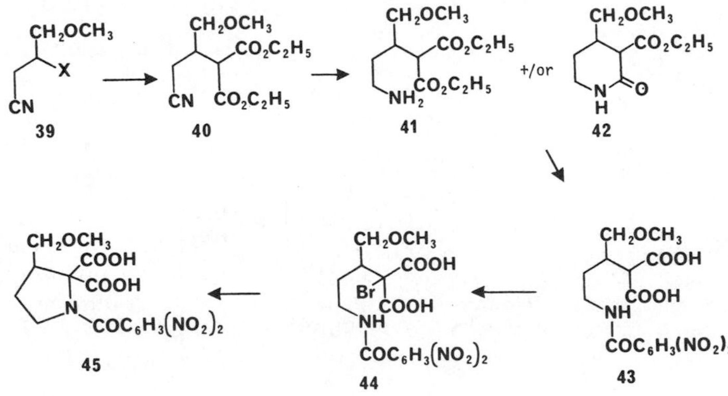

CH_2OCH_3 X CN **39** → CH_2OCH_3 $CO_2C_2H_5$ CN $CO_2C_2H_5$ **40** → CH_2OCH_3 $CO_2C_2H_5$ $CO_2C_2H_5$ NH_2 **41** +/or CH_2OCH_3 $CO_2C_2H_5$ N H O **42**

CH_2OCH_3 COOH COOH N $COC_6H_3(NO_2)_2$ **45** ← CH_2OCH_3 COOH Br COOH NH $COC_6H_3(NO_2)_2$ **44** ← CH_2OCH_3 COOH COOH NH $COC_6H_3(NO_2)_2$ **43**

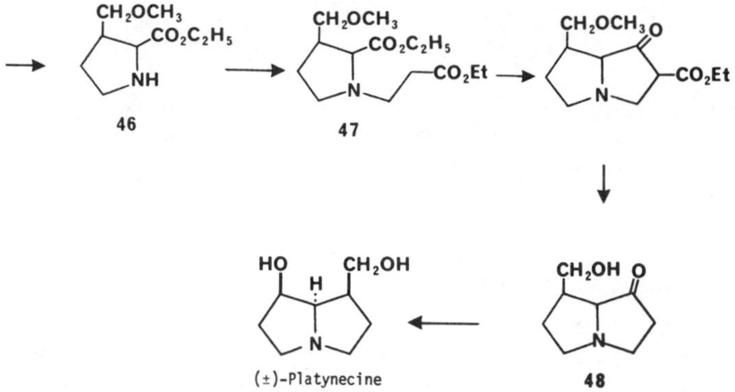

46

47

(±)-Platynecine

48

The readily available (±)-1-carbethoxy-2,3-dioxopyrrolizidine **35** (cf. Chart 5-7) was selected as the starting material. It was discovered that this substance exists almost exclusively in the enol form (**49**), as demonstrated by the fact that it cannot be extracted from the basic solution, is readily *O*-methylated by diazomethane, and exhibits no proton in its pmr spectrum attributable to the C_1 methine. Adams et al.[183] had reduced this substance catalytically over a rhodium catalyst followed by a further reduction with $LiAlH_4$ to a dihydroxypyrrolizidine **50** of undefined stereochemistry. Assuming the initial reduction proceeded by addition of hydrogen from the least hindered (convex) side, the stereochemistry of the diol should be that depicted in **50**. Physical and chemical data are reported in the present paper in support of this assignment. On the other hand, reduction of **49** with zinc in acetic acid provided a mixture of two diastereomers of which **51** predominated. Further reduction of this substance with lithium aluminum hydride yielded (±)-macronecine, which was resolved by fractional crystallization of the bromocamphorsulfonate salts.

Chart 5-11.[188]

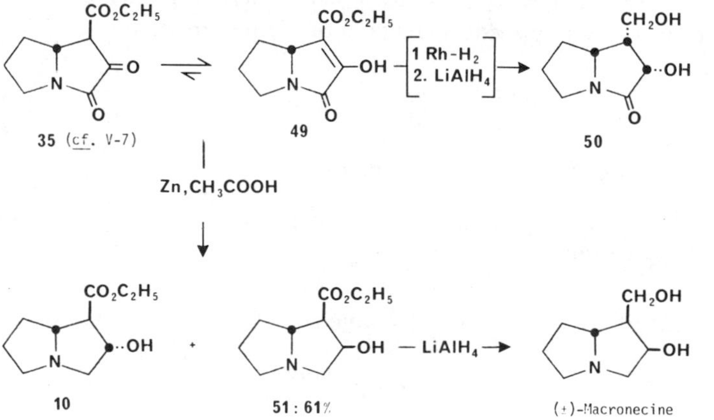

35 (cf. V-7)

49

1 Rh-H_2
2. $LiAlH_4$

50

Zn, CH_3COOH

10

51 : 61%

$LiAlH_4$

(±)-Macronecine

Chart 5-12.[189]

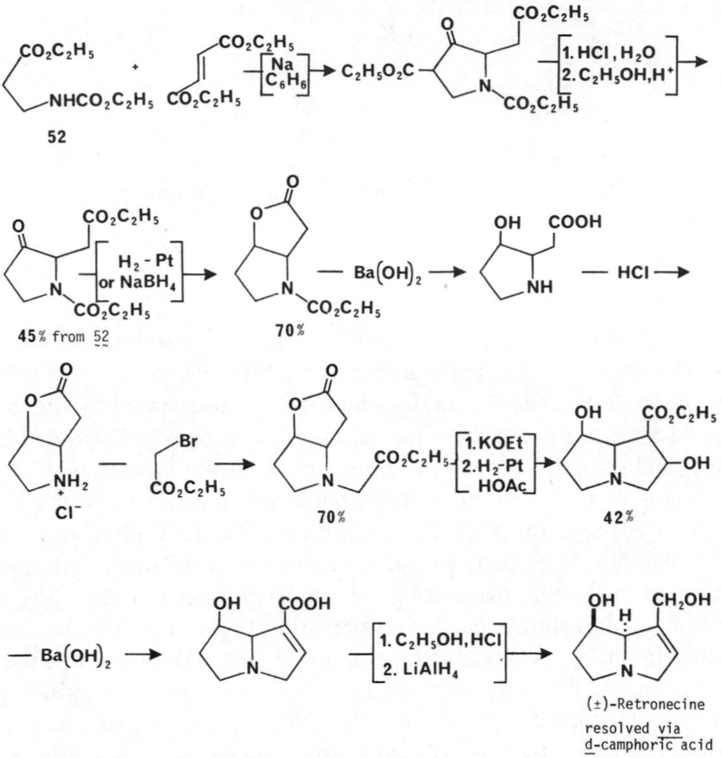

(±)-Retronecine resolved via d-camphoric acid

The key step in the recent elegant approach to pyrrolizidine alkaloids invovles the 1,3-dipolar cycloaddition of Δ'-pyrroline oxide to methyl γ-hydroxycrotonate. It had been established previously that methyl crotonate itself reacts with nitrones of this type of provide adducts of the desired orientation.

Chart 5-13.[190]

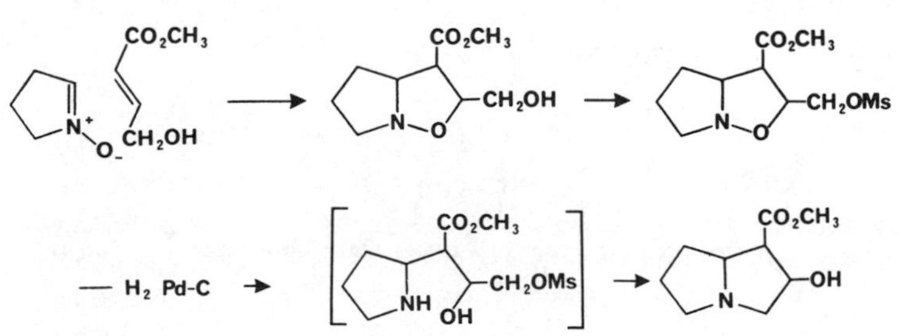

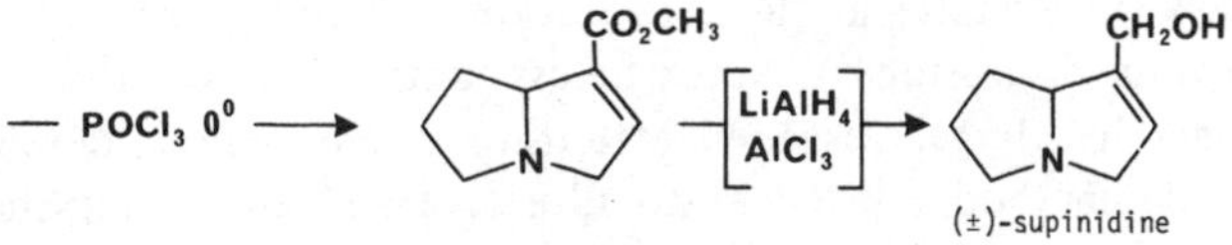

— $POCl_3$ 0^0 → [CO_2CH_3 pyrrolizine] —[$LiAlH_4$, $AlCl_3$]→ [CH_2OH pyrrolizine]

(±)-supinidine

6. GENERAL METHODS OF ALKALOID SYNTHESIS

The following compilation of nitrogenous plant products is at least representative of the structural diversity that abounds in the alkaloid realm. Many of these natural bases, or close relatives thereof, have made a significant impact on the science as a whole. But in spite of their apparently diverse nature, two structural units—the pyrrolidine and piperidine rings, usually fused to one or more other rings and not infrequently in alternate oxidation states—appear as common denominators in these and an impressively large number of other naturally occurring bases. This feature, perhaps more than any other, has made alkaloid

Pyridine Alkaloids

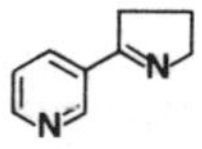

Myosmine

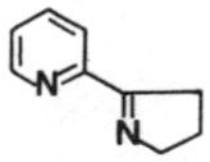

Apoferrorosamine

Mesembrine Alkaloids

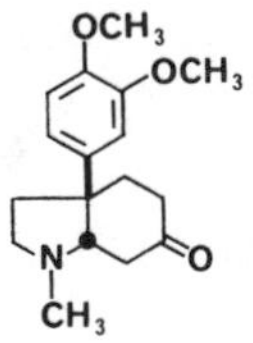

Mesembrine

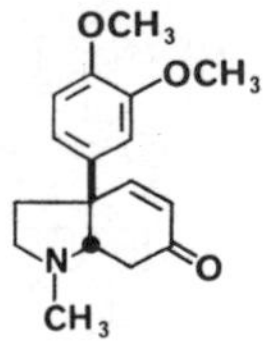

Mesembrenine

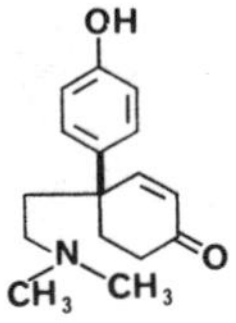

Joubertiamine

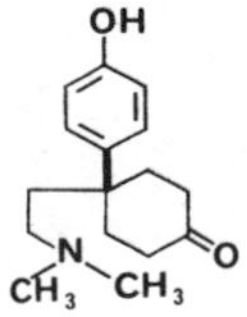

Dihydrojoubertiamine

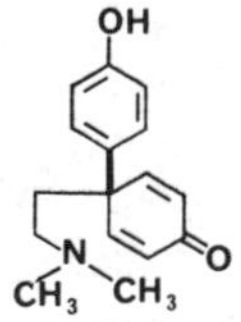

Dehydrojoubertiamine

synthesis largely an exercise in the elaboration of fused pyrrolidines or piperidines (see previous discussions!). Nevertheless, until quite recently this perhaps too obvious fact had largely escaped attention in any serious development of general synthetic methods. In order to illustrate and emphasize the types of considerations that have been important, and no doubt will continue to play an

Amaryllidaceae Alkaloids

Elwesine

Crinine

Narwedine

R = $-CH_2-$ Caranine
R = CH_3 Pluviine

Erythrina Alkaloids

Erysotrine

Erythratine

Aspidosperma Alkaloids

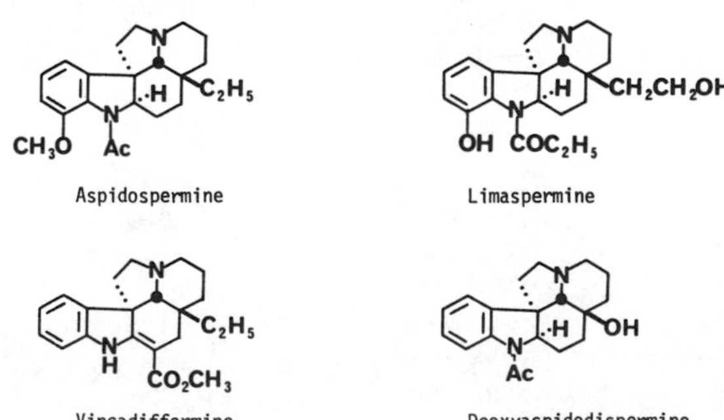

Aspidospermine

Limaspermine

Vincadifformine

Deoxyaspidodispermine

Lycopodium and Senecio Bases

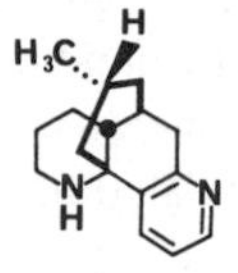

H$_3$C H N H N

Lycodine

CH$_2$OH N

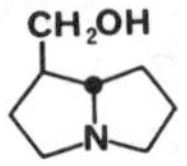

Trachelanthamidine

important role in the development of such methods, the author has intentionally emphasized in the following discussion only those studies that he has been intimately concerned with and had a hand in developing. In this way, it is hoped that the student will be provided with a first-hand narrative account of how such methods do, in fact, evolve.

A. The Pyridine Alkaloids

For reasons that become obvious as this theme is developed, we reasoned that any potentially general synthetic adventure into the vast array of structurally diverse alkaloid systems such as those already portrayed should, in principle, benefit from the nucleophilic properties associated with an endocyclic enamine of general structure (1), that is, a Δ^2-pyrroline. Thus electrophilic attack at the β-carbon of such an intermediate would proceed with simultaneous creation of an electrophilic center at the α-position and offers, as we now demonstrate, a simple and effective general approach to the synthesis of a variety of these natural bases. The initial synthetic task was, therefore, reduced to this central character, and a reliable, hopefully general synthetic procedure was sought. Among a number of methods of synthesis of Δ^2-pyrrolines **2** that were considered, the thermally induced rearrangement of cyclopropyl imines **1** by analogy with the rather extensively studied vinyl cyclopropane rearrangement (**5**) seemed admirably suited for the assignments we envisaged.

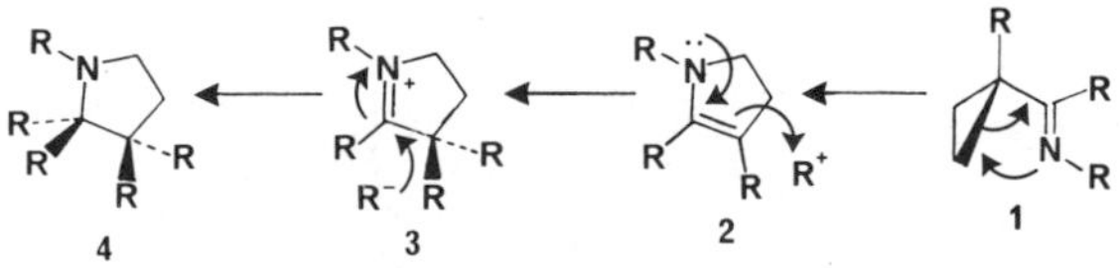

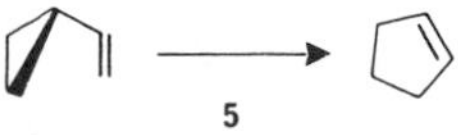

4 3 2 1

5

A literature search revealed that an example of the proposed rearrangement (**1** to **2**) had been reported as early as 1929 by J. B. Cloke.[191] The procedure employed by Cloke involved reacting cyclopropanecarbonitrile with phenylmagnesium bromide and subsequent decomposition of the magnesium salt with water. The desired ketimine **6a** was separated from neutral components by precipitation as the hydrochloride salt **7** and the free base regenerated by treating a hot chloroform solution of the salt with excess ammonia, thus precipitating ammonium chloride. Further purification of the thus partially purified imine **6b** was attempted by distillation. However, only the rearranged pyrroline **8** was obtained in high yield. The product of this rearrangement was later[192] reformulated as the Δ^1-isomer **9**, a revision in consonance with a growing body of information concerning these potentially tautomeric substances.[193]

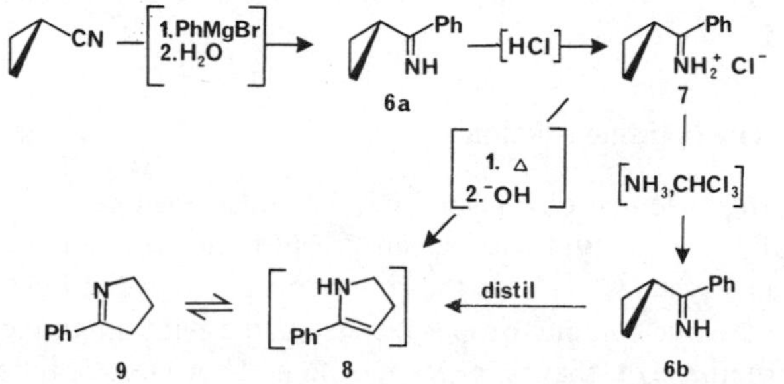

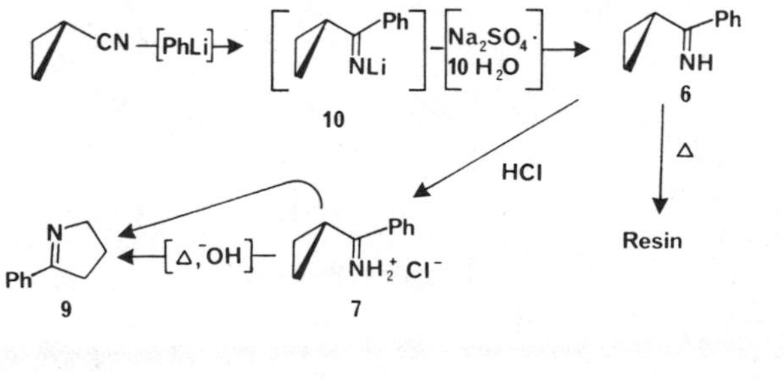

In order to gain experience with the propsed rearrangement, we decided to repeat the Cloke experiment.[194,195] The details of this process seemed unduly elaborate, and we decided upon a more efficient method. By employing phenyllithium and decomposing the intermediate lithium salt **10** with $Na_2SO_4 \cdot 10H_2O$, we were able to isolate, in high yield, a water-white liquid that distilled at 56 to 58°/0.3 mm. The pmr spectrum of this oil immediately informed us that in contrast to the Cloke experience, we had succeeded in isolating the presumed thermally labile cyclopropyl ketimine **6**. Since no information concerning the temperature at which Cloke had attempted the distillation of this substance was available, and in view of the rather low temperature and pressure we employed, a thermal study of the rearrangmenet step was initiated. Somewhat to our surprise, we discovered that heating sealed samples of the pure ketimine at various temperatures and time periods did not induce rearrangement. Indeed, even after 2 hr at 200°, only bearly detectable traces of pyrroline **9** were found. Higher temperatures only increased resinification. *We have been unable in this and a variety of other cases to induce smooth purely thermal rearrangements of cyclopropyl imines either in the neat or gas phases.*[194,195,*]

Our inability to effect a purely thermal rearrangement of **6** was not a serious cause for concern in our synthetic planning, since Cloke had also observed that substitution of the corresponding hydrochloride salt **7** in the thermal process is also effective, and we find this observation is completely reproducible (cf. **6** to **7** to **9**). The additional observation that smooth rearrangement could be induced by employing catalytic amounts of the hydrochloride salt [**6** + **7** (catalyst) → **9**] or even ammonium chloride (the presumably inocuous inorganic by-product of Cloke's purification procedure) greatly facilitated our subsequent work. Armed with this information, we began a systematic program to exploit this rearrangement as a devise for alkaloid synthesis. Naturally, we limited our initial efforts to simple systems such as the pyridine alkaloids myosmine and apoferrorosamine.

3-Lithio-pyridine was prepared by metal-halogen exchange according to the method of Gilman and Spatz[197] and treated with cyclopropanecarbonitrile at -78°. The intermediate lithio-salt **11** was decomposed with $Na_2SO_4 \cdot 10H_2O$ and distilled without rearrangement to yield pure ketimine **12**. Admixture of this base and a catalytic amount of its own hydrochloride and heating at 110° for 15 min, followed by distillation, gave nearly pure myosmine **13** in a 68% yield.[194]

CN + Li (pyridine) → [NLi] 11 —Na_2SO_4 $10H_2O$→ NH 12
BuLi -78° (from Br-pyridine)
HCl cat. 100°/15min.
Myosmine
13

The synthesis of ketimine **14** was accomplished in 35% of the yield in an analogous manner. Conversion of this base to its hydrochloride salt and admixture of a catalytic amount thereof to freshly prepared **14** followed by a 20-min

*The implication[196] that the thermally induced rearrangement of cyclopropyl imines is analogous to the well-documented vinyl cyclopropane rearrangement appears on the basis of our experience to be subjected to considerable doubt.

heat treatment at 110° gave, upon direct distillation from the reaction vessel, a 75% yield of apoferrorosamine **15**.[194]

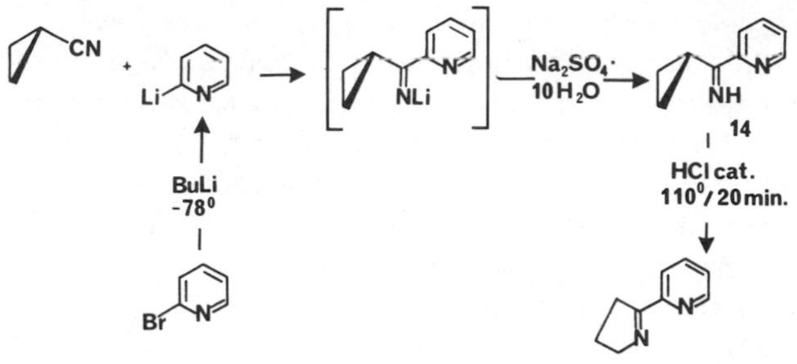

Apoferrorosamine
15

B. The Mesembrine Alkaloids

In order to expand our base of support for further synthetic designs, additional examples of the rearrangement were investigated. As soon becomes apparent, the substrates and products of this stage of the study were selected for potential deployment in the synthesis of the mesembrine and then ultimately *Amaryllidaceae* alkaloids.

Thus 1-phenylcyclopropane carbonitrile **16**, prepared from the sodium amide induced *bis*-alkylation of phenylacetonitrile with ethylene dibromide,[198] was exposed to an ethereal solution of methyllithium and the resultant lithio-salt decomposed via the sodium sulfate decahydrate procedure. In this manner, ketimine **17** was obtained in 73% of the yield. A purely thermal study of the rearrangement of this substrate was initiated but soon abandoned in view of previous results (see previous discussion) coupled with the observation that only starting material was recovered from a 450° gas phase thermolysis. By contrast, the thermal reorganization of this base to pyrroline **18** proceeded with ease by prior conversion to the hydrochloride salt. The formulation of this pyrroline as the Δ^1-isomer (in spite of the fact that the Δ^2-tautomer would be conjugated) is in agreement with the previously mentioned studies and is corroborated by its infrared, pmr, and ultraviolet spectra. Pyrroline **18** was conveniently converted to the extremely labile enamine **19** by alkylation with methyl iodide and subsequent basification. Alternatively, and perhaps more interestingly, the latter base could be secured by methyl iodide-induced alkylation of ketimine **17** to the quaternary immonium salt **20**, which is at the same time also a hydroiodide salt and as such would be expected to catalyze its own rearrangement. Indeed, when **17** was refluxed in methyl iodide as solvent followed by distillation of the excess methyl iodide and a brief 155° heat treatment, subsequent pasification provided the Δ^2-pyrroline **19** in 54% of the yield.

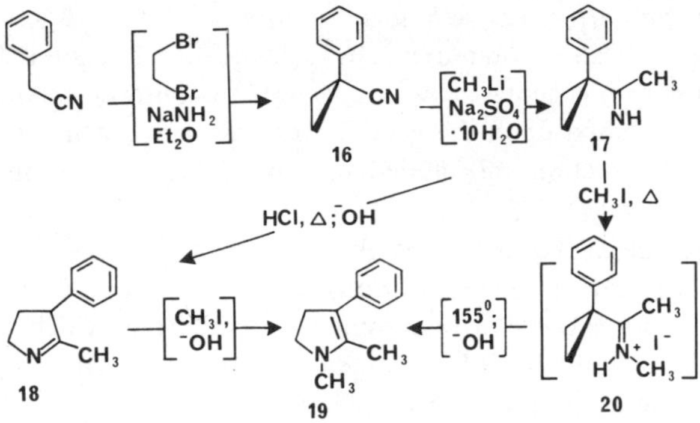

Our interest in securing endocyclic enamine **19** was based upon it potential deployment in the total synthesis of the mesembrine model 22. We had hoped that this base could be invited to participate in a combination of alkylation, intramolecular acylation with methyl acrylate. The product (21) of this transformation remains only a stereoselective reduction step away from the model 22. Analogy for this proposition can be found in the recent literature[193,199] not the least interesting of which is the fate of enamine 23 when exposed to methyl acrylate.[193] However, in spite of what would appear to be ample precedence, pyrroline **19** failed to undergo an analogous transformation. Indeed, only a complex mixture of products were obtained the major components of which were isolated and found devoid of the rather characteristic infrared absorbtion characteristic of β-acyl enamine chromophores.[200] Although it is difficult to assign any one factor as being responsible for the failure of this experiment, we are inclined to believe that the extreme lability of this particular enamine is an important factor. In spite of this momentary disappointment, we were now convinced that *the acid-catalyzed, thermally induced rearrangement of cyclo-*

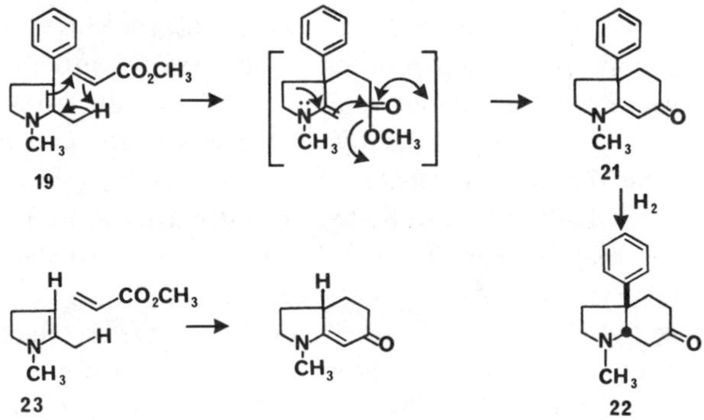

propyl imines is an effective method of synthesis of Δ^1- or Δ^2-pyrrolines. This assertion is further corroborated in the subsequent discussion. However, the problem of how to incorporate this knowledge into a useful general approach to alkaloid synthesis based upon the general principles set out in the introductory paragraphs of this section still eluded us. This delicate situation was soon to change dramatically.

As a consequence of the aforementioned studies, we had on hand a substantial supply of 1-phenylcyclopropane carbonitrile **16**. Selective hydride reduction of this material yielded the corresponding aldehyde from which the *N*-methyl imine **24** was readily obtained. Rearrangement of this substance proceeded without incident to the Δ^2-pyrroline **25** in high yield.

CN; [$LiAlH_4$, H_3O^+]; CHO; 16; CH_3NH_2, $MgSO_4$; H; N; CH_3; 24; [HBr, Δ, ^-OH]; N; CH_3; 25

With the obtention of endocyclic enamine **25**, we were now in a position to test our assumptions concerning the mode of attack of electrophilic reagents at the β-carbon of such an intermediate and to take advantage of the simultaneous creation of an electrophilic center at the α-position. Of particular interest was the manner in which methyl vinyl ketone might enter into reaction with this endocyclic enamine. We had envisaged the Michael addition of **25** with methyl vinyl ketone to proceed by any one, or possibly all, the stages outlined here. Should the initial alkylation step proceed at all, then we anticipate a possible menacing divergence between competing 1,2- and 1,4-addition processes. Since the 1,2-addition process and dihydropyran formation are known, except in special cases, to be thermally reversible,[201] we anticipated the proton transfer step required for the 1,4-addition to be crucial. Although our initial experiments were inconclusive, they were instrumental in the ultimate satisfactory solution of this problem. Pyrroline **25** was refluxed in a nitrogen-purged ethanolic solution with methyl vinyl ketone until it could no longer be detected by thin layer chromatography (tlc). The reaction mixture was then concentrated, and further purification attempted by distillation. We were at first surprised to discover that a major portion of the distillate was the starting enamine **25** since this had been judged by tlc to be absent. This result suggested that under the conditions employed the reaction had proceeded largely via either 1,2-addition or dihydropyran formation. The attempted distillation at a relatively higher temperature had apparently served only to reverse this process. Indeed, methyl vinyl ketone was secured from the vacuum trap. Although low yields of the desired product **26** could be isolated from this reaction by preparative layer chromatography, this interpretation suggested that employment of higher reaction temperatures might accelerate the decomposition of the (postulated) thermally labile addition products, thus freeing the starting materials for the desired 1,4-combination. No effort was made to further corroborate these assumptions when it was discovered that employment of hot ethylene glycol as solvent resulted in an effective yield of **26**. Although the annelation of exocyclic enamines with methyl vinyl ketone finds a prominent role in organic synthesis, its employment here with an endocyclic enamine is unique.*

The stereochemical course of the annelation had been anticipated. Thus inspection of models reveals that maximum overlap of the termini of the π-orbital system is most readily achieved via perpendicular attack as illustrated in expression **27a** (as opposed to **27b**). The maintenance of similar geometry in the sterically more demanding intermediate **28** has been recently reported,[204b] and the application of this principle to intramolecular conjugate addition has been emphasized.[204b] The experience gained in these experiments was next applied to the total synthesis of racemic mesembrine.[203b]

In contrast to phenylacetonitrile itself, the corresponding dimethoxy isomer **29** (R = OCH_3) could not be induced to react with ethylene dibromide under the conditions employed for the former substrate ($NaNH_2$, Et_2O). Although electronically understandable, this result was, nevertheless, annoying. While searching for an alternative solution, a timely communication appeared[205]

*Almost simultaneous with our initial report[203a] of this important reaction two communicaitons from other laboratories described analogous results.[202a,b]

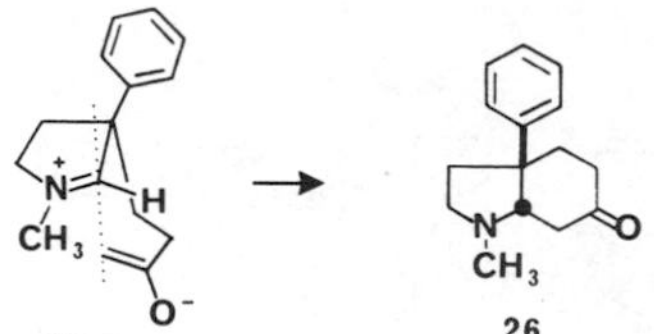

27a

26

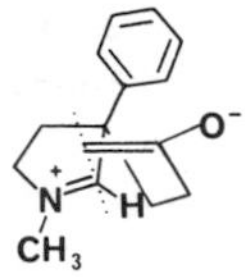

27b

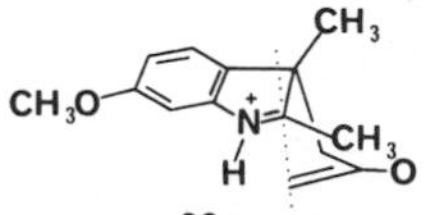

28

wherein the formation of the dilithiated species **30** (R = H) was demonstrated (the exact structure of this species remains unknown). The expected greater covalent character revealed by this experiment immediately attracted our attention as a logical solution to the problem at hand. Indeed, exposure of **29** (R = OCH_3) to two equivalents of *n*-butyllithium and rapid quenching in D_2O gave greater than 97% α-dideuteriophenylacetonitrile. Alkylation of the dilithiosalt **30** with ethylene dibromide yielded the desired cyclopropane **31**. With this intermediate now in hand and with the experience gained in model studies, we turned our attention to the ultimate goal of the present investigation, the *Aizoaceae* alkaloid mesembrine.

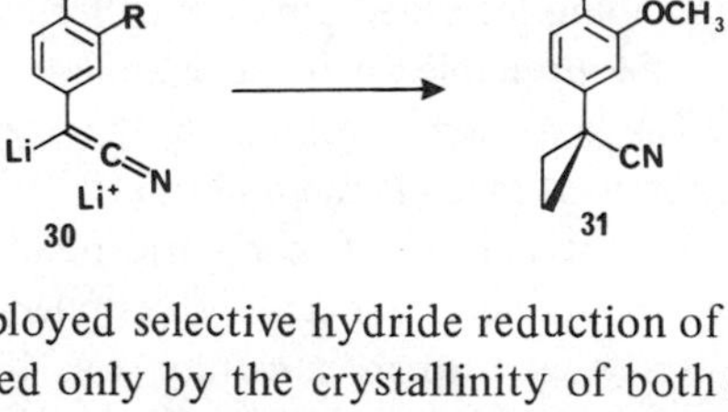

29

2BuLi

$NaNH_2$ Et_2O

30

31

The previously employed selective hydride reduction of **31** to the aldehyde **32** was, amusingly, marred only by the crystallinity of both substances, thus creating a separation problem. This was easily overcome by treating the crude reduction mixture of **31** and **32** with saturated sodium bisulfite and extraction of the neutral nitrile. Pure crystalline aldehyde was then regenerated directly from the bisulfite solution (later in our work it was discovered that diisobutylaluminum hydride can be employed in this selective reduction and generally gives excellent yields; this fact has been taken advantage of in our subsequent work). A 91% yield of analytically pure aldimine **33** was obtained by treating a benzene solution of aldehyde with a tenfold excess of methylamine in the presence of suspended magnesium sulfate for 20 hr at room temperature. As noted previously, the acid-catalyzed, thermally induced rearrangement of cyclopropyl imines is best achieved by employing only catalytic amounts of acid. In the present case, a 76% yield of analytically pure pyrroline **34** was obtained from a 20-min, 148° run using the hydrobromide salt as catalyst. Racemic mesembrine was secured via the methyl vinyl ketone annelation procedure already described as a 56% yield. Thus the combination of acid-catalyzed, thermally induced rearrangement of cyclopropyl imines and annelation of the resultant endocyclic enamine with methyl vinyl ketone provided a simple five-step total synthesis of this alkaloid.

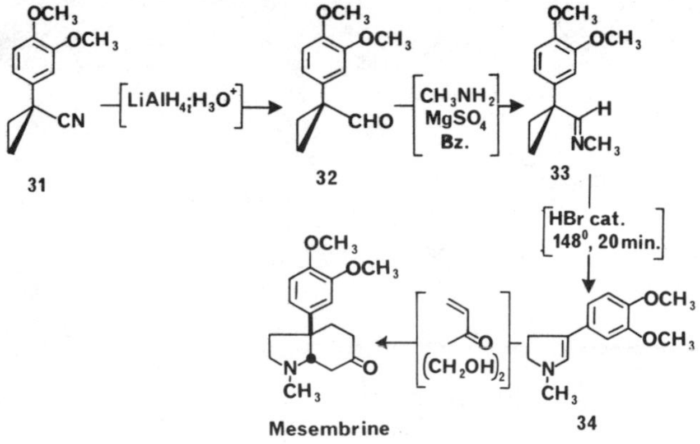

Pyrroline **34** has also been annelated with methyl β-chlorovinyl ketone as the final step in the synthesis of mesembrenine.

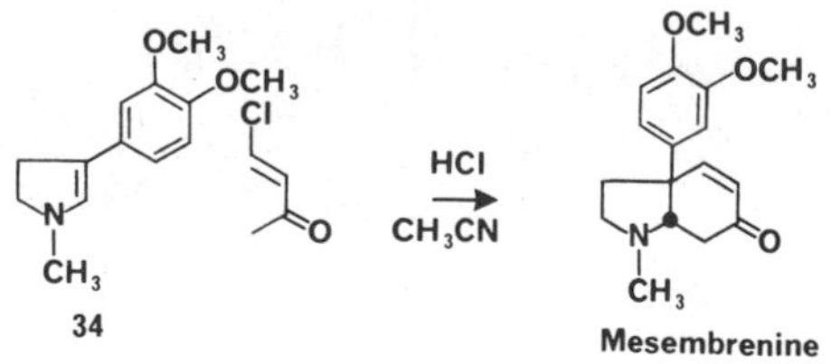

The discovery[206] of the *seco*-mesembrine alkaloids joubertiamine, dihydrojoubertiamine, and dehydrojoubertiamine provided us with an excellent opportunity to further corroborate and expand upon the synthetic principles and procedures just outlined. Thus cyclopropanation of *p*-methoxyphenyl acetonitrile was proceeded smoothly and in a 75% yield by employing $LiNH_2$ base in glyme. Selective reduction of the nitrile (**35**) with DIBAL-H in benzene provided the aldehyde (86%) whose transformation into the corresponding aldimine simply required stirring a benzene solution with excess CH_3NH_2 for 2 days at room temperature in the presence of suspended $MgSO_4$ (90% yield). The crucial rearrangement of this aldimine proceeded in virtually quantitative yield to the pyrroline **36** by heating to 130° in the presence of NH_4Cl as the acidic catalyst. Acid-catalyzed MVK annelation completed the carbon skeleton (83%). Conversion to joubertiamine methyl ether was accomplished in a 62% yield by refluxing in neat CH_3I and direct β-elimination of the intermediate methiodide salt upon aqueous base workup. Finally, demethylation with hot hydrobromic acid completed the synthesis of joubertiamine whose reduction also provided dihydrojoubertiamine.

OCH_3 — $LiNH_2$, $(CH_2Br)_2$ → Ar, CN **35** — i-Bu_2AlH →

Ar, CHO — CH_3NH_2, $MgSO_4$ → Ar, N, CH_3 — NH_4Cl 130° →

Ar, N, CH_3 **36** — HCl; MVK → Ar, N, CH_3, O — CH_3I →

Ar, I^-, N^+, CH_3, CH_3, O — ^-OH → Ar, H_3C, N, CH_3, O — H_2 Pd → Ar, N, CH_3, CH_3, O

Joubertiamine

Dihydro-joubertiamine

C. The *Amaryllidaceae* Alkaloids

The *Amaryllidaceae* family has been known for some time now to be a rich source of complex and intriguing alkaloids. The structural diversity that has been revealed among these bases is truly remarkable and necessitates their subdivision into several skeletally homogeneous groups. One such major group includes those alkaloids that incorporate the 5,10b-ethanophenanthridine nucleus and is most often referred to as the crinine group after the parent natural product. A recent review of these alkaloids lists 35 closely related members of this family, and from a careful inspection of their structures, we conceived of a number of potentially general synthetic approaches that, perhaps with only minor modification, could be employed in the synthesis of a number of these alkaloids. We selected for our initial efforts elwesine (dihydrocrinine) a minor alkaloid of *Galanthus elwesii* Hook, f. The somewhat deceiving similar structural features of mesembrine and this base had, from the very beginning of our investigation,[194] not escaped our attention.

Thus the now familiar approach to the synthesis of endocyclic enamine **39** was attempted and not found lacking.[207] The additional observations that lithium amide in dimethoxyethane is even more efficient than butyllithium in the *bis*-alkylation step and that diisobutylaluminum hydride (DIBAL) is a more selective and convenient reducing agent in the **37** to **38** transformation further strengthens the utility of the method. MVK annelation of the thus prepared pyrroline proceeded as expected to **40**. A number of attempts to demethylate this intermediate or to induce incorporation of this methyl function into the last ring met with failure. This was not a serious cause for concern since substitution of the adamant *N*-methyl function by the more labile benzyl group required only a slight variation of the scheme and no compromise whatever in efficiency.

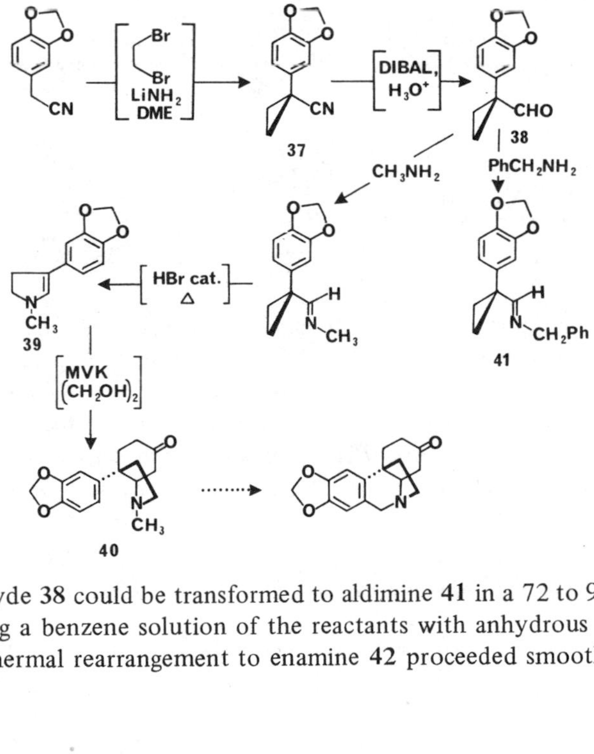

Br
Br
$LiNH_2$
DME
CN
37
DIBAL,
H_3O^+
CHO
38
CH_3NH_2
$PhCH_2NH_2$
HBr cat.
Δ
N
H
CH_3
CH_2Ph
41
39
MVK
$(CH_2OH)_2$
40

Thus aldehyde **38** could be transformed to aldimine **41** in a 72 to 90% yield by simply stirring a benzene solution of the reactants with anhydrous $CaCl_2$ for 2 to 3 days. Thermal rearrangement to enamine **42** proceeded smoothly in 72 to

80% yield by employing NH_4Cl as the acidic catalyst. We were somewhat surprised to observe that the MVK annelation of this intermediate gave only complex unstable mixtures containing little if any of the desired product since the same procedure had been so successfully employed with other closely related substrates. However, 55 to 65% yields of the pure *cis*-octahydroindole **43** could be secured by employing acidic catalysis. Sodium borohydride reduction of **43** provided a 3:1 mixture of two epimeric alcohols, which were easily separated by preparative layer chromatography. Debenzylation of the major isomer **44** (*vida infra*) yielded **46** whose HCHO induced Pictet-Spengler cyclization completed the synthesis of 3-*epi*-elwesine **47**. In order to increase the efficiency of this procedure, it was necessary to reverse the **44** to **45** ratio. This was accomplished in quantitative yield by catalytic hydrogenation of **43**. Debenzylation and Pictet-Spengler cyclization then provided racemic elwesine (dihydrocrinine).

Perhaps it is appropriate now to assert that *the acid-catalyzed, thermally induced rearrangement of cyclopropyl imines appears to be a general method of approach to a variety of useful Δ^2-pyrrolines and that subsequent methyl vinyl ketone (or close equivalent thereof) annelation of these and (as we see later) other endocyclic enamines offer convenient and efficient stereoselective approaches to a wide variety of alkaloid system.* We now try to justify this assertion further.

D. The *Erythrina* Alkaloids

Attention will now be focused on a distinctly different class of nitrogeneous plant products: The pharmacollogically interesting *Erythrina* alkaloids[208] of which erysotrine and erythratrine are perhaps the most representative. Our initial efforts in this field have been most encouraging[209] and were a consequence of a desire to further expand the scope of the annelation sequence under development as a general approach to alkaloid synthesis.

The endocyclic enamine **48** required to test the crucial annelation step was prepared in two steps by modification of existing procedures.[210] We were pleased to observe that admixture of this enamine and methyl vinyl ketone in refluxing ethanol did, in fact, yield the desired ketone **49**, which had previously been reported as a degradation product of erysotrine. The stereochemical course of this important result had been anticipated from our previous work.

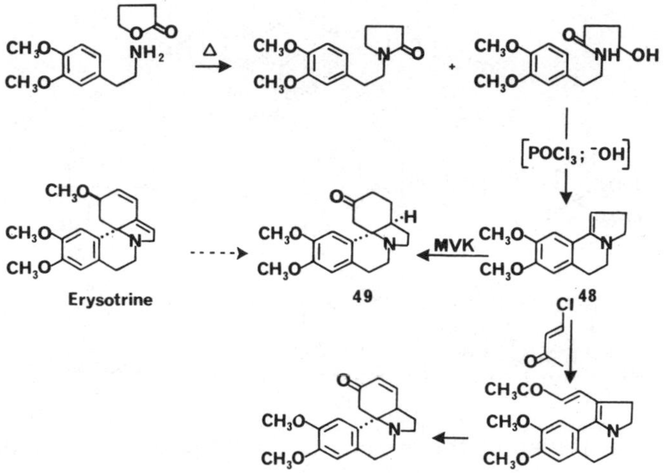

E. The *Aspidosperma* and Related Alkaloids

Thus far in our discussion the crucial annelation step has been applied or suggested only in the case of substituted Δ^2-pyrroline, which in turn have usually been provided by acid-catalyzed thermal rearrangements of cyclopropyl imines. However, in principle at least, *the annelation of any properly selected endocyclic enamine with methyl vinyl ketone or related derivatives thereof provides even broader possible applications.* For example, the biologically important *Aspidosperma* alkaloids, of which aspidospermine is the parent base, provides us with an interesting test of this assertion. Although our studies in this area are only now commencing, enough information has been gathered to fairly precisely outline our method of attacking this problem.

Malonic ester has been converted by a series of trivial steps to lactam **50**.[211] Benzylation of this intermediate and selective reduction by means of diisobutylaluminum hydride yielded the desired endocyclic enamine **51**. Methyl vinyl ketone annelation of this enamine proceeded very smoothly in refluxing ethanol providing yet another example of the utility of this process. Subsequent debenzylation yields amino ketone **52**. In principle, two possible stereoisomers could have been produced in this transformation; however, only the *cis* isomer was obtained, a fact that is in consonance with the stereochemical considerations already discussed (cf. **27**). This result was particularly gratifying since it offered a new, potentially general method for the synthesis of angularly substituted hydroquinolones, which in turn had previously been established as valuable intermediates in the synthesis of hydrolulolidines **53** and ultimately aspidospermine itself.[212]

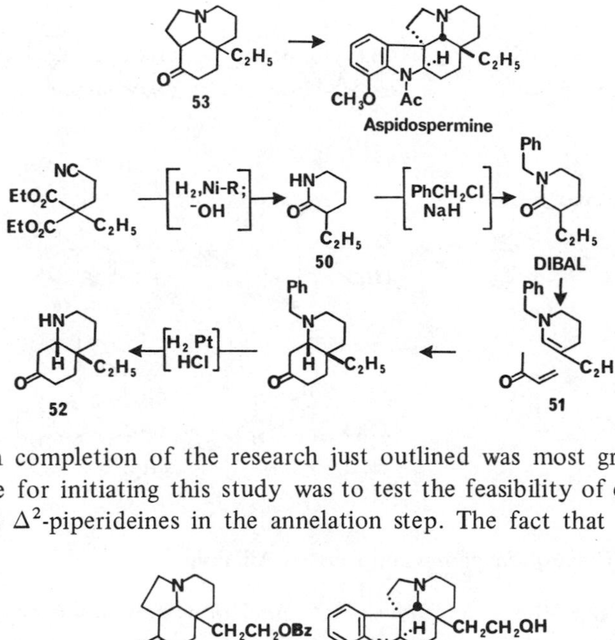

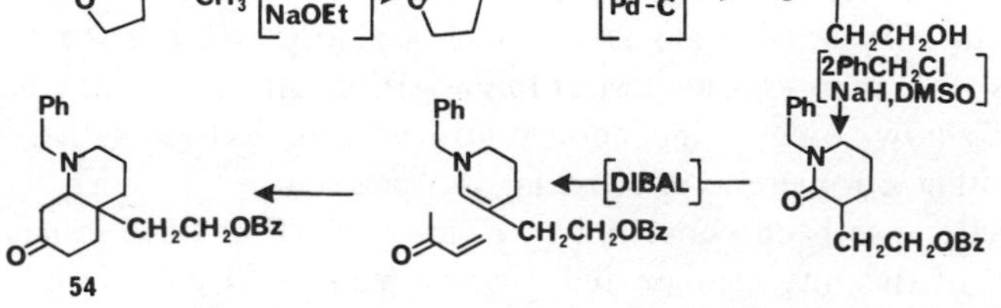

Although completion of the research just outlined was most gratifying, our real motive for initiating this study was to test the feasibility of employing β-substituted Δ^2-piperideines in the annelation step. The fact that the ultimate product, **52** in this case, was a known compound justifies its selection, and its previous conversion to aspidospermine served as a delightful, but not fortuitous, bonus. A further example of the implied generality of this method of approach to synthesis of angularly substituted hydroquinolones was achieved[211] as outlined here. The conversion of **54** to limaspermine via **55** can now be anticipated.

Having at least partially established the utility of this type of enamine in the annelation process, now even more novel and efficient approaches to the *Aspidosperma* alkaloids could be envisaged.[213] Thus cyclopropane carboxaldehyde **56** and ketal amine **57** (prepared as outlined here) have been condensed to cyclopropyl imine **58**. The acid-catalyzed thermal rearrangement of this substance was investigated and not found lacking. In this case, ammonium chloride proved to be the catalyst of choice.

$CO_2C_2H_5$ CN — [Br Br; NaOEt, EtOH] → $CO_2C_2H_5$ CN — [HO HO; H_2SO_4] → $CO_2C_2H_5$ 64%

[$NaBH_4$; H_3O^+]

59 : 82% ← [NH_4Cl, Δ] — 58 : 86% ← [$-H_2O$] — 56 : 61% CHO + H_2N 57 : 87%

[H_2, Ni-R, NH_3]

O Et NC — [Na] → O Et NC — [OH OH; TsOH] → Et NC

Upon exposure to anhydrous HCl gas in ether[214] **59** cyclized to the bicyclic ketal of **60** the ketal group being readily hydrolyzed off during workup. Treatment of **60** with methoxide and subsequent acidification of the basic methanolic solution with HCl gas yield the two crystalline tricyclic enolethers **61** and **62** in a 3:1 ratio. Conversion of **61** to **63** was accomplished by the steps indicated.

The total synthesis of Sceletium A-4 has now appeared and includes an improved synthesis of the key 2-pyrroline intermediate.

The cyclopropylimine rearrangement has been extended to the synthesis of novel *N*-substituted 3-phenylthio-2-pyrrolines. The latter substances have been employed in the total synthesis of the pyrrolizidine alkaloid–isoretronecanol–and the indolizidine alkaloids–δ-coneceine, ipalbidine, and septicine.

59

[HCl, Et_2O; K_2CO_3, H_2O]

60 : 86 %

[$NaOCH_3$]

Na^+

[CH_3OH, HCl]

61

+

62

[LiAlH$_4$]

96%

[H_3O^+]

70%

[H_2 Pd-C]

63

Senecio A-4

NH_4

isoretronecanol

δ-coniceine

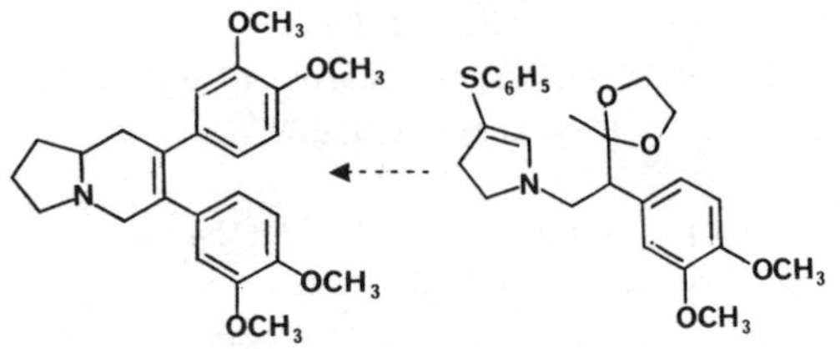

septicine

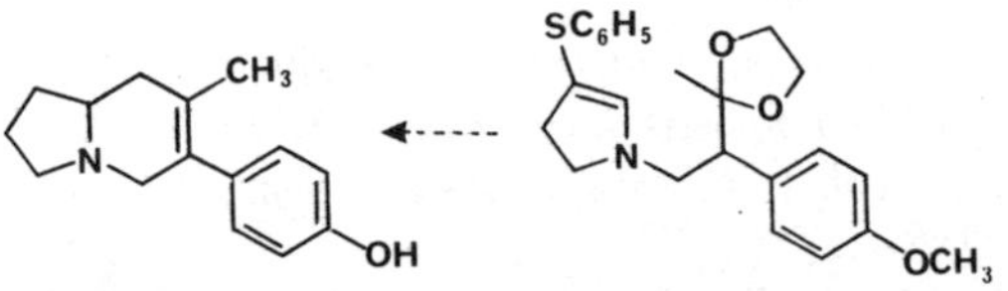

ipalbidine

ACKNOWLEDGMENTS

A portion of this manuscript was written at Iowa State University during a summer as Visiting Professor. The author expresses his gratitude to Mrs. Millie Allen Clarke who assisted in the collection of many of the references cited in this review and to Mrs. Jean Long for her skill and care in typing the manuscript. Finally, the able assistance of those collaborators who devoted their time and talent in executing the experiments outlined in Section 6 of this review is gratefully acknowledged.

REFERENCES

1. E. J. Corey and T. Wipke, *Science,* **166**, 179 (1969).
2. A. Popelak and G. Lettenbauer, *The Alkaloids,* **9**, 467 (1967).
3. P. W. Jeffs, R. L. Hawks, and D. S. Farrier, *J. Amer. Chem. Soc.*, **91**, 3831 (1969).
4. P. W. Jeffs, G. Ahmann, H. F. Campbell, D. S. Farrier, G. Ganguli, and R. L. Hawks, *J. Org. Chem.,* **35**, 3512 (1970).
5. R. R. Arndt and P. E. J. Kruger, *Tetrahedron Lett.*, 3237 (1970).
6. A. Popelak, G. Lettenbauer, E. Haack, and H. Spingler, *Naturwissenschaften,* **47**, 231 (1960).
7. M. Shamma and H. R. Rodriguez, *Tetrahedron,* **24**, 6583 (1968); *Tetrahedron Lett.*, 4847 (1965).
8. W. C. Wildman, *J. Amer. Chem. Soc.*, **78**, 4180 (1956).
9. H. W. Whitlock, Jr., and G. L. Smith, *J. Amer. Chem. Soc.,* **89**, 3600 (1967); *Tetrahedron Lett.,* 1389 (1965); 2711 (1966).

10. L. Raiford and D. Fox, *J. Org. Chem.*, **9**, 170 (1944).
11. W. C. Wildman and R. B. Wildman, *J. Org. Chem.*, **17**, 581 (1952).
12. R. Pappo, D. S. Allen, Jr., R. V. Lemieux, and W. S. Johnson, *J. Org. Chem.*, **28**, 250 (1963).
13. A. Marwuet and J. Jacques, *Tetrahedron Lett.*, **24** (1959).
14. S. Marmor, *J. Org. Chem.*, **28**, 250 (1963).
15. P. L. Julian, W. Cole, E. W. Meyer, and B. M. Regan, *J. Amer. Chem. Soc.*, **77**, 4601 (1955).
16. R. V. Stevens and M. P. Wentland, *J. Amer. Chem. Soc.*, **90**, 5580 (1968); *Tetrahedron Lett.*, 2613 (1968).
17. C. Dupin and R. Fraisse-Julien, *Bull. Soc. Chim. France*, 1993 (1964).
18. D. I. Schuster and J. D. Roberts, *J. Org. Chem.*, **27**, 51 (1962).
19. S. L. Keely, Jr., and F. C. Tahk, *J. Amer. Chem. Soc.*, **90**, 5584 (1968); *Chem. Comm.*, 441 (1968).
20. T. J. Curphey and H. L. Kim, *Tetrahedron Lett.*, 1441 (1968).
21. E. A. Prill and S. M. McElvain, *J. Amer. Chem. Soc.*, **55**, 1233 (1933).
22. J. M. Bruce, *J. Chem. Soc.*, 2366 (1959).
23. F. H. Howell and D. H. Taylor, *J. Chem. Soc.*, 4252 (1956).
24. T. Oh-ishi and H. Kugita, *Chem. Pharm. Bull. (Japan)*, **18**, 291, 299 (1970); *Tetrahedron Lett.*, 5445 (1968).
25. C. F. Koelsch and H. M. Walker, *J. Amer. Chem. Soc.*, **72**, 346 (1950).
26. C. F. Koelsch and D. L. Ostercamp, *J. Org. Chem.*, **26**, 1104 (1961).
27. Y. Ban, Y. Sato, I. Inoue, M. Nagai, T. Oishi, M. Terashima, O. Yonemitsu, and K. Kanaoka, *Tetrahedron Lett.*, 2261 (1965).
28. H. Taguchi, T. Oh-ishi, and H. Kugita, *Chem. Pharm. Bull. (Japan)*, **18**, 1008 (1970); *Tetrahedron Lett.*, 5763 (1968).
29. R. V. Stevens and J. T. Lai, *J. Org. Chem.*, **37**, 2138 (1972).
30. W. C. Wildman, *The Alkaloids*, **6**, 289 (1960); **11**, 307 (1968).
31. W. C. Wildman and L. H. Mason, *J. Amer. Chem. Soc.*, **76**, 6194 (1954).
32. H. Muxfeldt, R. S. Schneider, and J. B. Mooberry, *J. Amer. Chem. Soc.*, **88**, 3670 (1966).
33. T. Kametani and H. Ida, *J. Pharm. Soc. Japan*, **73**, 681 (1953).
34. H. Muxfeldt, *Angew. Chem.*, **74**, 825 (1962).
35. H. Meerwein, W. Florian, N. Schön, and G. Stopp, *Annalen*, **641**, 1 (1961).
36. A. E. Wick, D. Felix, K. Steen, and A. Eschenmose, *Helv. Chim. Acta*, **47**, 2425 (1964).
37. W. C. Wildman, *J. Amer. Chem. Sc.*, **80**, 2567 (1958).
38. H. Born, R. Pappo, and J. Szmuszkovicz, *J. Chem. Soc.*, 1779 (1953).
39 R. O. Clinton and S. C. Laskowski, *J. Amer. Chem. Soc.*, **70**, 135 (1948).
40. H. W. Heine, *J. Amer. Chem. Soc.*, **85**, 2743 (1963).
41. H. M. Fales and W. C. Wildman, *J. Org. Chem.*, **26**, 881 (1961).
42. H. Goering, *Rec. Chem. Progr.*, **21**, 109 (1960).
43. S. Uyeo, H. Irie, A. Yoshitake, and A. Ito, *Chem. Pharm. Bull. (Tokyo)*, **13**, 427 (1965).

44. H. Irie, Y. Tsuda, and S. Uyeo, *J. Chem. Soc.*, 1446 (1959).
45. W. S. Wadsworth, Jr., and W. D. Emmons, *J. Amer. Chem. Soc.*, **83**, 1773 (1961).
46. H. Irie, S. Uyeo, and A. Yoshitake, *J. Chem. Soc. (C)*, 1802 (1968); *Chem. Comm.*, 635 (1966).
47. J. B. Hendrickson, C. Foote, and N. Yoshimura, *Chem. Comm.*, 165 (1965).
48. J. B. Hendrickson, T. L. Bogard, and M. E. Fisch, *J. Amer. Chem. Soc.*, **92**, 5538 (1970).
49. R. V. Stevens, L. E. DuPree, Jr., and P. L. Lowenstein, *J. Org. Chem.*, **37**, in press; *Chem. Comm.*, 1585 (1970).
50. I. Ninomaya, T. Naito, and T. Kiguchi, *Chem. Comm.*, 1669 (1970).
51. O. L. Chapman and P. G. Cleveland, *Chem. Comm.*, 1064 (1967).
52. O. L. Chapman and G. L. Eian, *J. Amer. Chem. Soc.*, **90**, 5329 (1968).
53. T. Kametani, T. Hohno, S. Shibuya, and K. Fukumoto, *Chem. Comm.*, 774 (1971).
54. S. Minami and S. Uyeo, *Chem. Pharm. Bull. (Tokyo)*, **12**, 1012 (1964).
55. M. Gates and G. Tschudi, *J. Amer. Chem. Soc.*, 78, 1380 (1956).
56. M. Tomita and S. Minami, *Yakugako Zasshi*, **33**, 1022 (1963).
57. T. Takahashi, M. Yori, and A. Kanbara, *Chem. Pharm. Bull (Tokyo)*, **7**, 917 (1959).
58. J. Koizumi, S. Kobayashi, and S. Uyeo, *Chem. Pharm. Bull. (Tokyo)*, **12**, 696 (1964).
59. N. Hakama, H. Irie, T. Mizutani, T. Shingu, M. Takada, S. Uyeo, and A. Yoshitake, *J. Chem. Soc. (C)*, 2947 (1968).
60. Y. Misaka, T. Mizutani, M. Sekido, and S. Uyeo, *J. Chem. Soc. (C)*, 2954 (1968); *Chem. Comm.*, 1258 (1967).
61. R. K. Hill, J. A. Joule, and L. J. Loeffler, *J. Amer. Chem. Soc.*, **84**, 4951 (1962).
62. R. B. Kelly, W. I. Taylor, and K. Wiesner, *J. Chem. Soc.*, 2094 (1953).
63. G. Humber, H. Kondo, K. Kotera, S. Takaga, K. Takeda, W. I. Taylor, B. R. Thomas, Y. Tsuda, K. Tsukamoto, S. Uyeo, H. Yajima, and N. Yanihara, *J. Chem. Soc.*, 4622 (1954).
64. S. Takagi and S. Uyeo, *J. Chem. Soc.*, (1961).
65. H. M. Fales, E. W. Warnhoff, and W. W. Wildman, *J. Amer. Chem. Soc.*, **77**, 5885 (1955).
66. N. Ueda, T. Tokuyama, and T. Sakan, *Bull. Chem. Soc. (Japan)*, **39**, 2012 (1966).
67. R. G. Naik and T. S. Wheeler, *J. Chem. Soc.*, 1780 (1938).
68. C. B. Clark and A. R. Pinder, *J. Chem. Soc.*, 1967 (1958).
69. J. F. Bunnett and B. F. Hautfiord, *J. Amer. Chem. Soc.*, **83**, 1691 (1961).
70. J. F. Bunnett and J. A. Skorcz, *J. Org. Chem.*, **27**, 3836 (1962).
71. J. F. Bunnett, and T. Kato, R. R. Flyn, and J. A. Skorcz, *J. Org. Chem.*, **28**, 1 (1963).
72. K. Takeda, K. Kotera, S. Mizujami, and M. Kobayashi, *Chem. Pharm. Bull. (Japan)*, 8, 483 (1960).
73. K. Kotera, *Tetrahedron*, **12**, 240 (1961).
74. K. Kotera, *Tetrahedron*, **12**, 248 (1961).
75. H. Irie, Y. Nishitani, M. Sugita, and S. Uyeo, *Chem. Comm.*, 1313 (1970).
76. R. T. Arnold and E. C. Coyner, *J. Amer. Chem. Soc.*, **66**, 1542 (1944).
77. R. Quelet, R. Duran, and G. Lukacs, *Compt. Rend.*, **258**, 1826 (1964).
78. R. Dran and T. Prange, *Compt. Rend.*, **262**, 02 (1966).

79. K. Kotera, *Tetrahedron,* **12**, 240 (1961).

80. N. Uneda, T. Tokuyama, and T. Dakan, *Bull. Chem. Soc. Japan,* **39**, 2012 (1966).

81. R. K. Hill and R. M. Carlson, *J. Org. Chem.,* **30**, 1571 (1965); *Tetrahedron Lett.,* 1157 (1964).

82. D. H. R. Barton and G. W. Kirby, *J. Chem. Soc.,* 806 (1962); *Proc. Chem. Soc.,* 392 (1960).

83. Ladenburg, Folkers, and Major, *J. Amer. Chem. Soc.,* **58**, 1292 (1936).

84. Czapliski, Kostanecki, and Lampe, *Berichte,* **42**, 831 (1909).

85. Späth, Orechoff, and Kuffper, *Berichte,* **67**, 1214 (1934).

86. Surrey, Mooradian, Cutler, Suter, and Buck, *J. Amer. Chem. Soc.,* **71**, 2421 (1949).

87. A. R. Battersby and W. I. Taylor, Eds., *Oxidative Coupling of Phenols,* Dekker, New York, 1967.

88. R. A. Abramovitch and S. Takahashi, *Chem. Ind.,* 1039 (1963).

89. B. Franck and H. J. Lubs, *Justus Liebigs Ann. Chem.,* **720**, 131 (1968).

90. B. Franck and H. J. Lubs, *Angew. Chem. Int. Ed.,* 7, 223 (1968).

91. M. Schiebel, Dissertation, Universität Göttingen, 1965.

92. T. Kametani, K. Yamaki, H. Tagi, and K. Fukumoto, *J. Chem. Soc. (C),* 2602 (1969).

93. D. D. Vaghani and J. R. Merchant, *J. Chem. Soc.,* 1066 (1961).

94. A. H. Jackson and J. A. Martin, *J. Chem. Soc. (C),* 2061 (1966).

95. M. A. Schwartz and R. A. Holton, *J. Amer. Chem. Soc.,* **92**, 1090 (1970); **91**, 2800 (1969).

96. B. Franck and H. J. Lubs, *Angew. Chem. Int. Ed.,* 7, 223 (1968).

97. A. Goosen, E. V. O. John, F. L. Warren, and K. C. Yates, *J. Chem. Soc.,* 4028 (1961).

98. W. Rolionow, *Bull. Soc. Chim. France,* **39**, 305 (1926).

99. M. W. Whitlock and G. L. Smith, *J. Amer. Chem. Soc.,* **89**, 8600 (1967).

100. D. B. McLean, *The Alkaloids,* **10**, 305 (1968); R. H. F. Manske, *The Alkaloids,* **5**, 295 (1955); K. Wiesner, *Fortsch. Chem. Org. Naturstoffe,* **20**, 271 (1962).

101. K. Wiesner, W. A. Ayer, L. R. Fowler, and Z. Valenta, *Chem. Ind.,* 564 (1957).

102. W. A. Ayer, D. A. Law, and K. Piers, *Tetrahedron Lett.,* 2959 (1964).

103. F. A. L. Anet, *Tetrahedron Lett.,* 13 (1960).

104. W. A. Ayer, D. A. Law, and J. Berezowski, *Can. J. Chem.,* **41**, 649 (1963).

105. W. A. Ayer, J. A. Berezowsky, and G. G. Iverach, *Tetrahedron,* **18**, 567 (1962).

106. W. A. Ayer and G. G. Iverach, *Can. J. Chem.,* **38**, 1823 (1960)

107. W. A. Harrison, M. Curcumelli-Rodostamo, D. F. Carson, L. R. C. Barclay, and D. B. MacLean, *Can. J. Chem.,* **39**, 2086 (1961).

108. F. A. L. Anet and M. V. Rao, *Tetrahedron, Lett.,* 9 (1960).

109. R. H. Burnell and D. R. Taylor, *Tetrahedron,* **15**, 473 (1961).

110. R. H. Burnell and D. R. Taylor, *Tetrahedron,* **18**, 1467 (1962).

111. J. C. F. Yang and D. B. MacLean, *Can. J. Chem.,* **41**, 2731 (1963).

112. A. Reissert, *Chem. Ber.,* **24**, (1891).

113. W. L. Mosby, in *The Chemistry of Heterocyclic Compounds,* Vol. 15B (A. Weissberger, Ed.), Interscience, New York.

114. F. G. Mann and B. B. Smith, *J. Chem. Soc.,* 1898 (1951).

115. F. G. Mann and P. I. Ittyerah, *J. Chem. Soc.*, 467 (1958).
116. P. A. S. Smith and T. Yu, *J. Amer. Chem. Soc.*, **74**, 1096 (1952).
117. D. B. Glass and A. Weissberger, *Org. Synthesis,* **26**, 40 (1946).
118. R. E. Rindfusz and V. L. Harnack, *J. Amer. Chem. Soc.*, **42**, 1720 (1920).
119. Z. Vanlenta, P. Deslongchamps, R. Ellison, and J. K. Wiesner, *J. Amer. Chem. Soc.*, **86**, 2533 (1964).
120. V. Boekelheide and G. P. Quinn, *J. Amer. Chem. Soc.*, **70**, 2830 (1948).
121. M. Protiva and V. Prelog, *Helv. Chim. Acta,* **32**, 621 (1949).
122. N. J. Leonard and M. J. Middleton, *J. Amer. Chem. Soc.*, **74**, 5114 (1952).
123. F. Bohlmann and C. Arndt, *Chem. Ber.*, **91**, 2167 (1958).
124. F. Bohlmann, *Chem. Ber.*, **91**, 2157 (1958).
125. E. Wenkert,
126. N. J. Leonard, L. A. Miller, and P. D. Thomas, *J. Amer. Chem. Soc.*, 78, 3463 (1956) and references cited therein.
127. K. Tsuda and S. Saeki, *Chem. Pharm. Bull. (Tokyo),* **6**, 391 (1958).
128. L. Mandell, J. U. Piper, and K. P. Singh, *J. Org. Chem.*, **28**, 3440 (1963); L. Mandell, B. A. Hall, and K. P. Singh, *J. Org. Chem.*, **29**, 3067 (1964).
129. E. Wenkert, K. G. Dave, F. Haglid, R. G. Lewis, T. Oishi, R. V. Stevens, and M. Terashima, *J. Org. Chem.*, **33**, 747 (1968).
130. E. Wenkert and B. Wickberg, *J. Amer. Chem. Soc.*, 87, 1580 (1965).
131. E. Wenkert, *Acct. Chem. Res.*,
132. E. Wenkert, K. G. Dave, and R. V. Stevens, *J. Amer. Chem. Soc.*, **96**, 6177 (1968).
133. E. Wenkert, K. G. Dave, and R. V. Stevens, *J. Amer. Chem. Soc.*, **90**, 6177 (1968).
134 E. P. Anderson, J. V. Crawford, and M. L. Sherrill, *J. Amer. Chem. Soc.*, **68**, 1294 (1946).
135. E. Colvin, J. Martin, W. Parker, and R. A. Raphael, *Chem. Comm.*, 596 (1966).
136. Z. Horii, S. W. Khim, T. Imanishi, and T. Momose, *Chem. Pharm. Bull.*, **18**, 2235 (1970) and references cited therein.
137. A. C. Cope and M. E. Synerholm, *J. Amer. Chem. Soc.*, **72**, 5228 (1950).
138. E. Colvin and W. Parker, *J. Chem. Soc.*, 5764 (1965).
139. Z. Valenta, P. Deslongchamps, R. A. Ellison, and K. Wiesner, *J. Amer. Chem. Soc.*, **86**, 2533 (1964).
140. G. Pinkus, *Chem. Ber.*, **25**, 2798 (1892).
141. N. J. Leonard and J. A. Adanick, *J. Amer. Chem. Soc.*, **81**, 595 (1959).
142. H. Dugas, R. A. Ellison, Z. Valenta, K. Wiesner, and C. M. Wong, *Tetrahedron Lett.*, 1279 (1965).
143. E. H. Böhme, Z. Valenta, and K. Wiesner, *Tetrahedron Lett.*, 2441 (1965).
144. I. N. Nazarov and S. I. Zar'valov, *Zhur, Obshchei Khim.*, **24**, 469 (1954).
145. K. Wiesner, I. Jirkovský, M. Fishman, and C. A. J. Williams, *Tetrhaedron Lett.*, 1523 1967).
146. C. A. Grob and H. J. Wilkens, *Helv. Chim. Acta,* **48**, 808 (1965).
147. K. Wiesner and I. Jirkovsky, *Tetrahedron Lett.*, 2077 (1967).
148. W. Nigata, S. Hirai, H. Itazaki, and K. Takeda, *J. Org. Chem.*, **26**, 2413 (1961).

149. K. Wiesner, Z. Valenta, W. A. Ayer, L. R. Fowler, and J. E. Francis, *Tetrahedron,* **4**, 87 (1958).

150 Z. Valenta, F. W. Stonner, C. Bankiewicz, and K. Wiesner, *J. Amer. Chem. Soc.*, 78, 2867 (1956).

151. D. B. MacLean and H. C. Prime, *Can. J. Chem.*, **31**, 543 (1953).

152. H. Dugas, M. E. Hazenberg, Z. Valenta, and K. Wiesner, *Tetrahedron Lett.*, 4931 (1967).

153. H. Reinshagen, *Annalen,* **681**, 84 (1965).

154. K. Wiesner and L. Poon, *Tetrahedron Lett.*, 4937 (1967); *Can J. Chem.*, **47**, 433 (1969).

155. E. E. Betts and D. B. MacLean, *Can. J. Chem.*, **35**, 211 (1957).

156. K. Wiesner, V. Musil, and K. J. Wiesner, *Tetrahedron Lett.*, 5643 (1968).

157. G. Stork, R. A. Kretchmer, and R. H. S. Schlessinger, *J. Amer. Chem. Soc.*, **90**, 1647 (1968).

158. N. L. Allinger and C. K. Riew, *Tetrahedron Lett.*, 1269 (1966).

159. G. Stork and P. Hudrlik, *J. Amer. Chem. Soc.*, **90**, 4462, 4464 (1968).

160. W. A. Ayer, W. R. Bowman, T. C. Joseph, and P. Smith, *J. Amer. Chem. Soc.*, **90**, 1648 (1968).

161. W. A. Ayer, W. R. Bowman, G. A. Cooke, and A. C. Soper, *Tetrahedron Lett.*, 2021 (1966).

162. G. F. H. Green and A. G. Long, *J. Chem. Soc.*, 2332 (1961).

163a. N. J. Leonard, in *The Alkaloids,* Vol. I, (R. H. F. Manske and H. L. Holmes, Ed.), Academic, New York, 1950, pp. 108-164.

163b. N. J. Leonard, in *The Alkaloids,* Vol. VI, (R. H. F. Manske and H. L. Holmes, Eds.), Academic, New York, 1960, pp. 37-121.

163c. F. L. Warren, in *The Alkaloids* Vol. XII, (R. H. F. Manske and H. L. Holmes, Eds.), Academic, New York, 1970, pp. 246-331.

164. N. K. Kochetkov and A. M. Likhosherstov, in *Advances in Heterocyclic Chemistry* Vol. 5 (A. R. Katritzky, Ed.), Academic, New York, 1965, pp. 315-367.

165. F. L. Warren, in *Progress in the Chemistry of Organic Natural Products,* Vol. XII (E. Zechmeister, Ed.), Springer-Verlan Wien, 1955, pp. 198-269.

165b. D. F. L. Warren, in *Progress in the Chemistry of Organic Natural Products,* Vol. XXIV, (E. Zechmeister, Ed.), Springer-Verlag, Wein, 1966, pp.329-406.

166. A. S. Labenskii and G. P. Men'shikov, *Zh. Obshch. Khim.*, **18**, 1936 (1948); *Chem. Abstr.*, **43**, 3827 (1949).

167. R. Adams and K. E. Hamlin, Jr., *J. Amer. Chem. Soc.*, **64**, 2597 (1942).

168. N. K. Hart, S. R. Johns, J. A. Lmaberton, *Austral. J. Chem.*, **21**, 1393 (1968).

169. G. P. Men'shikov and G. M. Borodina, *Zh. Obshch. Khim.*, **15**, 225 (1945); *Chem. Abstr.*, **44**, 1484 (1950).

170. F. Galinovsky, H. Goldberger, and M. Pöhm, *Monatsh. Chem.*, **80**, 550 (1949).

171. A. S. Labenskii, N. A. Serova, and G. P. Men'shikov, *Dokl. Akad. Nauk SSSR,* 88, 467 (1953); *Chem. Abstr.*, 48, 2721 (1954).

172. A. M. Likhosherstov, V. N. Kulakov, and N. K. Kochetkov, *Zh. Obshch. Khim.*, **34**, 2798 (1964); *Chem. Abstr.*, **61**, 14734 (1964).

173. F. L. Warren and M. E. von Klemperer, *J. Chem. Soc.*, 4574 (1958).

174. R. A. Konovalova and A. Orekhov, *Chem. Ber.*, **69**, 1908 (1936).

175. N. J. Leonard and D. L. Felley, *J. Amer. Chem. Soc.*, 72, 2537 (1950).

176. N. K. Kochetkov, A. M. Likhosherstov, and E. I. Budovskii, *Zh. Obshch. Khim.*, **30**, 2077 (1960); *Chem. Abstr.*, **55**, 7386i (1961).

177. A. M. Likhosherstov, L. M. Likhosherstov, and N. K. Kochetkov, *Zh. Obshch. Khim.*, **33**, 1801 (1963).

178. N. J. Leonard and S. W. Blum, *J. Amer. Chem. Soc.*, **82**, 503 (1960).

179. K. Babor, I. Jezo, V. Kalác, and M. Karvas, *Chem. Zvesti*, **13**, 163 (1969); *Chem. Abstr.*, **53**, 20107 (1959).

180. O. Cervinka, K. Pelz, and I. Jirkovský, *Collection Czech. Chem. Comm.*, **26**, 3116 (1961).

181. L. Mandell and W. A. Blanchard, *J. Amer. Chem. Soc.*, **79**, 2343, 6198 (1957).

182. N. K. Kochetkov, A. M. Likhosherstov, and A. S. Labedeva, *Zh. Obshch. Khim.*, **31**, 3461 (1961); *Chem. Abstr.*, **57**, 3490 (1962).

183. M. D. Nair and R. Adams, *J. Org. Chem.*, **26**, 3059 (1961); R. Adams, S. Miyano, and M. D. Nair, *J. Amer. Chem. Soc.*, **83**, 3323 (1961).

184. N. J. Leonard and T. Sato, *J. Org. Chem.*, **34**, 1066 (1969).

185. I. Jezo and V. Kalác, *Chem. Zvesti*, **11**, 696 (1957); *Chem. Abstr.*, **52**, 10052 (1958).

186. A. M. Likhosherstov, A. M. Kritsyn, and N. K. Kochetkov, *Zh. Obshch. Khim.*, **32**, 2377 (1962); *Chem. Abstr.*, **58**, 9154 (1963).

187. K. Babor, J. Jezo, V. Kalac, M. Karvas, and K. Tihlarik, *Chem. Zvesti*, **14**, 679 (1960); *Chem. Abstr.*, **55**, 17620 (1961).

188. A. J. Aasen and C. C. J. Culvenor, *J. Org. Chem.*, **34**, 4143 (1969).

189. T. A. Geissman and A. C. Waiss, Jr., *J. Org. Chem.*, **27**, 139 (1962).

190 J. J. Tufariello and J. P. Tette, *Chem. Comm.*, 469 (1971).

191. J. B. Cloke, *J. Amer. Chem. Soc.*, **51**, 1174 (1929); see also J. B. Cloke, L. H. Baer, J. M. Robbins, and G. E. Smith, *J. Amer. Chem. Soc.*, **67**, 2155 (1945) and references cited therein.

192. P. M. Maginnity with J. B. Cloke, *J. Amer. Chem. Soc.*, **73**, 49 (1951).

193. K. Blaha and O. Cervinka, in *Advances in Heterocycilc Chemistry*, Vol. 6, Academic, New York, 1966, p. 1949.

194. R. V. Stevens and M. C. Ellis, *Tetrahedron Lett.*, 5185 (1967).

195. R. V. Stevens, M. C. Ellis, and M. P. Wentland, *J. Amer. Chem. Soc.*, **90**, 5576 (1968).

196. R. Breslow, in *Molecular Rearrangments* (P. deMayo, Ed.), Interscience, New York, London, 1963, cf. p. 239.

197. H. Gilman and S. M. Spatz, *J. Org. Chem.*, **16**, 1485 (1951).

198. C. Dupin and R. Fraisse-Jullien, *Bull. Soc. Chim. France*, 1993 (1964).

199. F. Bohlmann and O. Schmidt, *Chem. Ber.*, 97, 1354 (1964); M. E. Kuehne and C. Bayha, *Tetrahedron Lett.*, 1311 (1966).

200. E. Wenkert, K. G. Dave, R. G. Lewis, T. Uishi, R. V. Stevens, and M. Terashima, *J. Org. Chem.*, **33**, 747 (1968).

201. J. Szmuszkovicz, in *Advances in Organic Chemistry*, Vol. 4, Interscience, New York, 1963; I. Fleming and M. N. Karger, *J. Chem. Soc. (C)*, 226 (1967).

202a. F. C. Tahk and S. L. Keely, Jr., *Chem. Comm.*, 441 (1968).

202b. T. J. Curphey and H. L. Kim, *Tetrahedron Lett.*, 1441 (1968).

203a. R. V. Stevens and M. P. Wentland, *Tetrahedron Lett.*, 2613 (1968).

203b. R. V. Stevens and M. P. Wentland, *J. Amer. Chem. Soc.*, **90**, 5580 (1968).

204a. R. B. Woodward, Abstracts, 20th National Organic Symposium of the American Chemical Society, Burlington, Vt., June, 1967, p. 104.

204b. E. L. Eliel, N. L. Allinger, S. J. Angyal, and G. A. Morrison, *Conformational Analysis*, Interscience, New York, 1965, p. 314.

205. E. Kaiser and C. Hauser, *J. Amer. Chem. Soc.*, 88, 2348 (1966).

206. R. R. Arndt and P. E. J. Kruger, *Tetrahedron Lett.*, 3237 (1970).

207. R. V. Stevens and L. E. DuPree, *Chem. Comm.*, 1585 (1970); R. V. Stevens, L. E. DuPree, and P. L. Loewstein, *J. Org. Chem.*, in press (1972).

208. For recent reviews on *Erythrina* alkaloids, see V. Voekelheide, *Alkaloids*, 7, 201 (1960) and R. K. Hill, *Alkaloids*, 9, 483 (1967).

209. R. V. Stevens and M. P. Wentland, *Chem. Comm.*, 1104 (1968).

210. K. Wiesner, Z. Valenta, A. J. Manson, and F. W. Stonner, *J. Amer. Chem. Soc.*, **77**, 675 (1955); see also V. Prelog, A. Langemann, O. Rodig, and M. Ternbah, *Helv. Chim. Acta*, **42**, 1301 (1959).

211. R. V. Stevens, R. K. Mehra, and R. L. Zimmerman, *Chem. Comm.*, 877 (1969).

212a. G. Stork and J. E. Dolfini, *J. Amer. Chem. Soc.*, **85**, 2872 (1963).

212b. Y. Ban, Y. Sato, I. Inove, M. Nagi, T. Oishi, M. Terashima, O. Yonemitsu, and Y. Kanoako, *Tetrahedron Lett.*, 2261 (1965).

213. R. V. Stevens, J. M. Fitzpatrick, M. Kaplan, and R. L. Zimmerman, *Chem. Comm.*, 857 (1971).

214. E. Wenkert, K. G. Dave, and R. V. Stevens, *J. Amer. Chem. Soc.*, **90**, 6177 (1968).

Selected Additional References—Crinine-type Alkaloids

1. H. A. Lloyd, E. A. Kielar, R. J. Highet, S. Uyeo, H. M. Fales, and W. C. Wildman, *J. Org. Chem.*, **27**, 373 (1962). Contains the synthesis of a degradation product of powellane and several references to interconversions between alkaloids of the series.
2. W. C. Wildman, *J. Amer. Chem. Soc.*, **80**, 2567 (1958). The structures of crinine, powelline, buphanidrine, and buphasine proved by degradation and partial synthesis.
3. H. M. Fales and W. C. Wildman, *J. Amer. Chem. Soc.*, **82**, 197 (1960).
4. H.-G. Boit and H. Ehmke, *Chem. Ber.*, 89, 2093 (1956). 6-Hydroxycrinamine → criwelline.
5. H. Irie, Y. Tsuda, and S. Uyeo, *J. Chem. Soc.*, 1446 (1959).
6. T. Ikeda, W. I. Taylor, Y. Tsuda, S. Uyeo, and H. Yajima, *J. Chem. Soc.*, 4749 (1956).
7. C. F. Murphy and W. C. Wildman, *Tetrahedron Lett.*, 3863 (1964).
8. W. Döpke, *Arch. Pharm.*, **298**, 704 (1965). The conversion of tubispacin into powellane, epoxypowellane, and epiepoxypowellane.
9. S. Uyeo, H. M. Fales, R. J. Highet, and W. C. Wildman, *J. Amer. Chem. Soc.*, **80**, 2590 (1958). Haemanthidine → apohaemanthidine → dihydroapohaemanthamine ← haemanthamine.
10. H. M. Fales and W. C. Wildman, *J. Org. Chem.*, **26**, 181 (1961). Partial synthesis.
11. H. M. Fales and W. C. Wildman, *J. Amer. Chem. Soc.*, 82, 3368 (1960). A summary of the interconversions of alkaloids related to crinine.

12. E. W. Warnhoff and W. C. Wildman, *Chem. Ind. (London),* 1293 (1958). Undulatine →powellane.

13. H.-G. Boit and W. Döpke, *Chem. Ber.,* **91**, 1965 (1958). Crinamine and haemanthamine → haemultine.

14. H.-G. Boit and W. Stender, *Chem. Ber.,* **89**, 161 (1956). Haemanthidine → tazettine.

15. W. C. Wildman, *Chem. Ind. (London),* 123 (1956). Haemanthidine → tazettine.

Selected Additional References—Galanthamine-type Alkaloids

16. L. Bubewa-lawanoa, *Chem. Ber.,* **95**, 1348 (1962). Galanthamine $\xrightarrow{H^+}$ (±)-epigalanthamine $\xrightarrow{MnO_2}$ (±)-Narwedine.

17. G. P. Volpp and L. Budewa-lwanowa, *Chem. Ber.,* **97**, 563 (1964).

18. J. Koizumi, S. Kobayashi, and S. Uyeo, *Chem. Pharm. Bull. (Tokyo),* **12**, 696 (1964).

19. S. Uyeo and S. Kobayashi, *Chem. Pharm. Bull. (Tokyo),* **1**, 139 (1953). Galanthamine → dihydrogalanthamine = lycoramine.

Selected Additional References—Lycorine and Lycorenine-type Alkaloids

20. H. M. Fales and W. C. Wildman, *J. Amer. Chem. Soc.,* **80**, 4395 (1958). The conversion of lycorine → caranine, methyl pseudolycorine → pluviine, powelline → dihydroepicrinine, and buphanidrine → buphanisine is described.

21. L. G. Humber, H. Kondo, K. Kotera, S. Takagi, K. Takeda, W. I. Taylor, B. R. Thomas, Y. Tsuda, K. Tsukamoto, S. Uyeo, H. Yajima, and N. Yanaihara, *J. Chem. Soc.,* 4622 (1954).

22. F. Benington and R. D. Morin, *J. Org. Chem.,* **27**, 143 (1962).

23. K. Kotera, *Tetrahedron,* **12**, 240 (1961).

24. K. Kotera, *Tetrahedron,* **12**, 248 (19610'

25. S. Mizukami, *Tetrahedron,* **11**, 89 (1960).

26. R. B. Kelly, W. I. Taylor, and K. Wiesner, *J. Chem. Soc.,* 2094 (1953).

27. W. Döpke, *Arch. Pharm.,* **295**, 920 (1962). Partial synthesis of hippamine = 2-*O*-methyllycorine.

28. K. Takeda, K. Kotera, et al., *Chem. Pharm. Bull. (Tokyo),* **8**, 483 (1960).

29. K. Takeda and K. Kotera, *Chem. Pharm. Bull. (Tokyo),* **5**, 234 (1957). Partial synthesis of aulamine = 2-*O*-acetyl lycorine.

30. H.-G. Boit and H. Ehmke, *Chem. Ber.,* **90**, 57 (1957). Oxidation of nerinine to albomaculine.

31. B. Mehlis, *Naturwissenschaften,* **52**, 33 (1965). Conversion of clivimine to clivonine.

32. T. Kitagawa, S. Uyeo, and N. Yokoyama, *J. Chem. Soc.,* 3741 (1959). Chemically interrelated hydrogenation products of lycorenine, homolycorine, and pluviine.

33. D. F. C. Garbutt, P. W. Jeffs, and F. L. Warren, *J. Chem. Soc.,* 5011 (1962). Krigenamine converted to falcatine methiodide.

34. T. Kitagawa, W. I. Taylor, S. Uyeo, and H. Yajima, *J. Chem. Soc.,* 1066 (1955). Lycorenine → homolycorine.

35. H.-G. Boit. L. Paul, an W. Stender, *Chem. Ber.*, 88, 133 (1955). Lycorenine → homolycorine.

36. S. Uyeo and H. Yajima, *J. Chem. Soc.*, 3393 (1955). Related via their Emde bases.

37. H.-G. Boit and H. Ehmke, *Chem. Ber.*, **90**, 57 (1957). Nerinine → albomaculine.

38. R. B. Kelly, W. I. Taylor and K. Wiesner, *J. Chem. Soc.*, 2094 (1953).

39. L. G. Humber, et al., *J. Chem. Soc.*, 4622 (1954). Synthesis of anhydrolycorine methiodide.

40. H. M. Fales, L. D. Giuffrida, and W. C. Wildman, *J. Amer. Chem. Soc.*, 78, 4145 (1956). Narcissamine → galanthamine methiodide.

41. C. C. J. Culvenor, G. M. O'Donovan, R. S. Sawhney, and L. W. Smith, *Austral. J. Chem.*, **23**, 347 (1970). Monocrotaline is not epoxidized by the usual peracid reagents. However, this transformation can be achieved with CF_3CO_3H in excess trifluoroacetic anhydride in which it is postulated the *N*-oxidation is suppressed by formation of an acyl-ammonium complex.

42. A. J. Aasen and C. C. J. Culvenor, *Austral. J. Chem.*, **22**, 2657 (1969).

43. A. R. Mattocks, *J. Chem. Soc. (C)*, 1155 (1969); see also Ref. 44. Methods are given for the conversion of various dehydrpyrrolizidine alkaloids and their *N*-oxides into dihydropyrrolizine derivatives. The latter substances have been suggested as the toxic metabolites of these alkaloids in animals.

44. C. C. J. Culvenor, J. A. Edgar, L. W. Smith, and H. J. Tweeddale, *Tetrahedron Lett.*, 3599 (1969); see also Ref. 34.

45. A. J. Aasen, C. C. J. Culvenor, and L. W. Smith, *J. Org. Chem.*, **34**, 4137 (1969).

46. B. Luning and H. Trankner, *Acta Chem. Scand.*, **22**, 2324 (1968). The structure of (+)-1-(methoxycarbonyl)pyrrolizidine was proved by degradation into pyrrolizidine, by reduction to lindelofidine, and by synthesis.

47. L. B. Bull, C. C. J. Culvenor, and A. T. Dick, *Frontiers of Biology*, Vol. 9, Interscience, New York, 1968.

48. C. C. J. Culvenor, G. M. O'Donovan, and L. W. Smith, *Austral. J. Chem.*, **20**, 757 (1967). Two new alkaloids (1β,2β-epoxy-1α-hydroxymethyl-8α-pyrrolizidine and 7β-acetoxy-1-methoxymethyl-1,2-dehydro-8α-pyrrolizidine) were isolated and characterized, and their partial synthesis was executed.

49. R. Schoental, *Potential Carcinogenic Hazards Drugs, Proc. Symp.*, (R. Truhaut, Ed.), 1965, pp. 152-61.

50. O. Cervinka, *Chem. Listy.*, **52**, 307 (1958).

51. F. Micheel and W. Flitsch, *Chem. Ber.*, 88, 509 (1955).

52. L. J. Dry, M. J. Koekemoer, and F. L. Warren, *J. Chem. Soc.*, 59 (1955).

53. M. S. Koekemoer and F. L. Warren, *J. Chem. Soc.*, 63 (1955).

54. C. C. J. Culvenor, *Austral. J. Chem.*, 7, 287 (1954). The conversion of supinidine to (−)-isoretronecanol is described.

55. R. Adams and B. L. van Duren, *J. Amer. Chem. Soc.*, **76**, 6379 (1954). Conversion of platynecine to anhydroplatynecine.

56. A. D. Kuzovkov and G. P. Men'shikov, *Zh. Obshch. Khim.*, **21**, 2245 (1951); *Chem. Abstr.*, **46**, 8130a (1952).

57. A. V. Danilova and R. A. Konovalova, *Zh. Obshch. Khim.*, **20**, 1921 (1950); *Chem. Abstr.* **45**, 2960a (1951). Conversion of renardine into renarcine.

58. G. P. Men'shikov and E. L. Gurevich, *Zh. Obshch. Khim.*, **19**, 1382 (1949); *Chem. Abstr.*, **44**, 3486a (1950). Reduction of supinidine to isoretronecanol.

59. G. P. Men'shikov and A. D. Kuzovkov, *Zh. Obshch. Khim.*, **19**, 1702 (1949); *Chem. Abstr.*, **44**, 1113g (1950). Chromic acid oxidation of hydroxyheliotridane to retronecanol.
60. N. J. Leonard and D. L. Felley, *J. Amer. Chem. Soc.*, **71**, 1758 (1949).
61. N. J. Leonard and G. L. Shoemaker, *J. Amer. Chem. Soc.*, **71**, 1760 (1949).
62. N. J. Leonard and G. L. Shoemaker, *J. Amer. Chem. Soc.*, **71**, 1762 (1949).
63. N. J. Leonard and K. M. Beck, *J. Amer. Chem. Soc.*, **70**, 2504 (1948).
64. N. J. Leonard, L. R. Hruda, and F. W. Long, *J. Amer. Chem. Soc.*, **69**, 690 (1947).
65. R. Adams and N. J. Leonard, *J. Amer. Chem. Soc.*, **66**, 257 (1944). The synthesis of retronecanone is described.
66. G. Menschikoff, *Chem. Ber.*, **69**, 1802 (1936).

APPENDIX

Since this review was first submitted, several important advances in the total synthesis of these alkaloids have appeared. The following list of titles was compiled from *Chemical Abstracts* (through April 26, 1976). Excellent accounts of progress in the chemistry of alkaloids are to be found in all five volumes of the *Chemical Society's Specialist Reports.*

The Mesembrine Alkaloids

1. J. B. P. Wynberg and W. N. Speckamp, *Tetrahedron Lett.*, 3963 (1975).
2. R. V. Stevens, P. M. Lesko, and R. Lapalme, *J. Org. Chem.*, **40**, 3495 (1975).
3. G. Otani and S. Yamada, *Chem. Pharm. Bull.*, **21**, 2130 (1973).

The *Amaryllidaceae* Alkaloids—Activity in this area has been intense since the original survey.

4. S. Yamada, K. Tomioka, and K. Koga, *Tetrahedron Lett.*, 57 (1976).
5. S. Yamada, K. Tomioka, and K. Koga, *Tetrahedron Lett.*, 61 (1976). Accompanies previous paper.
6. H. Lida, S. Aoyagi, and C. Kibayashi, *J. Chem. Soc. Perkin Trans.*, **1**, 2502 (1975).
7. Y. Tsuda, T. Sano, J. Taga, K. Isobe, J. Toda, H. Irie, H. Tanaka, S. Takagi, M. Yamaki, and M. Murata, *Chem. Comm.*, 933 (1975).
8. S. Ohta and S. Kimoto, *Tetrahedron Lett.*, 2279 (1975).
9. S. Tobinaga, *Bioorg. Chem.*, **4**, 110 (1975). Oxidation with Iron-DMF and Iron-DMSO complexes.
10. J. B. Hendrickson, T. L. Bogard, M. E. Fisch, S. Grossert, and N. Yoshimura, *J. Amer. Chem. Soc.*, **96**, 7781 (1974).
11 T. Onaka, Y. Kanda, and M. Natsume, *Tetrahedron Lett.*, 1179 (1974).
12. K. Torsell, *Tetrahedron Lett.*, 623 (1974).
13. H. Muxfeldt, J. P. Bell, J. A. Baker, and U. Cuntze, *Tetrahedron Lett.*, 4587 (1973).

14. P. W. Jeffs, *MTP (Med. Tech. Publ. Co.) Int. Rev. Sci. Org. Chem. Ser. One,* 9, 273 (1973). A review with 103 references.
15. I. Niomiya, T. Naito, and T. Kiguchi, *J. Chem. Soc. Perkin Trans.,* **1**, 2261 (1973). Full account of Ref. 50.
16. E. Wenkert, H. P. S. Chawla, and F. M. Schell, *Syn. Comm.,* **3**, 381 (1973).
17. E. Kotani, N. Takeuchi, and S. Tobinaga, *Tetrahedron Lett.,* 2735 (1973).
18. E. Kotani, N. Takeuchi, and S. Tobinago, *Chem. Comm.,* 550 (1973).
19. H. Irie, Y. Nagai, K. Tamoto, and H. Tanaka, *Chem. Comm.,* 302 (1973).

The Lycopodium Alkaloids

20. T. Harayama, M. Ohtani, M. Oki, and Y. Inubushi, *Chem. Pharm. Bull.,* **23**, 1511 (1975).
21. Y. Inubushi and T. Harayama, *Farumashia,* **11**, 126 (1975). A review.
22. Y. Ban, M. Kimura, and T. Oishi, *Heterocycles,* **2**, 323 (1974).
23. W. A. Ayer, *MTP (Med. Tech. Publ. Co.) Int. Rev. Sci. Org. Chem., Ser. One,* 9, 1-25 (1973). A review with 69 references.

The Pyrrolizidine Alkaloids

24. J. J. Tufariello and J. P. Tette, *J. Org. Chem.,* **40**, 3866 (1975).
25. A. Klasek and O. Weinbergova, *Chem. Nat. Carbon Comp.,* **6**, 35 (1975). A review with 107 references.
26. M. T. Pizzorno and S. M. Albonico, *J. Org. Chem.,* **39**, 731 (1974).
27. R. V. Stevens, Y. Luh, and J. Sheu, *Tetrahedron Lett.,* 3799 (1976).

Compound Index

Reaction Index